Chronic Inflammation

Masayuki Miyasaka • Kiyoshi Takatsu
Editors

Chronic Inflammation

Mechanisms and Regulation

Springer

Editors
Masayuki Miyasaka
Interdisciplinary Program for Biomedical Sciences
Institute for Academic Initiatives
Osaka University
Suita, Japan

WPI Immunology Frontier Center
Osaka University
Suita, Japan

MediCity Research Laboratory
University of Turku
Turku, Finland

Kiyoshi Takatsu
Department of Immunobiology and Pharmacological Genetics
Graduate School of Medicine and Pharmaceutical Sciences
University of Toyama
Toyama, Japan

Toyama Prefectural Institute for Pharmaceutical Research
Toyama, Japan

ISBN 978-4-431-56066-1 ISBN 978-4-431-56068-5 (eBook)
DOI 10.1007/978-4-431-56068-5

Library of Congress Control Number: 2016945256

Printed on acid-free paper

This Springer imprint is published by Springer Nature
The registered company is Springer Japan KK

Preface

Inflammation, a reaction characterized by redness, fever, swelling, and pain, has been considered a homeostatic tissue repair mechanism, which is evoked by the body in response to infections and/or tissue injury. However, accumulating evidence indicates that, when inflammation becomes chronic, it acts as a strong disease-promoting factor in a number of pathological disorders including arteriosclerosis, obesity, cancer, and Alzheimer disease. Chronic inflammation also promotes aging. Despite such importance, the dismaying fact is that we know very little about why inflammatory reactions that would usually subside continue and become chronic. More specifically, we do not know precisely what type of factors induce chronic inflammation and promote its prolongation. Also we have little knowledge about how chronic inflammation causes tissue degeneration and other disorders. Furthermore, we have no effective treatment against chronic inflammation at present.

Realizing these situations, a key funding body of the Government of Japan, the Japan Science and Technology Agency (JST), launched two major research programs (CREST and PRESTO) on chronic inflammation in 2010; CRESTO is a funding program for team-oriented research, whereas PRESTO is for independent research by young investigators. From 2010 until now, in the research area of chronic inflammation, 17 teams were selected for CRESTO and conducted research for 5 years (each team receiving 150–500 million yen in total), and 37 researchers were selected and conducted research for 3–5 years in PRESTO (each scientist receiving 30–40 million yen for 3-year research and 50–100 million yen for 5-year research).

This book represents a compendium of such research efforts. Members of the CREST and PRESTO projects contributed a chapter on their own work, and research supervisors of the CRESTO and PRESTO projects (M.M. and K.T., respectively) edited the book. As you see in this book, thanks to the painstaking and persistent hard work by the CRESTO and PRESTO members, we are now beginning to understand what induces and maintains the chronicity of inflammation, and what kinds of mechanisms chronic inflammation utilizes to induce specific diseases including cancer, degenerative neurological disorders, and arteriosclerotic diseases. We have also succeeded in creating novel technologies that allow for the early detection and quantitative assessment of chronic inflammation.

Producing this book required the efforts of many people who deserve credit and thanks. First, we would like to thank all the CRESTO and PRESTO investigators, who worked strenuously on the subject of chronic inflammation and contributed a nice chapter for the book. Second, our special thanks go to research officers of JST and AMED (Japan Agency for Medical Research and Development) (CREST "Chronic Inflammation" is now under the supervision of AMED since 2015), particularly to Shinichi Kato (JST-CREST), Akihiko Kasahara (AMED-CREST), and Isao Serizawa (PRESTO), who kept the projects organized and meticulously prepared a number of research meetings for the members. Third, we are indebted to the editorial assistance by Yuko Matsumoto and Yasutaka Okazaki of Springer Japan. Fourth, we wish to acknowledge the constant support and understanding of our wives, Chieko Takatsu and Etsuko Miyasaka. Finally, we thank you, the reader, for your interest in this research field. We will be more than happy if our efforts are successful in providing you with useful and stimulating information that will lead to new developments in the field of chronic inflammation.

Suita, Japan
Toyama, Japan
September 2015

Masayuki Miyasaka
Kiyoshi Takatsu

Contents

Part I Basic Mechanisms Underlying Induction, Progression, and Resolution of Chronic Inflammation

1 **Prostaglandins in Chronic Inflammation** 3
Tomohiro Aoki and Shuh Narumiya

2 **Cellular and Molecular Mechanisms of Chronic Inflammation-Associated Organ Fibrosis** 19
Tatsuya Tsukui, Shigeyuki Shichino, Takeshi Shimaoka, Satoshi Ueha, and Kouji Matsushima

3 **Sema4A and Chronic Inflammation** 37
Daisuke Ito and Atsushi Kumanogoh

4 **MicroRNAs in Chronic Inflammation** 49
Y. Ito, S. Mokuda, K. Miyata, T. Matsushima, and H. Asahara

5 **Genetic Dissection of Autoinflammatory Syndrome** 63
Koji Yasutomo

6 **Structural Biology of Chronic Inflammation-Associated Signalling Pathways: Toward Structure-Guided Drug Development** 77
Reiya Taniguchi and Osamu Nureki

7 **Lipid Signals in the Resolution of Inflammation** 89
Makoto Arita

8 **Regulation of Chronic Inflammation by Control of Macrophage Activation and Polarization** 97
Junko Sasaki and Takehiko Sasaki

9 **Clarification of the Molecular Mechanisms That Negatively Regulate Inflammatory Responses** 109
Takashi Tanaka

10 The Drosophila Toll Pathway: A Model of Innate Immune Signalling Activated by Endogenous Ligands 119
Takayuki Kuraishi, Hirotaka Kanoh, Yoshiki Momiuchi, Hiroyuki Kenmoku, and Shoichiro Kurata

Part II Imaging Analyses of Chronic Inflammation

11 Macrophage Dynamics During Bone Resorption and Chronic Inflammation 133
Junichi Kikuta, Keizo Nishikawa, and Masaru Ishii

12 Visualization of Localized Cellular Signalling Mediators in Tissues by Imaging Mass Spectrometry 147
Yuki Sugiura, Kurara Honda, and Makoto Suematsu

13 Tracking of Follicular T Cell Dynamics During Immune Responses and Inflammation 161
Takaharu Okada

Part III Chronic Inflammation and Cancer

14 The Role of Chronic Inflammation in the Promotion of Gastric Tumourigenesis 173
Hiroko Oshima, Kanae Echizen, Yusuke Maeda, and Masanobu Oshima

15 Cellular Senescence as a Novel Mechanism of Chronic Inflammation and Cancer Progression 187
Naoko Ohtani

16 Establishment of Diagnosis for Early Metastasis 201
Sachie Hiratsuka

17 Non-autonomous Tumor Progression by Oncogenic Inflammation 211
Shizue Ohsawa and Tatsushi Igaki

18 Inflammation-Associated Carcinogenesis Mediated by the Impairment of microRNA Function in the Gastroenterological Organs 223
Motoyuki Otsuka

19 Roles of Epstein–Barr Virus Micro RNAs in Epstein–Barr Virus-Associated Malignancies 235
Ai Kotani

Part IV Chronic Inflammation and Obesity/Environmental Stress

20 Chronicity of Immune Abnormality in Atopic Dermatitis: Interacting Surface Between Environment and Immune System . . . 249
Takanori Hidaka, Eri H. Kobayashi, and Masayuki Yamamoto

21 Role of Double-Stranded RNA Pathways in Immunometabolism in Obesity 277
Takahisa Nakamura

22 Molecular Mechanisms Underlying Obesity-Induced Chronic Inflammation 291
Takayoshi Suganami, Miyako Tanaka, and Yoshihiro Ogawa

23 Roles of Mitochondrial Sensing and Stress Response in the Regulation of Inflammation 299
Kohsuke Takeda, Daichi Sadatomi, and Susumu Tanimura

24 Oxidative Stress Regulation by Reactive Cysteine Persulfides in Inflammation 309
Tomohiro Sawa

Part V Chronic Inflammation and Innate Immunity

25 Posttranscriptional Regulation of Cytokine mRNA Controls the Initiation and Resolution of Inflammation 319
Osamu Takeuchi

26 Roles of C-Type Lectin Receptors in Inflammatory Responses 333
Shinobu Saijo

27 Elucidation and Control of the Mechanisms Underlying Chronic Inflammation Mediated by Invariant Natural Killer T Cells 345
Hiroshi Watarai

28 Understanding of the Role of Plasmacytoid Dendritic Cells in the Control of Inflammation and T-Cell Immunity 357
Katsuaki Sato

29 Mechanisms of Lysosomal Exocytosis by Immune Cells 369
Ji-hoon Song and Rikinari Hanayama

30 Potential Therapeutic Natural Products for the Treatment of Obesity-Associated Chronic Inflammation by Targeting TLRs and Inflammasomes 379
Yoshinori Nagai, Hiroe Honda, Yasuharu Watanabe, and Kiyoshi Takatsu

Part VI Chronic Inflammation and Adaptive Immunity

31 Human and Mouse Memory-Type Pathogenic Th2 (Tpath2) Cells in Airway Inflammation 401
Yusuke Endo, Kiyoshi Hirahara, Kenta Shinoda, Tomohisa Iinuma, Heizaburo Yamamoto, Shinichiro Motohashi, Yoshitaka Okamoto, and Toshinori Nakayama

32 Controlling the Mechanism Underlying Chronic Inflammation Through the Epigenetic Modulation of CD4 T Cell Senescence . . . 417
Masakatsu Yamashita, Makoto Kuwahara, Junpei Suzuki, and Takeshi Yamada

33 Adrenergic Control of Lymphocyte Dynamics and Inflammation . . . 429
Kazuhiro Suzuki

34 The Multifaceted Role of PD-1 in Health and Disease 441
Mohamed El Sherif Gadelhaq Badr, Kikumi Hata, Masae Furuhata, Hiroko Toyota, and Tadashi Yokosuka

35 The Role of Lysophospholipids in Immune Cell Trafficking and Inflammation 459
Masayuki Miyasaka, Akira Takeda, Erina Hata, Naoko Sasaki, Eiji Umemoto, and Sirpa Jalkanen

Part VII Chronic Inflammation and Autoimmune Diseases

36 Devising Novel Methods to Control Chronic Inflammation Via Regulatory T Cells 475
James B. Wing, Atsushi Tanaka, and Shimon Sakaguchi

37 Control of Chronic Inflammation Through Elucidation of Organ-Specific Autoimmune Disease Mechanisms 489
Mitsuru Matsumoto

38 Lysophosphatidylserine as an Inflammatory Mediator 501
Kumiko Makide, Asuka Inoue, and Junken Aoki

39 Aberrant Activation of RIG-I–Like Receptors and Autoimmune Diseases 511
Hiroki Kato and Takashi Fujita

40 Elucidation of the Exacerbation Mechanism of Autoimmune Diseases Caused by Disruption of the Ion Homeostasis 525
Masatsugu Oh-hora

Part VIII Chronic Inflammation and Ageing

41 Pathophysiological Role of Chronic Inflammation in Ageing-Associated Diseases 541
Yuichi Ikeda, Hiroshi Akazawa, and Issei Komuro

42 Uterine Cellular Senescence in the Mouse Model of Preterm Birth 555
Yasushi Hirota

Part IX Chronic Inflammation and Bowel Diseases

43 Physiological and Pathological Inflammation at the Mucosal Frontline 567
Yosuke Kurashima and Hiroshi Kiyono

44 Control of Intestinal Regulatory T Cells by Human Commensal Bacteria 591
Koji Atarashi

45 Roles of the Epithelial Autophagy in the Intestinal Mucosal Barrier 603
Koji Aoki and Manabu Sugai

46 Development of Sentinel-Cell Targeted Therapy for Inflammatory Bowel Diseases 617
Kenichi Asano and Masato Tanaka

47 Identification of Long Non-Coding RNAs Involved in Chronic Inflammation in *Helicobacter Pylori* Infection and Associated Gastric Carcinogenesis 627
Reo Maruyama

Part X Chronic Inflammation and Central Nervous System Disease

48 The Research for the Mechanism of Chronically Intractable Pain Based on the Functions of Microglia as Brain Immunocompetent Cell 641
Kazuhide Inoue and Makoto Tsuda

49 The Role of Innate Immunity in Ischemic Stroke 649
Takashi Shichita, Minako Ito, Rimpei Morita, and Akihiko Yoshimura

50 Chronic Neuroinflammation Underlying Pathogenesis of Alzheimer's Disease 661
Takashi Saito

Part XI Chronic Inflammation and Cardiovascular Diseases

51 The Roles of Hypoxic Responses During the Pathogenesis of Cardiovascular Diseases 675
Norihiko Takeda

52 Prevention and Treatment of Heart Failure Based on the Control of Inflammation 685
Motoaki Sano

Index 697

Part I
Basic Mechanisms Underlying Induction, Progression, and Resolution of Chronic Inflammation

Chapter 1
Prostaglandins in Chronic Inflammation

Tomohiro Aoki and Shuh Narumiya

Abstract Chronic inflammation underlies various chronic diseases including autoimmune diseases, cancer, neurodegenerative diseases, vascular diseases, and metabolic syndrome. Inasmuch as aspirin-like nonsteroidal anti-inflammatory drugs exert their effects by inhibiting prostaglandin (PG) biosynthesis, PGs have been traditionally thought to function only as mediators of acute inflammation by regulating short-lived events such as vasodilation, pain and fever. However, recent studies using mice deficient in PG receptor in various models of chronic inflammation have demonstrated that, in addition to their short-lived actions in acute inflammation, PGs exert long-term inflammatory actions by acting on mesenchymal, epithelial and immune cells and critically regulating gene expression at the transcription level. In these actions, PGs often cooperate with various cytokines and innate immunity molecules and amplify their actions. Through these studies, evidence now accumulates that PGs function in various aspects of chronic inflammation such as conversion to immune inflammation, amplification of inflammation by a positive feedback loop, sustained inflammatory cell infiltration, and tissue remodelling. Here we review these findings and discuss their relevance to human disease.

Keywords Prostaglandin • Cyclooxygenase (COX) • Cytokine • Pathogen-associated molecular pattern (PAMP) • NFκB • Helper T cell (Th) subset • Autoimmune disease • Intracranial aneurysm • Colorectal cancer • Angiogenesis

T. Aoki • S. Narumiya (✉)
AMED-CREST, Kyoto 606-8501, Japan

Center for Innovation in Immunoregulation Technologies and Therapeutics, Kyoto University Graduate School of Medicine, Kyoto 606-8501, Japan
e-mail: snaru@mfour.med.kyoto-u.ac.jp

M. Miyasaka, K. Takatsu (eds.), *Chronic Inflammation*,
DOI 10.1007/978-4-431-56068-5_1

1.1 Introduction

Chronic inflammation is inflammation of prolonged duration (weeks to months to years) in which active inflammation, tissue injury, and healing proceed simultaneously. It is histologically characterised by infiltration of mononuclear cells including macrophages, lymphocytes and plasma cells, tissue destruction by products of the inflammatory cells, and repair involving angiogenesis and fibrosis (Kumar et al. 2007). Given that chronic inflammation underlies various chronic diseases including autoimmune diseases, cancer, neurodegenerative diseases, vascular diseases, and metabolic syndrome (Ben-Neriah and Karin 2011; Libby et al. 2011), understanding mechanisms of chronic inflammation is important not only for human health but also for social economy. Supposedly, there are distinct mechanisms to make inflammatory response long-lasting and maintained chronically, and they include (1) conversion of acute inflammation to immune inflammation by acquired immunity, (2) amplification and continuation of inflammatory processes by positive feedback mechanism or suppression of negative feedback mechanism, (3) progression of inflammation by a chain of changes in active cell populations at the inflammatory site, (4) tissue remodelling, and (5) epigenetic changes associated with the above processes to sustain inflammation.

Prostaglandins (PGs) including PGD_2, PGE_2, $PGF_{2\alpha}$, PGI_2, and thromboxane (TX) A_2 are cyclooxygenase (COX) metabolites of C20-unsaturated fatty acids such as arachidonic acid, which are produced and released in response to extrinsic, often noxious, stimuli. PGs exert their actions through a family of G protein-coupled receptors (GPCRs) specific for each PG, PGD receptor (DP), EP1 to EP4 subtypes of PGE receptor, PGF receptor (FP), PGI receptor (IP), and TXA receptor (TP), and CRTH2/DP2 for PGD, another GPCR in a different family (Hirata and Narumiya 2011). Because COX, an enzyme initiating PG biosynthesis, is the target of aspirin-like nonsteroidal anti-inflammatory drugs (NSAIDs) with anti-inflammatory, antipyretic and analgesic actions, PGs have been traditionally thought as mediators of acute inflammation. Recent studies, however, have revealed that PGs play important roles in the above-mentioned mechanisms of chronic inflammation, its transition from acute inflammation, progression, and maintenance. Here, we summarise these findings and discuss their implications in chronic inflammation in humans.

1.1.1 PGs as an Amplifier of Cytokines and Innate Immunity Molecules (Fig. 1.1)

Innate immunity molecules such as pathogen- or damage-associated molecular patterns (PAMPs and DAMPs) are now recognised as a trigger of inflammation. Because PAMPs such as lipopolysaccharide (LPS) and proinflammatory cytokines

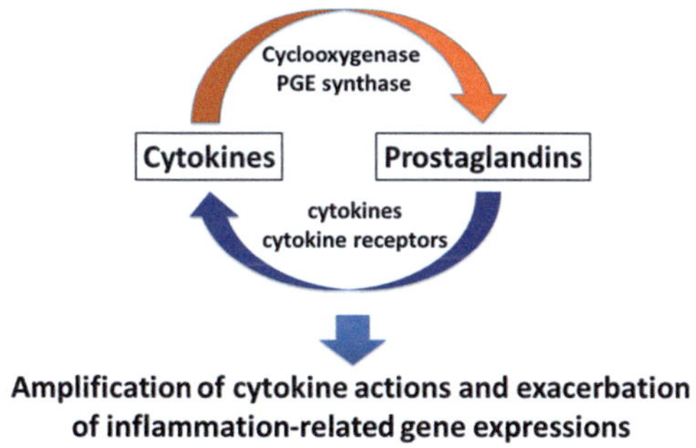

Fig. 1.1 Amplification of inflammatory signalling by the crosstalk between PGs and cytokines. Cytokines induce expression of cyclooxygenase and PGE synthase, and PGs thus formed induce expression of cytokines and cytokine receptor, therefore these two groups of inflammatory mediators form an amplification loop to exacerbate inflammation

induced by these molecules such as interleukin (IL)-1β and IL-6 can induce an inducible isoform of COX, COX-2, and initiate PG biosynthesis, it is generally thought that PGs function as terminal mediators of inflammation to elicit acute inflammation symptoms such as vasodilation and fever downstream of innate immunity. However, PGs can amplify the actions of PAMPs and cytokines, and there is bidirectional crosstalk between the two. For example, Honda et al. reported that PGI_2-IP signalling amplifies actions of IL-1β in collagen-induced arthritis (CIA) of mice (Honda et al. 2006). They found that IP deficiency significantly reduced the severity of arthritis assessed by synovial cell proliferation, inflammatory cell infiltration, and joint destruction in this model, which were accompanied by significant reduction in the content of IL-6 in arthritic paws. They then showed that indomethacin, a COX inhibitor, significantly reduced the IL-1β-induced IL-6 production in cultured synovial fibroblasts and the addition of an IP agonist, cicaprost, restored the IL-6 production. Microarray analysis revealed that in addition to IL-6, PGI_2-IP signalling amplified expression of various genes induced by IL-1β in these cells, including those related to inflammation such as IL-11 and CXCL7, those related to cell proliferation such as fibroblast growth factor (FGF), and vascular and endothelial growth factor (VEGF), and those related to tissue remodelling such as RANKL and the ADAM family molecules. Inasmuch as PGI_2 alone did not induce expression of these genes, these results suggest that PGI_2 can function as an amplifier of IL-1β signalling. Intriguingly, PGI_2-IP signalling augmented expression of the IL-1 receptor (IL1R1) itself. Therefore, this study revealed induction of the receptor for relevant cytokine as one mechanism of PG-mediated amplification of cytokine action. This mechanism combined with cytokine-induced COX-2 expression makes a self-amplification loop for inflammation (Fig. 1.1). As described below, induction of the relevant cytokine receptor is seen also in PG-mediated facilitation of differentiation and expansion of Th subsets in acquired immunity (see Sect. 1.1.2). Amplification by PGs is not limited either to actions of cytokines or to receptor induction. For example, Oshima et al. found that LPS induced expression of COX-2, IL-1β, and IL-6 in cultured macrophages, and that this induction was ameliorated by the addition of celecoxib, a selective COX-2 inhibitor, or RQ-00015986, an EP4 antagonist (Oshima et al. 2011). These findings suggest that endogenously formed PGE_2 acts on the EP4 receptor and amplifies actions by LPS. Similar amplification of TLR actions by PGE_2 was reported for induction of the p19 subunit of IL-23 (*Il23a*) in dendritic cells (DCs; see Sect. 1.1.2). As reviewed below, the action of PGs as an amplifier of cytokines and PAMPs/DAMPs constitutes one of the basic mechanisms whereby PGs function in chronic inflammation.

1.1.2 PGs in Conversion of Acute Inflammation to Immune Inflammation (Fig. 1.2)

One possible mechanism by which inflammation becomes chronic is conversion of acute inflammation to immune inflammation by acquired immunity. Acquired immunity is initiated by presentation of antigen to naïve T cells by DCs and activated T cells differentiate to distinct helper T cell (Th) subsets under the influence of specific cytokine milieu. Among Th subsets, Th1 and Th17, characterised by production of interferon-γ (IFN-γ) and IL-17, respectively, play

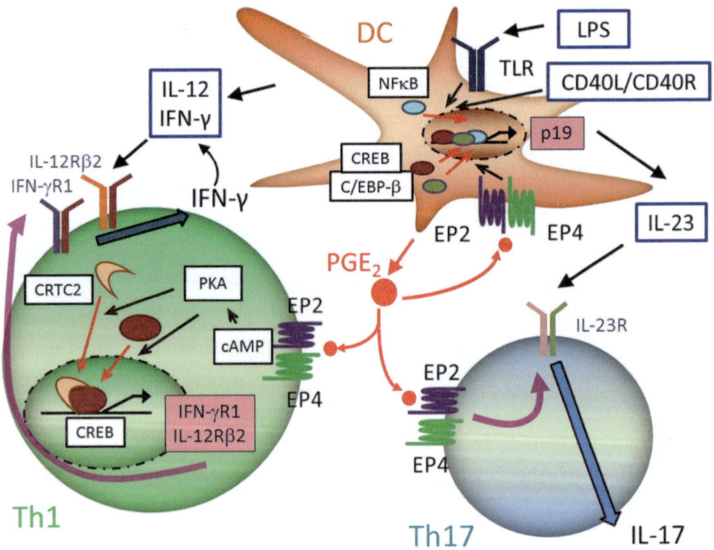

Fig. 1.2 PGE_2 in conversion of acute inflammation to immune inflammation. PGE_2 facilitates Th1 differentiation and Th17 expansion via EP2 and EP4 synergistically with respective cytokines through induction of their receptor IFN-γR1, IL-12Rβ2, and IL-23R PGE_2 also promote IL-23 production from dendritic cells (DCs) synergistically with TLR ligands and CD40 stimulation to further facilitate Th17 cell expansion

the crucial role in immune inflammation. Th1 differentiation is induced by IL-12 and facilitated by IFN-γ. Th17 differentiation and expansion are induced by TGF-β/IL-6 and IL-23, respectively. Although PGs were previously considered as an immunosuppressor (Harris et al. 2002), there is now substantial in vitro and in vivo evidence that PGs act as an immune-activator under many circumstances. Yao et al. found that under the Th1 skewing condition and with strengthened TCR stimulation, PGE_2 enhanced IL-12–mediated Th1 differentiation from mouse naïve T cells in a concentration-dependent manner (Yao et al. 2009). Facilitation of Th1 differentiation by PGE_2 was mimicked by an EP2 or EP4 selective agonist and abolished in T cells deficient in EP2 or EP4, suggesting that PGE_2-EP2/EP4 signalling enhances Th1 differentiation. They further clarified the underlying mechanism that PGE_2-EP2/4 signalling activates the cAMP-PKA pathway, induces expression of IL-12Rβ2 and IFN-γR1 genes via activating CREB and its coactivator CRTC2, and amplifies the actions of IL-12 and IFN-γ on Th1 differentiation (Yao et al. 2013; Fig. 1.2). Notably, in addition to Th1 differentiation, PGE_2 also facilitates IL-23-induced Th17 expansion. This is mediated redundantly via EP2 and EP4 receptors in mouse (Yao et al. 2009), and preferentially via EP2 in humans (Chizzolini et al. 2008; Boniface et al. 2009; Napolitani et al. 2009). In human Th17 cells, PGE_2-EP2 signalling exerts this effect apparently by upregulating expression of IL-23 receptor and IL-1 receptor (Boniface et al. 2009; Fig. 1.2). These studies thus provide further examples for the cytokine amplifying action of PGE_2 through receptor induction. Intriguingly, PGE_2-EP2/4 signalling facilitates not only IL-23-induced Th17 expansion but also enhances production of IL-23 from DCs. Ganea's group showed that PGE_2 potently enhances expression of IL-23 p19 induced by various TLR ligands such as LPS, Poly-I-C, CpG, and proteoglycan from DCs, and that this action is via EP2 and EP4 (Sheibanie et al. 2004; Khayrullina et al. 2008). They further demonstrated that this PGE_2 action is exerted by interaction of NFκB activated by TLR pathway and CREB and C/EBP-β activated by PGE_2-EP4-cAMP signalling (Kocieda et al. 2012). Yao et.al found this PGE_2-mediated enhancement of IL-23 production also in DCs stimulated with anti-CD40 antibody and further found that the treatment with indomethacin or an EP4 antagonist almost completely suppressed the IL-23 production (Yao et al. 2009), suggesting the presence of the PGE_2-mediated self-amplification cycle for IL-23 production in activated DCs. Interestingly, whereas the PGE_2-enhanced IL-23 mRNA expression by TLR ligands or TNF-α is transient, peaking at 1 h, that by CD40 stimulation is of long duration lasting over 36 h, in which the early phase is mediated canonical and the late phase is mediated by non-canonical NFκB signalling, both being similarly enhanced by the PGE_2-EP4 signalling (Ma et al. 2016).

Consistently with these in vitro findings on Th1 and Th17 cells, genetic loss or pharmacological antagonism of EP2 and EP4 significantly ameliorated disease progression in mouse contact hypersensitivity (CHS), experimental autoimmune encephalomyelitis (EAE), and transfer colitis, which are all Th1- and Th17-mediated disease conditions (Yao et al. 2009, 2013). This amelioration was accompanied by remarkable suppression of antigen-induced proliferation and IFN-γ and

IL-17 production of lymph node cells. Pharmacological antagonism of EP4 also ameliorated CIA progression with concomitant suppression of IFN-γ and IL-17 production from lymph node cells (Chen et al. 2010). Conversely, Sheibanie et al. demonstrated that intraperitoneal administration of PGE_2 or an EP3/EP4 agonist, misoprostol, exacerbated CIA and 2,4,6-trinitrobenzene sulfonic acid (TNBS)-induced colitis accompanied by the increase of IL-23 p19 and IL-17 in the lesions (Sheibanie et al. 2007a, b). Such an immune activating function of PGs is not limited to PGE_2 but also seen in PGI_2-IP signalling, which couples to cAMP elevation like EP2/EP4. Nakajima et al. demonstrated that PGI_2-IP signalling facilitates Th1 differentiation in vitro and that IP deficiency impairs CHS (Nakajima et al. 2010). These results suggest that PG signalling functions to facilitate Th1 differentiation and Th17 expansion in vivo in mouse models of various autoimmune diseases. Consistently with these mouse studies, recent genome-wide association studies (GWAS) have identified *Ptger4* (EP4) as a susceptible locus in a number of autoimmune diseases including Crohn's disease (Glas et al. 2012), multiple sclerosis (International Multiple Sclerosis Genetics Consortium; Wellcome Trust Case Control Consortium2 2011), and allergy (Hinds et al. 2013). Furthermore, causal single nucleotide polymorphism (SNPs) related to these autoimmune diseases are enriched in active enhancer region labelled with acetylated H3K27 in *Ptger4* locus and well correlated with EP4 expression (Farh et al. 2015), supporting the clinical relevance of the findings on PGE_2-EP4 signalling in mouse. In addition, Kofler et al. recently reported that EP2 undergoes RORC-dependent silencing in T cells of healthy individuals, but is overexpressed in T cells of patients of multiple sclerosis and simulation of EP2 in these patients' T cells induces a highly pathogenic phenotype expressing both IL-17 and IFN-γ (Kofler et al. 2014), providing clinical relevance of the finding on EP2 and Th17 in mouse studies.

1.1.3 PGs in the Positive Feedback Loop to Amplify Inflammatory Responses (Fig. 1.3)

The formation of a positive feedback loop to amplify and sustain inflammatory responses can be another mechanism whereby inflammation becomes chronic. Indeed, several studies have demonstrated the formation of such positive feedback loops involving PG and their contribution to chronic inflammation. Intracranial aneurysm (IA) is a regional bulging of intracranial arteries at bifurcation sites and a major cause of subarachnoid hemorrhage (van Gijn et al. 2007). IA is chronic inflammation of the artery histologically characterised by degenerative changes of arterial walls and inflammatory cell infiltration mainly of macrophages (Chyatte et al. 1999). High wall shear stress loaded on bifurcation sites of intracranial arteries by blood flow is believed to trigger IA formation (Turjman et al. 2014). Aoki et al. demonstrated induction of COX-2 and EP2 in endothelial cells at a

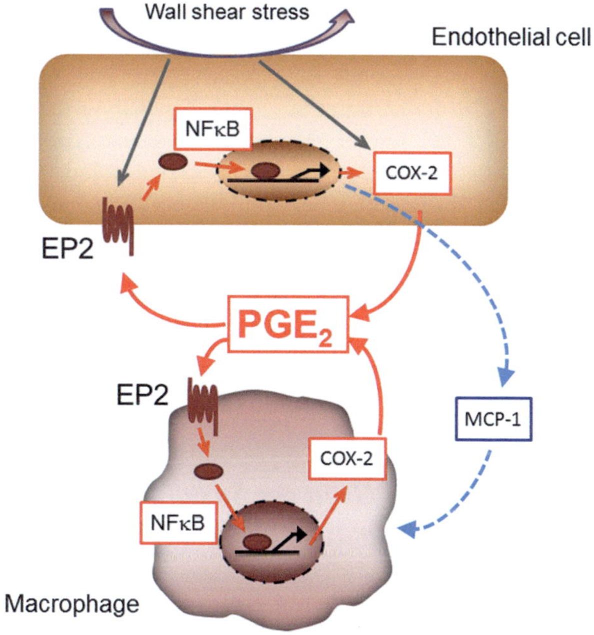

Fig. 1.3 PGE_2 in positive feedback loop for inflammation. In IA, a positive feedback loop consisting of COX-2-PGE_2-EP2-NFκB is formed in arterial endothelial cells upon high wall shear stress. Macrophages are recruited by NFκB-dependent MCP-1 induction in this loop and also form a similar loop for further amplification of inflammation

prospective site of IA formation in an animal model of IA, which was mimicked in vitro in cultured endothelial cells under high shear stress (Aoki et al. 2011). It is important to note that COX-2 inhibition by celecoxib significantly suppressed EP2 expression and EP2 deficiency suppressed COX-2 induction, and both suppressed inflammation in IA walls and prevented IA formation in vivo. PGE_2-EP2 signalling activates NFκB and stimulates NFκB-mediated expression of various proinflammatory genes including MCP-1 in cultured endothelial cells in vitro (Aoki et al. 2011), which is consistent with the previous finding that IA formation is dependent on NFκB (Aoki et al. 2007b). These findings together with the finding that NFκB transcriptionally regulates COX-2 expression (Newton et al. 1997) suggest that high wall shear stress triggers COX-2 expression through NFκB activation in endothelial cells, which triggers a positive feedback loop of COX-2-PGE_2-EP2-NFκB to amplify inflammatory responses (Fig. 1.3). The same feedback loop is formed in macrophages recruited by MCP-1 in IA walls for further amplification (Aoki et al. 2009, 2011; Kanematsu et al. 2011).

Chronic inflammation also underlies cancer development, and is typically characterised by COX-2 expression in tumor lesion (Chulada et al. 2000). Pharmacological inhibition of COX by NSAIDs is long known to reduce incidence of colorectal cancer (CRC) in humans (Rothwell et al. 2010; Janne and Mayer 2000), and genetic deletion of COX-2 in mice reduced intestinal adenoma formation in an animal model of human familial adenomatous polyposis coli (Oshima et al. 1996). Sonoshita et al. used mice deficient in each EP subtype with $Apc^{\Delta 716}$ mutation and found that EP2 deficiency selectively reduced the number and size of adenomas in this model (Sonoshita et al. 2001). They also demonstrated that COX-2 and EP2 were strongly expressed in the stromal region of adenomas and further that EP2 deficiency almost completely abolished COX-2 expression (Sonoshita et al. 2001), suggesting the presence of a positive feedback loop between PGE_2, EP2, and COX-2, as in IA (Aoki et al. 2011). To analyse the mode and the role of inflammation in CRC further, Ma et al. used azoxymethane-dextran sodium sulfate treatment as a model of colitis-associated colon cancer (Ma et al. 2015). They found that EP2 deficiency remarkably reduced inflammatory infiltrates and suppressed the number of colon tumors in this model (Ma et al. 2015). Notably, EP2 was expressed in both neutrophils, the major infiltrating cells in lesions, and tumor-associated fibroblasts surrounding tumor cells, and functioned synergistically with TNF-α to produce various cytokines and chemokines through the self-amplification loop of COX-2-PGE_2-EP2 to promote tumorigenesis.

1.1.4 Role of PGs in the Sustained Infiltration of Inflammatory Cells to Affected Sites (Fig. 1.4)

Although inflammatory cell infiltration is transient in acute inflammation, chronic inflammation exhibits sustained infiltration of inflammatory cells, which is crucial for progression and maintenance of inflammation in various diseases. For example, sustained infiltration of macrophages plays a crucial role in the pathogenesis of IA, as administration of chlodronate liposome to deplete macrophages, gene deletion of MCP-1, a macrophage chemokine, or expression of its dominant negative form all significantly suppressed macrophage infiltration and prevented IA formation (Aoki et al. 2009; Kanematsu et al. 2011). PG signalling plays a critical role in this process, because MCP-1 expression is induced and amplified by a positive feedback loop of the COX-2-PGE_2-EP2-NFκB pathway formed in endothelial cells at the prospective site of IA formation in the cerebral artery (Aoki et al. 2009, 2011). Recruited macrophages then form this amplification loop, and produce MCP-1 by themselves in addition to various cytokines and tissue-destructive proteinases, thus making an autocrine loop for sustained macrophage accumulation and further exacerbation of inflammation in the lesion (Aoki et al. 2007a, 2009; Fig. 1.4). Oshima et al. (2011) reported a similar augmentation of MCP-1-mediated macrophage recruitment by PG signalling. They used *Helicobacter pylori*-infected gastric tumor as a model and found that bacterial colonisation and PGE_2-EP4 signalling cooperatively induced MCP-1 expression to recruit macrophages to promote gastric tumors (Oshima et al. 2011). On the other hand, Ma et al. (2015) found the involvement of PGE_2-EP2 signalling in sustained neutrophil recruitment in the AOM-DSS model of colitis-associated colon cancer. They found extensive neutrophil infiltration and significant expression of CXCL1, a neutrophil chemokine, in the tumor lesion in this model (Ma et al. 2015). Intriguingly, infiltrating neutrophils expressed EP2 and CXCL1 as well, and EP2 deficiency suppressed neutrophil

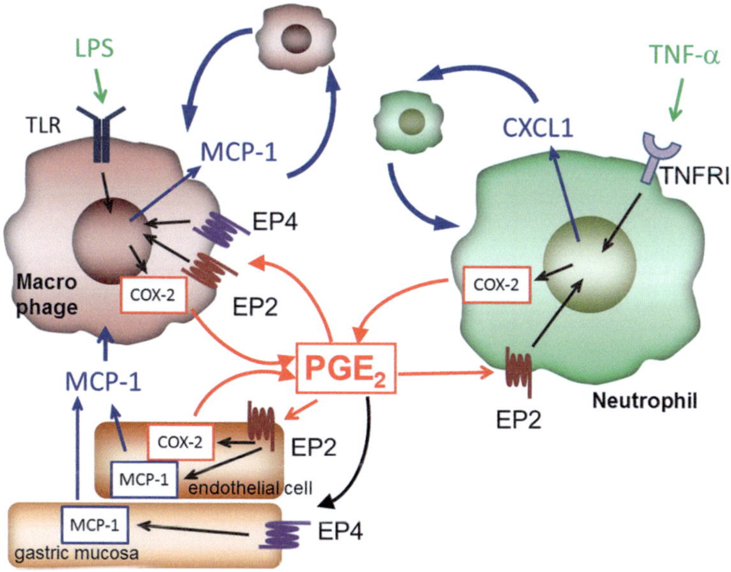

Fig. 1.4 PGs in sustained infiltration of inflammatory cells. PGE_2 induces production of chemokines such as MCP-1 and CXCL1 from various types of cells via EP2 or EP4 and recruits relevant inflammatory cells to affected sites. The recruited cells then produce the chemokines by their own, which forms an autocrine/paracrine loop (*blue colour*) in the affected region and sustains inflammatory cell infiltration

infiltration and CXCL1 expression. Furthermore, EP2 stimulation of primary culture of neutrophils augmented CXCL1 expression synergistically with TNF-α. These results suggest that neutrophils self-amplify their recruitment through the PGE_2 EP2-CXCL1 pathway, which critically contributes to tumorigenesis in their model.

These findings clearly show that PG signalling sustains infiltration of inflammatory cells under different inflammatory settings through induction of various chemoattractants in a positive feedback manner and makes inflammation long-lasting (Fig. 1.4).

1.1.5 Role of PGs in Tissue Remodelling (Fig. 1.5)

In chronic inflammation, destruction and repair of affected tissues simultaneously occur and these processes lead to tissue remodelling including tissue metaplasia, fibrosis, angiogenesis, and granulation. PGs either facilitate or suppress tissue remodelling in a context-dependent manner (Fig. 1.5). For example, the airway undergoes extensive remodelling in bronchial asthma. In the ovalbumin (OVA)-induced allergic asthma model, OVA challenge induced expression of genes

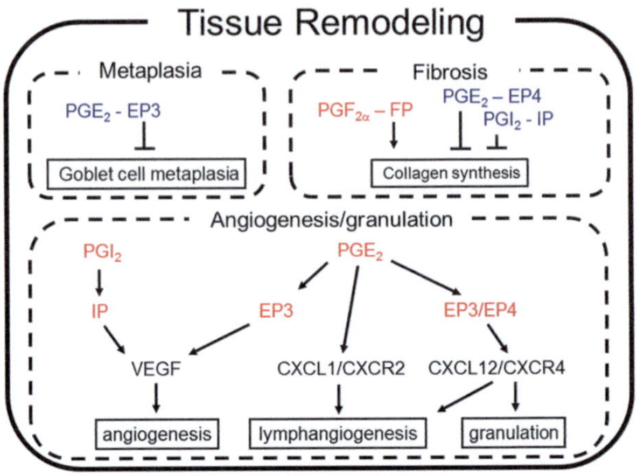

Fig. 1.5 PGs in tissue remodelling. PGs either promote or suppress tissue remodelling, including metaplasia, fibrosis, angiogenesis, or granulation in affected tissues depending on the microenvironment. *Red* or *blue colour* indicates examples of PG contribution to promotion or suppression of tissue remodelling, respectively

involved in tissue remodelling such as Gob-5, Munc5ac, MMP-12, and ADAM-8, and stimulation of PGE_2-EP3 potently suppresses this induction (Kunikata et al. 2005). On the other hand, the same PGE_2-EP3 signalling facilitates angiogenesis associated with chronic inflammation and tumor. Amano et al. implanted a Matrigel sponge or tumor cells in mice, and found induction of angiogenesis in these implants in wild-type mice (Amano et al. 2003). This angiogenesis was suppressed in $EP3^{-/-}$ mice with reduced VEGF expression in the stroma. Bone marrow transfer experiment by the same group indicates that bone marrow cells bearing EP3 is responsible for VEGF expression in the stroma around the implant, and recruitment of $VEGFR\text{-}1^{+}/VEGFR\text{-}2^{+}$ cells there (Ogawa et al. 2009). Perhaps in relation to these findings, Wang et al. found that PGE_2 induced expression of CXCL1, a chemokine for endothelial cells, from CRC cells in vitro, and that administration of PGE_2 in vivo augmented CXCL1 expression in LS-174 T cells transplanted in immune-compromised mice and enhanced angiogenesis around the xenograft, which was abolished by the injection of anti-CXCL1 antibody (Wang et al. 2006). They suggested clinical relevance of their findings by showing correlation between CXCL1 expression and PGE_2 content in specimens of human CRC tumors (Wang et al. 2006). For granulation, Katoh et al. (2010) used implants of tumor cells and micropore chamber and found expression of CXCL12 around the implants, which was sensitive to COX-2 inhibitor, augmented by PGE_2 and absent in $EP3^{-/-}$ or $EP4^{-/-}$ mice. These authors further found that this chemokine recruits $CXCR4^{+}S100A4^{+}$ fibroblasts from bone marrow to the site for granulation. This PGE_2-EP3/4 signalling at the site of implant functions not only in stroma formation but also in lymphangiogenesis (Katoh et al. 2010). Matrigel implants containing FGF-2 induced proliferative inflammation and associated lymphangiogenesis, which was suppressed by COX-2 inhibitor, augmented by PGE_2 and absent again in $EP3^{-/-}$ or $EP4^{-/-}$ mice (Katoh et al. 2010). Notably, agonists selective to EP3 or EP4 induced VEGF-3 and VEGF-4 from macrophages and/or fibroblast in culture, suggesting that PGE_2 induces growth factors for lymph endothelial cells for lymphangiogenesis (Hosono et al. 2011). Lastly, PGs also regulate tissue fibrosis. Tissue fibrosis is characterised by proliferation of fibroblasts and excessive deposition of extracellular matrix proteins, and disrupts normal tissue architecture and functions. Oga et al. (2009) used bleomycin-induced pulmonary fibrosis as a model of idiopathic pulmonary fibrosis in humans, and demonstrated that the fibrosis in this model was attenuated in $FP^{-/-}$ mice. Intriguingly, the loss of FP did not affect inflammatory responses in lesions but decreased collagen synthesis independently of TGF-β (Oga et al. 2009). Consistently, $PGF_{2\alpha}$ enhanced collagen synthesis in lung fibroblasts in vitro in an additive way to TGF-β. These results indicate that $PGF_{2\alpha}$-FP signalling exerts action on its own in fibrosis. In contrast to this profibrotic effect of $PGF_{2\alpha}$, Lovgren et al. (2006) demonstrated that the loss of IP augmented lung fibrosis in a bleomycin-induced pulmonary fibrosis model. Such anti-fibrotic action of PGI_2-IP signalling was also reported in the heart in mice subjected to pressure overload (Hara et al. 2005). Francois et al. (2005) also reported that $IP^{-/-}$ mice developed cardiac fibrosis, which was suppressed completely by coincidental deletion of TP, suggesting IP signalling and TP signalling antagonise in cardiac fibrosis. Protective action of PGs against tissue fibrosis was also reported for the PGE_2-EP4 signalling, which functions against tubulointerstitial fibrosis in the kidney of mice subjected to unilateral ureteral obstruction (Nakagawa et al. 2012).

1.2 Conclusions

As reviewed, substantial evidence now accumulates that PGs function in various aspects of chronic inflammation from immune inflammation to tissue remodelling, at least in animal models. Although we have not discussed the role of PGs in epigenetic changes associated with chronic inflammation in this review, we believe it will be uncovered soon. It is therefore important now to extrapolate the findings in animal experiments to human disease and to identify the context-dependent action of each PG and its receptor in chronic inflammation associated with various human diseases. Given development of agonists and antagonists selective to each subtype of PG receptors and given the potential and adverse effects of traditional NSAIDs and COX-2 inhibitors, it is the time to examine the potential of receptor-selective drugs to manipulate chronic diseases such as cancer, autoimmune, neurodegenerative, and vascular diseases.

References

Amano H, Hayashi I, Endo H, Kitasato H, Yamashina S, Maruyama T, Kobayashi M, Satoh K, Narita M, Sugimoto Y, Murata T, Yoshimura H, Narumiya S, Majima M (2003) Host prostaglandin E_2-EP3 signalling regulates tumor-associated angiogenesis and tumor growth. J Exp Med 197(2):221–232

Aoki T, Kataoka H, Morimoto M, Nozaki K, Hashimoto N (2007a) Macrophage-derived matrix metalloproteinase-2 and -9 promote the progression of cerebral aneurysms in rats. Stroke 38 (1):162–169. doi:10.1161/01.STR.0000252129.18605.c8

Aoki T, Kataoka H, Shimamura M, Nakagami H, Wakayama K, Moriwaki T, Ishibashi R, Nozaki K, Morishita R, Hashimoto N (2007b) NFκB is a key mediator of cerebral aneurysm formation. Circulation 116(24):2830–2840. doi:10.1161/CIRCULATIONAHA.107.728303

Aoki T, Kataoka H, Ishibashi R, Nozaki K, Egashira K, Hashimoto N (2009) Impact of monocyte chemoattractant protein-1 deficiency on cerebral aneurysm formation. Stroke 40(3):942–951. doi:10.1161/STROKEAHA.108.532556

Aoki T, Nishimura M, Matsuoka T, Yamamoto K, Furuyashiki T, Kataoka H, Kitaoka S, Ishibashi R, Ishibazawa A, Miyamoto S, Morishita R, Ando J, Hashimoto N, Nozaki K, Narumiya S (2011) PGE_2-EP2 signalling in endothelium is activated by haemodynamic stress and induces cerebral aneurysm through an amplifying loop via NFκB. Br J Pharmacol 163 (6):1237–1249. doi:10.1111/j.1476-5381.2011.01358.x

Ben-Neriah Y, Karin M (2011) Inflammation meets cancer, with NFκB as the matchmaker. Nat Immunol 12(8):715–723. doi:10.1038/ni.2060

Boniface K, Bak-Jensen KS, Li Y, Blumenschein WM, McGeachy MJ, McClanahan TK, McKenzie BS, Kastelein RA, Cua DJ, de Waal MR (2009) Prostaglandin E_2 regulates Th17 cell differentiation and function through cyclic AMP and EP2/EP4 receptor signalling. J Exp Med 206(3):535–548. doi:10.1084/jem.20082293

Chen Q, Muramoto K, Masaaki N, Ding Y, Yang H, Mackey M, Li W, Inoue Y, Ackermann K, Shirota H, Matsumoto I, Spyvee M, Schiller S, Sumida T, Gusovsky F, Lamphier M (2010) A novel antagonist of the prostaglandin E_2 EP4 receptor inhibits Th1 differentiation and Th17 expansion and is orally active in arthritis models. Br J Pharmacol 160(2):292–310. doi:10.1111/j.1476-5381.2010.00647.x

Chizzolini C, Chicheportiche R, Alvarez M, de Rham C, Roux-Lombard P, Ferrari-Lacraz S, Dayer JM (2008) Prostaglandin E_2 synergistically with interleukin-23 favors human Th17 expansion. Blood 112(9):3696–3703. doi:10.1182/blood-2008-05-155408

Chulada PC, Thompson MB, Mahler JF, Doyle CM, Gaul BW, Lee C, Tiano HF, Morham SG, Smithies O, Langenbach R (2000) Genetic disruption of *Ptgs-1*, as well as *Ptgs-2*, reduces intestinal tumorigenesis in Min mice. Cancer Res 60(17):4705–4708

Chyatte D, Bruno G, Desai S, Todor DR (1999) Inflammation and intracranial aneurysms. Neurosurgery 45(5):1137–1146

Farh KK, Marson A, Zhu J, Kleinewietfeld M, Housley WJ, Beik S, Shoresh N, Whitton H, Ryan RJ, Shishkin AA, Hatan M, Carrasco-Alfonso MJ, Mayer D, Luckey CJ, Patsopoulos NA, De Jager PL, Kuchroo VK, Epstein CB, Daly MJ, Hafler DA, Bernstein BE (2015) Genetic and epigenetic fine mapping of causal autoimmune disease variants. Nature 518(7539):337–343. doi:10.1038/nature13835

Francois H, Athirakul K, Howell D, Dash R, Mao L, Kim HS, Rockman HA, Fitzgerald GA, Koller BH, Coffman TM (2005) Prostacyclin protects against elevated blood pressure and cardiac fibrosis. Cell Metab 2(3):201–207. doi:10.1016/j.cmet.2005.08.005

Glas J, Seiderer J, Czamara D, Pasciuto G, Diegelmann J, Wetzke M, Olszak T, Wolf C, Muller-Myhsok B, Balschun T, Achkar JP, Kamboh MI, Franke A, Duerr RH, Brand S (2012) *PTGER4* expression-modulating polymorphisms in the 5p13.1 region predispose to Crohn's disease and affect NFκB and XBP1 binding sites. PLoS One 7(12):e52873. doi:10.1371/journal.pone.0052873

Hara A, Yuhki K, Fujino T, Yamada T, Takayama K, Kuriyama S, Takahata O, Karibe H, Okada Y, Xiao CY, Ma H, Narumiya S, Ushikubi F (2005) Augmented cardiac hypertrophy in response to pressure overload in mice lacking the prostaglandin I2 receptor. Circulation 112 (1):84–92. doi:10.1161/CIRCULATIONAHA.104.527077

Harris SG, Padilla J, Koumas L, Ray D, Phipps RP (2002) Prostaglandins as modulators of immunity. Trends Immunol 23(3):144–150

Hinds DA, McMahon G, Kiefer AK, Do CB, Eriksson N, Evans DM, St Pourcain B, Ring SM, Mountain JL, Francke U, Davey-Smith G, Timpson NJ, Tung JY (2013) A genome-wide association meta-analysis of self-reported allergy identifies shared and allergy-specific susceptibility loci. Nat Genet 45(8):907–911. doi:10.1038/ng.2686

Hirata T, Narumiya S (2011) Prostanoid receptors. Chem Rev 111(10):6209–6230. doi:10.1021/cr200010h

Honda T, Segi-Nishida E, Miyachi Y, Narumiya S (2006) Prostacyclin-IP signalling and prostaglandin E_2-EP2/EP4 signalling both mediate joint inflammation in mouse collagen-induced arthritis. J Exp Med 203(2):325–335. doi:10.1084/jem.20051310

Hosono K, Suzuki T, Tamaki H, Sakagami H, Hayashi I, Narumiya S, Alitalo K, Majima M (2011) Roles of prostaglandin E_2-EP3/EP4 receptor signalling in the enhancement of lymphangiogenesis during fibroblast growth factor-2-induced granulation formation. Arterioscler Thromb Vasc Biol 31(5):1049–1058. doi:10.1161/ATVBAHA.110.222356

International Multiple Sclerosis Genetics Consortium, Wellcome Trust Case Control Consortium2 (2011) Genetic risk and a primary role for cell-mediated immune mechanisms in multiple sclerosis. Nature 476(7359):214–219. doi:10.1038/nature10251

Janne PA, Mayer RJ (2000) Chemoprevention of colorectal cancer. N Engl J Med 342 (26):1960–1968. doi:10.1056/NEJM200006293422606

Kanematsu Y, Kanematsu M, Kurihara C, Tada Y, Tsou TL, van Rooijen N, Lawton MT, Young WL, Liang EI, Nuki Y, Hashimoto T (2011) Critical roles of macrophages in the formation of intracranial aneurysm. Stroke 42(1):173–178. doi:10.1161/STROKEAHA.110.590976

Katoh H, Hosono K, Ito Y, Suzuki T, Ogawa Y, Kubo H, Kamata H, Mishima T, Tamaki H, Sakagami H, Sugimoto Y, Narumiya S, Watanabe M, Majima M (2010) COX-2 and prostaglandin EP3/EP4 signalling regulate the tumor stromal proangiogenic microenvironment via CXCL12-CXCR4 chemokine systems. Am J Pathol 176(3):1469–1483. doi:10.2353/ajpath.2010.090607

Khayrullina T, Yen JH, Jing H, Ganea D (2008) In vitro differentiation of dendritic cells in the presence of prostaglandin E_2 alters the IL-12/IL-23 balance and promotes differentiation of Th17 cells. J Immunol 181(1):721–735

Kocieda VP, Adhikary S, Emig F, Yen JH, Toscano MG, Ganea D (2012) Prostaglandin E_2-induced IL-23p19 subunit is regulated by cAMP-responsive element-binding protein and C/AATT enhancer-binding protein beta in bone marrow-derived dendritic cells. J Biol Chem 287(44):36922–36935. doi:10.1074/jbc.M112.402958

Kofler DM, Marson A, Dominguez-Villar M, Xiao S, Kuchroo VK, Hafler DA (2014) Decreased RORC-dependent silencing of prostaglandin receptor EP2 induces autoimmune Th17 cells. J Clin Invest 124(6):2513–2522. doi:10.1172/JCI72973

Kumar V, Abbas A, Fausto N, Mitchell R (2007) Robbins basic pathology, 8th edn. Elsevier, Missouri

Kunikata T, Yamane H, Segi E, Matsuoka T, Sugimoto Y, Tanaka S, Tanaka H, Nagai H, Ichikawa A, Narumiya S (2005) Suppression of allergic inflammation by the prostaglandin E receptor subtype EP3. Nat Immunol 6(5):524–531. doi:10.1038/ni1188

Libby P, Ridker PM, Hansson GK (2011) Progress and challenges in translating the biology of atherosclerosis. Nature 473(7347):317–325. doi:10.1038/nature10146

Lovgren AK, Jania LA, Hartney JM, Parsons KK, Audoly LP, Fitzgerald GA, Tilley SL, Koller BH (2006) COX-2-derived prostacyclin protects against bleomycin-induced pulmonary fibrosis. Am J Physiol Lung Cell Mol Physiol 291(2):L144–L156. doi:10.1152/ajplung.00492.2005

Ma X, Aoki T, Tsuruyama T, Narumiya S (2015) Definition of prostaglandin E_2-EP2 signals in the colon tumor microenvironment which amplify inflammation and tumor growth. Cancer Res 75 (14):2822–2832

Ma X, Aoki T, Narumiya S (2016) Prostaglandin E_2-EP4 signalling persistently amplifies CD40-mediated induction of IL-23 p19 expression through canonical and non-canonical NFκB pathways. Cell Mol Immunol 13(2):240–250

Nakagawa N, Yuhki K, Kawabe J, Fujino T, Takahata O, Kabara M, Abe K, Kojima F, Kashiwagi H, Hasebe N, Kikuchi K, Sugimoto Y, Narumiya S, Ushikubi F (2012) The intrinsic prostaglandin E_2-EP4 system of the renal tubular epithelium limits the development of tubulointerstitial fibrosis in mice. Kidney Int 82(2):158–171. doi:10.1038/ki.2012.115

Nakajima S, Honda T, Sakata D, Egawa G, Tanizaki H, Otsuka A, Moniaga CS, Watanabe T, Miyachi Y, Narumiya S, Kabashima K (2010) Prostaglandin I_2-IP signalling promotes Th1 differentiation in a mouse model of contact hypersensitivity. J Immunol 184(10):5595–5603. doi:10.4049/jimmunol.0903260

Napolitani G, Acosta-Rodriguez EV, Lanzavecchia A, Sallusto F (2009) Prostaglandin E_2 enhances Th17 responses via modulation of IL-17 and IFN-gamma production by memory CD4+ T cells. Eur J Immunol 39(5):1301–1312. doi:10.1002/eji.200838969

Newton R, Kuitert LM, Bergmann M, Adcock IM, Barnes PJ (1997) Evidence for involvement of NFκB in the transcriptional control of COX-2 gene expression by IL-1β. Biochem Biophys Res Commun 237(1):28–32. doi:10.1006/bbrc.1997.7064

Oga T, Matsuoka T, Yao C, Nonomura K, Kitaoka S, Sakata D, Kita Y, Tanizawa K, Taguchi Y, Chin K, Mishima M, Shimizu T, Narumiya S (2009) Prostaglandin $F_{2\alpha}$ receptor signalling facilitates bleomycin-induced pulmonary fibrosis independently of transforming growth factor-beta. Nat Med 15(12):1426–1430. doi:10.1038/nm.2066

Ogawa Y, Suzuki T, Oikawa A, Hosono K, Kubo H, Amano H, Ito Y, Kitasato H, Hayashi I, Kato T, Sugimoto Y, Narumiya S, Watanabe M, Majima M (2009) Bone marrow-derived EP3-expressing stromal cells enhance tumor-associated angiogenesis and tumor growth. Biochem Biophys Res Commun 382(4):720–725. doi:10.1016/j.bbrc.2009.03.094

Oshima M, Dinchuk JE, Kargman SL, Oshima H, Hancock B, Kwong E, Trzaskos JM, Evans JF, Taketo MM (1996) Suppression of intestinal polyposis in $Apc^{\Delta 716}$ knockout mice by inhibition of cyclooxygenase 2 (COX-2). Cell 87(5):803–809

Oshima H, Hioki K, Popivanova BK, Oguma K, Van Rooijen N, Ishikawa TO, Oshima M (2011) Prostaglandin E signalling and bacterial infection recruit tumor-promoting macrophages to mouse gastric tumors. Gastroenterology 140(2):596–607 e597. doi:10.1053/j.gastro.2010.11.007

Rothwell PM, Wilson M, Elwin CE, Norrving B, Algra A, Warlow CP, Meade TW (2010) Long-term effect of aspirin on colorectal cancer incidence and mortality: 20-year follow-up of five randomised trials. Lancet 376(9754):1741–1750. doi:10.1016/S0140-6736(10)61543-7

Sheibanie AF, Tadmori I, Jing H, Vassiliou E, Ganea D (2004) Prostaglandin E_2 induces IL-23 production in bone marrow-derived dendritic cells. FASEB J 18(11):1318–1320. doi:10.1096/fj.03-1367fje

Sheibanie AF, Yen JH, Khayrullina T, Emig F, Zhang M, Tuma R, Ganea D (2007a) The proinflammatory effect of prostaglandin E_2 in experimental inflammatory bowel disease is mediated through the IL-23-->IL-17 axis. J Immunol 178(12):8138–8147

Sheibanie AF, Khayrullina T, Safadi FF, Ganea D (2007b) Prostaglandin E_2 exacerbates collagen-induced arthritis in mice through the inflammatory interleukin-23/interleukin-17 axis. Arthritis Rheum 56(8):2608–2619. doi:10.1002/art.22794

Sonoshita M, Takaku K, Sasaki N, Sugimoto Y, Ushikubi F, Narumiya S, Oshima M, Taketo MM (2001) Acceleration of intestinal polyposis through prostaglandin receptor EP2 in Apc $Apc^{\Delta 716}$ knockout mice. Nat Med 7(9):1048–1051. doi:10.1038/nm0901-1048

Turjman AS, Turjman F, Edelman ER (2014) Role of fluid dynamics and inflammation in intracranial aneurysm formation. Circulation 129(3):373–382. doi:10.1161/CIRCULATIONAHA.113.001444

van Gijn J, Kerr RS, Rinkel GJ (2007) Subarachnoid haemorrhage. Lancet 369(9558):306–318. doi:10.1016/S0140-6736(07)60153-6

Wang D, Wang H, Brown J, Daikoku T, Ning W, Shi Q, Richmond A, Strieter R, Dey SK, DuBois RN (2006) CXCL1 induced by prostaglandin E_2 promotes angiogenesis in colorectal cancer. J Exp Med 203(4):941–951. doi:10.1084/jem.20052124

Yao C, Sakata D, Esaki Y, Li Y, Matsuoka T, Kuroiwa K, Sugimoto Y, Narumiya S (2009) Prostaglandin E_2-EP4 signalling promotes immune inflammation through Th1 cell differentiation and Th17 cell expansion. Nat Med 15(6):633–640. doi:10.1038/nm.1968

Yao C, Hirata T, Soontrapa K, Ma X, Takemori H, Narumiya S (2013) Prostaglandin E_2 promotes Th1 differentiation via synergistic amplification of IL-12 signalling by cAMP and PI3-kinase. Nat Commun 4:1685. doi:10.1038/ncomms2684

Chapter 2
Cellular and Molecular Mechanisms of Chronic Inflammation-Associated Organ Fibrosis

Tatsuya Tsukui, Shigeyuki Shichino, Takeshi Shimaoka, Satoshi Ueha, and Kouji Matsushima

Abstract Fibrotic or scar tissue represents a condition where normal tissue architecture has become distorted and been replaced by extracellular matrix (ECM). ECM deposition in injured tissues is a natural part of the wound healing process that facilitates efficient restoration of tissue integrity. However, if the injury persists, the excessive accumulation of ECM leads to the loss of organ function and eventual organ failure. Persistent injuries arise due to various causes, depending on the organ. Some fibrotic diseases are associated with the chronic inflammation that accompanies infection or autoimmune conditions. Other fibrotic diseases are triggered by chemical or pathophysiological insults to epithelial cells, or by unknown causes, as is the case for idiopathic pulmonary fibrosis (IPF). Although the etiology of fibrosis varies between specific diseases, the fibrotic process that takes place in each organ shares a number of common characteristics. In particular, it is widely accepted that excessive amounts of ECM components are produced by activated fibroblasts that accumulate in injured tissue. In the first half of this chapter, we discuss the controversial origin of activated fibroblasts as well as the mechanisms of their activation. In the second half of this chapter, we describe the cellular and molecular mediators that regulate fibrotic responses in the specific example of pulmonary fibrosis.

Keywords Bleomycin • Fibroblast • Fibrosis • Intratracheal transfer • IPF • Macrophage • Monocyte • Myofibroblast • Silica • Silicosis

T. Tsukui • S. Shichino • T. Shimaoka • S. Ueha • K. Matsushima (✉)
Department of Molecular Preventive Medicine, Graduate School of Medicine, The University of Tokyo, 7-3-1, Hongo, Bunkyo-ku, Tokyo 113-0033, Japan

AMED-CREST, Tokyo 113-0033, Japan
e-mail: koujim@m.u-tokyo.ac.jp

M. Miyasaka, K. Takatsu (eds.), *Chronic Inflammation*,
DOI 10.1007/978-4-431-56068-5_2

2.1 The Origin of Collagen-Producing Fibroblasts and Myofibroblasts

Tissue injury often occurs at the epithelium. The precursors of activated fibroblasts become activated following exposure to inflammatory mediators produced upon injury, then proliferate and migrate into the injured area. Under certain conditions, such as in skin injury, these activated fibroblasts differentiate into myofibroblasts, which are contractile cells that often express α-smooth muscle actin (α-SMA; Tomasek et al. 2002). Myofibroblasts proliferate and produce ECM components such as collagen I (Col1) in a range of fibrotic diseases and organs (Hinz et al. 2012). Because therapeutic options for fibrotic diseases are very limited, the origin of myofibroblasts has remained an active area of research. Classically mesenchymal cells such as tissue resident fibroblasts were proposed to be the precursors of myofibroblasts, but over the years various cell types have been reported to differentiate into myofibroblasts, making the origin of these cells a controversial question (Fig. 2.1).

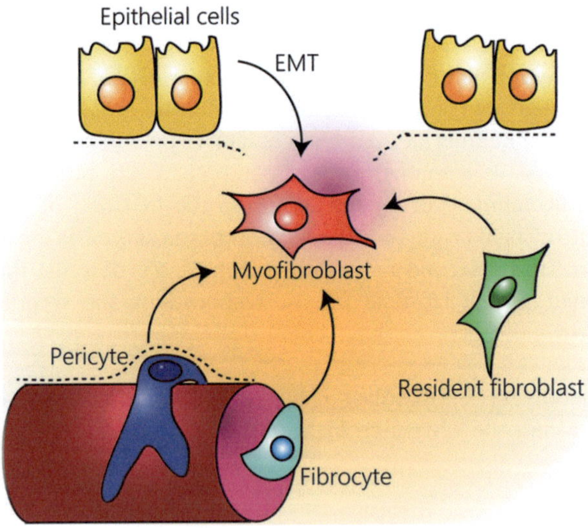

Fig. 2.1 The controversial origins of myofibroblasts. Multiple cell types have been reported to act as a source of myofibroblasts in organ fibrosis. Resident fibroblasts are classically considered to be the primary source of myofibroblasts. Epithelial cells may differentiate into myofibroblasts through EMT, although recent studies have failed to detect evidence of EMT. Bone-marrow–derived fibrocytes are supplied from the circulation and differentiate into myofibroblasts at sites of tissue injury tissue, but there is little evidence of collagen synthesis by these cells. Recent reports have demonstrated a pericyte origin of myofibroblasts in multiple organs

2.2 Myofibroblast Precursors from the Bone Marrow

Bone marrow-derived circulating cells have been reported to gain expression of Col1 in fibrotic organs. These cells are known as ‘fibrocytes’, and are defined as $CD45^+$ $Col1^+$ cells (Bucala et al. 1994). Recent studies using bone marrow chimeras or parabiosis have shown that only a small proportion of Col1-producing cells are supplied from the circulation in experimental models of fibrosis (Higashiyama et al. 2009; Kisseleva et al. 2006; Tsukui et al. 2013). Some myeloid cells internalise and degrade ECM components. These cells are detected as $Col1^+$ even though they may not synthesise Col1 themselves. A recent study revealed that haematopoietic cell-specific Col1 gene deletion does not reduce collagen deposition in bleomycin-induced pulmonary fibrosis (Kleaveland et al. 2014), suggesting that fibrocytes are not a significant source of Col1. Thus, previous studies reporting on $Col1^+$ $CD45^+$ fibrocytes may have been detecting myeloid cells that had internalised Col1 from their surroundings, rather than Col1-producing myeloid cells. However, it has also been reported that 35 % of myofibroblasts in the kidney come from the bone marrow, based on experiments using an α-SMA-RFP reporter in a unilateral ureteral obstruction model (LeBleu et al. 2013). Thus, further fibrocyte lineage tracing studies that address the synthesis and deposition of Col1 are necessary, and the extent to which bone-marrow–derived cells contribute to the activated fibroblast population remains unclear.

2.3 Myofibroblast Precursors from the Epithelium

It is widely accepted that epithelial cells can differentiate into mesenchymal cells via a process known as epithelial-to-mesenchymal transition (EMT) during development or cancer. It has also been reported that myofibroblasts differentiate from epithelial cells by EMT in experimental models of fibrosis. Most earlier studies suggesting a role for EMT in fibrosis were based on in vitro observations, whereas in vivo evidence implicating EMT in fibrosis consisted mainly of immunohistochemical staining showing coexpression of mesenchymal cell and epithelial cell markers. However, there were a number of technical limitations associated with these earlier studies (Kriz et al. 2011). Recent lineage tracing studies have failed to observe EMT in the lung (Rock et al. 2011), liver (Taura et al. 2010), or kidney (Humphreys et al. 2010), and thus raise doubt as to whether epithelial cells are in fact precursors of activated fibroblasts. However, upon tissue injury epithelial cells upregulate transcription factors associated with EMT in a manner similar to that which occurs during development, and these phenotypic changes may promote mesenchymal crosstalk (Rowe et al. 2011). Epithelial cells also play an important role in the initiation of fibrosis. Thus, epithelial cells remain an important target of research that aims to elucidate the underlying mechanisms of fibrosis.

2.4 A Mesenchymal Origin of Myofibroblasts

In recent years, pericytes and perivascular cells have attracted considerable attention as likely precursors for myofibroblasts. Pericytes are embedded in the endothelial basement membrane and are characterised by their direct attachment to endothelial cells. Perivascular cells typically represent interstitial cells that exist in close proximity to blood vessels but outside the endothelial basement membrane. The perivascular cell population may be heterogeneous, and differs in each organ. In the liver, hepatic stellate cells localise to perivascular regions known as the space of Disse, and have characteristics of both pericytes and resident fibroblasts. Hepatic stellate cells represent a major source of myofibroblasts in carbon tetrachloride (CCl_4)-induced liver fibrosis (Mederacke et al. 2013). In the quiescent state, hepatic stellate cells store lipid droplets and do not express Col1. Upon activation, these cells lose their lipid droplets and begin to express Col1 and α-SMA. However, in cholestatic fibrosis, portal fibroblasts have also been suggested to act as myofibroblast precursors, and it remains to be determined which cell type represents the major myofibroblast precursor (Iwaisako et al. 2014).

In the kidney, pericytes are considered to be the most likely precursor for myofibroblasts. Humphreys et al. showed that Foxd1-CreER-labeled pericytes are the major progenitor of myofibroblasts in the unilateral ureter obstruction model (Humphreys et al. 2010). More recently, the Gli1-CreER transgene was shown to label a specific subset of perivascular PDGFRβ^+ cells in the kidney, and ablation of Gli1-CreER-labeled myofibroblasts ameliorated kidney fibrosis (Kramann et al. 2015). Gli1-CreER-labeled cells also expressed PDGFRα, CD146, and Nestin, but not NG2. In addition, like mesenchymal stem cells, this population possessed trilineage differentiation capacity. Although PDGFRβ is a widely used marker for pericytes, this study suggests that pericyte and perivascular populations have a degree of heterogeneity, and that specific subsets may possess higher profibrotic potential. Accordingly, further characterisation of subpopulations of cells (such as Gli1$^+$ cells) and comparison with other kidney cell subsets is required.

In the spinal cord, Glast-expressing type A pericytes have been shown to differentiate into scar-forming myofibroblasts (Goritz et al. 2011). Although Goritz et al. showed that type A pericytes are normally surrounded by basal lamina, another group reported recently that type A pericytes in the spinal cord express Col1 even in the quiescent state, which is a characteristic of resident fibroblasts (Soderblom et al. 2013). Dulauroy et al. revealed that ADAM12-Cre-labelled cells are major myofibroblast precursors in models of dermis or muscle injury, and that these cells migrate into injured areas from the perivascular niche (Dulauroy et al. 2012). ADAM12-Cre-labelled cells localised in the perivascular region, surrounded by endothelial basement membrane. After injury, the labelled cells detached from the blood vessels and emerged from the basement membrane, although it was difficult to distinguish these cells from perivascular cells that acquired ADAM12 expression after the injury had occurred. ADAM12-Cre-labelled cells were PDGFRα$^+$, which is a marker used to identify resident fibroblasts in other organs. However, at present, the characteristics of mesenchymal populations such as fibroblasts and pericytes are poorly understood, and it is difficult to generalise subsets and markers across organs. Many studies have shown that myofibroblasts express PDGFRβ, but myofibroblast progenitors are not necessarily PDGFRβ$^+$, because PDGFRβ expression is induced in mesenchymal cells upon injury (Henderson et al. 2013). Because there is no single marker that defines these mesenchymal cell subsets, elucidation of the basic molecular signatures of pericytes and resident fibroblasts will facilitate further characterisation of these cells in different organs. The same principle applies to the characterisation of activated fibroblasts, and a range of different activation markers for Col1-producing cells has been reported, such as α-SMA, Fn1, S100a4, and Spp1. It remains unclear whether any of these fibroblast activation markers define subpopulations with specific functions in fibrosis.

2.5 The Origin of Col1-Producing Cells in the Lungs

For the last several years, we have been studying the origin and dynamics of activated fibroblasts in pulmonary fibrosis. Following bleomycin-induced lung injury, Col1-producing activated fibroblasts form massive clusters and destruction of the alveolar structure is observed (Tsukui et al. 2013). The precursors of these activated fibroblasts appear likely to be resident fibroblasts or pericytes. Rock et al. reported that NG2-CreER-labelled pericytes proliferate after injury, but that these cells do not express high levels of α-SMA (Rock et al. 2011). Instead, they observed the proliferation of PDGFRα$^+$ cells, suggesting that resident fibroblasts rather than NG2$^+$ pericytes are the major progenitors of activated fibroblasts in the lungs. Hung et al. used Foxd1-Cre mice to examine pericytes in pulmonary fibrosis (Hung et al. 2013). Foxd1 is expressed only during development and labels PDGFRβ$^+$ cells in the alveolar regions more efficiently than NG2-CreER. Among Foxd1-Cre-labelled cells, around 60 % are NG2$^+$ whereas greater than 80 % express PDGFRβ. This study showed that 68 % of α-SMA$^+$ cells in fibrotic lesions were derived from Foxd1-Cre-labelled cells at day 14 after bleomycin treatment. The discrepancy between the results of the studies of Rock et al. and Hung et al. may be related to the markers used. Because not all PDGFRβ$^+$ cells express NG2, it is possible that NG2$^-$ PDGFRβ$^+$ pericytes have the potential to differentiate into myofibroblasts. Moreover, both studies lacked in vivo evidence of Col1 production by proliferating or α-SMA$^+$ mesenchymal cells. Histological analysis has only limited capacity for the examination of multiple markers and functions. For future studies, it may be useful to combine histological analysis with flow cytometric analysis and ex vivo assays in order to characterise activated fibroblast populations.

Many reports suggest that myofibroblasts are the key player in fibrosis. It remains unclear the extent to which other profibrotic cell populations contribute to fibrosis, or even if other profibrotic cell populations exist. Our studies have demonstrated that only a small proportion of Col1-producing cells become α-SMA$^+$ after bleomycin-induced lung injury, whereas Col1-producing cells dramatically upregulate profibrotic genes and organise fibrotic lesions (Tsukui et al. 2013). The function of the α-SMA molecule is to mediate cell contraction. Although cell contraction and ECM production are closely associated elements of the wound healing process, α-SMA expression is not necessary for ECM production itself. Thus, direct evidence of excessive ECM deposition by myofibroblasts or other cell populations is necessary for understanding the contributions of profibrotic cell populations to pulmonary fibrosis.

The structure of the lungs is characterised by thin alveolar walls that enable efficient gas exchange. Because of these thin alveolar walls, interstitial cells including resident fibroblasts are very sparsely distributed throughout the alveolar walls in the quiescent state. The mechanism by which these interstitial cells generate clusters and fibrotic lesions following epithelial injury is important for understanding the origin of profibrotic cells. Our experiments using BrdU have confirmed that Col1-producing cells proliferate following bleomycin treatment (Tsukui et al. 2013). However, the proliferation of these cells was not robust enough to explain fully their formation of clusters, and histological analysis suggested that Col1-producing cell clusters might result from cell migration as well as proliferation. It has been reported previously that resident fibroblasts migrate into the alveolar airspaces following epithelial injury (Fukuda et al. 1985; Fig. 2.2a). The migration of fibroblasts after injury is well characterised in the skin (Tomasek et al. 2002), and the lungs may undergo a similar process of wound healing. Some reports suggest that fibroblastic foci in IPF patients are formed by the migration of resident fibroblasts into the alveolar airspaces (Noble 2005).

Our group conceived an experimental design to investigate whether activated fibroblasts that migrate into the alveolar airspaces can become interstitial fibroblasts that form fibroblastic foci. In this experiment, fibroblasts were delivered intratracheally into alveolar airspaces that were undergoing tissue remodelling following bleomycin-induced injury (Tsukui et al. 2015). We purified resident fibroblasts from Col1-GFP reporter mice, then intratracheally transferred the purified resident fibroblasts into wild-type mice that had been treated with bleomycin several days earlier. We found that the transferred fibroblasts dramatically upregulated profibrotic genes, displayed activated morphology, and formed fibroblastic foci (Fig. 2.2b), suggesting that activated fibroblasts in the alveolar airspaces eventually form fibrotic lesions. Interestingly, pericytes that were purified according to their expression of the NG2 reporter did not differentiate into Col1-producing activated fibroblasts following intratracheal transfer. Thus, as reported previously (Rock et al. 2011), NG2$^+$ pericytes in the lungs are unlikely to act as progenitors for myofibroblasts, or NG2$^+$ pericytes may be unable to respond to environmental changes in the alveolar airspaces in the same way as resident fibroblasts.

In our experiments, the intratracheal transfer of resident fibroblasts partly recapitulated the activation of Col1-producing cells, suggesting that exposure to

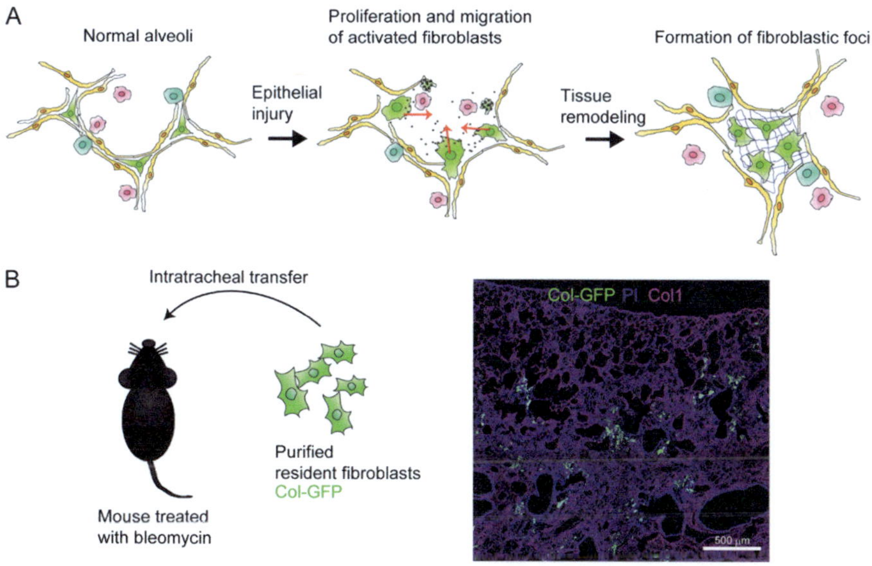

Fig. 2.2 Mechanisms of fibroblastic foci formation in pulmonary fibrosis. (**a**) Following epithelial injury, activated fibroblasts proliferate and migrate into injured alveolar airspaces. Activated fibroblasts form clusters and produce ECM after migration, generating fibroblastic foci at the site of injury. (**b**) Intratracheal cell transfer experiments have shown that fibroblasts in alveolar airspaces are able to become activated fibroblasts in fibroblastic foci. Purified resident fibroblasts from collagen I GFP (Col-GFP) reporter mice were transferred intratracheally into wild-type mice 10 days after intratracheal instillation of bleomycin. The lungs were analysed by immunohistochemistry 21 days after bleomycin instillation. Transferred fibroblasts (*green*) formed fibroblastic foci in the regions of collagen I (*magenta*) deposition. Nuclei were visualised by propidium iodide (PI, *blue*). Scale bar: 500 μm

alveolar airspaces may be pivotal to the process of fibroblastic foci formation by activated fibroblasts. Migratory fibroblasts may form clusters in injured alveolar airspaces, produce ECM components, and then contract, leading to coalescence of the alveolar walls and the formation of fibrotic lesions (Noble 2005). This hypothesis also explains the lack of robust proliferation of activated fibroblasts in bleomycin-induced pulmonary fibrosis, and the loss of lung volume that is observed in the later stages of IPF. However, many important questions remain about the mechanisms that underlie pulmonary fibrosis. If a subset of the pericyte population contributes to the formation of activated fibroblasts, what are the identifying features of that subset, and are those pericytes able to respond to changes in the alveolar airspace environment? The mechanisms of fibroblast migration and fibroblastic foci formation also remain poorly understood. For example, we found that osteopontin defines a subset of activated fibroblasts that form the leading edge of fibroblastic foci and encroach upon the alveolar airspaces (Tsukui et al. 2013), but the factors that elicit osteopontin$^+$ activated fibroblasts remain unknown. Further delineation of the process of fibroblastic foci formation may lead to the identification of novel therapeutic targets for pulmonary fibrosis.

2.6 Cellular and Molecular Mediators of Pulmonary Fibrosis

Wound healing following lung injury is often skewed towards fibrotic responses rather than towards the normal regeneration of the lung architecture (King 2005). Various cellular and molecular mediators, including activated tissue cells (epithelial cells, endothelial cells, and fibroblasts), leukocytes, and soluble mediators participate in a sequential cascade of wound healing responses. In the course of normal tissue regeneration, damaged cells release soluble mediators such as CC chemokine ligand 2, IL-1β, and IL-33 (Cavarra et al. 2004; Mercer et al. 2009; Pichery et al. 2012), which recruit and activate various leukocytes in the damaged tissues. These activated leukocytes play important roles in the clearance of the pathogens, debris, and foreign particles responsible for the injury (Forbes and Rosenthal 2014). In addition, these damaged tissue cells and activated leukocytes also promote tissue repair by secreting cytokines and growth factors such as IL-13, platelet-derived growth factors, and TGF-β1 (Bonner et al. 1991; Huaux et al. 2003). These profibrotic factors activate fibroblasts, inducing the production of temporary ECM that forms a scaffold for lung regeneration. During the resolution phase of the inflammatory response, the remaining leukocytes, particularly macrophages, clear the temporary ECM scaffolds and extracellular debris (Gibbons et al. 2011; Liang et al. 2012), and possibly also clear activated fibroblasts (Redente et al. 2014). This process of clearance is an essential step preceding tissue regeneration (Duffield et al. 2013). Pulmonary fibrosis arises as a result of the dysregulation of wound repair processes (Duffield et al. 2013). In the remainder of the chapter, we discuss the involvement of tissue cells, leukocytes, and inflammatory mediators in the development of pulmonary fibrosis.

2.7 The Role of Granulocyte Responses in the Induction of Lung Injury and Fibrosis

In the early stages of lung injury, granulocytes such as neutrophils and eosinophils infiltrate the injured lung (Wynn 2011). Neutrophils play a major role in the priming of acute inflammatory responses through the production of proinflammatory cytokines, reactive oxygen species, and reactive nitrogen species (Dostert et al. 2008; Hasan et al. 2013; Kolaczkowska and Kubes 2013). Neutrophils also promote tissue remodelling and directly degrade elastic fibers and the basement membrane through the production of proteases such as neutrophil elastase (Kang et al. 2001). Eosinophils enhance the recruitment of other inflammatory cells, such as effector T cells, and exacerbate inflammatory responses, particularly in allergic lung inflammation (Humbles et al. 2004). In addition, eosinophils promote fibrotic responses by producing the profibrotic cytokines IL-13 and TGF-β1 (Huaux et al. 2003; Minshall et al. 1997). Collectively, these granulocytes are likely to play important roles in the induction and progression of fibrotic responses following acute lung injury.

2.8 The Role of Macrophage Subsets in Pulmonary Fibrosis

Following the infiltration of granulocytes into the lung in acute lung injury, monocyte-derived macrophages (MMs) from the bone marrow infiltrate the lung (Forbes and Rosenthal 2014). In parallel with the accumulation of granulocytes and MMs, alveolar macrophages (AMs) become activated after phagocytosing the irritants and cellular debris that are present during lung injury (Hussell and Bell 2014). These macrophages have been shown to ameliorate or exacerbate pulmonary fibrosis in mice in a context-dependent manner (Gharaee-Kermani et al. 2003; Liang et al. 2012; Moore et al. 2001; Murray and Wynn 2011; Wynn 2011; Wynn and Barron 2010). In the bleomycin-induced acute model of pulmonary fibrosis, MMs and AMs promote lung inflammation and the activation of tissue cells in the early stages of pulmonary fibrosis through the secretion of proinflammatory mediators and growth factors such as TNF-α, IL-1β, IL-6, TGF-β, platelet derived growth factor, insulin-like growth factor-1, and vascular endothelial growth factor (Wynn et al. 2013). Because MMs and AMs are a major source of collagenolytic enzyme matrix metalloproteinases (MMPs), the infiltration of MMs and activation of AMs results in the disruption of lung architecture (Hussell and Bell 2014; Wynn and Barron 2010). On the other hand, MMs and AMs also suppress excessive damage and ECM deposition in the lungs through the secretion of anti-inflammatory mediators (Gibbons et al. 2011; Redente et al. 2014; Tighe et al. 2011), the clearance of fibrous connective tissue, and possibly through the clearance of proinflammatory extracellular debris, such as apoptotic cells (Liang et al. 2012). Indeed, depletion of AMs in the resolution phase of bleomycin-induced pulmonary fibrosis delays the clearance of fibrotic lesions (Gibbons et al. 2011). In a model of silica-induced pulmonary fibrosis, we recently demonstrated that MMs continuously infiltrate the silica-treated lungs through to the chronic phase of the disease. MMs selectively accumulated in the periphery of silica-induced fibrotic lesions, and notably, MMs limited the expansion of the fibrotic area and the induction of diffuse pulmonary fibrosis (Shichino et al. 2015). We also found that MMs suppress the expression of human IPF-related genes, particularly tissue remodelling-related genes, in the silica-treated lungs. In addition, the changes in the expression of IPF-related genes appear to be suppressed by MM-derived inflammatory mediators (unpublished observations). These observations suggest that MMs and AMs exert pathogenic effects at the onset of lung injury, but might exert protective effects during the development of chronic pulmonary fibrosis. These insights into the role of MMs in pulmonary fibrosis raise concerns about the potential adverse effects of MM-targeted therapies for pulmonary fibrosis, including anti-CCL2 therapy.

At least two possible hypotheses could explain the protective role of macrophages during chronic pulmonary fibrosis. One hypothesis is that macrophages undergo phenotypic conversion from a proinflammatory/profibrotic pattern to an anti-inflammatory/antifibrotic pattern. The anti-inflammatory phenotype is characterised by the increased expression of IL-10, arginase-1, and RELMα (Murray and Wynn 2011). The anti-fibrotic features of this phenotype include increased expression of MMP9, MMP12, and MMP13, and/or the increased capacity to phagocytose extracellular debris (Murray and Wynn 2011). Indeed, in the CCl_4-induced liver fibrosis model, MMs and resident Kupffer cells promote the resolution of fibrosis (Duffield et al. 2005; Mitchell et al. 2009) and switch from a proinflammatory/profibrotic phenotype to an anti-inflammatory/antifibrotic phenotype (Mitchell et al. 2009; Ramachandran et al. 2012). Another hypothesis is that there are dual roles for inflammatory mediators in lung fibrosis and repair. For example, TNF-α, one of the major proinflammatory mediators produced by MMs, not only promotes inflammation but also suppresses collagen production by fibroblasts (Inagaki et al. 1995; Siwik et al. 2000). In addition, a reduction in MM numbers or TNF-α inhibition leads to disrupted organisation of the granuloma in *M. tuberculosis*-infected mouse lungs (Chakravarty et al. 2008; Peters et al. 2001), consistent with our observations in the silica-induced pulmonary fibrosis model (Shichino et al. 2015). Moreover, the resolution of bleomycin-induced pulmonary fibrosis is delayed in TNF knockout mice (Redente et al. 2014). These observations suggest that overall, MMs may exert a protective effect on the pathological course of pulmonary fibrosis, possibly due to MM-associated 'beneficial inflammation' (Fig. 2.3). The molecular mechanism(s) that underlie the protective properties of MMs remain poorly understood. Further studies on the protective aspects of MMs may support the development of novel therapeutic strategies for chronic pulmonary fibrosis.

2.9 Adaptive Immune Cell Responses in Pulmonary Fibrosis

After the acute inflammatory phase of pulmonary fibrosis, effector T cells and B cells infiltrate into the injured lung and modulate inflammatory and fibrotic responses. $CD4^+$ T_H17 cells and T_H2 cells produce IL-17A and IL-13, respectively (Lo Re et al. 2013). Because IL-17A has proinflammatory properties and induces TGF-β1 expression (Wilson et al. 2010), and because IL-13 has profibrotic effects (Chiaramonte et al. 1999), these T-cell subsets are thought to promote lung injury and fibrosis. Conversely, $CD4^+$ T_H1 cells play a protective role in pulmonary fibrosis because they produce the antifibrotic cytokine IFN-γ (Giri et al. 1986). On the other hand, there are mixed reports regarding the effects of $CD4^+$ regulatory

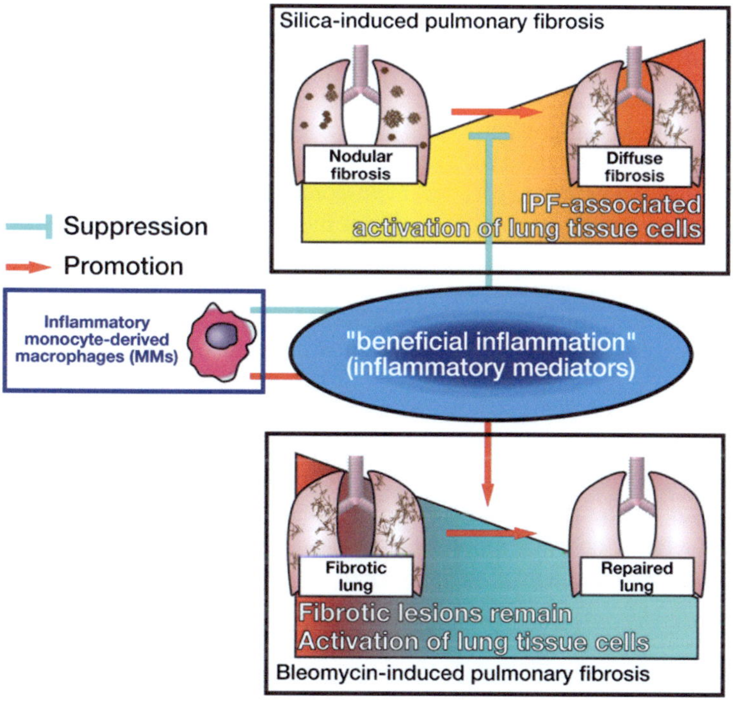

Fig. 2.3 Role of MMs in the progression/resolution of PF. MMs inhibit the development of diffuse PF and IPF-related transcriptomic signatures in silica-induced chronic PF. MMs also accelerate the resolution of bleomycin-induced self-limiting PF. Putative MM-mediated 'beneficial inflammation' in PF might mediate the protective effect of MMs by suppressing IPF-related tissue cell activation

T cells (T_{reg}) in pulmonary fibrosis. Because T_{reg} produce both the anti-fibrotic cytokine IL-10 (Kitani et al. 2003; Wilson et al. 2010) and the profibrotic cytokine TGF-β1 (Boveda-Ruiz et al. 2013), the mixed effects of T_{reg} might be explained by the predominance of these cytokines in different aspects of fibrotic responses. B cells also have mixed effects on the progression of pulmonary fibrosis. B cells accumulate in the lungs in bleomycin- or silica-induced pulmonary fibrosis, and fibrosis is alleviated in B-cell–deficient mice (Arras et al. 2006; Komura et al. 2008). On the other hand, the protective effect of IL-9 overexpression in silica-induced pulmonary fibrosis is negated in B-cell–deficient mice (Arras et al. 2006). These observations suggest a possible role for B cells in the progression of pulmonary fibrosis, the mechanisms of which remain largely unknown. Thus, overall, depending on their subset, T cells and B cells may influence the course of fibrotic responses. However, as discussed below, their involvement in the progression of irreversible pulmonary fibrosis is only minor in some experimental models.

2.10 Requirement for Leukocyte Subsets in Pulmonary Fibrosis: Lessons from Depletion Studies

Numerous studies using experimental models of pulmonary fibrosis have demonstrated that lung leukocyte subsets modulate lung injury, repair, and fibrosis (Wynn 2011). On the other hand, IPF, the most common type of idiopathic interstitial pneumonia, progresses even with less inflammatory cell infiltration than is observed in other types of interstitial pneumonia (Raghu et al. 2011). IPF is usually poorly responsive to various anti-inflammatory and immunosuppressive therapies including corticosteroids, anti-TNF biologics, and cyclosporin A (Raghu et al. 2011). The poor response of IPF to anti-inflammatory and immunosuppressive therapies raises doubts about the functional consequences of inflammatory leukocyte infiltration in the progression of chronic pulmonary fibrosis. Several types of immunodeficient mice, such as NOD/SCID and $Rag1^{-/-}$ mice lacking T and B cells and $Il5^{-/-}$ mice lacking eosinophils, still develop experimental pulmonary fibrosis (Hao et al. 2000; Helene et al. 1999; Hubbard 1989). In addition, the depletion of neutrophils or NK/NKT cells in the acute inflammatory phase of the disease does not hinder the development of experimental pulmonary fibrosis, and bleomycin-induced pulmonary fibrosis is not completely prevented by the depletion of AMs in the inflammatory or fibrotic phase (Beamer et al. 2010; Clark and Kuhn 1982; Gavett et al. 1992; Gibbons et al. 2011). Moreover, we recently found that the degree of infiltration into the lungs of $CD4^+$ T cells, $CD8^+$ T cells, eosinophils, and NK cells, and the number of AMs present in the lungs, were equivalent between a self-limiting bleomycin-induced pulmonary fibrosis model and a progressive silica-induced pulmonary fibrosis model, suggesting that these cells do not alter the course of the disease (Fig. 2.4, unpublished observations and Shichino et al. 2015). In contrast, MM infiltration increased to a greater extent in the silica model compared to the bleomycin model (Fig. 2.4; Shichino et al. 2015). However, $Ccr2^{-/-}$ mice, in which the infiltration of MMs into the lung is deficient, still displayed progression to silica- and bleomycin-induced pulmonary fibrosis (Shichino et al. 2015). The necessity for the development of pulmonary fibrosis of other leukocyte subsets such as CCR2-independent monocyte-derived macrophages, interstitial macrophages, and lung-resident dendritic cells remains unclear due to the lack of subset-specific depletion systems.

Recently, it was shown that fetal-liver–chimeric mice with CSF1R-deficient haematopoietic cells lack CCR2-independent monocyte-derived macrophages in their peripheral tissues (Alexander et al. 2014). In addition, long-term administration of an anti-CSF1R antibody also depletes tissue-resident macrophages (including macrophages located in the interstitium), but has minimal effect on dendritic cell subsets in most lymphoid tissues (MacDonald et al. 2010). These systems may present useful tools for investigating the role of CCR2-independent macrophages in the development of pulmonary fibrosis. Nevertheless, the experimental data available to date suggest that although adaptive immune cells, eosinophils, neutrophils, MMs, AMs, NK cells, and NKT cells in the lung may be involved in the

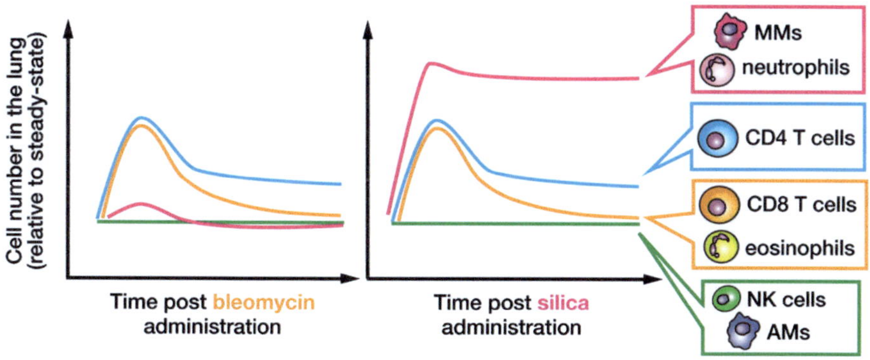

Fig. 2.4 Leukocyte subset kinetics differ between bleomycin-induced PF and silica-induced PF. The total number of $CD4^+$ T cells, $CD8^+$ T cells, eosinophils, NK cells, and AMs does not differ significantly between the two models. In contrast, MM and neutrophil numbers are dramatically elevated in the silica-treated lung, but not in the bleomycin-treated lung

pathogenesis of pulmonary fibrosis, these cells might not be essential for the progression of pulmonary fibrosis.

2.11 Conclusions and Future Perspectives

In recent years, the controversy surrounding the major source of myofibroblasts in fibrosis has been narrowed down to two candidates: tissue-resident fibroblasts and pericytes. However, the relative contributions of these potential precursors to the myofibroblast population remains elusive across a range of fibrosis models. In addition, there is a need for further characterisation of the functional heterogeneity within the activated fibroblast population, which is currently represented by α-SMA^+ myofibroblasts. Differences in inflamed tissue microenvironments might influence the differentiation or activation of fibroblasts and determine whether the wound healing process skews towards fibrosis or regeneration. Although interventions targeting specific subsets of inflammatory leukocytes have not been sufficient to ameliorate experimental pulmonary fibrosis, further elucidation of the spatio-temporal regulation of fibrotic responses by the various leukocyte subsets may reveal new points of therapeutic intervention for chronic inflammation-associated pulmonary fibrosis.

References

Alexander KA, Flynn R, Lineburg KE, Kuns RD, Teal BE, Olver SD, Lor M, Raffelt NC, Koyama M, Leveque L, Le Texier L, Melino M, Markey KA, Varelias A, Engwerda C, Serody JS, Janela B, Ginhoux F, Clouston AD, Blazar BR, Hill GR, MacDonald KP (2014)

CSF-1-dependant donor-derived macrophages mediate chronic graft-versus-host disease. J Clin Invest 124:4266–4280
Arras M, Louahed J, Simoen V, Barbarin V, Misson P, van den Brule S, Delos M, Knoops L, Renauld JC, Lison D, Huaux F (2006) B lymphocytes are critical for lung fibrosis control and prostaglandin E2 regulation in IL-9 transgenic mice. Am J Respir Cell Mol Biol 34:573–580
Beamer CA, Migliaccio CT, Jessop F, Trapkus M, Yuan D, Holian A (2010) Innate immune processes are sufficient for driving silicosis in mice. J Leukoc Biol 88:547–557
Bonner JC, Osornio-Vargas AR, Badgett A, Brody AR (1991) Differential proliferation of rat lung fibroblasts induced by the platelet-derived growth factor-AA, -AB, and -BB isoforms secreted by rat alveolar macrophages. Am J Respir Cell Mol Biol 5:539–547
Boveda-Ruiz D, D'Alessandro-Gabazza CN, Toda M, Takagi T, Naito M, Matsushima Y, Matsumoto T, Kobayashi T, Gil-Bernabe P, Chelakkot-Govindalayathil AL, Miyake Y, Yasukawa A, Morser J, Taguchi O, Gabazza EC (2013) Differential role of regulatory T cells in early and late stages of pulmonary fibrosis. Immunobiology 218:245–254
Bucala R, Spiegel LA, Chesney J, Hogan M, Cerami A (1994) Circulating fibrocytes define a new leukocyte subpopulation that mediates tissue repair. Mol Med 1:71–81
Cavarra E, Carraro F, Fineschi S, Naldini A, Bartalesi B, Pucci A, Lungarella G (2004) Early response to bleomycin is characterized by different cytokine and cytokine receptor profiles in lungs. Am J Physiol Lung Cell Mol Physiol 287:L1186–L1192
Chakravarty SD, Zhu G, Tsai MC, Mohan VP, Marino S, Kirschner DE, Huang L, Flynn J, Chan J (2008) Tumor necrosis factor blockade in chronic murine tuberculosis enhances granulomatous inflammation and disorganizes granulomas in the lungs. Infect Immun 76:916–926
Chiaramonte MG, Donaldson DD, Cheever AW, Wynn TA (1999) An IL-13 inhibitor blocks the development of hepatic fibrosis during a T-helper type 2-dominated inflammatory response. J Clin Invest 104:777–785
Clark JG, Kuhn C 3rd (1982) Bleomycin-induced pulmonary fibrosis in hamsters: effect of neutrophil depletion on lung collagen synthesis. Am Rev Respir Dis 126:737–739
Dostert C, Petrilli V, Van Bruggen R, Steele C, Mossman BT, Tschopp J (2008) Innate immune activation through Nalp3 inflammasome sensing of asbestos and silica. Science 320:674–677
Duffield JS, Forbes SJ, Constandinou CM, Clay S, Partolina M, Vuthoori S, Wu S, Lang R, Iredale JP (2005) Selective depletion of macrophages reveals distinct, opposing roles during liver injury and repair. J Clin Invest 115:56–65
Duffield JS, Lupher M, Thannickal VJ, Wynn TA (2013) Host responses in tissue repair and fibrosis. Annu Rev Pathol 8:241–276
Dulauroy S, Di Carlo SE, Langa F, Eberl G, Peduto L (2012) Lineage tracing and genetic ablation of ADAM12(+) perivascular cells identify a major source of profibrotic cells during acute tissue injury. Nat Med 18:1262–1270
Forbes SJ, Rosenthal N (2014) Preparing the ground for tissue regeneration: from mechanism to therapy. Nat Med 20:857–869
Fukuda Y, Ferrans VJ, Schoenberger CI, Rennard SI, Crystal RG (1985) Patterns of pulmonary structural remodelling after experimental paraquat toxicity. The morphogenesis of intraalveolar fibrosis. Am J Pathol 118:452–475
Gavett SH, Carakostas MC, Belcher LA, Warheit DB (1992) Effect of circulating neutrophil depletion on lung injury induced by inhaled silica particles. J Leukoc Biol 51:455–461
Gharaee-Kermani M, McCullumsmith RE, Charo IF, Kunkel SL, Phan SH (2003) CC-chemokine receptor 2 required for bleomycin-induced pulmonary fibrosis. Cytokine 24:266–276
Gibbons MA, MacKinnon AC, Ramachandran P, Dhaliwal K, Duffin R, Phythian-Adams AT, van Rooijen N, Haslett C, Howie SE, Simpson AJ, Hirani N, Gauldie J, Iredale JP, Sethi T, Forbes SJ (2011) Ly6Chi monocytes direct alternatively activated profibrotic macrophage regulation of lung fibrosis. Am J Respir Crit Care Med 184:569–581
Giri SN, Hyde DM, Marafino BJ Jr (1986) Ameliorating effect of murine interferon gamma on bleomycin-induced lung collagen fibrosis in mice. Biochem Med Metab Biol 36:194–197

Goritz C, Dias DO, Tomilin N, Barbacid M, Shupliakov O, Frisen J (2011) A pericyte origin of spinal cord scar tissue. Science 333:238–242

Hao HQ, Cohen DA, Jennings CD, Bryson JS, Kaplan AM (2000) Bleomycin-induced pulmonary fibrosis is independent of eosinophils. J Leukoc Biol 68:515–521

Hasan SA, Eksteen B, Reid D, Paine HV, Alansary A, Johannson K, Gwozd C, Goring KA, Vo T, Proud D, Kelly MM (2013) Role of IL-17A and neutrophils in fibrosis in experimental hypersensitivity pneumonitis. J Allergy Clin Immunol 131:1663–1673

Helene M, Lake-Bullock V, Zhu J, Hao H, Cohen DA, Kaplan AM (1999) T cell independence of bleomycin-induced pulmonary fibrosis. J Leukoc Biol 65:187–195

Henderson NC, Arnold TD, Katamura Y, Giacomini MM, Rodriguez JD, McCarty JH, Pellicoro A, Raschperger E, Betsholtz C, Ruminski PG, Griggs DW, Prinsen MJ, Maher JJ, Iredale JP, Lacy-Hulbert A, Adams RH, Sheppard D (2013) Targeting of alphav integrin identifies a core molecular pathway that regulates fibrosis in several organs. Nat Med 19:1617–1624

Higashiyama R, Moro T, Nakao S, Mikami K, Fukumitsu H, Ueda Y, Ikeda K, Adachi E, Bou–Gharios G, Okazaki I, Inagaki Y (2009) Negligible contribution of bone marrow-derived cells to collagen production during hepatic fibrogenesis in mice. Gastroenterology 137:1459–1466. e1451

Hinz B, Phan SH, Thannickal VJ, Prunotto M, Desmouliere A, Varga J, De Wever O, Mareel M, Gabbiani G (2012) Recent developments in myofibroblast biology: paradigms for connective tissue remodelling. Am J Pathol 180:1340–1355

Huaux F, Liu T, McGarry B, Ullenbruch M, Xing Z, Phan SH (2003) Eosinophils and T lymphocytes possess distinct roles in bleomycin-induced lung injury and fibrosis. J Immunol 171:5470–5481

Hubbard AK (1989) Role for T lymphocytes in silica-induced pulmonary inflammation. Lab Invest 61:46–52

Humbles AA, Lloyd CM, McMillan SJ, Friend DS, Xanthou G, McKenna EE, Ghiran S, Gerard NP, Yu C, Orkin SH, Gerard C (2004) A critical role for eosinophils in allergic airways remodelling. Science 305:1776–1779

Humphreys BD, Lin S-L, Kobayashi A, Hudson TE, Nowlin BT, Bonventre JV, Valerius MT, McMahon AP, Duffield JS (2010) Fate tracing reveals the pericyte and not epithelial origin of myofibroblasts in kidney fibrosis. Am J Pathol 176:85–97

Hung C, Linn G, Chow YH, Kobayashi A, Mittelsteadt K, Altemeier WA, Gharib SA, Schnapp LM, Duffield JS (2013) Role of lung pericytes and resident fibroblasts in the pathogenesis of pulmonary fibrosis. Am J Respir Crit Care Med 188:820–830

Hussell T, Bell TJ (2014) Alveolar macrophages: plasticity in a tissue-specific context. Nat Rev Immunol 14:81–93

Inagaki Y, Truter S, Tanaka S, Di Liberto M, Ramirez F (1995) Overlapping pathways mediate the opposing actions of tumor necrosis factor-alpha and transforming growth factor-beta on alpha 2(I) collagen gene transcription. J Biol Chem 270:3353–3358

Iwaisako K, Jiang C, Zhang M, Cong M, Moore-Morris TJ, Park TJ, Liu X, Xu J, Wang P, Paik YH, Meng F, Asagiri M, Murray LA, Hofmann AF, Iida T, Glass CK, Brenner DA, Kisseleva T (2014) Origin of myofibroblasts in the fibrotic liver in mice. Proc Natl Acad Sci U S A 111: E3297–E3305

Kang T, Yi J, Guo A, Wang X, Overall CM, Jiang W, Elde R, Borregaard N, Pei D (2001) Subcellular distribution and cytokine- and chemokine-regulated secretion of leukolysin/MT6-MMP/MMP-25 in neutrophils. J Biol Chem 276:21960–21968

King TE Jr (2005) Clinical advances in the diagnosis and therapy of the interstitial lung diseases. Am J Respir Crit Care Med 172:268–279

Kisseleva T, Uchinami H, Feirt N, Quintana-Bustamante O, Segovia JC, Schwabe RF, Brenner DA (2006) Bone marrow-derived fibrocytes participate in pathogenesis of liver fibrosis. J Hepatol 45:429–438

Kitani A, Fuss I, Nakamura K, Kumaki F, Usui T, Strober W (2003) Transforming growth factor (TGF)-beta1-producing regulatory T cells induce Smad-mediated interleukin 10 secretion that facilitates coordinated immunoregulatory activity and amelioration of TGF-beta1-mediated fibrosis. J Exp Med 198:1179–1188

Kleaveland KR, Velikoff M, Yang J, Agarwal M, Rippe RA, Moore BB, Kim KK (2014) Fibrocytes are not an essential source of type I collagen during lung fibrosis. J Immunol 193:5229–5239

Kolaczkowska E, Kubes P (2013) Neutrophil recruitment and function in health and inflammation. Nat Rev Immunol 13:159–175

Komura K, Yanaba K, Horikawa M, Ogawa F, Fujimoto M, Tedder TF, Sato S (2008) CD19 regulates the development of bleomycin-induced pulmonary fibrosis in a mouse model. Arthritis Rheum 58:3574–3584

Kramann R, Schneider RK, DiRocco DP, Machado F, Fleig S, Bondzie PA, Henderson JM, Ebert BL, Humphreys BD (2015) Perivascular Gli1+ progenitors are key contributors to injury-induced organ fibrosis. Cell Stem Cell 16:51–66

Kriz W, Kaissling B, Le Hir M (2011) Epithelial-mesenchymal transition (EMT) in kidney fibrosis: fact or fantasy? J Clin Investig 121:468–474

LeBleu VS, Taduri G, O'Connell J, Teng Y, Cooke VG, Woda C, Sugimoto H, Kalluri R (2013) Origin and function of myofibroblasts in kidney fibrosis. Nat Med 19:1047–1053

Liang J, Jung Y, Tighe RM, Xie T, Liu N, Leonard M, Gunn MD, Jiang D, Noble PW (2012) A macrophage subpopulation recruited by CC chemokine ligand-2 clears apoptotic cells in noninfectious lung injury. Am J Physiol Lung Cell Mol Physiol 302:L933–L940

Lo Re S, Lison D, Huaux F (2013) CD4+ T lymphocytes in lung fibrosis: diverse subsets, diverse functions. J Leukoc Biol 93:499–510

MacDonald KP, Palmer JS, Cronau S, Seppanen E, Olver S, Raffelt NC, Kuns R, Pettit AR, Clouston A, Wainwright B, Branstetter D, Smith J, Paxton RJ, Cerretti DP, Bonham L, Hill GR, Hume DA (2010) An antibody against the colony-stimulating factor 1 receptor depletes the resident subset of monocytes and tissue- and tumor-associated macrophages but does not inhibit inflammation. Blood 116:3955–3963

Mederacke I, Hsu CC, Troeger JS, Huebener P, Mu X, Dapito DH, Pradere JP, Schwabe RF (2013) Fate tracing reveals hepatic stellate cells as dominant contributors to liver fibrosis independent of its aetiology. Nat Commun 4:2823

Mercer PF, Johns RH, Scotton CJ, Krupiczojc MA, Konigshoff M, Howell DC, McAnulty RJ, Das A, Thorley AJ, Tetley TD, Eickelberg O, Chambers RC (2009) Pulmonary epithelium is a prominent source of proteinase-activated receptor-1-inducible CCL2 in pulmonary fibrosis. Am J Respir Crit Care Med 179:414–425

Minshall EM, Leung DY, Martin RJ, Song YL, Cameron L, Ernst P, Hamid Q (1997) Eosinophil-associated TGF-beta1 mRNA expression and airways fibrosis in bronchial asthma. Am J Respir Cell Mol Biol 17:326–333

Mitchell C, Couton D, Couty JP, Anson M, Crain AM, Bizet V, Renia L, Pol S, Mallet V, Gilgenkrantz H (2009) Dual role of CCR2 in the constitution and the resolution of liver fibrosis in mice. Am J Pathol 174:1766–1775

Moore BB, Paine R 3rd, Christensen PJ, Moore TA, Sitterding S, Ngan R, Wilke CA, Kuziel WA, Toews GB (2001) Protection from pulmonary fibrosis in the absence of CCR2 signaling. J Immunol 167:4368–4377

Murray PJ, Wynn TA (2011) Protective and pathogenic functions of macrophage subsets. Nat Rev Immunol 11:723–737

Noble PW (2005) Back to the future: historical perspective on the pathogenesis of idiopathic pulmonary fibrosis. Am J Respir Cell Mol Biol 33:113–120

Peters W, Scott HM, Chambers HF, Flynn JL, Charo IF, Ernst JD (2001) Chemokine receptor 2 serves an early and essential role in resistance to Mycobacterium tuberculosis. Proc Natl Acad Sci U S A 98:7958–7963

Pichery M, Mirey E, Mercier P, Lefrancais E, Dujardin A, Ortega N, Girard JP (2012) Endogenous IL-33 is highly expressed in mouse epithelial barrier tissues, lymphoid organs, brain, embryos, and inflamed tissues: in situ analysis using a novel Il-33-LacZ gene trap reporter strain. J Immunol 188:3488–3495

Raghu G, Collard HR, Egan JJ, Martinez FJ, Behr J, Brown KK, Colby TV, Cordier JF, Flaherty KR, Lasky JA, Lynch DA, Ryu JH, Swigris JJ, Wells AU, Ancochea J, Bouros D, Carvalho C, Costabel U, Ebina M, Hansell DM, Johkoh T, Kim DS, King TE, Kondoh Y, Myers J, Muller NL, Nicholson AG, Richeldi L, Selman M, Dudden RF, Griss BS, Protzko SL, Schunemann HJ, Comm AEJA (2011) An official ATS/ERSARS/ALAT statement: idiopathic pulmonary fibrosis: evidence-based guidelines for diagnosis and management. Am J Respir Crit Care Med 183:788–824

Ramachandran P, Pellicoro A, Vernon MA, Boulter L, Aucott RL, Ali A, Hartland SN, Snowdon VK, Cappon A, Gordon-Walker TT, Williams MJ, Dunbar DR, Manning JR, van Rooijen N, Fallowfield JA, Forbes SJ, Iredale JP (2012) Differential Ly-6C expression identifies the recruited macrophage phenotype, which orchestrates the regression of murine liver fibrosis. Proc Natl Acad Sci U S A 109:E3186–E3195

Redente EF, Keith RC, Janssen W, Henson PM, Ortiz LA, Downey GP, Bratton DL, Riches DW (2014) Tumor necrosis factor-alpha accelerates the resolution of established pulmonary fibrosis in mice by targeting profibrotic lung macrophages. Am J Respir Cell Mol Biol 50:825–837

Rock JR, Barkauskas CE, Cronce MJ, Xue Y, Harris JR, Liang J, Noble PW, Hogan BLM (2011) PNAS Plus: multiple stromal populations contribute to pulmonary fibrosis without evidence for epithelial to mesenchymal transition. Proc Natl Acad Sci 108:E1475–E1483

Rowe RG, Lin Y, Shimizu-Hirota R, Hanada S, Neilson EG, Greenson JK, Weiss SJ (2011) Hepatocyte-derived Snail1 propagates liver fibrosis progression. Mol Cell Biol 31:2392–2403

Shichino S, Abe J, Ueha S, Otsuji M, Tsukui T, Kosugi MK, Shand FHW, Hashimoto S, Suzuki HI, Morikawa T, Inagaki Y, Matsushima K (2015) Reduced supply of monocyte-derived macrophages leads to a transition from nodular to diffuse lesions and tissue cell activation in silica-induced pulmonary fibrosis in mice. Am J Pathol 185(11):2923–2938

Siwik DA, Chang DL, Colucci WS (2000) Interleukin-1beta and tumor necrosis factor-alpha decrease collagen synthesis and increase matrix metalloproteinase activity in cardiac fibroblasts in vitro. Circ Res 86:1259–1265

Soderblom C, Luo X, Blumenthal E, Bray E, Lyapichev K, Ramos J, Krishnan V, Lai-Hsu C, Park KK, Tsoulfas P, Lee JK (2013) Perivascular fibroblasts form the fibrotic scar after contusive spinal cord injury. J Neurosci 33:13882–13887

Taura K, Miura K, Iwaisako K, Osterreicher CH, Kodama Y, Penz-Osterreicher M, Brenner DA (2010) Hepatocytes do not undergo epithelial-mesenchymal transition in liver fibrosis in mice. Hepatology 51:1027–1036

Tighe RM, Liang J, Liu N, Jung Y, Jiang D, Gunn MD, Noble PW (2011) Recruited exudative macrophages selectively produce CXCL10 after noninfectious lung injury. Am J Respir Cell Mol Biol 45:781–788

Tomasek JJ, Gabbiani G, Hinz B, Chaponnier C, Brown RA (2002) Myofibroblasts and mechano-regulation of connective tissue remodelling. Nat Rev Mol Cell Biol 3:349–363

Tsukui T, Ueha S, Abe J, Hashimoto S, Shichino S, Shimaoka T, Shand FH, Arakawa Y, Oshima K, Hattori M, Inagaki Y, Tomura M, Matsushima K (2013) Qualitative rather than quantitative changes are hallmarks of fibroblasts in bleomycin-induced pulmonary fibrosis. Am J Pathol 183:758–773

Tsukui T, Ueha S, Shichino S, Inagaki Y, Matsushima K (2015) Intratracheal cell transfer demonstrates the profibrotic potential of resident fibroblasts in pulmonary fibrosis. Am J Pathol 185(11):2939–2948

Wilson MS, Madala SK, Ramalingam TR, Gochuico BR, Rosas IO, Cheever AW, Wynn TA (2010) Bleomycin and IL-1beta-mediated pulmonary fibrosis is IL-17A dependent. J Exp Med 207:535–552

Wynn TA (2011) Integrating mechanisms of pulmonary fibrosis. J Exp Med 208:1339–1350

Wynn TA, Barron L (2010) Macrophages: master regulators of inflammation and fibrosis. Semin Liver Dis 30:245–257

Wynn TA, Chawla A, Pollard JW (2013) Macrophage biology in development, homeostasis and disease. Nature 496:445–455

Chapter 3
Sema4A and Chronic Inflammation

Daisuke Ito and Atsushi Kumanogoh

Abstract Semaphorins were originally identified as axon guidance factors involved in the development of the nervous system. However, accumulating evidence indicates that several semaphorins, so-called 'immune semaphorins', have important functions in various phases of pathological and physiological immune responses, such as cell–cell interactions, immune-cell activation, differentiation, and trafficking/migration. For example, Sema4A, a class IV transmembrane semaphorin, plays crucial roles in dendritic cells, macrophages, and T cells. In addition, Sema4A is involved in various chronic inflammatory diseases including multiple sclerosis, asthma/allergic rhinitis/ atopic dermatitis, and retinitis pigmentosa.

Keywords Semaphorin • Sema4A • Multiple sclerosis • Asthma/allergic rhinitis/ atopic dermatitis • Retinitis pigmentosa

3.1 Introduction

Semaphorins, which are secretory and/or membrane proteins characterised by a conserved amino-terminal Sema domain, were originally identified as repulsive axon-guidance factors involved in neuronal development (Kolodkin et al. 1993; Tamagnone and Comoglio 2000; Pasterkamp and Kolodkin 2003). Vertebrate semaphorins, including immune semaphorins, are classified into classes III–VII, and invertebrate semaphorins are grouped into classes I and II. Semaphorins exert pleiotropic functions, playing roles in cardiogenesis (Toyofuku and Kikutani 2007; Toyofuku et al. 2008), angiogenesis (Toyofuku et al. 2007; Serini et al. 2003), tumor progression or suppression (Luchino et al. 2013), and bone homeostasis (Negishi-Koga et al. 2011; Hayashi et al. 2012).

D. Ito
Department of Respiratory Medicine, Allergy and Rheumatic Disease, Osaka University Graduate School of Medicine, Suita, Osaka, Japan

A. Kumanogoh (✉)
AMED-CREST, Suita, Osaka, Japan
e-mail: kumanogo@imed3.med.osaka-u.ac.jp

M. Miyasaka, K. Takatsu (eds.), *Chronic Inflammation*,
DOI 10.1007/978-4-431-56068-5_3

Recent findings indicate that several semaphorins, so-called 'immune semaphorins', are involved in various phases of immune responses, including immune-cell activation, differentiation, cell–cell interactions, and trafficking/migration (Kumanogoh and Kikutani 2013; Suzuki et al. 2008; Takamatsu et al. 2010). In particular, Sema4A, a class IV transmembrane semaphorin, plays crucial roles in dendritic cells (DCs), macrophages, and T cells. In addition, Sema4A has been implicated in various chronic inflammatory diseases. Here we review the current knowledge regarding Sema4A in the context of immune responses and chronic inflammation.

3.2 Sema4A in Immune Cells

3.2.1 *Sema4A and CD4⁺ T-Cell Function*

Sema4A is highly expressed in polarised T helper type 1 (Th1) cells and is important for helper T-cell differentiation (Fig. 3.1; Kumanogoh et al. 2005). In Sema4A-deficient mice, Th1 responses to heat-killed *Propionibacterium acne*, a Th1-inducing bacterium, are impaired in vivo. By contrast, T helper type 2 (Th2) responses against *Nippostrongylus brasiliensis*, a Th2-inducing intestinal

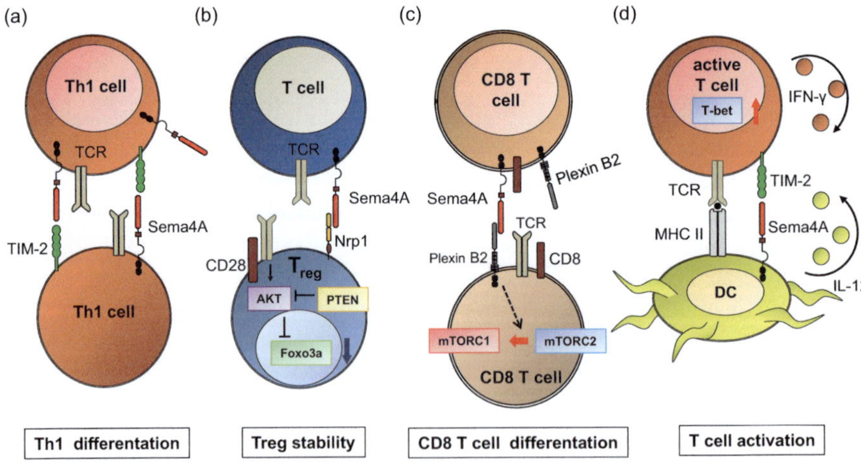

Fig. 3.1 (**a**) Sema4A promotes further Th1 differentiation via TIM-2. (**b**) Sema4A on conventional T cells ligates to Nrp-1 on Treg cells, leading to inhibition of Akt activation via recruitment of PTEN. Akt inactivation results in nuclear exclusion of FoxO molecules, which promotes the stability and increased function of Treg. (**c**) Sema4A on $CD8^+$ T cells promotes optimal $CD8^+$ T-differentiation via Plexin B2 by modulating mTOR-mediated signals. (**d**) When DCs encounter T cells, Sema4A on DCs directly stimulates T cells via TIM-2, leading to the optimal activation of antigen-specific T cells

nematode, are potentiated in these mice (Kumanogoh et al. 2005). TIM-2 appears to be the functional receptor for Sema4A, and the association between TIM-2 and Sema4A in Th1 cells negatively regulates Th2 cells (Kuchroo et al. 2003). However, there are some discrepancies between the phenotypes of Sema4A-deficient and Tim2-deficient mice, raising the possibility that there might be other binding partners of Sema4A. Indeed, Sema4A can bind plexin D1 and members of the plexin B subfamily (Toyofuku et al. 2007). Further studies are required to elucidate the details of the Sema4A-mediated pathway.

3.2.2 Sema4A/Nrp-1 and Regulatory T-Cell Stability

A recent study revealed a novel function of Sema4A associated with maintenance of regulatory T (Treg) cells (Fig. 3.1; Delgoffe et al. 2013). Sema4A directly interacts with the receptor neuropilin-1 (Nrp-1) expressed by Treg cells, and this interaction potentiates Treg-cell function and survival at inflammatory sites. In this pathway, Sema4A ligation of Nrp-1 restrains Akt phosphorylation in the cell body and at the immunologic synapse by recruiting phosphatase and tensin homologue (PTEN). This in turn promotes nuclear localisation of the transcription factor Foxo3a, which plays an important role in the development and programming of Treg cells (Fontenot et al. 2003; Hori et al. 2003). This Sema4A/Nrp-1-dependent pathway is crucial for maintenance of immune homeostasis, and is involved in inflammatory colitis and Treg-cell stability in tumor tissues in vivo. These findings suggest that Sema4A is a potential target for therapies aimed at limiting Treg-cell–mediated tumor-induced tolerance without inducing autoimmunity.

3.2.3 Sema4A and Dendritic-Cell Functions

Sema4A is expressed in dendritic cells, and Sema4A derived from both DCs and T cells is important for T-cell–mediated immunity (Fig. 3.1). Soluble Sema4A proteins and anti-Sema4A mAb bound to Sema4A on the surface of DCs enhance T-cell activation (Kumanogoh et al. 2002). Consistent with this, DCs derived from Sema4A-deficient mice stimulate allogeneic T cells poorly compared to wild-type DCs (Kumanogoh et al. 2002). By contrast, when $CD4^+$ T cells from Sema4A-deficient or wild-type mice are cultured with allogeneic DCs derived from wild-type mice, there are no differences in mixed lymphocyte reactions (Kumanogoh et al. 2002). These results suggest that DC derived Sema4A is directly and critically involved in the activation of T cells reactive to alloantigens on DCs.

3.2.4 *Sema4A and CD8$^+$ T-Cell Function*

Sema4A is abundantly expressed in naive CD8$^+$ T cells, but not in naive CD4$^+$ T cells. Indeed, Sema4A-deficient CD8$^+$ T cells exhibit impairments in cytokine production and induction of the effector molecules Granzyme B, Perforin, and FAS-L. Upon infection with ovalbumin-expressing *Listeria monocytogenes* (LM-OVA), which initiates an acute infection and activates CD8$^+$ T cells, pathogen-specific effector CD8$^+$ T-cell responses are significantly impaired in SEMA4A-deficient mice. Furthermore, Sema4A-deficient CD8$^+$ T cells exhibit reduced mTORC1 activity and elevated mTORC2 activity, suggesting that Sema4A is required for optimal activation of mTORC1 in CD8$^+$ T cells. Cytokine production and mTORC1 activity in Sema4A-deficient CD8$^+$ T cells can be restored by administration of recombinant Sema4A protein. In addition, Plexin B2 is the functional receptor for Sema4A in CD8$^+$ T cells (Fig. 3.1; Ito et al. 2015).

3.2.5 *Sema4A and Macrophage Function*

Soluble Sema4A protein stimulates macrophage migration in a dose-dependent manner. Moreover, the expression of Plexin D1 is enhanced in human macrophages stimulated with LPS and poly I:C, suggesting that Sema4A is potentially involved in macrophages during the inflammatory process. Consistent with this, blocking anti-Plexin D1 mAb inhibits the migratory effect of Sema4A in macrophages (Meda et al. 2012). This chemotactic effect, which is independent of inflammatory chemokines, might be direct. The role of Plexin D1 in mediating the promigratory effects of Sema4A is further corroborated by the finding that Sema4A inhibits VEGF-A–mediated endothelial cell migration via Plexin D1 expressed on the endothelium (Toyofuku et al. 2007). Of note, inhibition of Plexin D1 function strongly reduces the basal migration of macrophages, suggesting that endogenous Sema4A may be involved in promoting migration. Additional studies will be required to elucidate the signalling pathways triggered by Sema4A in macrophages and the possible role of Sema4A as an endogenous regulator of macrophage behavior.

3.3 Sema4A in Chronic Inflammation

3.3.1 *Sema4A and Multiple Sclerosis (MS)*

MS is a myelin-directed autoimmune disease and a leading cause of neurological disabilities. Both genetic elements and environmental factors are believed to participate in the pathogenesis of MS. When genetically predisposed individuals are exposed to an environmental trigger, myelin-specific T cells are activated, and MS is developed (McFarland and Martin 2007). Therefore, antigen presentation and activation and differentiation of subsequent CD4$^+$ T cells are important for the development of MS. Experimental autoimmune encephalomyelitis (EAE) is a widely used mouse model of MS (Mix et al. 2008). EAE is induced in susceptible animals by immunisation with myelin proteins such as myelin basic protein (MBP), myelin oligodendrocyte glycoprotein (MOG), and proteolipid protein (PLP) in combination with an adjuvant, and it can also be induced by the passive transfer of myelin antigen-reactive T cells. This model reproduces many of the clinical and histopathological features of MS.

Sema4A is expressed in dendritic and Th1 cells and plays important roles in the activation of Th cells and the differentiation of Th1 and Th17 cells (Kumanogoh et al. 2002; Kumanogoh et al. 2005). Consistent with this, the progression of MOG-induced EAE in wild-type mice can be ameliorated by injection of Sema4A monoclonal antibody at the same time as MOG immunisation. Infiltration of mononuclear inflammatory cells in the spinal cord is reduced in Sema4A antibody-treated mice. Moreover, CD4$^+$ T cells in draining lymph nodes exhibit markedly reduced responses to the MOG peptide (Kumanogoh et al. 2002). In patients with MS, serum Sema4A levels are significantly higher than in healthy subjects or those with other neurological diseases (ONDs), as determined by ELISA (Fig. 3.2; Nakatsuji et al. 2012).

Sema4A has multiple physiological sources. Monocytes and DCs derived from healthy control subjects express moderate amounts of Sema4A, whereas in

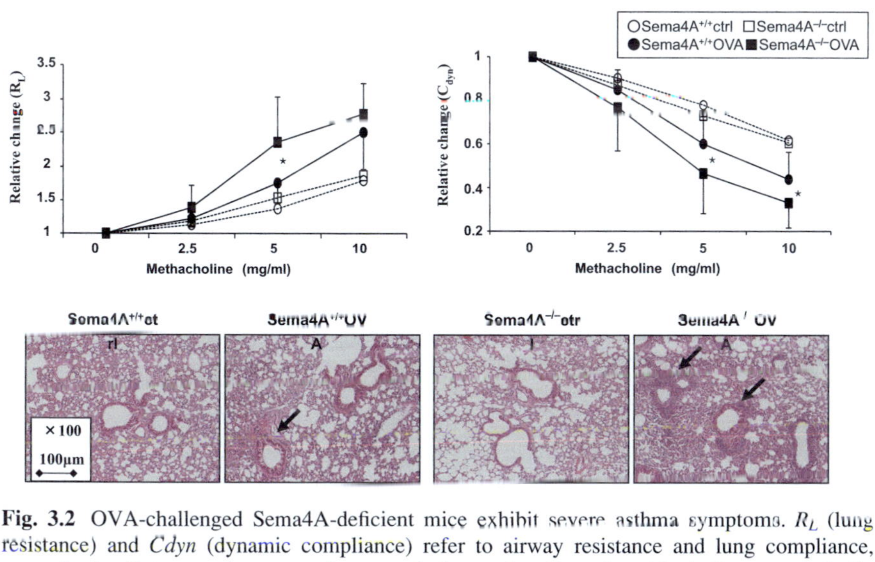

Fig. 3.2 OVA-challenged Sema4A-deficient mice exhibit severe asthma symptoms. R_L (lung resistance) and *Cdyn* (dynamic compliance) refer to airway resistance and lung compliance, respectively. Data are shown as changes relative to baseline. *Asterisks* indicate statistically significant differences (*$p < 0.05$). In H-E staining of asthma-induced lung tissue from Sema4A-deficient mice, infiltration of inflammatory cells around small airways and epithelium thickness are severe by *arrows* (100× magnification)

patients with MS these cells express significantly higher levels of the protein. The release of Sema4A can be inhibited by inhibitors of proteases such as matrix metalloproteinases (MMPs) and ADAM metalloproteinases, suggesting that these enzymes play an important role in releasing Sema4A from the cell surface. (Nakatsuji et al. 2012). Consistent with this, mRNA expression of metalloproteinases, such as MMPs and ADAM 10, is higher in peripheral blood mononuclear cells (PBMCs) from MS patients with high serum concentrations of Sema4A than those from healthy controls or MS patients with lower serum Sema4A levels. Collectively, these findings suggest that Sema4A, which is highly expressed in DCs and monocytes in patients with MS, is enzymatically shed in a subpopulation of the patients. There are several important hallmarks of MS patients with high Sema4A levels. One important feature is that MS patients with high serum Sema4A levels have a significantly higher proportion of IL-17–producing $CD4^+$ T cells than healthy subjects or patients with low serum Sema4A levels (Nakatsuji et al. 2012). In addition, high Sema4A patients have higher IL-2 levels. Therefore, Sema4A levels in patients with MS seem to be involved in Th17-mediated MS pathogenesis. Another important feature is that the disease course of MS patients with high Sema4A levels is significantly more severe. In addition, MS patients with high Sema4A levels are refractory to the first-line drug interferon IFN-β. These facts suggest that serum Sema4A may be useful as a biomarker of refractoriness to IFN-β, as well as a reliable aid for arriving at early diagnoses in the future.

3.3.2 Sema4A and Asthma/Allergic Rhinitis/Atopic Dermatitis

Allergic asthma, rhinitis, and atopic dermatitis are major allergic diseases, especially in developed countries, and the identification of effective therapies for these diseases is an important goal. These diseases are thought to be influenced by both genetic elements and environmental factors (Von Mutius 2009; Zhang et al. 2008), and they share several common features: eosinophil or mast cell infiltration in inflammatory sites, elevated levels of Th2-type cytokines, and elevated serum IgE levels (Cohn et al. 2004). Asthma develops due to inflammation in the lower airways in response to a wide variety of agents, such as airborne or dietary allergens, infectious agents, and chemical irritants. The features of allergic asthma are not limited to the typical allergic symptoms mentioned above, but also include mucus hypersecretion and airway hyperreactivity (AHR). Although antigen-presenting cells (APCs), Th17 cells, and epithelial cells play important roles in the pathogenesis of asthma, Th2 cells are thought to be essential inducers of allergic asthma (Kim et al. 2010; Lloyd and Hessel 2010). IL-4 produced by Th2 cells is required for immunoglobulin class-switching to IgE, as well as for the induction and maintenance of strong Th2 responses. IL-5 is also involved in the activation and recruitment of eosinophils, and IL-13 is critical for goblet cell hyperplasia and AHR.

As in allergic dermatitis, Sema4A is involved in the pathogenesis of allergic airway hyperreactivity. In a mouse model of OVA-specific experimental asthma, airway hyperreactivity in Sema4A-deficient mice is enhanced relative to that in wild-type mice. Further analysis of these mice revealed increased pulmonary eosinophil infiltration, and elevated levels of Th2 cytokines and IgE in bronchoalveolar lavage (BAL) fluid. These data demonstrate that Sema4A plays crucial roles in regulation of Th2-driven lung pathophysiology (Fig. 3.2; Nkyimbeng-Takwi et al. 2012; Morihana et al. 2013). Indeed, several reports have shown that immunological Th1/Th2 manipulation may be used to treat allergic diseases (Wegmann 2009). Many of these therapeutic effects were due to enhanced Th1 and suppressed Th2 immunity. Therefore, Sema4A may be useful for improvement of the diagnosis, prevention, and treatment of the allergic asthmatic diseases.

3.3.3 Sema4A and Inflammatory Bowel Disease

Inflammatory bowel disease (IBD), including Crohn's disease (CD) and ulcerative colitis (UC), is a group of inflammatory conditions of the intestinal tract that are considered to be autoimmune diseases. All of these diseases have typical clinical, histopathological, and endoscopic features. In addition, both diseases are characteried by T-cell overactivation because T-cell regulatory functions are downregulated and fail to suppress T effector cell responses (Geremia et al. 2014). In CD, upregulated activation of Th1 and Th17 cells is important for the induction of inflammation in bowel lamina propria. These cells react to bacterial bowel antigens when primary local immune responses fail to eliminate these antigens. This situation leads to the intensive and continuous production of proinflammatory cytokines.

Serum levels of Sema4A in patients with CD or UC were significantly lower than those in control subjects. In addition, Sema4A was abundantly expressed in lymphocytes of the lamina propria of CD and UC patients, although it was not expressed in patients with diverticulitis or in normal individuals (Vadasz et al. 2015). The association between Sema4A and IBDs has not been explained in detail. However, targeting Sema4A in IBD treatment might confer clinical benefits.

3.3.4 Sema4A and Retinal Degeneration

Retinitis pigmentosa (RP) is an inherited degenerative eye disease, caused by abnormalities of the photoreceptors (PRs) or the retinal pigment epithelium (RPE), that leads to severe vision impairment and often blindness. More than

100 associated mutations have been discovered in RP patients. Light exposure is the most obvious factor that makes PRs vulnerable to degeneration. Visible and ultraviolet light ionise biomolecules, and oxygen enhances this effect. Consequently, retinal damage can occur when light acts on photosensitising molecules including retinoids, and generates reactive oxygen and nitrogen species (RONS). Light damage to PRs is caused by the release of all-*trans* retinal from light-activated rhodopsin (Sun and Nathans 2001; Travis et al. 2007). Photoexcitation of all-*trans* retinal produces singlet dioxygen, which can lead to photo-oxidative damage. If gene mutations that affect the visual cycle block the recycling of all-*trans* retinal to 11-*cis* retinal, toxic bis-retinoids and adducts steadily accumulate over the course of aging.

In Sema4A-deficient mice, the outer segments of PRs are disrupted at P14 and completely lost by P28 (Fig. 3.3; Toyofuku et al. 2012). In response to illumination, Sema4A-deficient retinas exhibited a dramatic increase in the number of apoptotic cells in the outer nuclear layer before recovering to basal levels. Sema4A is expressed in RPE and bound to prosaposin, which is associated with procathepsin D in the Golgi membrane (Gopalakrishnan et al. 2004) and can be targeted to lysosomes (Kishimoto et al. 1992; Benes et al. 2008) or secreted into the extracellular space. Such secreted lysosomal precursor proteins are antiapoptotic in various neuronal populations (O'Brien et al. 1994; Benes et al. 2008). Oxidative stress caused by H_2O_2 treatment results in prosaposin transport to the cell periphery in wild-type RPE cells, but not in Sema4A-deficient RPE cells. Via its intercellular region, Sema4A binds to a complex of Rab11 and the adaptor protein FIP2 more

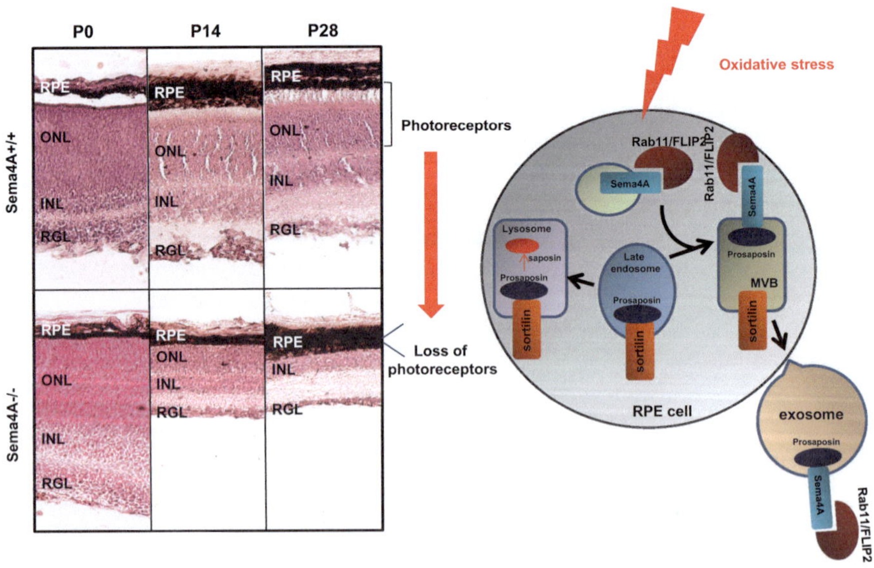

Fig. 3.3 (**a**) Sema4A$^{-/-}$ retinas exhibit progressive loss of the outer nuclear layer after exposure to light. (**b**) Sorting of prosaposin to the exosomal pathway is dependent on preferential binding of prosaposin to Sema4A and the Rab11/FIP2 endosomal sorting machinery

effectively than to FIP2 alone; Rab11 is involved in Sema4A-mediated prosaposin transport to the cell periphery under oxidative stress. In response to H_2O_2, prosaposin-containing vesicles are transported to the cell periphery via Sema4A/Rab11-mediated transport machinery (Fig. 3.3; Toyohuku et al. 2012). Thus, uncontrolled H_2O_2 production or defective H_2O_2 metabolism is likely to result in chronic inflammation (van der Vliet and Janssen-Heininger 2014). Sema4A plays a crucial role for protection against H_2O_2.

Three mutations in the Sema4A gene have been identified in patients with retinal degenerative diseases, including D345H, F350C, and R713Q (Abid et al. 2006). In addition, the expression of Sema4A (F350C) caused severe retinal degeneration in a series of knock-in mouse lines carrying mutated alleles of Sema4A (Nojima et al. 2013). In the RPE, Sema4A (F350C) tends to aggregate, and the resultant mislocalisation of Sema4A protein may lead to the impaired endosomal sorting of molecules such as prosaposin. Notably, virus-mediated gene transfer of Sema4A into RPE in neonatal Sema4A-deficient mice successfully prevents retinal degeneration for at least 4 months after injection. Thus, it is possible that Sema4A might be efficacious as a replacement gene therapy in patients with retinal degenerative diseases.

3.4 Perspective

As described above, Sema4A is shown to play crucial roles in the pathogenesis of chronic inflammatory disorders such as MS, IBD, and allergic diseases. Then, the question arises in which phases of chronic inflammation – induction, progression, or resolution phases - Sema4A is involved. Cumulative evidence indicates that Sema4A is involved in activation and differentiation of both $CD4^+$ and $CD8^+$ T cells. It thus appears that Sema4A is relevant to the induction phase of inflammation immune reactions. Indeed, Sema4A-deficent T cells showed impaired activation in response to TCR stimulation (Kumanogoh et al. 2005; Ito et al. 2015). On the other hand, several studies imply that Sema4A is also involved in the progression of chronic inflammation. For instance, the expression of Sema4A is shown to be upregulated in a later phase of $CD8^+$T cell activation (Ito et al. 2015). In addition, serum levels of Sema4A are elevated in patients with MS (Nakatsuji et al. 2012).

Notably, our recent finding indicates that Sema4A promotes a shift in mTOR-mediated signaling from mTORC2 to mTORC1 in $CD8^+$ T cells (Ito et al. 2015). In addition, our preliminary analysis found that the expression of Sema4A is dependent on mTOR-mediated signalling (unpublished data), showing a vicious cycle between Sema4A and mTOR-mediated signals in the course of chronic inflammation. Consistent with this scenario, it has been demonstrated that mTOR-mediated signals play a crucial role in chronic inflammatory disorders (Pollizzi and Powell 2014; Yoshizaki et al. 2010). To address these issues, careful and definitive evaluation by developing chronic inflammation experimental models and by using clinical samples will be required. Further studies will provide a clue to develop therapeutic targets for human chronic inflammatory diseases.

References

Abid A, Ismail M, Mehdi SQ, Khaliq S (2006) Identification of novel mutations in the SEMA4A gene associated with retinal degenerative diseases. J Med Genet 43:378–381

Benes P, Vetvicka V, Fusek M (2008) Cathepsin D-many functions of one aspartic protease. Crit Rev Oncol Hematol 68:12–28

Cohn L, Elias JA, Chupp GL (2004) Asthma: mechanisms of disease persistence and progression. Annu Rev Immunol 22:789–815

Delgoffe GM, Woo SR, Turnis ME, Gravano DM, Guy C, Overacre AE, Bettini ML, Vogel P, Finkelstein D, Bonnevier J, Workman CJ, Vignali DA (2013) Stability and function of regulatory T cells is maintained by a neuropilin-1-semaphorin-4a axis. Nature 501:252–256

Fontenot JD, Gavin MA, Rudensky AY (2003) Foxp3 programsthe development and function of CD41CD251 regulatory T cells. Nat Immunol 4:330–336

Geremia A, Biancheri P, Allan P, Corazza GR, Di Sabatino A (2014) Innate and adaptive immunity in inflammatory bowel disease. Autoimmun Rev 13:3–10

Gopalakrishnan MM, Grosch HW, Locatelli-Hoops S, Werth N, Smolenova E, Nettersheim M, Sandhoff K, Hasilik A (2004) Purified recombinant human prosaposin forms oligomers that bind procathepsin D and affect its autoactivation. Biochem J 383:507–515

Hayashi M, Nakashima T, Taniguchi M, Kodama T, Kumanogoh A, Takayanagi H (2012) Osteoprotection by semaphorin 3A. Nature 485:69–74

Hori S, Nomura T, Sakaguchi S (2003) Control of regulatory T cell development by the transcription factor Foxp3. Science 299:1057–1061

Ito D, Nojima S, Nishide M, Okuno T, Takamatsu H, Kang S, Kimura T, Yoshida Y, Morimoto K, Maeda Y, Hosokawa T, Toyofuku T, Ohshima J, Kamimura D, Yamamoto M, Murakami M, Morii E, Rakugi H, Isaka Y, Kumanogoh A (2015) mTOR complex signaling through the SEMA4A-plexin B2 axis is required for optimal activation and differentiation of CD8+ T cells. J Immunol 195(3):934–943

Kim HY, DeKruyff RH, Umetsu DT (2010) The many paths to asthma: phenotype shaped by innate and adaptive immunity. Nat Immunol 11:577–584

Kishimoto Y, Hiraiwa M, O'Brien JS (1992) Saposins: structure, function, distribution, and molecular genetics. J Lipid Res 33:1255–1267

Kolodkin AL, Matthes DJ, Goodman CS (1993) The semaphorin genes encode a family of transmembrane and secreted growth cone guidance molecules. Cell 31:1389–1399

Kuchroo VK, Umetsu DT, DeKruyff RH, Freeman GJ (2003) The TIM gene family: emerging roles in immunity and disease. Nat Rev Immunol 3:454–462

Kumanogoh A, Kikutani H (2013) Immunological functions of the neuropilins and plexins as receptors for semaphorins. Nat Rev Immunol 13:802–814

Kumanogoh A, Marukawa S, Suzuki K, Takegahara N, Watanabe C, Ch'ng E, Ishida I, Fujimura H, Sakoda S, Yoshida K, Kikutani H (2002) Class IV semaphorin Sema4A enhances Tcell activation and interacts with Tim-2. Nature 419:629–633

Kumanogoh A, Shikina T, Suzuki K, Uematsu S, Yukawa K, Kashiwamura S, Tsutsui H, Yamamoto M, Takamatsu H, Ko-Mitamura EP, Takegahara N, Marukawa S, Ishida I, Morishita H, Prasad DV, Tamura M, Mizui M, Toyofuku T, Akira S, Takeda K, Okabe M, Kikutani H (2005) Nonredundant roles of Sema4A in the immune system: defective T cell priming and Th1/Th2 regulation in Sema4A-deficient mice. Immunity 22:305–316

Lloyd CM, Hessel EM (2010) Functions of T cells in asthma: more than just T(H)2 cells. Nat Rev Immunol 10:838–848

Luchino J, Hocine M, Amoureux MC, Gibert B, Bernet A, Royet A, Treilleux I, Lécine P, Borg JP, Mehlen P, Chauvet S, Mann F (2013) Semaphorin 3E suppresses tumor cell death triggered by the plexin D1 dependence receptor in metastatic breast cancers. Cancer Cell 24:673–685

McFarland HF, Martin R (2007) Multiple sclerosis: a complicated picture of autoimmunity. Nat Immunol 8:913–919

Meda C, Molla F, De Pizzol M, Regano D, Maione F, Capano S, Locati M, Mantovani A, Latini R, Bussolino F, Giraudo E (2012) Semaphorin 4A exerts a proangiogenic effect by enhancing vascular endothelial growth factor-A expression in macrophages. J Immunol 188:4081–4092
Mix E, Meyer-Rienecker H, Zettl UK (2008) Animal models of multiple sclerosis for the development and validation of novel therapies – potential and limitations. J Neurol 255:7–14
Morihana T, Goya S, Mizui M, Yasui T, Prasad DV, Kumanogoh A, Tamura M, Shikina T, Maeda Y, Iwamoto Y, Inohara H, Kikutani H (2013) An inhibitory role for Sema4A in antigen-specific allergic asthma. J Clin Immunol 33:200–209
Nakatsuji Y, Okuno T, Moriya M, Sugimoto T, Kinoshita M, Takamatsu H, Nojima S, Kimura T, Kang S, Ito D, Nakagawa Y, Toyofuku T, Takata K, Nakano M, Kubo M, Suzuki S, Matsui-Hasumi A, Uto-Konomi A, Ogata A, Mochizuki H, Sakoda S, Kumanogoh A (2012) Elevation of sema4A implicates Th cell skewing and the efficacy of IFN-β therapy in multiple sclerosis. J Immunol 188:4858–4865
Negishi-Koga T, Shinohara M, Komatsu N, Bito H, Kodama T, Friedel RH, Takayanagi H (2011) Suppression of bone formation by osteoclastic expression of semaphorin 4D. Nat Med 17:1473–1480
Nkyimbeng-Takwi EH, Shanks K, Smith E, Iyer A, Lipsky MM, Detolla LJ, Kikutani H, Keegan AD, Chapoval SP (2012) Neuroimmune semaphorin 4A downregulates the severity of allergic response. Mucosal Immunol 5:409–419
Nojima S, Toyofuku T, Kamao H, Ishigami C, Kaneko J, Okuno T, Takamatsu H, Ito D, Kang S, Kimura T, Yoshida Y, Morimoto K, Maeda Y, Ogata A, Ikawa M, Morii E, Aozasa K, Takagi J, Takahashi M, Kumanogoh A (2013) A point mutation in Semaphorin 4A associates with defective endosomal sorting and causes retinal degeneration. Nat Commun 4:1406
O'Brien JS, Carson GS, Seo HC, Hiraiwa M, Kishimoto Y (1994) Identification of prosaposin as a neurotrophic factor. Proc Natl Acad Sci U S A 91:9593–9596
Pasterkamp RJ, Kolodkin AL (2003) Semaphorin junction: making tracks toward neural connectivity. Curr Opin Neurobiol 13:79–89
Pollizzi KN, Powell JD (2014) Integrating canonical and metabolic signalling programmes in the regulation of T cell responses. Nat Rev Immunol 14:435–446
Serini G, Valdembri D, Zanivan S, Morterra G, Burkhardt C, Caccavari F, Zammataro L, Primo L, Tamagnone L, Logan M, Tessier-Lavigne M, Taniguchi M, Püschel AW, Bussolino F (2003) Class 3 semaphorins control vascular morphogenesis by inhibiting integrin function. Nature 424:391–397
Sun H, Nathans J (2001) ABCR, the ATP-binding cassette transporter responsible for Stargardt macular dystrophy, is an efficient target of all-trans-retinal-mediated photooxidative damage in vitro. Implications for retinal disease. J Biol Chem 276:11766–11774
Suzuki K, Kumanogoh A, Kikutani H (2008) Semaphorins and their receptors in immune cell interactions. Nat Immunol 9:17–23
Takamatsu H, Takegahara N, Nakagawa Y, Tomura M, Taniguchi M, Friedel RH, Rayburn H, Tessier-Lavigne M, Yoshida Y, Okuno T, Mizui M, Kang S, Nojima S, Tsujimura T, Nakatsuji Y, Katayama I, Toyofuku T, Kikutani H, Kumanogoh A (2010) Semaphorins guide the entry of dendritic cells into the lymphatics by activating myosin II. Nat Immunol 11:594–600
Tamagnone L, Comoglio PM (2000) Signaling by semaphorin receptors: cell guidance and beyond. Trends Cell Biol 10:377–383
Toyofuku T, Kikutani H (2007) Semaphorin signaling during cardiac development. Adv Exp Med Biol 600:109–117
Toyofuku T, Yabuki M, Kamei J, Kamei M, Makino N, Kumanogoh A, Hori M (2007) Semaphorin-4A, an activator for T-cell-mediated immunity, suppresses angiogenesis via Plexin-D1. EMBO J 26:1373–1384

Toyofuku T, Yoshida J, Sugimoto T, Yamamoto M, Makino N, Takamatsu H, Takegahara N, Suto F, Hori M, Fujisawa H, Kumanogoh A, Kikutani H (2008) Repulsive and attractive semaphorins cooperate to direct the navigation of cardiac neural crest cells. Dev Biol 321:251–262

Toyofuku T, Nojima S, Ishikawa T, Takamatsu H, Tsujimura T, Uemura A, Matsuda J, Seki T, Kumanogoh A (2012) Endosomal sorting by Semaphorin 4A in retinal pigment epithelium supports photoreceptor survival. Genes Dev 26:816–829

Travis GH, Golczak M, Moise AR, Palczewski K (2007) Diseases caused by defects in the visual cycle: retinoids as potential therapeutic agents. Annu Rev Pharmacol Toxicol 47:469–512

Vadasz Z, Rainis T, Nakhleh A, Haj T, Bejar J, Halasz K, Toubi E (2015) The involvement of immune semaphorins in the pathogenesis of Inflammatory Bowel Diseases (IBDs). PLoS ONE 10:0125860

van der Vliet A, Janssen-Heininger YM (2014) Hydrogen peroxide as a damage signal in tissue injury and inflammation: murderer, mediator, or messenger? J Cell Biochem 115:427–435

Von Mutius E (2009) Gene-environment interactions in asthma. J Allergy Clin Immunol 123:3–11

Wegmann M (2009) Th2 cells as targets for therapeutic intervention in allergic bronchial asthma. Expert Rev Mol Diagn 9:85–100

Yoshizaki A, Yanaba K, Yoshizaki A, Iwata Y, Komura K, Ogawa F, Takenaka M, Shimizu K, Asano Y, Hasegawa M, Fujimoto M, Sato S (2010) Treatment with rapamycin prevents fibrosis in tight-skin and bleomycin-induced mouse models of systemic sclerosis. Arthritis Rheum 62:2476–2487

Zhang J, Paré PD, Sandford AJ (2008) Recent advances in asthma genetics. Respir Res 9:4

Chapter 4
MicroRNAs in Chronic Inflammation

Y. Ito, S. Mokuda, K. Miyata, T. Matsushima, and H. Asahara

Abstract MicroRNAs (miRNAs) are ~22-nucleotide single-stranded small non-coding RNAs that are highly conserved among eukaryotes. miRNAs play a central role in the posttranscriptional regulation of gene expression by inhibiting translation and/or by inducing mRNA degradation. Several recent studies have shown that miRNAs contribute to diverse physiological processes, including pathogenesis, through their role in gene suppression. In this chapter, we review the relationship between miRNAs and chronic inflammatory diseases, such as cancer and arthritis, and discuss the potential functions of miRNAs in chronic inflammation.

Keywords MicroRNA • Chronic inflammation • Osteoarthritis • Rheumatoid arthritis • miR-140 • miR-146a • miR-155

4.1 Introduction

miRNAs are 19–22-nucleotide long small noncoding RNAs that regulate many biological processes, including development and homeostasis, and are involved in the pathogenesis of various diseases (Esteller 2011; Kloosterman and Plasterk 2006; Ambros 2004). The first miRNA gene, *lin-4*, was identified in *Caenorhabditis elegans* by Ambros's research team (Lee et al. 1993). To date, more than 2000

Y. Ito • T. Matsushima
Department of Systems BioMedicine, Graduate School of Medical and Dental Sciences, Tokyo Medical and Dental University, Bunkyo, Tokyo 113-8519, Japan

S. Mokuda • K. Miyata
Department of Molecular and Experimental Medicine, The Scripps research Institute, San Diego, CA 92131, USA

H. Asahara (✉)
Department of Systems BioMedicine, Graduate School of Medical and Dental Sciences, Tokyo Medical and Dental University, Bunkyo, Tokyo 113-8519, Japan

Department of Molecular and Experimental Medicine, The Scripps research Institute, San Diego, CA 92131, USA

AMED-CREST, Tokyo 113-8519, Japan
e-mail: asahara.syst@tmd.ac.jp

M. Miyasaka, K. Takatsu (eds.), *Chronic Inflammation*,
DOI 10.1007/978-4-431-56068-5_4

miRNAs have been discovered within the human genome by RNA-sequencing and computational prediction (Kozomara and Griffiths-Jones 2014). Their function is to repress the expression of target genes by destabilisation and/or translation inhibition of target mRNAs (Bartel 2009). In order to do so, miRNAs act as RNA-induced silencing complexes (RISC), composed of miRNA and Argonaute (Ago) family protein. To target specific mRNAs, the intrinsic seed sequence of the respective miRNAs is used. After recruitment of a RISC complex (Bartel 2009), Ago inhibits the targeted mRNAs (Bartel 2009). This new paradigm will reveal novel aspects of the inflammation-related molecular network and should contribute to the development of new therapeutic tools for chronic inflammatory diseases.

4.2 MicroRNA Biogenesis

miRNAs have a unique synthesis process, as shown in Fig. 4.1 (Ceribelli et al. 2011). Many miRNAs have their own transcription initiation region, and some miRNAs are located on introns of other genes or reside as a cluster of transcriptional units. First, these are transcribed as primary miRNA (pri-miRNA), which is a long primary transcript containing one or more miRNAs. Nuclear RNase III Drosha then crops the pri-miRNA into a precursor miRNA (pre-miRNA), which has about 70 base pairs and a hairpin loop structure and is subsequently exported from the nucleus into the cytoplasm by Exportin 5 (Lee et al. 2003; Yi et al. 2003; Han et al. 2004). Finally, pre-miRNA is cleaved by cytoplasmic RNase III Dicer to form stem loops of short nucleotide duplexes, and either one of the single strands is

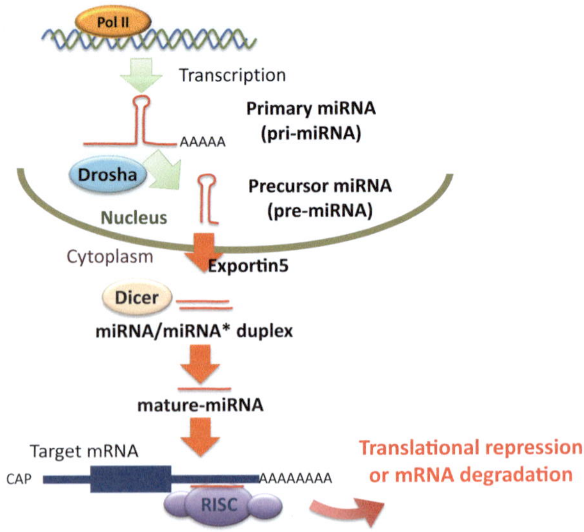

Fig. 4.1 miRNA biogenesis
The pri-miRNAs are transcribed as individual miRNA genes, and are cleaved into pre-miRNAs by Drosha. Pre-miRNAs are then exported to the cytoplasm by Exportin 5 and further processed into a miRNA/miRNA* duplex by Dicer. One of the strands is preferentially incorporated into RISC and guides it to target mRNAs

incorporated into an Ago protein complex, known as RISC. The single-strand RNA is a mature miRNA, which is able to bind a partial complementary sequence, mainly located on the 3'-untranslated region (3' UTR) of the target mRNAs, and induce its translation repression or degradation (Bernstein et al. 2001; Hutvagner et al. 2001; O'Connell et al. 2010).

Because Dicer knockout mice displayed severe growth arrest at the early embryonic stage (Bernstein et al. 2003), miRNAs are regarded as critical factors in several biological processes.

4.3 MicroRNAs in Cancer

To date, a number of reports have shown abnormalities in miRNA expression in cancer. The first report was made by Calin and his colleague in 2002. They indicated that miR-15 and miR-16 were located on chromosome 13q14, a region deleted in more than half the B-cell chronic lymphocytic leukemias (B-CLL), with both genes also deleted or downregulated in the majority of CLL cases (Calin et al. 2002).

The miR-17-92 cluster contains miR-17, miR-18a, miR-19a, miR-19b-1, miR-20a, and miR-92a-1, which are representative miRNAs considered as 'oncogenes'. *MYC* (v-myc avian myelocytomatosis viral oncogene homolog), one of the most famous oncogenes, promotes transcription of the miR-17-92 cluster, overexpressed in lymphoma or some solid tumours, such as those formed in lung cancer (Hayashita et al. 2005; He et al. 2005). Further analysis with miR-17-92 cluster transgenic mice showed an increase in lymphocytosis and lymphoma development. On the contrary, B-cell progenitor promoted apoptosis in knockout mice. These findings, acquired from in vivo analysis, confirmed the oncogenic function of the miR-17-92 cluster (Xiao et al. 2008; Sandhu et al. 2013). Subsequent reports identified its direct targets, such as tumour suppressing factor *PTEN* (phosphatase and tensin homologue deleted from chromosome 10), apoptotic inducer *BCL2L11* (BCL2-like 11), and cell cycle regulator *E2F1* (E2F transcription Factor 1), and described the mechanism through which the miR-17-92 cluster contributes to tumourigenesis (Xiao et al. 2008; Sylvestre et al. 2007; Woods et al. 2007).

Oncogenic factor *RAS* (rat sarcoma) is negatively regulated by the let-7 family, which is one of the most notorious miRNA families (Johnson et al. 2005). Analysis of cancerous and normal lung tissues indicated that expression of let-7 and *RAS* correlated negatively, also highlighting a potential oncogenic regulation of lung cancer through the control of their expression balance (Johnson et al. 2005).

Some reports showed that tumour suppressor *p53* regulates the expression of miR-34a (Chang et al. 2007; He et al. 2007; Raver-Shapira et al. 2007), contributing to *p53*-dependent apoptosis (Chang et al. 2007; Raver-Shapira et al. 2007) or induction of senescence (He et al. 2007). Overexpression of miR-34a reduced the cellular proliferation of HCT116 cells and several other cell lines, and its tumour-suppressive function was directly controlled via downregulation of E2F family or other cell-cycle regulators (Tazawa et al. 2007).

Because a considerably high number of reports focus on the role played by miRNA in cancer, only some of them are shown in this section. 'Circulating miRNA' has also been reported as a new concept describing how cell-free miRNA circulating in blood participates in crosstalks within cancer microenvironments (Kosaka et al. 2010; Srivastava et al. 2015). Further research on miRNAs will certainly lead to the establishment of novel strategies for the prevention and diagnosis of cancers as well as for the treatment of cancer patients.

4.4 MicroRNAs in Osteoarthritis

Osteoarthritis (OA) is the most common form of arthritis and degenerative joint disease, caused by chronic inflammation. OA is characterised by cartilage damage, which results in joint pain, swelling, and movement disorder. There are many reports on the alteration of miRNA expressions in OA chondrocytes.

It has been reported that miR-27a/b expressions decrease in OA chondrocytes (Tardif et al. 2009; Akhtar et al. 2010). The direct target of miR-27a/b is *MMP-13* (matrix metallopeptidase-13), the main proteinase responsible for aggrecan degradation in the articular cartilage. Therefore, they might contribute to maintain cartilage homeostasis (Tardif et al. 2009; Akhtar et al. 2010). The decreased expression of miR-199-3p was also detected in OA (Akhtar and Haqqi 2012). The direct target of miR-199a-3p is *COX-2* (cyclooxygenase-2), which might indicate its potential to suppress inflammations (Akhtar and Haqqi 2012). miR-558 expression was also shown to decrease in OA, its direct target being *COX-2* as well (Park et al. 2013). Significant reduction in the expression of miR-370 and miR-373 was also reported in OA chondrocytes (Song et al. 2015). This report indicated that miR-370 and miR-373 regulated OA pathogenesis through apoptosis inhibition and suppression of *MMP-13* expression by targeting *SHMT-2* (serine hydroxymethyltransferase-2) and *MECP-2* (methyl-CpG-binding protein-2), respectively (Song et al. 2015). Suppressed levels of miR-222 were also observed in OA chondrocytes (Song et al. 2014). miR-222 targets *HDAC-4*, that is involved in the regulation of *MMP-13* and chondrocyte apoptosis (Song et al. 2014).

On the contrary, elevated expression of miRNAs was also reported in OA. miR-145 was highly expressed in OA and its direct targets are *SOX9* (sex determining Y-box 9) and *SMAD3* (SMAD family member 3), significant factors for cartilage homeostasis (Yang et al. 2014; Martinez-Sanchez et al. 2012). miR-181b was also upregulated in OA chondrocytes, indirectly resulting in elevated expression level of *MMP-13* (Song et al. 2013). In addition, overexpression of

Table 4.1 Deregulated miRNA in osteoarthritis

Expression change	miRNA
Up-regulated	miR-145
	miR-181b
	miR-21
	miR-23b
	miR-22
	miR-29c
	miR-16
Down-regulated	miR-140
	miR-27a
	miR-27b
	miR-199a-3p
	miR-558
	miR-149
	miR-24
	miR-370
	miR-373
	miR-222

anti-miR-181b significantly reduced the cartilage destruction caused by surgical destabilisation of the medial meniscus (DMM) in mice (Song et al. 2013). These results suggest that miR-181b is an important miRNA in OA pathogenesis. Many additional reports on alteration of miRNA expressions in OA chondrocytes can be found (Table 4.1; Mirzamohammadi et al. 2014).

In our previous report, we detected a decreased expression of miR-140 in chondrocytes from OA patients' knee cartilages (Miyaki et al. 2009). We demonstrated considerable miRNA expression differences between human articular chondrocytes and mesenchymal stem cells. miR-140 exhibited the largest expression difference between these two cell types (Miyaki et al. 2009). miR-140 is induced by *SOX9*, the main chondrogenic factor in the cartilage growth layer, and controls cartilage homeostasis, by regulating *ADAMTS-5*, in order to protect against cartilage damage (Yamashita et al. 2012; Miyaki et al. 2010). In addition, miR-140$^{-/-}$ mice manifested a mild skeletal phenotype with a short stature and showed age-related OA-like changes characterised by proteoglycan loss and fibrillation of articular cartilage (Miyaki et al. 2010). In a surgical OA model with transection of medial meniscotibial ligament (MMTL), cartilage-specific miR-140 transgenic mice suppressed the cartilage degeneration (Miyaki et al. 2010). These results showed that miR-140 controls cartilage homeostasis and has potential as a therapeutic target of OA (Fig. 4.2).

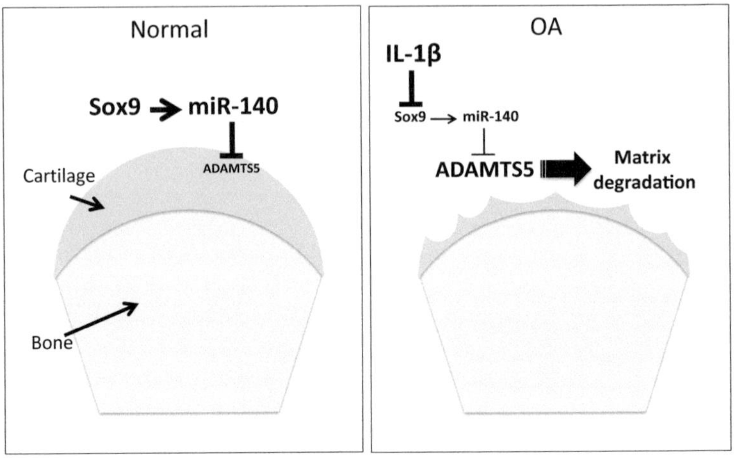

Fig. 4.2 Mechanism of cartilage homeostasis by miR-140
miR-140 is induced by *SOX9* and controls cartilage homeostasis by regulating *ADAMTS-5*, the proteinase responsible for aggrecan degradation in the articular cartilage

Table 4.2 Deregulated miRNA in rheumatoid arthritis

Expression change	miRNA
Up-regulated	miR-146a
	miR-155
	miR-133a
	miR-142
	miR-203
	miR-223
	miR-16
Down-regulated	miR-124a
	miR-34a-3p
	miR-363
	miR-498a

4.5 MicroRNAs in Rheumatoid Arthritis

Rheumatoid arthritis (RA) is a common chronic inflammatory disease characterised by hyperplastic synovial tissues and destruction of articular cartilages and adjacent bones. In RA, many previous works report on the alteration of miRNA expressions (Table 4.2; Ceribelli et al. 2011; Duroux-Richard et al. 2012). Therefore, miRNA might be involved in pathogenesis of RA.

In 2008, Stanczyk et al. reported for the first time that miR-146a and miR-155 were upregulated in RA synovial tissues and RA synovial fibroblasts (RASF), compared with OA (Stanczyk et al. 2008). miR-146a regulates inflammation by repressing *TRAF6* (tumour necrosis factor receptor associated factor 6) and *IRAK1* (IL-1 receptor-associated kinase 1), both downstream factors of Toll-like receptors

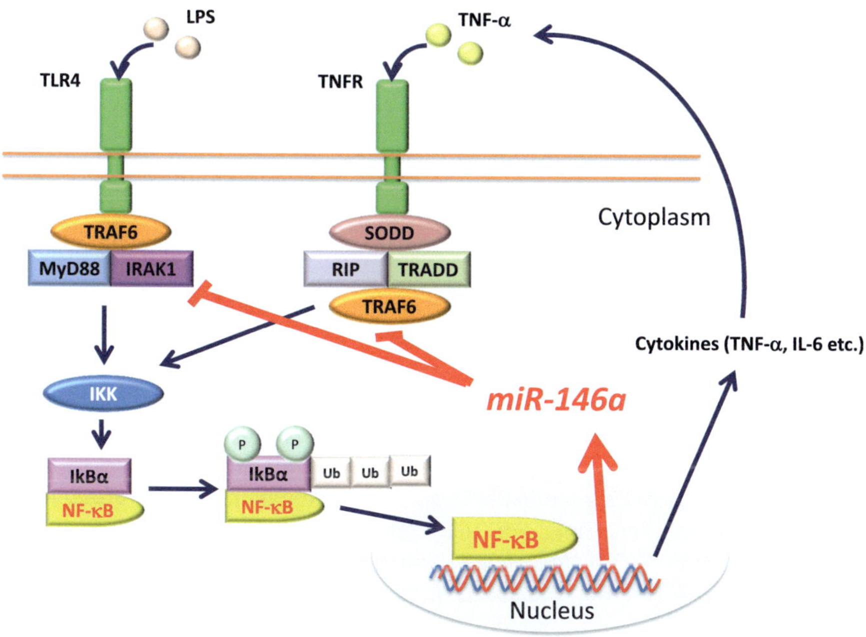

Fig. 4.3 Mechanism of inflammation repression by miR-146a
miR-146a regulates inflammation by repressing *TRAF6* and *IRAK1*, both downstream factors of TLR and inflammatory cytokine receptors

(TLR) and inflammatory cytokine receptors (Fig. 4.3; Taganov et al. 2006; Boldin et al. 2011; Zhao et al. 2011). In peripheral blood mononuclear cells (PBMC) and synovial tissues from RA patients, miR-146a was expressed in RASF, macrophage, CD4$^+$ T lymphocytes, IL 17-producing T lymphocytes, and B lymphocytes (Stanczyk et al. 2008; Pauley et al. 2008; Nakasa et al. 2008; Li et al. 2010). Furthermore, in collagen-induced arthritis (CIA) models, administration of double-stranded miR-146a prevented joint destruction in arthritic mice through inhibition of osteoclastogenesis (Nakasa et al. 2011). miR-155 might promote inflammation. miR-155$^{-/-}$ mice exhibited altered immune responses, through regulation of cytokines, chemokines, and transcription factors, in T and B lymphocytes and dendritic cells (Rodriguez et al. 2007; Thai et al. 2007). In RA, the expression of miR-155 was elevated in RASF and induced by TNF α (Tumour necrosis factor alpha) and LPS (lipopolysaccharide; Stanczyk et al. 2008). At first, researchers predicted that the role of miR-155 was protective against joint destruction. Indeed, overexpression of miR-155 in RASF induced the suppression of *MMP-3*, one of the cartilage destruction enzymes (Stanczyk et al. 2008). However, overexpression of miR-155 in monocytes/macrophages from RA patients promoted the production of TNF-α, interleukin-6 (IL-6), IL-1β, and IL-8 and inhibition of miR-155 suppressed the production of TNF-α (Kurowska-Stolarska et al. 2011). Furthermore, in CIA miR-155−/− mice models, suppression of inflammatory cytokines was observed and arthritis did not occur (Bluml et al. 2011). These results implied that miR-155 is involved in limiting arthritis pathogenesis.

miR-155 and miR-146a were also involved in other rheumatic diseases and inflammatory conditions. PBMC and synovial fluid MC (SFMC) from gouty arthritis showed strong expression of miR-155 (Jin et al. 2014). On the contrary, miR-146a expression was reduced during the acute flare period in PBMC from patients with gouty arthritis compared with the intercritical period (Dalbeth et al. 2015).

Decreased expression of miRNA in RA was also reported. It was shown that miR-23b is weakly expressed in autoimmune diseases, including RA and systemic lupus erythematosus (SLE) (Zhu et al. 2012). The expression of miR-23b was suppressed by IL-17. In addition, miR-23b inhibited cytokine expression as well as IL-17-, TNF-α-, and IL-1β-induced NF-kB activation, through the inhibition of *TAB2* (TGF-β activated kinase 1/MAP3K7 binding protein 2), *TAB3* and *IKK-α* (Inhibitor of nuclear factor kappa-B kinase subunit alpha), which are its direct targets. miR-23b holds a potential to suppress inflammation and, therefore, might become a therapeutic target for autoimmune diseases, including RA (Zhu et al. 2012). Additionally, the expression level of miR-124a was also low in RASF. The targets of miR-124a are *CDK2* (cyclin-dependent kinase 2) and *MCP-1* (monocyte chemoattractant protein-1). As a consequence, in RASF, suppression of miR-124a might lead to the promotion of proliferation via *CDK2* and result in facilitating inflammation via *MCP-1* (Nakamachi et al. 2009).

Starting from a report generated by Stanczyk et al. (ie, on miR-146a and miR-155), many reports on the relationship between miRNA and RA have been published. Further research is expected towards the elucidation of miRNA functions, their benefit as diagnostic markers and the development of novel therapeutic targets.

4.6 Relationship Between microRNAs and Chronic Inflammation

As mentioned above, many reports have shown abnormal miRNA expression in chronic inflammatory diseases, such as cancer, osteoarthritis, and rheumatoid arthritis. Therefore, miRNAs might be potential regulators of chronic inflammation. In chronic inflammation, the resolution phase of the inflammatory response does not occur, which can be attributed to a persisting stimulus and/or defective resolution mechanisms. Thus, normal immune responses can progress to chronic inflammatory states in instances where these events occur. The mechanisms underlying resolution of the inflammatory response have been a focus area of research in recent years, and many important steps and molecules involved in these mechanisms have been discovered. At the molecular level, the cytokine IL-10 (Fiorentino et al. 1991; Sakaguchi 2004), the signalling molecules, A20 (Lee et al. 2000), PD1 (Schwab et al. 2007), CTLA4 (Wing et al. 2008), and the secreted factor IL1RA (Ohlsson et al. 1990) are critical molecules for inhibition of immune cell activation. Furthermore, as described above, miR-146a regulates immune responses; it is induced in response to TLR signalling and forms a negative feedback loop by inhibiting *TRAF6* and *IRAK1*, which are critical upstream TLR-signalling mediators (Fig. 4.3). In recent years, an important role of this miRNA was confirmed by analysis of miR-146a knockout mice. These mice are hyperresponsive to LPS and demonstrate an exaggerated proinflammatory response when challenged with endotoxin (Boldin et al. 2011). Additionally, miR-146a-knockout mice develop a spontaneous autoimmune disorder characterised by splenomegaly, lymphadenopathy, and multiorgan inflammation (Boldin et al. 2011). Moreover, miR-146a knockout mice develop myeloid malignancies due to chronic activation of NF-kB signalling (Zhao et al. 2011). Together, these results indicate that lack of miR-146a expression leads to defects in the resolution of inflammation. Abnormal expression of miR-146a might be a cause of chronic inflammation in some cases.

Conversely, miR-155 functions as a proinflammatory miRNA. miR-155 is induced in activated myeloid cells and represses both SOCS1 and SHIP1 to enhance cytokine production by dendritic cells and macrophages (O'Connell et al. 2009). Overexpression of miR-155 in myeloid cells results in a myeloproliferative disorder due to reduced SHIP1 expression (O'Connell et al. 2009). Furthermore, it has been reported that overexpression of miR-155 in the haematopoietic compartment causes a chronic inflammatory disease (O'Connell et al. 2008) or leukemia (Costinean et al. 2006). These reports suggest that chronic dysregulation of miR-155 might be responsible for chronic inflammation.

Taken together, abnormal expression of miR-146a and miR-155 might be a cause of chronic inflammation, and hence, these miRNAs could serve as clinical targets for treatment of chronic inflammatory diseases. However, the precise functions of miRNAs in chronic inflammation remain largely unknown. For example, it is unclear how these miRNAs are deregulated in chronic inflammatory diseases. The current knowledge about the role of miRNAs in chronic inflammation is quite primitive and further investigations are required on this topic.

References

Akhtar N, Haqqi TM (2012) MicroRNA-199a* regulates the expression of cyclooxygenase-2 in human chondrocytes. Ann Rheum Dis 71:1073–1080

Akhtar N, Rasheed Z, Ramamurthy S, Anbazhagan AN, Voss FR, Haqqi TM (2010) MicroRNA-27b regulates the expression of matrix metalloproteinase 13 in human osteoarthritis chondrocytes. Arthritis Rheum 62:1361–1371

Ambros V (2004) The functions of animal microRNAs. Nature 431:350–355

Bartel DP (2009) MicroRNAs: target recognition and regulatory functions. Cell 136:215–233

Bernstein E, Caudy AA, Hammond SM, Hannon GJ (2001) Role for a bidentate ribonuclease in the initiation step of RNA interference. Nature 409:363–366

Bernstein E, Kim SY, Carmell MA, Murchison EP, Alcorn H, Li MZ, Mills AA, Elledge SJ, Anderson KV, Hannon GJ (2003) Dicer is essential for mouse development. Nat Genet 35:215–217

Bluml S, Bonelli M, Niederreiter B, Puchner A, Mayr G, Hayer S, Koenders MI, van den Berg WB, Smolen J, Redlich K (2011) Essential role of microRNA-155 in the pathogenesis of autoimmune arthritis in mice. Arthritis Rheum 63:1281–1288

Boldin MP, Taganov KD, Rao DS, Yang L, Zhao JL, Kalwani M, Garcia-Flores Y, Luong M, Devrekanli A, Xu J et al (2011) miR-146a is a significant brake on autoimmunity, myeloproliferation, and cancer in mice. J Exp Med 208:1189–1201

Calin GA, Dumitru CD, Shimizu M, Bichi R, Zupo S, Noch E, Aldler H, Rattan S, Keating M, Rai K et al (2002) Frequent deletions and down-regulation of micro- RNA genes miR15 and miR16 at 13q14 in chronic lymphocytic leukemia. Proc Natl Acad Sci U S A 99:15524–15529

Ceribelli A, Nahid MA, Satoh M, Chan EK (2011) MicroRNAs in rheumatoid arthritis. FEBS Lett 585:3667–3674

Chang TC, Wentzel EA, Kent OA, Ramachandran K, Mullendore M, Lee KH, Feldmann G, Yamakuchi M, Ferlito M, Lowenstein CJ et al (2007) Transactivation of miR-34a by p53 broadly influences gene expression and promotes apoptosis. Mol Cell 26:745–752

Costinean S, Zanesi N, Pekarsky Y, Tili E, Volinia S, Heerema N, Croce CM (2006) Pre-B cell proliferation and lymphoblastic leukemia/high-grade lymphoma in E(mu)-miR155 transgenic mice. Proc Natl Acad Sci U S A 103:7024–7029

Dalbeth N, Pool B, Shaw OM, Harper JL, Tan P, Franklin C, House ME, Cornish J, Naot D (2015) Role of miR-146a in regulation of the acute inflammatory response to monosodium urate crystals. Ann Rheum Dis 74:786–790

Duroux-Richard I, Jorgensen C, Apparailly F (2012) What do microRNAs mean for rheumatoid arthritis? Arthritis Rheum 64:11–20

Esteller M (2011) Non-coding RNAs in human disease. Nat Rev Genet 12:861–874

Fiorentino DF, Zlotnik A, Mosmann TR, Howard M, O'Garra A (1991) IL-10 inhibits cytokine production by activated macrophages. J Immunol 147:3815–3822

Han J, Lee Y, Yeom KH, Kim YK, Jin H, Kim VN (2004) The Drosha-DGCR8 complex in primary microRNA processing. Genes Dev 18:3016–3027

Hayashita Y, Osada H, Tatematsu Y, Yamada H, Yanagisawa K, Tomida S, Yatabe Y, Kawahara K, Sekido Y, Takahashi T (2005) A polycistronic microRNA cluster, miR-17-92, is overexpressed in human lung cancers and enhances cell proliferation. Cancer Res 65:9628–9632

He L, He X, Lim LP, de Stanchina E, Xuan Z, Liang Y, Xue W, Zender L, Magnus J, Ridzon D et al (2007) A microRNA component of the p53 tumour suppressor network. Nature 447:1130–1134

He L, Thomson JM, Hemann MT, Hernando-Monge E, Mu D, Goodson S, Powers S, Cordon-Cardo C, Lowe SW, Hannon GJ et al (2005) A microRNA polycistron as a potential human oncogene. Nature 435:828–833

Hutvagner G, McLachlan J, Pasquinelli AE, Balint E, Tuschl T, Zamore PD (2001) A cellular function for the RNA-interference enzyme Dicer in the maturation of the let-7 small temporal RNA. Science 293:834–838

Jin HM, Kim TJ, Choi JH, Kim MJ, Cho YN, Nam KI, Kee SJ, Moon JB, Choi SY, Park DJ et al (2014) MicroRNA-155 as a proinflammatory regulator via SHIP-1 down-regulation in acute gouty arthritis. Arthritis Res Ther 16:R88

Johnson SM, Grosshans H, Shingara J, Byrom M, Jarvis R, Cheng A, Labourier E, Reinert KL, Brown D, Slack FJ (2005) RAS is regulated by the let-7 microRNA family. Cell 120:635–647

Kloosterman WP, Plasterk RH (2006) The diverse functions of microRNAs in animal development and disease. Dev Cell 11:441–450

Kosaka N, Iguchi H, Ochiya T (2010) Circulating microRNA in body fluid: a new potential biomarker for cancer diagnosis and prognosis. Cancer Sci 101:2087–2092

Kozomara A, Griffiths-Jones S (2014) miRBase: annotating high confidence microRNAs using deep sequencing data. Nucleic Acids Res 42:D68–D73

Kurowska-Stolarska M, Alivernini S, Ballantine LE, Asquith DL, Millar NL, Gilchrist DS, Reilly J, Ierna M, Fraser AR, Stolarski B et al (2011) MicroRNA-155 as a proinflammatory regulator in clinical and experimental arthritis. Proc Natl Acad Sci U S A 108:11193–11198

Lee EG, Boone DL, Chai S, Libby SL, Chien M, Lodolce JP, Ma A (2000) Failure to regulate TNFinduced NF-kappaB and cell death responses in A20-deficient mice. Science 289:2350–2354

Lee RC, Feinbaum RL, Ambros V (1993) The C. elegans heterochronic gene lin-4 encodes small RNAs with antisense complementarity to lin-14. Cell 75:843–854

Lee Y, Ahn C, Han J, Choi H, Kim J, Yim J, Lee J, Provost P, Radmark O, Kim S et al (2003) The nuclear RNase III Drosha initiates microRNA processing. Nature 425:415–419

Li J, Wan Y, Guo Q, Zou L, Zhang J, Fang Y, Fu X, Liu H, Lu L, Wu Y (2010) Altered microRNA expression profile with miR-146a upregulation in CD4+ T cells from patients with rheumatoid arthritis. Arthritis Res Ther 12:R81

Martinez-Sanchez A, Dudek KA, Murphy CL (2012) Regulation of human chondrocyte function through direct inhibition of cartilage master regulator SOX9 by microRNA-145 (miRNA-145). J Biol Chem 287:916–924

Mirzamohammadi F, Papaioannou G, Kobayashi T (2014) MicroRNAs in cartilage development, homeostasis, and disease. Curr Osteoporos Rep 12:410–419

Miyaki S, Nakasa T, Otsuki S, Grogan SP, Higashiyama R, Inoue A, Kato Y, Sato T, Lotz MK, Asahara H (2009) MicroRNA-140 is expressed in differentiated human articular chondrocytes and modulates interleukin-1 responses. Arthritis Rheum 60:2723–2730

Miyaki S, Sato T, Inoue A, Otsuki S, Ito Y, Yokoyama S, Kato Y, Takemoto F, Nakasa T, Yamashita S et al (2010) MicroRNA-140 plays dual roles in both cartilage development and homeostasis. Genes Dev 24:1173–1185

Nakamachi Y, Kawano S, Takenokuchi M, Nishimura K, Sakai Y, Chin T, Saura R, Kurosaka M, Kumagai S (2009) MicroRNA-124a is a key regulator of proliferation and monocyte chemoattractant protein 1 secretion in fibroblast-like synoviocytes from patients with rheumatoid arthritis. Arthritis Rheum 60:1294–1304

Nakasa T, Miyaki S, Okubo A, Hashimoto M, Nishida K, Ochi M, Asahara H (2008) Expression of microRNA-146 in rheumatoid arthritis synovial tissue. Arthritis Rheum 58:1284–1292

Nakasa T, Shibuya H, Nagata Y, Niimoto T, Ochi M (2011) The inhibitory effect of microRNA-146a expression on bone destruction in collagen-induced arthritis. Arthritis Rheum 63:1582–1590

O'Connell RM, Chaudhuri AA, Rao DS, Baltimore D (2009) Inositol phosphatase SHIP1 is a primary target of miR-155. Proc Natl Acad Sci U S A 106:7113–7118

O'Connell RM, Rao DS, Chaudhuri AA, Baltimore D (2010) Physiological and pathological roles for microRNAs in the immune system. Nat Rev Immunol 10:111–122

O'Connell RM, Rao DS, Chaudhuri AA, Boldin MP, Taganov KD, Nicoll J, Paquette RL, Baltimore D (2008) Sustained expression of microRNA-155 in hematopoietic stem cells causes a myeloproliferative disorder. J Exp Med 205:585–594

Ohlsson K, Bjork P, Bergenfeldt M, Hageman R, Thompson RC (1990) Interleukin-1 receptor antagonist reduces mortality from endotoxin shock. Nature 348:550–552

Park SJ, Cheon EJ, Kim HA (2013) MicroRNA-558 regulates the expression of cyclooxygenase-2 and IL-1beta-induced catabolic effects in human articular chondrocytes. Osteoarthr Cart 21:981–989

Pauley KM, Satoh M, Chan AL, Bubb MR, Reeves WH, Chan EK (2008) Upregulated miR-146a expression in peripheral blood mononuclear cells from rheumatoid arthritis patients. Arthritis Res Ther 10:R101

Raver-Shapira N, Marciano E, Meiri E, Spector Y, Rosenfeld N, Moskovits N, Bentwich Z, Oren M (2007) Transcriptional activation of miR-34a contributes to p53-mediated apoptosis. Mol Cell 26:731–743

Rodriguez A, Vigorito E, Clare S, Warren MV, Couttet P, Soond DR, van Dongen S, Grocock RJ, Das PP, Miska EA et al (2007) Requirement of bic/microRNA-155 for normal immune function. Science 316:608–611

Sakaguchi S (2004) Naturally arising CD4+ regulatory t cells for immunologic self-tolerance and negative control of immune responses. Annu Rev Immunol 22:531–562

Sandhu SK, Fassan M, Volinia S, Lovat F, Balatti V, Pekarsky Y, Croce CM (2013) B-cell malignancies in microRNA Emu-miR-17~92 transgenic mice. Proc Natl Acad Sci U S A 110:18208–18213

Schwab JM, Chiang N, Arita M, Serhan CN (2007) Resolvin E1 and protectin D1 activate inflammationresolution programmes. Nature 447:869–874

Song J, Jin EH, Kim D, Kim KY, Chun CH, Jin EJ (2014) MicroRNA-222 regulates MMP-13 via targeting HDAC-4 during osteoarthritis pathogenesis. BBA Clin 3:79–89

Song J, Kim D, Chun CH, Jin EJ (2015) miR-370 and miR-373 regulate the pathogenesis of osteoarthritis by modulating one-carbon metabolism via SHMT-2 and MECP-2, respectively. Aging Cell 14:826–837

Song J, Lee M, Kim D, Han J, Chun CH, Jin EJ (2013) MicroRNA-181b regulates articular chondrocytes differentiation and cartilage integrity. Biochem Biophys Res Commun 431:210–214

Srivastava A, Filant J, Moxley KM, Sood A, McMeekin S, Ramesh R (2015) Exosomes: a role for naturally occurring nanovesicles in cancer growth, diagnosis and treatment. Curr Gene Ther 15:182–192

Stanczyk J, Pedrioli DM, Brentano F, Sanchez-Pernaute O, Kolling C, Gay RE, Detmar M, Gay S, Kyburz D (2008) Altered expression of MicroRNA in synovial fibroblasts and synovial tissue in rheumatoid arthritis. Arthritis Rheum 58:1001–1009

Sylvestre Y, De Guire V, Querido E, Mukhopadhyay UK, Bourdeau V, Major F, Ferbeyre G, Chartrand P (2007) An E2F/miR-20a autoregulatory feedback loop. J Biol Chem 282:2135–2143

Taganov KD, Boldin MP, Chang KJ, Baltimore D (2006) NF-kappaB-dependent induction of microRNA miR-146, an inhibitor targeted to signaling proteins of innate immune responses. Proc Natl Acad Sci U S A 103:12481–12486

Tardif G, Hum D, Pelletier JP, Duval N, Martel-Pelletier J (2009) Regulation of the IGFBP-5 and MMP- 13 genes by the microRNAs miR-140 and miR-27a in human osteoarthritic chondrocytes. BMC Musculoskelet Disord 10:148

Tazawa H, Tsuchiya N, Izumiya M, Nakagama H (2007) Tumor-suppressive miR-34a induces senescence-like growth arrest through modulation of the E2F pathway in human colon cancer cells. Proc Natl Acad Sci U S A 104:15472–15477

Thai TH, Calado DP, Casola S, Ansel KM, Xiao C, Xue Y, Murphy A, Frendewey D, Valenzuela D, Kutok JL et al (2007) Regulation of the germinal center response by microRNA-155. Science 316:604–608

Wing K, Onishi Y, Prieto-Martin P, Yamaguchi T, Miyara M, Fehervari Z, Nomura T, Sakaguchi S (2008) CTLA-4 control over Foxp3+ regulatory T cell function. Science 322:271–275

Woods K, Thomson JM, Hammond SM (2007) Direct regulation of an oncogenic micro-RNA cluster by E2F transcription factors. J Biol Chem 282:2130–2134

Xiao C, Srinivasan L, Calado DP, Patterson HC, Zhang B, Wang J, Henderson JM, Kutok JL, Rajewsky K (2008) Lymphoproliferative disease and autoimmunity in mice with increased miR-17-92 expression in lymphocytes. Nat Immunol 9:405–414

Yamashita S, Miyaki S, Kato Y, Yokoyama S, Sato T, Barrionuevo F, Akiyama H, Scherer G, Takada S, Asahara H (2012) L-Sox5 and Sox6 proteins enhance chondrogenic miR-140 microRNA expression by strengthening dimeric Sox9 activity. J Biol Chem 287:22206–22215

Yang B, Kang X, Xing Y, Dou C, Kang F, Li J, Quan Y, Dong S (2014) Effect of microRNA-145 on IL-1beta-induced cartilage degradation in human chondrocytes. FEBS Lett 588:2344–2352

Yi R, Qin Y, Macara IG, Cullen BR (2003) Exportin-5 mediates the nuclear export of pre-microRNAs and short hairpin RNAs. Genes Dev 17:3011–3016

Zhao JL, Rao DS, Boldin MP, Taganov KD, O'Connell RM, Baltimore D (2011) NF-kappaB dysregulation in microRNA-146a-deficient mice drives the development of myeloid malignancies. Proc Natl Acad Sci U S A 108:9184–9189

Zhu S, Pan W, Song X, Liu Y, Shao X, Tang Y, Liang D, He D, Wang H, Liu W et al (2012) The microRNA miR-23b suppresses IL-17-associated autoimmune inflammation by targeting TAB2, TAB3 and IKKalpha. Nat Med 18:1077–1086

Chapter 5
Genetic Dissection of Autoinflammatory Syndrome

Koji Yasutomo

Abstract Autoinflammatory syndromes are characterised by inflammatory responses without infection or autoimmune responses. As autoinflammatory syndromes cause both acute and chronic inflammation, the identification of the causative genes for this syndrome would help to reveal the molecular mechanisms underlying chronic inflammation. Recent studies have revealed that mutations in the *PSMB8* gene, which codes for an immunoproteasome component, cause an autoinflammatory disorder with lipodystrophy. The missense mutation in *PSMB8* disturbs the assembly of immunoproteasomes, which decreases immunoproteasome functions. The decreased immunoproteasome function increases the level of proinflammatory cytokines, including interleukin-6. Furthermore, dysfunction of the immunoproteasome suppresses adipocyte differentiation, which might explain the progressive lipodystrophy that is a characteristic symptom of these patients. The identification of immunoproteasome dysfunction-related autoinflammatory disorders highlights immunoproteasomes as crucial intracellular complexes in the regulation of inflammatory responses. Furthermore, analysis of this syndrome will provide new insights into the key molecular networks related to chronic inflammation.

Keywords Autoinflammatory disorders • Immunoproteasome • Interleukin-6 • Lipodystrophy • Inflammation • Adipocytes

5.1 Inflammatory Responses

Inflammation is a defensive physiological mechanism that protects our bodies from pathogens (Gordon and Trinchieri 2009). Inflammation is balanced between intrinsic stimulators and suppressive mechanisms in response to pathogen-derived molecules or cell-damage–associated molecules (Greenwald et al. 2005; Keir

K. Yasutomo (✉)
Department of Immunology and Parasitology, Graduate School of Medicine, Tokushima University, Tokushima, Japan

AMED-CREST, Tokushima, Japan
e-mail: yasutomo@tokushima-u.ac.jp

M. Miyasaka, K. Takatsu (eds.), *Chronic Inflammation*,
DOI 10.1007/978-4-431-56068-5_5

et al. 2008). Recent studies have revealed a variety of molecules that sense extracellular or intracellular pathogen invasion and transduce signals to initiate inflammatory responses (Takeuchi and Akira 2010). Although inflammation is required as a physiological response to protect our bodies, excessive responses or persistent stimulation induces irreversible cellular damage that leads to various chronic diseases. For instance, accumulating evidence has demonstrated that inflammation contributes to tumourigenesis and metabolic diseases in addition to classical chronic inflammatory diseases such as autoimmune diseases and inflammatory bowel diseases (Gregor and Hotamisligil 2011). Therefore, research into molecular mechanisms underlying inflammation has significantly expanded and has revealed many crucial molecules and molecular networks that regulate inflammatory responses. The mutual interactions among mechanisms that induce or suppress inflammation control the outcome of inflammation (Greenwald et al. 2005). Thus, it is important to understand the physiological roles of each molecule within the complex network of interactions.

5.2 Acute and Chronic Inflammation

Inflammatory responses are generally divided into two groups; acute and chronic. Acute inflammation is usually of sudden onset with redness, pain, swelling, and disability of organ functions, which are caused by increased vascular leakage and a predominance of exudative processes. In contrast, chronic inflammation is characterised by prolonged inflammation with global or scattered fibrotic regions. However, those are phenotypic definitions, and they raise the following question: how could we define acute and chronic inflammation on a molecular level? In other words, what is the molecular determinant to distinguish acute and chronic inflammation? Although it remains unclear how we define acute or chronic inflammation at a molecular level, there would be at least two explanations of how chronic inflammation is caused. First, long duration or high peak-strength acute inflammation could cause irreversible tissue damage with fibrotic regions. Even after clearance of acute inflammation, the tissue damage would leave ‘environments’ that cause persistent inflammation. The second explanation is that qualitatively altered cellular or molecular responses against acute inflammation could cause persistent inflammation or irreversible tissue fibrosis, which could not be corrected by their own cellular systems. Although there are many possible approaches to addressing the molecular basis underlying chronic inflammation, one way to search for key molecules or crucial cellular networks responsible for switching from acute to chronic inflammation is to identify causative genes for familial chronic inflammatory disorders. In this regard, I introduce the causative genes of autoinflammatory syndrome in this chapter, in particular, focusing on an autoinflammatory syndrome with lipodystrophy that has characteristics of both acute and chronic inflammation (Kitamura et al. 2011).

5.3 Autoinflammatory Syndromes

Autoinflammatory syndromes are defined by inflammatory symptoms that include fever and sterile serositis but lack infectious episodes (Aksentijevich and Kastner 2011). Research into autoinflammatory syndromes began with the analysis of familial autoinflammatory disorders and was expanded by discoveries of the causative genes. Familial Mediterranean fever (FMF) is caused by a mutation in the *MEFV* gene, and mutations in the p55 TNF receptor were discovered in familial Hibernian fever. After the discoveries of the two causative genes for these inherited autoinflammatory diseases, several genes that cause autoinflammatory diseases were reported (Aksentijevich and Kastner 2011). Many autoinflammatory syndromes are associated with IL-1β secretion caused by excessive NLRP3 inflammasome activation (Franchi et al. 2009). We and other groups have recently identified a mutation in *NLRC4* in several autoinflammatory syndromes. The mutation in *NLRC4* hyperactivated caspase-1, resulting in increased production of IL-1β (Canna et al. 2014; Romberg et al. 2014; Kitamura et al. 2014). Indeed, recent clinical trials have demonstrated that IL-1β blockade is an effective treatment for such diseases (Aksentijevich and Kastner 2011; Goldbach-Mansky et al. 2008; Stojanov et al. 2011). However, a significant proportion of autoinflammatory patients do not have genetic mutations in the NLRP3-mediated pathway, TNF receptor, or NLRC4, which has suggested the presence of other pathways for inducing autoinflammatory disorders.

A number of reports identified a previously unknown syndrome characterised by recurrent fever, nodular erythema, hepatosplenomegaly, and lipodystrophy without increased serum autoantibodies (Kitano et al. 1985). As with other symptoms, macroglossia and joint contraction with fibrotic changes were observed. Similar syndromes (CANDLE or JMP syndrome) have also been reported in the United States (Garg et al. 2010). Our group and others independently identified *PSMB8*, which codes for a subunit of the immunoproteasome, as the causative gene (Kitamura et al. 2011). This new type of autoinflammatory syndrome caused by defective immunoproteasomes highlights not only the important role of the immunoproteasome in inflammation, but also the broader spectrum of autoinflammatory diseases than were previously understood. In this chapter, we discuss this new type of autoinflammatory syndrome and the role of immunoproteasomes in inflammatory responses.

5.4 Immunoproteasomes

Proteasomes play an essential role in degrading proteins in cells and are crucial for removing misfolded proteins. Thus, they regulate the turnover of molecules to maintain cellular homeostasis (Fig. 5.1) (Ciechanover 2005). The 26S proteasome is a proteolytic enzyme complex that degrades polyubiquitinated proteins (Finley

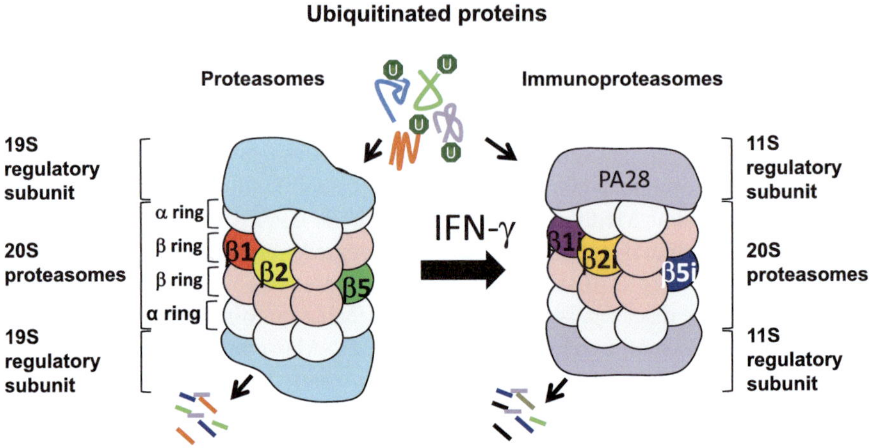

Fig. 5.1 Proteasomes and immunoproteasomes

2009). The 20S proteasome is composed of multiple subunits and is the catalytic core of the 26S proteasome. The 20S proteasome interacts with two 19S regulatory complexes (Tanaka and Ichihara 1988).

The 20S proteasome contains 14 subunits that organise four stacked rings of seven subunits (Murata et al. 2009) 5, (Finley 2009). Two outer α-rings are composed of seven α-subunits (α1–α7) and the two inner rings have seven different β-subunits (β1–β7). The 20S proteasome is an N-terminal nucleophilic hydrolase that generates the peptides presented by MHC class I by using its N-terminal threonine residues as the active site of β1, β2, and β5. The three subunits hydrolyse short peptide substrates through chymotrypsin-like (β5), trypsin-like (β2), or caspase-like (β1) activity. Three catalytic β-subunits (β1i/LMP2, β2i/MECL1, and β5i/LMP7) are upregulated by interferon-γ. Those three subunits are incorporated into the proteasome instead of β1, β2 and β5 to form a distinct 20S proteasome complex with an alternative subunit composition (Fig. 5.1) (Akiyama et al. 1994). These proteasomes are termed immunoproteasomes (Akiyama et al. 1994). The change of subunits with distinct activity results in immunoproteasomes having not only an altered cleavage site preference but also a different cleavage rate (Van den Eynde and Morel 2001). The proteasome activator-αβ (PA28-αβ) is also upregulated by interferon-γ. PA28 is a heptameric complex composed of α and β subunits and binds to the end of the 20S proteasome core (Fig. 5.1) (Hendil et al. 1998). It increases the cleavage rate (Preckel et al. 1999). Mice deficient in PA28-β produce altered peptide cleavage products compared with control mice and exhibit impaired immunoproteasome assembly (Preckel et al. 1999). Therefore PA28 is crucial to the organisation and functions of immunoproteasomes.

Immunoproteasomes process antigens and produce the peptides that are presented on MHC class I (Chen et al. 2001). LMP7-deficient mice showed reduced responses against two dominant influenza virus epitopes and increased responses to two subdominant epitopes (Chen et al. 2001), suggesting that the absence of LMP7 affected $CD8^+$ T cell repertoire formation and susceptibility to infectious diseases. Furthermore, mice that lacked three catalytic subunits showed that about 50 % of peptides were presented solely by cells expressing immunoproteasomes (Kincaid et al. 2012). Strikingly, the alteration of antigen presentation in mice with complete immunoproteasome deficiency led to the rejection of wild-type cells by $CD8^+$ T cells, indicating that immunoproteasome deficiency in mice results in significantly altered antigen processing and presentation that ultimately impacted the T cell repertoire (Kincaid et al. 2012).

Lymphoid tissues, in particular dendritic cells, macrophages, T cells, and B cells, constitutively express the immunoproteasome, whereas nonlymphoid peripheral tissues typically lack the immunosubunits. However, the expression of immunoproteasomes is not entirely restricted to haematopoietic cells. For example, primary hepatocytes express low amounts of β5i even in the absence of cytokine stimulation (Vasuri et al. 2010). We also found that β5i is expressed in subcutaneous regions even in the absence of extrinsic stimulation with IFN-γ (Kitamura et al. 2011). Therefore, immunoproteasome expression is likely regulated not only by IFN-γ but also by other stimuli, probably in a cell type-specific manner.

5.5 Immunoproteasomes and Inflammatory Diseases

The proteasome inhibitor MG-132 suppresses NF-kB activation by inhibiting the degradation of IkB, which inhibits the production of many proinflammatory cytokines including RANTES, MCP-1, and M-CSF (Wang et al. 1999; Ortiz-Lazareno et al. 2008). These data suggest that proteasomes are involved in regulating inflammatory responses through NF-kB activation. In a more specific manner, a PSMB8-specific inhibitor suppressed the secretion of inflammatory cytokines, including TNF-α, IL-23, and IL-6 from human peripheral mononuclear cells (Muchamuel et al. 2009). Furthermore, treatment with a PSMB8 inhibitor suppressed experimental arthritis, probably due to reduced production of inflammatory cytokines (Muchamuel et al. 2009). Additionally, LMP7 KO mice have lower susceptibility for experimental colitis (Basler et al. 2010; Schmidt et al. 2010). These data suggest that immunoproteasomes play important roles in ameliorating inflammatory responses. However, recent studies have provided evidence that immunoproteasomes are required for the reduction of inflammatory responses. For instance, a study revealed that IFN-γ stimulates the accumulation of oxidant-damaged polyubiquitylated proteins (van Deventer and Neefjes 2010), and the 26S immunoproteasome plays a crucial role in the elimination and efficient turnover of accumulated defective and ubiquitinated proteins. Furthermore, PSMB8-deficient mice have a reduced ability to eliminate accumulated oxidised proteins, which results in an increased susceptibility to experimental autoimmune encephalomyelitis (EAE) (Seifert et al. 2010). How can these conflicting findings on the role of the immunoproteasome in inflammatory responses be reconciled?

One possibility is that immunoproteasomes play a distinct role in inflammatory responses depending on the stage of inflammation. T cells from mice with EAE are characterised by hyperactivated Th17 and Th1 cells (Veldhoen et al. 2006). Therefore, INF-γ mediates the accumulation of oxidised proteins that cannot be efficiently removed by dysfunctional immunoproteasomes. In any case, clarifying the roles of the immunoproteasome in each inflammatory disease setting is necessary as clinicians employ immunoproteasome inhibitors for the treatment of immune-mediated diseases.

5.6 Clinical Characteristics of Inherited Inflammatory Diseases Caused by PSMB8 Dysfunction

Previous publications described a syndrome characterised by recurrent fever, progressive partial lipodystrophy, nodular erythema, and hepatosplenomegaly in consanguineous families in Japan (Kitano et al. 1985). Laboratory findings revealed high serum C-reactive protein and hypergamma globulinemia, although autoantibody titers were negative. Patients were treated with oral steroids that were effective in improving patients' inflammatory responses, including fever and nodular erythema, but not their progressive lipodystrophy. We termed this disease Japanese Autoinflammatory Syndrome with Lipodystrophy (JASL) (Kitamura et al. 2011). Our group attempted to identify the gene that caused JASL using patient samples from two Japanese consanguineous families by linkage analysis and exome resequencing. We found a homozygous missense mutation in *PSMB8* in JASL patients (Kitamura et al. 2011). Similar syndromes were also reported by other groups in Japan (Arima et al. 2011) as Nakajo–Nishimura syndrome, and it had the same mutation as JASL patients. Researchers in other countries described JMP and CANDLE syndromes (Agarwal et al. 2010), which had mutations in *PSMB8* that were distinct from the mutation seen in JASL patients. The phenotypes of JASL patients are similar to those of CANDLE and JMP syndromes, such as cutaneous eruption and lipodystrophy. The recurrent fever with acute phase reactant proteins seen in JASL is observed in CANDLE but not JMP syndrome patients. Differences in symptoms among these autoinflammatory syndromes are likely due to different *PSMB8* mutations or to background genetic differences. These issues should be clarified by comparing genotypes and phenotypes in a larger number of patients.

Mutations in *PSMB8* increased immunoproteasome assembly intermediates, which reduced the expression of mature immunoproteasomes and the final cleavage of PSMB8 (Fig. 5.2). Accordingly, chymotrypsin-like activity in immunoproteasomes was decreased in transformed B cells from JASL patients (Kitamura et al. 2011). Ubiquitin-coupled proteins accumulated in patients' skin and, as expected, PSMB8 expression was lower in JASL skin samples compared with healthy controls (Kitamura et al. 2011).

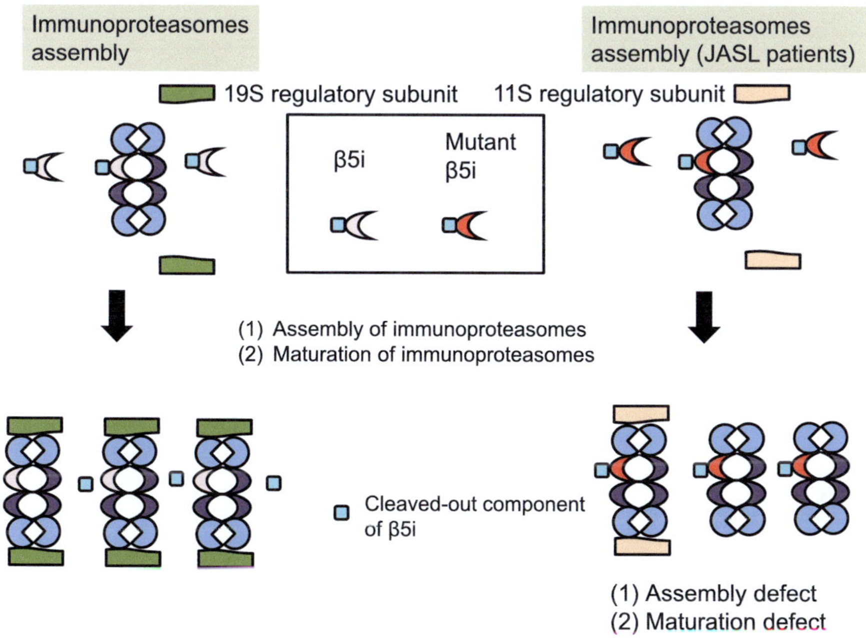

Fig. 5.2 Assembly defect of immunoproteasomes by mutant b5i

5.7 Immunoproteasome Defects Contribute to Autoinflammation

In order to understand the relationship between the reduced activity of immunoproteasomes and autoinflammation, we analysed the expression of IL-6 in transformed B cells (Kitamura et al. 2011). Another group has reported that serum IL-6 levels (Kasagi et al. 2008), but not IL-1β, were increased in JASL patients. We found that transformed B cells from JASL patients showed increased IL-6 expression, and it was inhibited by suppressing the p38 pathway. These data were corroborated by increased phosphorylation of p38 in JASL cells. Thus, p38 and IL-6 appeared to regulate, at least partially, autoinflammation (Fig. 5.3).

There are three important questions regarding immunoproteasome assembly defects and autoinflammation that should be addressed in future studies. First, what factors are responsible for p38 activation? Defects in immunoproteasome assembly caused by mutations in *PSMB8* decreased chymotrypsin-like activity and resulted in the accumulation of ubiquitinated proteins. Another group has demonstrated that *PSMB8* mutations result in the accumulation of oxidised proteins (Arima et al. 2011). One possibility is that the accumulation of oxidised proteins causes stress on cells, leading them to activate p38. This scenario is supported by one report which showed that PSMB8-deficient mice accumulated oxidised proteins, enhancing their susceptibility for experimental autoimmune

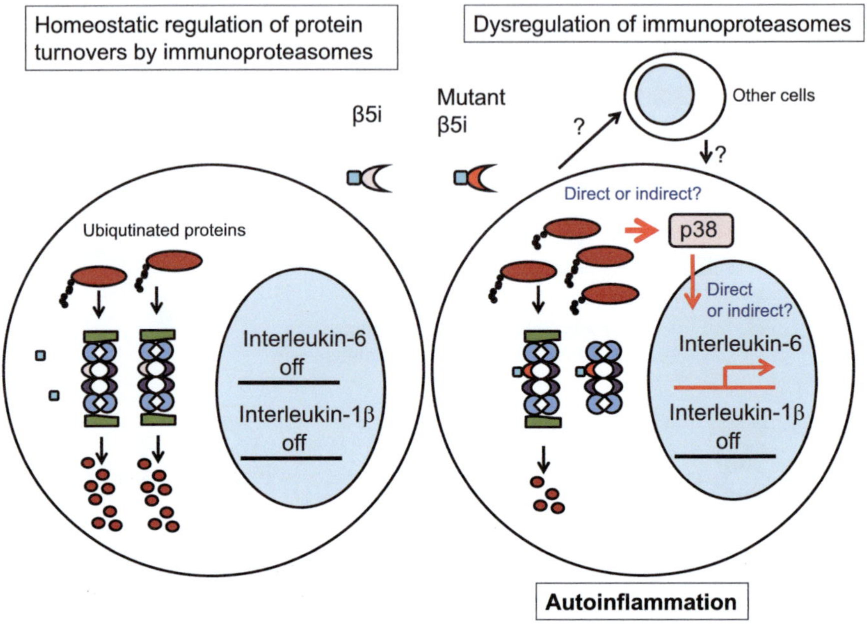

Fig. 5.3 Current model of immunoproteasomes and autoinflammation

encephalomyelitis (Seifert et al. 2010). The second question is whether IL-6 is really a causative factor for autoinflammation in JASL patients. We and other groups have demonstrated that IL-6 is highly upregulated in JASL patients (Kitamura et al. 2011). Furthermore, NF-kB was not activated and IL-1β production was not increased in JASL cells. In order to determine if IL-6 is really involved in the autoinflammation caused by mutant *PSMB8*, mice harboring the same *PSMB8* mutation should be established and analysed. The third question is whether the mutation in *PSMB8* affects the function of conventional proteasomes. Our analysis revealed a reduction in total proteasome activity in JASL cells. Another group demonstrated that the *PSMB8* mutation reduced not only chymotrypsin-like activity but also caspase- and trypsin-like activity (Arima et al. 2011), although PSMB8 regulates only chymotrypsin-like activity. These findings suggest that the JASL *PSMB8* mutation affects the assembly of immunoproteasomes, resulting in reduced total immunoproteasome activity. PSMB8-deficient mice do not develop spontaneous inflammatory diseases, although these mice show increased susceptibility for EAE (Seifert et al. 2010). Therefore, the inflammatory responses exhibited by JASL patients might represent mixed phenotypes due to reduced activity of both immunoproteasomes and standard proteasomes. This possibility should be clarified by analysing mice with the same *PSMB8* mutation found in JASL patients.

5.8 PSMB8 Controls Adipocyte Differentiation

Lipodystrophy is one of the characteristic symptoms of JASL. The lipodystrophy in JASL is evident in peripheral areas, such as distal arms and legs (Kitamura et al. 2011). A previous study suggested that lipodystrophy was associated with panniculitis-induced adipocyte death (Garg et al. 2010). However, the lipodystrophy in JASL is progressive despite treatment with steroids that are otherwise effective for inhibiting nodular erythema and high fever. This finding suggests that the progressive lipodystrophy is associated with an adipocyte-intrinsic dysfunction rather than a secondary effect of inflammation, such as panniculitis. Therefore, our group investigated whether *PSMB8* impacted preadipocyte differentiation. We found that reduced expression of *PSMB8* inhibited the differentiation of mouse and human preadipocytes to mature adipocytes in vitro. Furthermore, injection of siRNA against *PSMB8* into the back skin of mice reduced fat volume. These data suggest that PSMB8 expressed in the adipocyte cell lineage controls subcutaneous fat volume. In contrast, PSMB8-deficient mice do not exhibit lipodystrophy, although a detailed analysis of fat volume has not been conducted. In addition, we cannot completely eliminate the possibility that lipodystrophy in JASL patients at least partly results from inflammation of adipocyte tissue. Therefore, it will be necessary to perform a detailed analysis of fat metabolism in PSMB8-deficient mice (Fehling et al. 1994) and also establish JASL knock-in mice in future studies.

5.9 Future Prospects

Inflammatory responses are caused by a variety of signaling cascades, and immunoproteasome dysfunction is a recent addition to a growing list of inflammatory pathways. The biological importance of immunoproteasomes is evident from the autoinflammatory symptoms of humans with inherited deficiencies in *PSMB8*, but its clinical implications are still largely unexplored. There are many issues to be analysed. Specific areas of study include the following.

1. Analyse in detail the mechanism(s) by which mutations in *PSMB8* affect the biochemical and structural basis of PSMB8 incorporation into the immunoproteasome.
2. Develop animal models that more faithfully recapitulate human diseases and inflammation.
3. Analyse genetic variations of immunoproteasome components and their correlation with immunoproteasome function and inflammation.
4. Determine which stimuli elicit IL-6 secretion.
5. Determine the effects of oxidised protein accumulation and how that impacts the pathogenesis of autoinflammatory diseases by altering immunoproteasome function.

6. Study the conflicting roles of the immunoproteasome and how they impact inflammatory responses. One intriguing finding is that patients with *PSMB8* mutations show distinct cytokine signatures; for example, some show elevated IL-6 levels whereas other patients exhibit high IFN-γ production. These differences might be associated with the manner in which different *PSMB8* mutations impact PSMB8 function and immunoproteasome incorporation. In any case, it will be essential to analyse which cytokines are crucial for the development of inflammatory responses in JASL by establishing animal models.
7. Study how dysfunctions of JASL inflammation cause chronic inflammation in the joints. This is related to the preceding point. The identification of the molecular networks that cause inflammation in JASL would also reveal how dysfunctions of immunoproteasomes cause chronic inflammation with fibrotic regions.
8. Finally, an important question of biological and practical importance is whether immunoproteasome dysfunction is involved in the pathogenesis of common inflammatory diseases. Do immunoproteasomes have distinct and special roles in inhibiting inflammatory responses compared with conventional proteasomes? If the roles of the immunoproteasome in inflammatory responses and adipocyte differentiation generalise to other diseases, then modulating immunoproteasome function could be an attractive target for drug treatments aimed at inhibiting inflammation or fat accumulation. Delineating the remaining questions will broaden our understanding of the physiological roles of immunoproteasomes as well as inflammatory responses, and could open novel ways to develop new therapeutic approaches to treat inflammatory diseases.

References

Agarwal AK, Xing C, DeMartino GN, Mizrachi D, Hernandez MD, Sousa AB et al (2010) PSMB8 encoding the beta5i proteasome subunit is mutated in joint contractures, muscle atrophy, microcytic anemia, and panniculitis-induced lipodystrophy syndrome. Am J Hum Genet 87 (6):866–872. doi:10.1016/j.ajhg.2010.10.031; PubMed PMID: 21129723, PubMed Central PMCID: PMC2997366, Epub 2010/12/07

Akiyama K, Yokota K, Kagawa S, Shimbara N, Tamura T, Akioka H et al (1994) cDNA cloning and interferon gamma down-regulation of proteasomal subunits X and Y. Science 265 (5176):1231–1234; Epub 1994/08/26

Aksentijevich I, Kastner DL (2011) Genetics of monogenic autoinflammatory diseases: past successes, future challenges. Nat Rev Rheumatol 7(8):469–478. doi:10.1038/nrrheum.2011.94; Epub 2011/07/06

Arima K, Kinoshita A, Mishima H, Kanazawa N, Kaneko T, Mizushima T et al (2011) Proteasome assembly defect due to a proteasome subunit beta type 8 (PSMB8) mutation causes the autoinflammatory disorder, Nakajo-Nishimura syndrome. Proc Natl Acad Sci U S A 108 (36):14914–14919. doi:10.1073/pnas.1106015108; PubMed PMID: 21852578, PubMed Central PMCID: PMC3169106, Epub 2011/08/20

Basler M, Dajee M, Moll C, Groettrup M, Kirk CJ (2010) Prevention of experimental colitis by a selective inhibitor of the immunoproteasome. J Immunol 185(1):634–641. doi:10.4049/jimmunol.0903182; Epub 2010/06/08

Canna SW, de Jesus AA, Gouni S, Brooks SR, Marrero B, Liu Y et al (2014) An activating NLRC4 inflammasome mutation causes autoinflammation with recurrent macrophage activation syndrome. Nat Genet 46(10):1140–1146. doi:10.1038/ng.3089; PubMed PMID: 25217959, PubMed Central PMCID: PMC4177369

Chen W, Norbury CC, Cho Y, Yewdell JW, Bennink JR (2001) Immunoproteasomes shape immunodominance hierarchies of antiviral CD8(+) T cells at the levels of T cell repertoire and presentation of viral antigens. J Exp Med 193(11):1319–1326; PubMed PMID: 11390439, PubMed Central PMCID: PMC2193381, Epub 2001/06/08

Ciechanover A (2005) Proteolysis: from the lysosome to ubiquitin and the proteasome. Nat Rev Mol Cell Biol 6(1):79–87. doi:10.1038/nrm1552; Epub 2005/02/03

Fehling HJ, Swat W, Laplace C, Kuhn R, Rajewsky K, Muller U et al (1994) MHC class I expression in mice lacking the proteasome subunit LMP-7. Science 265(5176):1234–1237; Epub 1994/08/26

Finley D (2009) Recognition and processing of ubiquitin-protein conjugates by the proteasome. Annu Rev Biochem 78:477–513. doi:10.1146/annurev.biochem.78.081507.101607; Epub 2009/06/06

Franchi L, Eigenbrod T, Munoz-Planillo R, Nunez G (2009) The inflammasome: a caspase-1-activation platform that regulates immune responses and disease pathogenesis. Nat Immunol 10(3):241–247. doi:10.1038/ni.1703; PubMed PMID: 19221555, PubMed Central PMCID: PMC2820724, Epub 2009/02/18

Garg A, Hernandez MD, Sousa AB, Subramanyam L, de Martinez VL, dos Santos HG et al (2010) An autosomal recessive syndrome of joint contractures, muscular atrophy, microcytic anemia, and panniculitis-associated lipodystrophy. J Clin Endocrinol Metab 95(9):E58–E63. doi:10.1210/jc.2010-0488; PubMed PMID: 20534754, PubMed Central PMCID: PMC2936059, Epub 2010/06/11

Goldbach-Mansky R, Shroff SD, Wilson M, Snyder C, Plehn S, Barham B et al (2008) A pilot study to evaluate the safety and efficacy of the long-acting interleukin-1 inhibitor rilonacept (interleukin-1 Trap) in patients with familial cold autoinflammatory syndrome. Arthritis Rheum 58(8):2432–2442. doi:10.1002/art.23620; PubMed PMID: 18668591, PubMed Central PMCID: PMC2875198, Epub 2008/08/01

Gordon S, Trinchieri G (2009) Innate resistance and inflammation. Curr Opin Immunol 21(1):1–2. doi:10.1016/j.coi.2009.02.001; Epub 2009/02/19

Greenwald RJ, Freeman GJ, Sharpe AH (2005) The B7 family revisited. Annu Rev Immunol 23:515–548. doi:10.1146/annurev.immunol.23.021704.115611; Epub 2005/03/18

Gregor MF, Hotamisligil GS (2011) Inflammatory mechanisms in obesity. Annu Rev Immunol 29:415–445. doi:10.1146/annurev-immunol-031210-101322; Epub 2011/01/12

Hendil KB, Khan S, Tanaka K (1998) Simultaneous binding of PA28 and PA700 activators to 20 S proteasomes. Biochem J 332(Pt 3):749–754; PubMed PMID: 9620878, PubMed Central PMCID: PMC1219536, Epub 1998/06/11

Kasagi S, Kawano S, Nakazawa T, Sugino H, Koshiba M, Ichinose K et al (2008) A case of periodic-fever-syndrome-like disorder with lipodystrophy, myositis, and autoimmune abnormalities. Mod Rheumatol 18(2):203–207. doi:10.1007/s10165-008-0033-4; Epub 2008/03/15

Keir ME, Butte MJ, Freeman GJ, Sharpe AH (2008) PD-1 and its ligands in tolerance and immunity. Annu Rev Immunol 26:677–704. doi:10.1146/annurev.immunol.26.021607.090331; Epub 2008/01/05

Kincaid EZ, Che JW, York I, Escobar H, Reyes-Vargas E, Delgado JC et al (2012) Mice completely lacking immunoproteasomes show major changes in antigen presentation. Nat Immunol 13(2):129–135. doi:10.1038/ni.2203; PubMed PMID: 22197977, PubMed Central PMCID: PMC3262888, Epub 2011/12/27

Kitamura A, Maekawa Y, Uehara H, Izumi K, Kawachi I, Nishizawa M et al (2011) A mutation in the immunoproteasome subunit PSMB8 causes autoinflammation and lipodystrophy in humans. J Clin Invest 121(10):4150–4160. doi:10.1172/JCI58414; PubMed PMID: 21881205, PubMed Central PMCID: PMC3195477, Epub 2011/09/02

Kitamura A, Sasaki Y, Abe T, Kano H, Yasutomo K (2014) An inherited mutation in NLRC4 causes autoinflammation in human and mice. J Exp Med 211(12):2385–2396. doi:10.1084/jem.20141091; PubMed PMID: 25385754, PubMed Central PMCID: PMC4235634

Kitano Y, Matsunaga E, Morimoto T, Okada N, Sano S (1985) A syndrome with nodular erythema, elongated and thickened fingers, and emaciation. Arch Dermatol 121 (8):1053–1056; Epub 1985/08/01

Muchamuel T, Basler M, Aujay MA, Suzuki E, Kalim KW, Lauer C et al (2009) A selective inhibitor of the immunoproteasome subunit LMP7 blocks cytokine production and attenuates progression of experimental arthritis. Nat Med 15(7):781–787. doi:10.1038/nm.1978; Epub 2009/06/16

Murata S, Yashiroda H, Tanaka K (2009) Molecular mechanisms of proteasome assembly. Nat Rev Mol Cell Biol 10(2):104–115. doi:10.1038/nrm2630; Epub 2009/01/24

Ortiz-Lazareno PC, Hernandez-Flores G, Dominguez-Rodriguez JR, Lerma-Diaz JM, Jave-Suarez LF, Aguilar-Lemarroy A et al (2008) MG132 proteasome inhibitor modulates proinflammatory cytokines production and expression of their receptors in U937 cells: involvement of nuclear factor-kappaB and activator protein-1. Immunology 124(4):534–541. doi:10.1111/j.1365-2567.2008.02806.x; PubMed PMID: 18298552, PubMed Central PMCID: PMC2492945, Epub 2008/02/27

Preckel T, Fung-Leung WP, Cai Z, Vitiello A, Salter-Cid L, Winqvist O et al (1999) Impaired immunoproteasome assembly and immune responses in PA28−/− mice. Science 286 (5447):2162–2165; Epub 1999/12/11

Romberg N, Al Moussawi K, Nelson-Williams C, Stiegler AL, Loring E, Choi M et al (2014) Mutation of NLRC4 causes a syndrome of enterocolitis and autoinflammation. Nat Genet 46 (10):1135–1139. doi:10.1038/ng.3066; PubMed PMID: 25217960, PubMed Central PMCID: PMC4177367

Schmidt N, Gonzalez E, Visekruna A, Kuhl AA, Loddenkemper C, Mollenkopf H et al (2010) Targeting the proteasome: partial inhibition of the proteasome by bortezomib or deletion of the immunosubunit LMP7 attenuates experimental colitis. Gut 59(7):896–906. doi:10.1136/gut.2009.203554; Epub 2010/06/29

Seifert U, Bialy LP, Ebstein F, Bech-Otschir D, Voigt A, Schroter F et al (2010) Immunoproteasomes preserve protein homeostasis upon interferon-induced oxidative stress. Cell 142(4):613–624. doi:10.1016/j.cell.2010.07.036; Epub 2010/08/21

Stojanov S, Lapidus S, Chitkara P, Feder H, Salazar JC, Fleisher TA et al (2011) Periodic fever, aphthous stomatitis, pharyngitis, and adenitis (PFAPA) is a disorder of innate immunity and Th1 activation responsive to IL-1 blockade. Proc Natl Acad Sci U S A 108(17):7148–7153. doi:10.1073/pnas.1103681108; PubMed PMID: 21478439, PubMed Central PMCID: PMC3084055, Epub 2011/04/12

Takeuchi O, Akira S (2010) Pattern recognition receptors and inflammation. Cell 140(6):805–820. doi:10.1016/j.cell.2010.01.022; Epub 2010/03/23

Tanaka K, Ichihara A (1988) Involvement of proteasomes (multicatalytic proteinase) in ATP-dependent proteolysis in rat reticulocyte extracts. FEBS Lett 236(1):159–162; Epub 1988/08/15

Van den Eynde BJ, Morel S (2001) Differential processing of class-I-restricted epitopes by the standard proteasome and the immunoproteasome. Curr Opin Immunol 13(2):147–153; Epub 2001/03/03

van Deventer S, Neefjes J (2010) The immunoproteasome cleans up after inflammation. Cell 142 (4):517–518. doi:10.1016/j.cell.2010.08.002; Epub 2010/08/21

Vasuri F, Capizzi E, Bellavista E, Mishto M, Santoro A, Fiorentino M et al (2010) Studies on immunoproteasome in human liver. Part I: absence in fetuses, presence in normal subjects, and increased levels in chronic active hepatitis and cirrhosis. Biochem Biophys Res Commun 397 (2):301–306. doi:10.1016/j.bbrc.2010.05.104; Epub 2010/06/01

Veldhoen M, Hocking RJ, Flavell RA, Stockinger B (2006) Signals mediated by transforming growth factor-beta initiate autoimmune encephalomyelitis, but chronic inflammation is needed to sustain disease. Nat Immunol 7(11):1151–1156. doi:10.1038/ni1391; Epub 2006/09/26

Wang XC, Jobin C, Allen JB, Roberts WL, Jaffe GJ (1999) Suppression of NF-kappaB-dependent proinflammatory gene expression in human RPE cells by a proteasome inhibitor. Invest Ophthalmol Vis Sci 40(2):477–486; Epub 1999/02/09

Chapter 6
Structural Biology of Chronic Inflammation-Associated Signalling Pathways: Toward Structure-Guided Drug Development

Reiya Taniguchi and Osamu Nureki

Abstract Chronic inflammation is involved in the provocation, progression, and exacerbation of various diseases, such as cancer and fibrosis. To treat these diseases, it is essential to understand their pathogenesis and develop therapeutic compounds that target the underlying signalling pathways. Recent technical advances in structural biology have enabled us to determine the high-resolution structures of important drug-target proteins, including membrane receptors and transporters, and thus structure-guided drug development has now become a realistic approach. In this chapter, we focus on two clinically important pathways involved in diseases associated with chronic inflammation, and summarise the recent results of our structural studies. In the first section, we focus on the signalling mediated by the lipid mediator lysophosphatidic acid, and discuss the structural findings for its producing enzyme and receptor. In the second section, we focus on the anaemia associated with inflammatory conditions, and discuss the structural insights into the iron exporter ferroportin, which plays a key role in the anaemia of chronic inflammation. These examples provide ideas for using structural information as the basis for pharmacological analyses.

Keywords X-ray crystallography • Structural biology • Lysophosphatidic acid • Autotaxin • G protein-coupled receptors • Iron homeostasis • Anaemia of chronic inflammation • Ferroportin • Hepcidin

---

R. Taniguchi
Department of Biological Sciences, Graduate School of Science, The University of Tokyo, 2-11-16 Yayoi, Bunkyo-ku, Tokyo 113-0032, Japan

O. Nureki (✉)
Department of Biological Sciences, Graduate School of Science, The University of Tokyo, 2-11-16 Yayoi, Bunkyo-ku, Tokyo 113-0032, Japan

AMED-CREST, Tokyo 113-0032, Japan
e-mail: nureki@bs.s.u-tokyo.ac.jp

M. Miyasaka, K. Takatsu (eds.), *Chronic Inflammation*,
DOI 10.1007/978-4-431-56068-5_6

6.1 ATX-LPA Signalling

6.1.1 LPA Signalling

Various types of lipid mediators, such as prostaglandins and leukotrienes, provoke inflammatory responses. In addition to these well-known inflammatory lipid mediators, lysophospholipids are also involved in inflammatory responses (Gräler 2002). Lysophospholipids are membrane-derived phospholipids that lack one of the two acyl chains, and typically contain either a glycerol or a sphingosine backbone. Lysophosphatidic acid (LPA) is a representative lysolipid, and together with LPA-producing enzymes and LPA receptors, its biological functions have been extensively investigated.

LPA contains a phosphate group and a single acyl chain, with various lengths and saturations, attached to a glycerol backbone. LPA is mainly produced by the secreted lysophospholipase autotaxin (ATX), which hydrolyses lysophosphatidylcholine (LPC) into LPA and choline (Stracke et al. 1992; Umezu-Goto et al. 2002; Tokumura et al. 2002). The synthesised LPA subsequently activates the six G protein-coupled receptors (GPCRs) LPA_1–LPA_6, leading to various cellular responses, such as cell proliferation and migration (van Meeteren and Moolenaar 2007; Noguchi et al. 2009; Kihara et al. 2014). Accordingly, high expression levels of the ATX or LPA receptor promote tumour growth and metastasis, and are often associated with various types of cancers (Barbayianni et al. 2015). Moreover, the high expression or activity levels of ATX are also observed in nonmalignant tissues, such as fibrotic lungs (Oikonomou et al. 2012), chronically inflamed livers (Watanabe et al. 2007), and the venules of chronically inflamed tissues (Nakasaki et al. 2008), clearly indicating the involvement of ATX-LPA signalling in chronic diseases and immune responses. Thus, the ATX and LPA receptors are regarded as promising targets for drug development, and extensive studies have been conducted in order to generate clinically applicable compounds targeting these proteins.

Recently, our group and the Perrakis group reported the crystal structures of mouse and rat ATX (Nishimasu et al. 2011; Hausmann et al. 2011), which greatly advanced the understanding of the LPA production mechanism and facilitated the structure-guided development of ATX inhibitors. The crystal structures of LPA_1 were also reported last year (Chrencik et al. 2015), providing insights into its ligand recognition mechanism. The structural information will pave the way toward drug development targeting the ATX and LPA receptors.

6.1.2 ATX Structures

The structures of ATX were determined in the free form, the LPA-bound forms, and the inhibitor HA155-bound form (Nishimasu et al. 2011; Hausmann et al. 2011).

The ATX structure consists of four domains: the central catalytic domain, the nuclease-like domain and two somatomedin B-like domains, SMB1 and SMB2 (Fig. 6.1a). The nuclease-like domain tightly interacts with the central catalytic domain by disulfide bonds and salt bridges. These interactions are further reinforced by the Asn-linked glycan located at the domain interface. Our molecular dynamics simulations suggested that the lack of these tight interactions would destabilise the overall architecture of ATX and disrupt its catalytic activity (Koyama et al. 2012). On the opposite side of the nuclease-like domain, SMB1 and SMB2 interact with the central catalytic domain, further stabilising the overall structure. The central catalytic domain contains the active site of ATX, which consists of the catalytic threonine and two Zn^{2+} ions coordinated by conserved polar residues. The LPA-bound structures of ATX revealed that the head group of the LPA binds to the active site, whereas its acyl chain is accommodated in a deep hydrophobic pocket in close proximity to the catalytic center (Nishimasu et al. 2011) (Fig. 6.1b). The structure also showed that the ATX inhibitor HA155 occupies the same hydrophobic pocket (Hausmann et al. 2011), providing mechanistic insights into the ATX inhibition by chemical compounds. SMB1 interacts with the central catalytic domain near the active site, and a hydrophobic channel is formed between these two domains (Fig. 6.1a, b). This channel is connected to the entrance of the hydrophobic pocket, and residual electron density was observed within it (Nishimasu et al. 2011; Hausmann et al. 2011). Based on the structural information and the mass spectrometry analyses, we hypothesised that this hydrophobic channel contains the LPA molecule and is involved in the process of releasing the reaction product (Nishimasu et al. 2011). The other SMB domain, SMB2, probably mediates the interaction between ATX and the cell-surface integrin (Hausmann et al. 2011; Fulkerson et al. 2011), which anchors ATX on the cell membrane and thereby localises the LPA production near the cell surface.

The structural information about ATX revealed the detailed architecture of its active site, and accelerated the structure-guided optimisation of ATX inhibitors. Soon after the ATX structure determination, the Perrakis and Ovaa groups conducted the structure-based modification of the inhibitor HA155, and explored the structure–activity relationships of various HA155-derivatives (Albers et al. 2011). We also performed the screening and the structure-guided optimisation of novel ATX inhibitors. By determining the ATX structures in complexes with hit compounds or their derivatives, we successfully developed the potent ATX-selective inhibitor, 3-BoA, which exhibited inhibitory activity both in vitro and in vivo (Kawaguchi et al. 2013). These examples clearly demonstrate the power of structural information for drug development.

6.1.3 LPA Receptor Structure

Because ATX is the LPA-producing enzyme, the inhibition of ATX leads to the reduction of the overall LPA level, resulting in the nonselective inhibition of the

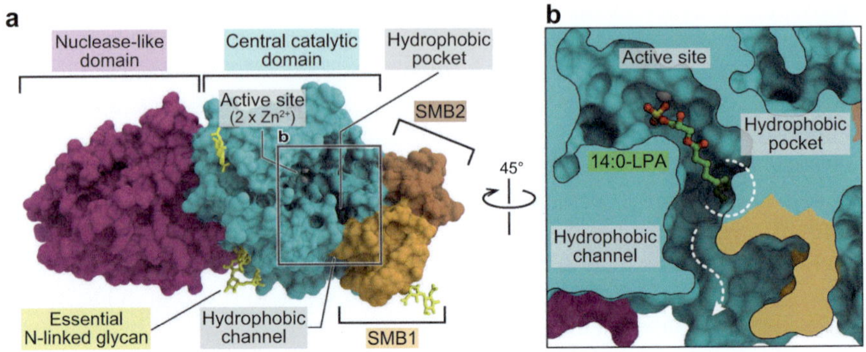

Fig. 6.1 (**a**) Surface representation of the mouse ATX structure in the free form (PDB ID: 3NKM). (**b**) Cut-away surface representation of the mouse ATX active site with bound 14:0-LPA (PDB ID: 3NKN)

downstream LPA signalling pathways. Thus, in order to treat specific diseases associated with particular downstream signalling, selective inhibition of certain LPA receptor subtypes would be advantageous. The LPA receptors are classified into two distinct families, the Edg family (LPA_1–LPA_3) and the non-Edg family (LPA_4–LPA_6), and they share high sequence similarity within each family (Kihara et al. 2014). This indicates that the receptors belonging to the same family also share similar ligand-binding sites, which makes the discovery of subtype-specific compounds extremely challenging. To overcome this difficulty, extensive structural studies of the LPA receptors have been conducted, and last year, the crystal structures of LPA_1 were reported by the Chun, Stevens, and Hanson groups (Chrencik et al. 2015).

LPA_1 is folded into the canonical seven transmembrane bundle, as observed in other GPCR structures, and the antagonist tool compound (ONO-9780307) is bound within the central ligand-binding pocket (Chrencik et al. 2015; Fig. 6.2a). This tool compound consists of a polar head and a bulky, relatively hydrophobic body. In the structure, the polar head of ONO-9780307 faces toward the extracellular side, and the hydrophobic body occupies the pocket formed by transmembrane helices 3, 4, 5, and 6. The N-terminus of the receptor forms a helical cap over the ligand-binding pocket, and provides polar residues for interactions with the head group of ONO-9780307 (Fig. 6.2b). Based on the ligand recognition mechanism observed in the structure, the authors further optimised the chemical structure of ONO-9780307, and validated their design strategy by determining the cocrystal structures (Chrencik et al. 2015). Their work exemplifies the utility of GPCR structures for the modification of tool compounds.

Prior to the structure determination of LPA_1, the crystal structure of sphingosine 1-phosphate (S1P) receptor 1 ($S1P_1$), another lysolipid GPCR, was reported in 2012 (Hanson et al. 2012; Fig. 6.2c, d). The endogenous ligand of this receptor, S1P, consists of a sphingosine backbone and a phosphate group, and has a quite similar molecular structure to LPA (Fig. 6.2e). Moreover, $S1P_1$ and four other S1P

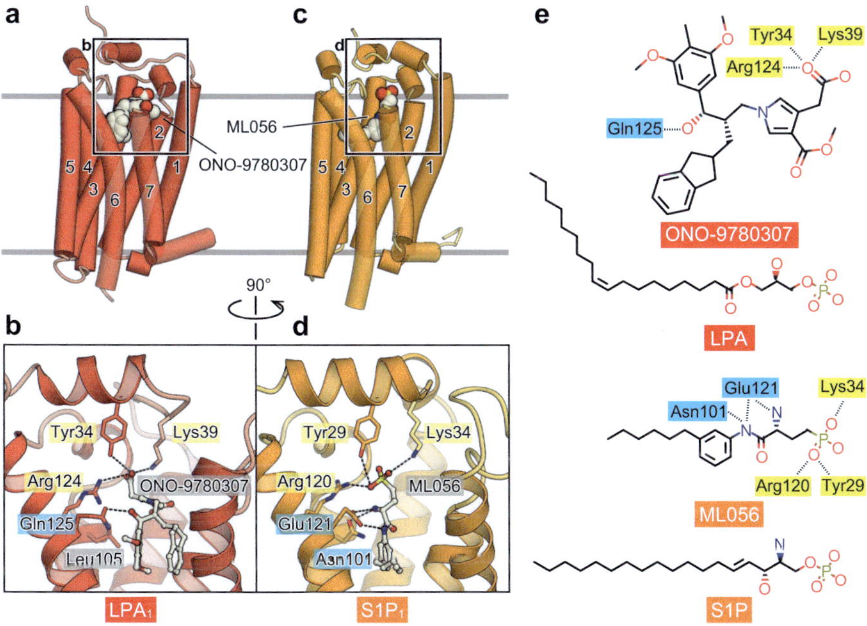

Fig. 6.2 (**a**) Overall structure of LPA_1 with the ligand ONO-9780307 (PDB ID: 4Z34). (**b**) Close-up view of the ligand-binding pocket of LPA_1. (**c**) Overall structure of $S1P_1$ with the ligand ML056 (PDB ID: 3V2Y). (**d**) Close-up view of the ligand-binding pocket of $S1P_1$. (**e**) Chemical structures of the ligands of LPA_1 and $S1P_1$. In (**a**) and (**c**), the seven TMs are numbered, and the loop between TM6 and TM7 is omitted for clarity. In (**b**) and (**d**), TM4 and TM5 are omitted for clarity. In (**b**), (**d**), and (**e**), the residues interacting with the polar heads of the ligands are highlighted in *yellow*, while those interacting with the ligand backbones are highlighted in *blue*

receptors ($S1P_2$–$S1P_5$) are all classified into the Edg family, and share sequence homology with the Edg family LPA receptors (LPA_1–LPA_3) (Kihara et al. 2014). This indicates that their overall structures and ligand recognition mechanisms are also comparable, in addition to the similarity in their endogenous ligands. Indeed, the structures of LPA_1 and $S1P_1$ superimposed well, with the conserved residues located at similar positions. The structure of $S1P_1$ was determined in complex with the S1P-mimicking antagonist ML056 (Hanson et al. 2012), and thus provided insights into the S1P-recognition mechanism.

Based on the functional and structural similarities between LPA_1 and $S1P_1$, we can rationally predict the LPA-recognition mechanism by LPA_1 from the putative S1P-binding mode deduced from the $S1P_1$ structure. In the structure of $S1P_1$, the phosphate head group of its ligand ML056 is recognised by Tyr29, Lys34, and Arg120, which are highly conserved among the Edg family receptors (Hanson et al. 2012; Fig. 6.2d, e). In the LPA_1 structure, the corresponding residues (Tyr34, Lys39, and Arg124, respectively) are located at the positions equivalent to those of Tyr29, Lys34 and Arg120 in the $S1P_1$ structure, and similarly interact with the polar head of ONO-9780307 (Chrencik et al. 2015; Fig. 6.2b, e). These observations indicated that LPA_1 recognises the phosphate group of LPA using these conserved residues, and explained the previous mutagenesis analyses (Parrill et al. 2000). The $S1P_1$ structure also revealed that the amine group of the ML056 head group, which mimics the amine group of the S1P sphingosine backbone, is recognised by Asn101 and Glu121 (Hanson et al. 2012; Fig. 6.2d, e). These two residues are strictly conserved among the five S1P receptors, whereas they are replaced with leucine and glutamine in LPA_1–LPA_3 (Leu105 and Gln125 in LPA_1). Inasmuch as the LPA molecule possesses a hydroxyl group at the position corresponding to the S1P amine group (Fig. 6.2e), these two residues of the Edg family LPA receptors, leucine and glutamine, are likely to determine the ligand selectivity of the receptors, as previously suggested by mutagenesis studies (Wang et al. 2001).

The structures of LPA_1 and $S1P_1$ provided insights into the recognition mechanism of the LPA head group. In addition, the determined LPA_1 structure can be utilised for various other analyses, including mutagenesis experiments, ligand docking simulations, and modelling of LPA_2 and LPA_3. Thus, the LPA_1 structure will not only advance the pharmacological analyses targeting LPA signalling, but also establish the basic framework for understanding the molecular mechanisms of the Edg family LPA receptors. However, many important questions still remain to be addressed. First, how the ligand-binding pockets of the LPA receptors accommodate diverse LPA molecules with different acyl-chain lengths and saturations is unclear. It is also unknown whether the lipidic ligands enter the ligand-binding pocket from the extracellular side or from within the lipid bilayer, as suggested from the $S1P_1$ structure (Hanson et al. 2012). Finally, the means by which the receptors distinguish LPA molecules from other similar lysolipids, such as LPCs and lysophosphatidylserines, remain elusive. To answer these questions, further biochemical and structural analyses are necessary.

6.1.4 Perspective

The ATX and LPA receptors are both highly important therapeutic targets for diverse diseases, such as cancer and fibrosis. Now that the ATX and LPA_1 structures have been reported, structure-guided drug development has become a realistic strategy. The structural studies of ATX and LPA_1 discussed in this chapter are examples of successful structure-guided compound optimisation, and similar approaches can be employed for other drug-target proteins to develop and modify potent chemical compounds. To pharmacologically target the LPA signalling pathway, many proteins remain to be studied, such as the non-Edg family LPA receptors (LPA_4–LPA_6), another LPA-producing enzyme, PA-$PLA_1\alpha$ (Sonoda et al. 2002; Inoue et al. 2011), and the LPA-degradation enzyme lipid-phosphate phosphatases (LPPs) (Pyne et al. 2004). The structural analyses of these proteins are also awaited to further clarify the LPA signalling mechanism and increase the possibility of drug discovery.

6.2 Hepcidin-Ferroportin Axis

6.2.1 *Inflammation and Iron Restriction*

Pathogen infection is the major cause of inflammation. The detection of pathogen-derived molecules by immune cells leads to the production and secretion of inflammatory mediators, such as cytokines and lipid mediators, which provoke various inflammatory responses in order to limit the expansion of pathogens, accelerate the removal of pathogens, and promote the healing of damaged tissues. Among the various reactions induced under inflammatory conditions, iron restriction is one of the simple but effective procedures for limiting the growth of pathogens. Iron is an essential nutrient for pathogenic microorganisms, and during infection, they obtain iron from their surrounding environment to proliferate and expand. Thus, the reduction of available iron in the host serum can prevent pathogen expansion, and thus facilitate the rapid removal of invading pathogens. Indeed, iron supplementation to the infected host leads to the exacerbation of pathogen infection (Holbein 1981; Schaible et al. 2002), indicating the significance of iron restriction for the host defense system.

When inflammation becomes chronic, however, this iron restriction system is constitutively activated without infection, leading to the continuous reduction of the serum iron level. Therefore, chronic inflammation and associated diseases, such as cancer and autoimmune disorders, cause the anaemic condition called ‘anaemia of chronic inflammation (ACI)’. To treat ACI, the identification and the resolution of the underlying diseases are straightforward approaches. However, in some cases the causes of the inflammatory conditions of ACI patients are unclear, and thus the treatment targeting ACI itself is still necessary. The molecular basis of ACI had long been elusive, but recent studies focusing on iron homeostasis regulation identified two key players in ACI, ferroportin and hepcidin.

6.2.2 *Ferroportin and Hepcidin*

Ferroportin (FPN) is the sole iron exporter in vertebrates, and it releases Fe^{2+} from the cytoplasm into serum, thereby maintaining the proper cytoplasmic and serum iron levels (McKie et al. 2000; Donovan et al. 2000; Abboud and Haile 2000). FPN is expressed on the basolateral membrane of duodenal enterocytes, and supplies the dietary absorbed iron to the blood. Thus, the level of FPN activity is a major determinant of the amount of iron absorption. In addition, macrophages and hepatocytes also express high levels of FPN (Donovan et al. 2000; Abboud and Haile 2000; Yang et al. 2002). In macrophages, FPN exports the iron retrieved from senescent erythrocytes, whereas in hepatocytes, it exports the iron stored in the liver.

The activity of FPN is downregulated by hepcidin, a small peptide hormone produced in the liver (Nemeth et al. 2004b; Jordan et al. 2009). Hepcidin directly binds to FPN, and induces ubiquitination-mediated internalisation and subsequent lysosomal degradation of FPN (Nemeth et al. 2004b; Qiao et al. 2012; Ross et al. 2012). This reaction reduces the cell surface expression level of FPN, thereby limiting the iron export rate. Hepcidin production is induced by interleukin-6 (IL-6), an inflammatory cytokine (Nemeth et al. 2004a). Therefore, during pathogen infection, the IL-6-induced increase of the hepcidin level leads to the subsequent transient inhibition of FPN-mediated iron export, resulting in the reduction of the serum iron level. In the case of ACI, the continuous downregulation of FPN by the high hepcidin expression level reduces the dietary iron absorption and the iron recycling, leading to a drop in the available iron level.

Thus, the hepcidin-FPN axis is an attractive therapeutic target for ACI, and the understanding of the molecular basis of this pathway is required in order to rationally develop potent chemical compounds. However, because no structural information about FPN or its homologous transporters had been reported, the overall architecture and the transport mechanism of FPN remained unclear. In addition, the hepcidin-binding site was predicted to be located on the extracellular loop of FPN (Wallace et al. 2010; Le Gac et al. 2013; Bonaccorsi di Patti et al. 2014; Callebaut et al. 2014), but this model could not explain the mechanism of the hepcidin-induced ubiquitination of FPN. Therefore, the detailed molecular mechanisms of the FPN-mediated iron export and the FPN downregulation by hepcidin remained largely unknown, which hampered the rational development of FPN-targeting therapeutic compounds.

6.2.3 BbFPN Structure and Insights into FPN Function

In 2015, our group determined the structures of a bacterial homologue of FPN from *Bdellovibrio bacteriovorus* (BbFPN; Bonaccorsi di Patti et al. 2015), in both the outward- and inward-facing states (Taniguchi et al. 2015; Fig. 6.3a). The structure consists of 12 transmembrane helices (TM), which can be divided into two 6-TM bundles related by twofold pseudosymmetry. A large hydrophilic cavity is formed between these two 6-TM bundles. This cavity opens toward the extracellular side in the outward-facing structure, and it opens toward the cytoplasmic side in the inward-facing structure. A conserved metal-binding site is present at the center of the N-terminal 6-TM bundle. Our biochemical analyses indicated that this site functions as the substrate transition metal-binding site (Taniguchi et al. 2015). Based on these results, we propose that FPN undergoes a conformational transition between the outward- and inward-facing states, during which the substrate is captured by the substrate-binding site in the N-terminal bundle and translocated across the cell membrane.

The hepcidin-binding residues of eukaryotic FPNs were previously identified by mutagenesis studies (Fernandes et al. 2009; Preza et al. 2011; Mayr et al. 2011;

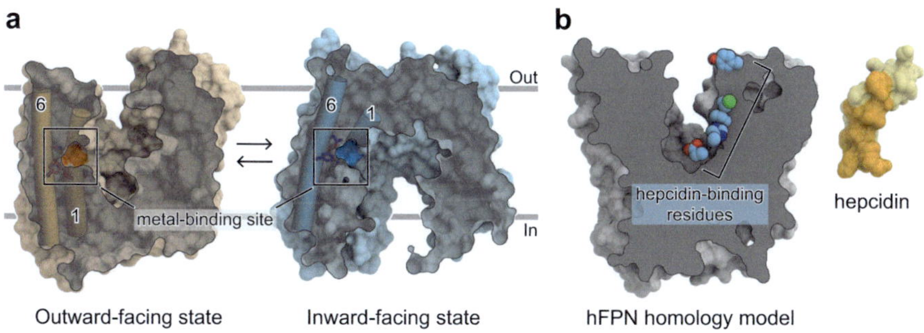

Fig. 6.3 (**a**) Cut-away surface representations of the outward-facing structure (PDB ID: 5AYN) and the inward-facing structure (PDB ID: 5AYO) of BbFPN, with semitransparent cross-sections. The metal-binding sites are highlighted, with TM1 and TM6, which form the metal-binding sites, shown as cylinders. (**b**) Surface representations of the hFPN homology model and the NMR structure of hepcidin (PDB ID: 2KEF), shown on the same scale. The cross-section view of the hFPN homology model is shown, with the hepcidin-binding residues highlighted as cpk models. In the hepcidin structure, the important region for the interaction with FPN is highlighted in *yellow*

Callebaut et al. 2014). Inasmuch as these residues are not conserved in BbFPN, we constructed a homology model of human FPN, based on the BbFPN outward-facing structure, to predict the hepcidin-binding site. In the homology model, the five hepcidin-binding residues are all located within the central cavity or at its entrance (Taniguchi et al. 2015; Fig. 6.3b), although they were predicted to be located on the extracellular loop. Taken together with the similarity between the sizes of hepcidin and the FPN central cavity, we propose that hepcidin enters the central cavity and induces a conformational change, which subsequently alters the architecture of the intracellular side of FPN and triggers its ubiquitination-dependent internalisation. Furthermore, this binding mode also suggests that hepcidin binding directly arrests the conformational transition between the outward- and inward-facing states. In addition to the internalisation-dependent downregulation, hepcidin binding may directly inhibit the transport activity of FPN, which is beneficial for the rapid regulation of the serum iron level.

The detailed structural information of BbFPN provided novel insights and ideas regarding the molecular basis of the hepcidin-FPN axis, such as the transport cycle of FPN and the approximate position of the hepcidin-binding site. These findings will substantially advance our understanding of this pathway, and provide the framework for further biochemical and structural analyses of these molecules. In addition, our structural analyses also suggested that the inside of the FPN central cavity can be utilised as the potential target region for drug development, which may open up new possibilities for the design of FPN-targeting therapeutic compounds for the treatment of ACI.

6.2.4 Perspective

In addition to membrane receptors and enzymes, membrane transporters have recently emerged as attractive drug targets. However, the structure determination of eukaryotic membrane transporters is still a challenging task, and the analysis of bacterial homologue proteins is currently the prevailing approach. Our structural analysis of BbFPN can be regarded as a successful example of such studies. Nevertheless, detailed information about the hepcidin-binding mode to FPN is definitely needed, and the structure determination of eukaryotic FPN in complex with hepcidin remains as an important challenge.

References

Abboud S, Haile DJ (2000) A novel mammalian iron-regulated protein involved in intracellular iron metabolism. J Biol Chem 275:19906–19912. doi:10.1074/jbc.M000713200

Albers HMHG, Hendrickx LJD, van Tol RJP et al (2011) Structure-based design of novel boronic acid-based inhibitors of autotaxin. J Med Chem 54:4619–4626. doi:10.1021/jm200310q

Barbayianni E, Kaffe E, Aidinis V, Kokotos G (2015) Autotaxin, a secreted lysophospholipase D, as a promising therapeutic target in chronic inflammation and cancer. Prog Lipid Res 58:76–96. doi:10.1016/j.plipres.2015.02.001

Bonaccorsi di Patti MC, Polticelli F, Cece G et al (2014) A structural model of human ferroportin and of its iron binding site. FEBS J 281:2851–2860. doi:10.1111/febs.12825

Bonaccorsi di Patti MC, Polticelli F, Tortosa V et al (2015) A bacterial homologue of the human iron exporter ferroportin. FEBS Lett 589:3829–3835. doi:10.1016/j.febslet.2015.11.025

Callebaut I, Joubrel R, Pissard S et al (2014) Comprehensive functional annotation of 18 missense mutations found in suspected hemochromatosis type 4 patients. Hum Mol Genet 23:4479–4490. doi:10.1093/hmg/ddu160

Chrencik JE, Roth CB, Terakado M et al (2015) Crystal structure of antagonist bound human lysophosphatidic acid receptor 1. Cell 161:1633–1643. doi:10.1016/j.cell.2015.06.002

Donovan A, Brownlie A, Zhou Y et al (2000) Positional cloning of zebrafish ferroportin1 identifies a conserved vertebrate iron exporter. Nature 403:776–781. doi:10.1038/35001596

Fernandes A, Preza GC, Phung Y et al (2009) The molecular basis of hepcidin-resistant hereditary hemochromatosis. Blood 114:437–443. doi:10.1182/blood-2008-03-146134

Fulkerson Z, Wu T, Sunkara M et al (2011) Binding of autotaxin to integrins localizes lysophosphatidic acid production to platelets and mammalian cells. J Biol Chem 286:34654–34663. doi:10.1074/jbc.M111.276725

Gräler M (2002) Lysophospholipids and their G protein-coupled receptors in inflammation and immunity. Biochim Biophys Acta - Mol Cell Biol Lipids 1582:168–174. doi:10.1016/S1388-1981(02)00152-X

Hanson MA, Roth CB, Jo E et al (2012) Crystal structure of a lipid G protein-coupled receptor. Science 335:851–855. doi:10.1126/science.1215904

Hausmann J, Kamtekar S, Christodoulou E et al (2011) Structural basis of substrate discrimination and integrin binding by autotaxin. Nat Struct Mol Biol 18:198–204. doi:10.1038/nsmb.1980

Holbein BE (1981) Enhancement of Neisseria meningitidis infection in mice by addition of iron bound to transferrin. Infect Immun 34:120–125

Inoue A, Arima N, Ishiguro J et al (2011) LPA-producing enzyme PA-PLA$_1\alpha$ regulates hair follicle development by modulating EGFR signalling. EMBO J 30:4248–4260. doi:10.1038/emboj.2011.296

Jordan JB, Poppe L, Haniu M et al (2009) Hepcidin revisited, disulfide connectivity, dynamics, and structure. J Biol Chem 284:24155–24167. doi:10.1074/jbc.M109.017764

Kawaguchi M, Okabe T, Okudaira S et al (2013) Screening and X-ray crystal structure-based optimization of autotaxin (ENPP2) inhibitors, using a newly developed fluorescence probe. ACS Chem Biol 8:1713–1721. doi:10.1021/cb400150c

Kihara Y, Maceyka M, Spiegel S, Chun J (2014) Lysophospholipid receptor nomenclature review: IUPHAR Review 8. Br J Pharmacol 171:3575–3594. doi:10.1111/bph.12678

Koyama M, Nishimasu H, Ishitani R, Nureki O (2012) Molecular dynamics simulation of Autotaxin: roles of the nuclease-like domain and the glycan modification. J Phys Chem B 116:11798–11808. doi:10.1021/jp303198u

Le Gac G, Ka C, Joubrel R et al (2013) Structure-function analysis of the human ferroportin iron exporter (SLC40A1): effect of hemochromatosis type 4 disease mutations and identification of critical residues. Hum Mutat 34:1371–1380. doi:10.1002/humu.22369

Mayr R, Griffiths WJH, Hermann M et al (2011) Identification of mutations in SLC40A1 that affect ferroportin function and phenotype of human ferroportin iron overload. Gastroenterology 140:2056–2063, 2063.e1. doi:10.1053/j.gastro.2011.02.064

McKie AT, Marciani P, Rolfs A et al (2000) A novel duodenal iron-regulated transporter, IREG1, implicated in the basolateral transfer of iron to the circulation. Mol Cell 5:299–309. doi:10.1016/S1097-2765(00)80425-6

Nakasaki T, Tanaka T, Okudaira S et al (2008) Involvement of the lysophosphatidic acid-generating enzyme autotaxin in lymphocyte-endothelial cell interactions. Am J Pathol 173:1566–1576. doi:10.2353/ajpath.2008.071153

Nemeth E, Rivera S, Gabayan V et al (2004a) IL-6 mediates hypoferremia of inflammation by inducing the synthesis of the iron regulatory hormone hepcidin. J Clin Invest 113:1271–1276. doi:10.1172/JCI200420945

Nemeth E, Tuttle MS, Powelson J et al (2004b) Hepcidin regulates cellular iron efflux by binding to ferroportin and inducing its internalization. Science 306:2090–2093. doi:10.1126/science.1104742

Nishimasu H, Okudaira S, Hama K et al (2011) Crystal structure of autotaxin and insight into GPCR activation by lipid mediators. Nat Struct Mol Biol 18:205–212. doi:10.1038/nsmb.1998

Noguchi K, Herr D, Mutoh T, Chun J (2009) Lysophosphatidic acid (LPA) and its receptors. Curr Opin Pharmacol 9:15–23. doi:10.1016/j.coph.2008.11.010

Oikonomou N, Mouratis M-A, Tzouvelekis A et al (2012) Pulmonary autotaxin expression contributes to the pathogenesis of pulmonary fibrosis. Am J Respir Cell Mol Biol 47:566–574. doi:10.1165/rcmb.2012-0004OC

Parrill AL, Wang D, Bautista DL et al (2000) Identification of Edg1 receptor residues that recognize sphingosine 1-phosphate. J Biol Chem 275:39379–39384. doi:10.1074/jbc.M007680200

Preza GC, Ruchala P, Pinon R et al (2011) Minihepcidins are rationally designed small peptides that mimic hepcidin activity in mice and may be useful for the treatment of iron overload. J Clin Invest 121:4880–4888. doi:10.1172/JCI57693

Pyne S, Kong K-C, Darroch PI (2004) Lysophosphatidic acid and sphingosine 1-phosphate biology: the role of lipid phosphate phosphatases. Semin Cell Dev Biol 15:491–501. doi:10.1016/j.semcdb.2004.05.007

Qiao B, Sugianto P, Fung E et al (2012) Hepcidin-induced endocytosis of ferroportin is dependent on ferroportin ubiquitination. Cell Metab 15:918–924. doi:10.1016/j.cmet.2012.03.018

Ross SL, Tran L, Winters A et al (2012) Molecular mechanism of hepcidin-mediated ferroportin internalization requires ferroportin lysines, not tyrosines or JAK-STAT. Cell Metab 15:905–917. doi:10.1016/j.cmet.2012.03.017

Schaible UE, Collins HL, Priem F, Kaufmann SHE (2002) Correction of the iron overload defect in −2-microglobulin knockout mice by lactoferrin abolishes their increased susceptibility to tuberculosis. J Exp Med 196:1507–1513. doi:10.1084/jem.20020897

Sonoda H, Aoki J, Hiramatsu T et al (2002) A novel phosphatidic acid-selective phospholipase A1 that produces lysophosphatidic acid. J Biol Chem 277:34254–34263. doi:10.1074/jbc.M201659200

Stracke ML, Krutzsch HC, Unsworth EJ et al (1992) Identification, purification, and partial sequence analysis of autotaxin, a novel motility-stimulating protein. J Biol Chem 267:2524–2529

Taniguchi R, Kato HE, Font J et al (2015) Outward- and inward-facing structures of a putative bacterial transition-metal transporter with homology to ferroportin. Nat Commun 6:8545. doi:10.1038/ncomms9545

Tokumura A, Majima E, Kariya Y et al (2002) Identification of human plasma lysophospholipase D, a lysophosphatidic acid-producing enzyme, as autotaxin, a multifunctional phosphodiesterase. J Biol Chem 277:39436–39442. doi:10.1074/jbc.M205623200

Umezu-Goto M, Kishi Y, Taira A et al (2002) Autotaxin has lysophospholipase D activity leading to tumour cell growth and motility by lysophosphatidic acid production. J Cell Biol 158:227–233. doi:10.1083/jcb.200204026

van Meeteren LA, Moolenaar WH (2007) Regulation and biological activities of the autotaxin-LPA axis. Prog Lipid Res 46:145–160. doi:10.1016/j.plipres.2007.02.001

Wallace DF, Harris JM, Subramaniam VN (2010) Functional analysis and theoretical modeling of ferroportin reveals clustering of mutations according to phenotype. Am J Physiol Cell Physiol 298:C75–C84. doi:10.1152/ajpcell.00621.2008

Wang DA, Lorincz Z, Bautista DL et al (2001) A single amino acid determines lysophospholipid specificity of the S1P1 (EDG1) and LPA1 (EDG2) phospholipid growth factor receptors. J Biol Chem 276:49213–49220. doi:10.1074/jbc.M107301200

Watanabe N, Ikeda H, Nakamura K et al (2007) Both plasma lysophosphatidic acid and serum autotaxin levels are increased in chronic hepatitis C. J Clin Gastroenterol 41:616–623. doi:10.1097/01.mcg.0000225642.90898.0e

Yang F, Liu X-B, Quinones M et al (2002) Regulation of reticuloendothelial iron transporter MTP1 (Slc11a3) by inflammation. J Biol Chem 277:39786–39791. doi:10.1074/jbc.M201485200

Chapter 7
Lipid Signals in the Resolution of Inflammation

Makoto Arita

Abstract Acute inflammation is an indispensable host defence mechanism, and its proper resolution is important in keeping tissue homeostasis. The resolution of inflammation involves active cellular and molecular programs that enable inflamed tissues to return to homeostasis. The mechanisms by which acute inflammation is resolved are of interest, and research in recent years has uncovered novel endogenous mechanisms that involve eosinophils, macrophages, and their lipid signals in controlling acute inflammation and resolution.

Keywords Omega-3 fatty acid • Resolution of inflammation • Lipidomics • Lipid mediator • Lipoxygenase • LC-MS/MS

7.1 Introduction

Inflammation is an indispensable host response to injury and infection, but excessive and uncontrolled inflammation is a key component of pathogenesis in many chronic diseases (Nathan and Ding 2010). Acute inflammation consists of two main phases: the initiation phase and the resolution phase. The initiation phase is characterised by the rapid infiltration of polymorphonuclear neutrophils (PMNs) and is followed by the infiltration of monocytes that mature into macrophages. PMNs constitute the first line of immune defence by migrating to sites of injury, and neutralising invading microorganisms or noxious materials by phagocytosis. The resolution phase is characterised by PMN apoptosis and their subsequent uptake by macrophages, which then emigrate rapidly from the inflamed site to the draining lymph nodes (DLNs). In healthy individuals, acute inflammation is self-limiting and has an active termination program (Serhan 2014). Therefore, the identification

M. Arita (✉)
Laboratory for Metabolomics, RIKEN Center for Integrative Medical Sciences, 1-7-22 Suehiro-cho, Tsurumi ku, Yokohama, Kanagawa 230-0045, Japan

Graduate School of Medical Life Science, Yokohama City University, 1-7-29 Suehiro-cho, Tsurumi-ku, Yokohama, Kanagawa 230-0045, Japan

PRESTO, Japan Science and Technology Agency, Kawaguchi 332-0012, Japan
e-mail: makoto.arita@riken.jp

M. Miyasaka, K. Takatsu (eds.), *Chronic Inflammation*,
DOI 10.1007/978-4-431-56068-5_7

of such endogenous anti-inflammatory and/or proresolution mechanisms is of wide interest.

7.2 Lipid Mediators That Regulate Inflammation

Polyunsaturated fatty acids (PUFAs) exhibit a range of biological effects, many of which are mediated through the formation and actions of lipid mediators (Serhan 2014; Funk 2001; Arita 2012) (Fig. 7.1). PUFA-derived lipid mediators are formed by enzymatic oxidation through the action of cyclooxygenases (COX), lipoxygenases (LOX), and cytochrome P450 monooxygenases (CYP). When cells are stimulated, arachidonic acid is released from membrane phospholipids, and unesterified arachidonic acid is immediately metabolised by fatty acid oxygenases such as COX, LOX, or CYP. The COX pathway leads to the formation of prostaglandins (PG) and thromboxanes (TX), the LOX pathway leads to leukotrienes (LT), lipoxins (LX), and the CYP pathway leads to hydroxy-eicosatetraenoic acids (HETE) and epoxy-eicosatrienoic acids (EET). COX-derived PGs are involved in pro-inflammatory processes, and are responsible for the hallmark signs of inflammation such as heat, redness, swelling, and pain. 5-LOX-derived LTs are involved

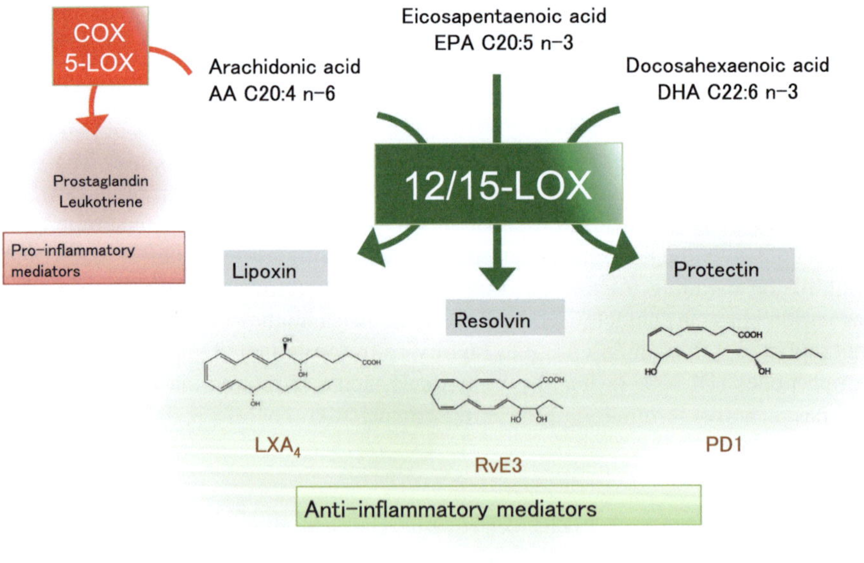

Fig. 7.1 Lipid mediators that regulate inflammatory responses. In addition to arachidonic acid-derived mediators such as COX-derived PGs, 5-LOX-derived LTs, 12/15-LOX-derived mediators are present such as LXs and n-3 PUFA-derived resolvins and protectins. These mediators are coordinately produced in the course of inflammation and resolution, and play important roles in regulating inflammatory responses

in proinflammatory processes such as PMN infiltration and increased vascular permeability. In contrast, 5- and 12/15-LOX-derived LXA_4 counterregulates the proinflammatory processes and may be important in the resolution of inflammation. The epoxygenation of arachidonic acid by CYP generates EETs, which may have roles in regulation of smooth muscle cells and vascular tone.

Eicosapentaenoic acid (EPA) and docosahexaenoic acid (DHA) are n-3 PUFAs that are widely held to be anti-inflammatory and tissue-protective. Dietary supplementation of n-3 PUFA has beneficial effects in many inflammatory disorders including cardiovascular disease, arthritis, colitis, and metabolic syndrome (Simopoulos 2002). Also, elevation of tissue n-3 PUFA levels in fatty acid n-3 desaturase (*fat-1*) transgenic mice is protective against inflammatory disease models (Kang 2007). N-3 PUFAs are thought to act via several mechanisms. One role is to serve as an alternative substrate for COX or 5-LOX, resulting in the production of less potent products. Another role is to be converted to endogenous anti-inflammatory and protective mediators. For example, 12/15-LOX-derived DHA product protectin D1 (PD1) displays bioactivity to promote the resolution of inflammation by reducing PMN influx and stimulating clearance of apoptotic cells and inflammatory debris (Schwab et al. 2007). Thus, endogenous cellular sources of PD1 and other proresolving mediators are of interest.

7.3 Lipid Mediator Class Switching in Acute Inflammation and Resolution

Mediator lipidomics concerns the simultaneous and quantitative analysis of bioactive lipid mediators and their metabolic pathways. A comprehensive analysis of fatty acid-derived mediators can be achieved by liquid chromatography tandem mass spectrometry (LC-MS/MS). We developed a comprehensive LC-MS/MS method that can simultaneously detect and quantify more than 500 of fatty acid metabolites (Arita 2012).

The inflammatory response is characterised by a sequential release of mediators and the recruitment of different types of inflammatory cells. Acute inflammation is characterised by oedema formation and the initial recruitment of PMNs followed by the recruitment of monocytes that differentiate into macrophages (Fig. 7.2a). Lipid mediators such as PGs and LTs, and cytokines and chemokines are temporarily induced and coordinately regulate vascular permeability and leukocyte infiltration in the initiation phase. Also, the resolution of inflammation involves the formation of resolution mediators that promote efficient clearance of inflammatory cells and exudates by lymphatic drainage into DLNs, and restoration of the inflamed tissue to homeostasis (Nathan and Ding 2010; Serhan 2014). To better understand the molecular and cellular mechanisms underlying the coordinated processes of

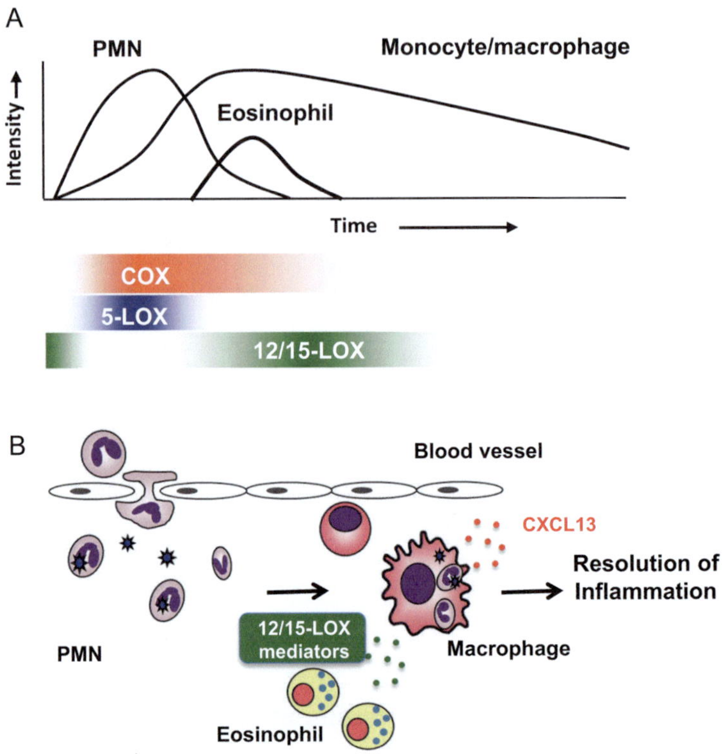

Fig. 7.2 Lipid signals in the resolution of acute inflammation. (**a**) Schematic sequence of events in the course of acute inflammation and resolution. 5-LOX and COX products were present in the initiation phase, and 12/15-LOX products were increased during the resolution phase. (**b**) Eosinophils are recruited in the early resolution phase, where they locally produce resolution mediators via a 12/15-LOX-initiated biosynthetic route, and play roles leading to the efficient clearance of inflammatory leukocytes and exudates

inflammation and resolution, we applied LC-MS/MS-based mediator lipidomics to the self-resolving acute inflammation model, namely mouse zymosan-induced peritonitis (Yamada et al. 2011). Temporal and quantitative differences in the lipid mediator profiles were observed in the course of acute inflammation and resolution (Fig. 7.2a). The maximum levels of 5-LOX-derived mediators such as LTB_4 were observed in the initiation phase, and they subsequently subsided during the resolution phase. In contrast, formation of 12/15-LOX-derived mediators such as PD1 was increased later in the resolution phase. A similar pattern of lipid mediator class switching was observed in other inflammatory models (Bannenberg et al. 2005; Blaho et al. 2009). Because 5-LOX-derived LTs are involved in proinflammatory processes such as oedema formation and PMN infiltration, we questioned whether 12/15-LOX-derived mediators are involved in the resolution processes.

To address this point, we first determined which cell type in the resolution phase express 12/15-LOX. As a result, we found that eosinophil express 12/15-LOX and was the major cell type producing 12/15-LOX-derived mediators such as PD1 in the resolution phase. In vivo depletion of eosinophils by the administration of anti-interleukin (IL)-5 antibody significantly reduced 12/15-LOX activity and its products in resolving exudates, and caused a resolution defect characterised by impaired DLN drainage along with sustained numbers of PMNs in inflamed tissues. Furthermore, the resolution deficit caused by eosinophil depletion was rescued by eosinophil restoration, and eosinophils deficient in 12/15-LOX were unable to rescue the resolution phenotype. These results clearly indicated that eosinophils are recruited to the inflamed site, where they locally produce lipid mediators such as PD1 via a 12/15-LOX-initiated biosynthetic route, and play roles in promoting resolution of acute inflammation (Fig. 7.2b) (Yamada et al. 2011).

7.4 Lipid Signals from Eosinophils Facilitate the Resolution of Inflammation

Macrophages play important roles in the resolution of inflammation by efficient clearance of apoptotic cells and/or tissue debris from inflamed sites, and facilitate tissue return to homeostasis. To further determine the molecular mechanisms by which eosinophils promote resolution of inflammation, we questioned whether 12/15-LOX-expressing eosinophils affect these macrophage functions (Tani et al. 2014).

We first examined gene expression patterns of resolution-phase macrophages by microarray analysis in wild-type and eosinophil-deficient ΔdblGATA mice. Microarray analysis revealed that a series of macrophage gene expressions was affected by the presence of eosinophils in the resolution phase. Among those, a chemokine CXCL13 was the gene on which eosinophils exert the most significant effect in a 12/15-LOX-dependent manner. CXCL13 expression in ΔdblGATA mice macrophages was only 20–30 % of that in wild-type mice, but could be restored by the adoptive transfer of wild–type eosinophils. In contrast, adoptive transfer of 12/15-LOX deficient eosinophils did not restore the CXCL13 expression. CXCL13 was found to be expressed in a subset of macrophages present in the resolution phase, and mice treated with anti-CXCL13 antibody displayed a resolution deficit accompanied by a significant increase in PMN number that was similar to eosinophil-deficient ΔdblGATA mice.

CXCL13 is a functional chemokine in the early development of peripheral lymph nodes. It is well characterised that positive feedback regulation of stromal CXCL13 by lymphotoxin released by CXCR5-expressing lymphoid tissue-inducer (LTi) cells is essential for early lymph node initiation (Mebius 2003). Inasmuch as activation and expansion of the DLNs are involved in promoting inflammatory cell clearance from the periphery, we next questioned whether 12/15-LOX-expressing eosinophils are involved in the regulation of inflamed DLNs in the resolution phase. Increase in DLN size and cellularity was significantly reduced in eosinophil-deficient or 12/15-LOX-deficient mice. Analysis of cellular composition of the inflamed DLNs revealed that eosinophil was present in inflamed DLNs, and its depletion reduced numbers of each cellular component, including lymphocytes, macrophages, PMNs, and DCs. Adoptive transfer of eosinophils or administration of recombinant CXCL13 dramatically restored the inflamed DLN hypertrophy with an increased number of cells. These results demonstrated that 12/15-LOX-expressing eosinophils control the resolution of inflammation and DLN hypertrophy through the CXCL13 pathway in mice (Fig. 7.2b) (Tani et al. 2014). Inflamed DLN expansion may be considered as a passive process after lymphatic drainage, but it is possible that eosinophils actively promote DLN remodeling and expansion that could in turn drive clearance of inflammatory components from inflammatory sites. Further study is needed to understand molecular mechanisms in more detail, but these results clearly demonstrate an active mechanism of endogenous cellular and molecular cascade that coordinately brings acute inflammation to a close.

7.5 Perspectives

Eosinophils are thought to be involved primarily in the innate immune response against parasitic infection and also in the allergic responses (Rothenberg and Hogan 2006). Recently, eosinophils were shown to increase the ratio of alternatively activated macrophages in adipose tissue, which in turn improved glucose metabolism (Wu et al. 2011). In addition to these activities, here we propose an active role for eosinophils as protective effector cells that contribute to the resolution of inflammation via the 12/15-LOX-mediated lipid signals. This is a novel endogenous cellular and molecular mechanism that coordinately brings acute inflammation to a close. The question remains of how eosinophils appear in a timely manner in the resolution phase. Also it is of interest whether resolution phase eosinophils represent a distinct functional subset.

In addition to eosinophils, 12/15-LOX is expressed in tissue-resident macrophages and airway epithelial cells (Kühn and O'Donnell 2006). 12/15-LOX-expressing macrophages are present in naïve mouse peritoneal cavity and also in the resolution phase of acute peritonitis, and display competent phagocytotic activity that potentially contributes to tissue homeostasis by efficient clearance of apoptotic cells (Uderhardt et al. 2012). Also 12/15-LOX expression levels can be upregulated by Th2 cytokines such as IL-4 and IL-13 (Kühn and O'Donnell 2006). Cells expressing 12/15-LOX might be involved in regulating inflammatory tones by locally producing lipid mediators. This is consistent with the findings of impaired host defence and inflammatory phenotypes in 12/15-LOX-deficient mice in several disease models (Morita et al. 2013; Gronert et al. 2005; Merched et al. 2008; Krönke et al. 2009), and potentially related to clinical finding of dysregulated 12/15-LOX activity in eosinophils from patients with severe asthma (Miyata et al. 2013). Identification of such endogenous anti-inflammatory and/or proresolving lipid signals (Serhan 2014; Isobe et al. 2012; Kubota et al. 2014; Yokokura et al. 2014; Endo et al. 2014) and elucidation of their molecular mechanisms could lead to the development of resolution agonist-based therapeutics when sustained inflammation and impaired resolution are suspected as key components of pathogenesis.

References

Arita M (2012) Mediator lipidomics in acute inflammation and resolution. J Biochem 152:313–319

Bannenberg GL, Chiang N, Ariel A et al (2005) Molecular circuits of resolution: formation and actions of resolvins and protectins. J Immunol 174:4345–4355

Blaho VA, Buczynski MW, Brown CR et al (2009) Lipidomic analysis of dynamic eicosanoid responses during the induction and resolution of lyme arthritis. J Biol Chem 284:21599–21612

Endo J, Sano M, Isobe Y et al (2014) 18-HEPE, an n-3 fatty acid metabolite released by macrophages, prevents pressure overload-induced cardiac remodeling. J Exp Med 211:1673–1687

Funk CD (2001) Prostaglandins and leukotrienes: advances in eicosanoid biology. Science 294:1871–1875

Gronert K, Maheshwari N, Khan N et al (2005) A role for the mouse 12/15-lipoxygenase pathway in promoting epithelial wound healing and host defense. J Biol Chem 280:15267–15278

Isobe Y, Arita M, Matsueda S et al (2012) Identification and structure determination of novel anti-inflammatory mediator resolvin E3, 17,18-dihydroxyeicosapentaenoic acid. J Biol Chem 287:10525–10534

Kang JX (2007) Fat-1 transgenic mice: a new model for omega-3 research. Prostaglandins Leukot Eessent Fat Acids 77:263–267

Krönke G, Katzenbeisser J, Uderhardt S et al (2009) 12/15-Lipoxygenase counteracts inflammation and tissue damage in arthritis. J Immunol 183:3383–3389

Kubota T, Arita M, Isobe Y et al (2014) Eicosapentaenoic acid is converted via ω-3 epoxygenation to the anti-inflammatory metabolite 12-hydroxy-17,18-epoxyeicosatetraenoic acid. FASEB J 28:586–593

Kühn H, O'Donnell VB (2006) Inflammation and immune regulation by 12/15-lipoxygenases. Prog Lipid Res 45:334–356

Mebius RE (2003) Organogenesis of lymphoid tissues. Nat Rev Immunol 3:292–303

Merched AJ, Ko K, Gotlinger KH et al (2008) Atherosclerosis. Evidence for impairment of resolution of vascular inflammation governed by specific lipid mediators. FASEB J 22:3595–3606

Miyata J, Fukunaga K, Iwamoto R et al (2013) Dysregulated synthesis of protectin D1 in eosinophils from patients with severe asthma. J Allergy Clin Immunol 131:353–360

Morita M, Kuba K, Ichikawa A et al (2013) The lipid mediator protectin D1 inhibits influenza virus replication and improves severe influenza. Cell 153:112–125

Nathan C, Ding A (2010) Nonresolving inflammation. Cell 140:871–882

Rothenberg ME, Hogan SP (2006) The eosinophil. Annu Rev Immunol 24:147–174

Schwab JM, Chiang N, Arita M et al (2007) Resolvin E1 and protectin D1 activate inflammation-resolution programmes. Nature 447:869–874

Serhan CN (2014) Pro-resolving lipid mediators are leads for resolution physiology. Nature 510:92–101

Simopoulos AP (2002) Omega-3 fatty acids in inflammation and autoimmune diseases. J Am Coll Nutr 21:495–505
Tani Y, Isobe Y, Imoto Y et al (2014) Eosinophils control the resolution of inflammation and draining lymph node hypertrophy through the proresolving mediators and CXCL13 pathway in mice. FASEB J 28:4036–4043
Uderhardt S, Herrmann M, Oskolkova OV et al (2012) 12/15-lipoxygenase orchestrates the clearance of apoptotic cells and maintains immunologic tolerance. Immunity 36:1–13
Wu D, Molofsky AB, Liang HE et al (2011) Eosinophils sustain adipose alternatively activated macrophages associated with glucose homeostasis. Science 332:243–247
Yamada T, Tani Y, Nakanishi H et al (2011) Eosinophils promote resolution of acute peritonitis by producing proresolving mediators in mice. FASEB J 25:561–568
Yokokura Y, Isobe Y, Matsueda S et al (2014) Identification of 14,20-dihydroxy-docosahexaenoic acid as a novel anti-inflammatory metabolite. J Biochem 156:315–321

Chapter 8
Regulation of Chronic Inflammation by Control of Macrophage Activation and Polarization

Junko Sasaki and Takehiko Sasaki

Abstract Heterogeneity and plasticity are hallmarks of macrophages. Activated macrophages can be classified into two groups: M1 (or classically activated) and M2 (or alternatively activated) macrophages. Various pathological conditions have been associated with dynamic changes in the phenotypes of macrophages. Impaired polarization of macrophages affects the progression of several diseases including inflammation, tumour progression, and obesity-associated insulin resistance. Identification of the activation status of macrophages and regulations of macrophage polarization from M1 to M2 or vice versa, might serve as novel diagnostic or therapeutic approaches for various diseases. Many signalling molecules, transcription factors, epigenetic modifiers, and miRNAs (microRNAs) are known to regulate macrophage polarization and activation. Here we show that phosphatase and PTEN (phosphatase and tensin homologue deleted on chromosome 10) and SHIP1 (Src homology 2-containing inositol phosphatase 1), which dephosphorylate second messenger PtdIns(3,4,5)P_3 (phosphatidylinositol 3,4,5-trisphosphate), can also act as regulators of macrophage polarization and activation. We compared *Pten*$^{-/-}$ macrophages and *Ship1*$^{-/-}$ macrophages under similar conditions in vitro and in vivo and found that they regulate macrophage polarization by different mechanisms.

Keywords Macrophage • Activation • Polarization • PtdIns(3,4,5)P_3 • PTEN • SHIP1

J. Sasaki (✉)
Department of Medical Biology, Akita University Graduate School of Medicine, Akita, Japan

PRESTO, Japan Science and Technology Agency, Kawaguchi 332-0012, Japan
e-mail: sasakij@med.akita-u.ac.jp

T. Sasaki
Department of Medical Biology, Akita University Graduate School of Medicine, Akita, Japan

Research Center for Biosignal, Akita University, Akita, Japan

M. Miyasaka, K. Takatsu (eds.), *Chronic Inflammation*,
DOI 10.1007/978-4-431-56068-5_8

8.1 Molecular Mechanisms of Macrophage Activation and Polarization

Macrophages are key players in innate immunity and act as critical effectors and regulators of inflammation. These cells also play a role in homeostasis, including tissue remodelling and metabolism (Murray and Wynn 2011; Gordon et al. 2014). Many studies reveal that heterogeneity and plasticity are hallmarks of macrophages (Mosser and Edwards 2008; Sica and Mantovani 2012). Macrophages can be differentially activated by the microenvironment to initiate specific and varied functions. Activated macrophages can be broadly classified into two groups: M1 macrophages (or classically- activated macrophages) induced by IFN (interferon)-γ and TLR (Toll-like receptor) ligands, and M2 macrophages (or alternatively-activated macrophages) induced by IL (interleukin)-4/IL-13, IL-10, glucocorticoid hormones, molecules released from apoptotic cells, and immune complexes. Macrophage polarization, thus, mirrors the Th1/Th2 polarization of T cells (Gordon 2003; Mantovani et al. 2002). M1 macrophages produce high levels of proinflammatory cytokines and chemokines such as TNF (tumour necrosis factor)-α, IL-12, IL-6, and CXCL1–3, as well as nitric oxide and reactive oxygen intermediates. They also exhibit potent microbicidal and tumouricidal properties and promote Th1 responses. In contrast, the phenotype of M2 macrophages is characterised by the upregulation of MR (mannose receptor), CD163, YM1 (a member of the chitinase family) and FIZZ1 (found in inflammatory zone 1). M2 macrophages produce anti-inflammatory cytokines such as IL-10 and produce ornithine and polyamines through the arginase pathway. M2 macrophages are involved in parasite containment and promote Th2 responses, tissue remodelling, and tumour progression (Mantovani et al. 2013; Sica et al. 2008). In pathological conditions, dynamic changes in the phenotype of macrophages have been observed. For example, M1 macrophages initiate and sustain inflammation, and M2 macrophages are associated with resolution or smoldering chronic inflammation (Martinez et al. 2009). Impaired polarization of macrophages accelerates inflammation, tumour progression, and obesity-associated insulin resistance. Moreover, several studies indicate that identification of the activation status of macrophages and regulation of macrophage polarization from M1 to M2, or vice versa, might serve as novel diagnostic or therapeutic approaches for various diseases (Jinushi and Komohara 2015; Lee et al. 2011; Waidmann et al. 2013).

Macrophage polarization and activation are regulated by a network of signalling molecules, transcription factors, epigenetic modifiers, and miRNAs (Fig. 8.1). IRF (interferon regulatory factor)/STAT (signal transducer and activator of transcription) signalling is a pivotal pathway in modulating M1/M2 polarization of macrophages (Lawrence and Natoli 2011). Activation of IRF/STAT signalling pathways by IFNs and TLR ligands skews macrophage function toward the M1 phenotype (via STAT1), wheras activation of IRF/STAT (via STAT3/STAT6) signalling pathways by IL-4, IL-13, and IL-10 skews macrophage function toward the M2 phenotype. TLR4 signalling stimulated by LPS (lipopolysaccharide) leads to

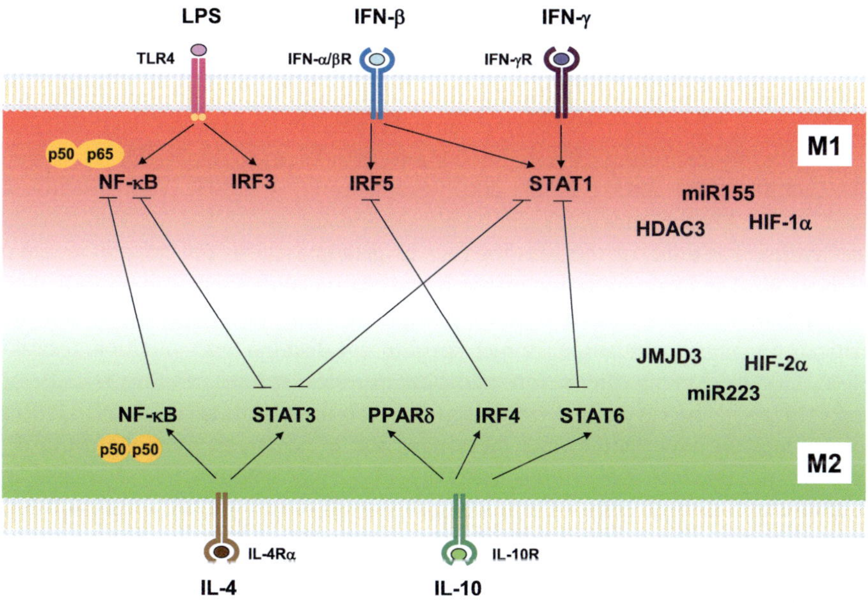

Fig. 8.1 Signalling pathways of macrophage polarization. This figure illustrates major pathways and molecules of macrophage polarization and shows the crosstalk between M1 and M2 macrophage polarization. The balance between STAT1 and STAT3/6 activation tightly controls macrophage polarization. In addition, activation of p50 and p65 NF-kB heterodimers, IRF3, IRF5, HIF-1α, HDAC3, and miR155 promotes M1 macrophage polarization. In contrast, activation of IRF4, PPARδ, p50 NF-kB homodimer, HIF-2α, JMJD3, and miR223 promotes M2 macrophage polarization

NF-kB (nuclear factor kappa B) p65/p50 heterodimer activation, which induces the expressions of many proinflammatory genes associated with M1 macrophages (Bonizzi and Karin 2004). In contrast, NF-kB p50 homodimer activation by IL-10 induces M2 polarization (Porta et al. 2009). Nuclear receptors PPARγ and PPARδ also induce M2 macrophage polarization (Chawla 2010). Hypoxia-induced factors HIF-1α and HIF-2α, epigenetic modifiers HDAC3 (histone deacetylase 3) and JMJD3 (Jumonji domain containing-3), and miRNAs miR155 and miR223 differentially induce M1 and M2 macrophage polarization (Takeda et al. 2010; Neele et al. 2015; Liu and Abraham 2013). In addition to these known regulators of macrophage polarization, we propose that PtdIns(3,4,5)P_3 phosphatases PTEN and SHIP1 regulate M1/M2 macrophage polarization.

8.2 Phosphoinositides

Phosphoinositides are membrane-signalling lipids that regulate numerous biological functions, including proliferation, survival, adhesion, migration, and vesicular trafficking, by recruiting effector proteins to cellular membranes (Di Paolo and De Camilli 2006; Sasaki et al. 2007; Stephens et al. 2008; Vanhaesebroeck et al. 2010). Phosphoinositides can be phosphorylated at the 3, 4, or 5 position of the hydroxyl residue of the inositol ring to generate seven phosphatidylinositol derivatives. The interconversion of seven phosphoinositide species is controlled by 19 phosphoinositide kinases and 28 phosphoinositide phosphatases, which are critical for phosphoinositide isomer-specific localisation and function (Sasaki et al. 2009). Among the seven phosphoinositides species, PtdIns(3,4,5)P_3 is reported to be involved in macrophage activation (Ooms et al. 2009). PtdIns (3,4,5)P_3 is an important second messenger and is generated from PtdIns(4,5)P_2 (phosphatidylinositol 4,5-bisphosphate) by PI3K (phosphoinositide 3-kinase) in response to a diverse array of extracellular stimuli (Cantley 2002). PtdIns(3,4,5) P_3 recruits signalling proteins, which contain lipid-binding domains such as the PH (pleckstrin homology) domain, to the cytoplasmic leaflet of the plasma membrane, leading to their activation (Itoh and Takenawa 2002). These downstream effectors include serine/threonine kinase, Akt (also known as protein kinase B), and its activating kinase PDK1 (phosphoinositide-dependent kinase 1), Btk (Bruton's tyrosine kinase), Rac-GEFs (guanine-nucleotide-exchange factors) such as P-Rex1 and P-Rex2, and ARF (ADP-ribosylation factor) -GEFs such as ARNO (ARF nucleotide-binding-site operator), and Grp1 (general receptor for phosphoinositides 1) (Wymann and Pirola 1998; Cullen and Venkateswarlu 1999; Franke 2008; Welch 2015). PtdIns(3,4,5)P_3 levels are tightly controlled by regulating its synthesis by PI3K and its degradation by phosphoinositide phosphatases. In macrophages, two major PtdIns(3,4,5)P_3 phosphatases, PTEN and SHIP1 control the PtdIns(3,4,5)P_3 levels (Kozicky and Sly 2015).

8.2.1 PTEN

PTEN hydrolyses the 3-position phosphate bound of PtdIns(3,4,5)P_3, forming PtdIns(4,5)P_2, thereby terminating the PI3K signalling pathway (Maehama 2007). PTEN is one of the most frequently mutated, deleted, or silenced tumour suppressor genes in human cancers. The most important function of PTEN is to regulate the key cell survival kinase Akt negatively, and thus the loss of PTEN protein function leads to enhanced cell proliferation, survival, migration, and cell size. Homozygous deletion of PTEN causes embryonic lethality. Studies in cell-type–specific PTEN knockout mice revealed the importance of PTEN not only for tumourigenesis but also for the maintenance of a variety of physiological functions to prevent conditions such as autoimmune diseases, heart failure, nonalcoholic steatohepatitis, insulin hypersensitivity, respiratory distress syndrome, and macroencephaly (Suzuki et al. 2008). Recently, several reports revealed that PTEN contributes to cell fate decisions of macrophages. Macrophages from myeloid-specific *Pten*$^{-/-}$ mice (LysMcrePTEN$^{flox/flox}$) are hyporesponsive to LPS, as shown by decreased proinflammatory cytokines and NO (nitric oxide) release (Cao et al. 2004; Kuroda et al. 2008). In addition, several M2 markers, such as Arg1, are upregulated in *Pten*$^{-/-}$ macrophages through increased expression and activation of the transcription factors C/EBPβ (CCAAT/enhancer-binding protein β) and STAT3, suggesting *Pten*$^{-/-}$ macrophages are M2 macrophages (Sahin et al. 2014). Moreover, PTEN deficiency in myeloid cells protects the liver from ischemia reperfusion injury by facilitating M2 macrophage polarization (Yue et al. 2014).

8.2.2 *SHIP1*

SHIP1 hydrolyses the 5-position phosphate bound of PtdIns(3,4,5)P_3, forming PtdIns(3,4)P_2 (phosphatidylinositol 3,4-bisphosphate) and is exclusively expressed in haematopoietic cells (Krystal 2000). This protein contains an N-terminal SH2 (Src homology 2) domain and a C-terminal proline-rich region that allows it to interact with a variety of proteins such as kinases, adaptor proteins, and membrane receptors with ITIM (immunoreceptor tyrosine inhibitory motif) or ITAM (immunoreceptor tyrosine activation motif) sequences (Fig. 8.2a). *Ship1*$^{-/-}$ mice are viable but have shortened lifespans (Helgason et al. 1998; Liu et al. 1999). This failure of *Ship1*$^{-/-}$ mice to thrive is likely because of progressive myeloid cell infiltration in the lungs and intestine, systemic mast cell hyperplasia, and elevated numbers of hyperresorptive osteoclasts. Polarization and activation of *Ship1*$^{-/-}$ macrophages are complicated. Bone-marrow–derived *Ship1*$^{-/-}$ macrophages produce higher levels of proinflammatory cytokines and NO in response to LPS compared to wild-type macrophages (Sly et al. 2009). Thus, *Ship1*$^{-/-}$ macrophages differentiated in vitro have characteristics of M1 macrophages. In agreement with this finding, it is reported that SHIP1 negatively regulates proinflammatory IL-12 secretion in response to *Francisella novicida* and *Trichuris* (Hadidi et al. 2012). In contrast, macrophages from *Ship1*$^{-/-}$ mice are hyporesponsive to LPS, secrete lower levels of TNF-α, IL-6, and IL-12, and express high levels of Arg1 and Ym1 compared to wild-type macrophages (Rauh et al. 2005). Thus, *Ship1*$^{-/-}$ macrophages differentiated in situ are M2 macrophages. Interpretations of these discrepancies require further studies, but the fact that macrophages in *Ship1*$^{-/-}$ mice only develop an M2 phenotype after 5 weeks of age suggests that this M2 skewing may be due to other factors such as TGF (transforming growth factor)-β or IL-4 from basophils (Kuroda et al. 2011).

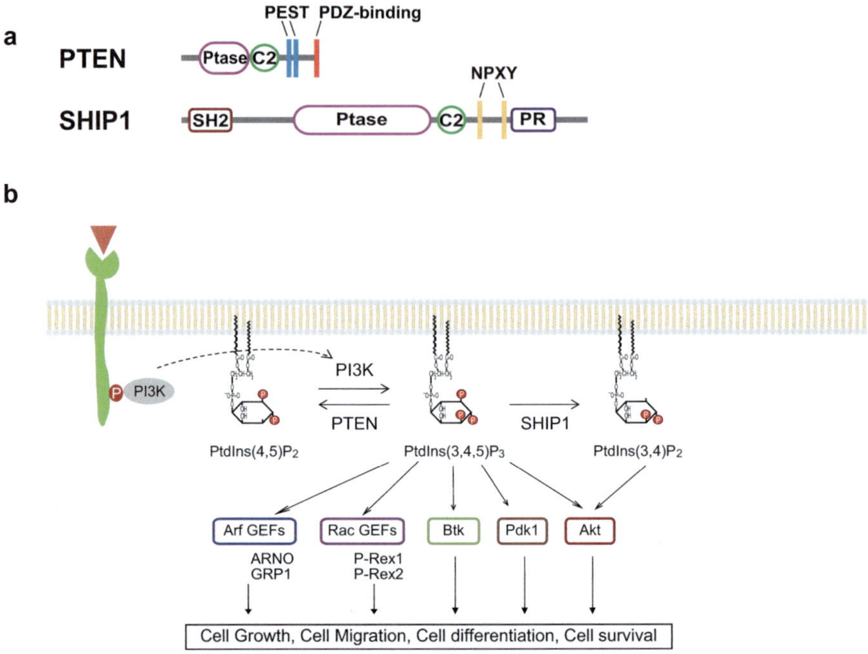

Fig. 8.2 Structure and signalling of PTEN and SHIP1. (**a**) The PTEN protein consists of a phosphatase domain, a lipid-binding C2 domain, proteolytic-recognition PEST (proline, glutamine, serine, threonine) sequence, and a PDZ domain, which is thought to be important for its stability and protein–protein interactions. Most PTEN mutations found in cancer patients are located in the phosphatase domain. The SHIP1 protein contains an SH2 domain, a phosphatase domain, a C2 domain, an NPXY (asparagine, proline, any amino acid, tyrosine) motif, and a proline-rich region. SHIP1 can interact with a variety of proteins through the SH2 domain, NPXY motif, and proline-rich region. (**b**) Activated cell-surface receptors recruit and activate PI3K, which generates PtdIns(3,4,5)P_3 from PtdIns(4,5)P_2 transiently. Then PtdIns(3,4,5)P_3 recruits and activates PH domain-containing proteins including Akt, PDK1, Btk, P-Rex, and ARNO, leading to a variety of cellular processes. PtdIns (3,4,5)P_3 signalling is terminated by PTEN and SHIP1, which convert PtdIns (3,4,5)P_3 to PtdIns(4,5)P_2 or PtdIns(3,4)P_2 respectively

8.3 Role of PtdIns(3,4,5)P_3 Phosphatases in Macrophage Polarization

8.3.1 Bone-Marrow–Derived Macrophages

Although many research articles show the role of PTEN or SHIP1 in macrophage polarization, no reports clarify their function concurrently in vitro and in vivo. Moreover, macrophages from mice are known to be affected by factors from many other cells. To clarify the roles of PTEN and SHIP1 in macrophage polarization and activation, we compared bone-marrow–derived $Pten^{-/-}$ macrophages and $Ship1^{-/-}$ macrophages under the same conditions. Macrophages can be differentiated from bone marrow cells by M-CSF. As shown in Fig. 8.3, *Pten*$^{-/-}$ macrophages from CD11bcrePTEN$^{flox/flox}$ (PTEN cKO) mice express higher levels of Arg1, MR, CD163, and IL-10, but not Ym1 or IL-12 compared to control macrophages. Moreover, TNF-α production in response to LPS is less in *Pten*$^{-/-}$ macrophages, which is a characteristic of M2 macrophages. On the other hand, *Ship1*$^{-/-}$ macrophages express a higher level of Arg1 as do *Pten*$^{-/-}$ macrophages, but lower levels of Ym1, MR, and CD163 compared to control macrophages, suggesting that *Ship1*$^{-/-}$ macrophages have the M1/M2 mixed phenotype. Because IL-12 expression and LPS-induced TNF-α production are enhanced in *Ship1*$^{-/-}$ macrophages, *Ship1*$^{-/-}$ macrophages are M1 macrophages functionally. As it has been shown that TGF-β skews *Ship1*$^{-/-}$ macrophages toward the M2 phenotype in vitro (Rauh et al. 2005), we analysed expression levels of M2 macrophage markers and cytokine profile of *Ship1*$^{-/-}$ macrophages differentiated by M-CSF and TGF-β. TGF-β-induced *Ship1*$^{-/-}$ macrophages expressed higher level of Arg1 and Ym1, but MR and CD163 expression remained low compared to control macrophages. Moreover, their cytokine profiles are IL-10^{low} IL-12^{high}, suggesting that TGF-β-induced *Ship1*$^{-/-}$ macrophages have M1 phenotype (Fig. 8.3c). Collectively, our results indicate that PTEN and SHIP1 control macrophage polarization and activation by different mechanisms.

8.3.2 Obesity and Metabolism

Chronic low-grade inflammation in adipose tissue contributes to obesity-related insulin resistance and diabetes (Hotamisligil 2006). In obesity, adipocytes can release proinflammatory mediators such as CCL2 (CC chemokine ligand 2), TNF-α, or free fatty acids, which induce the recruitment and activation of adipose tissue macrophages (ATMs). The activated ATMs secrete proinflammatory cytokines including TNF-α, IL-6, and IL-1β, which block the insulin-sensitising action of adiponectin and leptin, causing insulin resistance (Hill et al. 2014).

ATMs from obese mice show the M1 phenotype with upregulation of TNF-α and iNOS (inducible NO synthase). In contrast, ATMs from lean mice exhibit the M2 phenotype, which may control lipolysis and tissue homeostasis by preventing inflammation and promoting insulin sensitivity (Lumeng et al. 2007). Mice with macrophage-selective genetic inactivation of several molecules, which polarise their ATMs toward M2 macrophages, show inhibition of obesity-induced insulin resistance. Thus obesity-induced insulin resistance can be modulated by polarization and activation of ATMs (Odegaard and Chawla 2011).

To compare the in vivo functions of PTEN and SHIP1 in macrophage polarization, we generated CD11bcreSHIP1$^{flox/flox}$ (SHIP1 cKO) mice Because conventional SHIP1 KO mice die within 3 or 4 months. Male mice were fed a high fat diet (HFD) for 28–32 weeks, starting at 4 weeks of age. We isolated the stromal vascular fraction (SVF) of epididymal adipose tissue from PTEN cKO mice, SHIP1 cKO mice, and their littermate controls and quantified the proportions of M1 (F4/80$^+$ CD11b$^+$CD11c$^+$) and M2 (F4/80$^+$CD11b$^+$CD11c$^-$) macrophages by FACS

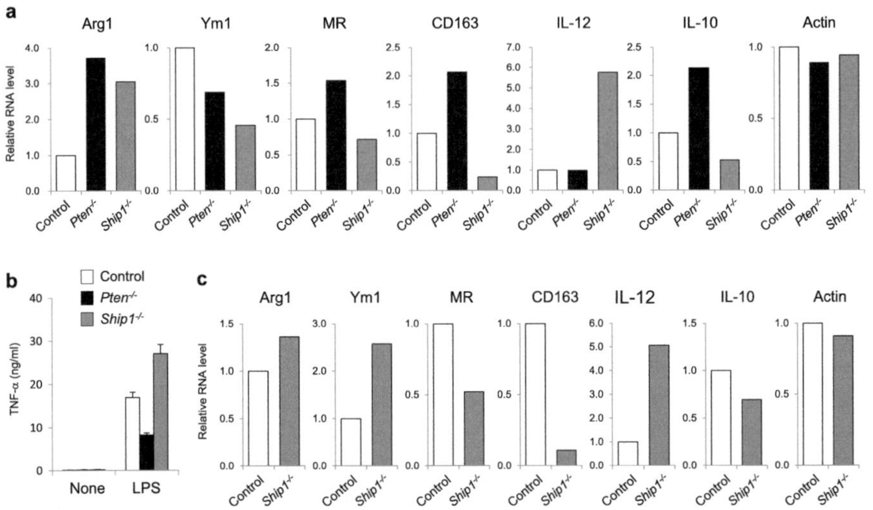

Fig. 8.3 Effects of PTEN or SHIP1 on macrophage polarization. (**a**) RT-PCR of M1 and M2 macrophage markers. Total RNA was prepared from bone-marrow–derived macrophages differentiated in the presence of M-CSF for 7 days. The data show relative expression levels of Arg1, Ym1, MR, CD163, IL-12, and IL-10 compared to control. (**b**) Bone-marrow–derived macrophages were stimulated with LPS (100 ng/mL) for 4 h. TNF-α production was measured by ELISA. (**c**) Expressions of Arg1, Ym1, MR, CD163, IL-12, and IL-10 (relative to control) in *Ship1*$^{-/-}$ macrophages differentiated with M-CSF or M-CSF plus TGF-β

analyses. As shown in Fig. 8.4a, adipose tissue of PTEN cKO mice showed a significantly reduced proportion of F4/80^{+}CD11b^{+}CD11c^{+} macrophages and an increased proportion of F4/80^{+}CD11b^{+}CD11c^{-} cells compared to adipose tissue of control mice. In contrast, analysis of ATM polarization in SVF showed a higher proportion of F4/80^{+}CD11b^{+}CD11c^{+} cells and a lower proportion of F4/80^{+}CD11b $^{+}$CD11c^{-} cells in SHIP1 cKO mice than in the control mice. These data suggest that *Pten*$^{-/-}$ ATMs are polarised toward M2 macrophages, whereas *Ship1*$^{-/-}$ ATMs are polarised toward M1 macrophages in vivo. Next, we investigated the effect of PTEN- or SHIP1-deleted macrophages on glucose homeostasis and insulin sensitivity. PTEN cKO mice were significantly more glucose tolerant and insulin sensitive compared to control mice. In contrast, SHIP1 cKO mice were more glucose intolerant and resistant to the glucose-lowering effects of exogenous insulin. These data suggest that PTEN and SHIP1 differentially regulate ATM polarization in vivo and modulate obesity-induced insulin resistance.

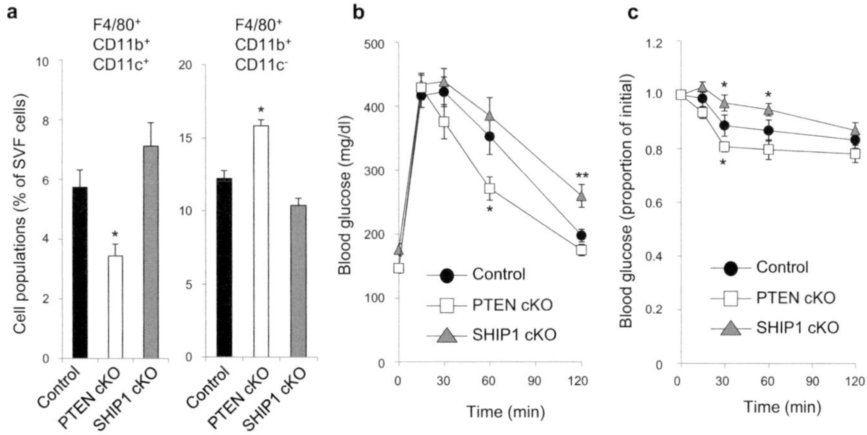

Fig. 8.4 Effects of deletion of PTEN or SHIP1 on glucose homeostasis in macrophages. (**a**) The stromal vascular fractions (SVFs) were prepared from epididymal adipose tissues of male mice fed a high fat diet. The isolated SVFs were stained for F4/80, CD11b, and CD11c and analysed by FACS. The number of $F4/80^{+}CD11b^{+}CD11c^{+}$ and $F4/80^{+}CD11b^{+}CD11c^{-}$ cells is indicated as percentage of total SVFs. (**b**) Oral glucose tolerance test. Obese mice were fasted for 16 h before intraperitoneal injection of glucose (2 g/kg). (**c**) Insulin (0.75U/kg) tolerance test in male mice after 30 weeks of feeding on high fat diet. Data are expressed as mean ± SEM; $*p < 0.05$, $**p < 0.01$, when compared to control counterpart,; n = 7–10 per group

8.4 Concluding remarks

Considerable progress has been made in understanding molecular networks that define polarised activation of macrophages. We present the PtdIns(3,4,5)P_3 phosphatases PTEN and SHIP1 as regulators of M1/M2 macrophage polarization. In vitro and in vivo studies show that PTEN regulates M2 macrophage polarization, whereas SHIP1 regulates M1 polarization. However, this raises the question of how PTEN and SHIP1 differentially regulate macrophage polarization. Their enzymatic products PtdIns(4,5)P_2 or PtdIns(3,4)P_2 might govern macrophage polarization, or the quantity or quality of PtdIns(3,4,5)P_3 might be different in PTEN- and SHIP1-deficient macrophages. Indeed it has been recently reported that there are many PtdIns(3,4,5)P_3 species depending on different acyl chains, although little is known about the physiological consequences of each PtdIns(3,4,5)P_3 molecule (Clark et al. 2011). Further studies are needed to clarify molecular mechanisms underlying macrophage activation by PTEN and SHIP1.

Acknowledgments This work was supported by Japan Science and Technology Agency (JST).

References

Bonizzi G, Karin M (2004) The two NF-kappaB activation pathways and their role in innate and adaptive immunity. Trends Immunol 25:280–288

Cantley LC (2002) The phosphoinositide 3-kinase pathway. Science 296:1655–1657

Cao X et al (2004) The inositol 3-phosphatase PTEN negatively regulates Fc gamma receptor signaling, but supports Toll-like receptor 4 signaling in murine peritoneal macrophages. J Immunol 172:4851–4857

Chawla A (2010) Control of macrophage activation and function by PPARs. Circ Res 106:1559–1569

Clark J et al (2011) Quantification of PtdInsP3 molecular species in cells and tissues by mass spectrometry. Nat Methods 8:267–272

Cullen PJ, Venkateswarlu K (1999) Potential regulation of ADP-ribosylation factor 6 signalling by phosphatidylinositol 3,4,5-trisphosphate. Biochem Soc Trans 27:683–689

Di Paolo G, De Camilli P (2006) Phosphoinositides in cell regulation and membrane dynamics. Nature 443:651–657

Franke TF (2008) Intracellular signaling by Akt: bound to be specific. Sci Signal 1:pe29

Gordon S (2003) Alternative activation of macrophages. Nat Rev Immunol 3:23–35

Gordon S, Pluddemann A, Martinez Estrada F (2014) Macrophage heterogeneity in tissues: phenotypic diversity and functions. Immunol Rev 262:36–55

Hadidi S et al (2012) Myeloid cell-specific expression of Ship1 regulates IL-12 production and immunity to helminth infection. Mucosal Immunol 5:535–543

Helgason CD et al (1998) Targeted disruption of SHIP leads to hemopoietic perturbations, lung pathology, and a shortened life span. Genes Dev 12:1610–1620

Hill AA, Reid Bolus W, Hasty AH (2014) A decade of progress in adipose tissue macrophage biology. Immunol Rev 262:134–152

Hotamisligil GS (2006) Inflammation and metabolic disorders. Nature 444:860–867

Itoh T, Takenawa T (2002) Phosphoinositide-binding domains: functional units for temporal and spatial regulation of intracellular signalling. Cell Signal 14:733–743

Jinushi M, Komohara Y (2015) Tumor-associated macrophages as an emerging target against tumors: creating a new path from bench to bedside. Biochim Biophys Acta 1855:123–130

Kozicky LK, Sly LM (2015) Phosphatase regulation of macrophage activation. Semin Immunol 27:276–285

Krystal G (2000) Lipid phosphatases in the immune system. Semin Immunol 12:397–403

Kuroda S et al (2008) Effective clearance of intracellular Leishmania major in vivo requires Pten in macrophages. Eur J Immunol 38:1331–1340

Kuroda E et al (2011) SHIP represses Th2 skewing by inhibiting IL-4 production from basophils. J Immunol 186:323–332

Lawrence T, Natoli G (2011) Transcriptional regulation of macrophage polarization: enabling diversity with identity. Nat Rev Immunol 11:750–761

Lee S et al (2011) Distinct macrophage phenotypes contribute to kidney injury and repair. J Am Soc Nephrol 22:317–326

Liu G, Abraham E (2013) MicroRNAs in immune response and macrophage polarization. Arterioscler Thromb Vasc Biol 33:170–177

Liu Q et al (1999) SHIP is a negative regulator of growth factor receptor-mediated PKB/Akt activation and myeloid cell survival. Genes Dev 13:786–791

Lumeng CN, Bodzin JL, Saltiel AR (2007) Obesity induces a phenotypic switch in adipose tissue macrophage polarization. J Clin Invest 117:175–184

Maehama T (2007) PTEN: its deregulation and tumorigenesis. Biol Pharm Bull 30:1624–1627

Mantovani A, Sozzani S, Locati M, Allavena P, Sica A (2002) Macrophage polarization: tumor-associated macrophages as a paradigm for polarized M2 mononuclear phagocytes. Trends Immunol 23:549–555

Mantovani A, Biswas SK, Galdiero MR, Sica A, Locati M (2013) Macrophage plasticity and polarization in tissue repair and remodelling. J Pathol 229:176–185
Martinez FO, Helming L, Gordon S (2009) Alternative activation of macrophages: an immunologic functional perspective. Annu Rev Immunol 27:451–483
Mosser DM, Edwards JP (2008) Exploring the full spectrum of macrophage activation. Nat Rev Immunol 8:958–969
Murray PJ, Wynn TA (2011) Protective and pathogenic functions of macrophage subsets. Nat Rev Immunol 11:723–737
Neele AE, Van den Bossche J, Hoeksema MA, de Winther MP (2015) Epigenetic pathways in macrophages emerge as novel targets in atherosclerosis. Eur J Pharmacol 763:79–89
Odegaard JI, Chawla A (2011) Alternative macrophage activation and metabolism. Annu Rev Pathol 6:275–297
Ooms LM et al (2009) The role of the inositol polyphosphate 5-phosphatases in cellular function and human disease. Biochem J 419:29–49
Porta C et al (2009) Tolerance and M2 (alternative) macrophage polarization are related processes orchestrated by p50 nuclear factor kappaB. Proc Natl Acad Sci U S A 106:14978–14983
Rauh MJ et al (2005) SHIP represses the generation of alternatively activated macrophages. Immunity 23:361–374
Sahin E et al (2014) Macrophage PTEN regulates expression and secretion of arginase I modulating innate and adaptive immune responses. J Immunol 193:1717–1727
Sasaki T, Sasaki J, Sakai T, Takasuga S, Suzuki A (2007) The physiology of phosphoinositides. Biol Pharm Bull 30:1599–1604
Sasaki T et al (2009) Mammalian phosphoinositide kinases and phosphatases. Prog Lipid Res 48:307–343
Sica A, Mantovani A (2012) Macrophage plasticity and polarization: in vivo veritas. J Clin Invest 122:787–795
Sica A et al (2008) Macrophage polarization in tumour progression. Semin Cancer Biol 18:349–355
Sly LM et al (2009) SHIP prevents lipopolysaccharide from triggering an antiviral response in mice. Blood 113:2945–2954
Stephens L, Milne L, Hawkins P (2008) Moving towards a better understanding of chemotaxis. Curr Biol 18:R485–R494
Suzuki A, Nakano T, Mak TW, Sasaki T (2008) Portrait of PTEN: messages from mutant mice. Cancer Sci 99:209–213
Takeda N et al (2010) Differential activation and antagonistic function of HIF-{alpha} isoforms in macrophages are essential for NO homeostasis. Genes Dev 24:491–501
Vanhaesebroeck B, Guillermet-Guibert J, Graupera M, Bilanges B (2010) The emerging mechanisms of isoform-specific PI3K signalling. Nat Rev Mol Cell Biol 11:329–341
Waidmann O et al (2013) Diagnostic and prognostic significance of cell death and macrophage activation markers in patients with hepatocellular carcinoma. J Hepatol 59:769–779
Welch HC (2015) Regulation and function of P-Rex family Rac-GEFs. Small GTPases 6:49–70
Wymann MP, Pirola L (1998) Structure and function of phosphoinositide 3-kinases. Biochim Biophys Acta 1436:127–150
Yue S et al (2014) Myeloid PTEN deficiency protects livers from ischemia reperfusion injury by facilitating M2 macrophage differentiation. J Immunol 192:5343–5353

Chapter 9
Clarification of the Molecular Mechanisms That Negatively Regulate Inflammatory Responses

Takashi Tanaka

Abstract Inflammation is an important host defence mechanism to fight against invading microbial pathogens. However, an excess of these responses can damage normal tissues and may cause autoimmune or allergic diseases. To prevent the onset of these diseases, inflammatory responses should be tightly regulated by the negative regulatory system. Dysfunction of this negative regulatory system is suggested to be the cause of these diseases, but little is known about the details. Our research goal is to identify key regulators of inflammation-related signal transduction pathways, and to clarify the molecular mechanisms for regulating inflammatory responses. We recently found that LIM domain-containing proteins constitute a new family of negative regulators of inflammatory responses. In this review, we summarise a series of our studies about the role of PDLIM2 (PDZ and LIM-domain protein-2) and related LIM proteins in the negative regulation of inflammatory responses. These studies should contribute to the development of new therapeutic tools to control the exaggerated inflammation seen in certain human diseases.

Keywords NF-κB • STAT4 • STAT3 • HSP70 • Inflammation • PDLIM2 • PDLIM1 • LIM protein • Ubiquitination • Ubiquitin E3 ligase

9.1 Introduction

Our immune system is classified into two distinct systems, innate and adaptive immunity. Innate immune cells, such as macrophages and dendritic cells, detect invading microbial pathogens by their sensors (e.g. Toll-like receptors [TLR]). The transcription factor nuclear factor κB (NF-κB) has a central role in TLR-mediated

T. Tanaka (✉)
Laboratory for Inflammatory Regulation, RIKEN Center for Integrative Medical Sciences (IMS), RIKEN Research Center for Allergy and Immunology (RCAI), Yokohama, Kanagawa 230-0045, Japan

PRESTO, Japan Science and Technology Agency, Kawaguchi 332-0012, Japan
e-mail: takashi.tanaka@riken.jp

M. Miyasaka, K. Takatsu (eds.), *Chronic Inflammation*,
DOI 10.1007/978-4-431-56068-5_9

activation of these cells. In resting state, heterodimer of p65/p50 subunits of NF-κB is sequestrated in the cytoplasm by binding to IκBα, an inhibitor of NF-κB. A signal through TLR leads to the proteasomal degradation of IκBα, which enables NF-κB to enter the nucleus and induce the expressions of a series of inflammation-related genes, including those encoding proinflammatory cytokines, such as interleukin-6 (IL-6) and IL-12 (Hayden and Ghosh 2004). These inflammatory responses then activate naïve T-helper (Th) lymphocytes, through the activation of transcription factors, signal transducer and activator of transcription 4 (STAT4), STAT6, and STAT3, and direct these cells differentiated into distinct subsets, such as Th1, Th2, and Th17 cells, respectively. These Th subsets are characterised by the cytokines they produce and the immunological functions they induce. Th1 and Th2 cells produce IFN-γ and IL-4, respectively, whereas Th17 cells secrete IL-17, IL-21, and IL-22 as effector cytokines. These effector T-cell subsets are important for fighting against different types of microbial infections, which are viral or intracellular bacteria infections by Th1, helminth infections by Th2, and fungal or extracellular bacteria infections by Th17 cells. On the other hand, these inflammatory responses should be terminated at appropriate time points. Otherwise, excessive and prolonged activation of these inflammation-related signal transduction pathways can cause massive damage to the host and result in human autoimmune or allergic diseases, including rheumatoid arthritis, inflammatory bowel diseases, and asthma (Gregersen and Olsson 2009). Indeed, NF-κB and STATs transcription factors are highly activated at sites of inflammation in these human diseases. Constitutive activation of NF-κB and STATs may contribute to the pathogenesis of these diseases. Thus a system to regulate the inflammatory signaling negatively is essential for preventing immunopathology. We therefore attempted to identify key molecules for regulating inflammatory signaling, and clarify the molecular mechanisms of how inflammatory responses are negatively regulated. We then examined if dysfunction of these regulators may cause autoimmune or allergic diseases. Eventually we expect that these molecules can be the targets for new treatments of these diseases.

9.1.1 Identification of PDLIM2 as a Ubiquitin E3 Ligase Negatively Regulating STAT Signaling

Among T-cell subsets described above, Th17 cells were newly identified about 10 years ago. Before the discovery of Th17 cells, $CD4^{+}$T cells were classically divided into two major populations, Th1 and Th2 cells (Wurster et al. 2000). It was thought that the polarised Th1 response causes inflammatory diseases, whereas excessive Th2 responses lead to allergic diseases, which led us to focus on Th1 cells to isolate negative regulators of inflammatory responses. In cytokine signaling, signals from several cytokines converge into the activation of one common STAT transcription factor, and then diverge by inducing a wide variety of immunological responses, suggesting that controlling the activity of the STAT transcription factor will be an efficient strategy for regulating inflammatory responses. We therefore sought to identify the molecules that interact with STAT4, which is essential for Th1 cell development, and affect STAT4 activity using a yeast two-hybrid screening. We isolated several STAT4-interacting molecules and one of them is a novel PDZ-LIM protein named SLIM (STAT4-interacting LIM protein) (Tanaka et al. 2005). SLIM has now been renamed PDLIM2 (PDZ and LIM domain protein-2), according to the international HUGO nomenclature. PDLIM2 is a nuclear protein that contains one PDZ domain at its N-terminus and one LIM domain at its C-terminus, and belongs to a large family of LIM proteins. The terms PDZ and LIM are derived from three proteins in which these domains were first identified; Postsynaptic density 65-Discs large-Zonula occludens-1 and Lin-1-Isl-1-Mec-3, respectively. Inasmuch as both PDZ and LIM domains are involved in protein–protein interactions, previous studies of these molecules have been focused on the identification of their binding partners, and have also shown that they are implicated in the regulation of various biological processes, including cytoskeleton organisation and oncogenesis (te Velthuis and Bagowski 2007). However, their functions in the immune system remain completely unknown. Based on the structural similarity of the LIM domain to the RING finger domain, a well-known domain responsible for ubiquitin E3 ligase, we found that the LIM domain has ubiquitin E3 ligase activity, and demonstrated for the first time that LIM domain-containing protein functions as an ubiquitin E3 ligase (Tanaka et al. 2005).

Several factors were reported to associate with STAT proteins and suppress their activity. For example, protein tyrosine phosphatases, such as PTP1B, TC-PTP, directly dephosphorylate activated STATs, whereas the protein inhibitor of the activated STAT (PIAS) family of proteins blocks STAT-DNA binding. In addition, the ubiquitin/proteasome-dependent degradation of STAT was also shown to regulate its activity. Ubiquitin is a small protein composed of 76 amino acids. The proteins tagged with the chain of ubiquitin molecules are degraded by 26S proteasome, a huge complex of proteolytic enzymes. Ubiquitin conjugation requires three enzymes: ubiquitin-activating enzyme (E1), ubiquitin-conjugating enzyme (E2), and ubiquitin ligase (E3). E3 selectively binds to both target proteins and E2, and promotes the transfer of the polyubiquitin chain from E2 to the substrates, implying that E3 confers the specificity to the reactions. However, the responsible endogenous ubiquitin E3 ligase for STAT proteins has not yet been identified. We found that PDLIM2 functions as a nuclear ubiquitin E3 ligase that targets STAT4. Upon IL-12 stimulation of T cells, STAT4 is tyrosine-phosphorylated and translocated into the nucleus to induce target gene expression. PDLIM2 interacts with phosphorylated STAT4 in the nucleus, and then promotes polyubiquitination and subsequent proteasome-dependent degradation of STAT4 proteins, thereby terminating STAT4-mediated signaling (Tanaka et al. 2005) (Fig. 9.1). Overexpression of PDLIM2 leads to impaired STAT4 activity due to reduced STAT4 protein levels. Consistently, PDLIM2-deficient mice revealed increased STAT4 expression and enhanced IFNγ production by Th1 cells. These

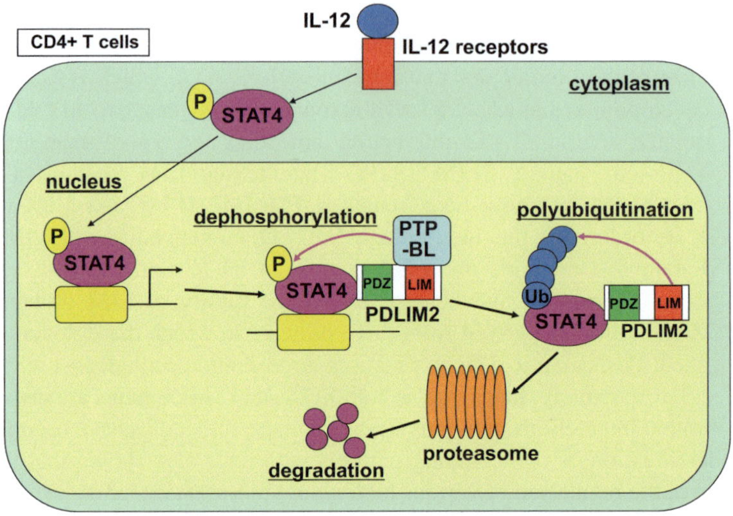

Fig. 9.1 PDLIM2 and PTPBL-mediated negative regulation of STAT4 signaling. PDLIM2 binds to STAT4 and promotes its ubiquitin/proteasome-dependent degradation, downregulating STAT4 activity. PTPBL also binds to STAT4 and enhances its dephosphorylation. PDLIM2 functions as an adaptor that bridges STAT4 and PTPBL, facilitating STAT4 dephosphorylation

data clearly showed that PDLIM2 negatively regulates STAT4-mediated Th1 cell differentiation as a ubiquitin E3 ligase.

9.1.2 Essential Role of a Tyrosine Phosphatase PTP-BL in the Negative Regulation of STAT Signaling

In addition, PDLIM2 also enhances dephosphorylation of activated STAT4, thereby downregulating STAT4 activity. PDLIM2 does not have a putative phosphatase domain, indicating the involvement of tyrosine phosphatases associated with PDLIM2. By a yeast two-hybrid screening, we isolated another STAT4-interacting molecule, protein tyrosine phopsphatase basophil-like (PTP-BL) (Nakahira et al. 2007). PTP-BL dephosphorylates STAT4, resulting in attenuation of STAT4-mediated gene activation, whereas PTP-BL deficiency leads to increased and prolonged activation of STAT4 and Th1 cell differentiation. Interestingly, PDLIM2 associated with both STAT4 and PTP-BL and enhanced STAT4 dephosphorylation synergistically with PTPBL, indicating that PDLIM2 functions as an adaptor that bridges STAT4 and PTPBL (our unpublished data).

9.1.3 PDLIM2 Inhibits Granulomatous Inflammation by Inhibiting STAT3-Mediated Th17 Cell Differentiation

Similar to Th1 cells, Th17 cells strongly induce inflammatory responses and eliminate microbial pathogens. However, Th17 cells are thought to be highly pathogenic, because an excess of Th17-mediated responses may cause autoimmune diseases. We have further shown that PDLIM2 also negatively regulated Th17 cell differentiation by acting as a nuclear ubiquitin E3 ligase targeting STAT3, the transcription factor critical for the commitment to the Th17 lineage (Tanaka et al. 2011). PDLIM2 promoted polyubiquitination and proteasome-dependent degradation of STAT3, thereby disrupting STAT3-mediated gene activation. PDLIM2 deficiency resulted in a larger amount of nuclear STAT3 and enhanced Th17 cell differentiation. To examine the role of PDLIM2 in the regulation of in vivo Th17 development, we used a mouse model for granulomatous inflammation. Granuloma formation is an important host defence mechanism against intracellular bacteria. However, uncontrolled granulomatous responses cause tissue damage and impair normal organ function in several human autoimmune diseases, such as Crohn's disease and sarcoidosis. Recent studies have implicated that exaggerated Th17 cell-mediated responses may contribute to the development of granulomatous inflammation. *Propionibacterium acnes*, a Gram-positive intracellular bacterium, is suspected to be the cause of human sarcoidosis. We demonstrated that PDLIM2-deficient mice revealed exacerbated *P. acnes*-induced liver granuloma formation, in which Th17 cells are suggested to have a pathogenic role. Consistently, *P. acnes*-induced in vivo and in vitro Th17 differentiation was augmented in PDLIM2-deficient mice. These findings delineate an essential role of PDLIM2 in negatively regulating Th17-mediated inflammatory responses and provide a potential therapeutic target for autoimmune diseases.

9.1.4 PDLIM2 Terminates NF-κB-Mediated Inflammatory Responses

As described above, in the innate immune cells, NF-κB is a key transcription factor for their activation (Kaisho and Tanaka 2008). On the other hand, NF-κB activation should be tightly regulated to prevent excessive inflammatory responses. One well-established mechanism of NF-κB suppression is the export of nuclear NF-κB by IκBα. IκBα is resynthesised in an NF-κB–dependent manner, enters the nucleus, and transports nuclear NF-κB back out to the cytoplasm, thereby downregulating NF-κB–mediated transcription. Recent studies disclosed another mechanism to shut down NF-κB activation by degrading nuclear NF-κB through the ubiquitin/proteasome-dependent pathway. However, ubiquitin E3 ligase that interacts with and polyubiquitinates nuclear NF-κB was not identified. Moreover, the fate of polyubiquitinated NF-κB in the nucleus was unclear.

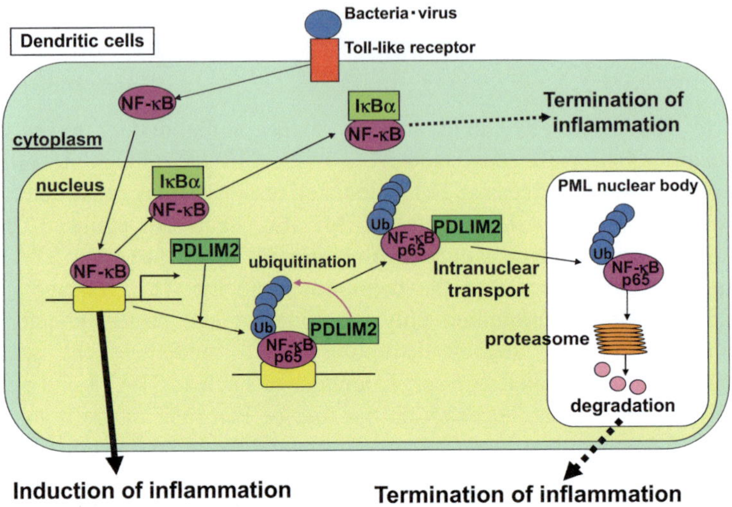

Fig. 9.2 PDLIM2-mediated termination of NF-κB activation. PDLIM2 binds to the p65 subunit of NF-κB and promotes p65 polyubiquitination in the nucleus. PDLIM2 then targets p65 into discrete intranuclear compartments, called PML nuclear bodies, where p65 is ultimately degraded by proteasome

We have demonstrated that PDLIM2 negatively regulates NF-κB activity, acting as a nuclear ubiquitin E3 ligase targeting the p65 subunit of NF-κB (Tanaka et al. 2007). PDLIM2 binds to p65 and promotes p65 polyubiquitination through its LIM domain. In addition, PDLIM2 targeted p65 to discrete intranuclear compartments, called PML nuclear bodies. PML nuclear bodies are nuclear proteolytic centers where proteasomal components are concentrated. Polyubiquitinated p65 is ultimately degraded by proteasome in these compartments (Fig. 9.2). A PDLIM2 mutant lacking the PDZ domain fails to target p65 to nuclear bodies and the PDZ domain of PDLIM2 can bind to α-actinin, an actin binding protein, suggesting that PDLIM2 mediates intranuclear trafficking of p65 in an α-actinin/actin-dependent manner through its PDZ domain. Consistently, PDLIM2 deficiency in dendritic cells results in larger amounts of nuclear p65, defective p65 ubiquitination, and augmented production of proinflammatory cytokines in response to TLR ligands, such as LPS and CpG-DNA. Moreover, targeted disruption of the PDLIM2 gene in mice enhances the sensitivity to the lethal effect of high-dose LPS treatment. These findings delineate a novel pathway by which PDLIM2 terminates NF-κB activation through intranuclear sequestration and subsequent degradation.

9.1.5 Molecular Mechanisms to Regulate PDLIM2-Mediated Termination of NF-κB Activation

As described above, PDLIM2 is essential for negatively regulating NF-κB–mediated inflammatory responses. However, it is still unclear how PDLIM2 activity itself is controlled in this signaling pathway. In dendritic cells, PDLIM2 suppresses NF-κB activation in the late phase but not in the early phase of TLR stimulation, although PDLIM2 is expressed even without TLR stimulation (Tanaka et al. 2014). These findings suggest the presence of the molecules that can bind to PDLIM2 and regulate PDLIM2 activity. We therefore attempted to identify PDLIM2-associated proteins by transfecting 293 T cells with Flag-tagged PDLIM2 and immunoprecipitation with anti-Flag antibody, followed by protein sequencing of PDLIM2-associated proteins by liquid chromatography–mass spectrometry. We then obtained 165 candidate proteins, including chaperone proteins (Heat shock protein [HSP] 70, HSP90, HSP40, HSP60, HSP105, and Agp-1), cytoskeletal proteins (α-actinin), and the components of ubiquitin ligase complex (our unpublished data). Among these candidate proteins identified in this assay, HSP70 was found to be the most reliable PDLIM2-associated protein based on False Discovery Rate (FDR) analysis. We therefore focused on the potential role of HSP70 in the regulation of PDLIM2-mediated NF-κB inactivation, and found that HSP70 is critical for PDLIM2 to degrade p65 and suppress NF-κB–mediated inflammatory responses. In dendritic cells, HSP70 expression is detected only in the cytoplasm without stimulation, but is translocated to the nucleus, where PDLIM2 is located, after 3–5 h of innate stimuli. HSP70 then associated with both PDLIM2 and BAG-1, a proteasome-associated protein, and promoted the transport of NF-κB–PDLIM2 complex to the proteasome, thereby facilitating p65 degradation (Fig. 9.3) (Tanaka et al. 2014). Consistently, either HSP70 deficiency or BAG-1 knockdown in dendritic cells resulted in enhanced nuclear p65 protein levels and thus enhanced

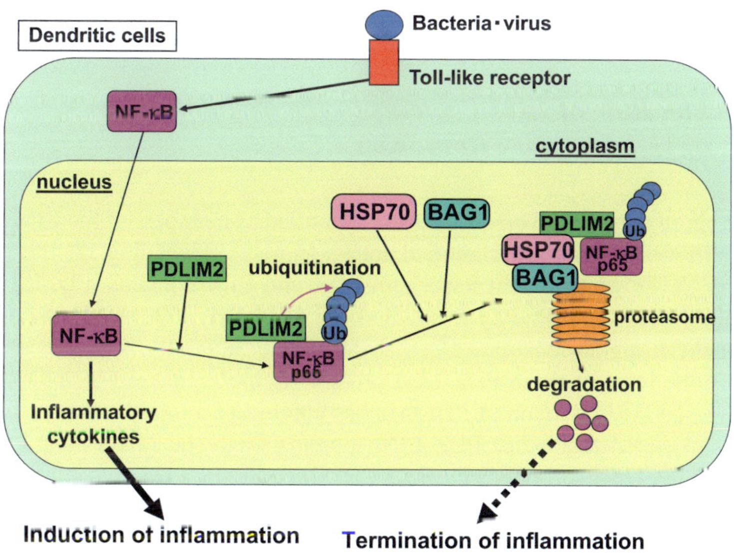

Fig. 9.3 HSP70 is essential for PDLIM2-mediated degradation of NF-κB p65. HSP70 binds to PDLIM2 and facilitates delivery of the NF-κB-PDLIM2 complex to the proteasome cooperatively with BAG1

production of proinflammatory cytokines in response to TLR stimuli. These data delineate a novel role of HSP70 in the regulation of NF-κB signaling, bridging between ubiquitin E3 ligase and the proteasome.

HSP70 is originally a chaperone protein that binds to and promotes the correct folding of substrate proteins by preventing their aggregation. However, if the folding is unsuccessful, HSP70 promotes ubiquitin/proteasome-dependent degradation of misfolded substrate proteins to protect cells from an accumulation of toxic aggregates of aberrant proteins (Amm et al. 2013). We speculate that HSP70 may also function as a molecular chaperone in the process of p65 degradation. Inasmuch as infection and inflammation can be environmental stresses that cause damage to proteins, a substantial amount of NF-κB, accumulated in the nucleus in response to microbial pathogens, may become damaged and need to be degraded. We assume that HSP70 binds to damaged p65 proteins and prevents their aggregation, which ultimately facilitates p65 degradation by the proteasome. It has been reported that failure of the protein quality control system in neurons is associated with the various neurodegenerative disorders, including Alzheimer's, Huntington's, and Parkinson's diseases (Amm et al. 2013). We therefore speculate that persistent inflammation may result in the accumulation of pathological protein aggregates in inflammation-related cells, such as dendritic cells, fibroblasts, and vascular endothelial cells, leading to degeneration of these cells. This pathology might be related to fibrinoid degeneration observed in autoimmune systemic vasculitis. Thus, HSP70, possibly in cooperation with other chaperone proteins identified as PDLIM2-associated proteins, may prevent the onset of these autoimmune diseases by eliminating damaged proteins during inflammation.

9.1.6 The Role of LIM Proteins in the Regulation of Inflammatory Responses

To date, more than 30 LIM proteins were identified as members of the LIM protein family. These proteins are classified into subgroups depending on the domain structure. Seven proteins that have both PDZ and LIM domains, PDLIM1/Elfin, PDLIM2, PDLIM3/ALP, PDLIM4/Ril, PDLIM5/ENH, PDLIM6/ZASP/Cypher, and PDLIM7/Enigma constitute a PDZ-LIM protein subfamily. We have attempted to clarify the roles of these PDZ-LIM proteins in the regulation of inflammatory responses. The LIM domain of PDLIM2 has ubiquitin E3 ligase activity, therefore we first predicted that other PDZ-LIM proteins might be ubiquitin E3 ligases. Unexpectedly, however, only PDLIM2, PDLIM6, and PDLIM7 have ubiquitin E3 ligase activity among the PDZ-LIM protein subfamily. We also checked the expression of these PDZ-LIM proteins. PDLIM1 and PDLIM7 are ubiquitously expressed in the immune cells at a high level as with PDLIM2, whereas PDLIM4 is exclusively expressed in $CD4^{+}T$ cells. In contrast, PDLIM3, PDLIM5, and PDLIM6 are expressed at high levels in skeletal and cardiac muscle, but at very low levels in the immune cells (our unpublished data). Based on these data, we focused on PDLIM1, PDLIM7, and PDLIM4 for subsequent analysis.

We have found that PDLIM7 and PDLIM1 are also negative regulators for NF-κB–mediated signalling in dendritic cells. Similar to PDLIM2, PDLIM7 was found to be a nuclear ubiquitin E3 ligase targeting p65 for proteasome-dependent degradation, terminating NF-κB activation. We further demonstrated that PDLIM7 and PDLIM2 could form heterodimers and synergistically promoted polyubiquitination and degradation of p65 protein (our unpublished data). On the other hand, PDLIM1 is located in the cytoplasm. As described above, PDLIM1 is not a ubiquitin E3 ligase, but instead PDLIM1 bound to and sequestered p65 subunit of NF-κB in the cytoplasm and suppressed its nuclear translocation, possibly by interaction with actin stress fibers through binding to α-actinin, an actin binding protein, via its PDZ domain. Interestingly, although p65 nuclear translocation is mainly regulated by IκBα, this effect of PDLIM1 on p65 nuclear translocation is independent of IκBα. Consistently, PDLIM1-deficient dendritic cells reveal enhanced p65 nuclear translocation and proinflammatory cytokine production in response to LPS stimulation (Ono et al. 2015).

We also demonstrated that PDLIM4 negatively regulated STAT4, STAT6, and STAT3-mediated signalling in $CD4^{+}T$ cells and suppressed Th1, Th2, and Th17 cell differentiation. PDLIM4 did not promote polyubiquitination of these STAT proteins, but associated with and recruited to a protein tyrosine phosphatase through its LIM domain and facilitated dephosphorylation of tyrosine residue of these STAT transcription factors, thereby terminating their activation (our unpublished data). Taking these data together, PDZ-LIM proteins appeared to be a new family of adaptors that can negatively regulate signal transduction pathways in the inflammation through different mechanisms.

9.2 Conclusion

In this study, we have shown that PDLIM2 and related LIM proteins, PDLIM1, PDLIM4, and PDLIM7, are negative regulators for inflammatory responses through different mechanisms. We first predicted that, as well as PDLIM2, other PDZ LIM proteins were all ubiquitin E3 ligase, but actually not all LIM proteins were ubiquitin E3 ligase. This may be because LIM domains have highly divergent sequences. Although LIM domains commonly have eight conserved residues of cysteine or histidine, the homology of the LIM domain between LIM proteins is only about 50 %. Because the functions of LIM proteins are defined by what kinds of proteins LIM domains can bind to, this divergence of LIM domains is considered to confer the different functions on each LIM protein. These findings suggest that LIM proteins might be a novel family of proteins that negatively regulate inflammatory responses through different mechanisms.

In the course of our study, we still have one more issue that should be addressed. This is a question of whether the dysfunction of LIM proteins is related to the onset of autoimmune diseases. To this end, we have analysed the association of human single nucleotide polymorphism (SNP) of LIM proteins with autoimmune diseases by genome-wide association study (GWAS). We have already found that nonsynonymous SNP of one PDZ-LIM protein is associated with human rheumatoid arthritis, suggesting that dysfunction of negative regulators of inflammatory signaling may contribute to the pathogenesis of autoimmune diseases. These studies should lead to the development of new treatments for these diseases.

References

Amm I, Sommer T, Wolf DH (2013) Protein quality control and elimination of protein waste: The role of the ubiquitin-proteasome system. Biochem Biophys Acta 1843:182–196

Gregersen PK, Olsson LM (2009) Recent advances in the genetics of autoimmune diseases. Annu Rev Immunol 27:363–391

Hayden MS, Ghosh S (2004) Signaling to NF-κB. Genes Dev 18:2195–2224

Kaisho T, Tanaka T (2008) Turning NF-κB and IRFs on and off in DCs. Trends Immunol 29:329–336

Nakahira M, Tanaka T, Robson BE, Mizgerd JP, Grusby MJ (2007) Regulation of signal transducer and activator of transcription signaling by the tyrosine phosphatase PTP-BL. Immunity 26:163–176

Ono R, Kaisho T, Tanaka T (2015) PDLIM1 inhibits NF-κB-mediated inflammatory signaling by sequestering the p65 subunit of NF-κB in the cytoplasm. Sci Rep 5:18327

Tanaka T, Soriano MA, Grusby MJ (2005) SLIM is a nuclear ubiquitin E3 ligase that negatively regulates STAT signaling. Immunity 22:729–736

Tanaka T, Grusby MJ, Kaisho T (2007) PDLIM2-mediated termination of transcription factor NF-κB activation by intranuclear sequestration and degradation of the p65 subunit. Nat Immunol 8:584–591

Tanaka T, Yamamoto Y, Muromoto R, Ikeda O, Sekine Y, Grusby MJ, Kaisho T, Matsuda T (2011) PDLIM2 inhibits T Helper 17 cell development and granulomatous inflammation through degradation of STAT3. Sci Signal 4:ra85

Tanaka T, Shibazaki A, Ono R, Kaisho T (2014) HSP70 mediates degradation of the p65 subunit of nuclear factor κB to inhibit inflammatory signaling. Sci Signal 7:ra119

te Velthuis AJ, Bagowski CP (2007) PDZ and LIM domain-encoding genes: molecular interactions and their role in development. Sci World J 7:1470–1492

Wurster AL, Tanaka T, Grusby MJ (2000) The biology of Stat4 and Stat6. Oncogene 19:2577–2584

Chapter 10
The Drosophila Toll Pathway: A Model of Innate Immune Signalling Activated by Endogenous Ligands

Takayuki Kuraishi, Hirotaka Kanoh, Yoshiki Momiuchi, Hiroyuki Kenmoku, and Shoichiro Kurata

Abstract The Drosophila Toll signalling pathway is partially homologous to the mammalian TLR innate immune pathway, but also has marked differences. The Drosophila Toll pathway is mainly activated by the endogenous ligand Spätzle, instead of by direct recognition of microbial molecules; and downstream signalling from adaptor molecules/kinases, MyD88/IRAKs, to the inhibitor of NF-κB (IκB) is a blackbox in Drosophila. These differences provide an interesting opportunity to decipher molecular mechanisms underlying microbial-independent activation of innate immunity. In our laboratory, we have performed ex vivo genome-wide RNAi screening and identified a HECT-type E3 ligase, Sherpa, as an essential component of intracellular Toll signalling; two protein kinases, Pitslre and Doa, as downstream kinases in the blackbox; and a Jumonji-like histone demethylase, Jarid2, as a transcription factor. Additionally, we found that the Drosophila larval peptide fraction has strong stimulatory activity on the Toll receptor. These findings and future analyses are likely to provide information on the as yet unclear molecular mechanisms of 'sterile inflammation' in mammals.

T. Kuraishi (✉)
Graduate School of Pharmaceutical Sciences, Tohoku University, Sendai, Japan

PRESTO, Japan Science and Technology Agency, Kawaguchi 332-0012, Japan

Faculty of Pharmacy, Institute of Medical, Pharmaceutical and Health Sciences, Kanazawa University, Ishikawa, Japan
e-mail: tkuraishi@staff.kanazawa-u.ac.jp

H. Kanoh (✉)
Graduate School of Pharmaceutical Sciences, Tohoku University, Sendai, Japan

Division of Glycopathology, Institute of Molecular Biomembrane and Glycobiology, Tohoku Pharmaceutical University, Sendai, Japan
e-mail: hirotaka.kanoh@gmail.com

Y. Momiuchi • H. Kenmoku • S. Kurata
Graduate School of Pharmaceutical Sciences, Tohoku University, Sendai, Japan

M. Miyasaka, K. Takatsu (eds.), *Chronic Inflammation*,
DOI 10.1007/978-4-431-56068-5_10

Keywords Drosophila • Innate immunity • Sterile inflammation • Toll • NF-kB • E3 ligase

10.1 Introduction

NF-kB signalling plays a central role in the initiation and persistence of inflammation. In mammals, many receptors are responsible for activation of NF-kB signalling, and the canonical and noncanonical intracellular pathways are complex, with a variety of innate and adaptive immune cells exhibiting different responses to NF-kB stimulation. Hence, although the molecular mechanisms and physiological and pathological functions of NF-kB signalling have been widely analysed, the complete picture is still unclear. NF-kB pathways are conserved in insects, including Drosophila melanogaster, a genetically tractable model animal, and there is partial similarity between the Drosophila Toll pathway and the mammalian Toll-like receptor (TLR) pathway. In this review, we describe lessons from the Drosophila Toll pathway that are applicable to mammalian pathways, and we discuss new components in Drosophila signalling that were recently identified in our laboratory.

10.2 Text

10.2.1 Drosophila Toll Pathway

10.2.1.1 Signalling Components

The Toll pathway was initially identified as a required pathway for dorsoventral patterning in Drosophila embryos (Anderson et al. 1985a, b). In 1996, the Toll pathway was shown to be essential for induction of an antifungal peptide, Drosomycin, and resistance to infection by pathogenic fungi (Lemaitre et al. 1996; Lemaitre and Hoffmann 2007; Ferrandon et al. 2007). Activation of the Toll receptor is induced by receptor dimerisation via binding of cleaved Spatzle (Spz) to the ectodomain of the Toll receptor (Jang et al. 2006; Weber et al. 2003). The dimerised Toll receptor recruits a plasma membrane-associated complex comprising dMyd88, Tube, and Pelle, a serine-threonine kinase similar to mammalian interleukin (IL)-1R–associated kinases (Tauszig-Delamasure et al. 2002; Towb et al. 1998; Sun et al. 2002, 2004). This complex formation is thought to lead to phosphorylation of Cactus, a Drosophila homologue of IkB. Phosphorylated IkB is degraded by the ubiquitin–proteasome system (Belvin et al. 1995; Reach et al. 1996), leading to release of the transcription factor Dif, a Drosophila NF-kB homologue that is responsible for expression of antimicrobial proteins (Ip et al. 1993; Manfruelli et al. 1999).

10.2.1.2 Comparison of Drosophila Toll Signalling with the Mammalian TLR Pathway

Recent studies have identified endogenous ligands for TLRs, but the nature of mammalian TLRs differs from that of the Drosophila Toll receptor in that mammalian TLRs are pattern-recognition receptors that directly sense microbial components (Takeuchi and Akira 2010; Valanne et al. 2011). TLRs and the Toll receptor both possess leucine-rich repeats in their extracellular regions, but have low similarity. TLRs and the Drosophila Toll receptor also contain a TIR intracellular domain that binds to dMyd88, and Drosophila Tube and Pelle are functional homologues of mammalian IRAK-4 and IRAK-1, respectively. Intriguingly, however, downstream components of mammalian IRAK-1, such as TRAF6, TAK1 and IKK, which function in degradation of IkB, are not involved in the Drosophila Toll pathway. Instead, Drosophila homologues of TRAF6, TAK1, and IKK are required for the Drosophila IMD pathway, which resembles the mammalian TNF-α pathway (Kleino and Silverman 2014). Drosophila cactus and Dif correspond to mammalian IkB and NF-kB, respectively. Thus, in the Drosophila Toll pathway, signalling from Pelle to Cactus is a blackbox (Fig. 10.1).

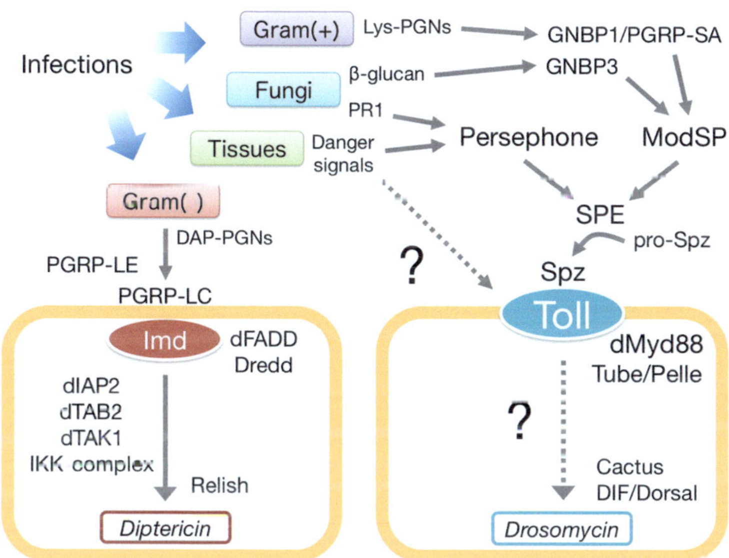

Fig. 10.1 Drosophila innate immune signalling pathway comprising the Toll and Imd pathways. Drosophila innate immune responses are induced through activation of the Toll and Imd pathways. The Imd pathway mainly recognises Gram-negative bacterial infections. The Toll pathway recognises a wide range of infections and tissue damage via an extensive extracellular signalling pathway. The molecular basis of the intracellular Toll pathway remains unclear. Direct signalling from danger signals to the cell-surface receptor has not been identified

10.2.2 Factors That Activate the Drosophila Toll Pathway

Except for developmental cues, the Toll pathway is activated by three factors: microbial-derived peptidoglycans, proteases from pathogenic fungi, and unidentified endogenous factors from dead cells. Peptidoglycans initiate ModSP-dependent Toll activation, and the proteases and endogenous factors promote psh-mediated Toll activation (Fig. 10.1). A complex of pattern-recognition receptors comprising PGRP-SA and GNBP1 senses lysine-type peptidoglycans derived from Gram-positive bacteria, and GNBP1 recognises β-glucans derived from fungi, both of which activate ModSP (Michel et al. 2001; Gobert et al. 2003; Gottar et al. 2006; Buchon et al. 2009). DAP-type peptidoglycans derived from Gram-negative bacteria are recognised by other PGRPs, resulting in activation of the IMD pathway (Kurata 2014). In addition, a recent report implied that Toll-7, rather than Drosophila Toll (also known as Toll-1), directly recognises viral infections and induces autophagy to degrade virus particles in lysosomes (Nakamoto et al. 2012). However, viral recognition by Toll-7 does not activate the canonical Toll pathway, which suggests the presence of other signalling pathways for antiviral responses.

So-called 'danger signals' activate psh (El Chamy et al. 2008): exogenous danger signals such as PR1 secreted from pathogenic fungi and virulence factors of Gram-positive bacteria, and endogenous danger signals (Gottar et al. 2006). In 2014, Ming et al. found induction of psh-, spz-, and Toll-dependent *Drosomycin* expression in larvae with a mutated caspase, Dronc, which is important for executing apoptosis (Ming et al. 2014). The same group subsequently reported that mutation of Dark, the Drosophila homologue of Apaf-1, causes spz- and Toll-dependent *Drosomycin* expression in adults (Igaki et al. 2002). It is likely that secondary necrotic cells originating from apoptosis-inhibited dead cells produce danger signals, but the molecular nature of the signal has yet to be identified.

10.2.3 Genome-Wide Screening for Factors That Play a Role in Activation of the Toll Pathway

To determine the details of the intracellular Toll pathway, we established an ex vivo experimental system for studying this pathway. S2 cells, a common Drosophila melanogaster cell line, are poorly responsive to stimulation of the Toll pathway (Goto et al. 2010). Thus, we tested several cell lines with a reporter assay and found that Toll pathway activation was induced 100-fold more strongly in DL1 cells, a cell line established from embryos, compared to other cell lines. In addition, the components and signalling hierarchy of the known Toll pathway in adult flies are rigidly conserved in DL1 cells. Therefore, we concluded that DL1 cells are suitable for studying the Toll pathway (Kanoh et al. 2015a).

To identify novel signalling components in the intracellular Toll pathway, we performed comprehensive genome-wide RNAi screening (Fig. 10.2) (Kanoh

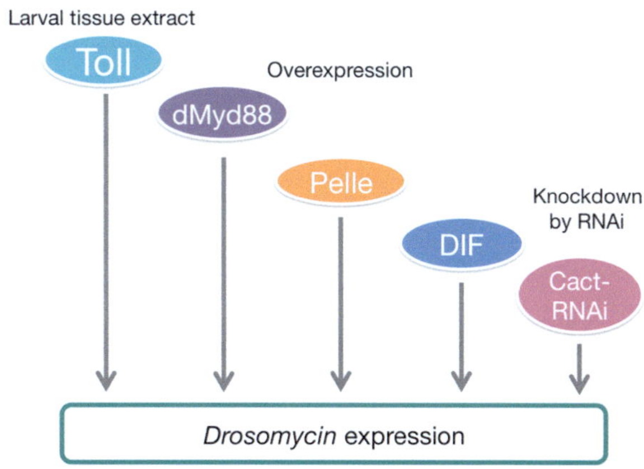

Fig. 10.2 Concept of comprehensive genome-wide RNAi screening for the intracellular Toll pathway. The intracellular Toll pathway was searched for unknown components using comprehensive genome-wide RNAi screening. Five distinct genome-wide RNAi screens efficiently covered a wide range of intracellular signalling events

et al. 2015a). In this screening, the Toll pathway in DL1 cells was stimulated with larval tissue extract that activates the Toll receptor, by overexpression of dMyD88, Pelle, or Dif, and by knockdown of Cactus to cover a wide range of intracellular signalling events. This approach identified known Toll pathway components, which verified the validity of this screening method. Based on bioinformatics analysis, candidate genes were categorised into functional groups of protein kinases, E3 ligases, and transcriptional regulators. Several of these candidate genes were shown to be essential for activation of the Toll pathway in vivo. Of these, we focus on Sherpa, an E3 ligase, in this review.

10.2.4 Sherpa, a Novel Component of the Toll Pathway

In our comprehensive genome-wide RNAi screening, a HECT domain-containing E3 ligase, Sherpa (SUMO-related HECT-domain and RCC-repeat Protein for Toll pathway Activation) was identified as a gene required for Toll activation initiated by dMyd88 overexpression, but not by Pelle overexpression (Kanoh et al. 2015a). A dsRNA against Sherpa exhibited repression of Toll pathway activation, which was rescued by overexpression of Sherpa. These results suggest that Sherpa is a novel component of the Toll pathway that functions genetically upstream of Pelle. Then, using this in vitro rescue system, we determined the domain structure required for the function of Sherpa (Fig. 10.3). The HECT domain, which was expected to possess E3 ligase activity, was found to be essential for activation of the Toll pathway. In the comprehensive genome-wide RNAi screening, Smt3/SUMO (small ubiquitin-like modifier) was also identified as one of the strongest hits. SUMOylation is known to regulate the subcellular localisation and stability of target proteins (Flotho and Melchior 2013) and RNAi-mediated knockdown of SUMO, SUMO E1, and E2 enzymes (Aso1, Uba2, and Uba9) reduced Toll activation induced by dMyd88 overexpression, but not that induced by Pelle

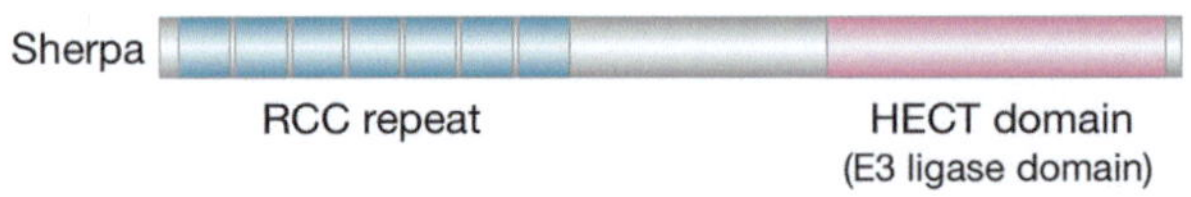

Fig. 10.3 Domain organisation of the novel E3-ligase protein Sherpa. Sherpa consists of two domains: an RCC repeat domain and a HECT domain. The HECT domain possesses E3 ligase activity and is conserved in the HECT-type E3 ligase family

overexpression. This indicates that SUMO-related factors work genetically upstream of Pelle, similarly to Sherpa.

These results suggested that dMyd88, a factor upstream of Pelle, is a possible substrate of Sherpa. To test this idea, we performed a coimmunoprecipitation assay of Sherpa and dMyd88 in DL-1 cells. Sherpa and dMyd88 strongly interacted when coexpressed in DL-1 cells and multiple modification signals were observed, including K63-linked polyubiquitin on dMyd88. Furthermore, polyubiquitination signals, including K63-linked polyubiquitin, were also observed on Sherpa itself. These posttranscriptional modifications on dMyd88 and Sherpa were dependent on the HECT domain. These results suggest that ubiquitination of dMyd88 is dependent on Sherpa.

To explore the function of ubiquitination and SUMOylation in the Toll pathway, the subcellular localisation of dMyd88 was examined by immunostaining. The adaptor complex of overexpressed dMyd88 and Tube was localised at the plasma membrane, as previously reported (Towb et al. 1998; Marek and Kagan 2012). This membrane localisation was abrogated and aggregates in the cytosol appeared following RNAi-mediated knockdown of Sherpa or SUMO. Moreover, the HECT domain was required for Sherpa localisation at the plasma membrane. These results suggest that Sherpa and SUMO mediate Toll pathway signalling by regulating the membrane localisation of the dMyd88-Tube complex via posttranslational modifications (Fig. 10.4).

Finally, we examined the function of Sherpa in adult flies. Sherpa mutant flies with a *sherpa* mRNA level less than 20 % that of wild-type flies rapidly succumbed to infection with Gram-positive bacteria. We next examined whether this susceptibility to Gram-positive bacterial infection was due to improper Toll signalling by monitoring expression of the antifungal peptide gene *Drosomycin* (*Drs*), a target of the Toll pathway. The Sherpa mutant flies had significantly reduced induction of *Drs* in response to Gram-positive bacterial infection, as observed in spz^{rm7} mutants. These experiments clearly demonstrate that *sherpa* is required for activation of the Toll pathway by bacterial infection in Drosophila adults (Kanoh et al. 2015a). Intriguingly, the Sherpa mutant flies showed no apparent developmental defects in the embryonic stage, which implies that Sherpa is specifically required for acute propagation of inflammatory signals in disease conditions.

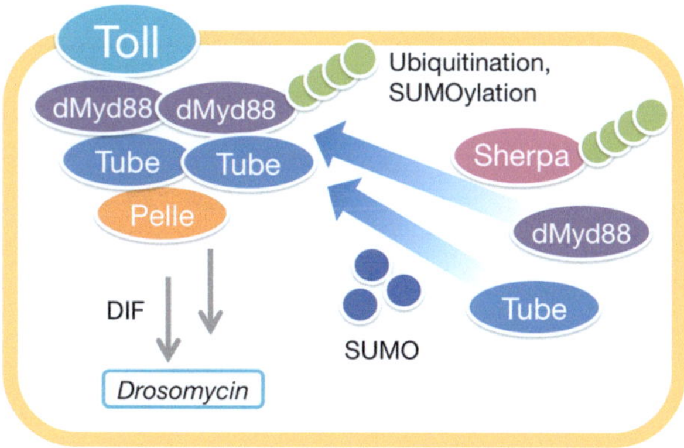

Fig. 10.4 Signalling mechanisms of the Toll pathway regulated by Sherpa. Drosophila Sherpa and SUMO lead adaptor proteins, dMyd88 and Tube, to the plasma membrane via ubiquitination or SUMOylation. This adaptor complex formation around the membrane is crucial for signal transduction and production of antimicrobial peptides

10.2.5 Novel Downstream Protein Kinases and Transcription Factors in the Toll Pathway

Our genome-wide RNAi screening also identified candidate genes for protein kinases and transcription factors (Kanoh et al. 2015a, 2015b). Pitslre and Doa, a cyclin-dependent kinase (CDK) homologue and a CDK-like kinase homologue, respectively, may mediate signal transduction between Pelle and Dif, in the blackbox of the Drosophila Toll pathway (Kanoh et al. 2015a). RNAi-mediated knockdown of Pitslre and Doa in adult fat body diminished antimicrobial peptide expression in response to bacterial infection. RNAi-mediated knockdown of all isoforms of Doa completely abrogated Toll pathway activation induced by overexpression of dMyd88 or Pelle in DL1 cells (Fig. 10.5). A recent report indicated that human CDK6 phosphorylates the NF-kB p65 subunit and is required for inflammatory cytokine production (Buss et al. 2012). Our study also suggested that some cyclin-dependent protein kinases contribute to NF-kB activation in insects and humans. In addition, Jarid2, a Jumonji-like transcription factor possessing histone demethylase activity, is required for Toll pathway signalling and resistance to bacterial infection (Kanoh et al. 2015b). In mammalians, the Jumonji-like transcription factor Jmjd3 regulates macrophage polarisation (Satoh et al. 2010) and inflammatory cytokine production (Kruidenier et al. 2012). These analyses imply that the regulatory mechanisms in Toll and TLR signalling by histone demethylases are evolutionarily conserved.

10.2.6 The Larval Peptide Fraction Activates the Toll Pathway

As described in ex vivo genome-wide RNAi screening (Sect. 10.2.3), our research found that the larval peptide fraction strongly potentiates Toll pathway activation

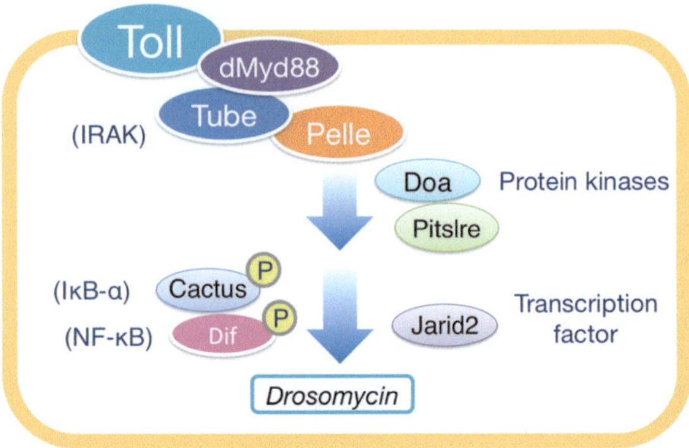

Fig. 10.5 Novel downstream components of the Drosophila Toll pathway. A cdc2-related kinase Pitslre and a cdc2-like kinase Doa mediate Toll pathway signalling between Pelle and Dif. A Jumonji-like transcription factor Jarid2 contributes to antimicrobial peptide expression and resistance to bacterial infection

(Kanoh et al. 2015b). The larval peptide fraction was purified from lysate of whole Drosophila larvae that were immediately dissected under acidic conditions. Using a *Drosomycin* reporter assay, we showed that the larval peptide fraction is a strong inducer of Toll pathway activation in DL1 cells (Kanoh et al. 2015b). This activity was not present in an organic solvent extract from the peptide fraction, and remained entirely in the aqueous phase. Protease treatment completely abolished the activity of the larval peptide fraction. These results suggest that this activity is dependent on hydrophilic peptides in the fraction. Genome-wide RNAi screening in DL1 cells showed that the larval peptide fraction induces Toll pathway activation that is dependent on the Toll receptor (Toll-1) and known intracellular components such as dMyd88, Tube, Pelle, and Dif. Shotgun mass spectrometry analysis indicated that the larval peptide fraction does not contain a detectable amount of Spz, a ligand for Toll-1 required for immune response; or Drosophila neurotrophins, which are ligands for Toll-6 and Toll-7 that are required for development of the nervous system (McIlroy et al. 2013; Sutcliffe et al. 2013). These analyses suggest that the larval peptide fraction contains novel endogenous ligands for the Toll receptor (Kanoh et al. 2015b).

10.3 Conclusion

In this review, we described the Drosophila Toll pathway and new members of this pathway: Sherpa, Pitslre, Doa, and Jarid2. Sherpa is a HECT-type E3 ligase that may target dMyd88, which is essential for activation of the Toll pathway.

NF-kB signalling is activated by exogenous microbial components such as peptidoglycans, β-glucans, and RP-1. Recently, it has been reported that endogenous components can also function as 'danger signals' and activate NF-kB signalling. In Drosophila larvae, mutation of the caspase Dronc may lead to secondary necrosis and constitutive activation of NF-kB signalling that is dependent on psh-, spz-, and Toll receptor. In addition, mutant larvae for oncogene Ras85D have a high expression level of *Drosomycin* (Hauling et al. 2014). In mammals, a nonhistone nuclear protein, HMGB1 (High Mobility Group Box-1), and a cold-inducible RNA binding protein, CIRP, are known to act as endogenous danger signals (Scaffidi et al. 2002; Qiang et al. 2013). These proteins are passively released by necrotic cells and activate NF-kB signalling via the TLR4/MD-2 pathway, and the signals are selectively repressed via the CD24–Siglec–axis (Chen et al. 2009).

It is important to note, recent reports suggest that endogenous ligands can activate TLR-mediated NF-kB signalling, but TLR-independent effects remain to be clarified. For example, RP105/MD-1 (TLR4/MD-2–like complex)-dependent NF-kB activation in obese mice cannot be explained by a known signaling pathway (Watanabe et al. 2012). Our ex vivo comprehensive RNAi screening, which identified Sherpa as a novel NF-kB signalling component, is likely to shed light on the blackbox of Drosophila Toll signalling, and contribute to identification of new mechanisms involving the mammalian NF-kB pathway in chronic inflammation.

Funding This work was supported by grants from the Japan Science and Technology Agency (JST).

Author contributions T. K., H. K, Y. M., and H. K. contributed equally to this work. T. K., H. K, Y. M., and H. K. wrote the draft. H. K. drew the figures. T. K. managed the writing. T. K., H. K, Y. M., H. K., and S. K. completed the manuscript.

Competing Interests The authors declare no conflict of interest.

Acknowledgments We are grateful to all members of the CREST/PRESTO Chronic Inflammation research group for discussions.

References

Anderson KV, Jürgens G, Nüsslein-Volhard C (1985a) Establishment of dorsal-ventral polarity in the Drosophila embryo: genetic studies on the role of the Toll gene product. Cell 42:779–789

Anderson KV, Bokla L, Nüsslein-Volhard C (1985b) Establishment of dorsal-ventral polarity in the drosophila embryo: the induction of polarity by the Toll gene product. Cell 42:791–798

Belvin MP, Jin Y, Anderson KV (1995) Cactus protein degradation mediates Drosophila dorsal-ventral signaling. Genes Dev 9:783–793

Buchon N, Poidevin M, Kwon HM, Guillou A, Sottas V, Lee BL, Lemaitre B (2009) A single modular serine protease integrates signals from pattern-recognition receptors upstream of the Drosophila Toll pathway. Proc Natl Acad Sci U S A 106:12442–12447

Buss H, Handschick K, Jurrmann N, Pekkonen P, Beuerlein K, Müller H, Wait R, Saklatvala J, Ojala PM, Schmitz ML, Naumann M, Kracht M (2012) Cyclin dependent kinase 6 phosphorylates NF-kB P65 at serine 536 and contributes to the regulation of inflammatory gene expression. PLoS One 7:e51847

Chen G-Y, Tang J, Zheng P, Liu Y (2009) CD24 and Siglec-10 selectively repress tissue damage-induced immune responses. Science 323:1722–1725

El Chamy L, Leclerc V, Caldelari I, Reichhart JM (2008) Sensing of 'danger signals' and pathogen-associated molecular patterns defines binary signaling pathways 'upstream' of Toll. Nat Immunol 9:1165–1170

Ferrandon D, Imler JL, Hetru C, Hoffmann JA (2007) The Drosophila systemic immune response: sensing and signalling during bacterial and fungal infections. Nat Rev Immunol 7:862–874

Flotho A, Melchior F (2013) Sumoylation: a regulatory protein modification in health and disease. Annu Rev Biochem 82:357–385

Gobert V, Gottar M, Matskevich AA, Rutschman S, Royet J, Belvin M, Hoffmann JA (2003) Dominique Ferrandon, Dual activation of the Drosophila toll pathway by two pattern recognition receptors. Science 302:2126–2130

Goto A, Yano T, Terashima J, Iwashita S, Oshima Y, Kurata S (2010) Cooperative regulation of the induction of the novel antibacterial Listericin by peptidoglycan recognition protein LE and the JAK-STAT pathway. J Biol Chem 285:15731–15738

Gottar M, Gobert V, Matskevich AA, Reichhart JM, Wang C, Butt TM, Belvin M, Hoffmann JA, Ferrandon D (2006) Dual detection of fungal infections in Drosophila via recognition of glucans and sensing of virulence factors. Cell 127:1425–1437

Hauling T, Krautz R, Markus R, Volkenhoff A, Kucerova L, Theopold U (2014) A Drosophila immune response against Ras-induced overgrowth. Biol Open 3:250–260

Igaki T, Yamamoto-Goto Y, Tokushige N, Kanda H, Miura M (2002) Down-regulation of DIAP1 triggers a novel Drosophila cell death pathway mediated by Dark and DRONC. J Biol Chem 277:23103–23106

Ip YT, Reach M, Engstrom Y, Kadalayil L, Cai H, Gonzalez-Crespo S, Tatei K, Levine M (1993) Dif, a dorsal-related gene that mediates an immune response in Drosophila. Cell 75:753–763

Jang IH, Chosa N, Kim SH, Nam HJ, Lemaitre B, Ochiai M, Kambris Z, Brun S, Hashimoto C, Ashida M, Brey PT, Lee WJ (2006) A Spatzle-processing enzyme required for toll signaling activation in Drosophila innate immunity. Dev Cell 10:45–55

Kanoh H, Tong L-L, Kuraishi T, Suda Y, Momiuchi Y, Shishido F, Kurata S (2015a) Genome-wide RNAi screening implicates the E3 ubiquitin ligase Sherpa in mediating innate immune signaling by Toll in Drosophila adults. Sci Signal 8:ra107

Kanoh H, Kuraishi T, Tong LL, Watanabe R, Nagata S, Kurata S (2015b) Ex vivo genome-wide RNAi screening of the Drosophila Toll signaling pathway elicited by a larva-derived tissue extract. Biochem Biophys Res Commun 467:400–406

Kleino A, Silverman N (2014) The Drosophila IMD pathway in the activation of the humoral immune response. Dev Comp Immunol 42:25–35

Kruidenier L, Chung C, Cheng Z, Liddle J, Che K, Joberty G, Bantscheff M, Bountra C, Bridges A, Diallo H, Eberhard D, Hutchinson S, Jones E, Katso R, Leveridge M, Mander PK, Mosley J, Ramirez-Molina C, Rowland P, Schofield CJ, Sheppard RJ, Smith JE, Swales C, Tanner R, Thomas P, Tumber A, Drewes G, Oppermann U, Patel DJ, Lee K, Wilson DM (2012) A selective jumonji H3K27 demethylase inhibitor modulates the proinflammatory macrophage response. Nature 488:404–408

Kurata S (2014) Peptidoglycan recognition proteins in Drosophila immunity. Dev Comp Immunol 42:36–41

Lemaitre B, Hoffmann J (2007) The host defense of Drosophila melanogaster. Annu Rev Immunol 25:697–743

Lemaitre B, Nicolas E, Michaut L, Reichhart JM, Hoffmann JA (1996) The dorsoventral regulatory gene cassette spätzle/Toll/cactus controls the potent antifungal response in Drosophila adults. Cell 86:973–983

Manfruelli P, Reichhart JM, Steward R, Hoffmann JA, Lemaitre B (1999) A mosaic analysis in Drosophila fat body cells of the control of antimicrobial peptide genes by the Rel proteins Dorsal and DIF. EMBO J 18:3380–3391

Marek LR, Kagan JC (2012) Phosphoinositide binding by the Toll adaptor dMyD88 controls antibacterial responses in Drosophila. Immunity 36:612–622

McIlroy G, Istvan F, Aurikko J, Wentzell JS, Lim MA, Fenton JC, Gay NJ, Hidalgo A (2013) Toll-6 and Toll-7 function as neurotrophin receptors in the Drosophila melanogaster CNS. Nat Neurosci 16:1248–1256

Michel T, Reichhart JM, Hoffmann JA, Royet J (2001) Drosophila Toll is activated by Gram-positive bacteria through a circulating peptidoglycan recognition protein. Nature 414:756–759

Ming M, Obata F, Kuranaga E, Miura M (2014) Persephone/Spätzle Pathogen Sensors Mediate the Activation of Toll Receptor Signaling in Response to Endogenous Danger Signals in Apoptosis-deficient Drosophila. J Biol Chem 289:7558–7568

Nakamoto M, Moy RH, Xu J, Bambina S, Yasunaga A, Shelly SS, Gold B, Cherry S (2012) Virus recognition by Toll-7 activates antiviral autophagy in Drosophila. Immunity 36:658–667

Qiang X, Yang W-L, Wu R, Zhou M, Jacob A, Dong W, Kuncewitch M, Ji Y, Yang H, Wang H, Fujita J, Nicastro J, Coppa GF, Tracey KJ, Wang P (2013) Cold-inducible RNA-binding protein (CIRP) triggers inflammatory responses in hemorrhagic shock and sepsis. Nat Med 19:1489–1495

Reach M, Galindo RL, Towb P, Allen JL, Karin M, Wasserman SA (1996) A gradient of cactus protein degradation establishes dorsoventral polarity in the Drosophila embryo. Dev Biol 180:353–364

Satoh T, Takeuchi O, Vandenbon A, Yasuda K, Tanaka Y, Kumagai Y, Miyake T, Matsushita K, Okazaki T, Saitoh T, Honma K, Matsuyama T, Yui K, Tsujimura T, Standley DM, Nakanishi K, Nakai K, Akira S (2010) The Jmjd3-Irf4 axis regulates M2 macrophage polarisation and host responses against helminth infection. Nat Immunol 11:936–944

Scaffidi P, Misteli T, Bianchi ME (2002) Release of chromatin protein HMGB1 by necrotic cells triggers inflammation. Nature 418:191–195

Sun H, Bristow BN, Qu G, Wasserman SA (2002) A heterotrimeric death domain complex in Toll signaling. Proc Natl Acad Sci U S A 99:12871–12876

Sun H, Towb P, Chiem DN, Foster BA, Wasserman SA (2004) Regulated assembly of the Toll signaling complex drives Drosophila dorsoventral patterning. EMBO J 23:100–110

Sutcliffe B, Forero MG, Zhu B, Robinson IM, Hidalgo A (2013) Neuron-Type Specific Functions of DNT1, DNT2 and Spz at the Drosophila Neuromuscular Junction. PLoS One 8:e75902

Takeuchi O, Akira S (2010) Pattern recognition receptors and inflammation. Cell 14:805–820

Tauszig-Delamasure S, Bilak H, Capovilla M, Hoffmann JA, Imler JL (2002) Drosophila MyD88 is required for the response to fungal and Gram-positive bacterial infections. Nat Immunol 3:91–97

Towb P, Galindo RL, Wasserman SA (1998) Recruitment of Tube and Pelle to signaling sites at the surface of the Drosophila embryo. Development 125:2443–2450

Valanne S, Wang JH, Ramet M (2011) The Drosophila Toll signaling pathway. J Immunol 186:649–656

Watanabe Y, Nakamura T, Ishikawa S, Fujisaka S, Usui I, Tsuneyama K, Ichihara Y, Wada T, Hirata Y, Suganami T, Izaki H, Akira S, Miyake K, Kanayama H, Shimabukuro M, Sata M, Sasaoka T, Ogawa Y, Tobe K, Takatsu K, Nagai Y (2012) The radioprotective 105/MD-1 complex contributes to diet-induced obesity and adipose tissue inflammation. Diabetes 61:1199–1209

Weber AN, Tauszig-Delamasure S, Hoffmann JA, Lelievre E, Gascan H, Ray KP, Morse MA, Imler JL, Gay NJ (2003) Binding of the Drosophila cytokine Spatzle to Toll is direct and establishes signaling. Nat Immunol 4:794–800

Part II
Imaging Analyses of Chronic Inflammation

Chapter 11
Macrophage Dynamics During Bone Resorption and Chronic Inflammation

Junichi Kikuta, Keizo Nishikawa, and Masaru Ishii

Abstract Macrophages are located in all body tissues, contributing to tissue homeostasis by clearing foreign materials, dead cells, and debris. Macrophages are heterogeneous populations of immune cells that rapidly change their functions in response to local microenvironmental signals and also play various roles in both the induction and resolution of inflammation. In this review, we introduce our recent studies on the dynamics and functions of various cells of the macrophage lineage. We first focus on bone-resorbing macrophages (the osteoclasts). We recently found that S-adenosylmethionine-mediated DNA methylation catalysed by DNA methyltransferase 3a regulates osteoclastogenesis via epigenetic repression of antiosteoclastogenic genes. We also visualised the in vivo behavior of bone-resorbing macrophages in living bone using intravital two-photon microscopy, and found that sphingosine-1-phosphate, a lipid mediator, dynamically regulates both the migration and localisation of bone-resorbing macrophage precursor cells. Furthermore, we identified two distinct functional states of differentiated bone-resorbing macrophages, ('bone-resorptive [R]' and 'nonresorptive [N]'), and found that Th17 cells, a subset of $CD4^+$ T cells, stimulated osteoclastic bone destruction by directly contacting N-state bone-resorbing macrophages to convert them to the R-state; this is a critical step in bone erosion within arthritic joints. Second, we discuss the roles played by adipose tissue macrophages during obesity induced chronic inflammation. Using intravital imaging techniques, we have shown that the adipose tissue-expressed S100A8 gene plays a critical role in macrophage recruitment in the very early stages of obesity. This review provides new insights into the physiological and pathological roles of macrophages, and suggests potential novel therapies targeting aspects of macrophage dynamics in patients with chronic inflammatory diseases.

J. Kikuta • K. Nishikawa • M. Ishii (✉)
Department of Immunology and Cell Biology, Graduate School of Medicine and Frontier Biosciences, and WPI-Immunology Frontier Research Center, Osaka University, 2-2 Yamada-oka, Suita, Osaka 565-0871, Japan

AMED-CREST, Suita, Osaka 565-0871, Japan
e-mail: mishii@icb.med.osaka-u.ac.jp

M. Miyasaka, K. Takatsu (eds.), *Chronic Inflammation*,
DOI 10.1007/978-4-431-56068-5_11

Keywords Osteoclast • Dnmt3a • S1P • RANKL • Th17 • Obesity • S100A8 • Imaging

11.1 Introduction

Macrophages, found in nearly all tissues, play significant roles in tissue development, homeostasis, and repair, as well as in the immune response. Macrophages may be divided into subpopulations in terms of both phenotype and function. Specialised tissue-resident macrophages include bone-resorbing macrophages of the bone, alveolar macrophages of the lung, Kupffer cells of the liver, Langerhans cells of the skin, and microglial cells of the brain. These tissue-specific macrophage subpopulations ingest foreign materials and recruit additional macrophages from the circulation during acute and chronic inflammation.

In this chapter, we focus especially on the molecular mechanisms controlling macrophage dynamics during bone resorption, and during development of chronic inflammation of adipose tissue. In addition, we also discuss novel macrophage-targeted therapies that will improve the future treatment of bone-destructive diseases and obesity.

11.1.1 DNA Methylation Regulates Osteoclastogenesis

11.1.1.1 Metabolic Diversity in Cells of the Macrophage Lineage

Tissue-resident macrophages originate from peripheral blood monocytes and/or tissue-resident precursors, and vary considerably in terms of both phenotype and specialisation (Davies et al. 2013). By analogy to Th1/Th2 cells, macrophages have been divided into two types: proinflammatory macrophages (also known as M1 or classical macrophages) and anti-inflammatory macrophages (the so-called M2 or alternative macrophages), based on in vitro experiments (Osborn et al. 2012). Proinflammatory macrophages are induced in vitro by growing bone-marrow–derived haematopoietic cells with IFN-γ and lipopolysaccharide; anti-inflammatory macrophages can be generated by culture in the presence of IL-4 and IL-13. Proinflammatory macrophages exhibit enhanced microbiocidal capacities and produce high levels of pro-inflammatory cytokines (IL-1, IL-6, and TNF-α), and increased concentrations of nitrogen radicals, enhancing their capacity to kill. In contrast, anti-inflammatory macrophages reduce the levels of proinflammatory cytokines and secrete certain components of the extracellular matrix. Thus, the proinflammatory macrophage is part of the first line of defense of the innate immune system, whereas the anti-inflammatory macrophage plays a greater role during resolution of disease, and in tissue remodelling. In addition, pro- and anti-inflammatory macrophages exhibit clear metabolic differences, reflected by their metabolism. In proinflammatory macrophages, aerobic glycolysis is induced upon activation; this increases both glucose uptake and the conversion of pyruvate to lactate. Simultaneously, respiratory chain activity is attenuated, generating reactive-oxygen species (ROS). The pentose phosphate pathway is also activated; this pathway is key in terms of NADPH generation by NADPH oxidase; NADPH is required for the production of both ROS and nitric oxides. A recent report showed that that the TCA cycle of proinflammatory macrophages was broken in two places (Jha et al. 2015). One break was evident after succinate production and the other at isocitrate dehydrogenase (IDH), which converts citrate to alpha-ketoglutarate. At the postsuccinate breakpoint, succinate acts as an inflammatory signal inducing IL-1-beta synthesis via a mechanism involving Hif-1-alpha (Tannahil et al. 2013). The arginosuccinate shunt replenishes fumarate and malate to allow citrate production, and also contributes to nitric oxide production. In contrast, downregulation of IDH triggered a buildup in citrate levels such that citrate was redirected toward the generation of the antimicrobial itaconic acid. In addition, citrate was withdrawn for fatty acid biosynthesis; this is another hallmark of proinflammatory macrophages. On the other hand, anti-inflammatory macrophages derive much of their energy from fatty acid oxidation and oxidative metabolism. Pentose phosphate pathway activity is more limited in anti-inflammatory macrophages. Furthermore, the levels of amino sugars and nucleotide sugars are increased, creating high levels of UDP-N-acetyl-alpha-D-glucosamine (UDP-GlcNAc), which is required for N-glycosylation. Highly glycosylated lectin/mannose receptors are markers of M2 macrophages. Thus, studies of the intracellular metabolism of several types of macrophages has emphasised that tight links exist between the metabolic states and the phenotypes of such cells.

Bone-resorbing macrophages, also termed osteoclasts, are resident in bone and have abundant mitochondria (Ishii et al. 2009a). The ‘receptor activator of nuclear factor kB ligand’ (RANKL) is the key cytokine involved in differentiation of precursor cells into bone-resorbing macrophages, and also controls mitochondrial biogenesis in such macrophages via PGC1-beta–mediated transcriptional regulation. Mitochondria play a central role in the metabolic pathway that consumes oxygen (i.e. oxidative metabolism) and are important energy-producing organelles. Indeed, a metabolic shift toward the oxidative process is induced during differentiation of precursor cells into bone-resorbing macrophages. Although it has traditionally been assumed that metabolic changes in bone-resorbing macrophages are important in terms of adaptation to the increased energy demands associated with bone resorption, very little is actually known about this topic.

11.1.1.2 A New Role for Oxidative Metabolism in Bone-Resorbing Macrophages

To gain a detailed understanding of changes in intracellular metabolism during bone-resorbing macrophage differentiation, fluctuations in the levels of intracellular metabolites were analysed with the aid of capillary electrophoresis–mass spectrometry. Metabolome analysis revealed that S-adenosylmethionine (SAM), a metabolite of the methionine cycle, increased in level during bone-resorbing macrophage differentiation (Nishikawa et al. 2015). The addition of an inhibitor of oxidative metabolism (i.e., an inhibitor of the mitochondrial electron transport chain) revealed that oxidative metabolism by bone-resorbing macrophages was important in terms of SAM production. Next, addition of an inhibitor of methionine adenosyltransferase, the enzyme catalysing SAM synthesis, revealed the importance of enhanced SAM production during macrophage differentiation. The inhibitor decreased the intracellular SAM concentration and inhibited cell differentiation. Thus, SAM, produced via oxidative metabolism, may be an important regulator of bone-resorbing macrophage differentiation.

Bone marrow cells, including bone-resorbing macrophages, typically live under conditions of low oxygen tension (Spencer et al. 2014). Furthermore, the low-oxygen regions of the medullary cavity are unevenly distributed; the middle of the bone (i.e., the diaphysis) is exposed to much less oxygen than is the end of the bony tissue (i.e., the epiphysis) (Kusumbe et al. 2014). Interestingly, as bone-resorbing macrophages are located principally in epiphyseal tissue, it would seem at first glance that such macrophages simply cannot exist in oxygen-poor environments. Although further research on this topic is necessary, bone-resorbing macrophages are influenced by the oxygen level in vivo, and the idea that functional differentiation and maturation might be fine-tuned by cellular metabolic status in situ is reasonable.

11.1.1.3 Epigenetic Regulation Links Bone-Resorbing Macrophage Differentiation to Intracellular Metabolism

SAM has been previously shown to donate the methyl group to methyltransferases. Therefore, we studied methyltransferases in a search for clues as to how SAM regulates bone-resorbing macrophage differentiation. We measured the expression levels of methyltransferases induced during macrophage differentiation using transcriptome data and found that the expression level of *Dnmt3a*, an enzyme transferring a methyl group to cytosine residues in DNA, was markedly enhanced during differentiation. We next created bone-resorbing macrophage-specific *Dnmt3a* knockout mice (using a conditional knockout technique) to allow us to assess the significance of Dnmt3a in bone metabolism. We found that the numbers of bone-resorbing macrophages decreased, and the bone volume markedly increased, in bone-resorbing macrophage-specific *Dnmt3a* knockout mice, suggesting that Dnmt3a regulates macrophage differentiation. More detailed analysis revealed how Dnmt3a functioned during bone-resorbing macrophage differentiation. Dnmt3a (1) suppressed the expression of several genes in a DNA methyltransferase-dependent manner, and (2) specifically accelerated macrophage differentiation by suppressing expression of the transcription factor *Irf8*, a negative regulator of the differentiation of such cells (Zhao et al. 2009). High-level *Irf8* expression is a characteristic of the precursor cells of bone-resorbing macrophages. However, overexpression of both Dnmt3a and SAM decreased *Irf8* expression in such cells. The data thus suggested that the combined effects of Dnmt3a and SAM suppressed *Irf8* expression.

11.1.1.4 Osteoporosis Treatments Targeting Epigenetic Regulation of Bone-Resorbing Macrophages

Recent work on the development of anticancer drugs has shown that targeting of epigenetic regulation may be effective in this context (Rodríguez-Paredes et al. 2011). Therefore, we explored the possibility that regulation of the DNA methylation associated with bone-resorbing macrophage differentiation might identify potential drug targets. Osteoporosis is triggered by abnormalities in bone-resorbing macrophages. We used postmenopausal osteoporotic mice in which osteoporosis had been induced by oophorectomy. After oophorectomy, the number of bone-resorbing macrophages increased and the bone volume decreased. However, bone-resorbing macrophage-specific *Dnmt3a* knockout mice did not develop osteoporosis. Thus, Dnmt3a may be a potential target of antiosteoporosis drugs. We next searched for such drugs. The aflavin 3,3′-digallate (TF-3) exhibited strong antiosteoporosis activity. We gave TF-3 to postoophorectomy mice to explore whether the drug protected against osteoporosis. TF-3 significantly reduced the severity of osteoporosis. Thus, drugs that inhibit bone resorption, and that target bone-resorbing macrophages, may be useful in humans. Such drugs would affect epigenetic regulation in target cells.

11.1.2 Intravital Bone Imaging Reveals the Dynamics of Bone-Resorbing Macrophages

11.1.2.1 Intravital Two-Photon Imaging of Bone Tissue

Bone is the hardest bodily tissue; it is technically difficult to visualise the cellular interactions in play within the bone marrow cavities of living animals. The morphology and structure of bone tissues can be analysed using conventional methods including microcomputed tomography, histomorphological analyses, and flow cytometry. These methods yield information on cell shape and molecular expression patterns, but not on dynamic cell movements in living bone marrow. The (relatively) recent introduction of fluorescence microscopy imaging allows us to better understand the cellular dynamics of organs and tissues in vivo (Cahalan et al. 2002; Germain et al. 2006). To this end, we established an advanced imaging system allowing us visualise living bone tissue by using intravital two-photon microscopy (Ishii et al. 2009b, 2010; Kikuta et al. 2013a, b).

Two-photon excitation-based laser microscopy affords certain advantages over conventional (single-photon) confocal microscopy. In confocal microscopy, a fluorophore absorbs energy from a single photon (excitation), and subsequently releases the energy as an emitted photon. In contrast, upon two-photon excitation, a fluorophore simultaneously absorbs two photons. Two-photon excitation is rare, being evident only in the region of a focal plane in which the photon density is high; the images are bright and of high resolution. Excitation by a laser operating at a near-infrared wavelength reduces phototoxic tissue damage, which is essential if intravital imaging is to yield useful results. Furthermore, light of near-infrared wavelengths penetrates deep into tissue (to 100–1,000 μm) but confocal microscopy yields data to only depths of < 100 μm. Two-photon excitation microscopy affords efficient light detection, reduces phototoxicity, and penetrates deeper into tissues. The technique is becoming increasingly useful when it is desirable to visualise the dynamic cellular behavior of deep intravital tissues, and to quantitate cell motility and cellular interactions (Cahalan et al. 2002; Germain et al. 2006).

Access to the deep bone marrow is difficult; light of near-infrared wavelengths is readily scattered by calcium phosphate crystals of the bone matrix. In the mouse parietal bone, the distance from the bone surface to the bone marrow cavity is only ~80–120 μM, which is sufficiently thin to permit controlled fluorophore excitation within the cavity. Intravital two-photon imaging of skull bone tissue allows visualisation of the real-time in vivo behavior of many cell types of bone marrow cavities; such cells include macrophages, neutrophils, lymphocytes, and haematopoietic stem cells. Moreover, such imaging may be useful when it is desirable to evaluate the effects of novel drugs targeting skeletal disease.

11.1.2.2 S1P-Dependent Migratory Control of the Cellular Precursors of Bone-Resorbing Macrophages

Bone-resorbing macrophages develop from haematopoietic precursor cells of the mononuclear/monocyte lineage. Circulating monocytes migrate and attach to bone surfaces, fuse to form giant cells, and trigger bone resorption. However, the means by which bone-resorbing macrophage precursors migrate to bony surfaces, and the *in vivo* controls exerted on migratory behavior, remain elusive. Intravital two-photon imaging of skull bone tissue allowed us to first define the in vivo behavior of bone-resorbing macrophages of the bone marrow cavity; we found that sphingosine-1-phosphate (S1P), a lipid mediator enriched in blood, controlled the migratory behavior of the macrophage precursors by acting in concert with several chemokines (Ishii et al. 2009b, 2010).

S1P is a bioactive sphingolipid metabolite that regulates various biological activities, including cell proliferation, motility, and survival (Cyster 2005). S1P signalling involves five receptors, S1PR1 to S1PR5, all of which are members of the 7-transmembrane-G-protein-coupled receptor family (Rosen et al. 2005; Rivera et al. 2008). Bone-resorbing macrophage precursors express both S1PR1 and S1PR2. These proteins exert opposite effects on the migration of bone-resorbing macrophage precursors. S1PR1 is extremely sensitive to low S1P concentrations, and promotes cell movement toward higher S1P concentrations of circulatory fluids. In contrast, S1PR2 requires a higher S1P concentration for activation, and negatively regulates the S1PR1 response when activated. In a high-S1P environment such as the bloodstream, S1PR1 is both activated and rapidly internalised, allowing S1PR2-mediated actions to predominate. Bone-resorbing macrophage precursors enter the bone marrow when chemorepulsion is mediated by both S1PR2 and other chemokines attracting macrophages to bone surfaces. When macrophages enter a low-S1P environment such as the bone marrow, S1PR1 is transported back to the cell surface, and the bone-resorbing macrophage precursors move from bone tissue into the blood vessels; this reflects chemotaxis toward an S1P gradient. The number of bone-resorbing macrophages on bone surfaces is thus determined by the extent of exchange of macrophage precursors to and from the circulation. This affords a novel means by which osteoclastogenesis may be controlled in vivo.

11.1.2.3 S1P-Targeted Osteoporosis Therapy

Targeting of S1P-dependent control of bone-resorbing macrophage precursor migration is attractive in terms of osteoporosis treatment. FTY720, a superagonist of four of the five S1P receptors (excluding S1PR2), acting preferentially on S1PR1, relieved ovariectomy-induced osteoporosis in mice by reducing the numbers of bone-resorbing macrophages attached to bone surfaces (Ishii et al. 2009b). This clearly suggests that S1P-targeted therapy, for example, prescription of S1P-receptor modulators, might aid in osteoporosis treatment. Moreover, the mechanism of action of S1P is distinct from those of conventional treatments. Bisphosphonates suppress the resorptive activity of fully differentiated bone-resorbing macrophages. Synergistic therapeutic effects can be expected if drugs with different modes of action are simultaneously administered.

An anti bone resorptive drug active on the S1P receptor already exists; this is vitamin D, a well-known bone-protecting factor that significantly inhibits bone destruction. Active vitamin D analogues have been clinically prescribed for patients with bone and mineral disorders, although the detailed mechanisms underlying the pharmacological action remain elusive. Recently, we showed – for the first time – that vitamin D controlled the migratory behavior of circulatory precursors of bone-resorbing macrophages (Kikuta et al. 2013a). Both active vitamin D, calcitriol (1,25-D), and the clinically active vitamin D analogue eldecalcitol (ELD), significantly suppressed S1PR2 expression in monocytes (including bone-resorbing macrophage precursors) both in vitro and in vivo. Intravital two-photon microscopy of living bone revealed that the motility of bone-resorbing macrophage precursors was significantly increased in mice treated with active vitamin D derivatives including 1,25-D and ELD, suggesting that in vivo administration of active vitamin D might suppress both S1PR2 expression and mobilisation of bone-resorbing macrophage precursors from the blood to the bone marrow. This should limit osteoclastic bone resorption in vivo; this is the principal therapeutic effect of active vitamin D.

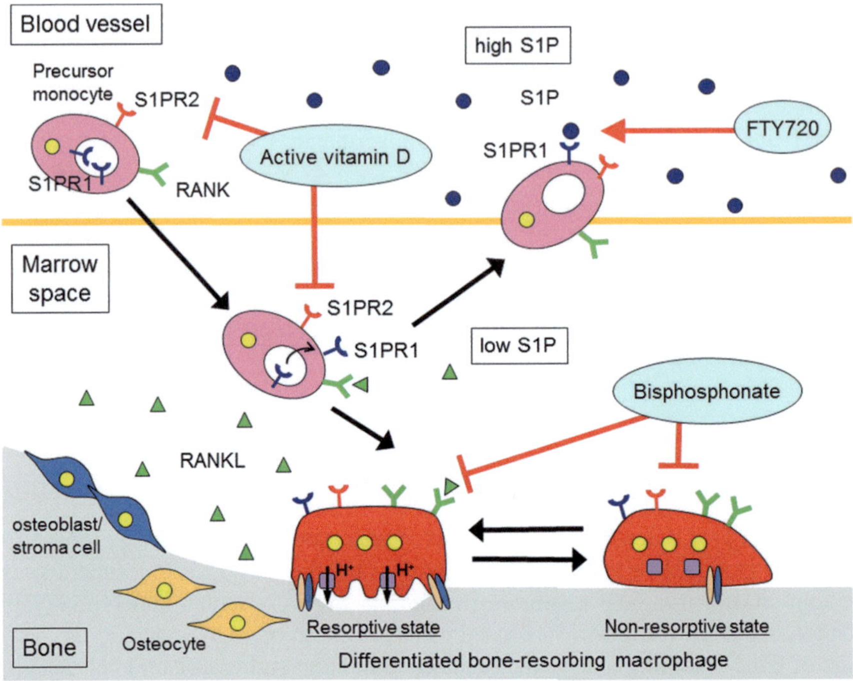

Fig. 11.1 Bone-resorbing macrophage dynamics as revealed by intravital bone imaging
Bone-resorbing macrophage precursors enter into, and exit from, the bone marrow cavity in a process regulated by sphingosine-1-phosphate (S1P). Fully differentiated bone-resorbing macrophages exist in two distinct functional states, bone-resorptive (R) and nonresorptive (N). An S1P receptor agonist, FTY720, facilitates the exit of bone-resorbing macrophage precursors, limiting the numbers of such precursors in the endosteum, and thus inhibiting osteoclastic bone resorption. Active vitamin D suppresses S1PR2 expression and mobilisation of bone-resorbing macrophage precursors from the blood to the bone marrow, thus limiting osteoclastic bone resorption. In contrast, bisphosphonate acts on fully differentiated bone-resorbing macrophages to inhibit bone resorption

We thus suggest that migration of bone-resorbing macrophage precursors to bone sites that will then be resorbed, and cell positioning at such sites, are critical in terms of regulation of bone destruction (Fig. 11.1).

11.1.2.4 RANKL- and TH17-Mediated Control of Bone-Resorbing Macrophages

We recently explored the in vivo activities of fully differentiated bone-resorbing macrophages via intravital two-photon microscopy. We identified two distinct functional states of differentiated macrophages; bone-resorbing (R) cells were firmly adherent to bones, and devoured the bone matrix by secreting acids; nonresorbing (N) cells were relatively loosely attached to bones and moved laterally along bone surfaces. Pathological conditions (e.g., osteoporosis) affected both the proportions of these populations and the total numbers of differentiated macrophages. Also, RANKL not only promoted macrophage differentiation, but also regulated the bone-resorptive function of fully differentiated macrophages.

Furthermore, we found that differentiated bone-resorbing macrophages interacted with $CD4^+$ T helper 17 (Th17) cells, a subset of $CD4^+$ helper T cells. Th17 cell numbers increase in the synovial fluid of rheumatoid arthritis (RA) patients, and enhance bone destruction by osteoclasts (Kong et al. 1999; Takayanagi et al. 2000). Th17 cells express surface RANKL (Sato et al. 2006); this has been suggested to be significant in terms of osteoclastic bone destruction of arthritic joints. However, any functional role for RANKL expressed on the surfaces of Th17 cells located in regions of bone erosion remains elusive. Intravital bone imaging has shown that RANKL-bearing Th17 cells stimulate osteoclastic bone destruction by directly contacting N-state bone-resorbing macrophages, converting such cells to the R-state. This may critically support bone erosion of arthritic joints, which would be a novel Th17 action; Th17 may thus be usefully targeted when it is sought to alleviate RA.

11.1.3 The Adipose Tissue Protein S100A8 Recruits Inflammatory Macrophages to Fatty Tissue of Obese Subjects

11.1.3.1 Macrophage Dynamics in Adipose Tissue During Chronic Inflammation

Obesity is a form of chronic low-grade adipose tissue inflammation. During the development of obesity, various immune system cells, including inflammatory macrophages, infiltrate adipose tissue and become involved in the creation and progression of a diseased state (Weisberg et al. 2003). Macrophages of inflamed adipose tissue (predominantly M1 macrophages) produce inflammatory cytokines and chemokines, and recruit other macrophages and immune cells from the circulation. Our intravital imaging system has recently revealed the macrophage dynamics of obese adipose tissue during development of inflammation (Sekimoto et al. 2015). In adipose tissues of animals on a normal diet, only a few migratory macrophages were evident. Eight weeks after commencing a high-fat high-sucrose (HF/HS) diet (Nishimura et al. 2008), the adipocytes were larger, and the number of infiltrating macrophages had increased. Dead adipocytes were surrounded by some of these macrophages, forming crown-like structures (CLSs) (Murano et al. 2008); both the body weight and plasma glucose concentration were significantly elevated. In fact, just 5 days after the HF/HS diet commenced, macrophages were mobilised to infiltrate adipose tissue, although the adipocytes were not obviously hypertrophic at this time, macrophage accumulation was not prominent. At this time, the body weight and plasma glucose concentration were similar to those of animals on a normal diet. These results indicate that the initial (very early) step of obesity-induced chronic inflammation involves highly mobile macrophages.

11.1.3.2 The Role Played by the Adipose Tissue Protein S100A8 in the Very Early Stages of Obesity

Several chemokines, including CCL2 (MCP-1) and CCL3 (MIP-1α), have been reported to be involved in macrophage infiltration of adipose tissues and progression of chronic inflammation in obesity (Kanda et al. 2006; Weisberg et al. 2006). The levels of these chemokines were elevated in the later stages of obesity (8 weeks after an HF/HS diet commenced), but were essentially unaltered at the very early stage of obesity (only 5 days after the HF/HS diet commenced). We recently found that the S100A8 level increased in mature adipocytes just 5 days after the HF/HS diet commenced, and rose further at 8 weeks (Sekimoto et al. 2015).

Two S100 protein family members, S100A8 and S100A9, are endogenous alarmins released from activated phagocytes; the proteins are recognised by the TLR4 protein of monocytes (Vogl et al. 2007; Foell et al. 2007). S100A8 and S100A9 form a heterodimer (calprotectin); the blood and extracellular body fluid levels of these proteins increase in many inflammatory conditions, including adiposity (Sekimoto et al. 2012), RA, and inflammatory bowel disease.

Intravital two-photon imaging of adipose tissue revealed that recombinant S100A8 increased adipose macrophage mobility 1 h after treatment, suggesting that S100A8 acts, in vivo, very early in obesity, to mobilise adipose tissue macrophages. Recombinant S100A8 induced the synthesis of proinflammatory molecules, including TNF-α and CCL2, by both macrophages and adipocytes. Furthermore, an antibody neutralising S100A8 efficiently suppressed the HF/HS diet-induced initial inflammatory changes, including the very early increase in adipose macrophage mobility. Anti-S100A8 antibody also significantly improved HF/HS diet-induced insulin sensitivity without changing either the body weight or fat mass, suggesting that the increase in macrophage mobility mediated by S100A8 affected the development of insulin resistance and diabetes as obese adipose tissue became chronically inflamed.

Thus, in a series of studies, we identified a very early pathological event (increased macrophage motility) in adipose tissue inflammation; such motility is triggered by an increase in S100A8 expression and progresses to chronic inflammation in situ. S100A8 will be a valuable target of novel therapies for obesity-induced chronic inflammation (Fig. 11.2).

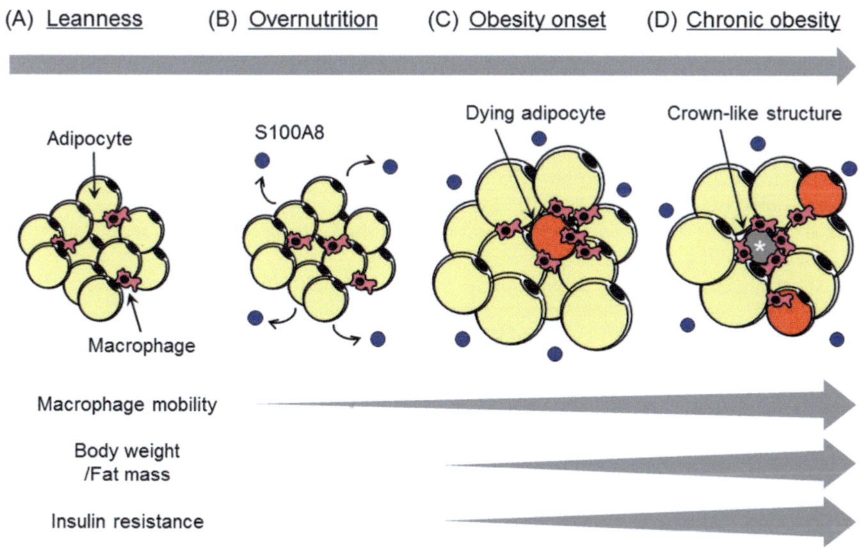

Fig. 11.2 The process of obesity-induced chronic inflammation in adipose tissues
In healthy adipose tissues, there are only a few migratory macrophages (**a**). During the development of obesity, the initial event is an increase of S100A8 expression in mature adipocytes (**b**). S100A8 enhances macrophage mobility in inflamed adipose tissues (**c**). As obese adipose tissues become chronic, both body weight and fat mass are elevated, and insulin resistance is induced (**d**). Dead adipocyte is surrounded by some macrophages, forming a crown-like structure (**d**, *asterisk*)

11.2 Conclusions

Macrophages are very dynamic and play critical roles not only during normal homeostasis, but also in the pathogenesis of several kinds of inflammatory disease. Intravital two-photon imaging yields spatiotemporal information from living organisms; this cannot be done using conventional methods. Such imaging has revealed, and continues to reveal, the dynamic nature of macrophages engaged in physiological and pathological processes. This technique can also be used to evaluate the effects of novel drugs targeting inflammatory diseases. Drugs that control macrophage recruitment and migration will be valuable in the treatment of macrophage-related inflammatory diseases.

References

Cahalan MD et al (2002) Two-photon tissue imaging: seeing the immune system in a fresh light. Nat Rev Immunol 2:872–880
Cyster JG (2005) Chemokines, sphingosine-1-phosphate, and cell migration in secondary lymphoid organs. Annu Rev Immunol 23:127–159
Davies LC et al (2013) Tissue-resident macrophages. Nat Immunol 14:986–995

Foell D et al (2007) S100 proteins expressed in phagocytes: a novel group of damage-associated molecular pattern molecules. J Leukoc Biol 81:28–37
Germain RN et al (2006) Dynamic imaging of the immune system: progress, pitfalls and promise. Nat Rev Immunol 6:497–507
Ishii KA et al (2009a) Coordination of PGC-1beta and iron uptake in mitochondrial biogenesis and osteoclast activation. Nat Med 15:259–266
Ishii M et al (2009b) Sphingosine-1-phosphate mobilizes osteoclast precursors and regulates bone homeostasis. Nature 458:524–528
Ishii M et al (2010) Chemorepulsion by blood S1P regulates osteoclast precursor mobilization and bone remodeling in vivo. J Exp Med 207:2793–2798
Jha AK et al (2015) Network integration of parallel metabolic and transcriptional data reveals metabolic modules that regulate macrophage polarization. Immunity 42:419–430
Kanda H et al (2006) MCP-1 contributes to macrophage infiltration into adipose tissue, insulin resistance, and hepatic steatosis in obesity. J Clin Invest 116:1494–1505
Kikuta J et al (2013a) Sphingosine-1-phosphate-mediated osteoclast precursor monocyte migration is a critical point of control in antibone-resorptive action of active vitamin D. Proc Natl Acad Sci U S A 110:7009–7013
Kikuta J et al (2013b) Dynamic visualization of RANKL and Th17-mediated osteoclast function. J Clin Invest 123:866–873
Kong YY et al (1999) Activated T cells regulate bone loss and joint destruction in adjuvant arthritis through osteoprotegerin ligand. Nature 402:304–309
Kusumbe AP et al (2014) Coupling of angiogenesis and osteogenesis by a specific vessel subtype in bone. Nature 507:323–328
Murano I et al (2008) Dead adipocytes, detected as crown-like structures, are prevalent in visceral fat depots of genetically obese mice. J Lipid Res 49:1562–1568
Nishikawa k et al (2015) DNA methyltransferase 3a regulates osteoclast differentiation by coupling to an S-adenosylmethionine-producing metabolic pathway. Nat Med 21:281–287
Nishimura S et al (2008) In vivo imaging in mice reveals local cell dynamics and inflammation in obese adipose tissue. J Clin Invest 11:710–721
Osborn O et al (2012) The cellular and signaling networks linking the immune system and metabolism in disease. Nat Med 6:363–374
Rivera J et al (2008) The alliance of sphingosine-1-phosphate and its receptors in immunity. Nat Rev Immunol 8:753–763
Rodríguez-Paredes M et al (2011) Cancer epigenetics reaches mainstream oncology. Nat Med 17: 330–339
Rosen H et al (2005) Sphingosine 1-phosphate and its receptors: an autocrine and paracrine network. Nat Rev Immunol 25:560–570
Sato K et al (2006) Th17 functions as an osteoclastogenic helper T cell subset that links T cell activation and bone destruction. J Exp Med 203:2673–2682
Sekimoto R et al (2012) High circulating levels of S100A8/A9 complex (calprotectin) in male Japanese with abdominal adiposity and dysregulated expression of S100A8 and S100A9 in adipose tissues of obese mice. Biochem Biophys Res Commun 419:782–789
Sekimoto R et al (2015) Visualized macrophage dynamics and significance of S100A8 in obese fat. Proc Natl Acad Sci U S A 112:E2058–E2066
Spencer JA et al (2014) Direct measurement of local oxygen concentration in the bone marrow of live animals. Nature 508:269–273
Takayanagi H et al (2000) T-cell-mediated regulation of osteoclastogenesis by signalling cross-talk between RANKL and IFN-gamma. Nature 408:600–605
Tannahil GM et al (2013) Succinate is an inflammatory signal that induces IL-1β through HIF-1α. Nature 496:238–242
Vogl T et al (2007) Mrp8 and Mrp14 are endogenous activators of Toll-like receptor 4, promoting lethal, endotoxin-induced shock. Nat Med 13:1042–1049

Weisberg SP et al (2003) Obesity is associated with macrophage accumulation in adipose tissue. J Clin Invest 112:1796–1808

Weisberg SP et al (2006) CCR2 modulates inflammatory and metabolic effects of high-fat feeding. J Clin Invest 116:115–124

Zhao B et al (2009) Interferon regulatory factor-8 regulates bone metabolism by suppressing osteoclastogenesis. Nat Med 15:1066–1071

Chapter 12
Visualization of Localized Cellular Signalling Mediators in Tissues by Imaging Mass Spectrometry

Yuki Sugiura, Kurara Honda, and Makoto Suematsu

Abstract Low-molecular–weight cellular signalling mediators orchestrate a programmed progression of a series of biological phenomena, including inflammation. In order to understand the biochemical basis of inflammatory diseases, it is important to first understand the spatiotemporal dynamics (i.e., production, diffusion, decomposition) of such inflammatory mediators in tissues; however, the lack of effective molecular imaging technology has made it difficult to determine their localisations in vivo. Because the in vivo concentrations of these mediators are maintained at low levels due to their strong ability to induce an inflammatory response, technological breakthroughs in molecular imaging methods toward a highly sensitive technology sufficient to detect inflammatory mediators in tissues are required.

We and other groups have attempted to fill this technical gap by developing highly sensitive imaging mass spectrometry (IMS) technologies. Owing to recent improvements in IMS regarding its sensitivity, molecular coverage, and spatial resolution, this imaging technology is expected to be a powerful and practical tool to reveal unexpected dynamics of inflammatory mediators in the tissues involved in chronic inflammation. Here, we review recent progress in the development of this technology and demonstrate how the highly sensitive IMS technique has

Y. Sugiura (✉)
Department of Biochemistry, Keio University School of Medicine, 35 Shinanomachi, Shinjuku-ku, Tokyo 160-8582, Japan

PRESTO, Japan Science and Technology Agency, Kawaguchi 332-0012, Japan
e-mail: yuki.sgi@keio.jp

K. Honda
Department of Biochemistry, Keio University School of Medicine, 35 Shinanomachi, Shinjuku-ku, Tokyo 160-8582, Japan

M. Suematsu
Department of Biochemistry, Keio University School of Medicine, 35 Shinanomachi, Shinjuku-ku, Tokyo 160-8582, Japan

Japan Science and Technology Agency, Exploratory Research for Advanced Technology (ERATO) Suematsu Gas Biology Project, Tokyo 160-8582, Japan

M. Miyasaka, K. Takatsu (eds.), *Chronic Inflammation*,
DOI 10.1007/978-4-431-56068-5_12

contributed to increasing our general understanding of the biological basis of disease mechanisms.

Keywords Imaging mass spectrometry • Postmortem degradation of metabolites • Lipid mediator

Abbreviations

ESI	electrospray ionisation
FMW	head-focused microwave irradiation
IMS	imaging mass spectrometry
ISF	in situ freezing
MALDI	matrix-assisted laser desorption/ionisation

12.1 Introduction

Identifying the molecular entities involved in specific biochemical processes and determining their spatial localisation among organs and within tissues are essential steps in biochemical research. In addition to hormones, low-molecular–weight signal mediators orchestrate the programmed progression of a series of biological phenomena, including inflammation. For example, bioactive lipids mediating inflammatory and anti-inflammatory signals are essential for mediating normal inflammation processes by regulating the function and localisation of lymphocytes, for example, recruitment to designated microregions of a tissue (Luster and Tager 2004; Wymann and Schneiter 2008; Spiegel and Milstien 2003; Takeda et al. 2016; Bai et al. 2013), as well as in the evacuation of immune cells during the resolution phase (Lawrence et al. 2002). In order to understand the biochemical basis of chronic inflammatory diseases, it is important to characterise the spatiotemporal dynamics (i.e., production, diffusion, decomposition) of inflammatory mediators in tissues. Various imaging technologies for large hormonal molecules have been developed using molecular probes. In general, the localisation of transcripts is visualised with the use of an oligonucleotide probe through in situ hybridisation, and the localisation of proteins is visualised using immunohistochemistry based on antibodies. However, the lack of imaging technology for small metabolites has made it difficult to determine their localisations in vivo. Moreover, in vivo concentrations of these mediators are maintained at low levels due to their strong ability to induce an inflammatory response; therefore, technological breakthroughs in molecular imaging methods are required for developing technology sensitive enough to detect inflammatory mediators in tissues.

Mass spectrometry (MS) has become a key analytical technique in broad areas of biology. Historically, the introduction of two ‘soft’ ionisation techniques, matrix-assisted laser desorption ionisation (MALDI) and electrospray ionisation (ESI), has made it possible to detect and quantify proteins,and peptides, as well as small molecules such as lipids and other metabolites, without requiring molecular probes. Recently, imaging mass spectrometry (IMS; also referred to as mass spectrometry imaging, MSI) has attracted the attention of researchers for visualising cellular metabolism *via* multiple-metabolite visualisation, and led to emergence of a high number of recent studies employing IMS in basic biological (Sugiura et al. 2011; Steinhauser et al. 2012; Girod et al. 2011; Moree et al. 2012) and medical research areas (Matsumoto et al. 2011; Yuki et al. 2014; Franck et al. 2009). Recent advances in IMS technology, which have been coupled with not only MALDI (the most versatile technique) (Stoeckli et al. 2001; Cornett et al. 2007) but also with secondary ion mass spectrometry (enabling nanometer-order spatial resolution) (Debois et al. 2009; Fletcher et al. 2007), desorption electrospray ionisation (realising ionisation under atmospheric pressure conditions) (Ifa et al. 2007; Wiseman et al. 2008), and other new desorption/ionisation techniques (Patti et al. 2010; Taira et al. 2008; Nemes and Vertes 2007), have expanded the capabilities of IMS, enabling visualisation of a wide variety of small molecules in a label-free manner. Given the enormous diversity of metabolite species and their distinct functional roles, simultaneous visualisation of multiple metabolites belonging to specific pathways has now made it possible to uncover the overall picture of metabolic systems in specific tissue regions where a cellular population of interest is localised.

However, the detection sensitivity and selectivity of IMS still need improvement for the imaging of cellular signalling mediators. For example, signal-transduction lipids (e.g., prostaglandins) are maintained at very low concentrations in vivo due to their strong ability to induce inflammation. By contrast, phospholipids, the major building blocks of cellular membranes, are much more abundant within tissue sections if compared to typical prostaglandins. In IMS, Because it is difficult to purify specific signalling molecules in tissue sections (in contrast to chromatography-related ESI-MS methods such as liquid chromatography-ESI-MS), innovation and effort are required to develop novel methods to effectively ionise and specifically detect such trace amounts of molecules that form complex tissues.

We and other research groups have approached these technical challenges by developing a highly sensitive IMS method. In particular, we have developed several key techniques, including (i) a sample preparation procedure that eliminates the common effect of the postmortem (PM) degradation of labile metabolites (Sugiura et al. 2014), and (ii) on-tissue derivatisation of metabolites that could enhance the analyte ionisation efficiency (Toue et al. 2014). These techniques have enabled the visualisation of diverse metabolite species, including abundant phospholipids (Beck et al. 2011; Sugiura et al. 2009; Hanada et al. 2012) as well as metabolites that are present in smaller amounts such as minor amino acids (Toue et al. 2014), TCA cycle intermediates (Kunisawa et al. 2015), nucleotides (Bao et al. 2013; Kim et al. 2013), neurotransmitters (Takenouchi et al. 2015; Takahashi et al. 2014; Sugiura et al. 2012), and signalling lipids (Bai et al. 2013) in various diseased model tissues (summarised in Fig. 12.1). We here review recent progress in IMS development, focusing on the fundamental technique (i) described above, that is, effective sample preparation to preserve labile metabolites from PM degradation,

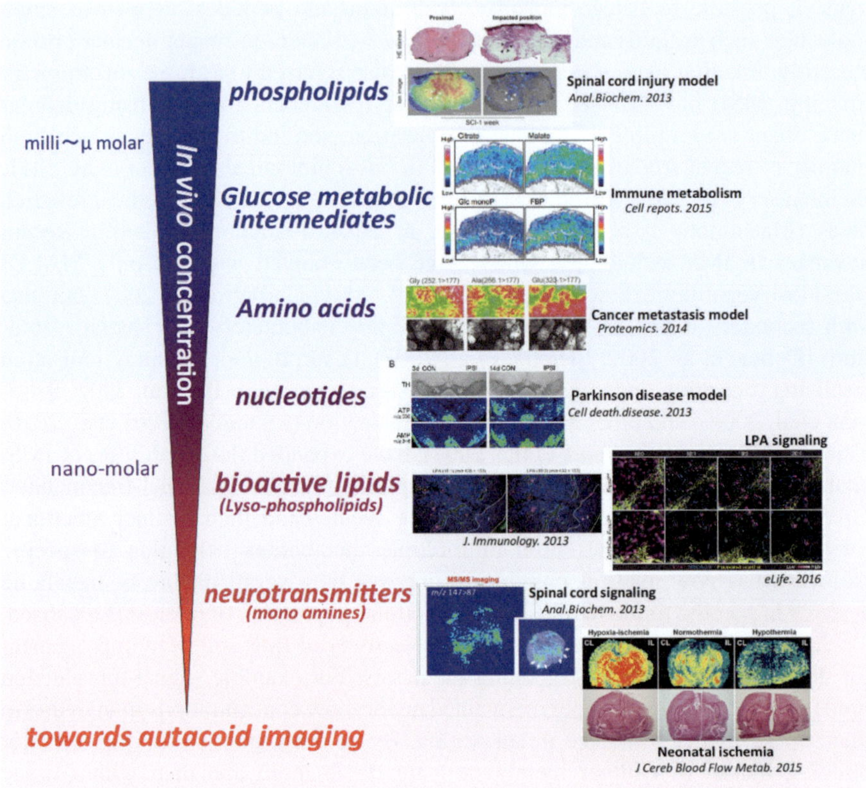

Fig. 12.1 Scheme summarising recent applications of highly sensitive IMS technology for visualising trace amounts of signalling molecules for analyses of various diseases models

which is in turn an important issue to achieve highly sensitive IMS for visualisation of cellular signalling mediators.

12.2 Prevention of PM Degradation of Labile Metabolites During Sample Preparation

A fundamental problem in quantifying in vivo metabolite concentrations with biochemical approaches is the requirement for invasive sampling. The PM degradation of labile metabolites during the organ sampling process remains a classical problem and a critical issue, because the degradation of major metabolites tends to occur in tissues within 1 min after death of the animal. In particular, small signal-transduction molecules, especially those containing organic phosphates (Hattori et al. 2010) such as sphingosine 1 phosphate (S1P) (Saigusa et al. 2014), are well known to be sensitive to PM degradation. To date, many classical studies, particularly those in neurochemistry, have described that high-energy phosphometabolites (Kerr 1935) as well as glucose intermediates (McGinty and Gesell 1925; Kinnersley and Peters 1929) are extremely sensitive to PM degradation in the brain.

Therefore, it is necessary to prevent the PM enzymatic degradation of metabolites during the sample preparation procedure. We have approached this problem by developing a method for the rapid 'fixation' of metabolites as they exist in vivo.

12.2.1 Head-Focused Microwave Irradiation (FMW)

For this purpose, we used a head-focused microwave irradiation method (FMW) (Moroji et al. 1977; Maruyama et al. 1978; Delaney and Geiger 1996; Schneider et al. 1981), in which the enzymes responsible for the decomposition of metabolites in the brain are inactivated within 1 s. We demonstrated that FMW treatment is an essential process for the visualisation of labile metabolites, especially for phosphometabolites, as they exist in vivo (Sugiura et al. 2014).

In our study, we showed that the FMW treatment preserved these metabolite concentrations as well as their distribution by halting the effect of PM degradation. As shown in Fig. 12.2a, when focusing on purine metabolic pathways, we revealed that the conventional decapitation protocol for brain extraction resulted in excessive autolytic reductions in ATP and ADP levels, and increased AMP levels. Subsequent quantification of these adenosine nucleotides with capillary electrophoresis-ESI-MS showed that 85 % of the ATP was degraded into downstream metabolites, and AMP showed a tenfold increase within 1 min after decapitation. FMW is more relevant for IMS than ESI-MS quantification, given that IMS is a longer procedure with additional steps, which could lead to more severe metabolite degradation. We further demonstrated that the imaging quality of high-energy adenosine nucleotides with FMW was substantially improved, with respect to the number of effective pixels and high image contrast, compared with that obtained using any other fixation method (Fig. 12.2).

From an animal welfare perspective, it is important to note that according to the American Veterinary Medical Association Recommendations (*AVMA Guidelines for the Euthanasia of Animals: 2013 Edition*) (Cima 2013), high-energy microwave irradiation is a humane method for euthanising small laboratory rodents. It is also worth noting that unconsciousness can be achieved in less than 100 ms with complete loss of brain function in less than 1 s.

12.2.2 Heat Stabilisation of Extracted Tissues

In contrast to the in vivo fixation nature of FMW, the heat stabilisation of surgically extracted and frozen tissues was also proven to be effective for preserving labile

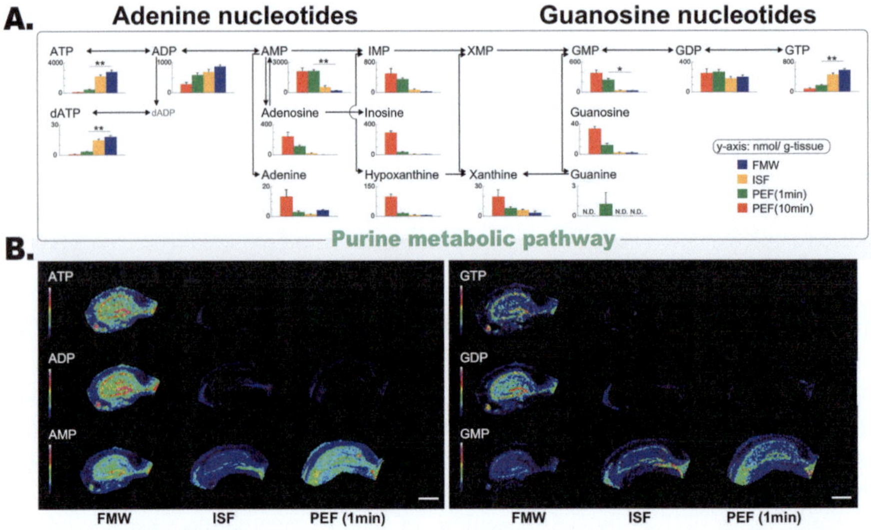

Fig. 12.2 FMW treatment minimises PM changes in labile metabolites of the brain (Sugiura et al. 2014). Focusing on the purine metabolic pathway, the absolute concentrations (nmol/mg tissue) of 15 purine metabolites were compared among the FMW, in situ freezing (ISF), and conventional decapitation (posteuthanised freezing, PEF) methods. Significant PM alterations were suppressed in the FMW-treated sample, as assessed by both quantitative (**a**) and imaging analyses (**b**)

metabolites. Using a commercially available instrument (stabiliser T1, Denator, Gothenburg, Sweden) (Svensson et al. 2009), which applies conductive heating to postextracted tissues under a mild vacuum, this ex vivo fixation method halts the PM enzymatic activity (Fig. 12.3a). This method has been employed for brain imaging of labile energy nucleotides (Blatherwick et al. 2013), as well as for the quantification of tissue contents of signalling lipids, including S1P and dihydrosphingosine 1-phosphate (dhS1P). Saigusa et al. (2014) demonstrated the severe degradation of S1P and dhS1P along the PM time course, especially within the first 15 min (Fig. 12.3b). To overcome this problem, the heat stabilising treatment was applied, which successfully fixed the concentrations of these phosphometabolites, thus demonstrating the utility of this method for accurate quantification of these signalling lipids by LC-MS/MS.

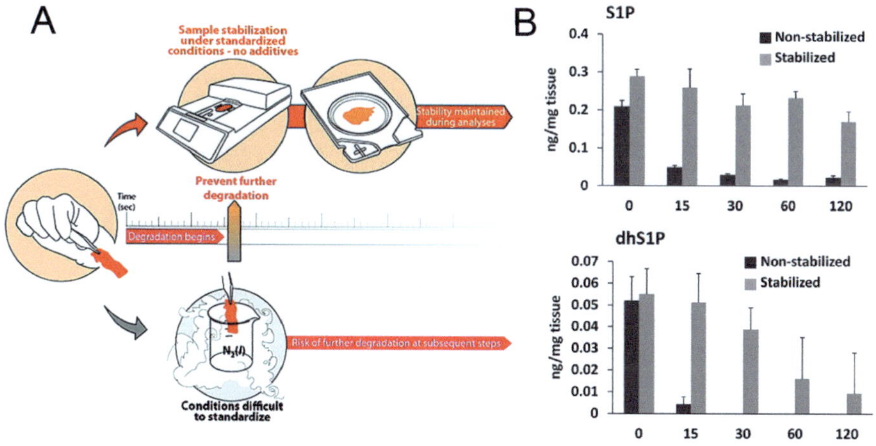

Fig. 12.3 The heat stabilisation of extracted tissues is effective for the protection of S1P and dhS1P from PM degradation. (**a**) The heat fixation method involves application of a combination of heat and mild vacuum to the surgically extracted and frozen tissue samples. (**b**) Results for the determination of sphingolipids in the liver at room temperature (25 °C) without ($n = 4$, dark gray, nonstabilised) or with ($n = 4$, clear gray, stabilised) the heat stabiliser. Values are mean ± SD (Modified from (Svensson et al. 2009) and (Saigusa et al. 2014))

12.3 Applications of IMS to Lysophosphatidic Acid (LPA) Imaging

12.3.1 Visualisation of LPA in the Lymph Node by Tandem MS Imaging

By employing an appropriate sample preparation procedure as described above, microscopic IMS observations have successfully provided molecular evidence for the production of specific lipid mediators within specific tissue structures. Bai et al. (2013) studied the signalling mechanisms involved in lymphocyte transmigration across the basal lamina of high endothelial venules (HEVs) in the lymph nodes. The results showed that lymphocyte transmigration is regulated by autotaxin (ATX) and its end-product, the signal-transducing lipid LPA. ATX is an HEV-associated ectoenzyme that produces LPA from lysophosphatidylcholine, which is abundant in the systemic circulation. In line with the selective expression of ATX in HEVs, LPA was constitutively and specifically visualised on the HEV structures (Fig. 12.4). Inasmuch as the effect of LPA was shown to be dependent of ATX in HEVs and deficiency of the LPA4 receptor expression in HEV ECs leads to impaired lymphocyte transmigration across HEVs (Hata et al. 2016), collectively, these results strongly suggest that LPA acts on HEV ECs to increase their motility, thereby promoting dynamic lymphocyte–HEV interactions and subsequent lymphocyte transmigration across the basal lamina of HEVs in steady state.

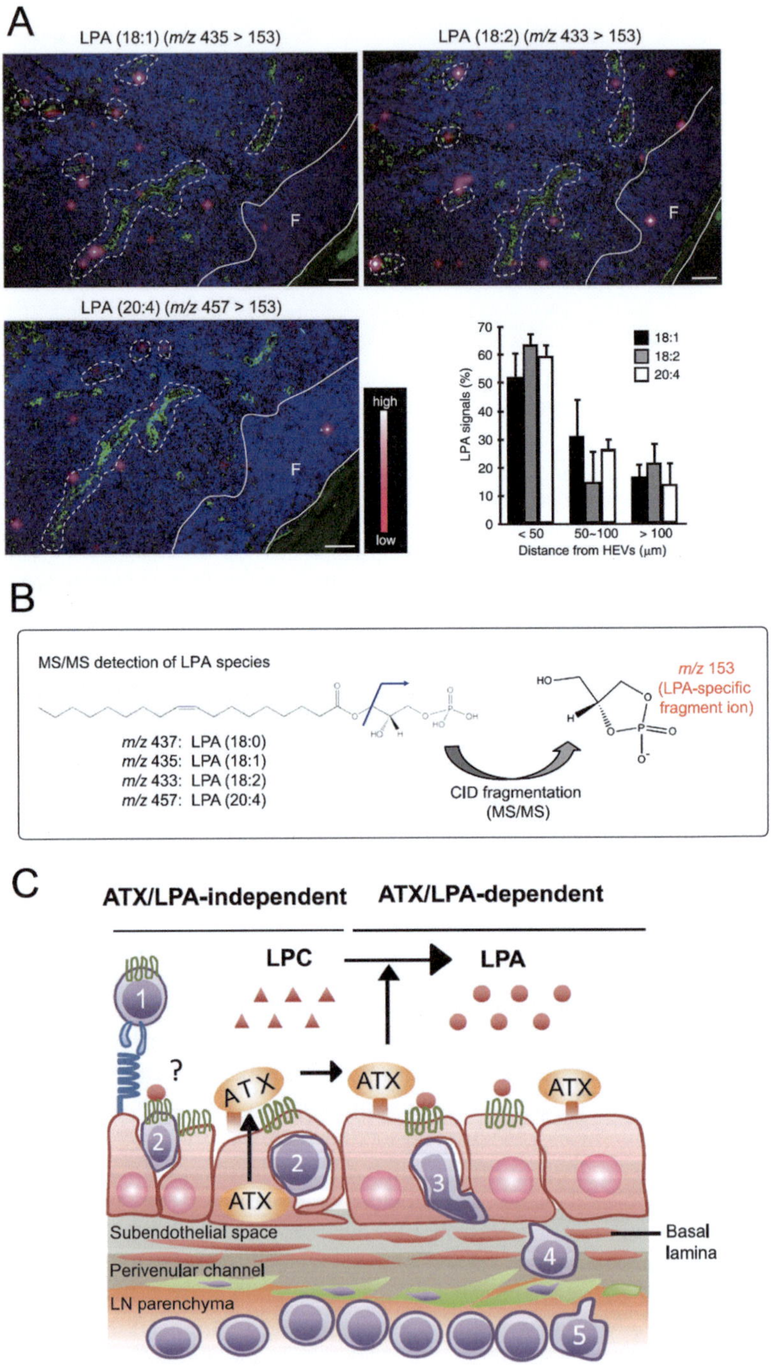

Fig. 12.4 LPA is detected in the vicinity of lymph node HEVs. (**a**) Tissue distribution of LPA species in lymph node sections by MALDI-MS/MS imaging. The LPA (18:1)-derived signal (ion transition from *m*/*z* 435 to 153), LPA (18:2)-derived signal (*m*/*z* 433 to 153), and LPA

It should be noted that this study utilised an ion trap-type mass spectrometer (iMScope, Shimadzu Corporation, Kyoto, Japan), which enabled tandem-MS imaging measurement and enhanced signal selectivity for the LPA molecular species. This analytical method utilises not only intact LPA ions but also their specific fragment ions generated by fragmentation of trapped LPA ions (Fig. 12.4b). Signal contamination from isobaric or molecules of similar molecular weight were eliminated by this analytical feature.

12.3.2 *Visualisation of LPA in the Spleen by Fourier Transform-MS Imaging*

Another strategy to obtain highly specific and sensitive molecular signals in IMS is through Fourier transform–ion cyclotron–resonance mass spectrometry (FT-ICR MS) Cornett et al. (2008), which provides the highest mass resolving power (mass resolution >100,000) and mass accuracy (a few ppm) for an IMS experiment. Accordingly, this type of instrument is capable of resolving small intense metabolite peaks with similar nominal masses of the most abundant molecules in a single-stage MS scan. The high-mass accuracy measurement in conjunction with isotopic profiles allows for the elemental composition of these particular ions to be determined (Marshall et al. 1998).

As an example, we obtained distribution images of LPA species in mouse spleen tissue sections using 7 T FT-ICR-MS (Solarix-XR Bruker Daltonik, Bremen,

Fig. 12.4 (continued) (20:4)-derived signal (*m/z* 457 to 153) overlapped with MECA-79 staining on a serial section of the popliteal lymph node (HEVs; *green*). Nuclei were counterstained with Hoechst 33342 (*blue*). The signals associated with HEVs are enclosed by dashed lines. Scale bars, 100 mm. The proportion of LPA signals reflects the distance from HEVs. Data are representative of two independent experiments. F, follicles
(**b**) The precursor ions at *m/z* 437, 435, 433, and 457, which include the signals of LPA (18:0), LPA (18:1), LPA (18:2), and LPA (20:4), respectively, were collected and fragmented by collision-induced dissociation in a mass spectrometer, and the generated signals were subsequently analysed. The signal at *m/z* 153 corresponds to the LPA- specific fragment ion
(**c**) Possible mode of action of the ATX/LPA axis in the regulation of lymphocyte trafficking across HEVs. Upper panel, lymphocytes migrating across the HEVs may follow multiple steps: (Luster and Tager 2004) tethering/adhesion to HEV ECs, (Wymann and Schneiter 2008) intra-EC infiltration, (Spiegel and Milstien 2003) retention in the sub-EC space, (Takeda et al. 2016) retention in the perivenular channel, and (Bai et al. 2013) egress from HEVs. The ATX/LPA axis is dispensable for lymphocyte infiltration into the EC layer (steps 1 and 2) but is indispensable for lymphocyte extravasation from the HEVs (steps 3, 4, and 5). ATX secreted by HEV ECs and neighbouring cells appears to be immobilised on the HEV EC surface, at least in part via glycosaminoglycan chains such as heparan sulfate. The EC-captured ATX converts circulating lysophosphatidylcholine (LPC) to LPA in situ. In turn, LPA induces their motility and permeability, as well as lymphocyte detachment from the EC, which may collectively drive directional lymphocyte movement from the basal laminal aspect of HEV ECs to the surrounding lymph node parenchymal compartment

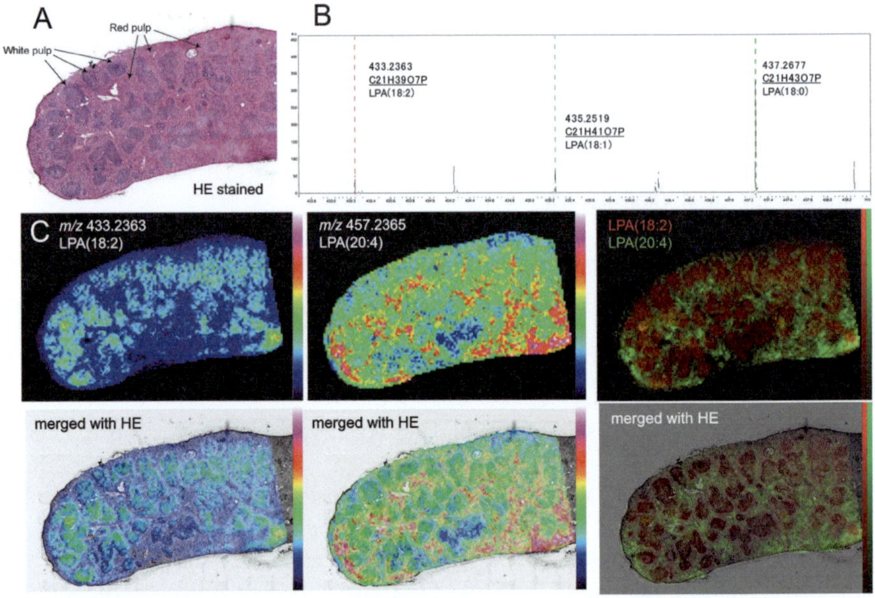

Fig. 12.5 Visualisation of the LPA molecular species on spleen tissue sections. IMS was performed using a Bruker Solarix 7 T FT-ICR-MS system (Bruker Daltonik, Bremen, Germany) in negative-ion mode, at a lateral resolution of 80 μm. Ions were detected over a mass range of *m*/*z* 400–500 with the continuous accumulation of selected ions (CASI) mode, which enhances the signal sensitivity over a selected mass region

(**a**) Haematoxylin and eosin (HE)-stained spleen tissue section after IMS measurement

(**b**) Averaged mass spectrum obtained from the whole spleen tissue section, in which specific signals for LPA species were detected. Metabolites were identified by matching to accurate masses with databases (METLIN, http://metlin.scripps.edu/;)

(**c**) IMS observation showing that two polyunsaturated fatty-acid–containing LPA species, namely LPA(18:2) and LPA(20:4), exhibited a complementary distribution pattern in the spleen tissue

Germany; Fig. 12.5a). Accurate mass measurement provides molecular specificity for the ion images on the basis of the elemental composition. The high mass accuracy allowed for selective ion signals for LPA species to be obtained within a mass window of 5 ppm (Fig. 12.5b). Although the signal specificity achieved in this example is high, it should nevertheless be acknowledged that the FT-ICR imaging strategy does not offer the same molecular specificity as achieved by acquiring an MS/MS spectrum at every pixel. In the presented case, however, the specific elemental composition of an ion could sufficiently identify LPA compounds by querying the highly accurate masses against databases.

By focusing on the distribution of LPAs, the IMS observation revealed an interesting pattern in that two polyunsaturated fatty-acid–containing LPA species, namely, LPA(18:2) and LPA(20:4), exhibited a complementary distribution pattern in the spleen tissue (Fig. 12.5c); LPA(18:2) showed splenic white pulp structure-specific localisation, whereas LPA(20:4) was distributed in the red pulp region. Because these LPA species with different fatty acid moieties, especially those with

PUFA moiety at sn-2 position, have distinct abilities to bind to LPA receptors (Cornett et al. 2008; Choi et al. 2010; Ishii et al. 2009), the visualised differences in LPA localisations might be a key observation toward elucidating the regulation mechanism of the specific migration patterns of immune cells expressing distinct LPA receptors.

12.4 Conclusion and Perspectives

For understanding the biochemical basis of disease progression and to develop new therapeutic strategies, it is important to characterise the spatiotemporal *in vivo* changes in strongly bioactive small molecules such as inflammatory mediators. This technical task was accomplished by developing a highly sensitive IMS technology. Though current MALDI-IMS already has powerful capabilities for visualising many metabolites in discrete areas, evidently further efforts should be made to improve the method to be sensitive enough to detect low levels of metabolites. Our group and others have tackled this challenge by optimisation of the sample preparation protocol. Several different approaches with efficient metabolic fixation have provided a unique platform to reveal in vivo metabolic status. We believe that these promising approaches will realise the highly sensitive IMS, which will in turn help to answer important unresolved questions in disease progression, such as determining the specific regions of organs or cells in vivo from which abnormal inflammatory mediators are produced. Resolving these issues is expected to provide the key to understanding the biological basis of chronic inflammatory diseases, discovering new diagnostic markers, and developing new therapies.

References

Bai ZB, Cai LJ, Umemoto E, Takeda A, Tohya K, Komai Y, Veeraveedu PT, Hata E, Sugiura Y, Kubo A, Suematsu M, Hayasaka H, Okudaira S, Aoki J, Tanaka T, Albers HMHG, Ovaa H, Miyasaka M (2013) Constitutive lymphocyte transmigration across the basal lamina of high endothelial venules is regulated by the autotaxin/lysophosphatidic acid axis. J Immunol 190:2036–2048

Bao Y, Mukai K, Hishiki T, Kubo A, Ohmura M, Sugiura Y, Matsuura T, Nagahata Y, Hayakawa N, Yamamoto T, Fukuda R, Saya H, Suematsu M, Minamishima YA (2013) Energy management by enhanced glycolysis in G(1) phase in human colon cancer cells in vitro and in vivo. Mol Cancer Res 11:973–985

Beck G, Sugiura Y, Shinzawa K, Kato S, Setou M, Tsujimoto Y, Sakoda S, Sumi-Akamaru H (2011) Neuroaxonal dystrophy in calcium-independent phospholipase A(2)beta deficiency results from insufficient remodeling and degeneration of mitochondrial and presynaptic membranes. J Neurosci 31:11411–11420

Blatherwick EQ, Svensson CI, Frenguelli BG, Scrivens JH (2013) Localisation of adenine nucleotides in heat-stabilised mouse brains using ion mobility enabled MALDI imaging. Int J Mass Spectrom 345:19–27

Choi JW, Herr DR, Noguchi K, Yung YC, Lee CW, Mutoh T, Lin ME, Teo ST, Park KE, Mosley AN, Chun J (2010) LPA receptors: subtypes and biological actions. Annu Rev Pharmacol Toxicol 50:157–186

Cima G (2013) AVMA guidelines for the euthanasia of animal, edition. Javma-J Am Vet Med A 242:715–716

Cornett DS, Reyzer ML, Chaurand P, Caprioli RM (2007) MALDI imaging mass spectrometry: molecular snapshots of biochemical systems. Nat Methods 4:828–833

Cornett DS, Frappier SL, Caprioli RM (2008) MALDI-FTICR imaging mass spectrometry of drugs and metabolites in tissue. Anal Chem 80:5648–5653

Debois D, Bralet MP, Le Naour F, Brunelle A, Laprevote O (2009) In Situ lipidomic analysis of nonalcoholic fatty liver by cluster TOF-SIMS imaging. Anal Chem 81:2823–2831

Delaney SM, Geiger JD (1996) Brain regional levels of adenosine and adenosine nucleotides in rats killed by high-energy focused microwave irradiation. J Neurosci Methods 64:151–156

Fletcher JS, Lockyer NP, Vaidyanathan S, Vickerman JC (2007) TOF-SIMS 3D biomolecular imaging of Xenopus laevis oocytes using buckminsterfullerene (C-60) primary ions. Anal Chem 79:2199–2206

Franck J, Arafah K, Elayed M, Bonnel D, Vergara D, Jacquet A, Vinatier D, Wisztorski M, Day R, Fournier I, Salzet M (2009) MALDI imaging mass spectrometry: state of the art technology in clinical proteomics. Mol Cell Proteomics 8:2023–2033

Girod M, Shi Y, Cheng JX, Cooks RG (2011) Mapping lipid alterations in traumatically injured rat spinal cord by desorption electrospray ionization imaging mass spectrometry. Anal Chem 83:207–215

Hanada M, Sugiura Y, Shinjo R, Masaki N, Imagama S, Ishiguro N, Matsuyama Y, Setou M (2012) Spatiotemporal alteration of phospholipids and prostaglandins in a rat model of spinal cord injury. Anal Bioanal Chem 403:1873–1884

Hata E, Sasakil N, Takeda A, Tohya K, Umemoto E, Akahoshi N, Ishii S, Bando K, Abe T, Kano K, Aoki J, H. Hayasaka H, Miyasaka M (2016) Lysophosphatidic acid receptors LPA4 and LPA6 differentially promote lymphocyte transmigration across high endothelial venules in lymph nodes. Int Immunol 28:283–292

Hattori K, Kajimura M, Hishiki T, Nakanishi T, Kubo A, Nagahata Y, Ohmura M, Yachie-Kinoshita A, Matsuura T, Morikawa T, Nakamura T, Setou M, Suematsu M (2010) Paradoxical ATP elevation in ischemic penumbra revealed by quantitative imaging mass spectrometry. Antioxid Redox Signal 13:1157–1167

Ifa DR, Wiseman JM, Song QY, Cooks RG (2007) Development of capabilities for imaging mass spectrometry under ambient conditions with desorption electrospray ionization (DESI). Int J Mass Spectrom 259:8–15

Ishii S, Noguchi K, Yanagida K (2009) Non-Edg family lysophosphatidic acid (LPA) receptors. Prostaglandins Other Lipid Mediat 89:57–65

Kerr SE (1935) Studies on the phosphorus compounds of brain. I. Phosphocreatine. J Biol Chem 110:625–635

Kim TW, Cho HM, Choi SY, Suguira Y, Hayasaka T, Setou M, Koh HC, Hwang EM, Park JY, Kang SJ, Kim HS, Kim H, Sun W (2013) (ADP-ribose) polymerase 1 and AMP-activated protein kinase mediate progressive dopaminergic neuronal degeneration in a mouse model of Parkinson's disease. Cell Death Dis 4:e919. doi:10.1038/cddis.2013.447

Kinnersley HW, Peters RA (1929) Observations upon carbohydrate metabolism in birds. I. The relation between the lactic acid content of the brain and the symptoms of opisthotonus in rice-fed pigeons. Biochem J 23:1126–1136

Kunisawa J, Sugiura Y, Wake T, Nagatake T, Suzuki H, Nagasawa R, Shikata S, Honda K, Hashimoto E, Suzuki Y, Setou M, Suematsu M, Kiyono H (2015) Mode of bioenergetic

metabolism during B cell differentiation in the intestine determines the distinct requirement for Vitamin B1. Cell Rep 13:122–131
Lawrence T, Willoughby DA, Gilroy DW (2002) Anti-inflammatory lipid mediators and insights into the resolution of inflammation. Nat Rev Immunol 2:787–795
Luster AD, Tager AM (2004) T-cell trafficking in asthma: lipid mediators grease the way. Nat Rev Immunol 4:711–724
Marshall AG, Hendrickson CL, Jackson GS (1998) Fourier transform ion cyclotron resonance mass spectrometry: a primer. Mass Spectrom Rev 17:1–35
Maruyama Y, Iida N, Horikawa A, Hosoya E (1978) A new microwave device for rapid thermal fixation of the murine brain. J Microw Power 13:53–57
Matsumoto J, Sugiura Y, Yuki D, Hayasaka T, Goto-Inoue N, Zaima N, Kunii Y, Wada A, Yang Q, Nishiura K, Akatsu H, Hori A, Hashizume Y, Yamamoto T, Ikemoto K, Setou M, Niwa S (2011) Abnormal phospholipids distribution in the prefrontal cortex from a patient with schizophrenia revealed by matrix-assisted laser desorption/ionization imaging mass spectrometry. Anal Bioanal Chem 400:1933–1943
McGinty DA, Gesell R (1925) On the chemical regulation of respiration. II. A quantitative study of the accumulation of lactic acid in the isolated brain during anaerobic conditions and the role of lactic acid as a continuous regulator of respiration. Am J Physiol 75:70–83
Moree WJ, Phelan VV, Wu CH, Bandeira N, Cornett DS, Duggan BM, Dorrestein PC (2012) Interkingdom metabolic transformations captured by microbial imaging mass spectrometry. Proc Natl Acad Sci U S A 109:13811–13816
Moroji T, Takahashi K, Ogura K, Toishi T, Arai S (1977) Rapid microwave fixation of rat brain. J Microw Power 12:273–286
Nemes P, Vertes A (2007) Laser ablation electrospray ionization for atmospheric pressure, in vivo, and imaging mass spectrometry. Anal Chem 79:8098–8106
Patti GJ, Shriver LP, Wassif CA, Woo HK, Uritboonthai W, Apon J, Manchester M, Porter FD, Siuzdak G (2010) Nanostructure-initiator mass spectrometry (Nims) imaging of brain cholesterol metabolites in Smith-Lemli-Opitz syndrome. Neuroscience 170:858–864
Saigusa D, Okudaira M, Wang J, Kano K, Kurano M, Uranbileg B, Ikeda H, Yatomi Y, Motohashi H, Aoki J (2014) Simultaneous quantification of sphingolipids in small quantities of liver by LC-MS/MS. Mass Spectrom (Tokyo) 3:S0046
Schneider DR, Felt BT, Goldman H (1981) Microwave radiation energy: a probe for the neurobiologist. Life Sci 29:643–653
Spiegel S, Milstien S (2003) Sphingosine-1-phosphate: an enigmatic signalling lipid. Nat Rev Mol Cell Biol 4:397–407
Steinhauser ML, Bailey AP, Senyo SE, Guillermier C, Perlstein TS, Gould AP, Lee RT, Lechene CP (2012) Multi-isotope imaging mass spectrometry quantifies stem cell division and metabolism. Nature 481:516–519
Stoeckli M, Chaurand P, Hallahan DE, Caprioli RM (2001) Imaging mass spectrometry: a new technology for the analysis of protein expression in mammalian tissues. Nat Med 7:493–496
Sugiura Y, Konishi Y, Zaima N, Kajihara S, Nakanishi H, Taguchi R, Setou M (2009) Visualization of the cell-selective distribution of PUFA-containing phosphatidylcholines in mouse brain by imaging mass spectrometry. J Lipid Res 50:1776–1788
Sugiura Y, Taguchi R, Setou M (2011) Visualization of spatiotemporal energy dynamics of hippocampal neurons by mass spectrometry during a Kainate-Induced Seizure. Plos One 6: e17952. doi:10.1371/journal.pone.0017952
Sugiura Y, Zaima N, Setou M, Ito S, Yao I (2012) Visualization of acetylcholine distribution in central nervous system tissue sections by tandem imaging mass spectrometry. Anal Bioanal Chem 403:1851–1861
Sugiura Y, Honda K, Kajimura M, Suematsu M (2014) Visualization and quantification of cerebral metabolic fluxes of glucose in awake mice. Proteomics 14:829–838

Svensson M, Boren M, Skold K, Falth M, Sjogren B, Andersson M, Svenningsson P, Andren PE (2009) Heat stabilization of the tissue proteome: a new technology for improved proteomics. J Proteome Res 8:974–981

Taira S, Sugiura Y, Moritake S, Shimma S, Ichiyanagi Y, Setou M (2008) Nanoparticle-assisted laser desorption/ionization based mass imaging with cellular resolution. Anal Chem 80:4761–4766

Takahashi T, Ohnishi H, Sugiura Y, Honda K, Suematsu M, Kawasaki T, Deguchi T, Fujii T, Orihashi K, Hippo Y, Watanabe T, Yamagaki T, Yuba S (2014) Non-neuronal acetylcholine as an endogenous regulator of proliferation and differentiation of Lgr5-positive stem cells in mice. Febs J 281:4672–4690

Takeda A, Kobayashi D, Aoi K, Sasaki N, Sugiura Y, Igarashi H, Tohya K, Inoue A, Hata E, Akahoshi N, Hayasaka H, Kikuta J, Scandella E, Ludewig B, Ishii S, Aoki J, Suematsu M, Ishii M, Takeda K, Jalkanen S, Miyasaka M, Umemoto E (2016) Fibroblastic reticular cell-derived lysophosphatidic acid regulates confined intranodal T-cell motility. Elife 5:e10561. doi:10.7554/eLife.10561

Takenouchi T, Sugiura Y, Morikawa T, Nakanishi T, Nagahata Y, Sugioka T, Honda K, Kubo A, Hishiki T, Matsuura T, Hoshino T, Takahashi T, Suematsu M, Kajimura M (2015) Therapeutic hypothermia achieves neuroprotection via a decrease in acetylcholine with a concurrent increase in carnitine in the neonatal hypoxia-ischemia. J Cerebr Blood F Met 35:794–805

Toue S, Sugiura Y, Kubo A, Ohmura M, Karakawa S, Mizukoshi T, Yoneda J, Miyano H, Noguchi Y, Kobayashi T, Kabe Y, Suematsu M (2014) Microscopic imaging mass spectrometry assisted by on-tissue chemical derivatization for visualizing multiple amino acids in human colon cancer xenografts. Proteomics 14:810–819

Wiseman JM, Ifa DR, Zhu YX, Kissinger CB, Manicke NE, Kissinger PT, Cooks RG (2008) Desorption electrospray ionization mass spectrometry: imaging drugs and metabolites in tissues. Proc Natl Acad Sci U S A 105:18120–18125

Wymann MP, Schneiter R (2008) Lipid signalling in disease. Nat Rev Mol Cell Biol 9:162–176

Yuki D, Sugiura Y, Zaima N, Akatsu H, Takei S, Yao I, Maesako M, Kinoshita A, Yamamoto T, Kon R, Sugiyama K, Setou M (2014) DHA-PC and PSD-95 decrease after loss of synaptophysin and before neuronal loss in patients with Alzheimer's disease. Sci Rep 4:7130. doi:10.1038/srep07130

Chapter 13
Tracking of Follicular T Cell Dynamics During Immune Responses and Inflammation

Takaharu Okada

Abstract Follicular helper T (Tfh) cells are a subtype of helper T cells critical for antibody responses, particularly germinal centre (GC) responses in the lymphoid organs. Tfh cells can become memory Tfh cells, which play important roles in memory antibody responses. Memory Tfh cells, after exiting the lymphoid organs into the circulation, possibly induce memory responses in lymphoid organs distal from the vaccination site. Circulating memory Tfh cells may also contribute to ectopic GC responses in inflamed nonlymphoid organs. Tfh cells and memory Tfh cells seem to be also involved in the development of autoimmune diseases. Human blood memory Tfh cells have begun to be characterised in detail to understand their function in protective or pathological antibody responses. Mouse studies, especially those using live imaging techniques, have illustrated Tfh cell dynamics within the lymphoid organs and identified important molecules that regulate them. In addition, imaging studies using photoactivatable or photoconvertible fluorescent probes have started to reveal how Tfh cells exit the lymphoid organs to become circulating memory cells. In future studies, in order to investigate directly the involvement of circulating memory Tfh cells in memory responses and inflammatory diseases, it is important to develop new methods to genetically and/or optogenetically manipulate Tfh cells for longitudinal tracking of their cell fate.

Keywords Antibody response • Inflammation • Autoimmune disease • Follicular helper T cell • In vivo imaging

---

T. Okada (✉)
RIKEN Center for Integrative Medical Sciences, Yokohama 230-0045, Japan

PRESTO, Japan Science and Technology Agency, Kawaguchi 332-0012, Japan

Graduate School of Medical Life Science, Yokohama City University, Yokohama 230-0045, Japan
e-mail: takaharu.okada@riken.jp

M. Miyasaka, K. Takatsu (eds.), *Chronic Inflammation*,
DOI 10.1007/978-4-431-56068-5_13

13.1 Follicular Helper T Cells

13.1.1 Roles for Tfh Cells in Immune Responses

B cell responses produce antibodies, which are the important arm of our immune system to fight against microbes. In most cases, B cell responses require help from T cells (MacLennan et al. 1997). Historically, based on in vitro experiments, the chief T cell type responsible for this help was thought to be type 2 helper T (Th2) cells secreting IL-4 (Mosmann and Coffman 1989). Within the last 20 years, knowledge about helper T cell heterogeneity was expanded in terms of not only their cytokine expression profiles but also their localization. It is now known that helper T cells that provide vital help to B cells are localized in B cell follicles in the lymphoid tissues and are therefore called follicular helper T (Tfh) cells (Crotty et al. 2010; Ueno et al. 2015). In order to achieve follicular localization, Tfh cells highly express CXCR5, the receptor for chemokine CXCL13 expressed in B cell follicles, and express little or no CCR7, the receptor for chemokine CCL21 and CCL19 expressed in the T cell zone in the lymphoid tissues. Tfh cells abundantly express IL-21, IL-4, and CD40 ligand, which promote proliferation and differentiation of antigen-engaged B cells (Okada et al. 2012).

Development of Tfh cells requires expression of the transcription factor Bcl6, which is not significantly expressed in naïve T cells or other types of helper T cells (Okada et al. 2012). Moderate expression of Bcl6 and CXCR5 together with downregulation of CCR7 is induced in $CD4^+$ T cells by interactions with dendritic cells presenting their specific antigen. The activated $CD4^+$ T cells are relocated to the border area between the T cell zone and B cell follicles, and are interacted with activated B cells presenting cognate antigen. The interaction with B cells induces further upregulation of Bcl6 and CXCR5 and further downregulation of CCR7 to drive them to the back of the follicles. Proliferating B cells and Tfh cells continue their interactions in the back of the follicles, leading to generation of short-lived plasma cells and development of germinal centres (GCs) in the centre of the follicles (Okada et al. 2012). GCs are important for high-affinity antibody production against pathogens and also for affinity diversification of gut IgA to regulate commensal microbiota (Kato et al. 2014; Allen et al. 2007; Victora and Nussenzweig 2012). The GC formation is severely impaired if activated $CD4^+$ T cells fail to express Bcl6 properly and fail to become Tfh cells (Crotty et al. 2010).

After GCs are formed, a fraction of Tfh cells are localized in the GCs. The ratio of GC Tfh cells to non-GC Tfh cells increases as the immune response progresses. GC Tfh cells express particularly high levels of CXCR5 and virtually no CCR7 (Okada et al. 2012). GC Tfh cells also highly express PD-1, which is expressed at slightly lower levels by non-GC Tfh cells. PD-1 in Tfh cells is reportedly important for the generation of long-lived plasma cells from GCs (Good-Jacobson et al. 2010). PD-1 signals seem to control Tfh cell numbers in Peyer's patch GCs, and contribute to shaping the IgA repertoire and microbiota in the gut (Kato et al. 2014).

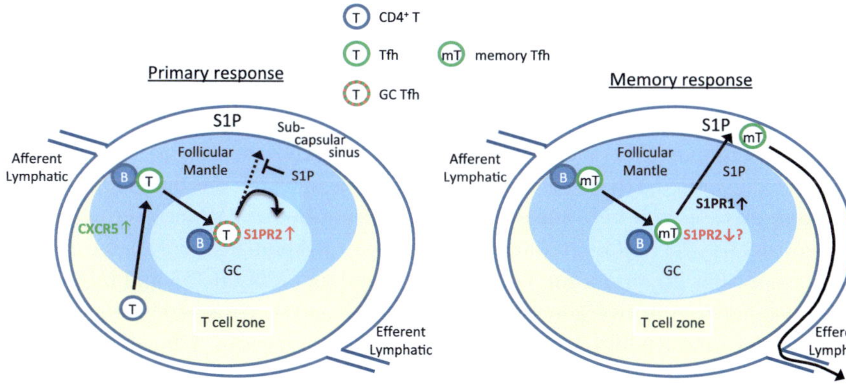

Fig. 13.1 Schematic diagrams of Tfh and memory Tfh cell dynamics in the lymphoid organ during primary and memory antibody responses. Tfh cells become GC Tfh cells by expressing S1PR2 to be retained in primary GCs during primary immune responses. Low amounts of S1P are thought to be present in the follicular mantle surrounding the GC and to act on S1PR2 for the GC Tfh cell retention. Memory Tfh cells are presumably generated from Tfh cells. Reactivation of memory Tfh cells induces secondary GC responses. Reactivated memory Tfh cells access GCs but are not retained in GCs. Some of them migrate into the subcapsular sinus to be carried away in the lymph in which high amounts of S1P exist. Downregulation of S1PR2 and upregulation of S1PR1 may be involved in this step

Until recently, little was known about localization mechanisms of GC Tfh cells except for a partial dependence on CXCR5. The author's group found that GC Tfh cells but not other T cells highly expressed transcripts of S1PR2, the type 2 receptor of sphingosine-1-phosophate (S1P; Moriyama et al. 2014). S1P was already known to be the critical lipid mediator for lymphocyte egress from the lymphoid tissue to circulation, which is driven by the type 1 receptor of S1P, S1PR1 (Cyster and Schwab 2012). However, the role for S1P in GC formation was not known. During the time when the author's group was investigating roles for S1PR2 in GC Tfh dynamics, the other group reported that S1PR2 contributed to localization of GC B cells (Green et al. 2011). The author's group showed by using two-photon excitation fluorescence microscopy of intact lymph nodes that S1PR2 was also important for retention of GC Tfh cells in GCs (Moriyama et al. 2014; Fig. 13.1). Strikingly, double deficiency of CXCR5 and S1PR2 in T cells led to very severe impairment of GC formation. When the double-deficient T cells coexist with wild-type T cells, which could support GC formation, they did not significantly infiltrate in GCs (Moriyama et al. 2014). Thus, S1PR2 plays a cooperative role with CXCR5 in GC Tfh cell biology.

13.1.2 Tfh Cells and Inflammatory Autoimmune Diseases

Several mouse studies have shown that dysregulated formation of Tfh cells is associated with inflammatory autoimmune diseases (Ueno et al. 2015). A single amino acid mutation (M199R) in the gene coding the RNA-binding protein Roquin-1 causes exaggerated Tfh cell and GC responses and spontaneous development of a lupus-like disease with high titres of antinuclear antibodies (Vinuesa et al. 2005). Mechanistically, the mutation impairs the ability of Roquin-1 to repress the expression of molecules important for Tfh cell formation, including ICOS and Interferon-γ. Because deletion of two genes coding Roquin-1 and its paralogue Roquin-2 but not deletion of the Roquin-1 gene only promoted spontaneous Tfh cell responses, the mutated Roquin may act in a dominant negative manner (Pratama et al. 2013; Vogel et al. 2013). BXSB-yaa mice with a duplication of X chromosome genes including Toll-like receptor 7 also develop aberrant Tfh cell responses and a lupus-like disease (Bubier et al. 2009). It is also important to note that genome-wide association studies of human single-nucleotide polymorphism have identified multiple Tfh-related genes as risk loci for autoimmune diseases, including Il21 (systemic lupus erythematosus, rheumatoid arthritis) and Cxcr5 (multiple sclerosis, Sjogren's syndrome; Ueno et al. 2015).

GCs are formed not only in the secondary lymphoid organs such as lymph nodes and spleen but also within lymphoid aggregates formed in other organs such as lung and brain under inflammatory conditions (Pitzalis et al. 2014). GCs formed in the inflamed organs (often called ectopic GCs) have been found in autoimmune diseases. Helper T cells that resemble Tfh cells are found in lymphoid aggregates in the inflamed organs and are thought to play key roles in development of ectopic GCs. Although precise roles for ectopic GCs in autoimmune disease pathogenesis are not known, patients with ectopic lymphoid aggregates in inflamed lesions tend to have worse disease outcomes (Pitzalis et al. 2014). It is possible that ectopic GCs become major sites for autoantibody production because of their advantageous location for self-antigen accessibility (Ueno et al. 2015).

13.2 Memory Tfh Cells

13.2.1 Mouse Memory Tfh Cells

There are a growing number of mouse studies that have shown that Tfh cells can become memory cells after primary immune responses (Hale and Ahmed 2015). These memory cells are called memory Tfh cells, which contain a circulating population. Memory Tfh cells retain expression of CXCR5, albeit at reduced levels compared to Tfh cells, but lack detectable expression of Bcl6. Upon secondary immunisation, however, memory Tfh cells efficiently and rapidly re-express Bcl6 and fully regain Tfh cell characteristics (Ise et al. 2014). Reactivated memory Tfh cells have been shown to be potent in helping memory B cell responses (Hale and Ahmed 2015). It is believed that memory Tfh cells can become a circulating population to contribute to memory antibody responses in various tissues. Circulating memory Tfh cells are likely involved in the formation of ectopic GCs in inflamed nonlymphoid organs.

The author's group has shown that Tfh cells gradually reduce Bcl6 protein expression during primary GC responses while keeping CXCR5 expression. Bcl6-low Tfh cells upregulate IL-7 receptor expression and quickly become quiescent, suggesting that they acquire memory cell–like characteristics. However, Bcl6-low Tfh cells express only small amounts of S1PR1 transcripts (Kitano et al. 2011). Thus, it is unclear whether circulating memory Tfh cells are generated during primary GC responses. As discussed later in this chapter, circulating memory Tfh cells may develop more efficiently during secondary antibody responses (Suan et al. 2015).

13.2.2 Human Memory Tfh Cells

It has been known for many years that human peripheral blood contains significant numbers of CXCR5-expressing $CD4^+$ T cells. Recent studies have shown that these cells are very potent in induction of B cell proliferation and plasma cell differentiation, establishing the circulating population of human memory Tfh cells (Ueno et al. 2015). Within the blood memory Tfh cells, there are several subpopulations, which differ in their capability to induce production of immunoglobulins of different isotypes. Interestingly, positive correlations have been reported between the frequency of blood memory Tfh cells and the severity of autoimmune diseases, including systemic lupus erythematosus, Sjogren's syndrome, and multiple sclerosis (Ueno et al. 2015).

13.3 Longitudinal Tracking of Tfh Cell Dynamics and Fate

13.3.1 Photoactivation and Photoconversion Methods to Track Tfh Cells

Exactly when and how do Tfh cells become memory Tfh cells recirculating in the body? This is an important question related to the mechanisms underlying the correlation between the human blood memory Tfh cell frequency and the activity of autoimmune diseases. In order to track long-range movements of Tfh cells, a recent study used a method of in vivo photoactivation-based fluorescent cell labelling with a two-photon laser (Shulman et al. 2013). This study reported that, within 20 h after photoactivation of GC Tfh cells that express photoactivatable

GFP, about 30 % of these cells moved out of original GCs and are found in other GCs and follicles in the same lymph node. Although this study did not assess egress of GC Tfh cells from lymph nodes to circulation, the data highlighted unrestricted movement of Tfh cells (Shulman et al. 2013) and might suggest their possible access to cortical, medullary, and subcapsular sinuses, the exits from the lymph node, near the follicles. However, this may seem inconsistent with the S1PR2-dependent retention of Tfh cells in GCs described above (Moriyama et al. 2014). One possible explanation is that the nature of Tfh cells was different because the photoactivation study used a prime-boost immunization protocol and their Tfh cells might be more like reactivated memory Tfh cells (Shulman et al. 2013).

A more recent report described an in vivo tracking method using the photoconvertible fluorescent protein Kaede (Suan et al. 2015). According to this study, 98 % of photoconverted GC Tfh cells in primary responses were found in the original GCs and follicles. In contrast, in secondary responses, reactivated memory Tfh cells freely move out of the original GCs and follicles. Note that some of these cells were found to enter the subcapsular sinus, and then were carried away in the lymph flow (Suan et al. 2015; Fig. 13.1). Thus the mouse studies point out a possibility that human blood memory Tfh cells may be developed through reactivation of already formed Tfh cells or memory Tfh cells in the lymphoid organs. Increased frequencies of human blood memory Tfh cells might be due to chronic exposure of autoreactive Tfh cells to self-antigens.

13.3.2 Genetic and Optogenetic Approaches for Future Studies

The maximum tracking time of Tfh cells expressing photoactivatable GFP or Kaede is about a day or two. Therefore, it is nearly impossible to track them after they go into the circulation and respond to a secondary antigen challenge in other tissues. In order to make this possible, genetic labelling methods are needed to track Tfh cell fate longitudinally. Previously, there were no Tfh cell–specific genes that could be reliably used to generate fate-mapping reporter mice for Tfh cells, as Bcl6 and CXCR5 are transiently expressed in acutely activated $CD4^+$ T cells that are not to become Tfh cells (Okada et al. 2012). However, the published work (Moriyama et al. 2014; Shinnakasu et al. 2016) and unpublished work by the author's group suggest that the *S1pr2* gene may suit the purpose of specifically tracking the fate of Tfh cells as well as GC B cells.

There are developing tools to induce gene expression by light illumination (Wang et al. 2012; Kennedy et al. 2010; Bacchus and Fussenegger 2012; Schindler et al. 2015). The author's group is making efforts to apply the light-activatable split-Cre system (Fig. 13.2) to optogenetically label Tfh cells in vivo. For this purpose, retroviral transduction is used for introducing the light-activatable split-Cre genes in haematopoietic stem cells, which are subsequently transplanted to irradiated

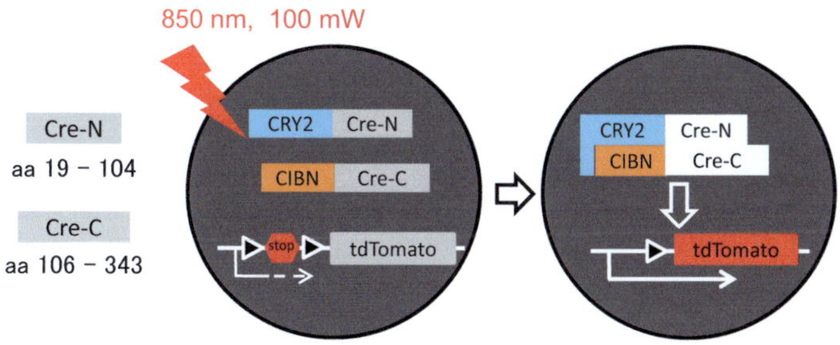

Fig. 13.2 Schematic diagram of the light-activatable split-Cre system to optogeneticaly label cells. Arabidopsis thaliana Cryptochrome 2 (CRY2) and CIB1 N-terminal fragment (CIBN) are fused to fragments of Cre recombinase (Cre-N and Cre-C) (Kennedy et al. 2010). Upon two-photon laser illumination, CRY2 binds to CIBN, and split Cre-fragments reconstitute the recombinase activity, which deletes floxed stop codons located before the tandem Tomato (tdTomato) gene inserted in the *Rosa26* locus (Madisen et al. 2010; Schindler et al. 2015). The resulting tdTomato expression is semi-irreversible and is used for longitudinal cell tracking

mice. Preliminary experiments gave results that 1–4 % of Tfh cells and 2–20 % of GC B cells were irreversibly labelled by repetitive illumination of GCs in the lymph nodes by two-photon laser (850 nm, 100 mW after objective lens, 100 ~ 500 times). In order to increase the labelling efficiency and its reproducibility, it will be important to generate germline transgenic animals in which all target cells express stoichiometrically optimal amounts of protein products of the light-activatable split-Cre genes. In addition, improvements of light-activatable modules (Taslimi et al. 2014) will help establish more powerful and versatile optogenetic methods for fate mapping of Tfh cells as well as other cells that are locatable by intravital microscopy. Establishment of fate mapping strategies will contribute to understanding of roles for memory Tfh cells in protective and pathological immune responses and understanding of mechanisms of ectopic GC development in inflamed tissues.

References

Allen CD, Okada T, Cyster JG (2007) Germinal-center organization and cellular dynamics. Immunity 27(2):190–202. doi:10.1016/j.immuni.2007.07.009

Bacchus W, Fussenegger M (2012) The use of light for engineered control and reprogramming of cellular functions. Curr Opin Biotechnol 23(5):695–702. doi:10.1016/j.copbio.2011.12.004

Bubier JA, Sproule TJ, Foreman O, Spolski R, Shaffer DJ, Morse HC 3rd, Leonard WJ, Roopenian DC (2009) A critical role for IL-21 receptor signaling in the pathogenesis of systemic lupus erythematosus in BXSB-Yaa mice. Proc Natl Acad Sci U S A 106(5):1518–1523. doi:10.1073/pnas.0807309106

Crotty S, Johnston RJ, Schoenberger SP (2010) Effectors and memories: Bcl-6 and Blimp-1 in T and B lymphocyte differentiation. Nat Immunol 11(2):114–120, doi:ni.1837[pii]10.1038/ni.1837

Cyster JG, Schwab SR (2012) Sphingosine-1-phosphate and lymphocyte egress from lymphoid organs. Annu Rev Immunol 30:69–94. doi:10.1146/annurev-immunol-020711-075011

Good-Jacobson KL, Szumilas CG, Chen L, Sharpe AH, Tomayko MM, Shlomchik MJ (2010) PD-1 regulates germinal center B cell survival and the formation and affinity of long-lived plasma cells. Nat Immunol 11(6):535–542. doi:10.1038/ni.1877

Green JA, Suzuki K, Cho B, Willison LD, Palmer D, Allen CD, Schmidt TH, Xu Y, Proia RL, Coughlin SR, Cyster JG (2011) The sphingosine 1-phosphate receptor S1P(2) maintains the homeostasis of germinal center B cells and promotes niche confinement. Nat Immunol 12 (7):672–680. doi:10.1038/ni.2047

Hale JS, Ahmed R (2015) Memory T follicular helper CD4 T cells. Front Immunol 6:16. doi:10.3389/fimmu.2015.00016

Ise W, Inoue T, McLachlan JB, Kometani K, Kubo M, Okada T, Kurosaki T (2014) Memory B cells contribute to rapid Bcl6 expression by memory follicular helper T cells. Proc Natl Acad Sci U S A 111(32):11792–11797. doi:10.1073/pnas.1404671111

Kato LM, Kawamoto S, Maruya M, Fagarasan S (2014) The role of the adaptive immune system in regulation of gut microbiota. Immunol Rev 260(1):67–75. doi:10.1111/imr.12185

Kennedy MJ, Hughes RM, Peteya LA, Schwartz JW, Ehlers MD, Tucker CL (2010) Rapid blue-light-mediated induction of protein interactions in living cells. Nat Methods 7(12):973–975. doi:10.1038/nmeth.1524

Kitano M, Moriyama S, Ando Y, Hikida M, Mori Y, Kurosaki T, Okada T (2011) Bcl6 protein expression shapes pre-germinal center B cell dynamics and follicular helper T cell heterogeneity. Immunity 34(6):961–972. doi:10.1016/j.immuni.2011.03.025

MacLennan IC, Gulbranson-Judge A, Toellner KM, Casamayor-Palleja M, Chan E, Sze DM, Luther SA, Orbea HA (1997) The changing preference of T and B cells for partners as T-dependent antibody responses develop. Immunol Rev 156:53–66

Madisen L, Zwingman TA, Sunkin SM, Oh SW, Zariwala HA, Gu H, Ng LL, Palmiter RD, Hawrylycz MJ, Jones AR, Lein ES, Zeng H (2010) A robust and high-throughput Cre reporting and characterization system for the whole mouse brain. Nat Neurosci 13(1):133–140. doi:10.1038/nn.2467

Moriyama S, Takahashi N, Green JA, Hori S, Kubo M, Cyster JG, Okada T (2014) Sphingosine-1-phosphate receptor 2 is critical for follicular helper T cell retention in germinal centers. J Exp Med 211(7):1297–1305. doi:10.1084/jem.20131666

Mosmann TR, Coffman RL (1989) TH1 and TH2 cells: different patterns of lymphokine secretion lead to different functional properties. Annu Rev Immunol 7:145–173. doi:10.1146/annurev.iy.07.040189.001045

Okada T, Moriyama S, Kitano M (2012) Differentiation of germinal center B cells and follicular helper T cells as viewed by tracking Bcl6 expression dynamics. Immunol Rev 247(1):120–132. doi:10.1111/j.1600-065X.2012.01120.x

Pitzalis C, Jones GW, Bombardieri M, Jones SA (2014) Ectopic lymphoid-like structures in infection, cancer and autoimmunity. Nat Rev Immunol 14(7):447–462. doi:10.1038/nri3700

Pratama A, Ramiscal RR, Silva DG, Das SK, Athanasopoulos V, Fitch J, Botelho NK, Chang PP, Hu X, Hogan JJ, Mana P, Bernal D, Korner H, Yu D, Goodnow CC, Cook MC, Vinuesa CG (2013) Roquin-2 shares functions with its paralog Roquin-1 in the repression of mRNAs controlling T follicular helper cells and systemic inflammation. Immunity 38(4):669–680. doi:10.1016/j.immuni.2013.01.011

Schindler SE, McCall JG, Yan P, Hyrc KL, Li M, Tucker CL, Lee JM, Bruchas MR, Diamond MI (2015) Photo-activatable Cre recombinase regulates gene expression in vivo. Scientific Reports 5:13627. doi:10.1038/srep13627

Shinnakasu R, Inoue T, Kometani K, Moriyama S, Adachi Y, Nakayama M, Takahashi Y, Fukuyama H, Okada T, Kurosaki T (2016) Regulated selection of germinal-center cells into the memory B cell compartment. Nat Immunol [Epub ahead of print]. doi:10.1038/ni.3460

Shulman Z, Gitlin AD, Targ S, Jankovic M, Pasqual G, Nussenzweig MC, Victora GD (2013) T follicular helper cell dynamics in germinal centers. Science 341(6146):673–677. doi:10.1126/science.1241680

Suan D, Nguyen A, Moran I, Bourne K, Hermes JR, Arshi M, Hampton HR, Tomura M, Miwa Y, Kelleher AD, Kaplan W, Deenick EK, Tangye SG, Brink R, Chtanova T, Phan TG (2015) T follicular helper cells have distinct modes of migration and molecular signatures in naive and memory immune responses. Immunity 42(4):704–718. doi:10.1016/j.immuni.2015.03.002

Taslimi A, Vrana JD, Chen D, Borinskaya S, Mayer BJ, Kennedy MJ, Tucker CL (2014) An optimized optogenetic clustering tool for probing protein interaction and function. Nat Commun 5:4925. doi:10.1038/ncomms5925

Ueno H, Banchereau J, Vinuesa CG (2015) Pathophysiology of T follicular helper cells in humans and mice. Nat Immunol 16(2):142–152. doi:10.1038/ni.3054

Victora GD, Nussenzweig MC (2012) Germinal centers. Annu Rev Immunol 30:429–457. doi:10.1146/annurev-immunol-020711-075032

Vinuesa CG, Cook MC, Angelucci C, Athanasopoulos V, Rui L, Hill KM, Yu D, Domaschenz H, Whittle B, Lambe T, Roberts IS, Copley RR, Bell JI, Cornall RJ, Goodnow CC (2005) A RING-type ubiquitin ligase family member required to repress follicular helper T cells and autoimmunity. Nature 435(7041):452–458. doi:10.1038/nature03555

Vogel KU, Edelmann SL, Jeltsch KM, Bertossi A, Heger K, Heinz GA, Zoller J, Warth SC, Hoefig KP, Lohs C, Neff F, Kremmer E, Schick J, Repsilber D, Geerlof A, Blum H, Wurst W, Heikenwalder M, Schmidt-Supprian M, Heissmeyer V (2013) Roquin paralogs 1 and 2 redundantly repress the Icos and Ox40 costimulator mRNAs and control follicular helper T cell differentiation. Immunity 38(4):655–668. doi:10.1016/j.immuni.2012.12.004

Wang X, Chen X, Yang Y (2012) Spatiotemporal control of gene expression by a light-switchable transgene system. Nat Methods 9(3):266–269. doi:10.1038/nmeth.1892

Part III
Chronic Inflammation and Cancer

Chapter 14
The Role of Chronic Inflammation in the Promotion of Gastric Tumourigenesis

Hiroko Oshima, Kanae Echizen, Yusuke Maeda, and Masanobu Oshima

Abstract Cyclooxigenase-2 (COX-2) is an inducible enzyme for prostaglandin biosynthesis which plays a key role in inflammatory responses. The expression of COX-2 is induced in most cancer tissues, and plays a key role in tumourigenesis through the production of inflammatory mediator, prostaglandin E_2 (PGE_2). We investigated the role of COX-2/PGE_2 pathway-associated inflammation in gastric tumourigenesis using *Gan* mice, which develop gastric tumours through the activation of both the COX-2/PGE_2 pathway and oncogenic Wnt signalling. Notably, the induction of the COX-2/PGE_2 pathway induces chronic inflammation in the stomach, which leads to the generation of the inflammatory tumour microenvironment. We found that macrophage-derived tumour necrosis factor-α (TNF-α) hyperactivates the Wnt activation level of the adjacent tumour epithelial cells. Wnt signalling is important for the maintenance of stemness in gastrointestinal stem cells, and activation of Wnt signalling causes tumour formation. Moreover, TNF-α induces the expression of Noxo1 in tumour cells (a component of the NADPH oxidase 1 [NOX1] complex), resulting in the activation of reactive oxygen species (ROS) signalling. NOX1-derived ROS have been shown to be important for cancer stem cell ability. Accordingly, in tumour tissues, the COX-2/PGE_2 pathway promotes gastric tumourigenesis through the generation of the inflammatory microenvironment; where TNF-α activates the NOX1 complex in tumour cells, resulting

H. Oshima
Division of Genetics, Cancer Research Institute, Kanazawa University, Kanazawa 920-1192, Japan

K. Echizen • M. Oshima (✉)
Division of Genetics, Cancer Research Institute, Kanazawa University, Kanazawa 920-1192, Japan

AMED-CREST, Kanazawa 920-1192, Japan
e-mail: oshimam@staff.kanazawa-u.ac.jp

Y. Maeda
Division of Genetics, Cancer Research Institute, Kanazawa University, Kanazawa 920-1192, Japan

Division of Gene Regulation, Institute for Advanced Medical Research, Keio University, Tokyo 160-8582, Japan

M. Miyasaka, K. Takatsu (eds.), *Chronic Inflammation*,
DOI 10.1007/978-4-431-56068-5_14

in the maintenance of stemness. Thus the inhibition of the inflammatory pathways from COX-2/PGE_2 to NOX1/ROS will be an effective strategy for preventing gastric cancer development.

Keywords Gastric cancer • Inflammation • COX-2 • PGE_2 • TNF-α • NOX1 • ROS

14.1 Introduction

Epidemiological studies have indicated that the regular use of nonsteroidal anti-inflammatory drugs (NSAIDs), such as aspirin, lowers the incidence of gastrointestinal cancer (Thun et al. 1991). The major target of NSAIDs is cyclooxygenase (COX)-1 and COX-2, which are rate-limiting enzymes for prostaglandin biosynthesis. COX-1, which is constitutively expressed in most tissues, performs housekeeping functions such as protecting the gastrointestinal mucosa. In contrast, COX-2 expression is induced in inflamed tissues by various stimuli and plays an important role in inflammatory responses through prostaglandin signalling (Fletcher et al. 1992). The induction of COX-2 expression is also found in a variety of cancer tissues, and the level of prostaglandin E_2 (PGE_2) is significantly increased in cancer tissues (Wang and DuBois 2010a, b). It has also been indicated that inflammation, which promotes tumourigenesis through a variety of mechanisms, is one of the important hallmarks of cancer (Grivennikov et al. 2010; Hanahan and Weinberg 2011). Accordingly, these results strongly suggest that the COX-2/PGE_2 pathway is linked to inflammation and cancer, and that the COX-2/PGE_2-dependent inflammatory responses play an important role in tumourigenesis.

In a mouse genetic study, we previously demonstrated that the disruption of the *Ptgs2* gene encoding COX-2 in $Apc^{\Delta716}$ mice (a model for intestinal polyposis) resulted in the significant suppression of intestinal tumour development (Oshima et al. 1996). Moreover, treatment with NSAIDs or COX-2 inhibitors significantly suppressed intestinal tumour development in chemically-induced mouse tumour models (Oshima and Taketo 2002). An inducible PGE_2 converting enzyme, microsomal PGE-synthase-1 (mPGES-1), has been shown to be functionally coupled with COX-2 for PGE_2 biosynthesis. Its expression is also induced in both inflamed and cancerous tissues. Notably, the disruption of the *Ptges* gene encoding mPGES-1 in *Apc* mutant mice resulted in the significant suppression of intestinal tumourigenesis (Nakanishi et al. 2008). Furthermore, the disruption of *Ptger2* encoding the PGE_2 receptor EP2 significantly suppressed intestinal polyp development in $Apc^{\Delta716}$ mice (Sonoshita et al. 2001). We have also shown that a PGE_2 receptor EP4 inhibitor suppressed gastric tumourigenesis in a genetically engineered gastric tumour mouse model (*Gan* mice; see below; Oshima et al. 2011). A 'multistep tumourigenesis' model demonstrated the development of malignant cancer from normal tissues via benign adenoma takes place over a period of 15–20 years (Fig. 14.1). Taken together, these genetic results indicate that the inflammatory microenvironment is required for cancer development, and that

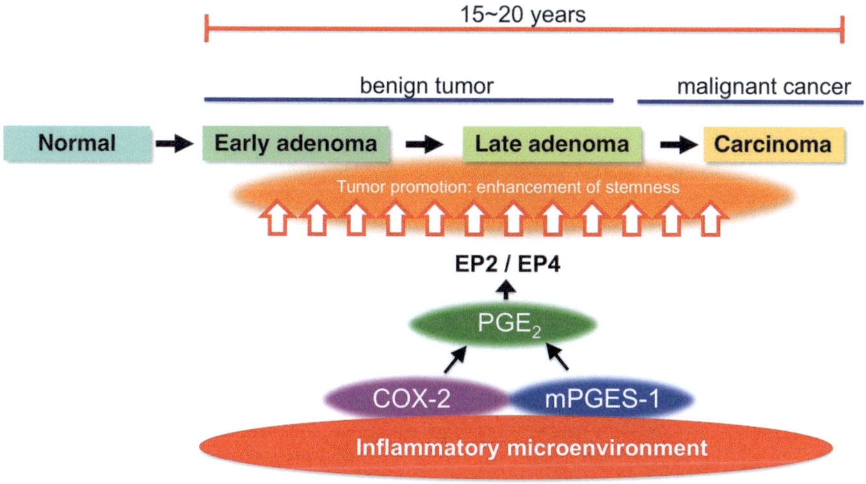

Fig. 14.1 A schematic drawing of multistep tumourigenesis. It takes approximately 15–20 years to develop malignant cancer from benign adenoma. Inflammatory responses are associated throughout the process of tumourigenesis. The expression of COX-2 and mPGES-1 is induced in the inflammatory microenvironment, resulting in the increase of the PGE_2 level. PGE_2 signalling through the EP2 and EP4 receptors has been shown to be important for tumour promotion

the COX-2/PGE_2 pathway plays a key role in tumour promotion through PGE_2 receptor EP2/EP4 signalling. In this chapter, we discuss the role of COX-2/PGE_2 pathway-associated chronic inflammation in gastric tumourigenesis, which we elucidated through mouse model studies.

14.2 '*Gan* mice' Develop Inflammation-Associated Gastric Tumours

14.2.1 The Induction of Gastric Inflammation and Hyperplasia by COX-2/PGE₂ Pathway

Infection with *H. pylori* causes chronic atrophic gastritis, which is strongly associated with gastric cancer development. *H. pylori* infection is therefore recognised as an important risk factor for gastric tumourigenesis. The COX-2/PGE_2 pathway is activated in the *H. pylori*-infected and inflamed gastric mucosa; the eradication of *H. pylori* results in decreased levels of COX-2 expression (McCarthy et al. 1999). Thus, it is possible that the COX-2/PGE_2 pathway plays an essential role in *H. pylori* infection-associated gastric tumourigenesis. To examine the role of the COX-2/PGE_2 pathway in gastric tumourigenesis, we constructed *K19-C2mE* mice, which expressed *Ptgs2* and *Ptges* (encoding COX-2 and mPGES-1, respectively) in gastric epithelial cells under the control of the *Krt19* gene promoter. In the gastric mucosa of *K19-C2mE* mice, the PGE_2 level was significantly increased and infiltrating macrophages were activated in mucosa, where they induced inflammatory responses (Oshima et al. 2004, 2009). Consistently, the expression levels of proinflammatory cytokines, TNF-α, IL-1β, and IL-11, and chemokines, CCL2, CXCL1, and CXCL2 were increased to a significant extent in the *K19-C2mE* mouse stomach. Accordingly, it is possible that the induction of the COX-2/PGE_2 pathway was responsible for inflammatory responses (Fig. 14.2). Considering that the COX-2/PGE_2 pathway is induced in most solid cancer tissues (Wang and DuBois 2010a, b), it is possible that the COX-2/PGE_2 pathway leads to the generation of the inflammatory microenvironment in most types of solid cancers.

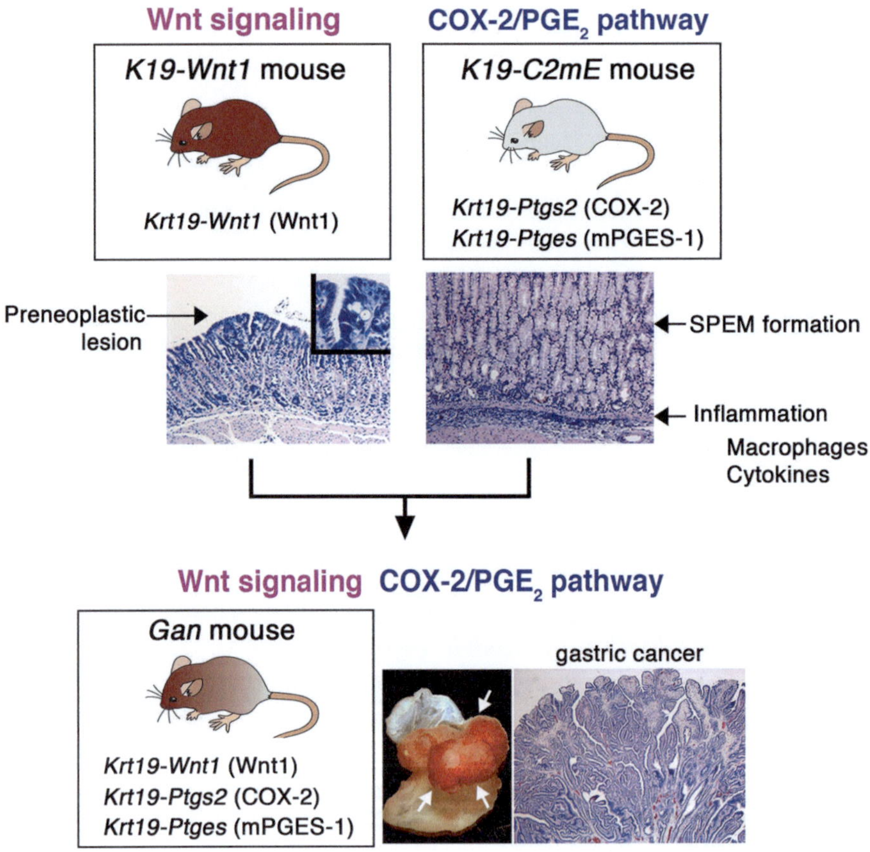

Fig. 14.2 Mouse models of gastric tumourigenesis. *K19-Wnt1* and *K19-C2mE* are transgenic mice that express *Wnt1* and *Ptgs1*/*Ptges*, respectively. These models are bred together to construct *Gan* mice. *K19-Wnt1* mice develop small preneoplastic lesions due to Wnt activation, whereas *K19-C2mE* mice show inflammation and SPEM formation, which are caused by the induction of PGE_2 signalling. *Gan* mice develop gastric tumours following the simultaneous activation of both pathways (Reproduced from Oshima et al. 2009)

It is important to note that *K19-C2mE* mice have been shown to develop metaplastic hyperplasia in the glandular stomach, which consists of mucous epithelial cells, which are histologically similar to 'spasmolytic polypeptide (trefoil factor 2: TFF2)-expressing metaplasia (SPEM)'; (Oshima et al. 2005; Fig. 14.2). SPEM lesions are frequently found adjacent to human gastric cancer. SPEM is therefore thought to be a precancerous lesion of gastric cancer (Weis and Goldenring 2009). These results indicate that COX-2/PGE_2-associated inflammation causes the formation of SPEM, which may lead to the development of gastric cancer. Although the underlying mechanism remains to be elucidated, it is possible that inflammatory responses disturb the normal differentiation process of epithelial cells through the induction of cytokines and growth factors, thereby leading to SPEM formation.

14.2.2 Tumour Promotion by Wnt Activation and COX-2/PGE_2-Associated Inflammation

Wnt signalling is required for the maintenance of stemness of normal intestinal stem cells. The genetic alteration of the *Apc* or *Ctnnb1* genes encoding APC and β-catenin, respectively, results in the constitutive activation of Wnt signalling, which causes the initiation of tumour development in the intestine (Gregorieff and Clevers 2005; Taketo 2006). Although genetic alterations in *APC* or *CTNNB1* are not common in human gastric cancer (Cancer Genome Atlas Network 2014), the accumulation of β-catenin in tumour cells, a hallmark of Wnt signalling activation, can be found by immunohistochemistry in approximately 50 % of gastric cancer tissue specimens (Oshima et al 2006). It is therefore possible that Wnt activation by mechanism(s) other than mutations in *APC* and *CTNNB1* may initiate human gastric cancer development. To assess this possibility, we constructed *K19-Wnt1* mice that expressed *Wnt1*, one of the canonical Wnt signalling ligands, in their gastric epithelial cells. The expression of *Wnt1* causes constitutive Wnt signalling activation in the gastric mucosa. Notably, the *K19-Wnt1* mice developed a number of small preneoplastic lesions (consisting of dysplastic epithelial cells), indicating that Wnt activation can initiate gastric tumourigenesis (Fig. 14.2).

However, the preneoplastic lesions of *K19-Wnt1* mice never progressed to gastric tumours, indicating that Wnt activation alone is not sufficient to initiate tumour development. We therefore crossed *K19-C2mE* mice and *K19-Wnt1* mice to generate compound transgenic mice (*Gan* mice) that expressed *Ptgs2*, *Ptges*, and *Wnt1* simultaneously in their gastric epithelial cells. This resulted in the activation of both Wnt signalling and the COX-2/PGE_2 pathway, which is similar to what occurs in human gastric cancer tissues. Note that all of the *Gan* mice developed inflammation-associated gastric tumours (Oshima et al. 2006, 2009). Taken together, these results indicate that the COX-2/PGE_2 pathway promotes tumourigenesis in Wnt-activated epithelial cells through the generation of the inflammatory microenvironment (Fig. 14.2).

14.3 The Mechanisms of Tumour Promotion by Inflammatory Responses

14.3.1 Hyperactivation of Wnt Signalling by TNF-α Signalling

Macrophage infiltration is present in the stroma of preneoplastic lesions of *K19-Wnt1* mouse stomach; such macrophage infiltration is not found in the normal gastric mucosa. Interestingly, the immunostaining intensity of β-catenin was found to be substantially increased in dysplastic epithelial cells from preneoplastic lesions of *K19-Wnt1* mice (Oguma et al. 2008). These staining results indicate that the Wnt signalling activation level is further increased (hyperactivated) in preneoplastic lesions of the gastric mucosa. From these results, we hypothesised that the activated macrophage-derived factors promote the Wnt signalling activity of the adjacent epithelial cells, and that such Wnt hyperactivation would lead to tumourous changes in the gastric mucosa.

In the $Apc^{\Delta 716}$ mouse model of intestinal polyposis, mice develop intestinal tumours through the activation of Wnt signalling, which occurs due to the loss of heterozygosity of the *Apc* gene (Oshima et al. 1995). Note that intestinal tumourigenesis was significantly suppressed when $Apc^{\Delta 716}$ mice were crossed with op/op mutant mice, in which macrophages are depleted due to a *Csf1* mutation (Oguma et al. 2008; Fig. 14.3). Taken together, these results indicate that macrophage infiltration in the tumour stroma is required for gastrointestinal tumour development, which possibly occurs through the hyperactivation of Wnt signalling activity in tumour cells.

14.3.2 Wnt Promotion by TNF-α Signalling in Gastric Cancer Cells

The level of Wnt signalling activity in cancer cells is indicated by the TOP-GFP reporter vector (Oguma et al. 2008). Notably, the GFP expression level was found to be significantly increased (Wnt signalling was activated) in gastric cancer cells, when these cells were treated with a conditioned medium (CM) of activated macrophages, indicating that macrophage-derived factor promotes the Wnt signalling of tumour cells (Fig. 14.3). Note that the promotion of Wnt signalling by CM was significantly suppressed by the treatment of cells with TNF-α neutralising

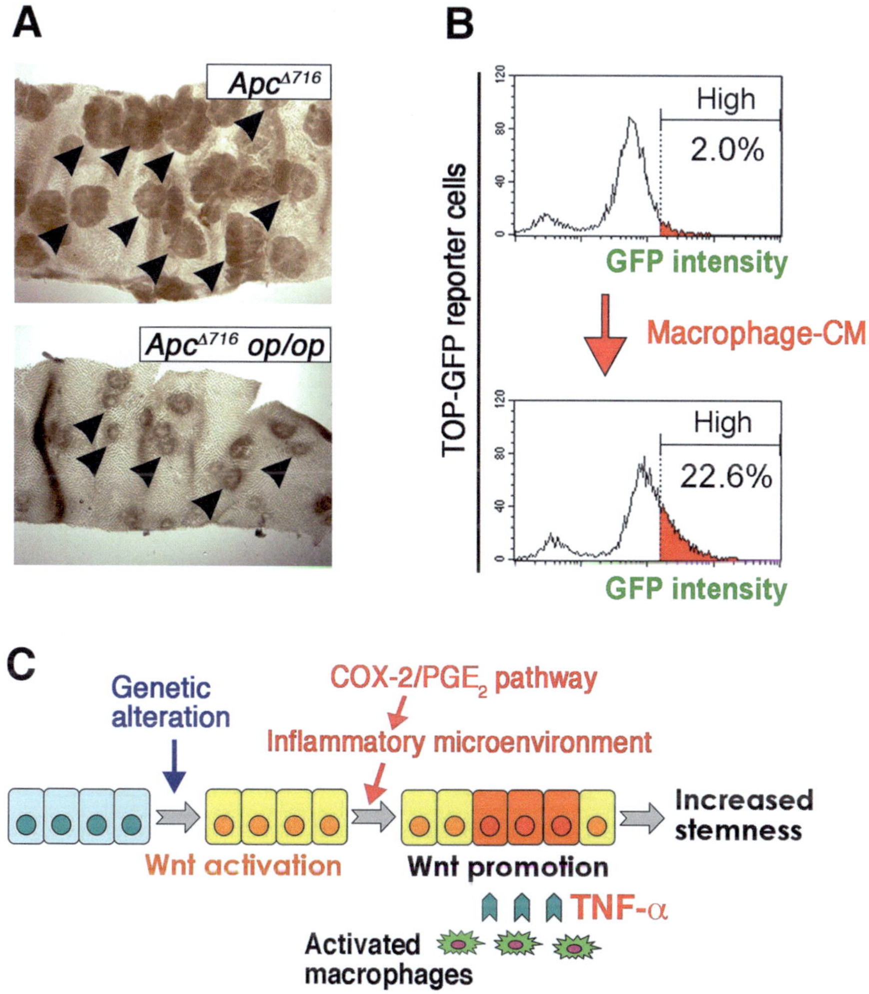

Fig. 14.3 Tumour promotion by macrophage-derived TNF-α. (**a**) The suppression of intestinal tumour development in *Apc*$^{\Delta 716}$ *op/op* mice by macrophage depletion. (**b**) The increase in the GFP High population (Wnt activation) of TOP-GFP reporter cells in response to macrophage CM treatment. (**c**) A schematic drawing of TNF-α–dependent tumour promotion. In the COX/PGE$_2$-induced inflammatory microenvironment, TNF-α derived from macrophages promotes the Wnt signalling of tumour cells, resulting in increased stemness, which contributes to tumour promotion (Reproduced from Oguma et al. 2008, 2009)

antibody, indicating that TNF-α signalling enhances Wnt signalling activity in gastric cancer cells (Oguma et al. 2008, 2010).

Because Wnt signalling is required for the maintenance of epithelial stem cells, it is possible that the inflammatory responses enhance the stemness of tumour cells through the activation of TNF-α signalling. TNF-α signalling activates transcription factor NF-kB, which plays a central role in inflammation through the induction of cytokine and chemokine expression. Recently, it has been shown that the simultaneous activation of Wnt signalling and NF-κB leads to the acquisition of stem cell properties in differentiated intestinal epithelial cells (Schwitalla et al. 2013). Accordingly, it is possible that the TNF-α/NF-κB pathway promotes gastric tumourigenesis through the enhancement of the stemness of tumour cells by the activation of Wnt signalling (Fig. 14.3). Interestingly, cell proliferation is significantly increased in the inflamed mucosa of the *H. pylori*-infected mouse stomach and the number of differentiated cells (such as parietal cells) is significantly decreased, suggesting that epithelial differentiation is suppressed by inflammatory responses (Oguma et al. 2008).

14.3.3 The Promotion of Gastric Tumourigenesis by TNF-α Signalling

Gastric tumour development was dramatically suppressed in *Gan* mice when the TNF-α gene (*Tnf*) was disrupted, and the mean tumour size in the *Gan Tnf−/−* mice was decreased to less than 20 % of that in the littermate control mice (Oshima et al. 2014) (Fig. 14.4). Although TNF-α signalling was suppressed in *Gan Tnf−/−* mouse tumours, macrophage infiltration was still present and other cytokines including IL-1β and IL-6 were induced at high levels in gastric tumours. Accordingly, it is possible that, among the various cytokines induced in the tumour microenvironment, TNF-α is indispensable for gastric tumourigenesis. Note that bone marrow transplantation from wild-type mice to *Gan Tnf−/−* mice significantly rescued the gastric tumour phenotype, indicating that TNF-α expressed by bone-marrow–derived cells, including macrophages, promotes gastric tumourigenesis.

To determine the TNF-α-dependent tumour-promoting factors, a microarray analysis was performed using gastric tumour tissues of *Gan Tnf−/−* mice and control *Gan* mice (Gene Expression Omnibus accession, GSE43145). One hundred fifty-seven (157) genes were selected as upregulated genes in tumour tissues in a TNF-α–dependent mechanism (Oshima et al. 2014). Interestingly, the marker genes for intestinal stem cells and progenitor cells, such as *Cd44*, *Prom1*, and *Sox9* (Itzkovitz et al. 2011) were significantly downregulated in *Gan Tnf−/−* mouse tumours. It is therefore possible that TNF-α signalling plays a role in the maintenance of stemness or the undifferentiated status of gastric tumour cells, which is consistent with the results of Sect. 14.3.2.

The TNF-α–dependent upregulated genes were further compared with gene the expression profile of normal gastric stem cells (Barker et al. 2011), and 11 genes from the list including *Aqp1*, *Azin1*, *Cd44*, *Cldn2*, *Gfpt1*, *Gna4*, *Noxo1*, *Prom1*, *Sgms1*, and *Sox9* were found to be upregulated in both normal gastric stem cells and tumour tissues in a TNF-α–dependent manner (Oshima et al. 2014). Interestingly,

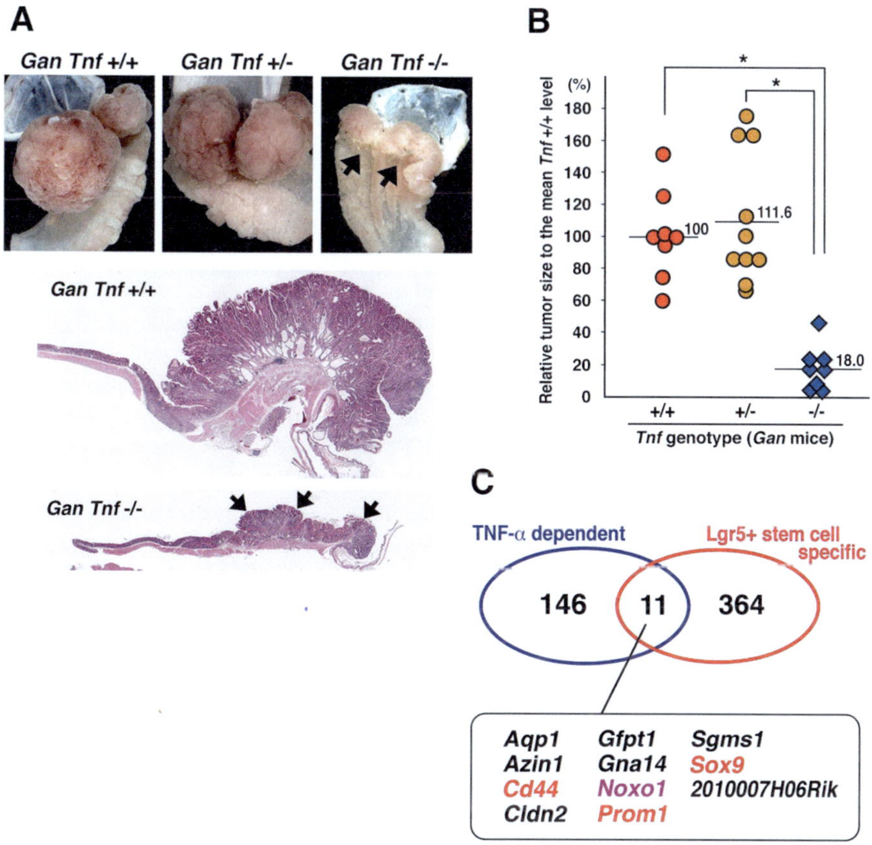

Fig. 14.4 The suppression of gastric tumourigenesis by the disruption of the TNF α gene (*Tnf*). (**a**) A representative macroscopic tumour phenotype (*top*) and whole view histology (H&E) (*bottom*) of gastric tumours of mice of the respective genotypes. Arrows indicate regressed gastric tumours in *Gan Tnf*−/− mice. (**b**) The gastric tumour sizes of *Gan Tnf*+/− and *Gan Tnf*−/− mice relative to the mean level of *Gan Tnf*+/+ mouse tumours (set at 100 %). *Asterisks* indicate statistical significance ($P < 0.05$). (**c**) A list of the genes that are upregulated in tumour tissues in a TNF-α–dependent manner and also in the Lgr5+ normal gastric stem cells. The stem-cell–related factors are indicated in *red*. Noxo1 is highlighted in *pink* (Reproduced from Oshima et al. 2014)

CD44, Prom1, and SOX9 are well-described stem-cell–related factors whose expression is induced by Wnt signalling (Fig. 14.4). Taken together, it is possible that TNF-α-induced Wnt activation is responsible for increased undifferenitated state of tumor cells.

14.3.4 The TNF-α–Induced Activation of NOX1-Dependent ROS Signalling

To identify the most important TNF-α–dependent tumour-promoting factors, a functional analysis was performed using a soft agar colony formation assay for the 11 selected genes (see Sect. 3.3). As expected, the inhibition of *Cd44*, *Prom1*, and *Sox9* resulted in the suppression of gastric cancer cell colony formation. However, the most marked suppression of colony formation was found when *Noxo1* expression was inhibited (Oshima et al. 2014).

Noxo1 is a cytosolic regulatory subunit of the NADPH oxidase1 (NOX1) complex, which produces reactive oxygen species (ROS). It has been shown that NOX1 expression is upregulated in a variety of cancers including colon cancer, and several mechanisms of NOX1-produced ROS have been reported to occur in tumourigenesis (Block and Gorin 2012). For example, NOX1-dependent ROS suppresses protein tyrosine phosphatases (PTPs), which results in the activation of tyrosine kinase receptor (RTK) signalling, increasing the proliferation of tumour cells. Moreover, NOX1 expression is upregulated by oncogenic RAS activation, and is required for cell transformation (Adachi et al. 2008). More recently, it has been demonstrated that the NOX1 complex is activated in intestinal tumour cells of Apc^{Min} mice, and the NOX1 complex-dependent ROS, together with NF-κB pathway, has been demonstrated to play a role in the maintenance of stemness in tumour epithelial cells (Myant et al. 2013). Note that we found that the ROS production in cancer cells induced by the NOX1 complex is regulated by Noxo1 expression, thus TNF-α can regulate the ROS level by the induction of Noxo1 expression in cancer cells (Echizen, unpublished results). Taken together, these results suggest that the inflammatory microenvironment supports gastric tumour promotion through the maintenance or acquisition of stemness in tumour cells via the induction of NOX1 complex-dependent ROS signalling, which is regulated by exogenous TNF-α stimulation (Fig. 14.5). It is therefore possible that targeting the NOX1/ROS pathway will be effective as a preventive strategy against gastric cancer development.

14.4 The Mechanism Underlying the Generation of the Inflammatory Microenvironment

14.4.1 The Innate Immunity and Stemness of Intestinal Epithelial Cells

It has been shown that commensal bacteria constitutively stimulate intestinal mucosa through Toll-like receptor (TLR)2 and TLR4, which is important for the homeostasis of the intestinal mucosa and is required for regeneration following mucosal damage (Greten et al. 2004). Moreover, the inhibition of MyD88

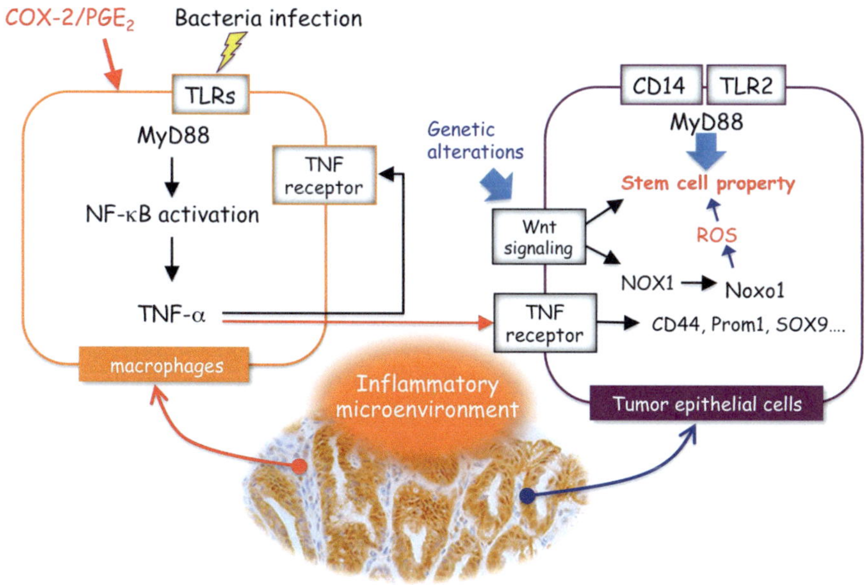

Fig. 14.5 A schematic drawing of the inflammatory microenvironment and gastric tumourigenesis. The COX-2/PGE$_2$ pathway and TLR signalling cooperatively induce the generation of the inflammatory microenvironment. TNF-α is expressed by activated macrophages. TNF-α stimulates tumour cells to induce the expression of stem-cell–related factors, such as CD44, Prom1, and Sox9. Noxo1 is also induced by TNF-α, which contributes to the activation of NOX1 complex to produce ROS. NOX1/ROS signalling is activated by the Wnt pathway, which enhances the stemness of tumour cells. TLR2 signalling has also been shown to be important for the stemness of tumour cells

(an adapter molecule of the TLRs) signalling resulted in the significant suppression of intestinal tumourigenesis in *Apc*^{Min} mice (Rakoff-Nahoum and Medzhitov 2007). We found that the expression level of TLR2 is significantly increased (in comparison to the other TLRs) in the tumour epithelial cells of the gastric tumour tissues of *Gan* mice and gp130$^{F/F}$ mice. gp130$^{F/F}$ mice are another gastric tumour model which develop inflammation-associated gastric tumours due to IL-11 signalling-induced hyper Stat3 activation. Notably, the disruption of the *Tlr2* gene significantly suppressed gastric tumourigenesis in gp130$^{F/F}$ mice, indicating the role of TLR2 signalling in tumourigenesis (Tye et al. 2012). More recently, it has been demonstrated that TLR2 signalling through MyD88 is important for the maintenance of stemness of intestinal and mammary gland epithelial stem cells (Scheeren et al. 2014). These results strongly suggest that the innate immune responses that occur via the TLR2/MyD88 pathway play a role in tumourigenesis.

14.4.2 Bacterial Infection and Gastric Tumourigenesis

The basal expression levels of cytokines and chemokines such as TNF-α, IL-1β, CXCL1, and CXCL2 were significantly decreased in the germfree mouse stomach in comparison to SPF mice, suggesting that indigenous bacteria contribute to the basal expression level of inflammatory factors in the gastric mucosa. Although it has been established that *H. pylori* infection is strongly associated with gastric cancer development, there is limited information on whether indigenous bacteria play a role in gastric tumourigenesis through the modulation of innate mucosal immunity.

We therefore examined the role of commensal bacteria in gastric tumourigenesis by the construction of germfree *Gan* mice (Oshima et al. 2011). Notably, gastric tumourigenesis was significantly suppressed in germfree *Gan* mice and tumour volume was decreased to approximately 30 % of that in the SPF control *Gan* mice. Interestingly, infection with *Helicobacter felis*, a close relative of *H. pylori*, in the stomach of germfree *Gan* mice caused the development of gastric tumours, indicating the role of bacterial infection in gastric tumourigenesis. An histological analysis revealed that macrophage infiltration was suppressed in germfree *Gan* mouse tumours, and that cytokine induction levels were significantly decreased, indicating a role of innate immunity in the COX-2/PGE_2-dependent inflammatory responses. Through genetic studies, we also found that signalling through the TLRs plays a role in gastric tumourigenesis in *Gan* mice (Maeda et al. 2016; Fig. 14.5). Accordingly, it is possible that the COX-2/PGE_2 pathway and innate immune signalling through the TLRs cooperatively generate the inflammatory tumour microenvironment, which further contributes to the promotion of gastric tumourigenesis.

14.5 Conclusions and Perspectives

It has been established that inflammation plays a role in cancer development. Using a *Gan* mouse model, we showed that the COX-2/PGE_2 pathway plays a tumour-promoting role through the generation of the inflammatory microenvironment. Our results also showed the role of TNF-α in tumourigenesis for the maintenance of undifferentiated state of tumour cells through the induction of the NOX1/ROS signalling pathway. Moreover, we showed that infectious stimuli to the mucosa by indigenous bacteria contribute to the generation of inflammatory responses, together with the COX-2/PGE_2 pathway, suggesting a role of innate immunity in tumour development. It has recently been reported that a *K-ras* mutation causes the activation of innate immune responses, which leads to cytokine circuit activation (Zhu et al. 2014), and that a p53 mutation prolongs the TNF-α-dependent NF-κB activation in cancer cells (Cooks et al. 2013). Accordingly, it is conceivable that microbiota and oncogenic mutations cooperatively activate innate immunity, thereby contributing to the generation of the inflammatory microenvironment, which promotes tumourigenesis and malignant progression through the enhancement of the proliferation and stemness of transformed epithelial cells.

References

Adachi Y, Shibai Y, Mitsushita J et al (2008) Oncogenic Ras upregulates NADPH oxidase 1 gene expression through MEK-ERK-dependent phosphorylation of GATA-6. Oncogene 27: 4921–4932

Barker N, Huch M, Kujala P et al (2011) Lgr5^{+ve} stem cells drive self-renewal in the stomach and build long-lived gastric units in vitro. Cell Stem Cell 6:25–36

Block K, Gorin Y (2012) Aiding and abetting roles of NOX oxidases in cellular transformation. Nat Rev Cancer 12:627–637

Cancer Genome Atlas Network (2014) Comprehensive molecular characterization of gastric adenocarcinoma. Nature 513:202–209

Cooks T, Pateras I, Tarcic O et al (2013) Mutant p53 prolongs NF-kB activation and promotes chronic inflammation and inflammation-associated colorectal cancer. Nat Cell Biol 23: 634–646

Fletcher BS, Kujubu DA, Perrin DM et al (1992) Structure of the mitogen-inducible TIS10 gene and demonstration that the TIS10-encoded protein is a functional prostaglandin G/H synthase. J Biol Chem 267:4338–4344

Gregorieff A, Clevers H (2005) Wnt signaling in the intestinal epithelium: from endoderm to cancer. Genes Dev 19:877–890

Greten FR, Eckmann L, Greten TF et al (2004) IKKβ links inflammation and tumorigenesis in a mouse model of colitis-associated cancer. Cell 118:285–296

Grivennikov SI, Greten FR, Karin M (2010) Immunity, inflammation and cancer. Cell 140: 883–899

Hanahan D, Weinberg RA (2011) Hallmarks of cancer: the next generation. Cell 144:646–674

Itzkovitz S, Lyubimova A, Blat IC et al (2011) Single-molecule transcript counting of stem-cell markers in the mouse intestine. Nat Cell Biol 14:106–114

Maeda Y, Echizen K, Oshima H et al (2016) Myeloid differentiation factor 88 signaling in bone marrow-derived cells promotes gastric tumorigenesis by generation of inflammatory microenvironment. Cancer Prev Res 9:253–263

McCarthy CJ, Crofford LJ, Greenson J et al (1999) Cyclooxygenase-2 expression in gastric antral mucosa before and after eradiation of *Helicobacter pylori* infection. Am J Gastroenterol 94: 1218–1223

Myant KB, Cammareri P, McGhee EJ et al (2013) ROS production and NF-kB activation triggered by RAC1 facilitates WNT-driven intestinal stem cell proliferation and colorectal cancer initiation. Cell Stem Cell 12:761–773

Nakanishi M, Montrose DC, Clark P et al (2008) Genetic deletion of mPGES-1 suppresses intestinal tumorigenesis. Cancer Res 68:3251–3259

Oguma K, Oshima H, Aoki M et al (2008) Activated macrophages promote Wnt signaling through tumor necrosis factor-a in gastric tumor cell. EMBO J 27:1671–1681

Oguma K, Oshima H, Oshima M (2010) Inflammation, tumor necrosis factor and Wnt promotion in gastric cancer development. Future Oncol 6:515–526

Oshima M, Taketo MM (2002) COX selectivity and animal models for colon cancer. Curr Pharm Des 8:1021–1034

Oshima M, Oshima H, Kitagawa K et al (1995) Loss of *Apc* heterozygosity and abnormal tissue building in nascent intestinal polyps in mice carrying a truncated *Apc* gene. Proc Natl Acad Sci U S A 92:4482–4486

Oshima M, Dinchuk JE, Kargman SL et al (1996) Suppression of intestinal polyposis in *Apc*$^{\Delta 716}$ knockout mice by inhibition of cyclooxygenase 2 (COX-2). Cell 87:803–809

Oshima H, Oshima M, Inaba K et al (2004) Hyperplastic gastric tumors induced by activated macrophages in COX-2/mPGES-1 transgenic mice. EMBO J 23:1669–1678

Oshima H, Oshima M, Matsunaga A et al (2005) Hyperplastic gastric tumors with spasmolytic polypeptide-expressing metaplasia caused by tumor necrosis factor-α-dependent inflammation in cyclooxygenase-2/microsomal prostaglandin E synthase-1 transgenic mice. Cancer Res 65: 9147–9151

Oshima H, Matsunaga A, Fujimura T et al (2006) Carcinogenesis in mouse stomach by simultaneous activation of the Wnt signaling and prostaglandin E_2 pathway. Gastroenterology 131: 1086–1095

Oshima H, Oguma K, Du YC et al (2009) Prostaglandin E_2, Wnt and BMP in gastric tumor mouse models. Cancer Sci 100:1779–1785

Oshima H, Hioki K, Popivanova BK et al (2011) Prostaglandin E_2 signaling and bacterial infection recruit tumor-promoting macrophages to mouse gastric tumors. Gastroenterology 140: 596–607

Oshima H, Ishikawa T, Yoshida GJ et al (2014) TNF-α/TNFR1 signaling promotes gastric tumorigenesis through induction on *NOXO1* and *Gna14* in tumor cells. Oncogene 33: 3820–3829

Rakoff-Nahoum S, Medzhitov R (2007) Regulation of spontaneous intestinal tumorigenesis though the adaptor protein MyD88. Science 317:124–127

Scheeren FA, Kuo AH, van Weele LJ et al (2014) A cell-intrinsic role for TLR2-MYD88 in intestinal and breast epithelia and oncogenesis. Nat Cell Biol 16:1238–1248

Schwitalla S, Fingerle AA, Cammareri P et al (2013) Intestinal tumorigenesis initiated by dedifferentiation and acquisition of stem-cell-like properties. Cell 152:25–38

Sonoshita M, Takaku K, Sasaki N et al (2001) Acceleration of intestinal polyposis through prostaglandin receptor EP2 in Apc(Delta716) knockout mice. Nat Med 7:1048–1051

Taketo MM (2006) Wnt signaling and gastrointestinal tumorigenesis in mouse models. Oncogene 25:7522–7530

Thun MJ, Namboodiri MM, Heath CW Jr et al (1991) Aspirin use and reduced risk of fatal colon cancer. N Engl J Med 325:1593–1596

Tye H, Kennedy CL, Najdovska M et al (2012) STAT3-driven upregulation of TLR2 promotes gastric tumorigenesis independent of tumor inflammation. Cancer Cell 22:466–478

Wang D, Dubois RN (2010a) Eicosanoids and cancer. Nat Rev Cancer 10:181–193

Wang D, Dubois RN (2010b) The role of COX-2 in intestinal inflammation and colorectal cancer. Oncogene 29:781–788

Weis VG, Goldenring JR (2009) Current understanding of SPEM and its standing in the preneoplastic process. Gastric Cancer 12:189–197

Zhu Z, Aref AR, Cohoon TJ et al (2014) Inhibition of KRAS-driven tumorigenicity by interruption of an autocrine cytokine circuit. Cancer Dicov 4:452–465

Chapter 15
Cellular Senescence as a Novel Mechanism of Chronic Inflammation and Cancer Progression

Naoko Ohtani

Abstract Cellular senescence is a state of irreversible cell proliferation arrest provoked by a persistent DNA damage induced by a variety of potentially oncogenic signals, and it was initially identified as a tumour-suppression mechanism. Recent studies, however, revealed that senescent cells have the potential to secrete numerous inflammatory cytokines, chemokines, growth factors, and matrix-remodeling factors, because unlike apoptotic cells, senescent cells are viable for a long period of time. This newly identified phenotype of cellular senescence, called senescence-associated secretory phenotype (SASP or senescence-associated secretome), could potentially provide beneficial effects, such as reinforcement of cellular senescence and tissue repair, but sometimes could induce deleterious side effects, such as chronic inflammation and cancer progression, depending on the biological context. Since obesity is associated with chronic inflammation and cancer, we thought that the senescence-associated secretome could be closely involved in the inflammation and tumourigenesis accompanying obesity. In this review, I first discuss the role and the mechanism of cellular senescence, and then introduce the role of senescence-associated secretome in vivo including our recent findings on the mechanism of obesity-associated liver cancer, promoted by the senescence-associated secretome in hepatic stellate cells, which could form a cancer-promoting microenvironment.

Keywords Cellular senescence • Senescence-associated secretome • Inflammatory cytokines • Obesity • Liver cancer • Deoxycholic acid

---

N. Ohtani (✉)
Department of Applied Bioscience, Faculty of Science and Technology, Tokyo University of Science, 2641 Yamazaki, Noda-shi, Chiba, Japan

PRESTO, Japan Science and Technology Agency, Kawaguchi 332-0012, Japan
e-mail: naoko.ohtani@rs.tus.ac.jp

M. Miyasaka, K. Takatsu (eds.), *Chronic Inflammation*,
DOI 10.1007/978-4-431-56068-5_15

15.1 Introduction

Cellular senescence is the state of irreversible cell-cycle arrest provoked by a persistent DNA damage, which is induced by a variety of potentially oncogenic stimuli, and has therefore long been considered to suppress tumourigenesis, playing a role as a guardian of tissue homeostasis (Finkel et al. 2007; Campisi 2005; Herbig and Sedivy 2006; Adams 2009; Collado and Serrano 2010; Kuilman et al. 2010; Muñoz-Espín and Serrano 2014). However, new evidence has revealed that senescent cells have a character to promote the secretion of inflammatory cytokines, chemokines, growth factors, matrix remodelling factors, growth factors and so on (Rodier and Campisi 2011). These factors alter the local tissue environment and contribute to some physiological effects, such as tissue repair (Rodier and Campisi 2011; Jun and Lau 2010; Demaria et al. 2014), but also to deleterious effects, such as chronic inflammation and cancer development (Rodier and Campisi 2011; Coppé et al. 2008; Kuilman and Peeper 2009), depending on the biological context. This newly identified senescence phenotype, termed the senescence-associated secretory phenotype (SASP; Coppé et al. 2008) or the senescence-messaging secretome (SMS; Kuilman and Peeper 2009; hereafter designated as the senescence-associated secretome), is known to be induced by persistent DNA damage (Rodier et al. 2009) that promotes the induction of cellular senescence. All of these senescence-associated secreted factors seem to be involved in homeostatic disorders, such as cancer. Our recent study focused on obesity-associated pathological conditions, and particularly on the link between obesity and liver cancer, and we discovered the role of the senescence-associated secretome in tumour-promoting microenvironments in obesity-associated liver cancer (Yoshimoto et al. 2013). In this review, we first introduce what cellular senescence is and its associated molecular mechanisms, and then discuss the role of the senescence-associated secretome especially in vivo including our recent findings on the mechanism of obesity-associated liver carcinogenesis (Yoshimoto et al. 2013).

15.2 Cellular Senescence: Irreversible Cell Proliferation Arrest

Most human somatic cells can initially proliferate well when cultured in vitro, but they stop dividing permanently after a finite number of cell divisions, and enter a state of irreversible cell proliferation arrest (Haff and Swim 1956; Hayflick 1965). This phenomenon of proliferation limit is called ‘cellular senescence’ or ‘replicative cellular senescence’. Unlike quiescent (G0 phase) cells, which are induced by low serum or contact inhibition conditions, senescent cells are irreversibly arrested predominantly in the G1 phase and are no longer able to divide even with proliferative stimuli, although they remain viable and metabolically active for long periods of time. In contrast, most cancer cells appear to have bypassed this proliferative limit and evaded cellular senescence. Thus, cellular senescence has been considered as a barrier to cancer, by playing an important role in preventing the extensive cell divisions required for malignant transformation (Muñoz-Espín and Serrano 2014; Finkel et al. 2007; Campisi 2005; Herbig and Sedivy 2006; Collado and Serrano 2010; Kuilman et al. 2010).

In human cells, the mechanism underlying replicative cellular senescence is thought to be telomere shortening. Telomeres are located at the ends of eukaryotic chromosomes, and they have special DNA structures with repetitive DNA elements to protect the DNA ends from degradation and chromosomal end-to-end fusion (Deng et al. 2008). The telomere length is maintained by a specific enzyme called telomerase, which is expressed in human germline cells and stem cells, but not in most normal somatic cells. Due to the incompletion of the DNA replication at the end of the telomere by the lack of telomerase, telomeres become shorter and shorter with each round of cell division (Harley et al. 1990). Eventually, progressive telomere erosion results in telomere dysfunction, and this is thought to initiate DNA damage response signals to activate the p53-dependent checkpoints that contribute to cellular senescence (de Lange 2005; Deng et al. 2008; Fumagalli et al. 2012).

In contrast to human cells, there is no strong evidence that rodent cells undergo replicative cellular senescence by telomere erosion (Wright and Shay 2000; Sherr and Depinho 2000), because the telomeres in rodent cells are known to be quite long, and many somatic rodent cells have telomerase activity. However, the senescence-like proliferative block in rodent cells occurs without detectable telomere shortening. This phenomenon suggests that a mechanism other than telomere shortening can cause cellular senescence in rodent cells. In this regard, it is interesting to note that primary mouse embryonic fibroblasts are known to proliferate indefinitely, if the cells are maintained under certain culture conditions, such as low oxygen conditions (Parrinello et al. 2003; Takeuchi et al. 2010). In addition, rat oligodendrocyte precursor cells and rat Schwann cells do not senesce in serum-free medium, but do senesce in the presence of serum (Lloyd 2002). These findings clearly demonstrated that cellular senescence can be induced without apparent telomere shortening, when cells are exposed to nonphysiological conditions in vitro. Notably moreover, the fact that antioxidant treatment delays cellular senescence induction strongly suggests that reactive oxygen species (ROS) can trigger cellular senescence (Finkel et al. 2007; Kuilman et al. 2010; Takahashi et al. 2006; Imai et al. 2014). Recent studies in human cells revealed that a similar irreversible proliferation block can be induced quite rapidly when normal cells are exposed to persistent DNA damage caused by a variety of potentially oncogenic stimuli, excessive levels of ROS, treatment with DNA damaging agents, activation of certain oncogenes, and so on (Finkel et al. 2007; Campisi 2005; Herbig and Sedivy 2006; Collado and Serrano 2010; Kuilman et al. 2010; Serrano et al. 1997; Serrano and Blasco 2001; d'Adda di Fagagna 2008). These types of oncogenic stress-induced senescence are now referred to as 'stress-induced senescence' (SIS) or 'oncogene-induced senescence' (OIS), in cases where oncogenic stimuli induce the cellular senescence.

15.3 Cellular Senescence as an Important Tumour Suppression Mechanism

The basic mechanisms for the induction of senescent cell-cycle arrest have been well documented. In mammalian cells, the RB and p53 tumour suppressor proteins play critical roles in the induction of cellular senescence (Shay et al. 1991; Hara et al. 1991). Particularly, RB and its family members, p107 and p130, are essential for the onset of senescent cell cycle-arrest (Dannenberg et al. 2000; Sage et al. 2000, 2003). The activities of RB-family proteins are precisely regulated by phosphorylation, protein–protein interactions, and other protein modifications (Cobrinik et al. 1992; Classon and Harlow 2002). A series of cyclin-dependent kinases (CDKs), including CDK2, CDK4, and CDK6, play key roles in regulating the activities of RB-family proteins (Sherr and Roberts 1999). When RB is phosphorylated by these CDKs, it loses its ability to bind to the E2F family of transcription factors and is unable to repress E2F functions that facilitate DNA replication and cell-cycle progression (Cobrinik et al. 1992; Classon and Harlow 2002; Sherr and Roberts 1999).

Senescent cells arrest themselves in the G1 phase of the cell cycle, and are no longer able to synthesize DNA. In many cases, this is because the CDKs are inactivated in senescent cells by CDK inhibitors (CDKI; Sherr and Roberts 1999). There are two different classes of CDKIs, the KIP/CIP family CDKIs ($p21^{Cip1}$, $p27^{Kip1}$, and $p57^{KipII}$) and the INK4 family CDKIs ($p16^{Ink4a}$, $p15^{Ink4b}$, $p18^{Ink4c}$, and $p19^{Ink4d}$). The KIP/CIP family members inhibit a broad range of CDKs, whereas the INK4 family proteins specifically bind and inactivate CDK4 and CDK6 (Sherr and Roberts 1999). In normal proliferating cells, the expression levels of CDKIs are very low (Noda et al. 1994; Hara et al. 1996; Alcorta et al. 1996). However, in response to irreparable DNA damage caused by a variety of oncogenic stimuli, the expression levels of the $p21^{Cip1}$ and $p16^{Ink4a}$ genes are significantly upregulated by p53-dependent and p53-independent mechanisms, respectively (Noda et al. 1994; Hara et al. 1996; Alcorta et al. 1996; Ohtani et al. 2001; el-Deiry et al. 1993). In addition, the simultaneous induction of $p21^{Cip1}$and $p16^{Ink4a}$ cooperatively inactivates all CDKs that phosphorylate RB-family proteins (Takeuchi et al. 2010; McConnell et al. 1999; Mitra and Dai 1999), thereby causing cellular senescence, a permanent proliferation arrest. It is noteworthy that the senescence pathway including $p16^{Ink4a}$, CDKs and RB is highly disregulated in nearly all human cancers (Gil and Peters 2006), illustrating the importance of the p16/RB pathway in tumour suppression through senescence induction.

15.4 Quiescence–Senescence Switch

We have recently identified the mechanism that switches cellular quiescence to cellular senescence (Imai et al. 2014). As discussed above, cell-cycle arrest in both quiescence and senescence is primarily regulated by the retinoblastoma (Rb) - tumour-suppressor activation. We have shown that in cultured fibroblasts, the molecular switch that determines whether cells undergo quiescence or senescence is mainly dependent on the level of intracellular ROS, which is determined by the expression of SOD2, a ROS-scavenging enzyme, evoked by the overlapping functions of the FoxO3a and FoxM1, fork-head transcription factors family (Lam et al. 2013). The activity of FoxO3a is regulated by its phosphorylation by AKT kinase, an important responder to mitogenic signals, which provokes nuclear export of FoxO3a thereby suppressing its activity (Naka et al. 2010). We found that FoxM1 is regulated by E2F transcription factor and strongly induced by mitogenic signals. In the absence of mitogenic signals, even when FoxM1 expression is repressed by the Rb pathway, FoxO3a transcription factor functions in the nuclear and prevents ROS production by maintaining SOD2 expression, leading to quiescence. However, if the Rb pathway is activated in the presence of mitogenic signals, in this situation FoxM1 expression is suppressed and FoxO3a is inactivated by AKT signalling, thus reducing SOD2 expression and consequently allowing ROS production. The accumulation of ROS elicits senescence through irreparable DNA damage (Imai et al. 2014).

15.5 Senescence-Associated Secretome

As mentioned above, there is no doubt that cellular senescence functions as an important tumour suppression mechanism. However, unlike apoptotic cells, senescent cells are known to remain viable for long periods of time. Indeed, we and others confirmed that senescent cells with high p16 expression accumulate in various aged organs (Yamakoshi et al. 2009; Krishnamurthy et al. 2004), with the accumulation of DNA damage markers. Of particular biological importance, it has recently become apparent that senescent cells exhibit increased expression of genes encoding a series of secreted proteins, such as inflammatory cytokines, chemokines, matrix remodeling factors, and growth factors, which can be beneficial or deleterious, depending on the biological context (Fig. 15.1; 8, 9; Acosta et al. 2008; Kuilman et al. 2008; Maier et al. 1990; Ohanna et al. 2011; Coppé et al. 2008). This newly identified senescent phenotype, the senescence-associated secretome, is observed in both replicative and oncogene-/stress-induced cellular senescence (Coppé et al. 2008) induced by DNA damage (Rodier et al. 2009). Although the mechanisms of the senescence-associated secretome have been initially investigated, mainly using normal fibroblasts (Acosta et al. 2008; Kuilman et al. 2008; Coppé et al. 2008), a series of papers reported that the secretome induced by

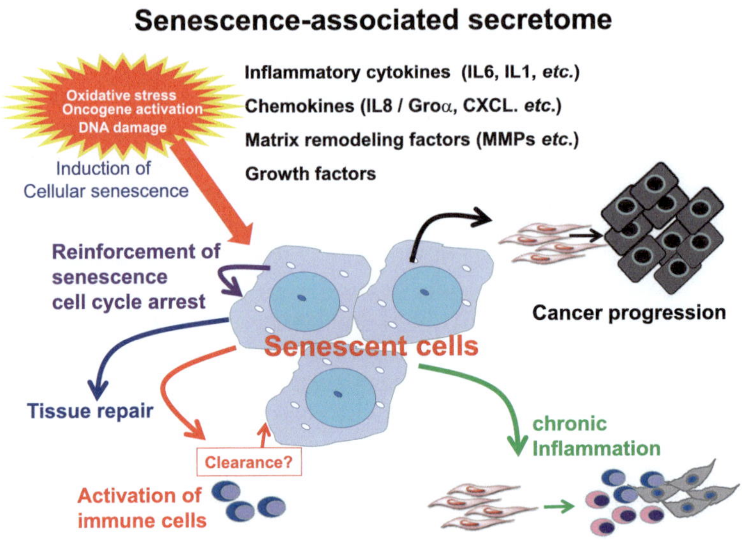

Fig. 15.1 The role of senescence-associated secretome. Cellular senescence is a state of irreversible cell proliferation arrest provoked by a variety of DNA damage-inducing signals. However, unlike apoptotic cells, senescent cells remain viable for a long time, and they eventually secrete a series of inflammatory cytokines, chemokines, and matrix-remodelling factors (senescence-associated secretory phenotype, SASP or senescence-associated secretome). Some of the secreted proteins can reinforce the senescence proliferation block. Other factors could promote the infiltration of immune cells, such as NK cells or macrophages, to clear the senescent cells. Moreover, these factors were recently found to accelerate tissue repair. However, some of the secreted factors, such as IL-6 and IL-8, are also known to promote malignant transformation, in cooperation with certain oncogene activation that may therefore contribute to form a cancer-promoting microenvironment. The sustained secretion of these factors could contribute to chronic inflammation

cellular senescence is observed in other types of cells, such as human melanocytes (Kuilman et al. 2008), human endothelial cells (Maier et al. 1990; Coppé et al. 2008), and cancerous cells (Ohanna et al. 2011; Chien et al. 2011).

15.5.1 Physiological Role of Senescence-Associated Secretome

The factors produced in the senescence-associated secretome reportedly suppress tumourigenesis by reinforcing the senescence proliferation arrest acting as a further tumour suppression mechanism (Acosta et al. 2008; Kuilman et al. 2008; Kortlever et al. 2006). Interesting moreover, are some recent reports demonstrating the physiological roles of senescence-associated secretome in vivo. Cellular senescence and senescence-associated secretome of hepatic stellate cells were induced in response to liver damage in carbon-tetrachloride–treated mouse liver

(Krizhanovsky et al. 2008). These senescent hepatic stellate cells seemed to limit the liver fibrosis by suppressing their proliferation, and secrete some extracellular matrix-degrading enzymes and NK cell-receptor ligands to enhance immune surveillance to clear the senescent cells (Krizhanovsky et al. 2008; Kang et al. 2011). Other reports showed the induction of senescence of skin fibroblasts in the process of wound healing. Skin wound rapidly induces matricellular protein CCN1 at the site of wound repair and provokes ROS production, thereby inducing cellular senescence of skin fibroblasts that express antifibrototic genes, to limit the skin fibrosis (Jun and Lau 2010). A recent paper also revealed the cellular senescence of fibroblasts on skin wound, and one of the senescence-associated secretome factors, PDGF-AA, could accelerate wound healing (Demaria et al. 2014). These reports suggest that physiologically, the senescence program seems to limit the fibrogenic response to acute tissue damage, and senescence-associated secretome factors promote normal tissue repair (Demaria et al. 2014). Furthermore, recently, two papers appeared describing developmental senescence in the apical ectodermal ridge and mesonephric tubes during physiological mammalian embryonic development (Muñoz-Espín et al. 2013; Storer et al. 2013) secreting some factors important for tissue development.

15.5.2 *Deleterious Role of Senescence-Associated Secretome*

In addition to these beneficial effects of senescence-associated secretome, there are some reports showing a dark side of senescence-associated secretome, because secreted factors such as IL-6 and IL-8 are also known to promote malignant transformation and proliferation of cancer cells in cooperation with certain oncogenes (Sparmann and Bar-Sagi 2004; Ancrile et al. 2007; Park et al. 2010). The factors secreted from senescent fibroblasts have been shown to induce an epithelial–mesenchymal transition (EMT), an important step in cancer progression and metastasis (Coppé et al. 2008). The senescence-associated secretome may also affect the microenvironment of cancer tissue. Indeed, the stromal fibroblasts of human ovarian cancer tissue were reportedly senescent cells that secreted a tumour-promoting chemokine (Yang et al. 2006), implying that the senescence-associated secretome might promote human ovarian cancer. The observations that the proteins secreted by senescent cells can promote degenerative hyperproliferative or metastatic changes in neighbouring cells (Coppé et al. 2008; Ohanna et al. 2011; Pribluda et al. 2013), indicate that the release of the senescence-associated secretome factors results in undesirable cancer-developing consequences. Recently the mechanism that switches beneficial or deleterious consequences was demonstrated. These secreted factors are known to act in a paracrine manner to reinforce the senescence cell-cycle arrest when p53 status of the target cells is normal, whereas on the other hand, they provide deleterious cell-nonautonomous side effects that promote tumourigenesis when surrounding cells harbour inactive p53 (Acosta et al. 2008; Kuilman et al. 2008; Chien et al. 2011; Kortlever et al. 2006;

Pribluda et al. 2013). Thus, the p53 status determines whether the senescence-associated secretome factors enhance the cellular senescence or the cancer progression (Pribluda et al. 2013).

15.5.3 The Mechanism to Induce the Expression of Senescence-Associated Secretome

Most of the genes encoding the senescence-associated secretome factors are known to be regulated by NFkB transcription factors (Lee et al. 2010; Ohanna et al. 2011; Chien et al. 2011). In addition to the role of NFkB, we identified an epigenetic mechanism to induce the expression of the senescence-associated secretome factors (Takahashi et al. 2012). DNA damage response signals in senescent cells induce proteasomal degradation of G9a and GLP, major histone H3K9 mono- and dimethyltransferases, through Cdc14B- and $p21^{Waf1/Cip1}$-dependent activation of APC/C^{Cdh1} ubiquitin ligase, thereby causing a global decrease in H3K9 dimethylation, an epigenetic mark for chromatic gene silencing. Interestingly, induction of IL-6 and IL-8, major players of the senescence-associated secretory phenotype, correlated with a decline of H3K9 dimethylation around the respective gene promoters and knockdown of Cdh1 abolished IL-6/IL-8 expression in senescent cells, suggesting that the APC/C^{Cdh1}–G9a/GLP axis plays crucial roles in aspects of senescent phenotype. These findings indicate a role for APC/C^{Cdh1} and reveal how the DNA damage response signals integrate with epigenetic processes to induce senescence-associated gene expression (Takahashi et al. 2012).

15.6 Obesity Promotes Cancer Progression Through Senescence-Associated Secretome

It has also been suggested that obesity could alter the character of the senescence-associated secretome, which might contribute to obesity-associated diseases, including cancer. In the following sections, our recent findings on the link between obesity and cancer are introduced as an example of the contribution of the senescence-associated secretome to cancer.

15.6.1 Obesity Promotes Liver Cancer and Senescence of Hepatic Stellate Cells

The population of obese individuals continues to rise worldwide (Calle et al. 2003), and obesity is known to increase the risk of not only diabetes and cardiovascular diseases, but also various cancers, including colorectal cancer, liver cancer, and prostate cancer (Samanic et al. 2004; Møller et al. 1994; Kaaks and Kühn 2014; Haslam and James 2005). Therefore, the prevention of obesity is clearly important for cancer prevention. Inflammatory cytokines such as IL-6 or other secreted proteins such as PAI-1, both of which are known to be the senescence-associated secretome factors, increased in obesity (Kortlever et al. 2006; Park et al. 2010; Yang et al. 2006), we thought that the senescence-associated secretome could be closely associated with obesity-induced inflammation and tumourigenesis.

We found that the long-term high-fat diet (HFD) accelerates development of cancers in mice treated with DMBA, a chemical carcinogen (Yoshimoto et al. 2013). In particular, all of the HFD-fed mice developed hepatocellular carcinomas (HCCs) after 30 weeks, whereas no tumours were found in the lean mice fed with normal diet (ND). Genetically obese mice fed the normal diet also developed HCCs, indicating that obesity itself, but not HFD, accelerated development of HCCs. Histological analysis revealed that activated hepatic stellate cells were observed in the stromal tissue around the lipid-rich hepatocellular carcinoma tumour areas. Interestingly, these hepatic stellate cells exhibited the signs of cellular senescence, the state of stable cell-cycle arrest that can be induced by DNA damage provoked by a variety of oncogenic stimuli. Interestingly, the senescent hepatic stellate cells were expressed multiple senescence-associated secretome factors, which in turn could promote the carcinogenesis of the surrounding hepatocytes. Indeed, depletion of senescent hepatic stellate cells or the deficiency of IL-1β, an upstream regulator of cytokine cascade, prevented the HCC development (Fig. 15.2; Yoshimoto et al. 2013).

15.6.2 Secondary Bile Acid from Gut Bacteria Promotes Liver Cancer Through Senescence-Associated Secretome

The important question is how obesity induced the senescence of hepatic stellate cells. We particularly investigated the alteration of the intestinal microbiota, which are known to alter the profile by obesity (Ley et al. 2006). Lipopolysaccharide, a component of Gram-negative bacteria in the intestinal microbiota, reportedly acts as a ligand of Toll-like receptor 4 (TLR4), to elicit inflammation and promote liver cancer formation, in a different liver cancer-inducing mouse model (administration of DEN plus carbon tetrachloride) (Dapito et al. 2012). The elimination of both the Gram-positive and Gram-negative intestinal microbiota by four antibiotic cocktails as well as the depletion of Gram-positive bacteria by vancomycin treatment significantly suppressed the development of HCC, suggesting that Gram-positive intestinal bacteria could play key roles in obesity-induced HCC development. Indeed, dietary or genetic obesity induces alterations of gut microbiota, thereby increasing the levels of deoxycholic acid (DCA), a gut bacterial metabolite known to cause

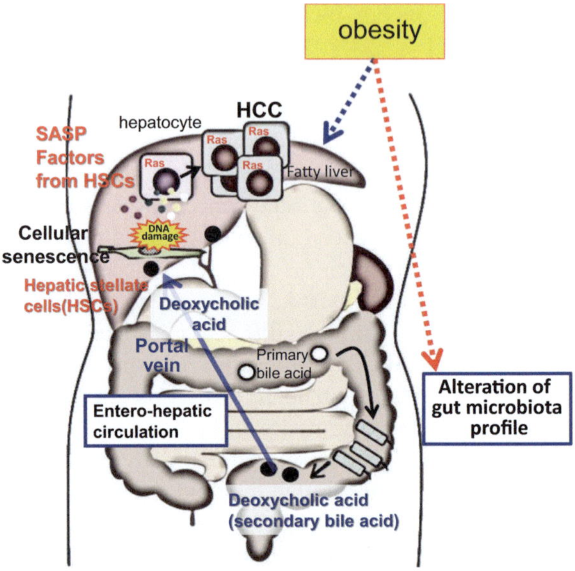

Fig. 15.2 Model for obesity-induced HCC development through senescence-associated secretome. Dietary or genetic obesity induces alteration of gut microbiota, thereby causing promotion of DCA production in the intestinal tract. Elevated levels of DCA provoke SASP factors in hepatic stellate cells through enterohepatic circulation, which in turn, secretes various inflammatory and tumour-promoting factors in the liver. This event, together with the activation of various oncogenic signalling pathways, results in the promotion of HCC development

DNA damage (Ridlon et al. 2006; Payne et al. 2007). The enterohepatic circulation of DCA provokes senescence-associated secretory phenotypes in hepatic stellate cells, which accelerate the secretion of various inflammatory and tumour-promoting factors in the liver, thus facilitating HCC development in mice after exposure to chemical carcinogen DMBA. Notably, reducing the serum DCA level efficiently prevents HCC development in obese mice, indicating that the DCA-SASP axis in HSCs has important roles in obesity-associated HCC development. Moreover, in humans, signs of senescence-associated secretome were also observed in the hepatic stellate cells in the area of HCC arising in patients with nonalcoholic steatohepatitis (NASH), indicating that a similar pathway may contribute to certain aspects of obesity-associated HCC development in humans as well (Yoshimoto et al. 2013). Interestingly, the histology of obesity-associated HCC accompanying the senescence-associated secretome in the hepatic stellate cells revealed less fibrotic in our mouse model, suggesting that our system could be a model of noncirrhotic NASH-associated HCC (Takuma and Nouso 2010).

15.7 Conclusion

In this review, I discussed the roles and the mechanisms of cellular senescence, and introduced our recent work as one of the examples of cancer progression promoted by the senescence-associated secretome. As described, senescence-associated secretome produced in the senescent cells influences various physiologies and pathologies. Indeed, the number of reports on the role of senescence-associated secretome is recently increasing. For example, the tumour-promoting effect of senescence-associated secretome on the prostate cancer microenvironment has been recently reported (Laberge et al. 2015). Considering the accumulation of senescent cells with age, it is possible that the role of the senescence-associated secretome could also contribute to the age-related increase in chronic inflammation and cancer.

References

Acosta JC, O'Loghlen A, Banito A et al (2008) Chemokine signaling via the CXCR2 receptor reinforces senescence. Cell 133:1006–1018

Adams PD (2009) Healing and hurting: molecular mechanisms, functions, and pathologies of cellular senescence. Mol Cell 36:2–14

Alcorta DA, Xiong Y, Phelps D et al (1996) Involvement of the cyclin-dependent kinase inhibitor p16 ^{INK4a} in replicative senescence of normal human fibroblasts. Proc Natl Acad Sci U S A 93:13742–13747

Ancrile B, Lim KH, Counter CM (2007) Oncogenic Ras-induced secretion of IL6 is required for tumorigenesis. Genes Dev 21:1714–1719

Calle EE, Rodriguez C, Walker-Thurmond K et al (2003) Overweight, obesity, and mortality from cancer in a prospectively studied cohort of U.S. adults. N Engl J Med 348:1625–1638

Campisi J (2005) Senescent cells, tumor suppression, and organismal aging: good citizens, bad neighbors. Cell 120:513–522

Chien Y, Scuoppo C, Wang X et al (2011) Control of the senescence-associated secretory phenotype by NF-kB promotes senescence and enhances chemosensitivity. Genes Dev 25:2125–2136

Classon M, Harlow E (2002) The retinoblastoma tumour suppressor in development and cancer. Nat Rev Cancer 2:910–917

Cobrinik D, Dowdy SF, Hinds PW et al (1992) The retinoblastoma protein and the regulation of cell cycling. Trends Biochem Sci 17:312–315

Collado M, Serrano M (2010) Senescence in tumours: evidence from mice and humans. Nat Rev Cancer 10:51–57

Coppé JP, Patil CK, Rodier F et al (2008) Senescence-associated secretory phenotypes reveal cell-nonautonomous functions of oncogenic RAS and the p53 tumor suppressor. PLoS Biol 6:2853–2868

d'Adda di Fagagna F (2008) Living on a break: cellular senescence as a DNA-damage response. Nat Rev Cancer 8:512–522

Dannenberg JH, van Rossum A, Schuijff L et al (2000) Ablation of the retinoblastoma gene family deregulates G(1) control causing immortalization and increased cell turnover under growth-restricting conditions. Genes Dev 14:3051–3064

Dapito DH, Mencin A, Gwak GY et al (2012) Promotion of hepatocellular carcinoma by the intestinal microbiota and TLR4. Cancer Cell 21:504–516
de Lange T (2005) Shelterin: the protein complex that shapes and safeguards human telomeres. Genes Dev 19:2100–2110
Demaria M, Ohtani N, Youssef SA et al (2014) An essential role for senescent cells in optimal wound healing through secretion of PDGF-AA. Dev Cell 31:722–733
Deng Y, Chan SS, Chang S (2008) Telomere dysfunction and tumour suppression: the senescence connection. Nat Rev Cancer 8:450–458
el-Deiry WS, Tokino T, Velculescu VE et al (1993) WAF1, a potential mediator of p53 tumor suppression. Cell 75:817–825
Finkel T, Serrano M, Blasco MA (2007) The common biology of cancer and ageing. Nature 448:767–774
Fumagalli M, Rossiello F, Clerici M et al (2012) Telomeric DNA damage is irreparable and causes persistent DNA-damage-response activation. Nat Cell Biol 14:355–365
Gil J, Peters G (2006) Regulation of the INK4b-ARF-INK4a tumour suppressor locus: all for one or one for all. Nat Rev Mol Cell Biol 7:667–677
Haff RF, Swim HE (1956) Serial propagation of 3 strains of rabbit fibroblasts; their susceptibility to infection with vaccinia virus. Proc Soc Exp Biol Med 93:200–204
Hara E, Smith R, Parry D, Tahara H, Stone S, Peters G (1996) Regulation of $p16^{CDKN2}$ expression and its implications for cell immortalization and senescence. Mol Cell Biol 16:859–867
Hara E, Tsurui H, Shinozaki A et al (1991) Cooperative effect of antisense-Rb and antisense-p53 oligomers on the extension of life span in human diploid fibroblasts, TIG-1. Biochem Biophys Res Commun 179:528–534
Harley CB, Futcher AB, Greider CW (1990) Telomeres shorten during ageing of human fibroblasts. Nature 345:458–460
Haslam DW, James WP (2005) Obesity. Lancet 366:1197–1209
Hayflick L (1965) The limited in vitro lifetime of human diploid cell strains. Exp Cell Res 37:614–636
Herbig U, Sedivy JM (2006) Regulation of growth arrest in senescence: telomere damage is not the end of the story. Mech Ageing Dev 127:16–24
Imai Y, Takahashi A, Hanyu A et al (2014) Crosstalk between the Rb pathway and AKT signaling forms a quiescence-senescence switch. Cell Rep 7:194–207
Jun JI, Lau LF (2010) The matricellular protein CCN1 induces fibroblast senescence and restricts fibrosis in cutaneous wound healing. Nat Cell Biol 12:676–685
Kaaks R, Kühn T (2014) Epidemiology: obesity and cancer-the evidence is fattening up. Nat Rev Endocrinol 10:644–645
Kang TW, Yevsa T, Woller N et al (2011) Senescence surveillance of pre-malignant hepatocytes limits liver cancer development. Nature 479:547–551
Kortlever RM, Higgins PJ, Bernards R (2006) Plasminogen activator inhibitor-1 is a critical downstream target of p53 in the induction of replicative senescence. Nat Cell Biol 8:877–884
Krishnamurthy J, Torrice C, Ramsey MR et al (2004) Ink4a/Arf expression is a biomarker of aging. J Clin Invest 114:1299–1307
Krizhanovsky V, Yon M, Dickins RA et al (2008) Senescence of activated stellate cells limits liver fibrosis. Cell 134:657–667
Kuilman T, Michaloglou C, Vredeveld LC et al (2008) Oncogene-induced senescence relayed by an interleukin-dependent inflammatory network. Cell 133:1019–1031
Kuilman T, Peeper DS (2009) Senescence-messaging secretome: SMS-ing cellular stress. Nat Rev Cancer 9:81–94
Kuilman T, Michaloglou C, Mooi WJ et al (2010) The essence of senescence. Genes Dev 24:2463–2479
Laberge RM, Sun Y, Orjalo AV et al (2015) MTOR regulates the pro-tumorigenic senescence-associated secretory phenotype by promoting IL1A translation. Nat Cell Biol 17:1049–1061

Lam EW, Brosens JJ, Gomes AR et al (2013) Forkhead box proteins: tuning forks for transcriptional harmony. Nat Rev Cancer 13:482–495

Lee MY, Wang Y, Vanhoutte PM (2010) Senescence of cultured porcine coronary arterial endothelial cells is associated with accelerated oxidative stress and activation of NFkB. J Vasc Res 47:287–298

Ley RE, Turnbaugh PJ, Klein S, Gordon JI (2006) Microbial ecology: human gut microbes associated with obesity. Nature 444:1022–1023

Lloyd AC (2002) Limits to lifespan. Nat Cell Biol 4:E25-7

Maier JA, Voulalas P, Roeder D et al (1990) Extension of the life-span of human endothelial cells by an interleukin-1 a antisense oligomer. Science 249:1570–1574

McConnell BB, Gregory FJ, Stott FJ et al (1999) Induced expression of $p16^{INK4a}$ inhibits both CDK4- and CDK2-associated kinase activity by reassortment of cyclin-CDK-inhibitor complexes. Mol Cell Biol 19:1981–1989

Mitra J, Dai CY (1999) Somasundaram K, El-Deiry WS, Satyamoorthy K, Herlyn M, Enders GH. Induction of $p21^{WAF1/CIP1}$ and inhibition of Cdk2 mediated by the tumor suppressor $p16^{INK4a}$. Mol Cell Biol 19:3916–3928

Møller H, Mellemgaard A, Lindvig K et al (1994) Obesity and cancer risk: a Danish record-linkage study. Eur J Cancer 30A:344–350

Muñoz-Espín D, Serrano M (2014) Cellular senescence: from physiology to pathology. Nat Rev Mol Cell Biol 15:482–496

Muñoz-Espín D, Cañamero M, Maraver A et al (2013) Programmed cell senescence during mammalian embryonic development. Cell 155:1104–1118

Naka K, Hoshii T, Muraguchi T et al (2010) TGF-beta-FOXO signalling maintains leukaemia-initiating cells in chronic myeloid leukaemia. Nature 463:676–680

Noda A, Ning Y, Venable SF et al (1994) Cloning of senescent cell-derived inhibitors of DNA synthesis using an expression screen. Exp Cell Res 211:90–98

Ohanna M, Giuliano S, Bonet C et al (2011) Senescent cells develop a PARP-1 and nuclear factor-kB-associated secretome (PNAS). Genes Dev 25:1245–1261

Ohtani N, Zebedee Z, Huot TJ et al (2001) Opposing effects of Ets and Id proteins on $p16^{INK4a}$ expression during cellular senescence. Nature 409:1067–1070

Park EJ, Lee JH, Yu GY et al (2010) Dietary and genetic obesity promote liver inflammation and tumorigenesis by enhancing IL-6 and TNF expression. Cell 140:197–208

Parrinello S, Samper E, Krtolica A et al (2003) Oxygen sensitivity severely limits the replicative lifespan of murine fibroblasts. Nat Cell Biol 5:741–747

Payne CM, Weber C, Crowley-Skillicorn C et al (2007) Deoxycholate induces mitochondrial oxidative stress and activates NF-kappaB through multiple mechanisms in HCT-116 colon epithelial cells. Carcinogenesis 28:215–222

Pribluda A, Elyada E, Wiener Z et al (2013) A senescence-inflammatory switch from cancer-inhibitory to cancer-promoting mechanism. Cancer Cell 24:242–256

Ridlon JM, Kang DJ, Hylemon PB (2006) Bile salt biotransformations by human intestinal bacteria. J Lipid Res 47:241–259

Rodier F, Coppe JP, Patil CK et al (2009) Persistent DNA damage signalling triggers senescence-associated inflammatory cytokine secretion. Nat Cell Biol 11:973–979

Rodier F, Campisi J (2011) Four faces of cellular senescence. J Cell Biol 192:547–556

Sage J, Mulligan GJ, Attardi LD et al (2000) Targeted disruption of the three Rb-related genes leads to loss of G(1) control and immortalization. Genes Dev 14:3037–3050

Sage J, Miller AL, Perez-Mancera PA et al (2003) Acute mutation of retinoblastoma gene function is sufficient for cell cycle re-entry. Nature 424:223–228

Samanic C, Gridley G, Chow WH et al (2004) Obesity and cancer risk among white and black United States veterans. Cancer Causes Control 15:35–43

Serrano M, Lin AW, McCurrach ME et al (1997) Oncogenic ras provokes premature cell senescence associated with accumulation of p53 and $p16^{INK4a}$. Cell 88:593–602

Serrano M, Blasco MA (2001) Putting the stress on senescence. Curr Opin Cell Biol 13:748–753

Shay JW, Pereira-Smith OM, Wright WE (1991) A role for both RB and p53 in the regulation of human cellular senescence. Exp Cell Res 196:33–39

Sherr CJ, Roberts JM (1999) CDK inhibitors: positive and negative regulators of G1-phase progression. Genes Dev 13:1501–1512

Sherr CJ, Depinho RA (2000) Cellular senescence: mitotic clock or culture shock? Cell 102:407–410

Sparmann A, Bar-Sagi D (2004) Ras-induced interleukin-8 expression plays a critical role in tumor growth and angiogenesis. Cancer Cell 6:447–458

Storer M, Mas A, Robert-Moreno A et al (2013) Senescence is a developmental mechanism that contributes to embryonic growth and patterning. Cell 155:1119–1130

Takahashi A, Imai Y, Yamakoshi K et al (2012) DNA damage signaling triggers degradation of histone methyltransferases through APC/C(Cdh1) in senescent cells. Mol Cell 45:123–131

Takahashi A, Ohtani N, Yamakoshi K et al (2006) Mitogenic signalling and the $p16^{INK4a}$-Rb pathway cooperate to enforce irreversible cellular senescence. Nat Cell Biol 8:1291–1297

Takeuchi S, Takahashi A, Motoi N et al (2010) Intrinsic cooperation between $p16^{INK4a}$ and $p21^{Waf1/Cip1}$ in the onset of cellular senescence and tumor suppression in vivo. Cancer Res 70:9381–9390

Takuma Y, Nouso K (2010) Nonalcoholic steatohepatitis-associated hepatocellular carcinoma: our case series and literature review. World J Gastroenterol 16:1436–1441

Wright WE, Shay JW (2000) Telomere dynamics in cancer progression and prevention: fundamental differences in human and mouse telomere biology. Nat Med 6:849–851

Yamakoshi K, Takahashi A, Hirota F et al (2009) Real-time in vivo imaging of $p16^{Ink4a}$ reveals cross talk with p53. J Cell Biol 186:393–407

Yoshimoto S, Loo TM, Atarashi K et al (2013) Obesity-induced gut microbial metabolite promotes liver cancer through senescence secretome. Nature 499:97–101

Yang G, Rosen DG, Zhang Z et al (2006) The chemokine growth-regulated oncogene 1 (Gro-1) links RAS signaling to the senescence of stromal fibroblasts and ovarian tumorigenesis. Proc Natl Acad Sci U S A 103:16472–16477

Chapter 16
Establishment of Diagnosis for Early Metastasis

Sachie Hiratsuka

Abstract Primary tumours secrete many cytokines and chemokines including CCL2, VEGF, TNFα, and TGF-β. In the natural course of primary tumour growth, these factors stimulate bone marrow cells and tissue-resident cells including immune cells and endothelial cells in the distant host tissues. Despite ubiquitously spreading in blood vessels, the secreted factors from primary tumours mediate discrete hyperpermeability foci by several overlapping cascades such as S100A8-SAA3-TLR4/MD-2 and VEGF-FAK-E-selectin. The former is that these secreted factors stimulate CCR2 in potentially hyperpermeable areas, further inducing the secretion of SAA3 and S100A8, which increase vascular permeability via TLR4/MD-2. The latter is that tumour-derived VEGF generates the permeable foci with E-selectin upregulation through endothelial FAK in the premetastatic lungs. Blocking the cascade of the permeability in hyperpermeability regions before metastasis results in the reduction of tumour cell homing. Thus, it is an advantage to target very early metastatic or premetastatic phase.

Keywords Premetastatic • Permeability • Innate immune • S100A8/A9 • SAA3 • Fibrinogen

16.1 Introduction

Distant metastasis is the major cause of cancer mortality, and the lung is a common site for metastatic disease (Fidler 2003; Hynes 2003; Gupta and Massague 2006; Gupta et al. 2007). The metastatic process includes a cascade of sequential events, including tumour cell intravasation from the primary site, tumour cell circulation, homing to distant organs, extravasation into target secondary organs, and subsequent colonisation, growth, and angiogenesis (Couzin 2003; Kang and Massague 2004; Fidler 2003).

S. Hiratsuka (✉)
Department of Pharmacology, Tokyo Women's Medical University School of Medicine, 8-1 Kawada-cho, Shinjuku-ku, Tokyo 162-8666, Japan

PRESTO, Japan Science and Technology Agency, Kawaguchi 332-0012, Japan
e-mail: shira@research.twmu.ac.jp

M. Miyasaka, K. Takatsu (eds.), *Chronic Inflammation*,
DOI 10.1007/978-4-431-56068-5_16

In an adaptation of Paget's 'seed and soil' concept (Paget 1889) – which states that tumour cells (the 'seeds') form metastatic nodules if they encounter hospitable 'soil' in a distant organ – several groups have proposed that tumours prepare the distant soil in a so-called 'premetastatic phase'. This term describes a period during which a distant organ is rendered more hospitable towards circulating tumour cells with metastatic potential, facilitated by the presence of a distant primary tumour. Several mechanisms may contribute towards creation of this state in the lung, many related to primary tumour production of cytokines such as vascular endothelial growth factor (VEGF), transforming growth factor-β (TGF-β), tumour necrosis factor α (TNFα), angiopoietin-2, and amine oxidases: all of which can modify the lung environment prior to tumour cell arrival (Hiratsuka et al. 2002, 2006, 2008; Kaplan et al. 2005; Erler et al. 2009; Kim et al. 2009; Huang et al. 2009). These factors in turn can promote pulmonary inflammation, vascular hyperpermeability, and recruitment of immune cells to the lung, which can encourage subsequent seeding of tumour cells.

When tumours spontaneously metastasise to the host lungs, they often present in an oligometastatic pattern or even as solitary metastatic nodules (Schlappack et al. 1988; Jaklitsch et al. 2001). In contrast, premetastatic events such as bone–marrow derived cell infiltration and activation of inflammatory pathways affect the lungs in a diffuse fashion (Hiratsuka et al. 2002, 2006, 2008; Huang et al. 2009; Kaplan et al. 2005; Kim et al. 2009; Erler et al. 2009), and all pulmonary microvessels are exposed to a steady stream of circulating tumour cells (Riethdorf and Pantel 2008; Mostert et al. 2009). The determinants of discrete oligometastasis formation remain unidentified. My hypothesis is that there may be a distinct series of mechanisms that facilitates homing of circulating tumour cells to specific, predetermined sites within the lungs.

My research group demonstrates that metastatic primary tumours, and soluble factors released by them, can induce distinct macroscopic regions of vascular hyperpermeability in the lungs. We also show that these regions serve as discrete fertile fields of premetastatic 'soil', demonstrating increased following tumour cell homing. In addition, the similar fibrinogen deposition pattern by permeability foci was observed in tumour-bearing mouse lung and human lung, indicating that the risk area might be predicted in the patients before metastasis.

16.1.1 Detection of Hyperpermeable Foci

In order to investigate the effects of primary tumours on the lungs, we implanted three murine tumour types with metastatic potential (LLC, Lewis lung carcinoma; E0771, mammary carcinoma; B16, melanoma), in immunocompetent C57BL/6 mice. Once tumours reached approximately 6–7 mm in diameter, we injected Evans Blue (EB) intravenously and then excised the lungs 3 h later. In all models, we found focal regions of increased EB leakage (Fig. 16.1).

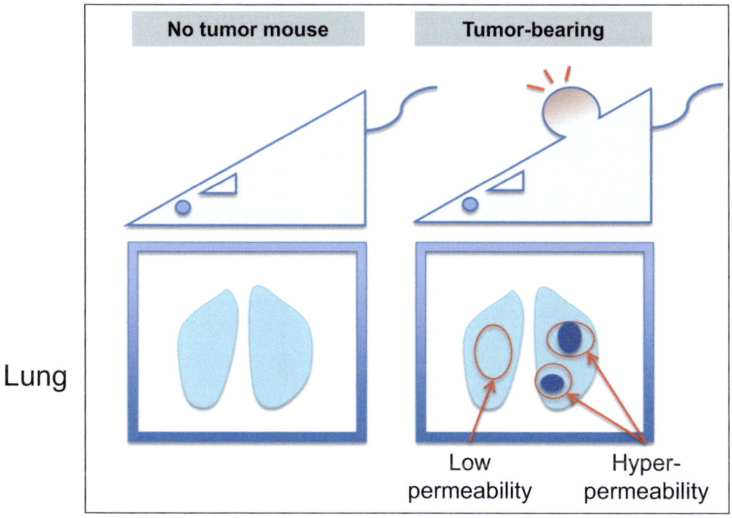

Fig. 16.1 Hyperpermeability spots in the premetastatic mouse lungs. Some hyperpermeability areas visualised by Evans Blue dye injection were found in LLC or E0771-bearing mouse lungs

Because we could not rule out that focal leakage of EB was a consequence of metastatic tumour cell seeding in these models, we repeated the experiments after intravenous injection of tumour-conditioned medium (TCM) from E0771 (ETCM) or LLC (LTCM) in nontumour-bearing mice. Both ETCM and LTCM injections induced hyperpermeable regions with focal EB leakage in the lungs.

16.1.2 Circulating Tumour Cells Accumulate the Foci

To test if hyperpermeable foci served as sites for tumour cell homing, we developed a three-step assay system (Fig. 16.2). First, we applied a permeability inducer (tumour implantation or intravenous injection of TCM). Second, we infused EB systemically to detect and measure hyperpermeability of the lung vasculature. Finally, we infused fluorescence-labeled metastatic tumour cells and measured the number of cells that homed to the lungs. We found increased numbers of fluorescent tumour cells homing in the lungs of tumour-bearing and TCM-stimulating mice. It is important to note that the tumour cells preferentially homed to areas of high vascular permeability (Fig. 16.2).

To determine the localisation of homing tumour cells about the permeable vessels, we infused pegylated PEG-microbeads instead of EB, after stimulation of focal hyperpermeability. Consistent with EB-leakage measurements, the higher density of tumour cells was observed in the region of the high density of extravasated PEG-microbeads, after TCM infusion.

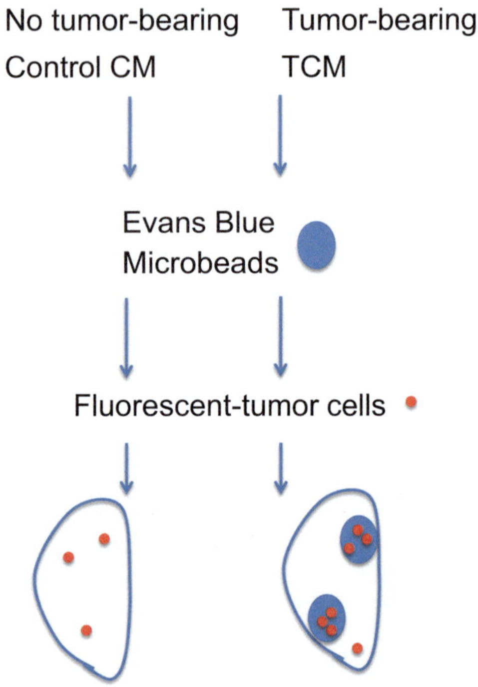

Fig. 16.2 Tumour cells prefer to accumulate in the hyperpermeability regions Acute assay system of circulating tumour cells accumulation to the hyperpermeability spots First TCM or Tumour-bearing situation induced an hyperpermeability area in normal mouse lungs. Next, labelled tumour cells were injected and the numbers of the cells in EB-high and EB-low areas were counted. Tumour cells prefer to accumulate in the EB-high area

16.1.3 Blocking Vascular Permeability Reduces Tumour Cell Homing – KO of Endothelial FAK

Src kinase-FAK complex activation mediates vascular permeability via TCM including VEGF (Eliceiri et al. 2002). To study the specific contribution of FAK in endothelial cells towards the formation of hyperpermeable foci and subsequent tumour cell homing, we developed an inducible transgenic mouse model. This model allowed in vivo regulation of endothelial cell-mediated vascular permeability. Doxycycline (Dox)-mediated endothelial-specific FRNK (the dominant negative form of FAK) overexpression significantly reduced ETCM- and LTCM-induced formation of the hyperpermeability foci in the lungs. Moreover, this reduction in endothelial cell-mediated hyperpermeability led to a significant reduction in tumour cell homing to the lungs after metastatic tumour cell infusion.

16.1.4 Molecular-Based Inflammation in the Foci

Because the inflammatory mediators all lie downstream of signalling factor NF-kB, the master regulator of inflammation, we next determined the activation of NF-kB by measuring phosphorylation of IkBα(p-IkBα). We found that ETCM and LTCM

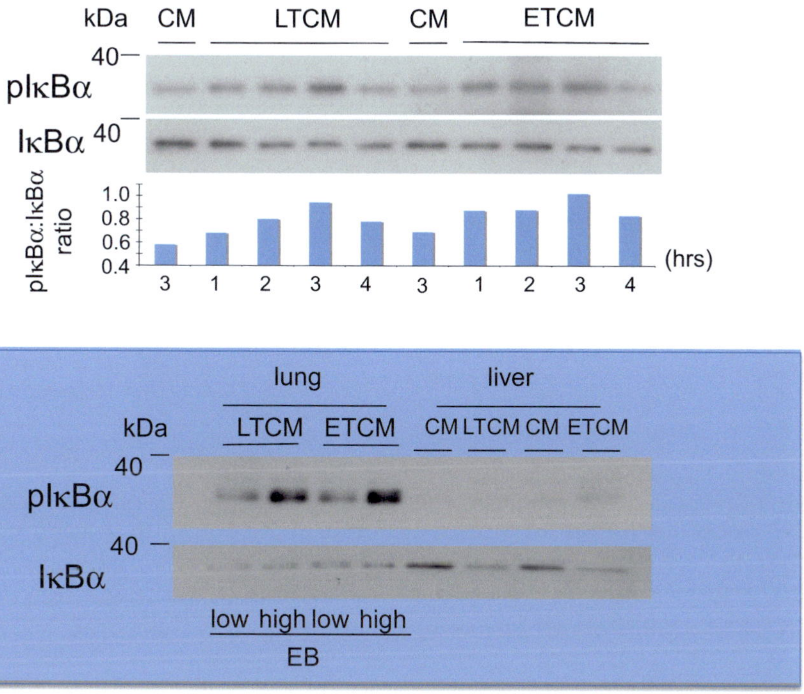

Fig. 16.3 Hyperpermeable regions in the lung display increased levels of inflammatory response The kinetics of phosphorylation of IkBα after stimulation by ETCM or LTCM in lungs (*upper*). The levels of IkBα phosphorylation were increased in the areas of high EB leakage in the lung 3 h after TCM stimulation (*lower*)

activate the NF-kB pathway in the lungs, and this NF-kB activation was largely restricted to the EB-high areas and was specific to the lungs rather than liver (Fig. 16.3).

16.1.4.1 VEGF-Mediating Permeability (Hiratsuka et al. 2011)

Using ELISA, we found that TCM contained high concentrations of vascular endothelial growth factor (VEGF) and placental growth factor (PlGF, data not shown). These factors are known to increase vascular permeability directly or indirectly (Nagy et al. 2008; Carmeliet 2005; Ferrara et al. 2003). Infusion of recombinant (r)VEGF or rPlGF also induced discrete foci of hyperpermeability with focal EB leakage in the lungs. The increase in EB leakage was specific to the lungs and was not seen in other organs after either intravenous or intracardiac injection of TCM. Treatment of tumour-bearing mice with an anti-VEGF monoclonal blocking antibody inhibited the lung vascular hyperpermeability. This result supports that VEGF mediates this effect in part. (Fig. 16.4)

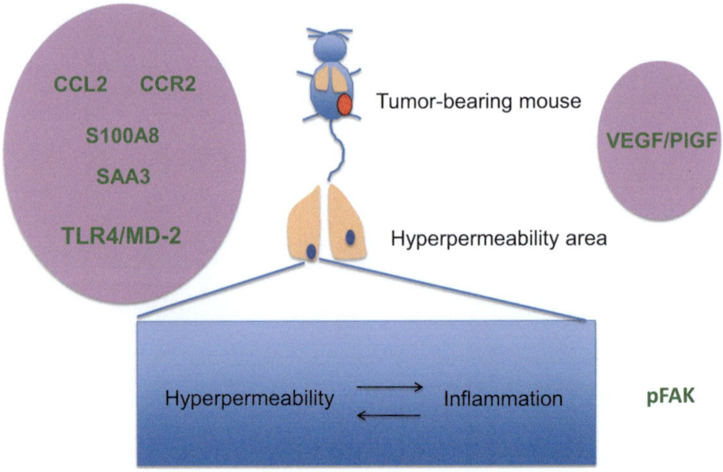

Fig. 16.4 Innate immune system-mediating hyperpermeability regions in the premetastatic lungs The permeability is regulated by S100A8-SAA3-mediating and VEGF-mediating signalling pathways. There may be crosstalk between these cascades

16.1.4.2 S100A8-SAA3-Mediating Permeability (Hiratsuka et al. 2013)

To find another pathway in addition to the VEGF cascade, we compared the gene expression levels of the receptors between hyperpermeable (H) and poorly permeable regions (L) in the tumour-bearing mouse lungs. C-C chemokine receptor type 2 (CCR2) showed the largest increase in the gene expression between the hyperpermeable area and the low permeable area in the tumour-bearing mice. To search a correlation between EB-leakage regions and the localisation of CCR2 and its ligand CCL2, we carried out immunohistochemical analyses. Among them, the CCR2-positive cells expressed CD45 (˜65 %), MECA32 (˜35 %), and Lyve-1-positive lymphatic cells (~1.5 %). To examine the functional role of this system, we used CCL2 and CCR2 knockout mice. In association with the lack of hyperpermeable foci in the tumour-bearing knockout mice, tumour cell homing was suppressed. Furthermore, we found that S100A8 and SAA3, which are downstream molecules of the CCL2-cascade, produced strong permeability responses. (Fig. 16.4)

16.1.5 The Possibility of Hyperpermeable Foci in Patients (Hiratsuka et al. 2013)

We tried to detect hyperpermeable areas by examining for regions in which fibrinogen expression was induced. Fibrinogen has been reported to be associated with microvascular leakage (Tyagi et al. 2008; Wygrecka et al. 2008). We found

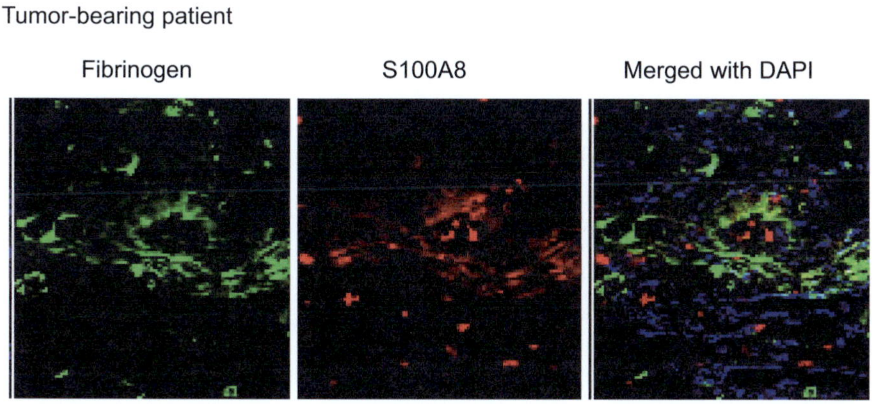

Fig. 16.5 Increased S100A8 and fibrinogen expression were observed in the lung endothelial cells of the tumour-bearing patient

that fibrinogen expression was increased in the hyperpermeable regions of the tumour-bearing mouse lungs compared with the low permeable regions and that its upregulation occurred in the same regions as the upregulation of CCR2. Based on these results, we applied the criteria for the analysis of human samples.

To search whether distant primary tumours stimulate the colocalised induction of fibrinogen and CCR2 expression in tumour-bearing human lungs, we examined the expression of both molecules in healthy lung lobes from patients who carried tumours in their extrapulmonary organs. It should be noted that we detected relatively low levels of fibrinogen and CCR2 in lungs from noncancer patients. In contrast, the colocalisation of upregulated fibrinogen and CCR2 expression was detected in the lungs of patients with tumours. We also observed that the tumour-bearing lungs exhibited significantly increased fibrinogen and S100A8 expression levels compared with the nontumour-bearing lungs (Fig. 16.5).

16.2 Discussion

16.2.1 Implications for Localised Lung Metastasis Formation and Therapy

The present study depicts a series of mechanisms by which primary tumours might give rise to focal lung metastases. We have observed that tumour-secreted factors, including VEGF, induce discrete foci of vascular hyperpermeability in the lungs, before metastasis has occurred, via FAK in lung endothelial cells. These areas also harbour an inflammation-like reaction that is mediated by the S100A8/A9-SAA3-TLR4/MD-2 cascade, which further exacerbates vascular permeability and tumour cell homing (Fig. 16.6). Our findings lead us to propose that primary-tumour–

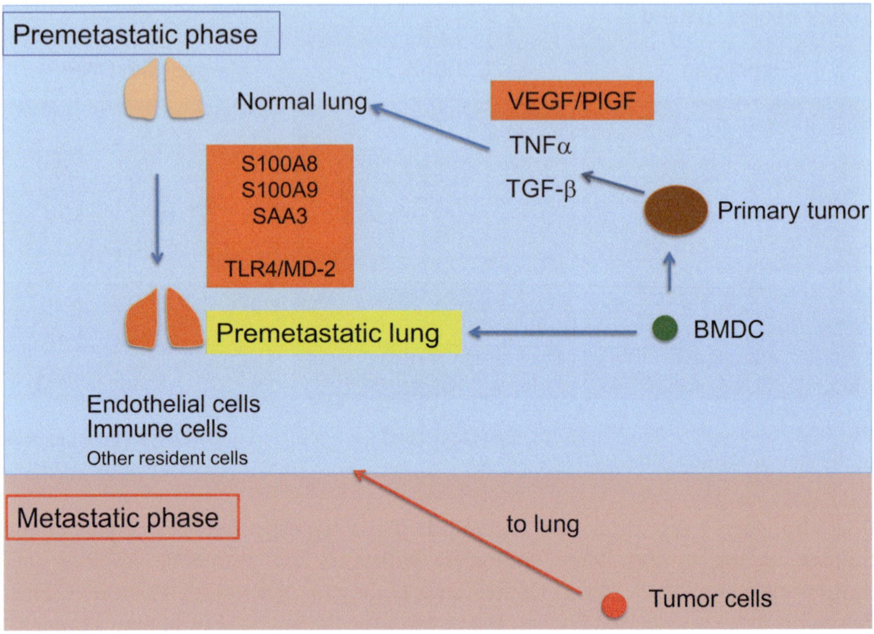

Fig. 16.6 Distant primary tumours make preferable soil in premetastatic phase. This scheme depicts molecules contributing to the premetastatic soil. Some cytokines and chemokines, derived from primary tumours stimulate distant target organs such as the lung and liver. Inflammatory factors related to the innate immune system are secreted from specific organs. Before metastasis, lungs become a soil with abundant inflammatory factors

induced pulmonary hyperpermeability is (a) a focal phenomenon and (b) critical to determining the sites for tumour cell homing.

Our results suggest a possible permeable area may exist in tumour-bearing patients. Further understanding of the cascade of pathophysiological changes in the metastatic site stroma and the molecular determinants of the early steps of metastatic cell dissemination to the lungs may greatly impact on future strategies for preventing and treating metastasis.

References

Carmeliet P (2005) Angiogenesis in life, disease and medicine. Nature 438:932–936

Couzin J (2003) Medicine. Tracing the steps of metastasis, cancer's menacing ballet. Science 299: 1002–1006

ELICEIRI BP, Puente XS, Hood JD, Stupack DG, Schlaepfer DD, Huang XZ, Sheppard D, Cheresh DA (2002) Src-mediated coupling of focal adhesion kinase to integrin alpha(v)beta5 in vascular endothelial growth factor signaling. J Cell Biol 157:149–160

Erler JT, Bennewith KL, Cox TR, Lang G, Bird D, Koong A, Le QT, Giaccia AJ (2009) Hypoxia-induced lysyl oxidase is a critical mediator of bone marrow cell recruitment to form the premetastatic niche. Cancer Cell 15:35–44

Ferrara N, Gerber HP, Lecouter J (2003) The biology of VEGF and its receptors. Nat Med 9: 669–676

Fidler IJ (2003) The pathogenesis of cancer metastasis: the 'seed and soil' hypothesis revisited. Nat Rev Cancer 3:453–458

Gupta GP, Massague J (2006) Cancer metastasis: building a framework. Cell 127:679–695

Gupta GP, Nguyen DX, Chiang AC, Bos PD, Kim JY, Nadal C, Gomis RR, Manova-Todorova K, Massague J (2007) Mediators of vascular remodelling co-opted for sequential steps in lung metastasis. Nature 446:765–770

Hiratsuka S, Nakamura K, Iwai S, Murakami M, Itoh T, Kijima H, Shipley JM, Senior RM, Shibuya M (2002) MMP9 induction by vascular endothelial growth factor receptor-1 is involved in lung-specific metastasis. Cancer Cell 2:289–300

Hiratsuka S, Watanabe A, Aburatani H, Maru Y (2006) Tumour-mediated upregulation of chemoattractants and recruitment of myeloid cells predetermines lung metastasis. Nat Cell Biol 8:1369–1375

Hiratsuka S, Watanabe A, Sakurai Y, Akashi-Takamura S, Ishibashi S, Miyake K, Shibuya M, Akira S, Aburatani H, Maru Y (2008) The S100A8-serum amyloid A3-TLR4 paracrine cascade establishes a pre-metastatic phase. Nat Cell Biol 10:1349–1355

Hiratsuka S, Goel S, Kamoun WS, Maru Y, Fukumura D, Duda DG, Jain RK (2011) Endothelial focal adhesion kinase mediates cancer cell homing to discrete regions of the lungs via E-selectin up-regulation. Proc Natl Acad Sci U S A 108:3725–3730

Hiratsuka S, Ishibashi S, Tomita T, Watanabe A, Akashi-Takamura S, Murakami M, Kijima H, Miyake K, Aburatani H, Maru Y (2013) Primary tumours modulate innate immune signalling to create pre-metastatic vascular hyperpermeability foci. Nat Commun 4:1853

Huang Y, Song N, Ding Y, Yuan S, Li X, Cai H, Shi H, Luo Y (2009) Pulmonary vascular destabilization in the premetastatic phase facilitates lung metastasis. Cancer Res 69:7529–7537

Hynes RO (2003) Metastatic potential: generic predisposition of the primary tumor or rare, metastatic variants-or both? Cell 113:821–823

Jaklitsch MT, Mery CM, Lukanich JM, Richards WG, Bueno R, Swanson SJ, Mentzer SJ, Davis BD, Allred EN, Sugarbaker DJ (2001) Sequential thoracic metastasectomy prolongs survival by re-establishing local control within the chest. J Thorac Cardiovasc Surg 121:657–667

Kang Y, Massague J (2004) Epithelial-mesenchymal transitions: twist in development and metastasis. Cell 118:277–279

Kaplan RN, Riba RD, Zacharoulis S, Bramley AH, Vincent L, Costa C, Macdonald DD, Jin DK, Shido K, Kerns SA, Zhu Z, Hicklin D, Wu Y, Port JL, Altorki N, Port ER, Ruggero D, Shmelkov SV, Jensen KK, Rafii S, Lyden D (2005) VEGFR1-positive haematopoietic bone marrow progenitors initiate the pre-metastatic niche. Nature 438:820–827

Kim S, Takahashi H, Lin WW, Descargues P, Grivennikov S, Kim Y, Luo JL, Karin M (2009) Carcinoma-produced factors activate myeloid cells through TLR2 to stimulate metastasis. Nature 457:102–106

Mostert B, Sleijfer S, Foekens JA, Gratama JW (2009) Circulating tumor cells (CTCs): detection methods and their clinical relevance in breast cancer. Cancer Treat Rev 35:463–474

Nagy JA, Benjamin L, Zeng H, Dvorak AM, Dvorak HF (2008) Vascular permeability, vascular hyperpermeability and angiogenesis. Angiogenesis 11:109–119

Paget S (1889) The distribution of secondary growths in cancer of the breast. Lancet 1:571–573

Riethdorf S, Pantel K (2008) Disseminated tumor cells in bone marrow and circulating tumor cells in blood of breast cancer patients: current state of detection and characterization. Pathobiology 75:140–148

Schlappack OK, Baur M, Steger G, Dittrich C, Moser K (1988) The clinical course of lung metastases from breast cancer. J Mol Med 66:790–795

Tyagi N, Roberts AM, Dean WL, Tyagi SC, Lominadze D (2008) Fibrinogen induces endothelial cell permeability. Mol Cell Biochem 307:13–22

Wygrecka M, Jablonska E, Guenther A, Preissner KT, Markart P (2008) Current view on alveolar coagulation and fibrinolysis in acute inflammatory and chronic interstitial lung diseases. Thromb Haemost 99:494–501

Chapter 17
Non-autonomous Tumor Progression by Oncogenic Inflammation

Shizue Ohsawa and Tatsushi Igaki

Abstract Defects in Mitochondrial respiratory function is frequently observed in human cancers. However, the mechanism by which mitochondrial dysfunction contributes to tumour growth and progression has been unclear. Recent studies in *Drosophila* epithelium have uncovered that mitochondrial defects induce tumour progression of surrounding tissue by cooperating with oncogenic Ras. Simultaneous Ras activation and mitochondrial dysfunction cause chronic 'oncogenic inflammation', which induces overgrowth and metastatic behavior in neighbouring benign tumours via upregulation of an IL-6-like inflammatory cytokine Upd. Further genetic analyses revealed that the nonautonomous tumour progression by oncogenic inflammation is caused through cellular senescence and senescence-associated secretary phenotype (SASP). These findings provide a novel mechanistic basis for tumour progression through cell–cell communication triggered by Ras activation and mitochondrial dysfunction, frequent alterations in human cancers.

Keywords Tumour progression • Mitochondrial dysfunction • Inflammatory cytokine • Cell–cell communication • *Drosophila*

17.1 Introduction

It is becoming increasingly clear that tumour progression is achieved not only by the sequential acquisition of genetic alterations but also by oncogenic cell–cell interactions. For instance, interactions between epithelial tumour cells and surrounding stromal cells, such as inflammatory cells, immune cells, and fibroblasts,

S. Ohsawa
Laboratory of Genetics, Graduate School of Biostudies, Kyoto University, Yoshida-Konoecho-cho, Sakyo-ku, Kyoto 606-8501, Japan

T. Igaki (✉)
Laboratory of Genetics, Graduate School of Biostudies, Kyoto University, Yoshida-Konoecho-cho, Sakyo-ku, Kyoto 606-8501, Japan

PRESTO, Japan Science and Technology Agency, Kawaguchi 332-0012, Japan
e-mail: igaki@lif.kyoto-u.ac.jp

M. Miyasaka, K. Takatsu (eds.), *Chronic Inflammation*,
DOI 10.1007/978-4-431-56068-5_17

as well as normal epithelial cells, are crucial for driving epithelial cancers (Bissell and Hines 2011; Jacks and Weinberg 2002). In addition, most cancers are comprised of heterogeneous cell populations with distinct genetic mutations (Going 2003; Novelli et al. 1996), suggesting the presence of interclonal oncogenic cooperation that promotes tumour growth and progression (Lyons et al. 2008). However, the mechanisms by which each oncogenic alteration cooperates with others to drive tumour progression through cell–cell interactions have remained elusive. The genetic mosaic techniques available in *Drosophila* allow us to analyse the behavior of cell clones with distinct mutations (Xu and Rubin 1993), making this the ideal model organism for studying cell–cell interactions and interclonal oncogenic cooperation during tumourigenesis (Brumby and Richardson 2005; Pastor-Pareja and Xu 2013; Wu et al. 2010). In this review, we summarise our recent studies in *Drosophila* revealing that mitochondrial dysfunction cooperates with Ras acti vation to drive nonautonomous tumour progression in the surrounding tissue via 'oncogenic inflammation'.

17.2 Mitochondrial Dysfunction in Ras-Activated Cells Triggers Nonautonomous Overgrowth of Surrounding Tissue

To understand the molecular basis of tumour progression through cell–cell communication, a genetic screen was conducted in *Drosophila* eye imaginal epithelium for mutations that cause Ras^{V12}-expressing cells to induce overproliferation of surrounding cells, which was named *nag* (nonautonomous growth) mutations (Fig.17.1a, b). Intriguingly, it was found that a series of *nag* mutants isolated from the screen had mutations in genes required for mitochondrial respiratory function, such as components of mitochondrial respiratory complexes (e.g., *Pdsw* in complex I, *CoVa* in complex IV) and mitochondrial ribosomal proteins (e.g., *mRpL4*) (Fig. 17.1c). Note that neither mitochondrial dysfunction alone nor Ras activation alone induces overproliferation of surrounding cells, indicating that mitochondrial dysfunction and Ras activation cooperatively induce noncell autonomous overgrowth.

It has been reported that mitochondrial respiratory function is frequently downregulated in many types of human cancers (Pedersen 1978). The high rates of the mitochondrial malfunctions in cancers are caused by both somatic and germline mutations in mitochondrial DNA (mtDNA; Brandon et al. 2006; Carew and Huang 2002; Modica-Napolitano et al. 2007; Penta et al. 2001), which encodes components of mitochondrial respiratory complexes. Although it had long been controversial whether mitochondrial dysfunction is responsible for oncogenicity, several studies have indicated that mtDNA mutation positively regulates tumourigenesis and malignant progression (Lee et al. 2010). However, the

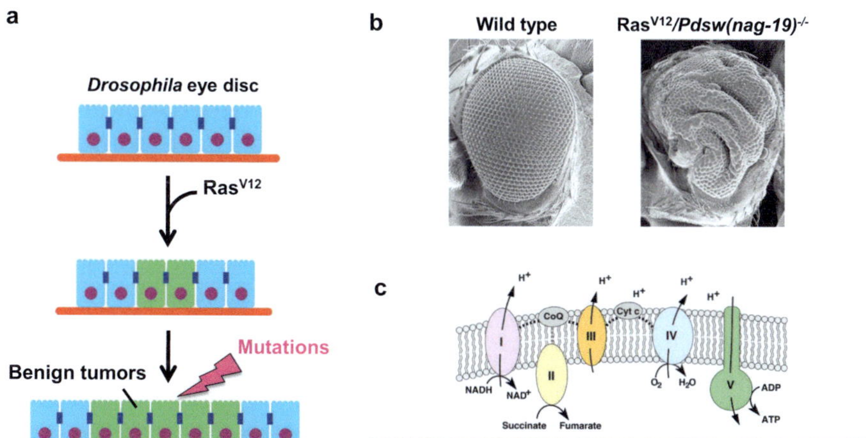

Allele	Affected gene	Function	Affected complex
nag-19	*Pdsw*	NADH dehydrogenase	I
nag-28	*CG9140*	NADH dehydrogenase	I
nag-37	*CG11455*	NADH dehydrogenase	I
nag-43	*CG13240*	NADH dehydrogenase	I
nag-20	*RFeSP*	ubiquinol-cytochrome c reductase	III
nag-8	*CG10664*	cytochrome c oxidase	IV
nag-46	*CG1746*	hydrogen-exporting ATPase	V
nag-24	*mRpL4*	Mitochondrial ribosomal protein	I, III, IV, V
nag-35	*mRpL13*	Mitochondrial ribosomal protein	I, III, IV, V
nag-36	*mRpL24*	Mitochondrial ribosomal protein	I, III, IV, V

Fig. 17.1 Mitochondrial dysfunction cooperates with Ras signalling to induce nonautonomous overgrowth. (**a**) A genetic screen for mutations that cause nonautonomous growth (*nag*) in conjunction with Ras activation. Clones of cells expressing Ras^{V12} result in benign tumours, and mutations that cause Ras^{V12}-tumours to induce non-autonomous overgrowth were screened. (**b**) Scanning electron microscopic images of adult eye bearing wild-type (*left*) or Ras^{V12}/*Pdsw*$^{nag\text{-}19-/-}$ (*right*) clones in the eye disc during development. (**c**) Schematic representation of the mitochondrial respiratory chain and a list of genes affected in *nag* mutants. Genes encoding the components of mitochondrial respiratory complexes (I, III, IV, or V) or mitochondrial ribosomal proteins were isolated as responsible genes for *nag* mutants in the screen (Modified from Ohsawa et al. 2012)

mechanism by which mitochondrial dysfunction contributes to tumour growth and progression has been elusive.

17.3 Ras^{V12}/*mito*$^{-/-}$ Cells Cause Nonautonomous Overgrowth via Induction of Inflammatory Cytokine

To dissect the mechanism by which Ras activation and mitochondrial dysfunction (Ras^{V12}/*mito*$^{-/-}$) cooperate to induce nonautonomous tissue overgrowth, a second ary genetic screen was conducted for dominant modifiers of *nag* phenotype caused by Ras^{V12}/*mito*$^{-/-}$ cells. This secondary screen identified a mutation of *stat92E*, a component of the JAK/STAT pathway, as a strong suppressor of the *nag* phenotype. This indicates that the JAK/STAT pathway is involved in Ras^{V12}/*mito*$^{-/-}$-induced nonautonomous overgrowth. The *Drosophila* JAK/STAT pathway is activated by the inflammatory cytokine Upd. Binding of Upd to its receptor Domeless (Dome) induces activation of the transcription factor STAT92E, thereby promoting cell proliferation (Hou et al. 2002). Indeed, Ras^{V12}/*mito*$^{-/-}$ cells upregulated Upd, however, neither Ras^{V12} alone nor *mito*$^{-/-}$ alone induced Upd expression. In addition, downregulation of Upd expression within Ras^{V12}/*mito*$^{-/-}$ cells or heterozygosity for the *upd* gene in the whole tissue suppressed nonautonomous tissue overgrowth. Thus, induction of the inflammatory cytokine Upd in Ras^{V12}/*mito*$^{-/-}$ cells plays a key role in the nonautonomous tissue overgrowth; we therefore called this phenomenon 'oncogenic inflammation'Enomoto et al. 2015).

17.4 Upd Is Induced by ROS-Dependent JNK Activation in Ras^{V12}/*mito*$^{-/-}$ Cells

Subsequent genetic analyses revealed that activation of c-Jun N-terminal kinase (JNK) signalling is required for Upd induction in Ras^{V12}/*mito*$^{-/-}$ cells. Furthermore, it was found that JNK activation is caused by overproduction of reactive oxygen species (ROS) in Ras^{V12}/*mito*$^{-/-}$ cells. In conjunction with Ras activation, defects in mitochondrial respiratory complex I (*Pdsw*) or complex IV (*CoVa*) overproduce O_2^- or H_2O_2, respectively, and reducing oxidative stress in Ras^{V12}/*mito*$^{-/-}$ cells by superoxide dismutase (SOD) or Catalase significantly suppressed JNK activation and subsequent nonautonomous overgrowth. Thus, Ras^{V12}/*mito*$^{-/-}$ cells cause ROS-mediated activation of JNK signalling, which induces upregulation of Upd expression. It is an interesting open question how mitochondrial defects in conjunction with Ras activation cause overproduction of ROS.

17.5 Ras^{V12}/*mito*$^{-/-}$ Cells Induce Upd via Inactivation of the Hippo Pathway

Although JNK signalling was found to be required for oncogenic inflammation, it is not sufficient for Upd induction. This suggests that an additional signalling in Ras^{V12}/*mito*$^{-/-}$ cells cooperates with JNK to induce oncogenic inflammation. Interestingly, it was found that coactivation of Ras and JNK signalling causes inactivation of the Hippo pathway. The Hippo pathway is an evolutionarily conserved tumour-suppressor pathway that limits tissue growth by inhibiting the transcriptional coactivator Yokie (Yki) (Halder and Johnson 2011; Meng et al. 2016; Pan 2010; Zhao et al. 2010). Inactivation of the Hippo pathway thus causes activation of Yki, leading to induction of its target genes including *upd* (Karpowicz et al. 2010; Ren et al. 2010; Shaw et al. 2010; Staley and Irvine 2010). Indeed, it was found that Upd and a secreted growth factor Wingless (Wg) are upregulated in Ras^{V12}/*mito*$^{-/-}$ cells through inactivation of the Hippo pathway. In addition, expression of *upd-RNAi* or *wg-RNAi* within Ras^{V12}/*mito*$^{-/-}$ cells suppressed noncell autonomous overgrowth. Thus, JNK activation in Ras^{V12}/*mito*$^{-/-}$ cells cooperates with Ras signalling to inactivate the Hippo pathway, which induces expression of Yki target genes *upd* and *wg*, leading to nonautonomous overgrowth of surrounding tissue.

17.6 Ras^{V12}/*mito*$^{-/-}$ Cells Cause Tumour Progression of Adjacent Benign Tumors via Oncogenic Inflammation

In human cancers, oncogenic cooperation among mutant cell clones with different mutations can induce clonal evolution. The fact that mitochondrial dysfunction in Ras-activated cells causes overgrowth of surrounding tissue suggests that such mutant cells may cause clonal evolution in nearby oncogenic cells. Strikingly, it was found that Ras^{V12}/*mito*$^{-/-}$ cells cause surrounding Ras-activated benign tumours to exhibit metastatic behavior; the surrounding tumour cells migrate out of the eye imaginal epithelium and invade adjacent tissue ventral nerve cord. Note that eye disc cells expressing Ras^{V12} alone never exhibit such invasive behavior. This indicates a specific interclonal oncogenic cooperation between Ras^{V12}/*mito*$^{-/-}$ and Ras^{V12} cells, which may indeed occur in human cancer tissues. The nonautonomous tumour progression caused by Ras^{V12}/*mito*$^{-/-}$ cells was blocked by reducing the Upd-JAK/STAT pathway activity, indicating that it is caused through oncogenic inflammation (Fig. 17.2; Ohsawa et al. 2012).

17.7 Ras^{V12}/*mito*$^{-/-}$ Cells Undergo Cellular Senescence

In analysing the nonautonomous overgrowth caused by Ras^{V12}/*mito*$^{-/-}$ cells, it was noticed that despite the secretion of growth factors Upd and Wg, Ras^{V12}/*mito*$^{-/-}$ cells grow poorly. This raises the possibility that cell-cycle progression is blocked in Ras^{V12}/*mito*$^{-/-}$ cells. Indeed, Ras^{V12}/*mito*$^{-/-}$ cells appeared to be undergoing cell-cycle arrest, which was reminiscent of cellular senescence.

Cellular senescence is an irreversible cell-cycle arrest in the G1 phase, which is triggered by various cellular stresses such as chromosomal instability, telomere shortening, oxidative stress, and DNA damage, as well as aberrant activation of oncogenes (Collado et al. 2007; Lowe et al. 2004; Ohtani and Hara 2013; Rodier and Campisi 2011). It has been shown that cells undergoing cellular senescence not only become quiescent but exert noncell autonomous effects by secreting a variety of cytokines, chemokines, and growth factors, which is known as the senescence-associated secretory phenotype (SASP) or the senescence-messaging secretome (SMS; Coppe et al. 2010; Davalos et al. 2010; Kuilman et al. 2010; Kuilman and

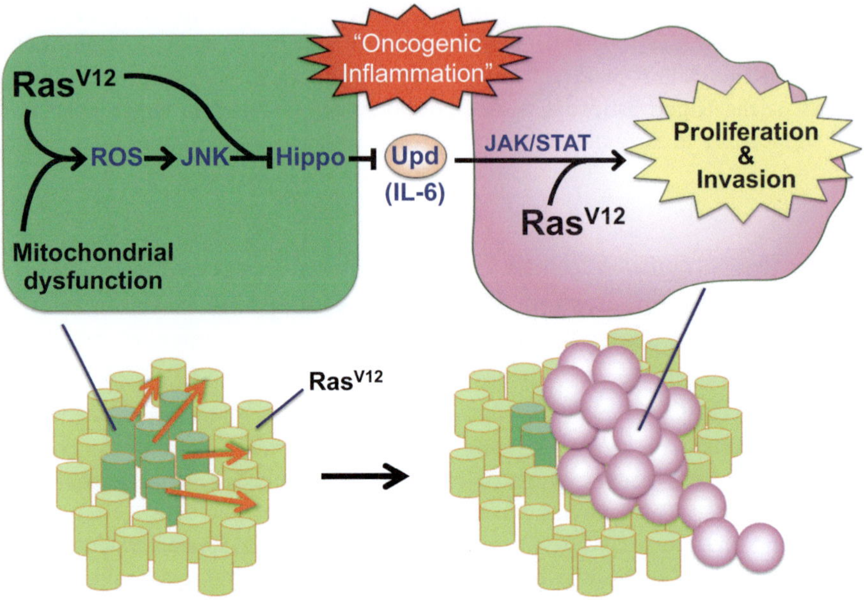

Fig. 17.2 A model for oncogenic inflammation triggered by Ras activation and mitochondrial dysfunction (See text for details)

Peeper 2009; Ohtani and Hara 2013; Rodier and Campisi 2011; Young and Narita 2009). Similar to Ras-induced cellular senescence observed in mammalian cell culture systems, it was found that clones of cells expressing Ras^{V12} in *Drosophila* imaginal epithelium exhibit several markers of cellular senescence, such as elevated SA-β-gal activity, upregulation of a Cdk inhibitor Dacapo (Dap, a p21/p27 homologue), elevation of Histone H3-K9 trimethylation (which is associated with the senescence-associated heterochromatin foci; SAHF), upregulation of heterochromatin protein-1 (HP-1; which binds to methylated K9 of Histone H3), and increased cell size (cellular hypertrophy). Intriguingly, however, despite showing such senescence markers, Ras^{V12}-expressing cells do not show cell-cycle arrest, indicating that aberrant Ras activation is not sufficient for the induction of cellular senescence in *Drosophila* epithelium. Strikingly, mitochondrial dysfunction in Ras-activated cells (Ras^{V12}/*mito*$^{-/-}$ cells) resulted in cell-cycle arrest in G1 phase, in addition to exhibiting other cellular senescence markers listed above, whereas mitochondrial dysfunction alone only slightly suppressed cell-cycle progression. Moreover, mitochondrial dysfunction strongly enhanced cellular hypertrophy of Ras-activated cells, whereas mitochondrial dysfunction alone did not affect cell size. Furthermore, as described earlier, Ras^{V12}/*mito*$^{-/-}$ cells appear to exhibit SASP, as these mutant cells upregulate an inflammatory cytokine Upd. Note that Ras^{V12}/*mito*$^{-/-}$ cells, but not Ras^{V12} cells or *mito*$^{-/-}$ cells, exhibit intranuclear foci of phosphorylated Histone H2A variant (γH2Av), an indicator of DNA damage response (DDR) that is also crucial for the induction of cellular senescence

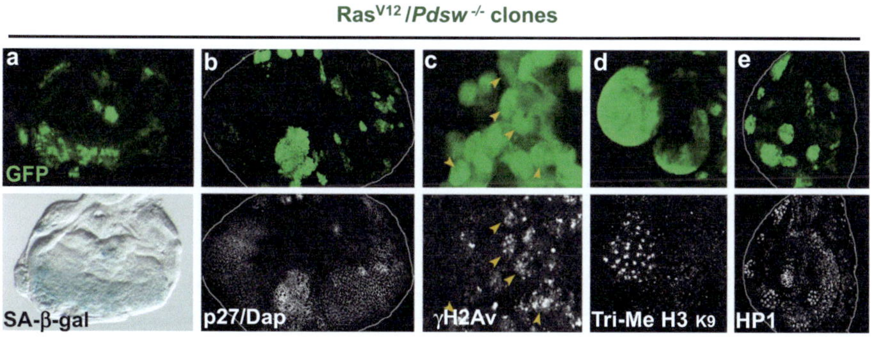

Fig. 17.3 Mitochondrial defect causes Ras-activated cells to undergo cellular senescence. (**a**) Eye-antennal disc bearing GFP-labeled $Ras^{V12}/Pdsw^{-/-}$ clones was subjected to SA-β-gal staining. (**b**)–(**d**) Eye-antennal disc bearing GFP-labeled $Ras^{V12}/Pdsw^{-/-}$ clones was stained with anti-Dacapo (**b**) anti-γ-H2Av (**c**) or anti-Histone-H3-trimethyl-K9 (**d**) antibodies. *Arrowheads* in (**c**) indicate representative clones upregulating γ-H2Av. (**e**) $Ras^{V12}/Pdsw^{-/-}$ clones were induced in eye-antennal disc bearing the RFP-HP1 reporter and stained with anti-RFP

(Bartkova et al. 2006; d'Adda di Fagagna et al. 2003; Di Micco et al. 2006; Rodier et al. 2009; Takahashi et al. 2012; Takai et al. 2003; Fig. 17.3). Thus, mitochondrial dysfunction causes Ras-activated cells to exhibit full aspects of cellular senescence, which induces oncogenic inflammation through SASP.

17.8 Cell-Cycle Arrest Is Required for SASP Induction

Although several studies in mammalian systems have reported the mechanisms of SASP induction (Coppe et al. 2010; Davalos et al. 2010; Kuilman et al. 2010; Kuilman and Peeper 2009; Ohtani and Hara 2013; Rodier and Campisi 2011; Young and Narita 2009), the molecular link between each senescence phenotype and SASP induction has been elusive. Interestingly, in *Drosophila* imaginal epithelium, it was found that cell-cycle arrest in the G1 phase is essential for the induction of SASP. Overexpression of Cyclin E (CycE), which counters the function of p21/p27/Dap and promotes the G1/S transition, in $Ras^{V12}/mito^{-/-}$ clones blocked Upd induction and nonautonomous overgrowth. In addition, overexpression of CycE significantly suppressed JNK activation, an upstream event of Upd induction. In contrast, CycE overexpression did not affect the induction of oxidative stress, an upstream event of JNK activation. These data indicate that cell-cycle arrest is required for SASP induction at the upstream of JNK activation and the downstream of oxidative stress. Indeed, forced induction of cell-cycle arrest in the G1 phase by introducing homozygous mutations of *cycE* in Ras^{V12} cells causes noncell autonomous overgrowth through JNK-dependent Upd induction. Furthermore, reducing one copy of the cell-cycle inhibitor *Rb* gene significantly suppressed nonautonomous overgrowth caused by $Ras^{V12}/mito^{-/-}$ clones. On the contrary, cellular hypertrophy is not required for SASP induction in $Ras^{V12}/mito^{-/-}$ cells, as blocking cellular hypertrophy by overexpressing *HP-1-RNAi* in $Ras^{V12}/mito^{-/-}$ cells does not affect the noncell autonomous growth phenotype.

17.9 Cell Cycle Arrest Amplifies JNK Signalling Activity

As described earlier, overexpression of CycE suppresses JNK activation in $Ras^{V12}/mito^{-/-}$ cells, which raises the possibility that JNK signalling is activated through cell-cycle arrest. Indeed, it was found that cell-cycle progression negatively regulates JNK activation; JNK-dependent cell death caused by *Drosophila* TNF ligand Eiger was significantly suppressed by overexpression of CycE. Conversely, JNK-dependent cell death was enhanced by overexpression of *cycE-RNAi*. Furthermore, overexpression of CycE suppressed JNK activation caused by a constitutively active form of JNK kinase Hemipterous (Hep^{CA}). These lines of evidence suggest that cell-cycle arrest enhances JNK activation. Intriguingly, it was found that activation of JNK signalling causes cell-cycle arrest. For instance, clones of cells expressing Hep^{CA} in the eye disc show no incorporation of BrdU. Together, these data show that cell-cycle arrest and JNK activation form a positive feedback loop that amplifies both. Thus, $Ras^{V12}/mito^{-/-}$ senescent cells induce SASP via cell-cycle arrest-induced amplification of JNK signalling. The mechanism by which JNK activation causes cell-cycle arrest is currently unknown.

17.10 p53 Triggers the Cell-Cycle Arrest-Mediated Amplification of JNK Signalling

How do $Ras^{V12}/mito^{-/-}$ cells cause cycle arrest? In mouse fibroblasts undergoing Ras-induced cellular senescence, the tumour suppressor gene p53 is required for the induction of cell-cycle arrest (Kamijo et al. 1997; Kilbey et al. 2008; Serrano et al. 1997). On the other hand, in *Drosophila* imaginal epithelia, p53 and JNK signalling form a positive feedback loop that amplifies both activities (Shlevkov and Morata 2012). Interestingly, $Ras^{V12}/mito^{-/-}$ cells significantly elevate the protein level of p53 and its activity (as assessed by the expression of p53 target gene *reaper*). Removal of *p53* genes in $Ras^{V12}/mito^{-/-}$ cells suppressed JNK activation as well as noncell autonomous overgrowth. These data show that elevated p53 activity in $Ras^{V12}/mito^{-/-}$ cells causes cell-cycle arrest-mediated amplification of JNK signalling and the following SASP induction. Indeed, overexpression of p53 in Ras^{V12}-expressing cells (Ras^{V12}/p53 cells) causes JNK activation, Upd induction, and nonautonomous overgrowth. In addition, the overgrowth phenotype caused by Ras^{V12}/p53 clones is suppressed by blocking

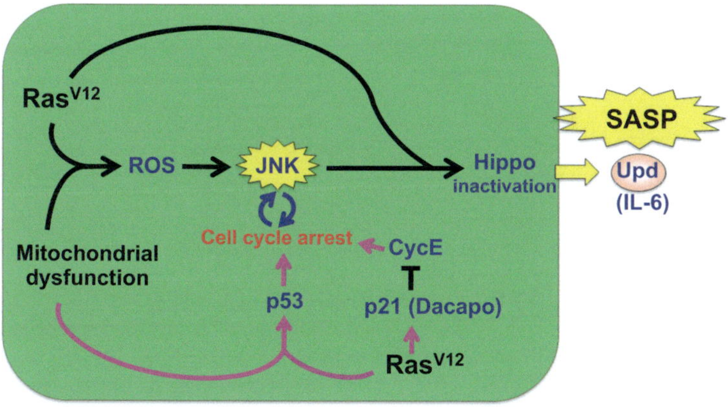

Fig. 17.4 A model for the genetic pathway to SASP by Ras activation and mitochondrial dysfunction (See text for details)

JNK activation, overexpressing CycE, or reducing Upd expression within the clones, suggesting that Ras^{V12}/p53 cells causes nonautonomous overgrowth through SASP. Altogether, these lines of evidence establish that coexistence of Ras activation and mitochondrial dysfunction in *Drosophila* epithelial cells causes ROS overproduction, p53 activation, and downregulation of CycE activity, all of which cooperatively induce cell-cycle arrest-mediated amplification of JNK signalling and subsequent oncogenic inflammation through SASP (Fig. 17.4; Nakamura et al. 2014).

17.11 Concluding Remarks

Recent genetic studies in *Drosophila* have uncovered a novel mechanistic basis for tumour progression through cell–cell communication; mitochondrial dysfunction with elevated Ras signalling causes cellular senescence, which leads to the induction of SASP factors including an inflammatory cytokine and nonautonomous tumour progression. In 1920s, Otto Warburg described that tumour cells utilise an excess of glucose for conversion of lactate through anaerobic respiration even in the presence of sufficient oxygen. This led to his 'hypothesis' that mitochondrial defects play an active role in the etiology of cancer (Warburg 1956). Genetic evidence in *Drosophila* described here not only supports Warburg's hypothesis but also provides a new concept that oncogenic inflammation plays a central role in nonautonomous tumour progression by Ras activation and mitochondrial dysfunction. Importantly, mtDNA mutations are frequently observed in pancreatic cancers (Modica-Napolitano et al. 2007), which mostly include (~90 %) Ras mutations (Downward 2003). Thus, similar oncogenic inflammation could have an important role in such aggressive cancer. Future studies should address the evolutionary conservation of oncogenic inflammation as well as its physiological roles in normal development and homeostasis.

References

Bartkova J, Rezaei N, Liontos M, Karakaidos P, Kletsas D, Issaeva N, Vassiliou LV, Kolettas E, Niforou K, Zoumpourlis VC et al (2006) Oncogene-induced senescence is part of the tumorigenesis barrier imposed by DNA damage checkpoints. Nature 444:633–637

Bissell MJ, Hines WC (2011) Why don't we get more cancer? A proposed role of the microenvironment in restraining cancer progression. Nat Med 17:320–329

Brandon M, Baldi P, Wallace DC (2006) Mitochondrial mutations in cancer. Oncogene 25: 4647–4662

Brumby AM, Richardson HE (2005) Using Drosophila melanogaster to map human cancer pathways. Nat Rev Cancer 5:626–639

Carew JS, Huang P (2002) Mitochondrial defects in cancer. Mol Cancer 1:9

Collado M, Blasco MA, Serrano M (2007) Cellular senescence in cancer and aging. Cell 130: 223–233

Coppe JP, Desprez PY, Krtolica A, Campisi J (2010) The senescence-associated secretory phenotype: the dark side of tumor suppression. Annu Rev Pathol 5:99–118

d'Adda di Fagagna F, Reaper PM, Clay-Farrace L, Fiegler H, Carr P, Von Zglinicki T, Saretzki G, Carter NP, Jackson SP (2003) A DNA damage checkpoint response in telomere-initiated senescence. Nature 426:194–198

Davalos AR, Coppe JP, Campisi J, Desprez PY (2010) Senescent cells as a source of inflammatory factors for tumor progression. Cancer Metastasis Rev 29:273–283

Di Micco R, Fumagalli M, Cicalese A, Piccinin S, Gasparini P, Luise C, Schurra C, Garre M, Nuciforo PG, Bensimon A et al (2006) Oncogene-induced senescence is a DNA damage response triggered by DNA hyper-replication. Nature 444:638–642

Downward J (2003) Targeting RAS signalling pathways in cancer therapy. Nat Rev Cancer 3: 11–22

Enomoto M, Vaughen J, Igaki T (2015) Non-autonomous overgrowth by oncogenic niche cells: cellular cooperation and competition in tumorigenesis. Cancer Sci 106:1651–1658

Going JJ (2003) Epithelial carcinogenesis: challenging monoclonality. J Pathol 200:1–3

Halder G, Johnson RL (2011) Hippo signaling: growth control and beyond. Development 138: 9–22

Hou SX, Zheng Z, Chen X, Perrimon N (2002) The Jak/STAT pathway in model organisms: emerging roles in cell movement. Dev Cell 3:765–778

Jacks T, Weinberg RA (2002) Taking the study of cancer cell survival to a new dimension. Cell 111:923–925

Kamijo T, Zindy F, Roussel MF, Quelle DE, Downing JR, Ashmun RA, Grosveld G, Sherr CJ (1997) Tumor suppression at the mouse INK4a locus mediated by the alternative reading frame product p19ARF. Cell 91:649–659

Karpowicz P, Perez J, Perrimon N (2010) The Hippo tumor suppressor pathway regulates intestinal stem cell regeneration. Development 137:4135–4145

Kilbey A, Terry A, Cameron ER, Neil JC (2008) Oncogene-induced senescence: an essential role for Runx. Cell Cycle 7:2333–2340

Kuilman T, Michaloglou C, Mooi WJ, Peeper DS (2010) The essence of senescence. Genes Dev 24:2463–2479

Kuilman T, Peeper DS (2009) Senescence-messaging secretome: SMS-ing cellular stress. Nat Rev Cancer 9:81–94

Lee HC, Chang CM, Chi CW (2010) Somatic mutations of mitochondrial DNA in aging and cancer progression. Ageing Res Rev 9(Suppl 1):S47–S58
Lowe SW, Cepero E, Evan G (2004) Intrinsic tumour suppression. Nature 432:307–315
Lyons JG, Lobo E, Martorana AM, Myerscough MR (2008) Clonal diversity in carcinomas: its implications for tumour progression and the contribution made to it by epithelial-mesenchymal transitions. Clin Exp Metastasis 25:665–677
Meng Z, Moroishi T, Guan KL (2016) Mechanisms of Hippo pathway regulation. Genes Dev 30: 1–17
Modica-Napolitano JS, Kulawiec M, Singh KK (2007) Mitochondria and human cancer. Curr Mol Med 7:121–131
Nakamura M, Ohsawa S, Igaki T (2014) Mitochondrial defects trigger proliferation of neighbouring cells via a senescence-associated secretory phenotype in Drosophila. Nat Commun 5:5264
Novelli MR, Williamson JA, Tomlinson IP, Elia G, Hodgson SV, Talbot IC, Bodmer WF, Wright NA (1996) Polyclonal origin of colonic adenomas in an XO/XY patient with FAP. Science 272:1187–1190
Ohsawa S, Sato Y, Enomoto M, Nakamura M, Betsumiya A, Igaki T (2012) Mitochondrial defect drives non-autonomous tumour progression through Hippo signalling in Drosophila. Nature 490:547–551
Ohtani N, Hara E (2013) Roles and mechanisms of cellular senescence in regulation of tissue homeostasis. Cancer Sci 104:525–530
Pan D (2010) The hippo signaling pathway in development and cancer. Dev Cell 19:491–505
Pastor-Pareja JC, Xu T (2013) Dissecting social cell biology and tumors using Drosophila genetics. Annu Rev Genet 47:51–74
Pedersen PL (1978) Tumor mitochondria and the bioenergetics of cancer cells. Prog Exp Tumor Res 22:190–274
Penta JS, Johnson FM, Wachsman JT, Copeland WC (2001) Mitochondrial DNA in human malignancy. Mutat Res 488:119–133
Ren F, Wang B, Yue T, Yun EY, Ip YT, Jiang J (2010) Hippo signaling regulates Drosophila intestine stem cell proliferation through multiple pathways. Proc Natl Acad Sci U S A 107: 21064–21069
Rodier F, Campisi J (2011) Four faces of cellular senescence. J Cell Biol 192:547–556
Rodier F, Coppe JP, Patil CK, Hoeijmakers WA, Munoz DP, Raza SR, Freund A, Campeau E, Davalos AR, Campisi J (2009) Persistent DNA damage signalling triggers senescence-associated inflammatory cytokine secretion. Nat Cell Biol 11:973–979
Serrano M, Lin AW, McCurrach ME, Beach D, Lowe SW (1997) Oncogenic ras provokes premature cell senescence associated with accumulation of p53 and p16INK4a. Cell 88: 593–602
Shaw RL, Kohlmaier A, Polesello C, Veelken C, Edgar BA, Tapon N (2010) The Hippo pathway regulates intestinal stem cell proliferation during Drosophila adult midgut regeneration. Development 137:4147–4158
Shlevkov E, Morata G (2012) A dp53/JNK dependant feedback amplification loop is essential for the apoptotic response to stress in Drosophila. Cell Death Differ 19:451–460
Staley BK, Irvine KD (2010) Warts and Yorkie mediate intestinal regeneration by influencing stem cell proliferation. Curr Biol 20:1580–1587
Takahashi A, Imai Y, Yamakoshi K, Kuninaka S, Ohtani N, Yoshimoto S, Hori S, Tachibana M, Anderton E, Takeuchi T et al (2012) DNA damage signaling triggers degradation of histone methyltransferases through APC/C(Cdh1) in senescent cells. Mol Cell 45:123–131
Takai H, Smogorzewska A, de Lange T (2003) DNA damage foci at dysfunctional telomeres. Curr Biol 13:1549–1556
Warburg O (1956) On the origin of cancer cells. Science 123:309–314
Wu M, Pastor-Pareja JC, Xu T (2010) Interaction between Ras(V12) and scribbled clones induces tumour growth and invasion. Nature 463:545–548

Xu T, Rubin GM (1993) Analysis of genetic mosaics in developing and adult Drosophila tissues. Development 117:1223–1237

Young AR, Narita M (2009) SASP reflects senescence. EMBO Rep 10:228–230

Zhao B, Li L, Lei Q, Guan KL (2010) The Hippo-YAP pathway in organ size control and tumorigenesis: an updated version. Genes Dev 24:862–874

Chapter 18
Inflammation-Associated Carcinogenesis Mediated by the Impairment of microRNA Function in the Gastroenterological Organs

Motoyuki Otsuka

Abstract Gastroenterological cancers are major causes of death in Japan. Most of the carcinomas occur subsequent to chronic inflammation, such as chronic gastritis, hepatitis and colitis. Emerging evidence supports the idea that microRNAs, non-coding RNAs regulating target gene expression through sequence homologies, are frequently involved in carcinogenesis. In particular, reduced expression levels of miRNAs are commonly observed in cancers. In fact, incomplete loss of the Dicer gene, or microRNAs, leads to carcinogenesis experimentally in mice, suggesting that impairment of microRNA expression or function is crucial for carcinogenesis. Chronic inflammatory stimuli impair microRNA function, possibly through reduced loading of microRNAs into the RNA-induced silencing complex (RISC) and the dissociation of the targeted mRNAs from processing bodies in the cells. Therefore, it is hypothesised that chronic inflammation impairs microRNA function, which, in turn, leads to carcinogenesis. This may be a mechanism for chronic inflammation leading to carcinogenesis, and interventional methods based on this idea should be developed for the prevention of inflammation-associated carcinogenesis.

Keywords Inflammation • Cancer • Gastritis • Hepatitis • Colitis • microRNA

Chronic inflammation, such as chronic gastritis, chronic hepatitis, and ulcerative colitis, is an important pathological factor in various kinds of gastroenterological diseases. Note that most are closely related to the subsequent occurrence of carcinoma in the affected organs. The current most frequent cause of mortality in Japan is cancer. Gastroenterological cancers, including gastric cancers, colon cancers, and liver cancers, all of which may arise subsequent to chronic

M. Otsuka (✉)
Department of Gastroenterology, Graduate School of Medicine, The University of Tokyo, 7-3-1 Hongo, Bunkyo-ku, Tokyo 113-8655, Japan

PRESTO, Japan Science and Technology Agency, Kawaguchi 332-0012, Japan
e-mail: otsukamo-tky@umin.ac.jp

M. Miyasaka, K. Takatsu (eds.), *Chronic Inflammation*,
DOI 10.1007/978-4-431-56068-5_18

inflammation, are the second, third, and fourth most common cancers, respectively. Therefore, elucidation of the disease pathogenesis from chronic inflammation to carcinoma is important for the prevention of these fatal diseases.

MicroRNAs (miRNAs) are short, single-stranded, noncoding RNAs that are expressed in most organisms, from plants to vertebrates (Carrington and Ambros 2003). Since the discovery of the first miRNA in *Caenorhabditis elegans* (Lee et al. 1993; Wightman et al. 1993), 1,881 miRNA precursors and 2,588 mature miRNA sequences from humans have been deposited in miRBase, a public database of the Sanger Institute, as of June 2015 (Kozomara and Griffiths-Jones 2011). It is suggested that miRNAs regulate more than 30 % of human protein-coding genes (John et al. 2004; Lewis et al. 2005). Through the regulation of gene expression, miRNAs are involved in various physiological and pathological processes, including cell proliferation, oncogenesis, and oncogenic suppression (Ambros 2004).

It is known that reduced global miRNA expression levels are closely involved in carcinogenesis (Calin and Croce 2006; Gaur et al. 2007; Lu et al. 2005), whereas deregulation of each miRNA may also be involved in oncogenesis (Calin and Croce 2006; Croce and Calin 2005; Zhang et al. 2006). In fact, experimental global miRNA reduction leads to oncogenesis in mice (Kumar et al. 2009). Therefore, it is important to examine the status of miRNA expression or function under chronic inflammation to elucidate the mechanisms of the subsequent oncogenesis. Because cellular stresses, such as inflammatory stimulation and starvation, have been reported to suppress miRNA function (Bhattacharyya et al. 2006), it may be that the inflammation-related impairment of global miRNA function is linked with subsequent carcinogenesis in corresponding tissues. These points are discussed below.

18.1 Chronic Inflammation-Associated Carcinomas in the Gastroenterological Organs

In Japan, the current most frequent cause of death is cancer, followed by cardiovascular diseases. Among deaths from cancer, the most frequent is lung. The second is gastric cancer, with an annual mortality of approximately 50,000. The third is colon cancer, with an annual mortality of 46,000. The fourth and fifth are pancreatic cancer and liver cancer, respectively, and the annual mortalities were both about 30,000 in 2013. The four organs among the top five related to cancer mortality are gastroenterological.

18.1.1 Gastric Cancer

Gastric cancer is known to be closely related with *Helicobacter pylori* (*H. pylori*) infection. *H. pylori* induces persistent inflammatory infiltration, which leads to atrophic and metaplastic changes in the gastric mucosa. Gastric cancers occur in the affected background gastric mucosa. The cytotoxin-associated gene A (CagA) protein of *H. pylori*, which is delivered into gastric epithelial cells via bacterial type IV secretion, is an oncoprotein that induces malignant transformation (Hatakeyama 2014). CagA affects multiple intracellular signalling pathways. It was reported that the oncogenic actions of CagA are successively taken over by a series of genetic and/or epigenetic alterations compiled in cancer-predisposed cells during long-standing infection (Hatakeyama 2014). Of these, *H. pyroli* virulence factor is important in the pathogenesis of gastric carcinomas, though the precise mechanisms involved in this process are still not clearly resolved. Although the majority of individuals infected with *H. pylori* remain asymptomatic throughout their life, essentially all develop chronic inflammation (Wang et al. 2014). Among infected individuals, only 1–3 % progress to gastric carcinoma (Warren 2000). During *H. pylori*-induced chronic inflammation and subsequent carcinogenesis, a variety of factors, including bacterial, host, and environmental factors, interact (Fig. 18.1). These changes may eventually lead to inflammation-associated oncogenesis, the precise mechanisms of oncogenesis, however are still unknown.

18.1.2 Colon Cancer

Patients with ulcerative colitis (UC) and Crohn's disease (CD) are at increased risk of colon cancers through mechanisms that remain incompletely understood (Pozza et al. 2011). An increased risk for the occurrence of colorectal carcinoma in up to 30 % of affected patients after 35 years of UC has been reported. The incidence is significantly increased compared to individuals without chronic colitis. Colitis-associated colorectal carcinoma does not display the adenoma–carcinoma sequence, which is typical for sporadic colorectal carcinoma (Rogler 2014), presenting a different sequence of tumourigenic events (Scarpa et al. 2014). However, recent findings have shown that inflammatory pathways not only are important in the development of colitis-associated colorectal carcinoma but are also involved in the pathogenesis of sporadic colorectal cancer (Moossavi and Bishehsari 2012). Therefore, chronic inflammation may also be closely related with sporadic oncogenesis in the colon (Fig. 18.1).

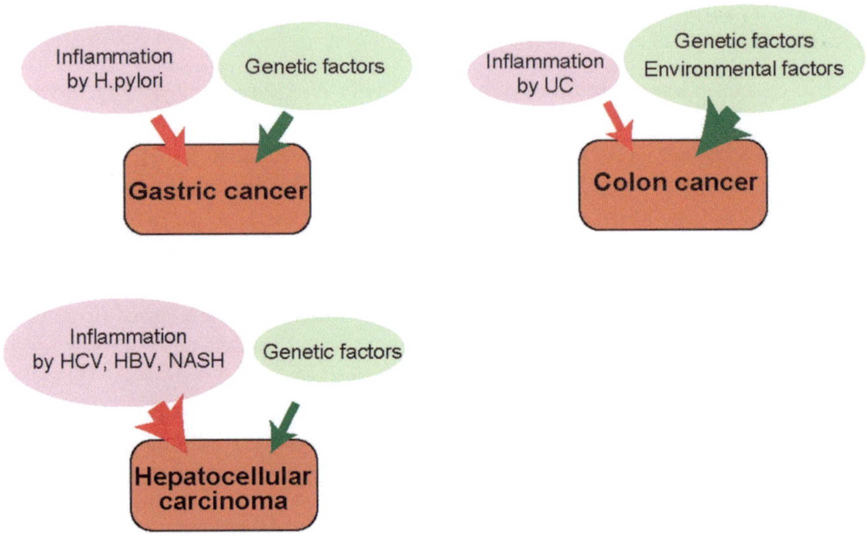

Fig. 18.1 Differential effects of inflammation and other factors in the carcinogenesis of different organs. Gastric carcinogenesis is affected by chronic inflammation, but genetic factors may also have similar effects. Colon carcinogenesis is affected by inflammation in some cases, such as UC, but sporadic carcinoma is mainly due to other genetic or environmental factors. Liver carcinogenesis is strongly affected by chronic inflammation. The size of the *arrows* and the *ovals* indicate the degree of the effects on carcinogenesis

18.1.3 Liver Cancer

Hepatocellular carcinoma (HCC) was the fourth leading cause of death in males and the sixth in females in 2013. The hepatitis C virus (HCV) is the major cause of HCC in Japan, with 70 % of cases being HCV related. Although the number of patients who die because of HCC has steadily increased over the last 50 years, the incidence of HCC is now decreasing, mainly because of the decreased prevalence of HCV-related HCC in Japan. In contrast, there has been a gradual increase in cases of nonviral chronic liver diseases, including nonalcoholic fatty liver disease (NAFLD), occurring with hepatocellular carcinoma (HCC). Among HCC-based nonviral diseases, alcoholic liver disease with HCC accounted for 7.2 % of all HCCs, followed by chronic liver disease of unknown etiology with HCC (5.1 %) and NAFLD with HCC (2.0 %). The clinical characteristics of these three HCC groups are clearly different (Tokushige et al. 2013).

HCC is a clear example of inflammation-related cancer as more than 90 % of HCCs arise in the context of hepatic inflammation (Bishayee 2014; Fig. 18.1). Chronic inflammation leads to sequential development of fibrosis, cirrhosis, and, eventually, HCC (Bishayee 2014). The identification of fundamental inflammatory signalling pathways causing transition from chronic liver injury to HCC could produce new targets for treating patients with chronic liver inflammation.

18.1.4 Pancreatic Cancer

Inflammatory conditions, such as pancreatitis, represent a risk factor for pancreatic cancer, although the precise mechanisms underlying the pathogenic role of inflammation need to be explored in detail (Zambirinis et al. 2014). Recognised risk factors for pancreatic cancer include cigarette smoking, chronic/hereditary pancreatitis, obesity, and type II diabetes (Greer and Whitcomb 2009). The recent explosion of investigations of the human microbiome have highlighted how perturbations of commensal bacterial populations can promote inflammation and promote disease processes, including carcinogenesis. However, elucidation of the precise mechanisms is still urgently needed.

18.2 Biogenesis and Functions of miRNAs

Transcription is the first step in miRNA expression. Similar to most protein-coding genes, transcriptional factors, enhancers, and silencers are involved in miRNA transcription (Cai et al. 2004; Lee et al. 2004). Epigenetic mechanisms, such as promoter methylation or histone modification, also regulate miRNA transcription.

Primary miRNAs, which possess stem-loop structures, are transcribed by RNA polymerase II (Ambros 2004). These pri-miRNAs are processed by a microprocessor complex comprised of Drosha (RNAase III; Lee et al. 2003) and DGCR8/Pasha (Han et al. 2004) in the nucleus (Denli et al. 2004). The processed products are approximately 65-nucleotide hairpin-shaped precursors (pre-miRNAs) that are transported to the cytoplasm via exportin-5 (Lund et al. 2004; Yi et al. 2003). Pre miRNAs are further cleaved into mature miRNAs by Drosha and Dicer RNA polymerase III. Mature miRNA duplexes are loaded onto an RNA-induced silencing complex (RISC) and are unwound into the single-stranded mature form (Gregory et al. 2005; Maniataki and Mourelatos 2005; Mourelatos et al. 2002). The resulting cocomplex directly targets the 3′-untranslated regions (3′UTRs) of target mRNAs, depending on sequence similarities, to regulate their expression negatively by enhancing mRNA cleavage or inhibiting translation (Ambros 2004). Because most miRNAs guide the recognition of imperfect matches of target mRNAs, individual miRNAs have multiple mRNA targets. In addition, multiple miRNAs can cooperate to regulate the expression of the same transcript (Krek et al. 2005).

Identifying functionally important miRNA target genes is crucial for understanding the impact of specific miRNAs on cellular function. However, this is still difficult because miRNAs usually have imperfect complementarity with their targets (Bartel 2009). In mammals, the most consistent requirement for miRNA–target interaction is a perfect pairing of the miRNA (nt 2–8), representing the ‘seed’ sequence (Bartel 2009). In many cases, the seed sequences determine this recognition, but not in all cases (Shin et al. 2010). Although public miRNA target prediction algorithms exist, candidates need to be validated experimentally.

18.3 miRNAs and Cancer

18.3.1 Individual miRNAs as onco-miR or Tumour-Suppressive miR

The involvement of miRNAs in cancer pathogenesis is well established. miRNAs can affect six hallmarks of malignant cells, which are (1) self-sufficiency in growth signals, (2) insensitivity to antigrowth signals, (3) evasion of apoptosis, (4) limitless replicative potential, (5) angiogenesis, and (6) invasion and metastasis (Hanahan and Weinberg 2011). miRNAs are frequently up- or downregulated in malignant tissues and can be considered as oncogenes or tumour suppressors, respectively. However, it is essential to experimentally test whether the deregulated miRNAs are actually causative to carcinogenesis, inasmuch as miRNAs have very restricted tissue-specific expression and the apparent miRNA modulation in cancer tissues may only reflect the different constituents of a cell population, as compared to normal tissues. Extensive analyses have confirmed the causative roles of miRNAs in cancer by using either human cancer cells or genetically engineered animal models, such as transgenic expression of miRNA-155, miRNA-21, and miRNA-15/16, which are sufficient to initiate lymphomagenesis in mice (Costinean et al. 2009; Klein et al. 2010; Medina et al. 2010). On the other hand, deletion of miRNA-122 induces liver cancers, indicating that miR-122 is a tumour-suppressive miRNA in the liver (Hsu et al. 2012; Tsai et al. 2012). These results suggest the potential role of miRNAs in the pathogenesis of carcinogenesis.

18.3.2 Deregulation of Global miRNA Expression or Function in Carcinogenesis

Although microRNAs (miRNAs) can function as both tumour suppressors and oncogenes in tumour development (Calin and Croce 2006; Croce and Calin 2005), widespread downregulated expression of miRNAs promotes cellular transformation and tumourigenesis and is commonly observed in human cancers (Calin and Croce 2006; Gaur et al. 2007; Lu et al. 2005). Similar to miRNAs, deregulated expression of miRNA processing pathway components can potentially modify miRNA expression profiles and may be associated with subsequent tumourigenesis. In fact, using mouse models with tissue-specific *Dicer1* ablation revealed that a single allele of *Dicer1* in mice leads to oncogenesis in a lung cancer model caused by a K-ras protooncogene mutation (Kumar et al. 2009), a retinoblastoma model caused by mutations in the tumour suppressor *RB* gene (Lambertz et al. 2010), and a lymphoma model caused by Eμ-myc (Arrate et al. 2010). Interestingly, in these models, monoallelic, but not complete, loss of *Dicer* alleles enhanced tumour formation. Therefore, it was proposed that Dicer may be unique among classical haploinsufficient tumour suppressor genes inasmuch as only partial, but not

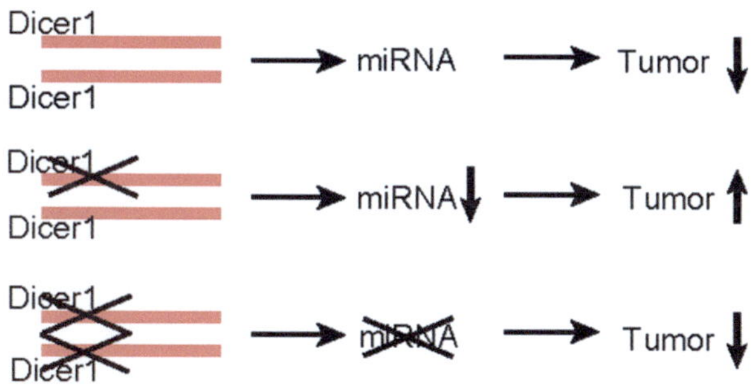

Fig. 18.2 Haploinsufficiency of the Dicer gene on carcinogenesis. Only partial, but not complete, loss of the Dicer gene and miRNAs promote tumour development in mouse experimental carcinogenesis models, suggesting that incomplete loss of global miRNA expression or function leaves a cell prone to carcinogenesis

complete, loss promotes tumour development (Davalos and Esteller 2012), which is consistent with the clinical phenomena showing downregulated miRNA expression, not complete loss of miRNA expression, in carcinogenesis (Fig. 18.2). We also detected similar phenomena in Dicer deletion in mouse colon epithelia in colon carcinogenesis (Yoshikawa et al. 2013), suggesting that incomplete loss of global miRNA expression or function may be related to carcinogenesis in corresponding organs.

18.4 Cellular Stresses Suppress miRNA Function

It was reported that the efficiency of miRNA-mediated gene suppression can be reduced when cells are exposed to stress (Bhattacharyya et al. 2006). In siRNA-mediated mRNA decay, which shares certain features with miRNA-mediated gene suppression, the effect was inhibited when cells underwent inflammatory stress, such as TNFα or IL-1β stimulation (Mols et al. 2008). Although the precise mechanisms of these phenomena are unknown, it was speculated that the target mRNA is released from cytoplasmic processing bodies to polysomes by cellular stresses (Bhattacharyya et al. 2006). We have confirmed similar phenomena in our experiments, as well as observed that the loading volume of miRNAs into the RISC is impaired by cellular inflammatory stimuli, which may account for the deregulated miRNA function under inflammatory stimuli.

18.5 Inflammation-Associated Carcinogenesis Due to Impaired miRNA Function

18.5.1 Screening Compounds for Enhancing miRNA Function

One may speculate that because inflammatory stress can also globally inhibit miRNA function as described above, such inhibition can promote tumourigenesis, as a global reduction in miRNA expression does. This hypothesis is intriguing for us and we have some experimental results that support this hypothesis. If this is the case, manipulation of global miRNA activity may be useful for preventing inflammation-associated tumourigenesis. To verify this, we screened a compound library which enhances miRNA function. Among 1280 compounds, Rho-associated, coiled-coil–containing protein kinase (ROCK) inhibitor was identified as one of the compounds which enhance miRNA function. The regulation of miRNA function by ROCK inhibitors was mediated, at least in part, by poly(A)-binding protein-interacting protein 2 (PAIP2), which enhances poly(A)-shortening of miRNA-targeted mRNAs and led to global upregulation of miRNA function. In the presence of a ROCK inhibitor, PAIP2 expression was enhanced by the transcription factor hepatocyte nuclear factor 4 alpha (HNF4A) through increased ROCK1 nuclear localisation and enhanced ROCK1 association with HNF4A (Yoshikawa et al. 2015). These data, revealing an unexpected role of ROCK1 as a cofactor of HNF4A in enhancing PAIP2 transcription, suggested that ROCK inhibitors may be useful for the various pathologies associated with the impairment of global miRNA function.

18.5.2 ROCK Inhibitor Prevents Inflammation-Associated Carcinogenesis

Based on the results above, we hypothesised that The effects of the ROCK inhibitor-induced increase in global miRNA activity, in addition to the inhibitory effects on cell-autonomous promotion of tumour cell invasion and metastasis (Rath and Olson 2012), may be useful for preventing carcinomas caused by a global reduction in, or insufficient activity of, miRNAs. To test this hypothesis, we applied the ROCK inhibitor to an AOM plus DSS-induced inflammation-associated colon tumour mouse model, expecting that the augmentation of miRNA function by ROCK inhibition would suppress tumourigenesis if the global impairment of miRNA function during chronic inflammation plays a role in tumour initiation and/or promotion. In fact, the inhibition of ROCK significantly reduced inflammation-associated colon tumourigenesis in a mouse experimental model. The effects werre miRNA-dependent, because the effects were lost when Dicer-deficient mice were used and ROCK inhibition did not have any effects on the spontaneous colon carcinogenesis in APC-disrupted mice. These data suggest that ROCK inhibitor treatment has suppressive effects on inflammation-associated tumours inasmuch as it releases miRNA function from suppression by inflammatory stimuli. Although some cytokines, including TNF-α, are generally believed to play major roles of inflammation-associated carcinogenesis, its downstream molecular mechanisms are largely unknown. Our results provide new information on the pathogenesis of inflammation-associated tumourigenesis and identify new potential methods for preventing such tumourigenesis. The ROCK inhibitor fasudil is already used clinically for the treatment of cerebral vasospasms; thus, its safety has already been demonstrated. Therefore, clinical trials are also planned to test its efficacy in the prevention of inflammation-associated tumourigenesis.

18.6 Conclusion

Although we described miRNA deregulation in the pathogenesis of inflammation-associated carcinogenesis as a possible mechanism, other mechanisms also exist. Because inflammation-associated tumourigenesis is closely linked with mortality worldwide, it is crucial to elucidate the precise mechanisms and develop preventative methods through continued research for improved global health.

References

Ambros V (2004) The functions of animal microRNAs. Nature 431:350–355

Arrate MP, Vincent T, Odvody J, Kar R, Jones SN, Eischen CM (2010) MicroRNA biogenesis is required for Myc-induced B-cell lymphoma development and survival. Cancer Res 70: 6083–6092

Bartel DP (2009) MicroRNAs: target recognition and regulatory functions. Cell 136:215–233

Bhattacharyya S, Habermacher R, Martine U, Closs E, Filipowicz W (2006) Relief of microRNA-mediated translational repression in human cells subjected to stress. Cell 125:1111–1124

Bishayee A (2014) The role of inflammation and liver cancer. Adv Exp Med Biol 816:401–435

Cai X, Hagedorn CH, Cullen BR (2004) Human microRNAs are processed from capped, polyadenylated transcripts that can also function as mRNAs. RNA 10:1957–1966

Calin GA, Croce CM (2006) MicroRNA signatures in human cancers. Nat Rev Cancer 6:857–866

Carrington J, Ambros V (2003) Role of microRNAs in plant and animal development. Science 301:336–338

Costinean S, Sandhu SK, Pedersen IM, Tili E, Trotta R, Perrotti D, Ciarlariello D, Neviani P, Harb J, Kauffman LR et al (2009) Src homology 2 domain-containing inositol-5-phosphatase and CCAAT enhancer-binding protein beta are targeted by miR-155 in B cells of Emicro-MiR-155 transgenic mice. Blood 114:1374–1382

Croce CM, Calin GA (2005) miRNAs, cancer, and stem cell division. Cell 122:6–7

Davalos V, Esteller M (2012) Rolling the dice to discover the role of DICER in tumorigenesis. Cancer Cell 21:717–719

Denli AM, Tops BB, Plasterk RH, Ketting RF, Hannon GJ (2004) Processing of primary microRNAs by the Microprocessor complex. Nature 432:231–235
Gaur A, Jewell DA, Liang Y, Ridzon D, Moore JH, Chen C, Ambros VR, Israel MA (2007) Characterization of microRNA expression levels and their biological correlates in human cancer cell lines. Cancer Res 67:2456–2468
Greer JB, Whitcomb DC (2009) Inflammation and pancreatic cancer: an evidence-based review. Curr Opin Pharmacol 9:411–418
Gregory RI, Chendrimada TP, Cooch N, Shiekhattar R (2005) Human RISC couples microRNA biogenesis and posttranscriptional gene silencing. Cell 123:631–640
Han J, Lee Y, Yeom KH, Kim YK, Jin H, Kim VN (2004) The Drosha-DGCR8 complex in primary microRNA processing. Genes Dev 18:3016–3027
Hanahan D, Weinberg RA (2011) Hallmarks of cancer: the next generation. Cell 144:646–674
Hatakeyama M (2014) Helicobacter pylori CagA and gastric cancer: a paradigm for hit-and-run carcinogenesis. Cell Host Microbe 15:306–316
Hsu SH, Wang B, Kota J, Yu J, Costinean S, Kutay H, Yu L, Bai S, La Perle K, Chivukula RR et al (2012) Essential metabolic, anti-inflammatory, and anti-tumorigenic functions of miR-122 in liver. J Clin Invest 122:2871–2883
John B, Enright AJ, Aravin A, Tuschl T, Sander C, Marks DS (2004) Human MicroRNA targets. PLoS Biol 2, e363
Klein U, Lia M, Crespo M, Siegel R, Shen Q, Mo T, Ambesi-Impiombato A, Califano A, Migliazza A, Bhagat G et al (2010) The DLEU2/miR-15a/16-1 cluster controls B cell proliferation and its deletion leads to chronic lymphocytic leukemia. Cancer Cell 17:28–40
Kozomara A, Griffiths-Jones S (2011) miRBase: integrating microRNA annotation and deep-sequencing data. Nucleic Acids Res 39:D152–157
Krek A, Grün D, Poy MN, Wolf R, Rosenberg L, Epstein EJ, MacMenamin P, da Piedade I, Gunsalus KC, Stoffel M et al (2005) Combinatorial microRNA target predictions. Nat Genet 37:495–500
Kumar MS, Pester RE, Chen CY, Lane K, Chin C, Lu J, Kirsch DG, Golub TR, Jacks T (2009) Dicer1 functions as a haploinsufficient tumor suppressor. Genes Dev 23:2700–2704
Lambertz I, Nittner D, Mestdagh P, Denecker G, Vandesompele J, Dyer MA, Marine JC (2010) Monoallelic but not biallelic loss of Dicer1 promotes tumorigenesis in vivo. Cell Death Differ 17:633–641
Lee RC, Feinbaum RL, Ambros V (1993) The C. elegans heterochronic gene lin-4 encodes small RNAs with antisense complementarity to lin-14. Cell 75:843–854
Lee Y, Ahn C, Han J, Choi H, Kim J, Yim J, Lee J, Provost P, Rådmark O, Kim S et al (2003) The nuclear RNase III Drosha initiates microRNA processing. Nature 425:415–419
Lee Y, Kim M, Han J, Yeom KH, Lee S, Baek SH, Kim VN (2004) MicroRNA genes are transcribed by RNA polymerase II. EMBO J 23(20):4051–4060
Lewis BP, Burge CB, Bartel DP (2005) Conserved seed pairing, often flanked by adenosines, indicates that thousands of human genes are microRNA targets. Cell 120:15–20
Lu J, Getz G, Miska EA, Alvarez-Saavedra E, Lamb J, Peck D, Sweet-Cordero A, Ebert BL, Mak RH, Ferrando AA et al (2005) MicroRNA expression profiles classify human cancers. Nature 435:834–838
Lund E, Güttinger S, Calado A, Dahlberg JE, Kutay U (2004) Nuclear export of microRNA precursors. Science 303:95–98
Maniataki E, Mourelatos Z (2005) A human, ATP-independent, RISC assembly machine fueled by pre-miRNA. Genes Dev 19:2979–2990
Medina PP, Nolde M, Slack FJ (2010) OncomiR addiction in an in vivo model of microRNA-21-induced pre-B-cell lymphoma. Nature 467:86–90
Mols J, van den Berg A, Otsuka M, Zheng M, Chen J, Han J (2008) TNF-alpha stimulation inhibits siRNA-mediated RNA interference through a mechanism involving poly-(A) tail stabilization. Biochim Biophys Acta 1779:712–719

Moossavi S, Bishehsari F (2012) Inflammation in sporadic colorectal cancer. Arch Iran Med 15: 166–170

Mourelatos Z, Dostie J, Paushkin S, Sharma A, Charroux B, Abel L, Rappsilber J, Mann M, Dreyfuss G (2002) miRNPs: a novel class of ribonucleoproteins containing numerous microRNAs. Genes Dev 16:720–728

Pozza A, Scarpa M, Ruffolo C, Polese L, Erroi F, Bridda A, Norberto L, Frego M (2011) Colonic carcinogenesis in IBD: molecular events. Ann Ital Chir 82:19–28

Rath N, Olson MF (2012) Rho-associated kinases in tumorigenesis: re-considering ROCK inhibition for cancer therapy. EMBO Rep 13:900–908

Rogler G (2014) Chronic ulcerative colitis and colorectal cancer. Cancer Lett 345:235–241

Scarpa M, Castagliuolo I, Castoro C, Pozza A, Kotsafti A, Angriman I (2014) Inflammatory colonic carcinogenesis: a review on pathogenesis and immunosurveillance mechanisms in ulcerative colitis. World J Gastroenterol 20:6774–6785

Shin C, Nam JW, Farh KK, Chiang HR, Shkumatava A, Bartel DP (2010) Expanding the microRNA targeting code: functional sites with centered pairing. Mol Cell 38:789–802

Tokushige K, Hashimoto E, Kodama K (2013) Hepatocarcinogenesis in non-alcoholic fatty liver disease in Japan. J Gastroenterol Hepatol 28(4):88–92

Tsai WC, Hsu SD, Hsu CS, Lai TC, Chen SJ, Shen R, Huang Y, Chen HC, Lee CH, Tsai TF et al (2012) MicroRNA-122 plays a critical role in liver homeostasis and hepatocarcinogenesis. J Clin Invest 122:2884–2897

Wang F, Meng W, Wang B, Qiao L (2014) Helicobacter pylori-induced gastric inflammation and gastric cancer. Cancer Lett 345:196–202

Warren JR (2000) Gastric pathology associated with Helicobacter pylori. Gastroenterol Clin North Am 29:705–751

Wightman B, Ha I, Ruvkun G (1993) Posttranscriptional regulation of the heterochronic gene lin-14 by lin-4 mediates temporal pattern formation in C. elegans. Cell 75:855–862

Yi R, Qin Y, Macara IG, Cullen BR (2003) Exportin-5 mediates the nuclear export of pre-microRNAs and short hairpin RNAs. Genes Dev 17:3011–3016

Yoshikawa T, Otsuka M, Kishikawa T, Takata A, Ohno M, Shibata C, Kang YJ, Yoshida H, Koike K (2013) Unique haploinsufficient role of the microRNA-processing molecule Dicer1 in a murine colitis-associated tumorigenesis model. PLoS One 8:e71969

Yoshikawa T, Wu J, Otsuka M, Kishikawa T, Ohno M, Shibata C, Takata A, Han F, Kang YJ, Chen CY et al (2015) ROCK inhibition enhances microRNA function by promoting deadenylation of targeted mRNAs via increasing PAIP2 expression. Nucleic Acids Research 43: 7577–7589.

Zambirinis CP, Pushalkar S, Saxena D, Miller G (2014) Pancreatic cancer, inflammation, and microbiome. Cancer J 20:195–202

Zhang B, Pan X, Anderson TA (2006) MicroRNA: a new player in stem cells. J Cell Physiol 209: 266–269

Chapter 19
Roles of Epstein–Barr Virus Micro RNAs in Epstein–Barr Virus-Associated Malignancies

Ai Kotani

Abstract Epstein–Barr virus (EBV) is an oncogenic human γ-herpes virus that causes various cancers such as Burkitt's lymphoma, Hodgkin's lymphoma, diffuse large B-cell lymphoma (DLBCL), gastric cancer, and nasopharyngeal carcinoma, one of the most common cancers in China.

miRNAs are small noncoding RNAs approximately 22 nucleotides long that posttranscriptionally regulate deadenylation, translation, and decay of their target messenger RNAs (mRNAs). The first viral miRNAs were discovered in human B-cells latently infected with EBV. EBV encodes at least 44 miRNAs. Forty of these miRNAs are transcribed from the BamH1 fragment A rightward transcript (*BART*) region. Several groups reported that EBV-encoded miRNAs target proapoptotic genes, preventing cells from entering the lytic phase. In this review, I discuss several recent findings centered on EBV-encoded miRNAs. In addition, I highlight the fact that small RNAs play important roles in inflammation.

Keywords EBV • miRNA • BART cluster • *BHRF1* • Lymphoma • Cancer

19.1 Introduction

Epstein–Barr virus (EBV) was discovered by examination of electron micrographs of cells isolated from Burkitt's lymphoma, a pediatric tumour common in sub-Saharan Africa, where its unusual geographic distribution indicated a viral etiology. EBV, which belongs to the γ-herpes virus family, has widespread distribution among humans and results in persistent asymptomatic infection of B cells.

A. Kotani (✉)
Division of Hematological Malignancy, Institute of Medical Science, Tokai University, Isehara, Japan

PRESTO, Japan Science and Technology Agency, Kawaguchi 332-0012, Japan
e-mail: aikotani@k-lab.jp

M. Miyasaka, K. Takatsu (eds.), *Chronic Inflammation*,
DOI 10.1007/978-4-431-56068-5_19

EBV usually results in clinically asymptomatic infections but may also cause infectious mononucleosis. In rare cases, EBV infection induces malignant transformation and development of cancers such as Burkitt's lymphoma, gastric carcinoma, and nasopharyngeal carcinoma (NPC), one of the most common cancers in China (Fang et al. 2008). In addition, EBV causes several lymphoid malignancies, including acquired immunodeficiency syndrome (AIDS)-associated lymphoma, Hodgkin's lymphoma, posttransplant lymphoma, age-associated B-cell lymphoma, and peripheral T- and NK-cell lymphomas (Parkin et al. 2005; Parkin 2006).

The primary site of EBV infection is the oropharyngeal cavity (Borza and Hutt-Fletcher 2002). EBV infection induces two distinct patterns of gene expression. During acute (lytic) EBV infection, the virus expresses all its genes sequentially. Linear, double-stranded viral genomes produced during the lytic state are packaged into virions. Shortly after the initial infection, EBV enters into a latent state and expresses only select 'latent' genes, which allows the virus to evade immune surveillance mechanisms of the host and establish a lifelong persistent infection (Ghosh et al. 2007). Serology analyses indicate that approximately 95 % adults worldwide are infected with EBV. After the primary infection, hosts remain lifelong carriers of the virus (Thorley-Lawson 2001).

19.2 EBV Micro-RNAs

19.2.1 Micro-RNAs

Micro-RNAs (miRNAs) are small, noncoding, single-stranded RNAs approximately 21–25 nucleotides in length. They posttranscriptionally regulate mRNA expression and are transcribed from the noncoding regions of genes in all multicellular organisms and in certain viruses (Wang et al. 2008; Chen and Rajewsky 2007). EBV was the first human virus found to encode miRNAs (Pfeffer et al. 2004). EBV encodes 44 miRNAs and a small RNA. EBV miRNAs are encoded by regions located within the *BHRF1* and BamHI-A rightward transcript (*BART*) loci of the EBV genome. The *BHRF1* cluster of miRNAs includes BHRF1-1, BHRF1-2, BHRF1-3, and BHRF1-4 (Chen and Rajewsky 2007; Pfeffer et al. 2004; Grundhoff et al. 2006). The remaining EBV miRNAs, except miR-BART2, are encoded by *BART* clusters 1 and 2. MiR-BART2 is encoded by a region outside the *BART* clusters (Chen and Rajewsky 2007; Pfeffer et al. 2004; Grundhoff et al. 2006; Griffiths-Jones et al. 2006). MiRNAs bind to the 3′ untranslated region (UTR) of mRNAs and interfere with their translation, thus downregulating protein expression. EBV miRNAs have been isolated from various EBV-associated carcinomas and lymphomas such as NPC, gastric carcinoma, diffuse large B-cell lymphoma, nasal T- and NK-cell lymphomas, and Hodgkin's lymphoma (Pfeffer et al. 2004; do Kim et al. 2007). Viral miRNAs play vital roles in immunogenesis, survival and proliferation of host cells, differentiation, lymphomagenesis, and regulation of viral infection and latency (Table 19.1) (Pfeffer et al. 2004; Rana 2007; Xia et al. 2008; Barth et al. 2011; Lu et al. 2012).

Table 19.1 Summary of EBV coding miRNAs

	Function	Target viral	Host target	Cluster
BLHF1-1	Transformation	BFLF2	LILRB-5,E2F1,p53, CBFA2T2	BHRF1
BLHF1-2	Transformation	BFLF2	PIK3R1	BHRF1
BLHF1-3	Transformation	BFLF2	CXCL11,PRF1,TGIF, NSEP1	BHRF1
BART1-5p	Cancer development	LMP1	CXCL12	BART Cluster1
BART2-5p	Viral replication	BALF5, LMP1	MIC B, Bim	
BART3		LMP1	IPO7, Bim	BART Cluster1
BART4		LMP1	Bim	BART Cluster1
BART5	Host cell survival	LMP1	PUMA, Bim	BART Cluster1
BART6	Maintain viral latency	EBNA2,LMP1, Rta,Zta	Dicer, Bim	BART Cluster1
BART7	EMT induction	LMP1	PTEN, Bim	BART Cluster2
BART8				BART Cluster2
BART9				BART Cluster2
BART10				BART Cluster2
BART11				BART Cluster2
BART12				BART Cluster2
BART13				BART Cluster2
BART14				BART Cluster2
BART15				BART Cluster1
BART16	Cancer development	LMP1	TOMM22	BART Cluster1
BART17	Cancer development	LMP1		BART Cluster1
BART18				BART Cluster2
BART19				BART Cluster2
BART20				BART Cluster2

(continued)

Table 19.1 (continued)

	Function	Target viral	Host target	Cluster
BART21				BART Cluster2
BART22		LMP2		BART Cluster2

19.2.2 Functions of EBV miRNAs

EBV miR-BART2 expression is low during latency. It prevents aberrant expression of *BALF5* mRNA, which is essential for viral DNA replication during lytic infection (Barth et al. 2008), The sequence of miR-BART2 is perfectly complementary to the 3′ UTR of *BALF5* mRNA. Thus, miR-BART2 inhibits viral DNA replication by degrading *BALF5* mRNA. MiRNA-guided cleavage of mRNAs requires association with AGO2 (Meister et al. 2004), a member of the Argonaute family of proteins and a component of RNA-induced silencing complex (RISC). MiR-BART2 associates with AGO2 and guides the sequence-specific cleavage of *BALF5* mRNA. MiR-BART2-guided cleavage of *BALF5* mRNA substantially decreases after induction of the lytic cycle in EBV-infected cells (Barth et al. 2008). The amount of miR-BART2 is reduced during lytic infection, which in turn derepresses BALF5 expression (Barth et al. 2008). However, it is unclear whether miR-BART2-mediated regulation of viral replication is completely controlled by BALF5.

MiR-BART6, which is regulated by RNA editing, is another regulator of the shift from latent EBV infection to lytic infection (Iizasa et al. 2010). Editing of wild-type primary (pre)-miR-BART6 dramatically decreases loading of miR-BART6-5p onto RISC without affecting the processing of precursor or mature miRNAs (Iizasa et al. 2010). Editing of pre-miRNA might affect selection and loading of the guide strand onto RISC (Khvorova et al. 2003). MiR-BART6-5p silences *DICER* by binding to multiple target sites in the 3′ UTR of *Dicer* mRNA. In contrast, miR-BART6-3p is unable to perform this function (Iizasa et al. 2010).

MiR-BART5 promotes host cell survival by regulating p53-upregulated modulator of apoptosis (PUMA) (Choy et al. 2008). PUMA is an apoptotic protein belonging to the BH3-only group of the Bcl-2 family of proteins and is encoded by *BBC3* (Han et al. 2001; Nakano and Vousden 2001; Yu et al. 2001). The 3′ UTR of *BBC3* is perfectly complementary to miR-BART5, and binding of miR-BART5 to this region suppresses PUMA expression.

Abundant expression of miR-BART5 significantly downregulates PUMA expression in 60 % of NPC tissues (Choy et al. 2008). BART5 uses this mechanism to promote survival of NPC cells, EBV-infected gastric carcinoma cells, and

EBV-infected epithelial cells (Choy et al. 2008). Therefore, miR-BART5 may be a good target for anticancer therapy in EBV-infected cancer cells.

LMP1 is a viral protein expressed during type III latent EBV infections (Hislop et al. 2007; Pagano et al. 2004). LMP1 promotes cell growth and B-cell transformation, resists serum deprivation-induced apoptosis, and induces phenotypic changes in epithelial cells. *BART1* cluster miRNAs negatively regulate LMP1 expression and limit its inappropriately high levels, thereby preventing apoptosis induced by LMP1-mediated changes in UPR. *BART1* cluster miRNAs, namely miR-BART16, miR-BART17-5p, and miR-BART1-5p, are recognized by target sites in the 3′ UTR of the mRNA expressing *LMP1* (Lo et al. 2007). These miRNAs regulate LMP1 expression at the posttranscriptional level.

BHRF1 is a latent protein expressed in growth-transformed cells that contributes to virus-associated lymphomagenesis (Kelly et al. 2009). MiR-BHRF1 downregulates BHRF1 expression, modulates cell transformation (Seto et al. 2010), and promotes B-cell proliferation following EBV infection. EBV-infected B cells lacking miR-BHRF1 progress into the cell cycle less efficiently and eventually die through apoptosis (Seto et al. 2010). MiR-BHRF1 is constitutively expressed in LCLs (Seto et al. 2010). The proportion of G1/G0 cells increases whereas that of S-phase cells decreases in the absence of miR-BHRF1 (Seto et al. 2010), indicating its key role in controlling proliferation of latently-infected cells. EBV-mediated differentiation of resting B cells into active B cells requires time. MiR-BHRF1 acts at the stage of the EBV life cycle during which multiple EBV oncogenes are activated.

EBV miR-BART7-3p enhances cell migration and invasion in vitro and cancer metastasis in vivo. EMT is characterized by the loss of epithelial markers and gain of mesenchymal features in NPC cells. Mechanistic studies indicate that EBV miR-BART7-3p targets PTEN, a major human tumour suppressor, and modulates PI3K/Akt/GSK-3β signalling, thus upregulating the expression and nuclear accumulation of Snail and β-catenin, which favor EMT (Lu et al. 2012; Cai et al. 2015).

19.3 EBV-Encoded Secretory miRNAs

19.3.1 Secretory miRNAs

Cellular and viral miRNAs control gene expression by repressing the translation of mRNAs into proteins, a process that is tightly regulated in healthy cells but that is deregulated in cancerous and virus-infected cells. Interestingly, miRNAs are not strictly intracellular; they are found in peripheral blood and are secreted into the culture media in small vesicles called exosomes (Kosaka et al. 2010). It has been suggested that exosome-associated miRNAs play important roles in intercellular communication (Kosaka et al. 2010); however, experimental evidence supporting this is not enough. Moreover, the dynamics and mechanism of miRNA secretion by exosomes are poorly understood. It is unclear whether miRNAs are secreted in physiologically relevant amounts and whether exogenous exosome-associated miRNAs can access the molecular machinery to undergo processing.

19.3.2 EBV-Encoded Secretory miRNAs

Pegtel et al. were the first to show that exosomes deliver viral miRNAs to noninfected cells (Pegtel et al. 2010). They used EBV B95.8-immortalised LCLs and showed that exosomes containing *BHRF1* cluster miRNAs targeted *CXCL11* mRNA in nearby uninfected cells. Furthermore, they showed that non-B cells obtained from EBV-infected patients with elevated viral loads contained EBV miRNAs, suggesting that exosomes transferred miRNAs to uninfected cells in vivo. These findings were confirmed by two studies reporting the release of exosomes from NPC cells. Gourzones et al. detected EBV miR-BART-containing exosomes in serum samples from patients with NPC and from mice xenografted with human NPC cells (Lu et al. 2012; Gourzones et al. 2010).

19.4 EBV miRNAs Regulate Inflammation and Immune Evasion

19.4.1 Inflammation and Oncogenesis

The potential of chronic inflammation to lead to oncogenesis has been established in malignancies. Chronic hepatitis caused by HCV can develop into hepatocarcinoma, autoimmune-mediated chronic colitis, colon cancer, *Helicobacter pilori*-induced chronic gastritis, and gastric cancer. These transitions require molecular and cellular interactions involving immune and nonimmune cells, cytokines, pathogens, and other factors (Grivennikov 2013; Shlomai et al. 2014).

Deficiency of proinflammatory cytokines such as TNF-alpha limits inflammation and suppresses oncogenesis in several models. Conversely, genetic deletion of immunosuppressive cytokines such as IL-10, and TGF-beta exacerbate inflammation and facilitate oncogenesis (Rickinson 2014). The mechanisms underlying inflammation-mediated regulation of oncogenesis have not been fully elucidated, but several studies are currently addressing this issue. Proinflammatory cytokines cause oxidative stress and production of reactive oxygen species (ROS) that induce DNA damage such as double strand breaks (DSB), and genomic instability (Lemercier 2015; Anuranjani 2014). Genomic instability can also be induced by active mutagenesis caused by activation-induced cytidine deaminase (AID). The minimal promoter of the gene encoding AID is controlled by NF-kB and gene expression is induced by several proinflammatory cytokines such as TNF-alpha (Okazaki et al. 2007). AID was originally discovered in the year 2000 as an essential component of the DNA modification step of class switch recombination (CSR) and somatic hypermutation (SHM) events that occur in B-cell immunoglobulin genes. AID expression is tightly regulated only in transient pre-B cells, in germinal center B cells (GC-B cells), and in activated mature B cells (Okazaki et al. 2007; Shimizu et al. 2012; Honjo et al. 2012). Interestingly, AID dysregulation induces SHM in genes other than immunogloblins. In B-cell lymphoma, MYC, BCL6, PIM1, and numerous other genes were found to be massively mutated (Kotani et al. 2005). Moreover, aberrant AID expression outside B cells has been reported to be involved in oncogenesis associated with cancers where inflammation has been linked to oncogenesis, such as gastric cancer, hepatocarcinoma, and others (Okazaki et al. 2007).

19.4.2 EBV-Related Cancer and Inflammation

EBV-related cancers are usually accompanied by severe inflammation. EBV-positive DLBCL of the elderly and EBV-positive Hodgkin's lymphoma show massive infiltration of tumours with lymphocytes, eosinophils, stromal cells, and fibroblasts. NKT lymphomas have poor prognoses and are characterised by severe inflammation. These observations strongly suggest that inflammation is involved in EBV oncogenesis (Lu et al. 2012). This hypothesis is supported by a study showing that in EBV-associated lymphoma, only tumour cells without bystander cells fail to be engrafted into immunodeficient mice. This observation suggests that bystander cells support tumour cell survival. EBV infection has been shown to activate NFkB, which as previously mentioned induces AID expression in mature B cells (Okazaki et al. 2007). However, whether aberrant AID expression during EBV infections is involved in EBV oncogenesis remains to be elucidated. Intriguingly, certain EBV-related cancers present with coinfection with malaria or HIV. For example, endemic Burkitt's lymphoma is common in sub-Saharan Africa, which also happens to be the endemic area for malaria. Several lines of evidence, including epidemiological studies, strongly indicate that malaria induces Burkitt's lymphoma through a mechanism that involves T-cell dysfunction, or other direct cofactors. Similarly, HIV infection has been reported to play certain roles in EBV-induced lymphomagenesis. The fact that T-cell maintenance in the HAART era is not associated with decreased incidence of EBV-related lymphomas such as Hodgkin's lymphoma, suggests that HIV infection and not the degree of immunocompromise, plays a role in these diseases (Rickinson 2014).

19.4.3 EBV miRNAs Regulate Inflammation and Immune Evasion

Several EBV miRNAs have been reported to be involved in inflammation and immune evasion. Major histocompatibility complex class I polypeptide-related sequence B (MICB) is a ligand of the NKG2D type II receptor, a stress-induced immune molecule (Bahram et al. 1994; Groh et al. 1996). Both B cells and endothelial cells, which are the targets of EBV, express MICB. Binding of MICB activates NK, $CD8^{+}\alpha\beta$, and $CD8^{+}\gamma\delta$ T cells (Suarez-Alvarez et al. 2009). MICB expression on the cell surface is upregulated in response to various insults such as viral infection, tumour formation, heat shock, and DNA damage. Therefore, EBV downregulates MICB expression to decrease immune detection by NK cells. Previous studies have shown that downregulated MICB expression decreases the lysis of infected cells by NK cells (Stern-Ginossar et al. 2007). The 3′ UTR of *MICB* mRNA has potential binding sites for EBV miR-BART2-5p (Nachmani et al. 2009). EBV downregulates MICB expression by employing miR-BART2-5p, thus decreasing NK-cell–mediated lysis to prevent detection by immune cells.

MiR-BART1-1 is expressed from the 5′ UTR and miR-BART1-2 and miR-BART1-3 are expressed from the 3′ UTR of *BHRF1* in EBV-infected cells (Xia et al. 2008). MiR-BART1-3 expression is markedly elevated in EBV-infected type III latent cells (Xia et al. 2008). In addition, miR-BART1-3 has been detected in cells isolated from EBV-positive primary effusion lymphoma and AIDS-associated diffuse large B-cell lymphoma (Xia et al. 2008). *BHRF1* cluster miRNAs are characteristically detected in EBV-infected type III latent cells (Xing and Kieff 2007). EBV miR-BHRF1-3 regulates host immunity by downregulating interferon (IFN)-inducible T-cell–attracting chemokine (I-TAC, also known as CXCL11). CXCL11 belongs to the CXC family of chemokines, and its expression is strongly induced by both IFN-β and IFN-γ (Rani et al. 1996). CXCL11 promotes cell-mediated immunity by attracting activated T cells. The 3′ UTR of *CXCL11* mRNA shows 100 % complementarity to miR-BART1-3 and therefore is a target of miR-BART1-3. MiR-BART1-3 inversely regulates CXCL11 expression whereas the antisense sequence of miR-BART1-3 has an opposite effect (Pfeffer et al. 2004). MiR-BART1-3 significantly reduces CXCL11 expression at both the mRNA and protein levels (Xia et al. 2008). Thus, because cellular chemokines are the targets of viral miRNAs, EBV may regulate antigen processing and presentation and downregulate CTL cytokine networks through this mechanism.

19.5 Concluding Remarks

EBV-associated cancers are generally difficult to cure. Despite extensive studies on well-known concepts and methods, the molecular mechanisms through which EBV induces tumourigenesis and eludes immune surveillance remain unclear. Recent studies using mouse models of EBV-mediated lymphoproliferative diseases have shown that EBV infection of B cells is necessary but not sufficient to induce tumourigenesis because all peripheral mononuclear cells are needed to generate tumours in these mice (Kuppers 2009). Immune cells are also indispensable for EBV-induced tumourigenesis. However, the detailed roles of inflammation in EBV-induced lymphomagenesis and the relationship between these cells and EBV-infected cells with respect to tumourigenesis remain unclear. Moreover, the mechanisms underlying drug resistance, which results in poor prognoses of EBV-related tumours, have not yet been elucidated. Therefore, it is important to study the biology of EBV-associated tumours from a new perspective such as that provided by investigations focused on EBV miRNAs.

References

Anuranjani BM (2014) Concerted action of Nrf2-ARE pathway, MRN complex, HMGB1 and inflammatory cytokines - implication in modification of radiation damage. Redox Biol 2:832–846

Bahram S, Bresnahan M, Geraghty DE, Spies T (1994) A second lineage of mammalian major histocompatibility complex class I genes. Proc Natl Acad Sci U S A 91(14):6259–6263

Barth S, Pfuhl T, Mamiani A et al (2008) Epstein-Barr virus-encoded microRNA miR-BART2 down-regulates the viral DNA polymerase BALF5. Nucleic Acids Res 36(2):666–675

Barth S, Meister G, Grässer FA (2011) EBV-encoded miRNAs. Biochim Biophys Acta 1809 (11–12):631–640

Borza CM, Hutt-Fletcher LM (2002) Alternate replication in B cells and epithelial cells switches tropism of Epstein-Barr virus. Nat Med 8(6):594–599

Cai LM, Lyu XM, Luo WR, Cui XF, Ye YF, Yuan CC, Peng QX, Wu DH, Liu TF, Wang E, Marincola FM, Yao KT, Fang WY, Cai HB, Li X (2015) EBV-miR-BART7-3p promotes the EMT and metastasis of nasopharyngeal carcinoma cells by suppressing the tumor suppressor PTEN. Oncogene 34(17):2156–66

Chen K, Rajewsky N (2007) The evolution of gene regulation by transcription factors and microRNAs. Nat Rev Genet 8(2):93–103

Choy EY, Siu KL, Kok KH et al (2008) An Epstein-Barr virus-encoded microRNA targets PUMA to promote host cell survival. J Exp Med 205(11):2551–2560

do Kim N, Chae HS, Oh ST et al (2007) Expression of viral microRNAs in Epstein-Barr virus-associated gastric carcinoma. J Virol 81(2):1033–1036

Fang W, Li X, Jiang Q et al (2008) Transcriptional patterns, biomarkers and pathways characterizing nasopharyngeal carcinoma of Southern China. J Transl Med 6:32

Ghosh SK, Forman LW, Akinsheye I, Perrine SP, Faller DV (2007) Short, discontinuous exposure to butyrate effectively sensitizes latently EBV-infected lymphoma cells to nucleoside analogue antiviral agents. Blood Cells Mol Dis 38(1):57–65

Gourzones C, Gelin A, Bombik I et al (2010) Extra-cellular release and blood diffusion of BART viral micro-RNAs produced by EBV-infected nasopharyngeal carcinoma cells. Virol J 7:271

Griffiths-Jones S, Grocock RJ, van Dongen S, Bateman A, Enright AJ (2006) miRBase: microRNA sequences, targets and gene nomenclature. Nucleic Acids Res 34(Database issue):D140–144

Grivennikov SI (2013) Inflammation and colorectal cancer: colitis-associated neoplasia. Semin Immunopathol 35(2):229–244

Groh V, Bahram S, Bauer S, Herman A, Beauchamp M, Spies T (1996) Cell stress-regulated human major histocompatibility complex class I gene expressed in gastrointestinal epithelium. Proc Natl Acad Sci U S A 93(22):12445–12450

Grundhoff A, Sullivan CS, Ganem D (2006) A combined computational and microarray-based approach identifies novel microRNAs encoded by human gamma-herpesviruses. RNA 12 (5):733–750

Han J, Flemington C, Houghton AB et al (2001) Expression of bbc3, a pro-apoptotic BH3-only gene, is regulated by diverse cell death and survival signals. Proc Natl Acad Sci U S A 98 (20):11318–11323

Hislop AD, Taylor GS, Sauce D, Rickinson AB (2007) Cellular responses to viral infection in humans: lessons from Epstein-Barr virus. Annu Rev Immunol 25:587–617

Honjo T, Kobayashi M, Begum N, Kotani A, Sabouri S, Nagaoka H (2012) The AID dilemma: infection, or cancer? Adv Cancer Res 113:1–44

Iizasa H, Wulff BE, Alla NR et al (2010) Editing of Epstein-Barr virus-encoded BART6 microRNAs controls their dicer targeting and consequently affects viral latency. J Biol Chem 285(43):33358–33370

Kelly GL, Long HM, Stylianou J et al (2009) An Epstein-Barr virus anti-apoptotic protein constitutively expressed in transformed cells and implicated in burkitt lymphomagenesis: the Wp/BHRF1 link. PLoS Pathog 5(3):e1000341

Khvorova A, Reynolds A, Jayasena SD (2003) Functional siRNAs and miRNAs exhibit strand bias. Cell 115(2):209–216

Kosaka N, Iguchi H, Ochiya T (2010) Circulating microRNA in body fluid: a new potential biomarker for cancer diagnosis and prognosis. Cancer Sci 101(10):2087–2092

Kotani A, Okazaki IM, Muramatsu M et al (2005) A target selection of somatic hypermutations is regulated similarly between T and B cells upon activation-induced cytidine deaminase expression. Proc Natl Acad Sci U S A 102(12):4506–4511

Kuppers R (2009) Molecular biology of Hodgkin lymphoma. Hematol Am Soc Hematol Educ Progr:1491–496

Lemercier C (2015) When our genome is targeted by pathogenic bacteria. Cell Mol Life Sci 72 (14):2665–2676

Lo AK, To KF, Lo KW et al (2007) Modulation of LMP1 protein expression by EBV-encoded microRNAs. Proc Natl Acad Sci U S A 104(41):16164–16169

Lu J, Chanda B, Kotani A (2012) EBV encoded miRNAs in EBV related malignancy. Hematol Sci Pract 409–425.

Meister G, Landthaler M, Patkaniowska A, Dorsett Y, Teng G, Tuschl T (2004) Human Argonaute2 mediates RNA cleavage targeted by miRNAs and siRNAs. Mol Cell 15 (2):185–197

Nachmani D, Stern-Ginossar N, Sarid R, Mandelboim O (2009) Diverse herpesvirus microRNAs target the stress-induced immune ligand MICB to escape recognition by natural killer cells. Cell Host Microbe 5(4):376–385

Nakano K, Vousden KH (2001) PUMA, a novel proapoptotic gene, is induced by p53. Mol Cell 7 (3):683–694

Okazaki IM, Kotani A, Honjo T (2007) Role of AID in tumorigenesis. Adv Immunol 94:245–273

Pagano JS, Blaser M, Buendia MA et al (2004) Infectious agents and cancer: criteria for a causal relation. Semin Cancer Biol 14(6):453–471

Parkin DM (2006) The global health burden of infection-associated cancers in the year 2002. Int J Cancer 118(12):3030–3044

Parkin DM, Bray F, Ferlay J, Pisani P (2005) Global cancer statistics, 2002. CA Cancer J Clin 55 (2):74–108
Pegtel DM, Cosmopoulos K, Thorley-Lawson DA et al (2010) Functional delivery of viral miRNAs via exosomes. Proc Natl Acad Sci U S A 107(14):6328–6333
Pfeffer S, Zavolan M, Grasser FA et al (2004) Identification of virus-encoded microRNAs. Science 304(5671):734–736
Rana TM (2007) Illuminating the silence: understanding the structure and function of small RNAs. Nat Rev Mol Cell Biol 8(1):23–36
Rani MR, Foster GR, Leung S, Leaman D, Stark GR, Ransohoff RM (1996) Characterization of beta-R1, a gene that is selectively induced by interferon beta (IFN-beta) compared with IFN-alpha. J Biol Chem 271(37):22878–22884
Rickinson AB (2014) Co-infections, inflammation and oncogenesis: future directions for EBV research. Semin Cancer Biol 26:99–115
Seto E, Moosmann A, Gromminger S, Walz N, Grundhoff A, Hammerschmidt W (2010) Micro RNAs of Epstein-Barr virus promote cell cycle progression and prevent apoptosis of primary human B cells. PLoS Pathog 6(8):e1001063
Shimizu T, Marusawa H, Endo Y, Chiba T (2012) Inflammation-mediated genomic instability: roles of activation-induced cytidine deaminase in carcinogenesis. Cancer Sci 103 (7):1201–1206
Shlomai A, de Jong YP, Rice CM (2014) Virus associated malignancies: the role of viral hepatitis in hepatocellular carcinoma. Semin Cancer Biol 26:78–88
Stern-Ginossar N, Elefant N, Zimmermann A et al (2007) Host immune system gene targeting by a viral miRNA. Science 317(5836):376–381
Suarez-Alvarez B, Lopez-Vazquez A, Baltar JM, Ortega F, Lopez-Larrea C (2009) Potential role of NKG2D and its ligands in organ transplantation: new target for immunointervention. Am J Transplant 9(2):251–257
Thorley-Lawson DA (2001) Epstein-Barr virus: exploiting the immune system. Nat Rev Immunol 1(1):75–82
Wang X, Gu J, Zhang MQ, Li Y (2008) Identification of phylogenetically conserved microRNA cis-regulatory elements across 12 Drosophila species. Bioinformatics 24(2):165–171
Xia T, O'Hara A, Araujo I et al (2008) EBV microRNAs in primary lymphomas and targeting of CXCL-11 by ebv-mir-BHRF1-3. Cancer Res 68(5):1436–1442
Xing L, Kieff E (2007) Epstein-Barr virus BHRF1 micro- and stable RNAs during latency III and after induction of replication. J Virol 81(18):9967–9975
Yu J, Zhang L, Hwang PM, Kinzler KW, Vogelstein B (2001) PUMA induces the rapid apoptosis of colorectal cancer cells. Mol Cell 7(3):673–682

Part IV
Chronic Inflammation and Obesity/ Environmental Stress

Chapter 20
Chronicity of Immune Abnormality in Atopic Dermatitis: Interacting Surface Between Environment and Immune System

Takanori Hidaka, Eri H. Kobayashi, and Masayuki Yamamoto

Abstract Atopic dermatitis is a pruritic inflammatory skin disease that is greatly increasing worldwide. As the disease can be divided into remission (acute) and nonremission (chronic) groups, atopic dermatitis is an appropriate model for exploration of the mechanisms underlying the chronicity of inflammation. The nonremission group is characterised by the subsequent complication with systemic allergic sequelae referred to as 'atopic march'. A central pathophysiology of atopic march has been revealed to be percutaneous sensitisation to environmental antigens, which results from memory Th2 cell induction. Characteristic pathophysiology of atopic dermatitis, including scratching, *Filaggrin*-null mutation, epidermal production of proinflammatory cytokines, such as *Tslp* and *Il33*, and induction of type2 innate lymphoid cells have been revealed to contribute to the maintenance and increment of memory Th2 cells, leading to the nonremission course of the disease. Recent rapid increment of the prevalence suggests that environmental factors make a significant contribution to the pathogenesis of atopic dermatitis. A number of recent studies have demonstrated a positive correlation between the prevalence of atopic dermatitis and air pollution levels. In addition, several studies also demonstrated that air pollutants contribute to the induction of atopic march and memory T-cell formation. These air pollutants appear to activate stress-responsive transcription factors, AhR and Nrf2, and these transcription factors elicit a response

T. Hidaka
Department of Medical Biochemistry, Tohoku University Graduate School of Medicine, 2-1 Seiryo-machi, Aoba-ku, Sendai 980-8575, Japan

Department of Dermatology, Tohoku University Graduate School of Medicine, 1-1 Seiryo-machi, Aoba-ku, Sendai 980-8574, Japan

E.H. Kobayashi
Department of Medical Biochemistry, Tohoku University Graduate School of Medicine, 2-1 Seiryo-machi, Aoba-ku, Sendai 980-8575, Japan

M. Yamamoto (✉)
Department of Medical Biochemistry, Tohoku University Graduate School of Medicine, 2-1 Seiryo-machi, Aoba-ku, Sendai 980-8575, Japan

AMED-CREST, Sendai 980-8575, Japan
e-mail: masiyamamoto@med.tohoku.ac.jp

M. Miyasaka, K. Takatsu (eds.), *Chronic Inflammation*,
DOI 10.1007/978-4-431-56068-5_20

associated with inflammation. However, the exact mechanisms underlying the air-pollutant–induced chronicity of atopic dermatitis remain to be elucidated.

Keywords Atopic Dermatitis • Sensitisation • Memory Th2 cell • Tslp • Il-33 • ILC2 • Air pollutant • AhR • Nrf2

20.1 Introduction: Atopic Dermatitis and Chronicity in Inflammation

The skin is an interface where the body and the environment interact. The skin functions as an immunological barrier that protects the body against environmental pathogens (Nestle et al. 2009). Indeed, the skin contains as many as 2×10^{10} T cells, which is approximately twice the number of circulating T cells (Clark et al. 2006b). These resident T cells are predominantly memory T cells (Clark et al. 2006b), which participate in long-lasting protection against specific pathogens. The breakdown of the cutaneous immune barrier leads to T-cell–mediated chronic inflammatory diseases, and T- cell–mediated excessive inflammation in the skin can cause a variety of diseases, including atopic dermatitis, psoriasis, alopecia areata, lichen planus, cutaneous graft-versus-host disease, and vitiligo.

Among these cutaneous diseases, atopic dermatitis is an appropriate model for investigating the mechanisms involved in chronicity because atopic dermatitis patients include both remission (acute) and nonremission (chronic) groups. Chronic atopic dermatitis patients develop characteristic complications that are related to immunological memory. Atopic dermatitis is a highly clinically significant disease because its prevalence has been rapidly increasing worldwide, and its incidence has reached 15–30% in children and 2–10% in adults over the last 30–40 years (Williams and Flohr 2006; Bieber 2008). Atopic dermatitis significantly impairs the quality of life as a result of intense pruritus and sleep disturbance (Beattie and Lewis-Jones 2006). Therefore, establishing a method to prevent atopic dermatitis is urgently needed. In this review, we describe the central pathophysiology that leads to the chronicity of atopic dermatitis, including antigen-sensitisation and memory Th2-cell formation. We also discuss the genetic, immunological, and environmental factors that are known to affect the chronicity of atopic dermatitis.

20.2 The Skin as an Immunological Barrier to External Insults

It is widely recognised that the pathophysiology of atopic dermatitis is caused by defects in the cutaneous immunological barrier that subsequently result in systemic immune abnormalities (Fig. 20.1). This concept is referred to as the 'outside-inside' model. Because the skin protects the body from environmental insults, such as

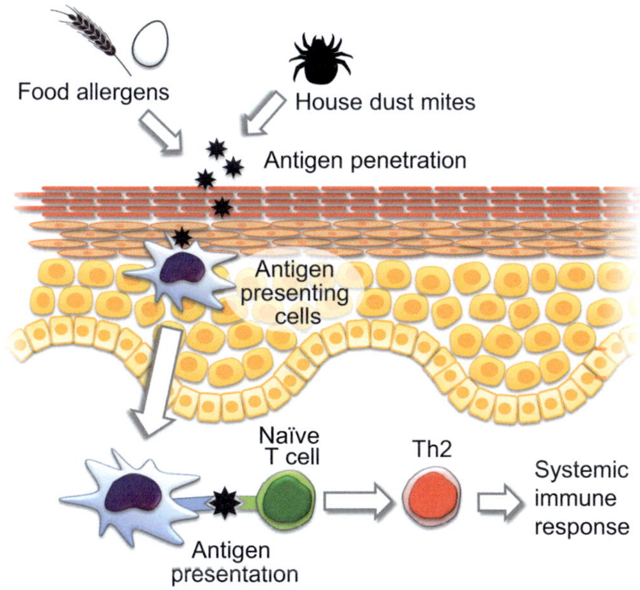

Fig. 20.1 The pathophysiology of atopic dermatitis. Atopic dermatitis leads to the percutaneous penetration of environmental antigens and subsequent immune responses. Antigens that exacerbate atopic dermatitis have been derived from foods, house dust mites, animal dander, and other environmental factors. Once antigens penetrate into the skin, they are engulfed and processed by antigen-presenting cells. The antigen-presenting cells then present the antigen to naïve helper T cells to provoke Th2 sensitisation, which leads to systemic abnormal immune responses

protein antigens (Nestle et al. 2009), disruption of the functions of the skin barrier can result in environmental antigens penetrating through the epidermis (Hogan et al. 2012). Penetrated antigens are engulfed and processed by antigen-presenting cells, such as Langerhans cells in the epidermis and by dendritic cells in the dermis. These antigen-presenting cells then migrate to the regional lymph nodes to present the antigen-derived epitopes to naïve T cells, which then express the T-cell receptor (TCR) that corresponds to the epitopes.

Antigen-presented naïve T cells then proliferate and differentiate into antigen-specific effector/memory T cells. Activated T cells migrate to sites that experience primary exposure to antigens, and once there, they act to remove the antigens. Finally, after an effector T cell has removed an antigen, the majority of these cells fall into apoptosis (contraction phase). However, a limited number of memory T cells are maintained for recurrent exposure to antigens (Murali-Krishna et al. 1998; Kupper and Fuhlbrigge 2004; Wherry and Ahmed 2004).

20.3 Sensitisation to Environmental Allergens Induces Chronicity in Atopic Dermatitis

A considerable proportion of the cases of atopic dermatitis result in a persistent and chronic condition (Fig. 20.2). Although a large number of atopic dermatitis cases spontaneously remit during childhood, approximately half of the cases of atopic dermatitis that have their onset in infancy persist into adulthood (Spergel and Paller 2003). Many atopic dermatitis patients subsequently develop other allergic diseases, including asthma, allergic rhinitis, and food allergies, which can become life-threatening symptoms. The development of these allergic disease sequelae is referred to as 'atopic march'. Approximately 30 and 60% of atopic dermatitis patients develop asthma and allergic rhinoconjunctivitis, respectively (Ricci et al. 2006). A number of studies have also confirmed that atopic dermatitis is associated with an increased risk of developing asthma (Illi et al. 2004; Lowe et al. 2008; Saunes et al. 2012; von Kobyletzki et al. 2012). It is therefore important to explore the mechanisms underlying chronicity in atopic dermatitis so that we can develop a comprehensive understanding of the events that lead to chronic inflammation.

One of the most influential factors that leads to a nonremission status in atopic dermatitis is sensitisation to environmental antigens (von Kobyletzki et al. 2015). A family history of atopic dermatitis and frequent scratching are also important factors (Illi et al. 2004; von Kobyletzki et al. 2015). Food sensitisation in atopic dermatitis patients increases the risk of developing asthma and contributes to the progress of atopic march (Ricci et al. 2006; Marenholz et al. 2009; Filipiak-Pittroff et al. 2011),

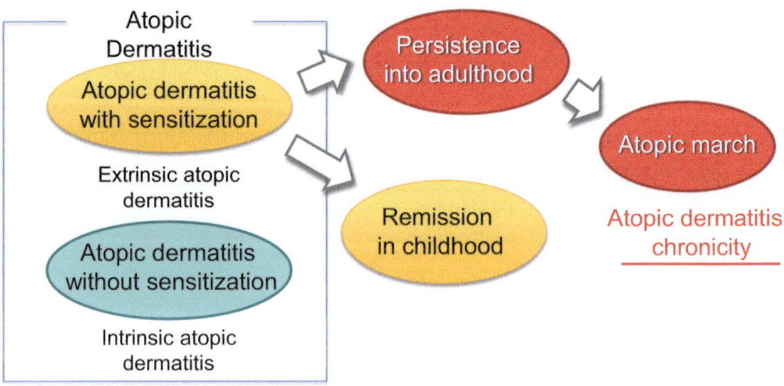

Fig. 20.2 The progression of chronic atopic dermatitis. Atopic dermatitis includes two subtypes. The major type of atopic dermatitis is accompanied by sensitisation to antigens and is called extrinsic atopic dermatitis. The minor type of atopic dermatitis, which is called intrinsic atopic dermatitis, lacks sensitisation and antigen-specific IgE production and is rarely accompanied by other allergic diseases. Although the majority of atopic dermatitis cases remit during childhood, a portion of atopic dermatitis cases result in chronic symptoms, including persistence into adulthood and coincidence with other allergic diseases, and this process is called atopic march

and antigen sensitisation is a significant predictor of asthmatic symptoms in later life (Illi et al. 2004). In support of the significance of antigen sensitisation in eliciting atopic march, 'intrinsic' or nonsensitised types of atopic dermatitis account for approximately 15% of such cases. These patients lack an antigen-specific IgE response to allergens (Ott et al. 2009; Roguedas-Contios and Misery 2011), and this disease rarely accompanies asthma (Wüthrich and Schmid-Grendelmeier 2002). These studies collectively indicate that sensitisation to antigens is a critical step in the chronicity of atopic dermatitis.

20.4 Th2 Cells Are Central Players in Atopic Dermatitis and Atopic March

Many types of cells are involved in sensitisation in atopic dermatitis, and T cells are central players. Inhibiting T-cell functions using calcineurin inhibitors had a significant therapeutic effect that included resolving atopic dermatitis lesions (Ashcroft et al. 2005). T helper type 2 (Th2) cells are a subset of T cells that are critical to the pathogenesis of atopic dermatitis. Atopic dermatitis lesions are characterised by the infiltration of IL-4–, IL-5–, and IL-13–producing Th2 cells, although IFN-γ and TNF-α–producing T helper type 1 (Th1) cells have been detected in chronic lesions (Leung et al. 2004). In relation to atopic march, Th2 cells are thought to be pathogenic in both atopic dermatitis and atopic-dermatitis–associated diseases (Spergel and Paller 2003; Cookson 2004; Locksley 2010). In support of this hypothesis, Th2 cells that were sensitised to allergens derived from house dust mites have been detected in the skin of atopic dermatitis patients (van Reijsen et al. 1992).

20.5 Skin-Homing Memory Th2 Cells Contribute to Chronicity in Atopic Dermatitis

Long-lived Th2 memory cells are essential to maintaining the chronicity of allergic inflammatory reactions and relapses upon re-exposure to allergens (Prescott et al. 1999; Mojtabavi et al. 2002). Memory T cells that possess the ability to home toward the skin are therefore considered to be key to chronicity in atopic dermatitis and to contribute to atopic march.

Cutaneous leukocyte-associated antigen (CLA) is often used as a marker for skin-homing T cells. CLA is a ligand for selectin, which is expressed on cutaneous postcapillary venules. CLA is essential to T cell homing toward the skin (Biedermann et al. 2002, 2006). Almost all T cells that infiltrate into the skin express CLA in combination with CD45RO, a marker of memory T cells (Clark et al. 2006a). Circulating $CLA^+/CD45RO^+$ skin-homing memory T cells obtained from atopic dermatitis patients produce higher levels of Th2 cytokines upon allergen stimulation than cells from healthy controls and/or CLA^- memory T cells (Seneviratne et al. 2007). The expression levels of activation markers, such as HLA-DR and CD25, are also enhanced in CLA^+ T cells obtained from atopic dermatitis patients (Santamaria Babi et al. 1995; Seneviratne et al. 2007). These data indicate an association between CLA^+ memory T cells and atopic dermatitis pathogenesis.

CLA^+ T cells also express the chemokine receptor CCR4 (Campbell et al. 1999; Biedermann et al. 2002; Clark et al. 2006a). CCR4 is the receptor for the chemokine ligands thymus and activation-regulated chemokine (TARC or CCL17) and macrophage-derived chemokine (MDC or CCL22). Both CCL17 and CCL22 are produced in atopic dermatitis lesions (Homey et al. 2006), and serum levels of CCL17 and CCL22 and the expression of CCR4 on peripheral blood $CD4^+$ T cells have been significantly correlated with disease activity (Kakinuma et al. 2001, 2002; Wakugawa et al. 2001).

Recently, CRTH2 (chemoattractant receptor–homologous molecule expressed on Th2 cells) was identified as a reliable marker for Th2-committed memory T cells (Cosmi et al. 2000; Wang et al. 2006). More $CRTH2^+$ T cells were observed in peripheral blood obtained from atopic dermatitis patients than in blood obtained from healthy subjects, and these cells were also detected in atopic dermatitis skin lesions (Cosmi et al. 2000; Iwasaki et al. 2002; Wang et al. 2006). A mouse model of epicutaneous allergen exposure also revealed that CRTH2 signalling is essential to antigen-induced allergic inflammation (Oiwa et al. 2008; He et al. 2010).

Based on these observations, circulating $CLA^+/CCR4^+/CRTH2^+$ memory Th2 cells are thought to be skin-homing memory T cells that are characteristic of atopic dermatitis. These memory Th2 cells are likely to be involved in the induction of atopic dermatitis chronicity, and they may migrate into other organs to provoke atopic march.

20.6 T-Cell Sensitisation Enhances the Maintenance of Immunological Memory

The chronicity of atopic dermatitis requires the continuous generation or maintenance of memory Th2 cells because the number of memory $CD4^+$ T cells would otherwise gradually decline over time (Homann et al. 2001; Kurtulus et al. 2013). Cytokines, such as IL-7 and IL-15, are critical to maintaining the homeostasis and survival of memory $CD4^+$ cells (Croft 2003; Kondrack et al. 2003; Lenz et al. 2004; Purton et al. 2007), and TCR signalling is also important for the persistence of memory T cells (Kassiotis et al. 2002; Bushar et al. 2010; Lees and Farber 2010).

Interestingly, prolonged antigen stimulation of $CD4^+$ T cells increases the number of memory T cells and promotes the production of cytokines upon recall by antigens (Moulton et al. 2006). Chronic exposure to antigens and the resulting amplification of Th2 sensitisation may therefore lead to the maintenance of memory

T cells, which is associated with chronic inflammation. In support of this hypothesis, recent studies have identified factors that promote Th2 sensitisation and the establishment of memory T cells by contributing to increased antigen infiltration and an exacerbated Th2-type immune response. In addition to these intrinsic factors, environmental factors also affect T-cell sensitisation. In the next section of this review, we discuss the factors that exacerbate allergic T-cell sensitisation and contribute to the development and persistence of atopic dermatitis.

20.7 Factors That Promote T-Cell Sensitisation During the Development of Atopic Dermatitis

20.7.1 Barrier Insufficiency Caused by Filaggrin Deficiency and Scratching

It has been shown that impaired skin barrier function results in the persistent infiltration of external antigens and the excess production of antigen-sensitised T cells (Fig. 20.3). Of the various structural proteins involved in maintaining the integrity of the skin barrier, filaggrin (FLG) is one of the most important molecules in the pathogenesis of atopic dermatitis. Terminal differentiation in keratinocytes results in the degradation of FLG. The products of this degradation perform multiple roles, including the agglutination of keratin filaments, which results in the flattening of keratinocytes and the formation of an effective barrier against external allergens, and acts as a natural moisturiser (Candi et al. 2005).

FLG-null mutations are frequently identified in atopic dermatitis patients, particularly in patients in whom the condition persists into adulthood (Palmer et al. 2006; Barker et al. 2007). Loss-of-function *FLG* mutations have been reported to be an important risk factor for food allergies (Brown et al. 2011), which can contribute to atopic march. Furthermore, the presence of an *FLG* mutation in either early onset atopic dermatitis or food sensitisation increases the risk of developing asthma later (Marenholz et al. 2009; Filipiak-Pittroff et al. 2011).

'Flaky tail' mice harbor a frame-shift mutation in the *Flg* gene and develop atopic dermatitis-like pruritic lesions. These mice exhibited an increase in the specific IgE against epicutaneously applied ovalbumin (OVA), a representative food antigen, indicating that the barrier insufficiency caused by the *Flg* mutation promoted T-cell sensitisation to external antigens (Fallon et al. 2009; Oyoshi et al. 2009; Moniaga et al. 2010).

Another common cause of barrier insufficiency is pruritus, which is commonly associated with scratching. Pruritus is the major diagnostic criterion for atopic dermatitis, and frequent scratching in early atopic dermatitis is a strong risk factor for atopic dermatitis chronicity (Illi et al. 2004). In atopic dermatitis patients, scratching provokes further pruritus and leads to recurrent scratching behaviour,

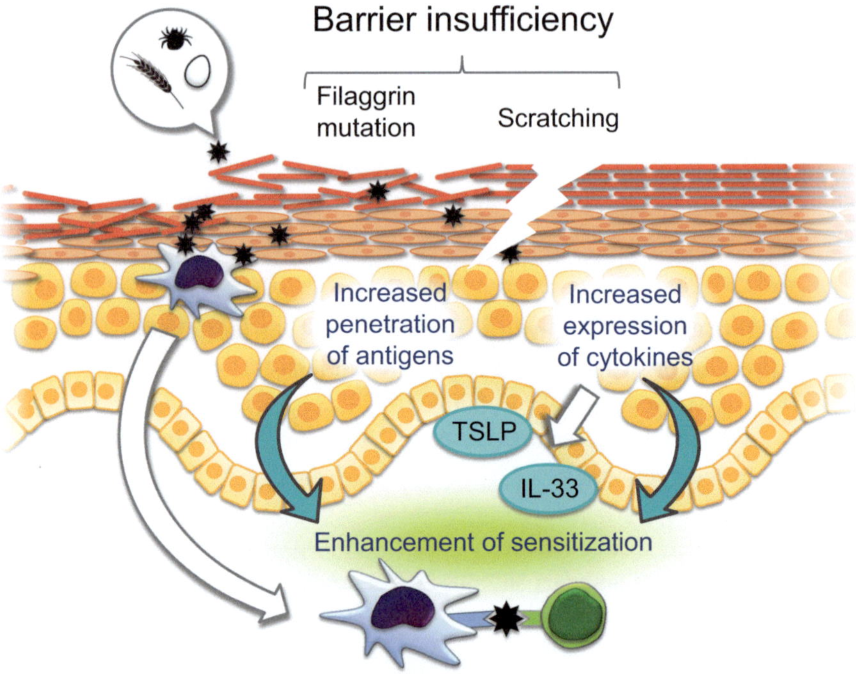

Fig. 20.3 Barrier insufficiency exacerbates inflammatory responses. Barrier insufficiency is mainly caused by *Filaggrin* mutations and scratching, which leads to an increase in the percutaneous penetration of antigens. Increased and prolonged penetration by antigens results in enhanced T-cell sensitisation. Mechanical injury caused by scratching also elevates the expression of the Th2 immune response-inducing cytokines TSLP and IL-33 in the epidermis, which enhances sensitisation

which is a characteristic symptom that is referred to as the 'itch–scratch cycle' (Hosogi et al. 2006; Bautista et al. 2014).

In addition to allowing an increase in antigen penetration, barrier insufficiency also enhances the expression of cytokines that support T-cell sensitisation and/or the formation of memory T cells. Mechanical injury also leads to an increase in inflammatory cytokine expression (Angelova-Fischer et al. 2010; Oyoshi et al. 2010; Savinko et al. 2012), which enhances T-cell sensitisation (discussed below). These data indicate that barrier insufficiency, which can be caused by *FLG* mutations, scratching, and a number of other factors, promotes the penetration of antigens through the barrier and the subsequent production of cytokines that support memory Th2 cell formation, resulting in chronicity in atopic dermatitis and an increased risk of developing atopic march.

20.7.2 Cytokines That Support Memory Th2 Cell Formation

20.7.2.1 TSLP

TSLP (Thymic stromal lymphopoietin) is a member of the IL-7 family of cytokines and is expressed mainly in the epidermis, lungs, and intestines (Liu et al. 2007). TSLP is expressed at higher levels in lesional keratinocytes in atopic dermatitis patients than in the same cells in nonatopic dermatitis patients or in nonlesional skin (Soumelis et al. 2002; Fig. 20.4). Serum TSLP levels are also higher in atopic dermatitis patients (Lee et al. 2010). Consistent with these findings, mice with skin-specific *Tslp* expression exhibit the hallmark phenotypes of human atopic dermatitis, including frequent scratching, dermal infiltration of T cells and eosinophils, and elevated serum levels of Th2 cytokines and IgE (Yoo et al. 2005; Li et al. 2005). Furthermore, epicutaneous application of a TSLP-inducing agent in combination with OVA caused dermatitis in wild-type mice but not in *Tslp*-deficient mice (Zhang et al. 2009). These data indicate that the excess production of TSLP in the epidermis promotes the Th2 immune response.

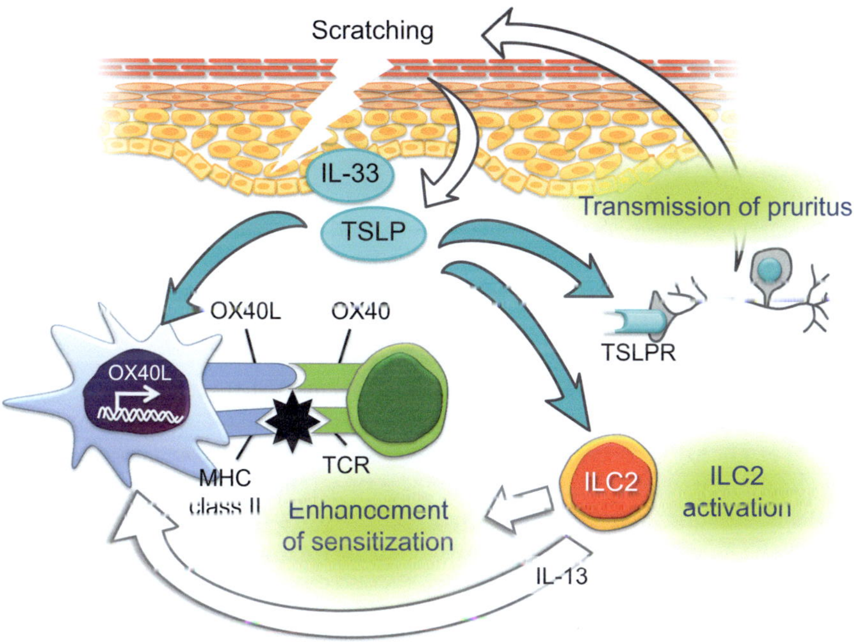

Fig. 20.4 Inflammatory cytokines produced in the skin mediate memory Th2 cell formation. Mechanical injury elevates the production of TSLP and IL-33 in the epidermis, leading to the enhancement of Th2 sensitisation and/or memory T-cell formation. TSLP elevates the expression of the costimulatory molecule OX40L in antigen-presenting cells and further induces Th2 sensitisation. TSLP also activates the TSLP-receptor (TSLPR)–expressing sensory nerves that innervate the skin, leading to the aggravation of pruritus. TSLP and IL-33 activate ILC2s, which activate DCs and enhance T-cell sensitisation

TSLP-treated human dendritic cells enhance the proliferation and development of naïve Th2-derived $CD4^+$ T cells (Soumelis et al. 2002; Watanabe et al. 2004). TSLP activates dendritic cells mainly by inducing the cell-surface expression of costimulatory molecules, including OX40L. OX40L induces Th2 polarization in naïve T cells through the formation of complexes with OX40, which is expressed on the surface of T cells. Neutralising OX40L abrogated the TSLP-mediated enhancement of Th2 development in dendritic cells (Ito et al. 2005). In cocultures of TSLP-expressing dendritic cells and memory T cells, blocking OX40L induced the expression of CDK (cyclin-dependent kinase) inhibitors, suggesting that the OX40L-OX40 interaction is required for the expansion in memory T cells that is activated by TSLP-treated dendritic cells (Wang et al. 2006). In experiments using memory Th2 cells that were collected from healthy donors, coculturing these cells with TSLP-treated $CD11c^+$ dendritic cells caused a robust expansion in memory Th2 cells (Wang et al. 2006), suggesting that TSLP is a key player in the maintenance of memory Th2 cells and chronic inflammation.

Data from experiments aimed at understanding the functions of TSLP in Th2 sensitisation and Th2 memory indicate that TSLP is involved in atopic march. In murine asthma models, inducing the skin-specific expression or epicutaneous application of TSLP promoted sensitisation to OVA and exacerbated the subsequent asthma-like airway inflammation that was induced by OVA challenge (Demehri et al. 2009; Han et al. 2012). However, specifically ablating the *Tslp* in keratinocytes alleviated the asthma-like phenotypes that were induced by epicutaneous sensitisation to OVA (Leyva-Castillo et al. 2013). OVA-specific IgE levels were therefore elevated when TSLP expression was increased and reduced when TSLP expression was ablated in mouse models, indicating that TSLP enhances allergen-specific T-cell sensitisation (Demehri et al. 2009; Han et al. 2012; Leyva-Castillo et al. 2013). Consistent with this hypothesis, polymorphisms in the human *TSLP* locus have been associated with the persistence of atopic dermatitis, asthma, and allergic rhinitis (Hunninghake et al. 2010; Harada et al. 2011; Ramasamy et al. 2011; Margolis et al. 2014).

A recent study showed that TSLP induces pruritus. TSLP activates the sensory neurons that innervate the skin because these cells express the receptor for TSLP (Wilson et al. 2013). Because TSLP expression is induced by mechanical injury (Angelova-Fischer et al. 2010), scratching itself might induce the expression of TSLP and provoke further pruritus. Controlling TSLP expression in the skin may therefore be an effective way to improve atopic dermatitis symptoms and prevent atopic dermatitis-related asthma and other allergic diseases.

20.7.2.2 IL-33

Another important cytokine that is associated with atopic dermatitis is IL-33, a member of the IL-1 family of cytokines. More cells are positive for IL-33 and ST2, a specific receptor for IL-33, in the lesional skin of atopic dermatitis patients than in nonlesional skin (Savinko et al. 2012; Fig. 20.4). These data indicate that IL-33-

ST2 signalling contributes to the development of atopic dermatitis. In addition, serum levels of IL-33 are elevated in atopic dermatitis patients and correlated with disease severity (Tamagawa-Mineoka et al. 2014), but IL-33 is not induced in nonatopic dermatitis patients or healthy controls. Furthermore, the mRNA expression levels of IL-33 and ST2 were found to be elevated in the skin of atopic dermatitis patients upon exposure to house dust mites (Savinko et al. 2012).

In agreement with these observations, single nucleotide polymorphisms in the human *ST2* gene were correlated with the incidence of atopic dermatitis (Shimizu et al. 2005). Inducing the keratinocyte-specific expression of IL-33 in mice using the Keratin 14 promoter also caused atopic dermatitis-like phenotypes, including spontaneous dermatitis with pruritus and the infiltration of eosinophils, mast cells, and IL-5–producing innate lymphoid cells (Imai et al. 2013; discussed below), supporting the hypothesis that IL-33 expression in the skin promotes allergic dermatitis.

Several studies have shown that the activation of the IL-33-ST2 axis enhances Th2-cell differentiation. Earlier studies showed that ST2 is expressed on the surface of Th2 cells but not on the surface of Th1 cells and that blocking ST2 signalling reduced the production of Th2 cytokines (Troehmann et al. 1998; Xu et al. 1998; Coyle et al. 1999). Treating human and murine $CD4^+$ cells with IL-33 enhanced Th2-cell polarization and the production of Th2-associated cytokines, such as IL-5 and IL-13 (Schmitz et al. 2005; Kurowska-Stolarska et al. 2008). IL-33 also activated dendritic cells (Besnard et al. 2011), and IL-33–treated dendritic cells showed an enhanced ability to promote Th2-cell differentiation (Rank et al. 2009). Thus, IL-33 is thought to activate dendritic cells at least in part by enhancing the OX40L-mediated activation of Th2 polarization (Murakami-satsutani et al. 2014).

Although the extent to which IL-33 is responsible for the development and persistence of atopic dermatitis has not been determined, a number of recent reports have shown that IL-33 is a key player in systemic Th2-type allergic inflammation. In an OVA-induced model of asthma, IL-33 treatment during OVA sensitisation exacerbated the observed increase in IL-5–producing cells and airway inflammation (Kurowska-Stolarska et al. 2008, 2009). Blocking IL-33-ST2 signalling with an anti-IL-33 antibody, an anti-ST2 antibody, and ST2-deficiency alleviated the severity of OVA-induced asthma (Kurowska-Stolarska et al. 2008, 2009; Kearley et al. 2009; Liu et al. 2009; Besnard et al. 2011). IL-33 enhanced the production of IL-5 in memory Th2 cells, and IL-33–deficiency ameliorated memory Th2-cell–mediated airway inflammation (Endo et al. 2015). These reports suggest that the IL-33-ST2 signalling axis is important in controlling atopic dermatitis and atopic-dermatitis–related allergic diseases.

20.7.3 Type 2 Innate Lymphoid Cell

Type 2 innate lymphoid cells (ILC2s) have been identified as the novel cell population that can induce allergic inflammation independently of adaptive immunity (Spits et al. 2013; McKenzie et al. 2014; Fig. 20.4). Innate lymphoid cells (ILCs) are members of innate immune cells derived from common lymphoid progenitor cells, which lack T- and B-cell receptors and conventional haematopoietic lineage markers, but express CD25, CD90, and CD127 (IL-7Rα; Spits and Cupedo 2012; Spits et al. 2013). In a similar way to classify helper T cells, ILCs are classified into three distinct subgroups in terms of their expression profiles of transcription factors and effector cytokines (Spits and Cupedo 2012; Sonnenberg et al. 2013). The first subgroup is Tbet-dependent IFNγ- and TNFα-producing ILC1s. The second subgroup is RORα-dependent, GATA3-expressing, and IL-5– and IL-13–producing ILC2s. The third subgroup is RORγt-dependent and IL-17A– and/or IL-22–producing ILC3s. In a murine asthma model, ILC2s produce a large amount of IL-5 and IL-13 equal to or greater than that done by T cells (Wolterink et al. 2012), whereas they produce a lesser amount of IL-4 than that done by T cells (Barlow et al. 2012). This capacity to produce such large amounts of Th2-type cytokines implies that ILC2s play an important role in the pathogenesis of allergic inflammation.

ILC2s are enriched in the interface between the environment and our body, such as intestine (Neill et al. 2010), lungs (Chang et al. 2011; Monticelli et al. 2011), and skin (Kim et al. 2013; Roediger et al. 2013). Those are organs where allergic inflammation occurs. ILC2s are actually found to play important roles in the induction of airway hyperreactivity (Chang et al. 2011) or infiltrate abundantly in the lesion of allergic inflammation, such as chronic rhinosinusitis (Mjösberg et al. 2011) and atopic dermatitis (Kim et al. 2013). Especially in skin, although all subgroups of ILCs have been identified, ILC2s are predominant in mice (Roediger et al. 2013). ILC2s constitute approximately 30% of total ILCs in humans (Dyring-Andersen et al. 2014; Villanova et al. 2014). In addition, the activation of ILC2s itself by IL-2 is sufficient to induce atopic dermatitis-like lesions via production of IL-5 and infiltration with eosinophils in murine skin (Roediger et al. 2013). These studies imply ILC2s play important roles not only in steady-state conditions but also at the lesional skin of atopic dermatitis.

Recently, the functional association between ILC2s and atopic dermatitis has been demonstrated. In both mouse and human systems, ILC2s express receptors for proinflammatory cytokines, such as TSLP, IL-33, and IL-25 (Mjösberg et al. 2012). Actually, keratinocyte-specific IL-33–overexpressing transgenic mice exhibit atopic dermatitis-like skin lesions with an accumulation of ILC2s (Imai et al. 2013). IL-33 activates ILC2s to produce IL-5 and IL-13 independently or with the combination of IL-2 and IL-25 (Furusawa et al. 2013; Salimi et al. 2013), leading to exacerbate Th2-type inflammation. Another epithelium-derived proinflammatory cytokine TSLP itself is also found to induce atopic dermatitis-like lesion with ILC2 infiltration in murine skin (Kim et al. 2013). Furthermore,

TSLP inhibits the anti-inflammatory effect of corticosteroid in the situation of IL-33–induced and ILC2s-mediated airway inflammation (Kabata et al. 2013). As both IL-33 and TSLP are expressed and ILC2s are enriched at the lesion of atopic dermatitis, these cytokines may also enhance Th2-type inflammation via ILC2s.

ILC2s enhance sensitisation, leading to chronicity of atopic dermatitis (Fig. 20.4). Indeed, although the exact role that ILC2s play in an IgE-dependent allergic response remains elusive, it has been shown that ILC2s are able to enhance IgE production from B cells in vitro (Fukuoka et al. 2013). In addition, as a mechanism to enhance sensitisation, ILC2s are demonstrated to enhance the priming step of Th2 cells directly in two different ways. First, in a murine lung inflammation model, activation of ILC2s is essential for dendritic cells to migrate to draining lymph nodes where antigen-presentation occurs (Halim et al. 2014). Second, ILC2s have been reported to express major histocompatibility complex class II and to stimulate $CD4^+$ T cells directly (Mirchandani et al. 2014). These results support the contention that ILC2s contribute to chronicity of atopic dermatitis via augmentation of sensitisation step.

Direct enrollment of ILC2s in the process of memory Th2 cell formation has been reported. In a papain application-induced atopic dermatitis model, Halim et al. demonstrated that IL-13 secreted from ILC2s induces memory Th2-cell–attracting chemokine, CCL17, expression by dermal $CD11b^+$ dendritic cells, which is a crucial step for memory Th2-cell formation (Halim et al. 2015). Taken together, these studies suggest that ILC2s may contribute to induce chronicity in atopic dermatitis by enforcement of the Th2-type sensitisation step and/or the direct induction of memory T cells via dendritic cells.

20.7.4 *Air Pollution*

The prevalence of atopic dermatitis is rapidly increasing, implying that environmental factors may also contribute to its pathogenesis (Fig. 20.5). A number of studies have suggested that air pollution is one of the risk factors for developing atopic dermatitis and allergic diseases. For example, the prevalence of allergic diseases is correlated with proximity to a major traffic road (McConnell et al. 2006; Morgenstern et al. 2008), and markers associated with exposure to cigarette smoke during the prenatal period are correlated with the prevalence of atopic dermatitis after birth (Wang et al. 2008). In a meta-analysis of published studies, increased exposure to particulate matter (PM), an important air pollutant, was found to be positively associated with the increased incidence of asthma (Gasana et al. 2012). A representative indoor air pollutant, cigarette smoke, is also positively correlated with the prevalence of atopic dermatitis, regardless of whether the exposure is mainstream or passive (Saulyte et al. 2014).

Although the mechanisms underlying these associations have not been determined, it is clear that air pollutants exacerbate allergic inflammation. In murine antigen-induced airway inflammation models, treatment with diesel exhaust

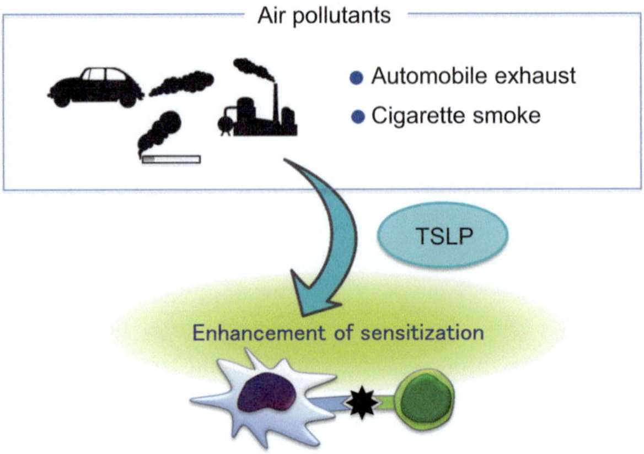

Fig. 20.5 Air pollution and atopic dermatitis. It has been suggested that air pollutants, including automobile exhaust, cigarette smoke, and factory smoke, exacerbate atopic dermatitis symptoms by inducing the expression of cytokines, such as TSLP. The prevalence of atopic dermatitis is correlated with air pollution levels, suggesting that exposure to air pollutants enhances the development of atopic dermatitis

particles (DEPs) in combination with OVA or other allergens promoted asthma-like phenotypes, including the production of Th2 and Th17 cytokines and antigen-specific IgE and IgG_1 and airway hyperresponsiveness (Takano et al.; Kim et al. 2011b; Brandt et al. 2013). Air pollutants may exert these inflammatory effects by enhancing TSLP expression. Administration of DEPs to human bronchial epithelial cells induced *TSLP* expression and subsequently Th2-supporting activity in dendritic cells (Bleck et al. 2008, 2010). Furthermore, one recent report showed that air pollutants enhanced the establishment of immunological memory. Following the induction of sensitisation to house dust mites, coexposure with DEPs led to enhanced immune responses against secondary allergen challenges even after a 7-week resting period (Brandt et al. 2015). Simultaneous exposure to DEPs and allergens therefore promoted immune responses and the accumulation of allergen-specific memory T cells.

Diverse compounds are included in DEPs. Among these, polycyclic aromatic hydrocarbons (PAHs) are important because they are generated during the burning of coal, oil, gas, other organic compounds, and cigarettes. Organic chemical components derived from DEPs often include PAHs, and PAH alone has been shown to enhance allergic responses, implying that PAHs may be the main mediators of the inflammation-exacerbating effects of air pollutants (Takenaka et al. 1995).

20.8 Transcription Factors Linking Pollution and Inflammatory Diseases

Pollutants usually exert their effects by activating transcription factors, which then mediate an environmental stress response. Although the association between stress response transcription factors and atopic dermatitis has not been investigated, some reports have suggested that these transcription factors alter inflammatory responses.

20.8.1 AhR

In addition to their inflammation-exacerbating effects, PAHs have also been known to exert carcinogenic effects by activating the transcription factor AhR (aryl hydrocarbon receptor or dioxin receptor; Fig. 20.6). PAHs act as ligands for AhR, through which they induce the expression of phase I detoxification enzymes, mainly including cytochrome P450s. Phase I enzymes are required for the elimination of PAHs from cells, but they also mediate the carcinogenicity of PAHs, which is induced by the electrophilic intermediates that result from PAH detoxification. Subcutaneous or topical application of benzo[a]pyrene, a PAH contained in cigarette smoke and automobile exhaust, produced skin tumours in wild-type mice,

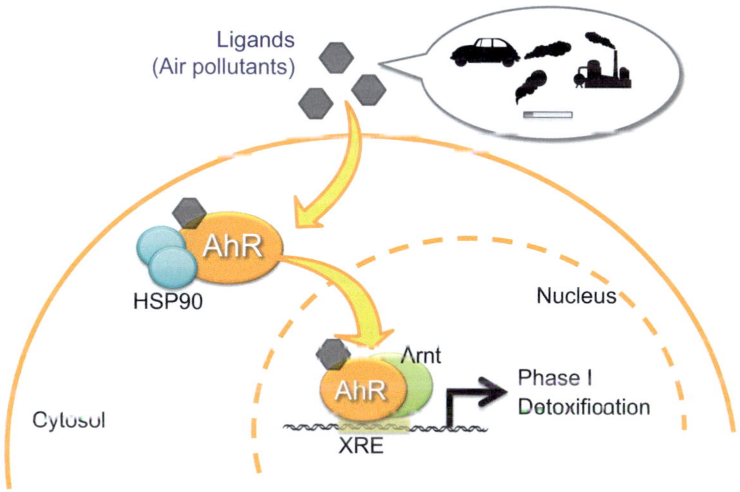

Fig. 20.6 The AhR-mediated induction of phase I detoxification enzymes. The transcription factor AhR is activated through its ligands, which include PAHs, and PAHs are the main inflammatory components of air pollutants. Upon ligand binding, AhR dislocates from a complex that includes heat shock protein 90 (HSP90) and translocates into the nucleus. In the nucleus, AhR forms heterodimers with Arnt and then binds to xenobiotic response elements (XREs) to upregulate the transcription of phase I detoxification enzymes, which are required to metabolise air pollutant molecules

whereas AhR deficiency resulted in a loss of the expression of phase I enzyme genes and a reduction in the incidence of tumours (Shimizu et al. 2000).

Ligands that can activate AhR include PAHs, dioxins, and other environmental pollutant molecules. When AhR is bound by a ligand, it disassociates from the cytoplasmic complex through a process involving heat shock protein 90 (HSP90), and it then translocates into the nucleus. In the nucleus, AhR forms heterodimers with aryl hydrocarbon receptor nuclear translocator (Arnt) and binds to xenobiotic response elements (XREs) in the *cis*-regulatory regions of its target genes to induce their transcription (Fujii-Kuriyama and Mimura 2005).

It has been suggested that AhR contributes to immune homeostasis, however, the role of AhR is complicated by the fact that its functions include both immune-promoting and immune-suppressive effects that depend on the cell type, the presence of particular ligands, and the experimental model being studied. Of the $CD4^+$ T cells, AhR is highly expressed in Th17 cells (Veldhoen et al. 2008). During Th17 differentiation in human and murine $CD4^+$ cells, the expression levels of the Th17 cytokine genes IL-17A, IL-17F, and IL-22 were increased by treatment with the AhR ligand 6-formylindolo[3,2-b]carbazole (FICZ). FICZ is generated via the UV (ultraviolet)-B irradiation of tryptophan and is thought to be an endogenous ligand of AhR. The Th17-enhancing effects of FICZ were shown to be dependent on AhR because they were reversed in cells obtained from AhR-deficient mice (Quintana et al. 2008; Veldhoen et al. 2008).

Interestingly, a well-investigated and highly toxic AhR ligand, 2,3,7,8-tetrachlorodibenzo-p-dioxin (TCDD), induced the differentiation of immunosuppressive regulatory T cells (Tregs) instead of Th17 (Quintana et al. 2008). The effects of AhR ligands have been examined in vivo in experimentally-induced autoimmune encephalomyelitis (EAE) in a murine model of multiple sclerosis. Similar to the results of in vitro experiments using $CD4^+$ T cells, FICZ exacerbated, but TCDD suppressed, EAE symptoms, and both of these effects were AhR-dependent (Quintana et al. 2008; Veldhoen et al. 2008). In contrast, other studies have reported that both FICZ and TCDD promoted Th17 differentiation in vitro, which contradicts the observation that some AhR ligands can repress the in vivo induction of Th17 cells during EAE, depending on the protocol being used (Duarte et al. 2013).

Collectively, the results of these studies indicate that the immune-related functions of AhR are regulated by numerous factors, including ligands, experimental conditions, and cell lineages. However, we have provided an intriguing line of evidence showing an association between AhR and inflammation. In these experiments, we used a transgenic mouse model (Tauchi et al. 2005) to show that inducing the expression of a constitutively active form of AhR in keratinocytes caused spontaneous skin inflammation that was accompanied by an increase in serum IgE levels. Although the mechanism underlying these effects remains to be determined, these results indicate that AhR is involved in the regulation of immune responses in the skin.

20.8.2 *Nrf2*

Nrf2 (Nuclear factor erythroid 2-related factor 2) is the master regulator of phase II detoxification responses, during which the metabolites produced in phase I detoxification reactions are modified to form less toxic and more easily excretable molecules (Fig. 20.7). Consistent with this role, Nrf2 loss-of-function increased the sensitivity of cells to various stresses, including those induced by toxic electrophilic and carcinogenic molecules (Taguchi et al. 2011; Suzuki et al. 2013). Recent studies also revealed that Nrf2 is a key player during the control of inflammation (Kobayashi et al. 2013).

The activation of Nrf2 is regulated by degradation processes that are mediated by Keap1 (Kelch-like ECH-associated protein 1), an adapter protein for E3 ubiquitin ligase (Itoh et al. 1999; Taguchi et al. 2011; Suzuki et al. 2013). In unstressed conditions, Nrf2 is constitutively degraded via a Keap1-mediated polyubiquitination and proteasome pathway. Upon exposure to oxidative and electrophilic stress, Keap1 senses the insult, and the degradation of Nrf2 is blocked. Newly synthesised Nrf2 proteins translocate into the nucleus, where they form heterodimers with sMaf proteins. Nrf2-sMaf heterodimers bind to antioxidant response elements/electrophile response elements (AREs/EpREs) in the regulatory regions of detoxification-related genes to upregulate their transcription (Friling et al. 1990; Rushmore et al. 1991).

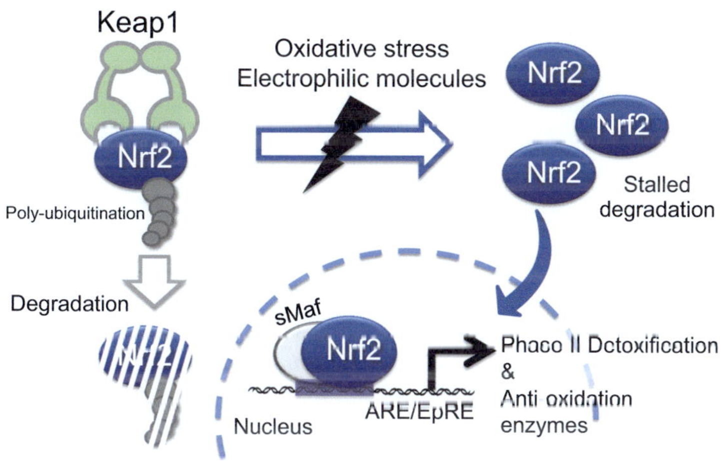

Fig. 20.7 The Nrf2-mediated induction of phase II detoxification enzymes and antioxidants. The transcription factor Nrf2 is required for the phase II detoxification of metabolites derived from air pollutants. Nrf2 is constitutively degraded in steady-state cells by binding to Keap1 proteins. However, Nrf2 accumulates under xenobiotic and oxidative stress conditions. Accumulated Nrf2 translocates into the nucleus, where it forms heterodimers with sMaf and then binds to ARE/EpRE sequences. Nrf2 induces the expression of various genes, including phase II detoxification enzymes and antioxidant genes. The ability to initiate an Nrf2-mediated antioxidative response is required for redox homeostasis, which prevents the development of excessive inflammation

In addition to detoxification, Nrf2 has been shown to regulate inflammation in diverse murine in vivo models (Itoh et al. 2004; Rangasamy et al. 2005; Ishii et al. 2005; Iizuka et al. 2005; Thimmulappa et al. 2006), including models of asthma. The Nrf2-inducing small chemical molecule Tecfidera® has been approved as a treatment for multiple sclerosis, supporting the efficacy of targeting the control of Nrf2 activity to treat inflammatory diseases (Gold et al. 2012; Burness and Deeks 2014). In relation to air pollution, cigarette smoke-induced emphysema is also exacerbated in Nrf2-deficient mice (Iizuka et al. 2005). Exposure to DEPs and ambient ultrafine particles was also shown to exacerbate asthma-like allergic responses against OVA, and these responses were stronger in Nrf2-deficient mice than in wild-type mice (Li et al. 2010; Li et al. 2013).

Nrf2 is thought to repress inflammation at least partially by its involvement in the elimination of reactive oxygen species (ROS) because Nrf2 upregulates a number of antioxidant genes (Hirotsu et al. 2012) to maintain redox homeostasis in cells. In some experimental models of inflammation, the elimination of ROS using N-acetyl cysteine (NAC) has been shown to alleviate the exacerbated inflammation observed in Nrf2-deficient mice, suggesting that the Nrf2-mediated control of ROS involves the repression of inflammatory phenotypes (Thimmulappa et al. 2006; Kong et al. 2010).

In an allergic dermatitis, ROS has been demonstrated to be an exacerbating factor. A mouse model for allergic dermatitis, an epicutaneous treatment of NC/Nga mice with 2,4-dinitrofluorobenzene, is alleviated by α-lipoic acid that functions as an antioxidant (Kim et al. 2011a). In blood samples from atopic dermatitis patients, an increase in oxidative stress and a decrease in antioxidants have been reported (Sivaranjani 2013), implying the relation of ROS to development of atopic dermatitis. Furthermore, basal and induced *IL4* gene expression in T cells from atopic dermatitis patients are attenuated by the inhibitor of mitochondrial respiratory complex I, because mitochondrial ROS upregulates expression of the *IL4* and *IL2* genes (Kaminski et al. 2010).

Consistent with these reports, loss-of-Nrf2-function has been reported to exacerbate a murine atopic dermatitis model (El Ali et al. 2013), implying that Nrf2-mediated elimination of ROS is important to control dermatitis. In addition to ROS elimination, some reports have identified novel Nrf2 target genes in addition to the previously identified antioxidative stress-response genes in immune cells (Ishii et al. 2004; Harvey et al. 2011), whereas the molecular mechanisms underlying the Nrf2-mediated repression of inflammation have not been fully explored.

20.9 Conclusion

The establishment of percutaneously sensitised memory Th2 cells may cause chronicity in atopic dermatitis, which manifests as nonremission and the subsequent development of atopic march. The characteristic pathophysiologies of atopic dermatitis include the epidermal production of TSLP and other proinflammatory cytokines, barrier insufficiency, and induction of ILC2s which leads to the maintenance and increased production of memory Th2 cells. Exposure to air pollutants also enhances atopic dermatitis and contributes to the subsequent development of atopic march. These air pollutants may exert their toxic effects via the activation of stress-responsive transcription factors, such as AhR and Nrf2, although only limited information is currently available to support this hypothesis. Future studies focused on the association between Th2 immunological memory and air pollution would be beneficial to the aims of preventing and treating atopic dermatitis and atopic march.

References

Angelova-Fischer I, Fernandez IM, Donnadieu MH et al (2010) Injury to the stratum corneum induces in vivo expression of human thymic stromal lymphopoietin in the epidermis. J Invest Dermatol 130:2505–2507. doi:10.1038/jid.2010.143

Ashcroft DM, Dimmock P, Garside R et al (2005) Efficacy and tolerability of topical pimecrolimus and tacrolimus in the treatment of atopic dermatitis: meta-analysis of randomised controlled trials. BMJ 330:516. doi:10.1136/bmj.38376.439653.D3

Barker JNWN, Palmer CN a, Zhao Y et al (2007) Null mutations in the filaggrin gene (FLG) determine major susceptibility to early-onset atopic dermatitis that persists into adulthood. J Invest Dermatol 127:564–567. doi:10.1038/sj.jid.5700587

Barlow JL, Bellosi A, Hardman CS et al (2012) Innate IL-13-producing nuocytes arise during allergic lung inflammation and contribute to airways hyperreactivity. J Allergy Clin Immunol 129:1367–1370. doi:10.1016/j.jaci.2011.09.041

Bautista DM, Wilson SR, Hoon M a (2014) Why we scratch an itch: the molecules, cells and circuits of itch. Nat Neurosci 17:175–182

Beattie PE, Lewis-Jones MS (2006) A comparative study of impairment of quality of life in children with skin disease and children with other chronic childhood diseases. Br J Dermatol 155:145–151. doi:10.1111/j.1365-2133.2006.07185.x

Besnard A-G, Togbe D, Guillou N et al (2011) IL-33-activated dendritic cells are critical for allergic airway inflammation. Eur J Immunol 41:1675–1686. doi:10.1002/eji.201041033

Bieber T (2008) Atopic dermatitis. N Engl J Med 358:1483–1494. doi:10.1056/NEJMra074081

Biedermann T, Schwärzler C, Lametschwandtner G et al (2002) Targeting CLA/E-selectin interactions prevents CCR4-mediated recruitment of human Th2 memory cells to human skin in vivo. Eur J Immunol 32:3171–3180. doi:10.1002/1521-4141(200211)32:11<3171::AID-IMMU3171>3.0.CO;2-4

Biedermann T, Lametschwandtner G, Tangemann K et al (2006) IL-12 instructs skin homing of human Th2 cells. J Immunol 177:3763–3770. doi:10.4049/jimmunol.177.6.3763

Bleck B, Tse DB, Curotto de Lafaille MA et al (2008) Diesel exhaust particle-exposed human bronchial epithelial cells induce dendritic cell maturation and polarization via thymic stromal lymphopoietin. J Clin Immunol 28:147–156. doi:10.1007/s10875-007-9149-0

Bleck B, Tse DB, Gordon T et al (2010) Diesel exhaust particle-treated human bronchial epithelial cells upregulate Jagged-1 and OX40 ligand in myeloid dendritic cells via thymic stromal lymphopoietin. J Immunol 185:6636–6645. doi:10.4049/jimmunol.1000719

Brandt EB, Kovacic MB, Lee GB et al (2013) Diesel exhaust particle induction of IL-17A contributes to severe asthma. J Allergy Clin Immunol 132:1194–1204.e2. doi:10.1016/j.jaci.2013.06.048

Brandt EB, Biagini Myers JM, Acciani TH et al (2015) Exposure to allergen and diesel exhaust particles potentiates secondary allergen-specific memory responses, promoting asthma susceptibility. J Allergy Clin Immunol 136:295. doi:10.1016/j.jaci.2014.11.043

Brown SJ, Asai Y, Cordell HJ et al (2011) Loss-of-function variants in the filaggrin gene are a significant risk factor for peanut allergy. J Allergy Clin Immunol 127:661–667. doi:10.1016/j.jaci.2011.01.031

Burness CB, Deeks ED (2014) Dimethyl fumarate: a review of its use in patients with relapsing-remitting multiple sclerosis. CNS Drugs 28:373–387. doi:10.1007/s40263-014-0155-5

Bushar ND, Corbo E, Schmidt M et al (2010) Ablation of SLP-76 signaling after T cell priming generates memory CD4 T cells impaired in steady-state and cytokine-driven homeostasis. Proc Natl Acad Sci U S A 107:827–831. doi:10.1073/pnas.0908126107

Campbell JJ, Haraldsen G, Pan J et al (1999) The chemokine receptor CCR4 in vascular recognition by cutaneous but not intestinal memory T cells. Nature 400:776–780. doi:10.1038/23495

Candi E, Schmidt R, Melino G (2005) The cornified envelope: a model of cell death in the skin. Nat Rev Mol Cell Biol 6:328–340. doi:10.1038/nrm1619

Chang Y-J, Kim HY, Albacker L a et al (2011) Innate lymphoid cells mediate influenza-induced airway hyper-reactivity independently of adaptive immunity. Nat Immunol 12:631–638. doi:10.1038/ni.2045

Clark R a, Chong BF, Mirchandani N et al (2006a) A novel method for the isolation of skin resident T cells from normal and diseased human skin. J Invest Dermatol 126:1059–1070. doi:10.1038/sj.jid.5700199

Clark RA, Chong B, Mirchandani N et al (2006b) The vast majority of CLA+ T cells are resident in normal skin. J Immunol 176:4431–4439. doi:10.4049/jimmunol.176.7.4431

Cookson W (2004) The immunogenetics of asthma and eczema: a new focus on the epithelium. Nat Rev Immunol 4:978–988. doi:10.1038/nri1500

Cosmi L, Annunziato F, Galli MIG et al (2000) CRTH2 is the most reliable marker for the detection of circulating human type 2 Th and type 2 T cytotoxic cells in health and disease. Eur J Immunol 30:2972–2979. doi:10.1002/1521-4141(200010)30:10<2972::AID-IMMU2972>3.0.CO;2-#

Coyle BAJ, Lloyd C, Tian J et al (1999) Crucial role of the interleukin 1 receptor family member T1/ ST2 in T helper cell type 2-mediated lung mucosal immune responses. J Exp Med 190: 895–902. doi:10.1084/jem.190.7.895

Croft M (2003) Co-stimulatory members of the TNFR family: keys to effective T-cell immunity? Nat Rev Immunol 3:609–620. doi:10.1038/nri1148

Demehri S, Morimoto M, Holtzman MJ, Kopan R (2009) Skin-derived TSLP triggers progression from epidermal-barrier defects to asthma. PLoS Biol 7:e1000067. doi:10.1371/journal.pbio.1000067

Duarte JH, Di Meglio P, Hirota K et al (2013) Differential influences of the aryl hydrocarbon receptor on Th17 mediated responses in vitro and in vivo. PLoS One 8:e79819. doi:10.1371/journal.pone.0079819

Dyring-Andersen B, Geisler C, Agerbeck C et al (2014) Increased number and frequency of group 3 innate lymphoid cells in nonlesional psoriatic skin. Br J Dermatol 170:609–616. doi:10.1111/bjd.12658

El Ali Z, Gerbeix C, Hemon P et al (2013) Allergic skin inflammation induced by chemical sensitizers is controlled by the transcription factor Nrf2. Toxicol Sci 134:39–48. doi:10.1093/toxsci/kft084

Endo Y, Hirahara K, Iinuma T et al (2015) The interleukin-33-p38 kinase axis confers memory T helper 2 cell pathogenicity in the airway. Immunity 42:294–308. doi:10.1016/j.immuni.2015.01.016

Fallon PG, Sasaki T, Sandilands A et al (2009) A homozygous frameshift mutation in the mouse Flg gene facilitates enhanced percutaneous allergen priming. Nat Genet 41:602–608. doi:10.1038/ng.358

Filipiak-Pittroff B, Schnopp C, Berdel D et al (2011) Predictive value of food sensitization and filaggrin mutations in children with eczema. J Allergy Clin Immunol 128:1235–1241.e5. doi:10.1016/j.jaci.2011.09.014

Friling RS, Bensimon A, Tichauer Y, Daniel V (1990) Xenobiotic-inducible expression of murine glutathione S-transferase Ya subunit gene is controlled by an electrophile-responsive element. Proc Natl Acad Sci U S A 87:6258–6262

Fujii-Kuriyama Y, Mimura J (2005) Molecular mechanisms of AhR functions in the regulation of cytochrome P450 genes. Biochem Biophys Res Commun 338:311–317. doi:10.1016/j.bbrc.2005.08.162

Fukuoka A, Futatsugi-Yumikura S, Takahashi S et al (2013) Identification of a novel type 2 innate immunocyte with the ability to enhance IgE production. Int Immunol 25:373–382. doi:10.1093/intimm/dxs160

Furusawa J, Moro K, Motomura Y et al (2013) Critical role of p38 and GATA3 in natural helper cell function. J Immunol 191:1818–1826. doi:10.4049/jimmunol.1300379

Gasana J, Dillikar D, Mendy A et al (2012) Motor vehicle air pollution and asthma in children: a meta-analysis. Environ Res 117:36–45. doi:10.1016/j.envres.2012.05.001

Gold R, Kappos L, Arnold DL et al (2012) Placebo-controlled phase 3 study of oral BG-12 for relapsing multiple sclerosis. N Engl J Med 367:1098–1107. doi:10.1056/NEJMoa1114287

Halim TYF, Steer CA, Mathä L et al (2014) Group 2 innate lymphoid cells are critical for the initiation of adaptive T helper 2 cell-mediated allergic lung inflammation. Immunity 40:425–435. doi:10.1016/j.immuni.2014.01.011

Halim TYF, Hwang YY, Scanlon ST et al (2015) Group 2 innate lymphoid cells license dendritic cells to potentiate memory TH2 cell responses. Nat Immunol 17:2–11. doi:10.1038/ni.3294

Han H, Xu W, Headley MB et al (2012) Thymic stromal lymphopoietin (TSLP)-mediated dermal inflammation aggravates experimental asthma. Mucosal Immunol 5:342–351. doi:10.1038/mi.2012.14

Harada M, Hirota T, Jodo AI et al (2011) Thymic stromal lymphopoietin gene promoter polymorphisms are associated with susceptibility to bronchial asthma. Am J Respir Cell Mol Biol 44:787–793. doi:10.1165/rcmb.2009-0418OC

Harvey CJ, Thimmulappa RK, Sethi S et al (2011) Targeting Nrf2 signaling improves bacterial clearance by alveolar macrophages in patients with COPD and in a mouse model. Sci Transl Med 3:78ra32. doi:10.1126/scitranslmed.3002042

He R, Oyoshi MK, Wang JYT et al (2010) The prostaglandin D2 receptor CRTH2 is important for allergic skin inflammation after epicutaneous antigen challenge. J Allergy Clin Immunol 126:784–790. doi:10.1016/j.jaci.2010.07.006

Hirotsu Y, Katsuoka F, Funayama R et al (2012) Nrf2-MafG heterodimers contribute globally to antioxidant and metabolic networks. Nucleic Acids Res 40:10228–10239. doi:10.1093/nar/gks827

Hogan MB, Peele K, Wilson NW (2012) Skin barrier function and its importance at the start of the atopic march. J Allergy 2012:901940. doi:10.1155/2012/901940

Homann D, Teyton L, Oldstone MB (2001) Differential regulation of antiviral T-cell immunity results in stable CD8+ but declining CD4+ T-cell memory. Nat Med 7:913–919. doi:10.1038/90950

Homey B, Steinhoff M, Ruzicka T, Leung DYM (2006) Cytokines and chemokines orchestrate atopic skin inflammation. J Allergy Clin Immunol 118:178–189. doi:10.1016/j.jaci.2006.03.047

Hosogi M, Schmelz M, Miyachi Y, Ikoma A (2006) Bradykinin is a potent pruritogen in atopic dermatitis: a switch from pain to itch. Pain 126:16–23. doi:10.1016/j.pain.2006.06.003

Hunninghake GM, Soto-Quirós ME, Avila L et al (2010) TSLP polymorphisms are associated with asthma in a sex-specific fashion. Allergy 65:1566–1575. doi:10.1111/j.1398-9995.2010.02415.x

Iizuka T, Ishii Y, Itoh K et al (2005) Nrf2-deficient mice are highly susceptible to cigarette smoke-induced emphysema. Genes Cells 10:1113–1125. doi:10.1111/j.1365-2443.2005.00905.x

Illi S, von Mutius E, Lau S et al (2004) The natural course of atopic dermatitis from birth to age 7 years and the association with asthma. J Allergy Clin Immunol 113:925–931. doi:10.1016/j.jaci.2004.01.778

Imai Y, Yasuda K, Sakaguchi Y et al (2013) Skin-specific expression of IL-33 activates group 2 innate lymphoid cells and elicits atopic dermatitis-like inflammation in mice. Proc Natl Acad Sci U S A 110:13921–13926. doi:10.1073/pnas.1307321110

Ishii T, Itoh K, Ruiz E et al (2004) Role of Nrf2 in the regulation of CD36 and stress protein expression in murine macrophages: activation by oxidatively modified LDL and 4-hydroxynonenal. Circ Res 94:609–616. doi:10.1161/01.RES.0000119171.44657.45

Ishii Y, Itoh K, Morishima Y et al (2005) Transcription factor Nrf2 plays a pivotal role in protection against elastase-induced pulmonary inflammation and emphysema. J Immunol 175:6968–6975

Ito T, Wang Y-H, Duramad O et al (2005) TSLP-activated dendritic cells induce an inflammatory T helper type 2 cell response through OX40 ligand. J Exp Med 202:1213–1223. doi:10.1084/jem.20051135

Itoh K, Wakabayashi N, Katoh Y et al (1999) Keap1 represses nuclear activation of antioxidant responsive elements by Nrf2 through binding to the amino-terminal Neh2 domain. Genes Dev 13:76–86. doi:10.1101/gad.13.1.76

Itoh K, Mochizuki M, Ishii Y et al (2004) Transcription factor Nrf2 regulates inflammation by mediating the effect of 15-deoxy-delta(12,14)-prostaglandin j(2). Mol Cell Biol 24:36–45. doi:10.1128/MCB.24.1.36

Iwasaki M, Nagata K, Takano S et al (2002) Association of a new-type prostaglandin D2 receptor CRTH2 with circulating T helper 2 cells in patients with atopic dermatitis. J Invest Dermatol 119:609–616. doi:10.1046/j.1523-1747.2002.01862.x

Kabata H, Moro K, Fukunaga K et al (2013) Thymic stromal lymphopoietin induces corticosteroid resistance in natural helper cells during airway inflammation. Nat Commun 4:2675. doi:10.1038/ncomms3675

Kakinuma T, Nakamura K, Wakugawa M et al (2001) Thymus and activation-regulated chemokine in atopic dermatitis: serum thymus and activation-regulated chemokine level is closely related with disease activity. J Allergy Clin Immunol 107:535–541. doi:10.1067/mai.2001.113237

Kakinuma T, Nakamura K, Wakugawa M et al (2002) Serum macrophage-derived chemokine (MDC) levels are closely related with the disease activity of atopic dermatitis. Clin Exp Immunol 127:270–273. doi:10.1046/j.1365-2249.2002.01727.x

Kaminski MM, Sauer SW, Klemke C-D et al (2010) Mitochondrial reactive oxygen species control T cell activation by regulating IL-2 and IL-4 expression: mechanism of ciprofloxacin-mediated immunosuppression. J Immunol 184:4827–4841. doi:10.4049/jimmunol.0901662

Kassiotis G, Garcia S, Simpson E, Stockinger B (2002) Impairment of immunological memory in the absence of MHC despite survival of memory T cells. Nat Immunol 3:244–250. doi:10.1038/ni766

Kearley J, Buckland KF, Mathie SA, Lloyd CM (2009) Resolution of allergic inflammation and airway hyperreactivity is dependent upon disruption of the T1/ST2-IL-33 pathway. Am J Respir Crit Care Med 179:772–781. doi:10.1164/rccm.200805-666OC

Kim G-D, Kim T-H, Jang A-H et al (2011a) α-lipoic acid suppresses the development of DNFB-induced atopic dermatitis-like symptoms in NC/Nga mice. Exp Dermatol 20:97–101. doi:10.1111/j.1600-0625.2010.01165.x

Kim J, Natarajan S, Vaickus LJ et al (2011b) Diesel exhaust particulates exacerbate asthma-like inflammation by increasing CXC chemokines. Am J Pathol 179:2730–2739. doi:10.1016/j.ajpath.2011.08.008

Kim BS, Siracusa MC, Saenz S a et al (2013) TSLP elicits IL-33-independent innate lymphoid cell responses to promote skin inflammation. Sci Transl Med 5:170ra16. doi:10.1126/scitranslmed.3005374

Kobayashi E, Suzuki T, Yamamoto M (2013) Roles nrf2 plays in myeloid cells and related disorders. Oxid Med Cell Longev 2013:529219. doi:10.1155/2013/529219

Kondrack RM, Harbertson J, Tan JT et al (2003) Interleukin 7 regulates the survival and generation of memory CD4 cells. J Exp Med 198:1797–1806. doi:10.1084/jem.20030735

Kong X, Thimmulappa R, Kombairaju P, Biswal S (2010) NADPH oxidase-dependent reactive oxygen species mediate amplified TLR4 signaling and sepsis-induced mortality in Nrf2-deficient mice. J Immunol 185:569–577. doi:10.4049/jimmunol.0902315

Kupper TS, Fuhlbrigge RC (2004) Immune surveillance in the skin: mechanisms and clinical consequences. Nat Rev Immunol 4:211–222. doi:10.1038/nri1310

Kurowska-Stolarska M, Kewin P, Murphy G et al (2008) IL-33 induces antigen-specific IL-5+ T cells and promotes allergic-induced airway inflammation independent of IL-4. J Immunol 181: 4780–4790. doi:10.4049/jimmunol.181.7.4780

Kurowska-Stolarska M, Stolarski B, Kewin P et al (2009) IL-33 amplifies the polarization of alternatively activated macrophages that contribute to airway inflammation. J Immunol 183: 6469–6477. doi:10.4049/jimmunol.0901575

Kurtulus S, Tripathi P, Hildeman DA (2013) Protecting and rescuing the effectors: roles of differentiation and survival in the control of memory T cell development. Front Immunol 3: 1–13. doi:10.3389/fimmu.2012.00404

Lee EB, Kim KW, Hong JY et al (2010) Increased serum thymic stromal lymphopoietin in children with atopic dermatitis. Pediatr Allergy Immunol 21:e457-60. doi:10.1111/j.1399-3038.2009.00919.x

Lees JR, Farber DL (2010) Generation, persistence and plasticity of CD4 T-cell memories. Immunology 130:463–470. doi:10.1111/j.1365-2567.2010.03288.x

Lenz DC, Kurz SK, Lemmens E et al (2004) IL-7 regulates basal homeostatic proliferation of antiviral CD4+T cell memory. Proc Natl Acad Sci U S A 101:9357–9362. doi:10.1073/pnas.0400640101

Leung DYM, Boguniewicz M, Howell MD et al (2004) New insights into atopic dermatitis. J Clin Invest 113:651–657. doi:10.1172/JCI21060

Leyva-Castillo JM, Hener P, Jiang H, Li M (2013) TSLP produced by keratinocytes promotes allergen sensitization through skin and thereby triggers atopic march in mice. J Invest Dermatol 133:154–163. doi:10.1038/jid.2012.239

Li M, Messaddeq N, Teletin M et al (2005) Retinoid X receptor ablation in adult mouse keratinocytes generates an atopic dermatitis triggered by thymic stromal lymphopoietin. Proc Natl Acad Sci U S A 102:14795–14800. doi:10.1073/pnas.0507385102

Li YJ, Takizawa H, Azuma A et al (2010) Nrf2 is closely related to allergic airway inflammatory responses induced by low-dose diesel exhaust particles in mice. Clin Immunol 137:234–241. doi:10.1016/j.clim.2010.07.014

Li N, Wang M, Barajas B et al (2013) Nrf2 deficiency in dendritic cells enhances the adjuvant effect of ambient ultrafine particles on allergic sensitization. J Innate Immun 5:543–554. doi:10.1159/000347060

Liu Y-J, Soumelis V, Watanabe N et al (2007) TSLP: an epithelial cell cytokine that regulates T cell differentiation by conditioning dendritic cell maturation. Annu Rev Immunol 25:193–219. doi:10.1146/annurev.immunol.25.022106.141718

Liu X, Li M, Wu Y et al (2009) Anti-IL-33 antibody treatment inhibits airway inflammation in a murine model of allergic asthma. Biochem Biophys Res Commun 386:181–185. doi:10.1016/j.bbrc.2009.06.008

Locksley RM (2010) Asthma and allergic inflammation. Cell 140:777–783. doi:10.1016/j.cell.2010.03.004

Lowe AJ, Carlin JB, Bennett CM et al (2008) Do boys do the atopic march while girls dawdle? J Allergy Clin Immunol 121:1190–1195. doi:10.1016/j.jaci.2008.01.034

Marenholz I, Kerscher T, Bauerfeind A et al (2009) An interaction between filaggrin mutations and early food sensitization improves the prediction of childhood asthma. J Allergy Clin Immunol 123:911–916. doi:10.1016/j.jaci.2009.01.051

Margolis DJ, Kim B, Apter AJ et al (2014) Thymic stromal lymphopoietin variation, filaggrin loss of function, and the persistence of atopic dermatitis. JAMA Dermatol 150:254–259. doi:10.1001/jamadermatol.2013.7954

McConnell R, Berhane K, Yao L et al (2006) Traffic, susceptibility, and childhood asthma. Environ Health Perspect 114:766–772. doi:10.1289/ehp.8594

McKenzie ANJ, Spits H, Eberl G (2014) Innate lymphoid cells in inflammation and immunity. Immunity 41:366–374. doi:10.1016/j.immuni.2014.09.006

Mirchandani AS, Besnard A-G, Yip E et al (2014) Type 2 innate lymphoid cells drive CD4+ Th2 cell responses. J Immunol 192:2442–2448. doi:10.4049/jimmunol.1300974

Mjösberg JM, Trifari S, Crellin NK et al (2011) Human IL-25- and IL-33-responsive type 2 innate lymphoid cells are defined by expression of CRTH2 and CD161. Nat Immunol 12:1055–1062. doi:10.1038/ni.2104

Mjösberg J, Bernink J, Golebski K et al (2012) The transcription factor GATA3 is essential for the function of human type 2 innate lymphoid cells. Immunity 37:649–659. doi:10.1016/j.immuni.2012.08.015

Mojtabavi N, Dekan G, Stingl G, Epstein MM (2002) Long-lived Th2 memory in experimental allergic asthma. J Immunol 169:4788–4796. doi:10.4049/jimmunol.169.9.4788

Moniaga CS, Egawa G, Kawasaki H et al (2010) Flaky tail mouse denotes human atopic dermatitis in the steady state and by topical application with dermatophagoides pteronyssinus extract. Am J Pathol 176:2385–2393. doi:10.2353/ajpath.2010.090957

Monticelli LA, Sonnenberg GF, Abt MC et al (2011) Innate lymphoid cells promote lung-tissue homeostasis after infection with influenza virus. Nat Immunol 12:1045–1054. doi:10.1038/ni.2131

Morgenstern V, Zutavern A, Cyrys J et al (2008) Atopic diseases, allergic sensitization, and exposure to traffic-related air pollution in children. Am J Respir Crit Care Med 177:1331–1337. doi:10.1164/rccm.200701-036OC

Moulton VR, Bushar ND, Leeser DB et al (2006) Divergent generation of heterogeneous memory CD4 T cells. J Immunol 177:869–876. doi:10.4049/jimmunol.177.2.869

Murakami-satsutani N, Ito T, Nakanishi T et al (2014) IL-33 promotes the induction and maintenance of Th2 immune responses by enhancing the function of OX40 ligand. Allergol Int 63:443–455. doi:10.2332/allergolint.13-OA-0672

Murali-Krishna K, Altman JD, Suresh M et al (1998) Counting antigen-specific CD8 T cells: a reevaluation of bystander activation during viral infection. Immunity 8:177–187. doi:10.1016/S1074-7613(00)80470-7

Neill DR, Wong SH, Bellosi A et al (2010) Nuocytes represent a new innate effector leukocyte that mediates type-2 immunity. Nature 464:1367–1370. doi:10.1038/nature08900

Nestle FO, Di Meglio P, Qin J-Z, Nickoloff BJ (2009) Skin immune sentinels in health and disease. Nat Rev Immunol 9:679–691. doi:10.1038/nri2622

Oiwa M, Satoh T, Watanabe M et al (2008) CRTH2-dependent, STAT6-independent induction of cedar pollen dermatitis. Clin Exp Allergy 38:1357–1366. doi:10.1111/j.1365-2222.2008.03007.x

Ott H, Stanzel S, Ocklenburg C et al (2009) Total serum IgE as a parameter to differentiate between intrinsic and extrinsic atopic dermatitis in children. Acta Derm Venereol 89:257–261. doi:10.2340/00015555-0627

Oyoshi MK, Murphy GF, Geha RS (2009) Filaggrin-deficient mice exhibit TH17-dominated skin inflammation and permissiveness to epicutaneous sensitization with protein antigen. J Allergy Clin Immunol 124:485–93, 493.e1. doi:10.1016/j.jaci.2009.05.042

Oyoshi MK, Larson RP, Ziegler SF, Geha RS (2010) Mechanical injury polarizes skin dendritic cells to elicit a T(H)2 response by inducing cutaneous thymic stromal lymphopoietin expression. J Allergy Clin Immunol 126:976–84, 984.e1–5. doi:10.1016/j.jaci.2010.08.041

Palmer CNA, Irvine AD, Terron-Kwiatkowski A et al (2006) Common loss-of-function variants of the epidermal barrier protein filaggrin are a major predisposing factor for atopic dermatitis. Nat Genet 38:441–446. doi:10.1038/ng1767

Prescott SL, Macaubas C, Smallacombe T et al (1999) Development of allergen-specific T-cell memory in atopic and normal children. Lancet 353:196–200. doi:10.1016/S0140-6736(98)05104-6

Purton JF, Tan JT, Rubinstein MP et al (2007) Antiviral CD4+ memory T cells are IL-15 dependent. J Exp Med 204:951–961. doi:10.1084/jem.20061805

Quintana FJ, Basso AS, Iglesias AH et al (2008) Control of T(reg) and T(H)17 cell differentiation by the aryl hydrocarbon receptor. Nature 453:65–71. doi:10.1038/nature06880

Ramasamy A, Curjuric I, Coin LJ et al (2011) A genome-wide meta-analysis of genetic variants associated with allergic rhinitis and grass sensitization and their interaction with birth order. J Allergy Clin Immunol 128:996–1005. doi:10.1016/j.jaci.2011.08.030

Rangasamy T, Guo J, Mitzner W a et al (2005) Disruption of Nrf2 enhances susceptibility to severe airway inflammation and asthma in mice. J Exp Med 202:47–59. doi:10.1084/jem.20050538

Rank MA, Kobayashi T, Kozaki H et al (2009) IL-33-activated dendritic cells induce an atypical TH2-type response. J Allergy Clin Immunol 123:1047–1054. doi:10.1016/j.jaci.2009.02.026

Ricci G, Patrizi A, Baldi E et al (2006) Long-term follow-up of atopic dermatitis: retrospective analysis of related risk factors and association with concomitant allergic diseases. J Am Acad Dermatol 55:765–771. doi:10.1016/j.jaad.2006.04.064

Roediger B, Kyle R, Yip KH et al (2013) Cutaneous immunosurveillance and regulation of inflammation by group 2 innate lymphoid cells. Nat Immunol 14:564–573. doi:10.1038/ni.2584

Roguedas-Contios A-M, Misery L (2011) What is intrinsic atopic dermatitis? Clin Rev Allergy Immunol 41:233–236. doi:10.1007/s12016-011-8276-9

Rushmore TH, Morton MR, Pickett CB (1991) The antioxidant responsive element activation by oxidative stress and identification of the DNA consensus sequence required for functional activity. J Biol Chem 266:11632–11639

Salimi M, Barlow JL, Saunders SP et al (2013) A role for IL-25 and IL-33-driven type-2 innate lymphoid cells in atopic dermatitis. J Exp Med 210:2939–2950. doi:10.1084/jem.20130351

Santamaria Babi LF, Picker LJ, Perez Soler MT et al (1995) Circulating allergen-reactive T cells from patients with atopic dermatitis and allergic contact dermatitis express the skin-selective homing receptor, the cutaneous lymphocyte-associated antigen. J Exp Med 181:1935–1940

Saulyte J, Regueira C, Montes-Martínez A et al (2014) Active or passive exposure to tobacco smoking and allergic rhinitis, allergic dermatitis, and food allergy in adults and children: a systematic review and meta-analysis. PLoS Med 11:e1001611. doi:10.1371/journal.pmed.1001611

Saunes M, Øien T, Dotterud CK et al (2012) Early eczema and the risk of childhood asthma: a prospective, population-based study. BMC Pediatr 12:168. doi:10.1186/1471-2431-12-168

Savinko T, Matikainen S, Saarialho-Kere U et al (2012) IL-33 and ST2 in atopic dermatitis: expression profiles and modulation by triggering factors. J Invest Dermatol 132:1392–1400. doi:10.1038/jid.2011.446

Schmitz J, Owyang A, Oldham E et al (2005) IL-33, an interleukin-1-like cytokine that signals via the IL-1 receptor-related protein ST2 and induces T helper type 2-associated cytokines. Immunity 23:479–490. doi:10.1016/j.immuni.2005.09.015

Seneviratne SL, Black AP, Jones L et al (2007) The role of skin-homing T cells in extrinsic atopic dermatitis. QJM 100:19–27. doi:10.1093/qjmed/hcl132

Shimizu Y, Nakatsuru Y, Ichinose M et al (2000) Benzo[a]pyrene carcinogenicity is lost in mice lacking the aryl hydrocarbon receptor. Proc Natl Acad Sci 97:779–782. doi:10.1073/pnas.97.2.779

Shimizu M, Matsuda A, Yanagisawa K et al (2005) Functional SNPs in the distal promoter of the ST2 gene are associated with atopic dermatitis. Hum Mol Genet 14:2919–2927. doi:10.1093/hmg/ddi323

Sivaranjani N (2013) Role of reactive oxygen species and antioxidants in atopic dermatitis. J Clin Diagnostic Res 7(12):2683–2685. doi:10.7860/JCDR/2013/6635.3732

Sonnenberg GF, Mjösberg J, Spits H, Artis D (2013) Snapshot: innate lymphoid cells. Immunity 39:622–622.e1. doi:10.1016/j.immuni.2013.08.021

Soumelis V, Reche PA, Kanzler H et al (2002) Human epithelial cells trigger dendritic cell mediated allergic inflammation by producing TSLP. Nat Immunol 3:673–680. doi:10.1038/ni805

Spergel JM, Paller AS (2003) Atopic dermatitis and the atopic march. J Allergy Clin Immunol 112: S118–S127. doi:10.1016/j.jaci.2003.09.033

Spits H, Cupedo T (2012) Innate lymphoid cells: emerging insights in development, lineage relationships, and function. Annu Rev Immunol 30:647–675. doi:10.1146/annurev-immunol-020711-075053

Spits H, Artis D, Colonna M et al (2013) Innate lymphoid cells – a proposal for uniform nomenclature. Nat Rev Immunol 13:145–149. doi:10.1038/nri3365

Suzuki T, Motohashi H, Yamamoto M (2013) Toward clinical application of the Keap1-Nrf2 pathway. Trends Pharmacol Sci 34:340–346. doi:10.1016/j.tips.2013.04.005

Taguchi K, Motohashi H, Yamamoto M (2011) Molecular mechanisms of the Keap1-Nrf2 pathway in stress response and cancer evolution. Genes Cells 16:123–140. doi:10.1111/j.1365-2443.2010.01473.x

Takano H, Yoshikawa T, Ichinose TI et al. Diesel exhaust particles enhance antigen-induced airway inflammation and local cytokine expression in mice. Am J Respir Crit Care Med 156: 36–42. doi:10.1164/ajrccm.156.1.9610054

Takenaka H, Zhang K, Diaz-Sanchez D et al (1995) Enhanced human IgE production results from exposure to the aromatic hydrocarbons from diesel exhaust: direct effects on B-cell IgE production. J Allergy Clin Immunol 95:103–115. doi:10.1016/S0091-6749(95)70158-3

Tamagawa-Mineoka R, Okuzawa Y, Masuda K, Katoh N (2014) Increased serum levels of interleukin 33 in patients with atopic dermatitis. J Am Acad Dermatol 70:882–888. doi:10.1016/j.jaad.2014.01.867

Tauchi M, Hida A, Negishi T et al (2005) Constitutive expression of aryl hydrocarbon receptor in keratinocytes causes inflammatory skin lesions. Mol Cell Biol 25(21):9360–9368. doi:10.1128/MCB.25.21.9360

Thimmulappa RK, Lee H, Rangasamy T et al (2006) Nrf2 is a critical regulator of the innate immune response and survival during experimental sepsis. J Clin Invest 116:984–995. doi:10.1172/JCI25790

Troehmann ARNES, Oyle ANJC, Rogan JANELG et al (1998) T1/ST2 is preferentially expressed on murine Th2 cells, independent of interleukin 4, interleukin 5, and interleukin 10, and important for Th2 effector function. Proc Natl Acad Sci U S A 95(12):6930–6935

Van Reijsen FC, Bruijnzeel-Koomen CA, Kalthoff FS et al (1992) Skin-derived aeroallergen-specific T-cell clones of Th2 phenotype in patients with atopic dermatitis. J Allergy Clin Immunol 90:184–193. doi:10.1016/0091-6749(92)90070-I

Veldhoen M, Hirota K, Westendorf AM et al (2008) The aryl hydrocarbon receptor links TH17-cell-mediated autoimmunity to environmental toxins. Nature 453:106–109. doi:10.1038/nature06881

Villanova F, Flutter B, Tosi I et al (2014) Characterization of innate lymphoid cells in human skin and blood demonstrates increase of NKp44+ ILC3 in psoriasis. J Invest Dermatol 134: 984–991. doi:10.1038/jid.2013.477

Von Kobyletzki LB, Bornehag C-G, Hasselgren M et al (2012) Eczema in early childhood is strongly associated with the development of asthma and rhinitis in a prospective cohort. BMC Dermatol 12:11. doi:10.1186/1471-5945-12-11

Von Kobyletzki L, Svensson Å, Apfelbacher C, Schmitt J (2015) Factors that predict remission of infant atopic dermatitis: a systematic review. Acta Derm Venereol 95:389–394. doi:10.2340/00015555-1941

Wakugawa M, Nakamura K, Kakinuma T et al (2001) CC chemokine receptor 4 expression on peripheral blood CD4+ T cells reflects disease activity of atopic dermatitis. J Invest Dermatol 117:188–196. doi:10.1046/j.0022-202x.2001.01430.x

Wang Y-H, Ito T, Wang Y-H et al (2006) Maintenance and polarization of human TH2 central memory T cells by thymic stromal lymphopoietin-activated dendritic cells. Immunity 24: 827–838. doi:10.1016/j.immuni.2006.03.019

Wang I-J, Hsieh W-S, Wu K-Y et al (2008) Effect of gestational smoke exposure on atopic dermatitis in the offspring. Pediatr Allergy Immunol 19:580–586. doi:10.1111/j.1399-3038.2008.00759.x

Watanabe N, Hanabuchi S, Soumelis V et al (2004) Human thymic stromal lymphopoietin promotes dendritic cell-mediated CD4+ T cell homeostatic expansion. Nat Immunol 5:426–434. doi:10.1038/ni1048

Wherry EJ, Ahmed R (2004) Memory CD8 T-cell differentiation during viral infection. J Virol 78: 5535–5545. doi:10.1128/JVI.78.11.5535-5545.2004

Williams H, Flohr C (2006) How epidemiology has challenged 3 prevailing concepts about atopic dermatitis. J Allergy Clin Immunol 118:209–213. doi:10.1016/j.jaci.2006.04.043

Wilson SR, Thé L, Batia LM et al (2013) The epithelial cell-derived atopic dermatitis cytokine TSLP activates neurons to induce itch. Cell 155:285–295. doi:10.1016/j.cell.2013.08.057

Wolterink RGJK, Kleinjan A, van Nimwegen M et al (2012) Pulmonary innate lymphoid cells are major producers of IL-5 and IL-13 in murine models of allergic asthma. Eur J Immunol 42: 1106–1116. doi:10.1002/eji.201142018

Wüthrich B, Schmid-Grendelmeier P (2002) Natural course of AEDS. Allergy 57:267–268. doi:10.1034/j.1398-9995.2002.1n3572.x

Xu BD, Chan WL, Leung BP et al (1998) Selective expression of a stable cell surface molecule on type 2 but not type 1 helper T cells. J Exp Med 187:787–794. doi:10.1084/jem.187.5.787

Yoo J, Omori M, Gyarmati D et al (2005) Spontaneous atopic dermatitis in mice expressing an inducible thymic stromal lymphopoietin transgene specifically in the skin. J Exp Med 202: 541–549. doi:10.1084/jem.20041503

Zhang Z, Hener P, Frossard N et al (2009) Thymic stromal lymphopoietin overproduced by keratinocytes in mouse skin aggravates experimental asthma. Proc Natl Acad Sci U S A 106: 1536–1541. doi:10.1073/pnas.0812668106

Chapter 21
Role of Double-Stranded RNA Pathways in Immunometabolism in Obesity

Takahisa Nakamura

Abstract During the pathogenesis of obesity, a broad array of inflammatory and stress responses are frequently evoked in insulin-targeted metabolic tissues such as liver and adipose tissue, leading to chronic, low grade, local inflammation which plays a central role in the disruption of systemic metabolic homeostasis. This atypical state engages immune response pathways, including recruitment of immune cells into metabolic tissues, activation of IkB kinase (IKK) and c-Jun N-terminal kinase (JNK) pathways, and elevated production of an array of immune mediators, which negatively impact on nutrient metabolism and insulin action. However, the molecular basis for the induction of metabolic inflammation and the vast network of pathological responses remains elusive. Recent evidence indicates that metabolic inflammation results from deregulated double-stranded RNA (dsRNA) processing/signaling in metabolic tissues, which adversely regulates systemic glucose metabolism in obesity. These findings suggest the involvement of altered RNA networks in the immunometabolic regulation of obesity. This review focuses on the regulation of endogenous dsRNA and protein networks and how their functional changes are associated with inflammatory responses, resulting in the metabolic sequelae of obesity.

Keywords Obesity • Inflammation • Double-Stranded RNA • PKR • JNK • IKK • TRBP • microRNA • RNA modification

21.1 Introduction

Dramatic changes in lifestyle and dietary trends have triggered a rapid worldwide epidemic of chronic metabolic diseases including nonalcoholic fatty liver disease (NAFLD/steatosis), glucose intolerance, type 2 diabetes (T2D), and cardiovascular disease (Perry et al. 2014; Saltiel 2000; Semenkovich 2006). Approximately 39 %

T. Nakamura (✉)
Divisions of Endocrinology and Developmental Biology, Department of Pediatrics, Cincinnati Children's Hospital Medical Center, Cincinnati, OH, USA

PRESTO, Japan Science and Technology Agency, Kawaguchi 332-0012, Japan
e-mail: takahisa.nakamura@cchmc.org

M. Miyasaka, K. Takatsu (eds.), *Chronic Inflammation*,
DOI 10.1007/978-4-431-56068-5_21

of the world's adult population was overweight (and 13 % obese) in 2014, predisposing them to a large array of comorbidities (World Health Organization 2015; Ogden et al. 2014). Despite this enormous impact on global public health, mechanisms underlying this cluster of pathologies remain unclear and effective preventive and/or therapeutic strategies are limited.

Over the past two decades, intensive research has identified that the obese state has strong inflammatory underpinnings (Donath and Shoelson 2011; Gregor and Hotamisligil 2011; Lumeng and Saltiel 2011; Olefsky and Glass 2010; Pedersen and Febbraio 2010; Sabio and Davis 2010). A broad array of changes in metabolite and protein networks are evoked in obese metabolic tissues, which are often associated with the induction of inflammatory responses (Jin et al. 2013). These atypical inflammatory events are considered a focal point in the initiation and progression of metabolic diseases including insulin resistance and T2D; however, a critical question concerning how these inflammatory events are initiated remains elusive. Recent studies have implicated the sensors for pathogen-associated molecular patterns (PAMPs) and danger-associated molecular patterns (DAMPs) in the induction of atypical inflammation in the pathogenesis of obesity. Toll-like receptors (TLRs) and NOD-like receptors (NLRs) can sense a variety of PAMPs and DAMPs, including lipopolysaccharide (LPS), saturated free fatty acids, and nucleic acids, and initiate immune responses toward homeostatic disturbances associated with obesity (Davis et al. 2008; Holland et al. 2011; Jin et al. 2013; Konner and Bruning 2011; Vandanmagsar et al. 2011; Wen et al. 2011). In such processes, stress and inflammatory signaling cascades, such as IkB kinase (IKK) and c-Jun N-terminal kinase (JNK) pathways, are activated and contribute to the production of inflammatory cytokines (Hotamisligil 2006, 2010). The processes lead to a feedforward amplification of inflammatory responses that further activate the signaling cascades in autocrine and paracrine manners. The activation of JNK and IKK, as well as protein kinase C, S6 kinase (S6K), mammalian target of rapamycin (mTOR), and extracellular signal-regulated kinase (ERK), results in the inhibition of insulin action, a condition of insulin resistance, in part through the serine phosphorylation of insulin receptor substrates (IRS) 1 and 2 (Arkan et al. 2005; Hirosumi et al. 2002; Kim et al. 2004; Ozcan et al. 2008; Tzatsos and Kandror 2006). The insulin-resistant condition is highly correlated with other metabolic disorders including hyperglycemia, hyperlipidemia, and hypertension (Reaven et al. 1996; Shulman 2014).

In contrast to the changes in metabolite and protein networks in obesity, there exists only limited information regarding the impact of obesity and/or excess nutrients on functional changes in RNA networks and their links to metabolism. There are multiple proinflammatory proteins that sense RNA species among the sensors for PAMPs and DAMPs, such as TLR3, the retinoic acid-inducible gene-I (RIG-I)-like receptor family members, and double-stranded RNA (dsRNA)-dependent protein kinase (PKR; Fig. 21.1). In general, these proteins have been known to recognize exogenous dsRNA exclusively and induce inflammatory responses (Ding 2010; Nallagatla et al. 2011; Takeuchi and Akira 2009). However, recent evidence has clearly shown that a variety of endogenous dsRNA exist and are involved in multiple cellular events including RNA silencing (Jinek and

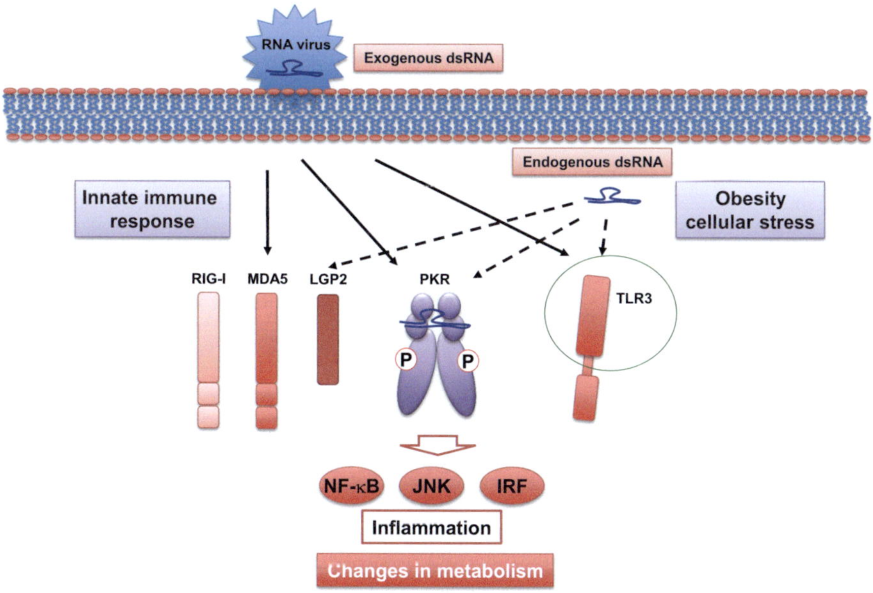

Fig. 21.1 Proposed molecular mechanisms of the involvement of proinflammatory dsRNA-binding proteins in recognition of both exogenous and endogenous dsRNAs. These dsRNA-binding proteins are known to recognize and interact with virus-derived exogenous RNAs upon virus infection and induce inflammatory responses. Under certain cellular stress conditions such as saturated fatty acid exposure, these dsRNA-binding proteins potentially sense changes in endogenous dsRNA networks through interacting, resulting in chronic inflammation and altered metabolism

Doudna 2009; Siomi and Siomi 2009; Wilson and Doudna 2013), which requires stringent regulation of dsRNA-to-protein interactions (Jinek and Doudna 2009; Siomi and Siomi 2009; Wilson and Doudna 2013). As a majority of dsRNA-binding proteins are intrinsically proinflammatory, regulatory disruption of endogenous dsRNA-protein networks may also cause an inflammatory response. We have hypothesized that such a disruption occurs in obesity, resulting in inflammation, insulin resistance, and metabolic dysfunction (Fig. 21.1). This hypothesis allowed us to identify PKR as a negative regulator of insulin action and glucose metabolism in obesity (Nakamura et al. 2010; Nakamura et al. 2014; Nakamura et al. 2015). Recent reports also showed the critical roles of TLR3 in the dysregulation of metabolism that occurs in obesity (Strodthoff et al. 2015; Wu et al. 2012).

PKR was originally identified as a pathogen sensor and a proposed regulator of the innate immune response against viral infections in higher eukaryotes (Meurs et al. 1990; Samuel 1993). A unique molecular aspect of PKR is its structure, that is, PKR is a serine/threonine kinase containing two dsRNA-binding domains (dsRBDs; Garcia et al. 2006; Williams 2001). Virus-derived dsRNA molecules are recognized and bound by PKR through its N-terminal dsRBDs, resulting in activation of the intramolecular kinase domain and autophosphorylation (Garcia et al. 2006; Williams 2001). Interestingly, in the context of infections, PKR can participate in two major cellular responses: inflammation and translation repression (Holcik and Sonenberg 2005; Ron and Walter 2007). These events are also implicated in the pathogenesis of obesity, as demonstrated by c-Jun N-terminal kinase (JNK) activation and eukaryotic initiation factor 2-α (eIF2 α) phosphorylation, both of which are highly-induced in obese metabolic tissues (Hotamisligil 2010). With these unique intrinsic molecular functions, we hypothesized that PKR might coordinately regulate the cellular events that disrupt glucose metabolism in obesity. If PKR influences the pathogenesis of obesity, it suggests the possibility that endogenous dsRNAs are involved in the process, and that PKR is a factor linking endogenous dsRNAs to the metabolic regulation in obesity.

21.2 PKR is Involved in Metabolic Inflammation and Regulates Glucose Metabolism

If PKR is involved in obesity-induced inflammatory responses, it would be anticipated that its activity is altered in conditions of nutrient surplus. Therefore, we first examined whether PKR activation is changed with nutrient-induced metabolic stress. As anticipated, there is marked activation of PKR in the liver and adipose tissue in both dietary and genetic obese mouse models (Nakamura et al. 2010). Additionally, despite the absence of exogenous dsRNA, PKR can respond to nutrient signals such as saturated fatty acids, and coordinate the activity of other critical inflammatory kinases such as JNK and the phosphorylation of eIF2a in cultured cells (Nakamura et al. 2010). Upon activation, PKR can also induce the inhibitory serine phosphorylation of insulin receptor substrate-1 (IRS-1), a critical insulin signaling component, by acting directly on IRS-1 or by indirectly activating JNK (Nakamura et al. 2010). These findings indicate that PKR, a sensor for dsRNA, is involved in the processes of obesity-induced inflammatory responses and metabolic disturbance and supports our hypothesis that the regulatory disruption of endogenous dsRNA-protein networks may also be associated with inflammatory responses in obesity (Fig. 21.2).

Given that PKR activity is linked to inflammatory responses in obesity, we then examined the functional significance of PKR in the pathogenesis of obesity, insulin resistance, and type 2 diabetes. To this end, we placed PKR-deficient mice on a high-fat diet (HFD) and analyzed their metabolic changes over the course of obesity. PKR-deficient mice have less weight gain and greater glucose tolerance and insulin sensitivity, accompanied by lower JNK activation and expression of inflammatory mediators in the HFD-feeding condition (Nakamura et al. 2010). Similar observations were made in an independently-derived strain of PKR-deficient mice (Carvalho-Filho et al. 2012). To explore the impact of PKR-deficiency on systemic glucose tolerance in a second setting, and without the potential confounding effects of body weight and adiposity, we employed a

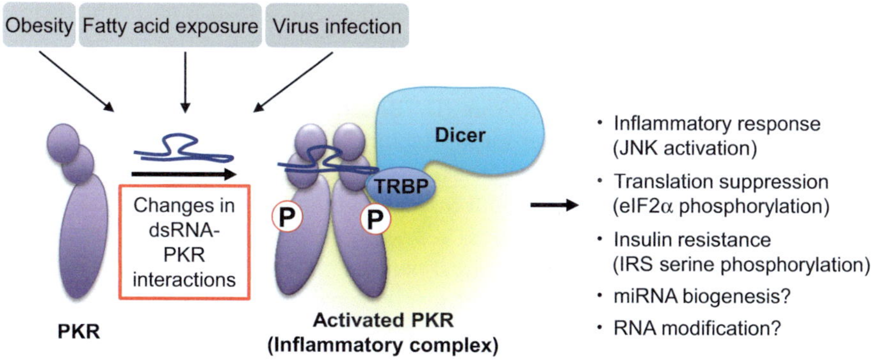

Fig. 21.2 Functional roles of PKR in virus infection and in the metabolic disorders. PKR was originally identified as an interferon-inducible gene and known to be involved in innate immune response. Recent studies have demonstrated that PKR is activated in obese metabolic tissues and by lipid-mediated stress and contributes to induction of metabolically-driven inflammatory responses, in which endogenous dsRNAs mediate PKR activation. When PKR is activated, PKR forms an inflammatory protein complex featuring components of RNA-induced silencing complex (RISC), TRBP, and Dicer. The inflammatory complex plays critical roles in the pathogenesis of metabolic disorders in obesity

mouse model of acute and severe insulin resistance induced by an infusion of lipids, in which a mixture of neutral triglycerides of predominantly linoleic, oleic, palmitic, linolenic, and stearic acids was intravenously administered to mice. Hyperinsulinemic–euglycemic clamp studies, a gold standard for assessing insulin action in vivo, revealed that PKR-deficient mice are protected from lipid-induced insulin resistance in a body-weight–independent manner (Nakamura et al. 2010). These results clearly implicate the involvement of PKR in the disruption of glucose metabolism in response to excess nutrient signals (Fig. 21.2).

As PKR-deficiency is protective against obesity- and excess nutrient-induced metabolic inflammation and dysfunction, pharmacologically targeting PKR could be a novel strategy for combating obesity-related metabolic diseases. However, in the HFD study, PKR-deficient mice were leaner than controls, which left the question of whether PKR might not be critical in inducing dysmetabolism in obesity, but may instead suppress energy expenditure. In order to target PKR for therapeutic purposes, there was a critical need to comprehend the significance of PKR in established obesity better. In an attempt to generate a new model that could provide necessary information for assessing PKR's effects on inflammation and glucose metabolism in obesity, we intercrossed PKR-deficient mice with the *Leptin*-deficient *ob/ob* model, which displays severe obesity with insulin resistance and hyperglycemia. In this setting, PKR-deficient mice exhibited similar body weights to controls in the C57BL/6 J-*ob/ob* genetic background. Nevertheless, PKR-deficiency improved glucose metabolism, and was accompanied by reduced JNK activation and inflammatory gene expression (Nakamura et al. 2015). Decreased levels of IRS1 serine phosphorylation and enhanced levels of insulin-induced Akt activation were also observed in this setting (Nakamura et al. 2015). Based on these observations, we then investigated the effects of two pharmacological inhibitors of PKR, an imidazolo–oxindole PKR inhibitor (Eley et al. 2007; Jammi et al. 2003) and 2-aminopurine (De Benedetti et al. 1985), as antidiabetic drugs. We assessed the effects of these structurally distinct small-molecule inhibitors of PKR on inflammatory responses, insulin sensitivity, and glucose metabolism in cells and in *ob/ob* mice. In both cases, inhibition of PKR resulted in reduced stress-induced JNK activation and IRS1 serine phosphorylation in vitro and in vivo (Nakamura et al. 2014). In addition, treatment with these PKR inhibitors reduced adipose tissue inflammation, improved insulin sensitivity, and improved glucose intolerance in mice after the establishment of obesity and insulin resistance (Nakamura et al. 2014). These findings suggest that pharmacologically targeting PKR may be an effective therapeutic strategy for the treatment of insulin resistance and type 2 diabetes.

21.3 Potential Role for Endogenous dsRNA in PKR Activation in Metabolic Stress

It is noteworthy that PKR responds to nutrients and plays a role in the mounting of inflammatory responses and inhibition of insulin action in the pathogenesis of obesity. However, a critical question was how PKR can sense nutrients. Because PKR is activated by exposure to palmitic acid, a proinflammatory saturated fatty acid, and TLR4 has been implicated in the response to saturated fatty acids (Shi et al. 2006), we asked if TLR4 is involved in PKR's activation by lipid exposure. However, in TLR4-deficient primary MEFs, palmitic-acid–induced PKR activation was still detected at levels similar to those in controls, indicating that PKR activation is not downstream of TLR4's lipid response (Nakamura et al. 2010). We then considered the novel idea that PKR might directly sense nutrients or other cellular molecules, including endogenous RNA species, in conditions of obesity-relevant metabolic stress. Inasmuch as PKR's defined RNA-binding activity is critical for its activation, we examined the role of PKR's RNA binding activity in such conditions. Intriguingly, a PKR RNA-binding defect variant (PKR-RM) with a point mutation at the lysine 64 residue loses the ability to respond to palmitate exposure (Nakamura et al. 2010; Nakamura et al. 2015). These data suggest that PKR's RNA-binding activity is indispensable to its nutrient response. In addition, these results further support our hypothesis that metabolic stress may alter dsRNA-to-dsRNA-binding protein networks, leading to the activation of dsRNA-binding proteins such as PKR, which in turn initiates inflammation and metabolic dysfunction.

21.4 Identification of PKR's Endogenous RNA Ligands in Metabolic Stress

Recent technical advances now allow us to analyze RNA expression profiles easily and determine changes in specific RNA expression levels under many physiological conditions. However, similar to DNA, RNA is subjected to various modifications, such as RNA methylation (Lee et al. 2014; Meyer and Jaffrey 2014). In addition, RNA can alter its cellular localization (nuclear to cytoplasm and vice versa), and have higher-order structure, which can be modified depending on the cellular environment. These changes in RNA are likely mediated by altered interactions with RNA-binding proteins, and may have influence on multiple cellular events, including the regulation of metabolism. The analysis of the RNA expression profile is unable to reveal these functional changes in RNA; however, by using a specific RNA-binding protein as a RNA sensor, changes in RNA under certain physiological conditions can be captured. In other words, by profiling RNAs specifically interacting with a RNA binding protein that is involved in the pathogenesis of diseases, we may be able to identify pathogenic RNAs critical for the development of such diseases. Thus, in order to understand the functional changes in dsRNA in obesity, PKR can be used as a RNA sensor to investigate the effects of excess nutrients on dsRNA species, to examine how PKR is activated in response to metabolic stress, and to determine the pathogenic RNAs responsible for inducing metabolic disturbances through PKR activation.

To shed light on dsRNA networks in the pathogenesis of metabolic diseases, we then studied the potential role of altered dsRNA-to-PKR interactions in metabolic inflammation. To use PKR as a dsRNA sensor, we prepared PKR-deficient MEFs reconstituted with wild-type PKR (PKR-WT) or PKR-RM, a PKR RNA-binding defect variant, and examined the effect of palmitate exposure on dsRNA-to-PKR interactions in conditions of cellular metabolic stress (Youssef et al. 2015). We performed a RNA immunoprecipitation (RIP)-seq analysis, in which both PKR-WT and RM were immunoprecipitated with a specific antibody, followed by purification and sequence analysis of bound RNAs. The RIP-seq showed that the majority of enriched RNAs that interacted with PKR-WT were small nucleolar RNAs (snoRNAs). The snoRNAs ubiquitously express noncoding RNAs of about 60–300 nucleotides and are mainly localized in the nucleolus. They can be classified into two groups, box C/D and H/ACA snoRNAs, based on their sequence elements. One of the established roles of these snoRNAs is the modification of ribosomal RNA (rRNA; Lafontaine and Tollervey 1998). Note that H/ACA box snoRNAs have a common secondary structure consisting of two hairpins and two single-stranded regions termed a hairpin–hinge-hairpin-tail structure, which contains dsRNA-like structures (Bachellerie et al. 2002). The selective interaction between these snoRNAs and PKR-WT (but not RM) was highly induced by palmitate exposure, implicating the involvement of snoRNAs in PKR activation and metabolic regulation. Indeed, we also confirmed that a subset of identified snoRNAs have the ability to bind and activate PKR in vitro. In addition, PKR could be activated in vitro by expressing the snoRNA through transfection. These results suggest an unprecedented and unexpected model whereby snoRNAs play a role in the activation of PKR under lipid exposure and support our hypothesis that changes in RNA-to-RNA-binding protein networks play a critical role in the response to metabolic stress.

It is worthy to note that snoRNAs are known to be involved in lipotoxic responses (Michel et al. 2011; Scruggs et al. 2012). Three conserved box C/D snoRNAs, U32a, U33, and U35a, were identified to be involved in the palmitate-induced oxidative stress and ER stress response. Palmitate exposure to the cells led to induction of these snoRNAs in the cytoplasm and not in nucleoli, suggesting a cytoplasmic role for these snoRNAs. Knockdown of the three snoRNAs with antisense oligonucleotides protected cells from palmitate-induced reactive oxygen species (ROS) production, ER stress, and subsequent cell death (Michel et al. 2011). Note that a spliceosomal protein SmD3 plays a critical role in generating these snoRNAs. The same research group also showed that disruption of one allele of the gene encoding SmD3, which results in reduction of snoRNA expression, confers resistance to palmitate-induced cell death (Scruggs et al. 2012). These results implicate SmD3 as a critical determinant in the processing of intronic noncoding RNAs including snoRNAs, and through this regulation, SmD3 can mediate the metabolic stress-response pathways (Fig. 21.3). As a broad range of intronic noncoding RNAs are processed through SmD3-mediated splicing (Scruggs et al. 2012), it is possible that RNAs bound to PKR may also be generated through this pathway, and ultimately activate PKR-induced inflammation and contribute to subsequent cell death in metabolic stress (Fig. 21.3).

21.5 Raveling a Novel Function for TRBP in Glucose Metabolism in Obesity

The fact that metabolic stress induces changes in dsRNA-PKR interactions raises the question of how altered dsRNA-to-dsRNA-binding protein networks impact metabolic homeostasis. dsRNA and dsRNA-binding protein interactions likely affect the function of dsRNA-binding protein to trigger downstream events, leading to the eventual modification of cellular homeostasis. In the case of PKR, interaction with endogenous dsRNAs induced by excess nutrients may cause conformational changes, leading to altered PKR kinase activity and protein complex formation.

To investigate this possibility, we identified a unique complex containing PKR forms and examined if the complex component (s) might impact PKR activity and adversely regulate glucose metabolism in obesity. Through mass spectrometry analyses, followed by biochemical and molecular validation experiments, we determined that PKR forms a distinct complex with components of RNA-induced silencing complex (RISC) and RISC-loading complex (RLC), featuring TAR-RNA binding protein (TRBP; Chendrimada et al. 2005; Gregory et al. 2005; Kok

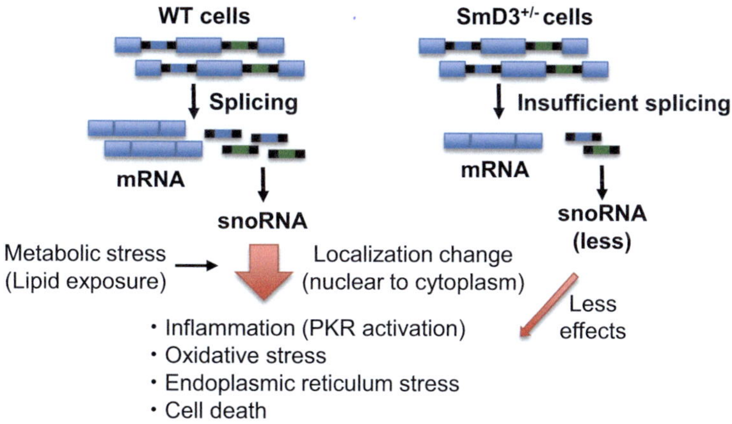

Fig. 21.3 Roles of snoRNA that have a dsRNA-like RNA structure in response to metabolic stress. A majority of snoRNAs are generated from intronic sequences, in which SmD3, a component of Spliceosome, plays a critical role. In metabolic stress conditions, the levels of cytoplasmic snoRNAs are increased, causing harmful effects including induction of oxidative stress, ER stress, and cell death, although the distinct molecular mechanisms are unclear. Given that PKR interacts with snoRNAs in lipid exposure, the SmD3–snoRNA axis may be involved in the pathogenesis of metabolic disorders in obesity

et al. 2007; Nakamura et al. 2015). The complex preferentially assembles in states of PKR activation, including the obese liver, polyinosinic-polycytidylic acid (PolyI:C) treatment, and palmitate exposure (Nakamura et al. 2015). TRBP is a unique dsRNA-binding protein that has three RNA-binding domains without any other typical motifs (Daniels and Gatignol 2012). Despite its simple protein structure, TRBP is known to have at least two molecular functions: enhancing microRNA (miRNA) maturation as a main component of the RLC (Chendrimada et al. 2005; Gregory et al. 2005; Paroo et al. 2009), and regulating inflammatory responses by modifying PKR activity (Daniels and Gatignol 2012; Nakamura et al. 2015). Note that the palmitate-induced interaction between PKR and TRBP requires PKR's RNA-binding activity, implicating involvement of endogenous RNAs in the interaction. In addition, we have also demonstrated that TRBP is required for palmitate induced PKR and JNK activation and eIF2α phosphorylation; these events are drastically diminished in the TRBP-deficient condition (Nakamura et al. 2015). These data suggest that TRBP function is regulated by metabolic stress, that PKR-TRBP complex formation occurs in response to nutrient signals, and that this complex may be critical for key downstream events, such as JNK activation (Fig. 21.2).

To assess the role of TRBP in physiological and pathophysiological conditions, we designed a short hairpin RNA (shRNA)-mediated TRBP knock down system and used adenovirus to acutely knockdown TRBP in the liver of *ob*/*ob* mice. Compared to controls, TRBP knockdown *ob*/*ob* mice exhibited improved glucose tolerance, accompanied by reduced expression levels of genes related to fatty acid synthesis, gluconeogenesis, and inflammation in the liver (Nakamura et al. 2015). In addition, suppression of hepatic TRBP resulted in significant reduction of JNK activity and eIF2α phosphorylation in obese liver, suggesting that TRBP is critical for regulation of these molecules in metabolic stress (Nakamura et al. 2015). These alterations are accompanied by decreased levels of IRS1 serine phosphorylation and a significant restoration of insulin receptor signaling as assessed by insulin-induced AKT phosphorylation levels, all of which phenocopy the loss of PKR (Nakamura et al. 2010; Nakamura et al. 2015).

It was previously reported that in the obese liver, global miRNA expression levels were changed, of which the vast majority were increased (Kornfeld et al. 2013). Given that TRBP plays a central role in miRNA biogenesis, a process wherein precursor miRNA (pre-miRNA) is processed to become mature miRNA (Chendrimada et al. 2005; Gregory et al. 2005), we hypothesized that the metabolic phenotypes observed in TRBP knockdown might also be due to changes in the miRNA expression profile. Microarray-based miRNA expression profiling (Chan et al. 2009; Chen et al. 2009; Iliopoulos et al. 2009) revealed a significant reduction of selective miRNAs including those known to be involved in the development of insulin resistance, such as the Let-7 family (Frost and Olson 2011; Zhu et al. 2011). Thus, altered miRNA expression regulated by TRBP may also be a key event in the pathogenesis of obesity. The critical question of whether TRBP-mediated inflammation or miRNA biogenesis is a dominant pathway in the metabolic dysregulation of obesity remains to be addressed and will provide important insight into the significance of dsRNA-mediated regulation of RNA silencing in metabolic stress.

21.6 Future Perspective

In the past, it was not widely recognized that there is a variety of endogenous dsRNA/dsRNA-like RNAs in the cell. Therefore, it was generally accepted that dsRNA-binding proteins exclusively sensed dsRNA from infectious organisms, which then triggered defensive inflammatory responses. However, it is now obvious that endogenous dsRNAs/dsRNA-like RNAs exist and play critical roles in the multiple cellular processes (Piatek and Werner 2014; Whipple et al. 2015; White et al. 2014). Given that PKR can recognize endogenous dsRNAs/dsRNA-like RNAs in response to metabolic stress to induce inflammation (Nakamura et al. 2010; Youssef et al. 2015), structural and functional changes in endogenous RNA species are an attractive core mechanism. However, is it still unclear how cellular stresses modify the RNA characteristics. Functional changes in RNA processing machinery, such as the spliceosome and RNA silencing, could be implicated as a cause of altered RNA networks (Scruggs et al. 2012; White et al. 2014), resulting in deranged RNA-to-RNA binding protein networks and dysmetabolism. RNAs are subjected to RNA modifications, including methylation, uridylation, and adenosine-to-inosine (A-I) editing, which can affect an RNA's secondary structure (Fu et al. 2014; Lee et al. 2014; Zipeto et al. 2015).

Investigating the roles of enzymes that mediate these modifications would provide us with a unique approach for better understanding the molecular mechanism of the pathogenesis of chronic inflammatory diseases.

There are various proteins involved in RNA processing and biology in the PKR complex we identified in the obese liver. Enriched Gene Ontology analysis with high-confidence PKR interactors has revealed that this PKR complex is closely integrated with endogenous RNA-processing pathways, particularly RNA splicing, RNA metabolic process, and mRNA processing in the obese liver (Nakamura et al. 2015). In addition, this complex contains proteins mediating proinflammatory signaling and RNA methylation (Nakamura et al. 2015). The functional and physiological roles of the majority of these components remain to be elucidated, especially in the pathogenesis of inflammatory metabolic disease. Investigation of RNA networks by using these RNA-modifying proteins as RNA sensors will be of vital interest and lead us to a better understanding of the effects of metabolic stress on RNA networks. As TRBP plays critical roles in the regulation of PKR function and RNA silencing, TRBP might be the signaling node that senses metabolic stress levels through the recognition of a variety of RNA networks, and may link to miRNA outputs and PKR-mediated inflammatory signaling networks. If changes in endogenous RNA dynamics is a key event that influences chronic inflammation through the modulation of TRBP-PKR functions in obesity, it would pave the way for novel therapeutic strategies targeting not only obesity-induced metabolic diseases but also other chronic inflammatory diseases.

Acknowledgment I am grateful to the members of my laboratory, particularly Celvie L. Yuan and Elise Bernhard for helpful discussions and comments. I thank Vivian Hwa and Jason Fan for thoughtful comments. I acknowledge support from the American Heart Association, Digestive Disease Research Core Center in Cincinnati (DK078392), the Diabetes Action Research and Education Foundation, and PRESTO from the Japan Science and Technology Agency.

References

Arkan MC, Hevener AL, Greten FR, Maeda S, Li ZW, Long JM, Wynshaw-Boris A, Poli G, Olefsky J, Karin M (2005) IKK-beta links inflammation to obesity induced insulin resistance. Nat Med 11.191–198

Bachellerie JP, Cavaille J, Huttenhofer A (2002) The expanding snoRNA world. Biochimie 84:775–790

Carvalho-Filho MA, Carvalho BM, Oliveira AG, Guadagnini D, Ueno M, Dias MM, Tsukumo DM, Hirabara SM, Reis LF, Curi R et al (2012) Double-stranded RNA-activated protein kinase is a key modulator of insulin sensitivity in physiological conditions and in obesity in mice. Endocrinology 153:5261–5274

Chan SY, Zhang YY, Hemann C, Mahoney CE, Zweier JL, Loscalzo J (2009) MicroRNA-210 controls mitochondrial metabolism during hypoxia by repressing the iron-sulfur cluster assembly proteins ISCU1/2. Cell Metab 10:273–284

Chen Y, Gelfond JA, McManus LM, Shireman PK (2009) Reproducibility of quantitative RT-PCR array in miRNA expression profiling and comparison with microarray analysis. BMC Genomics 10:407

Chendrimada TP, Gregory RI, Kumaraswamy E, Norman J, Cooch N, Nishikura K, Shiekhattar R (2005) TRBP recruits the dicer complex to Ago2 for microRNA processing and gene silencing. Nature 436:740–744

Daniels SM, Gatignol A (2012) The multiple functions of TRBP, at the hub of cell responses to viruses, stress, and cancer. Microbiol Mol Biol Rev 76:652–666

Davis JE, Gabler NK, Walker-Daniels J, Spurlock ME (2008) Tlr-4 deficiency selectively protects against obesity induced by diets high in saturated fat. Obesity 16:1248–1255

De Benedetti A, Williams GJ, Baglioni C (1985) Inhibition of binding to initiation complexes of nascent reovirus mRNA by double-stranded RNA-dependent protein kinase. J Virol 54:408–413

Ding SW (2010) RNA-based antiviral immunity. Nat Rev Immunol 10:632–644

Donath MY, Shoelson SE (2011) Type 2 diabetes as an inflammatory disease. Nat Rev Immunol 11:98–107

Eley HL, Russell ST, Tisdale MJ (2007) Attenuation of muscle atrophy in a murine model of cachexia by inhibition of the dsRNA-dependent protein kinase. Br J Cancer 96:1216–1222

Frost RJ, Olson EN (2011) Control of glucose homeostasis and insulin sensitivity by the Let-7 family of microRNAs. Proc Natl Acad Sci U S A 108:21075–21080

Fu Y, Dominissini D, Rechavi G, He C (2014) Gene expression regulation mediated through reversible m(6)a RNA methylation. Nat Rev Genet 15:293–306

Garcia MA, Gil J, Ventoso I, Guerra S, Domingo E, Rivas C, Esteban M (2006) Impact of protein kinase PKR in cell biology: from antiviral to antiproliferative action. Microbiol Mol Biol Rev 70:1032–1060

Gregor MF, Hotamisligil GS (2011) Inflammatory mechanisms in obesity. Annu Rev Immunol 29:415–445

Gregory RI, Chendrimada TP, Cooch N, Shiekhattar R (2005) Human RISC couples microRNA biogenesis and posttranscriptional gene silencing. Cell 123:631–640

Hirosumi J, Tuncman G, Chang L, Gorgun CZ, Uysal KT, Maeda K, Karin M, Hotamisligil GS (2002) A central role for JNK in obesity and insulin resistance. Nature 420:333–336

Holcik M, Sonenberg N (2005) Translational control in stress and apoptosis. Nat Rev Mol Cell Biol 6:318–327

Holland WL, Bikman BT, Wang LP, Yuguang G, Sargent KM, Bulchand S, Knotts TA, Shui G, Clegg DJ, Wenk MR et al (2011) Lipid-induced insulin resistance mediated by the proinflammatory receptor TLR4 requires saturated fatty acid-induced ceramide biosynthesis in mice. J Clin Invest 121:1858–1870

Hotamisligil GS (2006) Inflammation and metabolic disorders. Nature 444:860–867

Hotamisligil GS (2010) Endoplasmic reticulum stress and the inflammatory basis of metabolic disease. Cell 140:900–917

Iliopoulos D, Polytarchou C, Hatziapostolou M, Kottakis F, Maroulakou IG, Struhl K, Tsichlis PN (2009) MicroRNAs differentially regulated by Akt isoforms control EMT and stem cell renewal in cancer cells. Sci Signal 2:ra62

Jammi NV, Whitby LR, Beal PA (2003) Small molecule inhibitors of the RNA-dependent protein kinase. Biochem Biophys Res Commun 308:50–57

Jin C, Henao-Mejia J, Flavell RA (2013) Innate immune receptors: key regulators of metabolic disease progression. Cell Metab 17:873–882

Jinek M, Doudna JA (2009) A three-dimensional view of the molecular machinery of RNA interference. Nature 457:405–412

Kim JK, Fillmore JJ, Sunshine MJ, Albrecht B, Higashimori T, Kim DW, Liu ZX, Soos TJ, Cline GW, O'Brien WR et al (2004) PKC-theta knockout mice are protected from fat-induced insulin resistance. J Clin Invest 114:823–827

Kok KH, Ng MH, Ching YP, Jin DY (2007) Human TRBP and PACT directly interact with each other and associate with dicer to facilitate the production of small interfering RNA. J Biol Chem 282:17649–17657

Konner AC, Bruning JC (2011) Toll-like receptors: linking inflammation to metabolism. Trends Endocrinol Metab TEM 22:16–23

Kornfeld JW, Baitzel C, Konner AC, Nicholls HT, Vogt MC, Herrmanns K, Scheja L, Haumaitre C, Wolf AM, Knippschild U et al (2013) Obesity-induced overexpression of miR-802 impairs glucose metabolism through silencing of Hnf1b. Nature 494:111–115

Lafontaine DL, Tollervey D (1998) Birth of the snoRNPs: the evolution of the modification-guide snoRNAs. Trends Biochem Sci 23:383–388

Lee M, Kim B, Kim VN (2014) Emerging roles of RNA modification: m(6)a and U-tail. Cell 158:980–987

Lumeng CN, Saltiel AR (2011) Inflammatory links between obesity and metabolic disease. J Clin Invest 121:2111–2117

Meurs E, Chong K, Galabru J, Thomas NS, Kerr IM, Williams BR, Hovanessian AG (1990) Molecular cloning and characterization of the human double-stranded RNA-activated protein kinase induced by interferon. Cell 62:379–390

Meyer KD, Jaffrey SR (2014) The dynamic epitranscriptome: N6-methyladenosine and gene expression control. Nat Rev Mol Cell Biol 15:313–326

Michel CI, Holley CL, Scruggs BS, Sidhu R, Brookheart RT, Listenberger LL, Behlke MA, Ory DS, Schaffer JE (2011) Small nucleolar RNAs U32a, U33, and U35a are critical mediators of metabolic stress. Cell Metab 14:33–44

Nakamura T, Furuhashi M, Li P, Cao H, Tuncman G, Sonenberg N, Gorgun CZ, Hotamisligil GS (2010) Double-stranded RNA-dependent protein kinase links pathogen sensing with stress and metabolic homeostasis. Cell 140:338–348

Nakamura T, Arduini A, Baccaro B, Furuhashi M, Hotamisligil GS (2014) Small-molecule inhibitors of PKR improve glucose homeostasis in obese diabetic mice. Diabetes 63:526–534

Nakamura T, Kunz RC, Zhang C, Kimura T, Yuan CL, Baccaro B, Namiki Y, Gygi SP, Hotamisligil GS (2015) A critical role for PKR complexes with TRBP in immunometabolic regulation and eIF2alpha phosphorylation in obesity. Cell rep 11:295–307

Nallagatla SR, Toroney R, Bevilacqua PC (2011) Regulation of innate immunity through RNA structure and the protein kinase PKR. Curr Opin Struct Biol 21:119–127

Ogden CL, Carroll MD, Kit BK, Flegal KM (2014) Prevalence of childhood and adult obesity in the United States, 2011–2012. Jama 311:806–814

Olefsky JM, Glass CK (2010) Macrophages, inflammation, and insulin resistance. Annu Rev Physiol 72:219–246

Ozcan U, Ozcan L, Yilmaz E, Duvel K, Sahin M, Manning BD, Hotamisligil GS (2008) Loss of the tuberous sclerosis complex tumor suppressors triggers the unfolded protein response to regulate insulin signaling and apoptosis. Mol Cell 29:541–551

Paroo Z, Ye X, Chen S, Liu Q (2009) Phosphorylation of the human microRNA-generating complex mediates MAPK/Erk signaling. Cell 139:112–122

Pedersen BK, Febbraio MA (2010) Diabetes: treatment of diabetes mellitus: new tricks by an old player. Nat Rev Endocrinol 6:482–483

Perry RJ, Samuel VT, Petersen KF, Shulman GI (2014) The role of hepatic lipids in hepatic insulin resistance and type 2 diabetes. Nature 510:84–91

Piatek MJ, Werner A (2014) Endogenous siRNAs: regulators of internal affairs. Biochem Soc Trans 42:1174–1179

Reaven GM, Lithell H, Landsberg L (1996) Hypertension and associated metabolic abnormalities – the role of insulin resistance and the sympathoadrenal system. N Engl J Med 334:374–381

Ron D, Walter P (2007) Signal integration in the endoplasmic reticulum unfolded protein response. Nat Rev Mol Cell Biol 8:519–529

Sabio G, Davis RJ (2010) CJun NH2-terminal kinase 1 (JNK1): roles in metabolic regulation of insulin resistance. Trends Biochem Sci 35:490–496

Saltiel AR (2000) Series introduction: the molecular and physiological basis of insulin resistance: emerging implications for metabolic and cardiovascular diseases. J Clin Invest 106:163–164

Samuel CE (1993) The eIF-2 alpha protein kinases, regulators of translation in eukaryotes from yeasts to humans. J Biol Chem 268:7603–7606
Scruggs BS, Michel CI, Ory DS, Schaffer JE (2012) SmD3 regulates intronic noncoding RNA biogenesis. Mol Cell Biol 32:4092–4103
Semenkovich CF (2006) Insulin resistance and atherosclerosis. J Clin Invest 116:1813–1822
Shi H, Kokoeva MV, Inouye K, Tzameli I, Yin H, Flier JS (2006) TLR4 links innate immunity and fatty acid-induced insulin resistance. J Clin Invest 116:3015–3025
Shulman GI (2014) Ectopic fat in insulin resistance, dyslipidemia, and cardiometabolic disease. N Engl J Med 371:1131–1141
Siomi H, Siomi MC (2009) On the road to reading the RNA-interference code. Nature 457:396–404
Strodthoff D, Ma Z, Wirstrom T, Strawbridge RJ, Ketelhuth DF, Engel D, Clarke R, Falkmer S, Hamsten A, Hansson GK et al (2015) Toll-like receptor 3 influences glucose homeostasis and beta-cell insulin secretion. Diabetes 64(10):3425–3438
Takeuchi O, Akira S (2009) Innate immunity to virus infection. Immunol Rev 227:75–86
Tzatsos A, Kandror KV (2006) Nutrients suppress phosphatidylinositol 3-kinase/Akt signaling via raptor-dependent mTOR-mediated insulin receptor substrate 1 phosphorylation. Mol Cell Biol 26:63–76
Vandanmagsar B, Youm YH, Ravussin A, Galgani JE, Stadler K, Mynatt RL, Ravussin E, Stephens JM, Dixit VD (2011) The NLRP3 inflammasome instigates obesity-induced inflammation and insulin resistance. Nat Med 17:179–188
Wen H, Gris D, Lei Y, Jha S, Zhang L, Huang MT, Brickey WJ, Ting JP (2011) Fatty acid-induced NLRP3-ASC inflammasome activation interferes with insulin signaling. Nat Immunol 12:408–415
Whipple JM, Youssef OA, Aruscavage PJ, Nix DA, Hong C, Johnson WE, Bass BL (2015) Genome-wide profiling of the C. elegans dsRNAome. RNA 21:786–800
White E, Schlackow M, Kamieniarz-Gdula K, Proudfoot NJ, Gullerova M (2014) Human nuclear dicer restricts the deleterious accumulation of endogenous double-stranded RNA. Nat Struct Mol Biol 21:552–559
Williams BR (2001) Signal integration via PKR. Sci STKE 2001:RE2
Wilson RC, Doudna JA (2013) Molecular mechanisms of RNA interference. Annu Rev Biophys 42:217–239
World Health Organization (2015) Obesity and overweight. http://www.who.int/mediacentre/factsheets/fs311/en
Wu LH, Huang CC, Adhikarakunnathu S, San Mateo LR, Duffy KE, Rafferty P, Bugelski P, Raymond H, Deutsch H, Picha K et al (2012) Loss of toll-like receptor 3 function improves glucose tolerance and reduces liver steatosis in obese mice. Metab Clin Exp 61:1633–1645
Youssef OA, Safran SA, Nakamura T, Nix DA, Hotamisligil GS, Bass BL (2015) Potential role for snoRNAs in PKR activation during metabolic stress. Proc Natl Acad Sci U S A 112:5023–5028
Zhu H, Shyh-Chang N, Segre AV, Shinoda G, Shah SP, Einhorn WS, Takeuchi A, Engreitz JM, Hagan JP, Kharas MG et al (2011) The Lin28/let-7 axis regulates glucose metabolism. Cell 147:81–94
Zipeto MA, Jiang Q, Melese E, Jamieson CH (2015) RNA rewriting, recoding, and rewiring in human disease. Trends Mol Med 21(9):549–559

Chapter 22
Molecular Mechanisms Underlying Obesity-Induced Chronic Inflammation

Takayoshi Suganami, Miyako Tanaka, and Yoshihiro Ogawa

Abstract Obesity is a chronic low-grade inflammatory disease as well as a metabolic disease. Indeed, adipose tissue in obesity exhibits chronic inflammatory changes characterised by inflammatory cell infiltration and proinflammatory cytokine overproduction. In addition to lipid-laden mature adipocytes, adipose tissue is composed of various stromal cells such as preadipocytes, endothelial cells, fibroblasts, and immune cells that may be involved in adipose tissue functions. Accumulating evidence has suggested that adipocytes and stromal cells in adipose tissue change dramatically in number and cell type during the development of obesity, which is termed 'adipose tissue remodeling'. We have provided evidence suggesting that crosstalk between adipocytes and macrophages aggravates obesity-induced adipose tissue inflammation. Our recent data also demonstrate that adipose tissue fibrosis is a novel mechanism underlying obesity-induced ectopic lipid accumulation in nonadipose tissues such as the liver, where lipotoxicity impairs their metabolic functions. Thus, understanding the molecular mechanism underlying adipose tissue inflammation may lead to the identification of novel therapeutic strategies to prevent or treat obesity-induced adipose tissue inflammation.

T. Suganami (✉)
Department of Molecular Medicine and Metabolism, Research Institute of Environmental Medicine, Nagoya University, Nagoya, Japan

Department of Molecular Endocrinology and Metabolism, Graduate School of Medical and Dental Sciences, Tokyo Medical and Dental University, Tokyo, Japan

PRESTO, Japan Science and Technology Agency, Kawaguchi 332-0012, Japan
e-mail: suganami@riem.nagoya-u.ac.jp

M. Tanaka
Department of Molecular Medicine and Metabolism, Research Institute of Environmental Medicine, Nagoya University, Nagoya, Japan

Department of Molecular Endocrinology and Metabolism, Graduate School of Medical and Dental Sciences, Tokyo Medical and Dental University, Tokyo, Japan

Y. Ogawa
Department of Molecular Endocrinology and Metabolism, Graduate School of Medical and Dental Sciences, Tokyo Medical and Dental University, Tokyo, Japan

AMED-CREST, Tokyo, Japan

M. Miyasaka, K. Takatsu (eds.), *Chronic Inflammation*,
DOI 10.1007/978-4-431-56068-5_22

Keywords Metabolic syndrome • Obesity • Adipocytes • Macrophages • Chronic inflammation

22.1 Obesity-Induced Adipose Tissue Dysfunction and Metabolic Syndrome

The metabolic syndrome is a constellation of visceral obesity, impaired glucose and lipid metabolism, and blood pressure elevation, which increases a risk of atherosclerotic diseases, nonalcoholic steatohepatitis (NASH), and chronic kidney disease. Substantial evidence suggests that obesity induces adipose tissue dysfunction, which contributes to the pathogenesis of the metabolic syndrome. Adipose tissue functions as an endocrine organ that secretes a large number of adipocytokines including leptin, adiponectin, and proinflammatory cytokines and chemokines, which are involved in a variety of physiologic and pathologic processes (Matsuzawa et al. 1999; Kadowaki et al. 2006; Berg and Scherer 2005; Schenk et al. 2008; Hotamisligil 2006). Thus, adipocytokines are important mediators in organ crosstalk to maintain energy homeostasis (Fig. 22.1). In addition, unbalanced production of pro- and anti-inflammatory adipocytokines impairs glucose and lipid metabolism in obesity (Schenk et al. 2008; Hotamisligil 2006; Berg and Scherer 2005; Rocha and Libby 2009; Matsuzawa et al. 1999).

To store excessive energy as triglyceride is also a fundamental function of adipose tissue. In response to nutritional changes such as fasted and fed conditions,

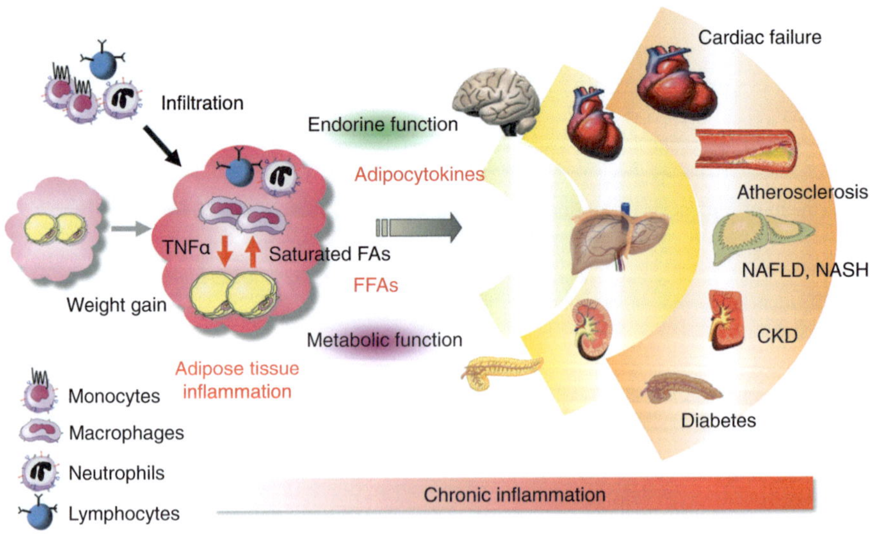

Fig. 22.1 Molecular mechanisms underlying obesity-induced adipose tissue inflammation *NAFLD* nonalcoholic fatty liver disease, *NASH* nonalcoholic steatohepatitis, *CKD* chronic kidney disease, *FFAs* free fatty acids, *FAs* fatty acids, *TNFα* tumour necrosis factor-α

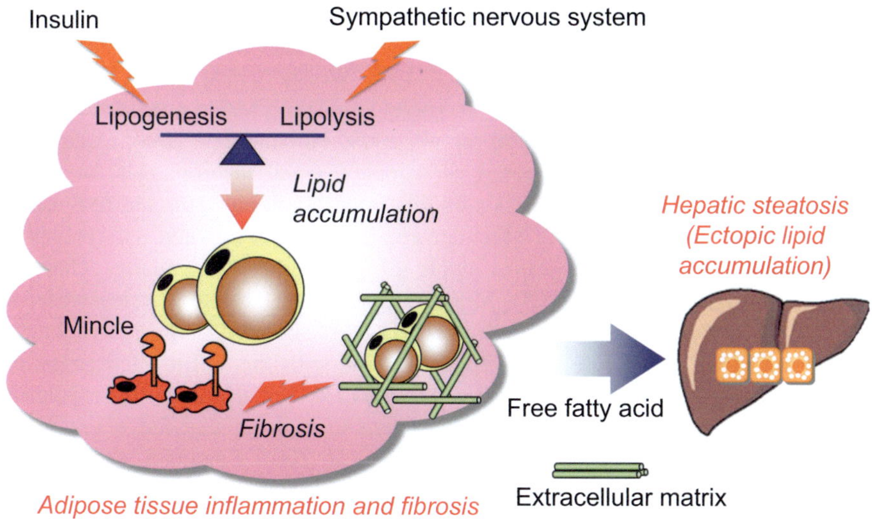

Fig. 22.2 Role of Mincle in obesity-induced adipose tissue fibrosis and ectopic lipid accumulation *Mincle* macrophage-inducible C-type lectin

lipid metabolism in adipose tissue is tightly regulated by hormones and the sympathetic nervous system (Fig. 22.2). For instance, insulin suppresses lipolysis and increases lipogenesis in the fed state. In contrast, catecholamines induce lipolysis in the fasted state to deliver free fatty acids as fuel to other organs. Moreover, adipose tissue controls its lipid storage capacity by changing its size (hypertrophy) and number (hyperplasia) during the development of obesity (Sun et al. 2011). When adipose tissue cannot meet the demand of storing excessive energy, triglyceride is also accumulated in nonadipose tissues as ectopic fat, thereby leading to tissue dysfunction such as insulin resistance in liver and skeletal muscle and insufficient insulin secretion in pancreas (lipotoxicity; Fig. 22.1; Olefsky and Glass 2010; Sun et al. 2011).

22.2 Role of Adipose Tissue Inflammation in Adipocytokine Production

Evidence has accumulated indicating that obesity is a state of chronic, low-grade inflammation, which may cause obesity-induced metabolic derangement (Schenk et al. 2008; Hotamisligil 2006; Berg and Scherer 2005; Rocha and Libby 2009). Indeed, a variety of immune cells including macrophages changes in number in adipose tissue during the development of obesity, thereby regulating adipose tissue function (Weisberg et al. 2003; Xu et al. 2003; Clement et al. 2004; Nishimura et al. 2009, 2013; Moro et al. 2010; Talukdar et al. 2012). Notably, macrophage infiltration in adipose tissue precedes or associates with the development of systemic insulin resistance, adipocytokine dysregulation, and ectopic lipid accumulation in obese animals and humans (Xu et al. 2003; Weisberg et al. 2003). In addition to mature adipocytes, adipose tissue macrophages also represent a major source of proinflammatory cytokines (Schenk et al. 2008; Hotamisligil 2006). Using an in vitro coculture system composed of adipocytes and macrophages, we have provided evidence suggesting that a paracrine loop involving saturated fatty acids and tumour necrosis factor-α (TNFα) derived from adipocytes and macrophages, respectively, establishes a vicious cycle of persistent inflammatory changes (Fig. 22.1; Suganami et al. 2005). Moreover, saturated fatty acids released from adipocytes activate Toll-like receptor (TLR4) signalling in macrophages (Suganami et al. 2007; Lee et al. 2001). In line with this, TLR4-deficiency or haematopoietic cell-specific deletion of TLR4 ameliorates dysregulated adipocytokines and systemic insulin resistance in high-fat diet-fed obese mice (Shi et al. 2006; Poggi et al. 2007; Tsukumo et al. 2007; Suganami et al. 2007; Saberi et al. 2009). Thus, such an intimate crosstalk between adipocytes and macrophages may be a molecular mechanism underlying obesity-induced adipose tissue inflammation.

22.3 Role of Mincle in Adipose Tissue Remodeling

In various tissues and organs, chronic inflammation is characterised by sustained interaction between parenchymal and stromal cells in response to exogenous and endogenous stresses, thereby leading to tissue dysfunction and remodeling (Serhan and Savill 2005; Medzhitov 2008). Adipose tissue in obesity also exhibits adipocyte hypertrophy, immune cell infiltration, angiogenesis, and extracellular matrix overproduction during the progression of chronic inflammation (Nishimura et al. 2007, 2008; Hotamisligil 2006; Schenk et al. 2008). Such dynamic histological changes observed in obese adipose tissue can be termed 'adipose tissue remodeling' (Sun et al. 2011; Itoh et al. 2011; Suganami and Ogawa 2010), which reminds us of 'vascular remodeling', the histological feature of atherosclerotic lesions (Rocha and Libby 2009). However, the molecular mechanisms underlying obesity-induced adipose tissue remodeling have been unclear.

Recently, we have demonstrated that macrophage-inducible C-type lectin (Mincle), a pathogen sensor for *Mycobacterium tuberculosis*, is involved in obesity-induced adipose tissue remodeling, in which Mincle activates myofibroblasts to induce interstitial fibrosis (Fig. 22.2; Tanaka et al. 2014). There is an unique histological structure termed crown-like structure (CLS) in obese adipose tissue, where macrophages scavenge the residual lipid droplets of dead adipocytes and adipocyte–macrophage crosstalk may occur in close proximity. Mincle is localised to the proinflammatory M1 macrophages constituting CLS, which are surrounded by myofibroblasts and collagen deposition. Interestingly, Mincle activation in macrophages is sufficient to induce adipose tissue fibrosis in lean mice.

On the other hand, recent evidence suggests that transcription factors in adipocytes such as hypoxia-inducible factor-1α and peroxisome proliferator-activated receptor-γ positively and negatively regulate obesity-induced interstitial fibrosis, respectively (Halberg et al. 2009; Jonker et al. 2012). Therefore, it is interesting to know the molecular mechanisms of the crosstalk between parenchymal and stromal cells in obesity-induced adipose tissue remodeling.

22.4 Role of Adipose Tissue Remodeling in Ectopic Lipid Accumulation

As discussed above, adipose tissue functions as an energy reservoir that stores excessive fatty acids in the form of triglyceride. Similarly to the liver and skeletal muscle, adipose tissue inflammation induces insulin resistance in adipose tissue, which disturbs lipid storage function of adipose tissue (Olefsky and Glass 2010). The crosstalk between adipocytes and macrophages increases production of proinflammatory cytokines such as TNFα, thereby inducing lipolysis (Olefsky and Glass 2010; Suganami et al. 2005). In addition to these functional alterations, histological changes are related to the reduction of lipid-storing function in adipose tissue in obesity (Sun et al. 2011). Recent studies have pointed to the overproduction of extracellular matrix components in adipose tissue from obese animals and humans (Khan et al. 2009; Liu et al. 2009; Mutch et al. 2009; Vila et al. 2014; Guglielmi et al. 2015). It is also known that adipose tissue fibrosis is negatively correlated with adipocyte diameters in human adipose tissue (Divoux et al. 2010), suggesting that increased interstitial fibrosis may limit adipose tissue expandability. In this regard, mice lacking collagen VI, which is expressed abundantly in adipose tissue, show uninhibited adipose tissue expansion, reduced fatty liver, and significant improvements in insulin sensitivity on a high-fat diet (Khan et al. 2009). These findings suggest that obesity-induced adipose tissue fibrosis may contribute to ectopic lipid accumulation in nonadipose tissues such as the liver, skeletal muscle, and pancreas, where lipotoxicity impairs their metabolic functions (Fig. 22.2; Duval et al. 2010; Wang et al. 2008; Strissel et al. 2007; Sun et al. 2011).

Interestingly, Mincle KO mice are protected against obesity-induced adipose tissue fibrosis and ectopic lipid accumulation in the liver. In addition, serum FFA concentrations are significantly decreased in Mincle KO mice relative to wild-type mice. These data support the notion that increased lipid-storage capacity of adipose tissue in Mincle KO mice may cause less efflux of FFA from adipose tissue to the liver (Fig. 22.2). Therefore, it is conceivable that Mincle is a novel regulator of ectopic lipid accumulation in obesity. Because it is known that Mincle can sense dead cells (Yamasaki et al. 2008), Mincle may be activated by an endogenous ligand released from adipocytes within CLS. From the therapeutic viewpoint, it is important to understand the difference in the intracellular signalling pathway of Mincle activated by endogenous and exogenous ligands.

22.5 Concluding Remarks

During the past decade, numerous studies have shown the molecular mechanism underlying obesity-induced chronic inflammation in adipose tissue, in which a variety of immune cells including macrophages crosstalk with adipocytes to impair adipocytokine production. Our recent data also demonstrated that the adipocyte–macrophage interaction induces adipose tissue fibrosis, which may limit lipid storage capacity of adipose tissue and increase ectopic lipid accumulation in the liver. Thus, chronic inflammation originated from adipose tissue may spread to various organs and tissues via adipocytokines and FFA (Fig. 22.1). These observations suggest that intercellular communication in CLS may be key to understanding the molecular mechanism of obesity-induced chronic inflammation. In this regard, it is interesting to know how macrophages aggregate to surround the dying adipocytes. Better understanding the molecular mechanisms underlying adipose tissue inflammation may lead to novel therapeutic strategies to prevent or treat obesity-induced insulin resistance and metabolic derangement.

Acknowledgments This work was supported by Japan Science and Technology Agency, Japan Agency for Medical Research and Development, and the Ministry of Education, Culture, Sports, Science and Technology of Japan.

References

Berg AH, Scherer PE (2005) Adipose tissue, inflammation, and cardiovascular disease. Circ Res 96(9):939–949

Clement K, Viguerie N, Poitou C, Carette C, Pelloux V, Curat CA, Sicard A, Rome S, Benis A, Zucker JD, Vidal H, Laville M, Barsh GS, Basdevant A, Stich V, Cancello R, Langin D (2004) Weight loss regulates inflammation-related genes in white adipose tissue of obese subjects. FASEB J 18(14):1657–1669

Divoux A, Tordjman J, Lacasa D, Veyrie N, Hugol D, Aissat A, Basdevant A, Guerre-Millo M, Poitou C, Zucker JD, Bedossa P, Clement K (2010) Fibrosis in human adipose tissue: composition, distribution, and link with lipid metabolism and fat mass loss. Diabetes 59(11): 2817–2825. doi:db10-0585

Duval C, Thissen U, Keshtkar S, Accart B, Stienstra R, Boekschoten MV, Roskams T, Kersten S, Muller M (2010) Adipose tissue dysfunction signals progression of hepatic steatosis towards nonalcoholic steatohepatitis in C57BL/6 mice. Diabetes 59(12):3181–3191. doi:db10-0224

Guglielmi V, Cardellini M, Cinti F, Corgosinho F, Cardolini I, D'Adamo M, Zingaretti MC, Bellia A, Lauro D, Gentileschi P, Federici M, Cinti S, Sbraccia P (2015) Omental adipose tissue fibrosis and insulin resistance in severe obesity. Nutr Diabetes 5:e175. doi:10.1038/nutd.2015.22

Halberg N, Khan T, Trujillo ME, Wernstedt-Asterholm I, Attie AD, Sherwani S, Wang ZV, Landskroner-Eiger S, Dineen S, Magalang UJ, Brekken RA, Scherer PE (2009) Hypoxia-inducible factor 1α induces fibrosis and insulin resistance in white adipose tissue. Mol Cell Biol 29(16):4467–4483. doi:10.1128/MCB.00192-09

Hotamisligil GS (2006) Inflammation and metabolic disorders. Nature 444(7121):860–867

Itoh M, Suganami T, Hachiya R, Ogawa Y (2011) Adipose tissue remodeling as homeostatic inflammation. Int J Inflam 2011:720926. doi:10.4061/2011/720926

Jonker JW, Suh JM, Atkins AR, Ahmadian M, Li P, Whyte J, He M, Juguilon H, Yin YQ, Phillips CT, Yu RT, Olefsky JM, Henry RR, Downes M, Evans RM (2012) A PPARγ-FGF1 axis is required for adaptive adipose remodelling and metabolic homeostasis. Nature 485(7398): 391–394. doi:10.1038/nature10998

Kadowaki T, Yamauchi T, Kubota N, Hara K, Ueki K, Tobe K (2006) Adiponectin and adiponectin receptors in insulin resistance, diabetes, and the metabolic syndrome. J Clin Invest 116(7):1784–1792

Khan T, Muise ES, Iyengar P, Wang ZV, Chandalia M, Abate N, Zhang BB, Bonaldo P, Chua S, Scherer PE (2009) Metabolic dysregulation and adipose tissue fibrosis: role of collagen VI. Mol Cell Biol 29(6):1575–1591. doi:MCB.01300-08

Lee JY, Sohn KH, Rhee SH, Hwang D (2001) Saturated fatty acids, but not unsaturated fatty acids, induce the expression of cyclooxygenase−2 mediated through toll-like receptor 4. J Biol Chem 276(20):16683–16689

Liu J, Divoux A, Sun J, Zhang J, Clement K, Glickman JN, Sukhova GK, Wolters PJ, Du J, Gorgun CZ, Doria A, Libby P, Blumberg RS, Kahn BB, Hotamisligil GS, Shi GP (2009) Genetic deficiency and pharmacological stabilization of mast cells reduce diet-induced obesity and diabetes in mice. Nat Med 15(8):940–945. doi:nm.1994

Matsuzawa Y, Funahashi T, Nakamura T (1999) Molecular mechanism of metabolic syndrome X: contribution of adipocytokines adipocyte-derived bioactive substances. Ann N Y Acad Sci 892:146–154

Medzhitov R (2008) Origin and physiological roles of inflammation. Nature 454(7203):428–435. doi:nature07201

Moro K, Yamada T, Tanabe M, Takeuchi T, Ikawa T, Kawamoto H, Furusawa J, Ohtani M, Fujii H, Koyasu S (2010) Innate production of T(H)2 cytokines by adipose tissue-associated c-Kit(+)Sca−1(+) lymphoid cells. Nature 463(7280):540–544. doi:nature08636

Mutch DM, Tordjman J, Pelloux V, Hanczar B, Henegar C, Poitou C, Veyrie N, Zucker JD, Clement K (2009) Needle and surgical biopsy techniques differentially affect adipose tissue gene expression profiles. Am J Clin Nutr 89(1):51–57. doi:ajcn.2008.26802

Nishimura S, Manabe I, Nagasaki M, Hosoya Y, Yamashita H, Fujita H, Ohsugi M, Tobe K, Kadowaki T, Nagai R, Sugiura S (2007) Adipogenesis in obesity requires close interplay between differentiating adipocytes, stromal cells, and blood vessels. Diabetes 56(6): 1517–1526. doi:db06-1749

Nishimura S, Manabe I, Nagasaki M, Seo K, Yamashita H, Hosoya Y, Ohsugi M, Tobe K, Kadowaki T, Nagai R, Sugiura S (2008) In vivo imaging in mice reveals local cell dynamics and inflammation in obese adipose tissue. J Clin Invest 118(2):710–721. doi:10.1172/JCI33328

Nishimura S, Manabe I, Nagasaki M, Eto K, Yamashita H, Ohsugi M, Otsu M, Hara K, Ueki K, Sugiura S, Yoshimura K, Kadowaki T, Nagai R (2009) CD8+ effector T cells contribute to macrophage recruitment and adipose tissue inflammation in obesity. Nat Med 15(8):914–920. doi:nm.1964

Nishimura S, Manabe I, Takaki S, Nagasaki M, Otsu M, Yamashita H, Sugita J, Yoshimura K, Eto K, Komuro I, Kadowaki T, Nagai R (2013) Adipose natural regulatory B cells negatively control adipose tissue inflammation. Cell Metab. doi:10.1016/j.cmet.2013.09.017

Olefsky JM, Glass CK (2010) Macrophages, inflammation, and insulin resistance. Annu Rev Physiol 72:219–246. doi:10.1146/annurev-physiol-021909-135846

Poggi M, Bastelica D, Gual P, Iglesias MA, Gremeaux T, Knauf C, Peiretti F, Verdier M, Juhan Vague I, Tanti JF, Burcelin R, Alessi MC (2007) C3H/HeJ mice carrying a toll-like receptor 4 mutation are protected against the development of insulin resistance in white adipose tissue in response to a high-fat diet. Diabetologia 50(6):1267–1276

Rocha VZ, Libby P (2009) Obesity, inflammation, and atherosclerosis. Nat Rev Cardiol 6(6): 399–409. doi:nrcardio.2009.55

Saberi M, Woods NB, de Luca C, Schenk S, Lu JC, Bandyopadhyay G, Verma IM, Olefsky JM (2009) Hematopoietic cell-specific deletion of toll-like receptor 4 ameliorates hepatic and adipose tissue insulin resistance in high-fat-fed mice. Cell Metab 10(5):419–429. doi:S1550-4131(09)00294-0

Schenk S, Saberi M, Olefsky JM (2008) Insulin sensitivity: modulation by nutrients and inflammation. J Clin Invest 118(9):2992–3002. doi:10.1172/JCI34260

Serhan CN, Savill J (2005) Resolution of inflammation: the beginning programs the end. Nat Immunol 6(12):1191–1197. doi:ni1276

Shi H, Kokoeva MV, Inouye K, Tzameli I, Yin H, Flier JS (2006) TLR4 links innate immunity and fatty acid-induced insulin resistance. J Clin Invest 116(11):3015–3025

Strissel KJ, Stancheva Z, Miyoshi H, Perfield JW 2nd, DeFuria J, Jick Z, Greenberg AS, Obin MS (2007) Adipocyte death, adipose tissue remodeling, and obesity complications. Diabetes 56(12):2910–2918. doi:db07-0767

Suganami T, Ogawa Y (2010) Adipose tissue macrophages: their role in adipose tissue remodeling. J Leukoc Biol 88(1):33–39. doi:jlb.0210072

Suganami T, Nishida J, Ogawa Y (2005) A paracrine loop between adipocytes and macrophages aggravates inflammatory changes: role of free fatty acids and tumor necrosis factor α. Arterioscler Thromb Vasc Biol 25(10):2062–2068

Suganami T, Tanimoto-Koyama K, Nishida J, Itoh M, Yuan X, Mizuarai S, Kotani H, Yamaoka S, Miyake K, Aoe S, Kamei Y, Ogawa Y (2007) Role of the toll-like receptor 4/NF-kB pathway in saturated fatty acid-induced inflammatory changes in the interaction between adipocytes and macrophages. Arterioscler Thromb Vasc Biol 27(1):84–91. doi:10.1161/01.ATV.0000251608.09329.9a

Sun K, Kusminski CM, Scherer PE (2011) Adipose tissue remodeling and obesity. J Clin Invest 121(6):2094–2101. doi:10.1172/JCI45887

Talukdar S, da Oh Y, Bandyopadhyay G, Li D, Xu J, McNelis J, Lu M, Li P, Yan Q, Zhu Y, Ofrecio J, Lin M, Brenner MB, Olefsky JM (2012) Neutrophils mediate insulin resistance in mice fed a high-fat diet through secreted elastase. Nat Med 18(9):1407–1412. doi:10.1038/nm.2885

Tanaka M, Ikeda K, Suganami T, Komiya C, Ochi K, Shirakawa I, Hamaguchi M, Nishimura S, Manabe I, Matsuda T, Kimura K, Inoue H, Inagaki Y, Aoe S, Yamasaki S, Ogawa Y (2014) Macrophage-inducible C-type lectin underlies obesity-induced adipose tissue fibrosis. Nat Commun 5:4982. doi:10.1038/ncomms5982

Tsukumo DM, Carvalho-Filho MA, Carvalheira JB, Prada PO, Hirabara SM, Schenka AA, Araujo EP, Vassallo J, Curi R, Velloso LA, Saad MJ (2007) Loss-of-function mutation in toll-like receptor 4 prevents diet-induced obesity and insulin resistance. Diabetes 56(8): 1986–1998

Vila IK, Badin PM, Marques MA, Monbrun L, Lefort C, Mir L, Louche K, Bourlier V, Roussel B, Gui P, Grober J, Stich V, Rossmeislova L, Zakaroff-Girard A, Bouloumie A, Viguerie N, Moro C, Tavernier G, Langin D (2014) Immune cell toll-like receptor 4 mediates the development of obesity- and endotoxemia-associated adipose tissue fibrosis. Cell Rep 7(4): 1116–1129. doi:10.1016/j.celrep.2014.03.062

Wang MY, Grayburn P, Chen S, Ravazzola M, Orci L, Unger RH (2008) Adipogenic capacity and the susceptibility to type 2 diabetes and metabolic syndrome. Proc Natl Acad Sci U S A 105(16):6139–6144. doi:0801981105

Weisberg SP, McCann D, Desai M, Rosenbaum M, Leibel RL, Ferrante AW Jr (2003) Obesity is associated with macrophage accumulation in adipose tissue. J Clin Invest 112(12):1796–1808

Xu H, Barnes GT, Yang Q, Tan G, Yang D, Chou CJ, Sole J, Nichols A, Ross JS, Tartaglia LA, Chen H (2003) Chronic inflammation in fat plays a crucial role in the development of obesity-related insulin resistance. J Clin Invest 112(12):1821–1830

Yamasaki S, Ishikawa E, Sakuma M, Hara H, Ogata K, Saito T (2008) Mincle is an ITAM-coupled activating receptor that senses damaged cells. Nat Immunol 9(10):1179–1188. doi:ni.1651

Chapter 23
Roles of Mitochondrial Sensing and Stress Response in the Regulation of Inflammation

Kohsuke Takeda, Daichi Sadatomi, and Susumu Tanimura

Abstract Mitochondria play critical roles in a wide variety of cellular functions beyond their intrinsic role as the energy powerhouse of the cell. Accumulating evidence has shown that mitochondria are deeply involved in innate immune responses to a variety of pathogen-associated molecular patterns (PAMPs) and damage-associated molecular patterns (DAMPs). Mitochondria serve as a signalling platform for antiviral innate immunity and are involved in the regulation of the NLRP3 inflammasome, an intracellular signalling complex required for the secretion of the proinflammatory cytokines interleukin (IL)-1β and IL-18. These findings suggest that mitochondrial conditions and integrity profoundly influence the extent and duration of inflammation. However, exposure of cells to various cytotoxic stressors easily causes damage to mitochondria and destroys mitochondrial integrity. If the damage is not severe, mitochondrial functions are maintained depending on the mitochondrial quality control system; however, once the damage exceeds capacity, cell-death–inducing signals are elicited by the mitochondria. Thus, mitochondrial sensing and stress response, as well as mitochondrial quality control, may be critically involved not only in the overall cellular response but also in the regulation of inflammation.

Keywords Inflammasome • Innate immunity • Mitochondria • Mitophagy • NLRP3 • Reactive oxygen species (ROS)

List of Abbreviations

ASC	Apoptosis-associated speck-like protein containing a C-terminal CARD
BMDM	Bone-marrow–derived macrophage

K. Takeda (✉)
Division of Cell Regulation, Graduate School of Biomedical Sciences, Nagasaki University, 1-14 Bunkyo-machi, Nagasaki 852-8521, Japan

PRESTO, Japan Science and Technology Agency, Kawaguchi 332-0012, Japan
e-mail: takeda-k@nagasaki-u.ac.jp

D. Sadatomi • S. Tanimura
Division of Cell Regulation, Graduate School of Biomedical Sciences, Nagasaki University, 1-14 Bunkyo-machi, Nagasaki 852-8521, Japan

M. Miyasaka, K. Takatsu (eds.), *Chronic Inflammation*,
DOI 10.1007/978-4-431-56068-5_23

CARD	Caspase recruitment domain
Cardif	CARD adaptor inducing IFN-β
CCCP	Carbonyl cyanide *m*-chlorophenylhydrazone
DAMPs	Damage-associated molecular patterns
DC	Dendritic cell
$\Delta\psi_m$	Membrane potential across the inner mitochondrial membrane
DRP1	Dynamin-related protein 1
ECSIT	Evolutionarily conserved signalling intermediate in Toll pathways
gC1qR	Receptor for globular head domain of complement component C1q
IL	Interleukin
IPS-1	Interferon-β promoter stimulator 1
IRF	Interferon regulatory factor
IRGM	Immunity-related GTPase family M protein
MAVS	Mitochondria-associated viral sensor
MDA5	Melanoma-differentiation-associated gene 5
MFN	Mitofusin
mtDNA	mitochondrial DNA
NLR	Nucleotide oligomerisation domain (NOD)-like receptor
NLRP3	NOD-like receptor family, pyrin domain containing 3
NLRX1	NOD-like receptor family member X1
NFP	*N*-formyl peptide
OPA1	Optic atrophy protein 1
PAMPs	Pathogen-associated molecular patterns
PINK1	PTEN-induced putative kinase 1
RIG-I	Retinoic acid inducible gene-I
ROS	Reactive oxygen species
TLR	Toll-like receptor
VISA	Virus-induced signalling adaptor

23.1 Introduction

The role of mitochondria in innate immunity has received much attention since MAVS (also called IPS-1, Cardif, and VISA), a crucial innate immune molecule, was first identified as a mitochondria-anchored protein (Arnoult et al. 2011). It was later shown that mitochondria play an important role in the regulation of the NLRP3 inflammasome, an intracellular signalling complex that mediates IL-1β and IL-18 secretion triggered by various PAMPs and DAMPs (Gurung et al. 2015). Maintaining mitochondrial integrity is crucial to support such roles of the mitochondria; however, mitochondrial integrity can easily be disturbed by various cytotoxic stressors. The membrane potential across the inner mitochondrial membrane ($\Delta\psi_m$) is normally maintained by the activity of the electron transport chain, but loss of $\Delta\psi_m$ is a common outcome of mitochondrial damage. Loss of $\Delta\psi_m$ is sensed by the mitochondrial stress-sensing machinery that appears to mediate mitochondrial quality control, or cell-death–inducing signalling, depending on the severity of the mitochondrial damage.

Here, we describe the roles of mitochondria in the regulation of inflammation, focusing on the importance of mitochondrial quality control and mitochondrial stress response.

23.2 Mitochondrial Quality Control System

Mitochondrial quality control is intrinsically indispensable for the maintenance of mitochondrial functions. However, it is also important to cope with dysfunctional mitochondria that may disturb cellular functions by excessive generation of reactive oxygen species (ROS) from the unregulated electron transport chain. The mitochondrial quality control system consists mainly of mitochondrial fission/fusion and mitophagy, of which the mechanisms of both have recently been extensively characterised.

23.2.1 Mitochondrial Fission and Fusion

Mitochondria are known to divide and fuse constantly. An individual mitochondrion divides by fission to generate two or more smaller mitochondria. Conversely, mitochondria undergo fusion, resulting in the fusion of the double membranes and the mixing of lipid membranes and intramitochondrial content. This dynamic nature of mitochondria is a major contributor to mitochondrial quality control (Westermann 2010). Damaged mitochondria are divided once and then fused to the intact mitochondria, thereby allowing the damaged mitochondria to recover their functions.

Most of the master regulator molecules of mitochondrial fission and fusion possess GTPase activity. In mammals, DRP1 and mitofusin (MFN) isoforms (MFN1 and MFN2) play central roles in fission and fusion, respectively. OPA1 is another GTPase that exists in the inner membrane and is required for fusion. OPA1 induces inner membrane fusion, whereas MFNs induce outer membrane fusion. Interestingly, OPA1 also acts as an intermediate between mitochondrial stress sensing and fission/fusion. Loss of $\Delta\psi_m$ induces the cleavage of OPA1 by the inner membrane-resident metalloendopeptidase OMA1 and thus suppresses the fusion-inducing activity of OPA1. This machinery prevents fusion of severely and irreversibly damaged mitochondria to intact mitochondria, maintaining mitochondrial quality as a whole in the cell.

23.2.2 Mitophagy

Mitochondrial fission/fusion is not sufficient for maintaining cellular homeostasis when a large portion of the mitochondria in the cell are severely impaired. Under such adverse conditions, dysfunctional mitochondria further disturb cellular functions by excess generation of ROS from the electron transport chain and, therefore, should be eliminated by mitophagy, the autophagy-based degradation machinery selective to mitochondria (Okamoto and Kondo-Okamoto 2012). Although mitophagy and autophagy basically share the same set of regulatory molecules, much focus has been placed on understanding the molecular mechanisms of how mitochondria are selectively degraded by mitophagy. Currently, the ubiquitin ligase Parkin and the mitochondrial serine/threonine kinase PINK1, both genes originally identified as the causative genes for Parkinson's disease, are considered to be the central regulators of mitophagy (Vives-Bauza and Przedborski 2011).

23.3 Mitochondrial Functions and Inflammation

23.3.1 Mitochondria and Innate Immunity

As mentioned above, the identification of the mitochondria-anchored protein MAVS as an important innate immune molecule was a clue used to elucidate the roles of mitochondria in innate immunity. The viral sensor proteins RIG-I and MDA5 recognise viral RNA in the cytosol and bind to MAVS on the outer mitochondrial membrane, provoking cytokine production through the transcription factor NF-kB and type I interferon production through IRF3 and IRF7. Several other innate immune molecules have been shown to exist in or physically associate with mitochondria, although it is largely unknown how mitochondria are involved in signalling through these molecules (Table 23.1). It has recently been shown that MAVS is functionally associated with MFNs and that antiviral signalling through MAVS requires mitochondria to maintain their $\Delta\psi_m$ (Koshiba et al. 2011). However, whether MFN-induced mitochondrial fusion is involved in antiviral signalling has not been clarified because MFN1 and MFN2 similarly induce mitochondrial fusion but facilitate or suppress MAVS-mediated signalling, respectively.

Table 23.1 Representative mitochondrial innate immune proteins

	Location	Function	References
ECSIT	IMM, OMM	Binds to TRAF6 and induces ROS generation	West et al. (2011)
gC1qR	OMM	Binds to MAVS and suppresses antiviral signalling	Xu et al. (2009)
IRGM	IMM, Matrix	Induces mitochondrial fission, autophagy, and cell death	Singh et al. (2010)
NLRX1	Matrix	Inhibits MAVS	Allen et al. (2011)
MAVS	OMM	See text	Reviewed in Arnoult et al. (2011)
MFN1	OMM	Promotes mitochondrial fusion and MAVS-mediated signalling	Onoguchi et al. (2010)
MFN2	OMM	Promotes mitochondrial fusion and inhibits MAVS-mediated signalling	Yasukawa et al. (2009)

IMM inner mitochondrial membrane, *OMM* outer mitochondrial membrane

It is known that a metabolic shift from mitochondrial oxidative phosphorylation to increased glycolysis occurs during activation and maturation of dendritic cells (DCs) and macrophages (Krawczyk et al. 2010). Various inflammatory stimuli potentially lead to loss of $\Delta\psi_m$ and subsequent apoptosis of these cells; however, ATP produced by the increased glycolysis appears to support the activated population of immune cells by maintaining mitochondrial functions. These findings strongly suggest that the maintenance of mitochondrial conditions and functions is a prerequisite for proper regulation of innate immunity.

23.3.2 Mitochondria Trigger Inflammation

When cells are severely damaged or mitochondrial damage exceeds the capacity of the mitochondrial quality control system or the control system itself is impaired, mitochondria can be sources of DAMPs, such as cytochrome *c*, mitochondrial DNA (mtDNA), and *N*-formyl peptide (NFP). Cytochrome *c* exists in the intermembrane space of mitochondria, but it is released into the cytosol in response to various stimuli and it triggers caspase-dependent apoptosis (Wang and Youle 2009). Whereas necrosis is known to accelerate inflammation, apoptosis generally suppresses inflammation by eliminating whole cells before they release various DAMPs, including mitochondrial DAMPs. Thus, the strict regulation of cytochrome *c* is required for the appropriate regulation of inflammation.

Similar to the role of PAMPs as sepsis inducers in microbial infection, mitochondrial DAMPs are known to cause systemic inflammation upon traumatic injury or tissue damage that does not involve infection, so-called sterile inflammation. It has been shown that mtDNA and NFP that are released from damaged cells induce inflammation by activating neutrophils through their Toll like receptor (TLR) 9 and formyl peptide receptor-1, respectively (Zhang et al. 2010). Mitochondrial DAMP-induced inflammation is negatively regulated not only by apoptosis but also by mitophagy. For instance, in mice deficient in DNase II, which is supposed to digest mtDNA during autophagy, the pressure overload-induced inflammatory response in the heart was accelerated, concomitant with mtDNA accumulation in autolysosomes in the myocardium (Oka et al. 2012). Thus, the tight regulation of mitochondrial DAMPs by mitophagy is one of the key mechanisms of preventing excess inflammation.

23.3.3 Mitochondria Regulate the NLRP3 Inflammasome

NLRP3 belongs to the NOD-like receptor (NLR) family, members of which generally recognise intracellular PAMPs and DAMPs, but NLRP3 responds to a wide variety of stimuli including external DAMPs and environmental irritants, as shown in Fig. 23.1. In response to these stimuli, NLRP3 binds to the adaptor protein

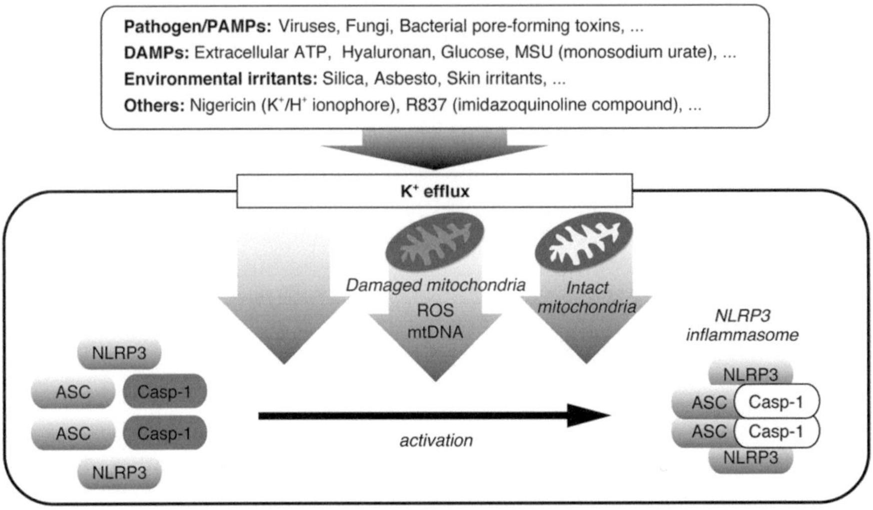

Fig. 23.1 The roles of mitochondria in the regulation of the NLRP3 inflammasome. The NLRP3 inflammasome, which consists of NLRP3, ASC, and caspase-1 (Casp-1), is activated by various PAMPs, DAMPs, and other stimuli. There appear to be mitochondria-independent and mitochondria-dependent activation mechanisms of the NLRP3 inflammasome. In the latter, damaged or intact mitochondria facilitate NLRP3 inflammasome activation in a context-dependent manner

ASC and the protease caspase-1; the resulting protein complex, called the NLRP3 inflammasome, induces secretion of IL-1β and IL-18 through caspase-1 activation mainly in DCs and macrophages. Among a wide variety of activation mechanisms of the NLRP3 inflammasome proposed so far, K^+ efflux and ROS have been identified as the main activators.

Zhou et al. reported, for the first time, the involvement of mitochondria in the regulation of the NLRP3 inflammasome (Zhou et al. 2011). They have shown that ROS generated from mitochondria contribute considerably to NLRP3 activation. In addition, the suppression of mitophagy increases mitochondrial ROS and concomitantly NLRP3 activation, suggesting that ROS generated from dysfunctional or damaged mitochondria principally participate in the activation of the NLRP3 inflammasome. Immediately after that report, Nakahira et al. reported that mtDNA released from dysfunctional mitochondria into the cytosol induces NLRP3 activation (Nakahira et al. 2011). It has also been reported that oxidised mtDNA binds directly to NLRP3, leading to the formation and activation of the NLRP3 inflammasome (Shimada et al. 2012). Moreover, cardiolipin, the phospholipid distributed exclusively in the mitochondrial inner membrane, has been shown to be exposed to the cytosol by an unknown mechanism and to bind directly to and activate NLRP3 (Iyer et al. 2013). These findings highlighting various mitochondrial factors as NLRP3 activators strongly suggest that mitochondria play a central role in the regulation of the NLRP3 inflammasome.

It has recently been shown that the microtubules, one of the major components of the cytoskeletons, mediate the activation of the NLRP3 inflammasome through mitochondria (Misawa et al. 2013). NLRP3 activators decrease the amount of NAD^+ by compromising mitochondrial functions and thus suppress Sirtuin 2, which deacetylates the microtubule component α-tubulin in an NAD^+-dependent manner. The resulting accumulation of acetylated α-tubulin facilitates the transport of mitochondria and endoplasmic reticulum (ER) on the microtubules and decreases the distance between these organelles. Under resting conditions, NLRP3 and ASC mainly exist on the ER and mitochondria, respectively; therefore, NLRP3 activators appear to induce the proximity of NLRP3 and ASC, forming the active NLRP3 inflammasome.

The above findings clearly demonstrate that mitochondrial dysfunction is a common trigger of activation of the NLRP3 inflammasome. However, it has also been found that mitochondrial integrity is required for NLRP3 activation in the case of infection with RNA viruses, such as influenza, measles, and encephalomyocarditis virus. NLRP3 appears to be activated by binding to MFN2 and MAVS on intact mitochondria that maintain their $\Delta\psi_m$ (Ichinohe et al. 2013). On the other hand, it has been reported that mitochondrial ROS and mitochondrial dysfunction and integrity are not necessarily required for NLRP3 activation, but K^+ efflux is the only common trigger for the activation of the NLRP3 inflammasome (Munoz-Planillo et al. 2013). Thus, mitochondria may not be a general mediator, but rather a context-dependent mediator of NLRP3 inflammasome activation.

In fact, we have found that not all of the experimentally used NLRP3 activators affect mitochondria, but even those affecting mitochondria induce different mitochondrial responses in mouse bone-marrow–derived macrophages (BMDMs), in terms of mitochondrial morphology, $\Delta\psi_m$ and ROS generation. Consistent with this, the uncoupler CCCP that abolishes $\Delta\psi_m$ does not generally affect the NLRP3 inflammasome, but only suppresses its activation induced by a quite narrow range of stimuli. Moreover, the profile of the requirement of mitochondrial integrity for NLRP3 activation in response to various stimuli differs between BMDMs and human monocyte THP-1 cells differentiated into macrophage-like cells, although both cells are frequently used in inflammasome research (Sadatomi et al., manuscript in preparation).

As a molecule that may be involved in the regulation of inflammation by regulating mitochondrial stress response, we have been focusing on phosphoglycerate mutase family member 5 (PGAM5). PGAM5 lacks phosphoglycerate mutase activity but instead acts as an atypical serine/threonine-specific protein phosphatase (Takeda et al. 2009; Sadatomi et al. 2013). PGAM5 is localised to the inner mitochondrial membrane through its N-terminal transmembrane domain and is cleaved within the domain upon loss of $\Delta\psi_m$ (Sekine et al. 2012). Our recent analysis of PGAM5-deficient BMDMs has suggested that PGAM5 is not a general regulator of the NLRP3 inflammasome but is required for its activation induced by a certain stimulus (unpublished data). Thus, the elucidation of PGAM5 functions will provide important insights into the roles of mitochondria in the regulation of inflammasomes.

Taken together, these findings indicate that the mechanisms of activation of the NLRP3 inflammasome are much more diverse than was previously expected and that the involvement of mitochondria is not a general requirement for the activation of the NLRP3 inflammasome. However, the ways and extent to which mitochondria influence the NLRP3 inflammasome are quite diverse and dependent on the types of stimuli and cells.

23.4 Conclusions and Perspectives

The evidence presented here indicates that mitochondria serve as an important signalling platform for innate immunity, whereas their disruption releases various mitochondrial DAMPs into the cytosol and extracellular space, which in turn triggers a strong inflammatory response. This two-sided nature of mitochondria implies that mitochondrial conditions and the capacity to respond to various stimuli and insults are critical determinants of inflammation. A similar point of view can also be applied to the regulation of the inflammasome, such that certain types of activation of the NLRP3 inflammasome require mitochondrial dysfunction or intact function in a context-dependent manner, suggesting that the extent and duration of activation of the inflammasome differ depending on the mitochondrial conditions. Thus, the mitochondrial stress-sensing system, which constantly monitors how mitochondria respond to, as well as to what extent mitochondria are damaged by, various stimuli and insults, is required for the precise regulation of inflammation. The elucidation of such a system will contribute to a better understanding of the cause of chronic inflammation and its association with various diseases.

References

Allen IC, Moore CB, Schneider M, Lei Y, Davis BK, Scull MA, Gris D, Roney KE, Zimmermann AG, Bowzard JB, Ranjan P, Monroe KM, Pickles RJ, Sambhara S, Ting JP (2011) NLRX1 protein attenuates inflammatory responses to infection by interfering with the RIG-I-MAVS and TRAF6-NF-kappaB signaling pathways. Immunity 34(6):854–865. doi:10.1016/j.immuni.2011.03.026

Arnoult D, Soares F, Tattoli I, Girardin SE (2011) Mitochondria in innate immunity. EMBO Rep 12(9):901–910. doi:10.1038/embor.2011.157

Gurung P, Lukens JR, Kanneganti TD (2015) Mitochondria: diversity in the regulation of the NLRP3 inflammasome. Trends Mol Med 21(3):193–201. doi:10.1016/j.molmed.2014.11.008

Ichinohe T, Yamazaki T, Koshiba T, Yanagi Y (2013) Mitochondrial protein mitofusin 2 is required for NLRP3 inflammasome activation after RNA virus infection. Proc Natl Acad Sci U S A 110(44):17963–17968. doi:10.1073/pnas.1312571110

Iyer SS, He Q, Janczy JR, Elliott EI, Zhong Z, Olivier AK, Sadler JJ, Knepper-Adrian V, Han R, Qiao L, Eisenbarth SC, Nauseef WM, Cassel SL, Sutterwala FS (2013) Mitochondrial cardiolipin is required for Nlrp3 inflammasome activation. Immunity 39(2):311–323. doi:10.1016/j.immuni.2013.08.001

Koshiba T, Yasukawa K, Yanagi Y, Kawabata S (2011) Mitochondrial membrane potential is required for MAVS-mediated antiviral signaling. Sci Signal 4(158):ra7. doi:10.1126/scisignal.2001147

Krawczyk CM, Holowka T, Sun J, Blagih J, Amiel E, DeBerardinis RJ, Cross JR, Jung E, Thompson CB, Jones RG, Pearce EJ (2010) Toll-like receptor-induced changes in glycolytic metabolism regulate dendritic cell activation. Blood 115(23):4742–4749. doi:10.1182/blood-2009-10-249540

Misawa T, Takahama M, Kozaki T, Lee H, Zou J, Saitoh T, Akira S (2013) Microtubule-driven spatial arrangement of mitochondria promotes activation of the NLRP3 inflammasome. Nat Immunol 14(5):454–460. doi:10.1038/ni.2550

Munoz-Planillo R, Kuffa P, Martinez-Colon G, Smith BL, Rajendiran TM, Nunez G (2013) K(+) efflux is the common trigger of NLRP3 inflammasome activation by bacterial toxins and particulate matter. Immunity 38(6):1142–1153. doi:10.1016/j.immuni.2013.05.016

Nakahira K, Haspel JA, Rathinam VA, Lee SJ, Dolinay T, Lam HC, Englert JA, Rabinovitch M, Cernadas M, Kim HP, Fitzgerald KA, Ryter SW, Choi AM (2011) Autophagy proteins regulate innate immune responses by inhibiting the release of mitochondrial DNA mediated by the NALP3 inflammasome. Nat Immunol 12(3):222–230. doi:10.1038/ni.1980

Oka T, Hikoso S, Yamaguchi O, Taneike M, Takeda T, Tamai T, Oyabu J, Murakawa T, Nakayama H, Nishida K, Akira S, Yamamoto A, Komuro I, Otsu K (2012) Mitochondrial DNA that escapes from autophagy causes inflammation and heart failure. Nature 485 (7397):251–255. doi:10.1038/nature10992

Okamoto K, Kondo-Okamoto N (2012) Mitochondria and autophagy: critical interplay between the two homeostats. Biochim Biophys Acta 1820(5):595–600. doi:10.1016/j.bbagen.2011.08.001

Onoguchi K, Onomoto K, Takamatsu S, Jogi M, Takemura A, Morimoto S, Julkunen I, Namiki H, Yoneyama M, Fujita T (2010) Virus-infection or 5′ppp-RNA activates antiviral signal through redistribution of IPS-1 mediated by MFN1. PLoS Pathog 6(7):e1001012. doi:10.1371/journal.ppat.1001012

Sadatomi D, Tanimura S, Ozaki K, Takeda K (2013) Atypical protein phosphatases: emerging players in cellular signaling. Int J Mol Sci 14(3):4596–4612. doi:10.3390/ijms14034596

Sekine S, Kanamaru Y, Koike M, Nishihara A, Okada M, Kinoshita H, Kamiyama M, Maruyama J, Uchiyama Y, Ishihara N, Takeda K, Ichijo H (2012) Rhomboid protease PARL mediates the mitochondrial membrane potential loss-induced cleavage of PGAM5. J Biol Chem 287(41):34635–34645. doi:10.1074/jbc.M112.357509

Shimada K, Crother TR, Karlin J, Dagvadorj J, Chiba N, Chen S, Ramanujan VK, Wolf AJ, Vergnes L, Ojcius DM, Rentsendorj A, Vargas M, Guerrero C, Wang Y, Fitzgerald KA, Underhill DM, Town T, Arditi M (2012) Oxidized mitochondrial DNA activates the NLRP3 inflammasome during apoptosis. Immunity 36(3):401–414. doi:10.1016/j.immuni.2012.01.009

Singh SB, Ornatowski W, Vergne I, Naylor J, Delgado M, Roberts E, Ponpuak M, Master S, Pilli M, White E, Komatsu M, Deretic V (2010) Human IRGM regulates autophagy and cell-autonomous immunity functions through mitochondria. Nat Cell Biol 12(12):1154–1165. doi:10.1038/ncb2119

Takeda K, Komuro Y, Hayakawa T, Oguchi H, Ishida Y, Murakami S, Noguchi T, Kinoshita H, Sekine Y, Iemura SI, Natsume T, Ichijo H (2009) Mitochondrial phosphoglycerate mutase 5 uses alternate catalytic activity as a protein serine/threonine phosphatase to activate ASK1. Proc Natl Acad Sci U S A 106(30):12301–12305. doi:0901823106 [pii]

Vives-Bauza C, Przedborski S (2011) Mitophagy: the latest problem for Parkinson's disease. Trends Mol Med 17(3):158–165. doi:10.1016/j.molmed.2010.11.002

Wang C, Youle RJ (2009) The role of mitochondria in apoptosis. Annu Rev Genet 43:95–118. doi:10.1146/annurev-genet-102108-134850

West AP, Brodsky IE, Rahner C, Woo DK, Erdjument-Bromage H, Tempst P, Walsh MC, Choi Y, Shadel GS, Ghosh S (2011) TLR signalling augments macrophage bactericidal activity through mitochondrial ROS. Nature 472(7344):476–480. doi:10.1038/nature09973

Westermann B (2010) Mitochondrial fusion and fission in cell life and death. Nat Rev Mol Cell Biol 11(12):872–884. doi:10.1038/nrm3013

Xu L, Xiao N, Liu F, Ren H, Gu J (2009) Inhibition of RIG-I and MDA5-dependent antiviral response by gC1qR at mitochondria. Proc Natl Acad Sci U S A 106(5):1530–1535. doi:10.1073/pnas.0811029106

Yasukawa K, Oshiumi H, Takeda M, Ishihara N, Yanagi Y, Seya T, Kawabata S, Koshiba T (2009) Mitofusin 2 inhibits mitochondrial antiviral signaling. Sci Signal 2(84):ra47. doi:10.1126/scisignal.2000287

Zhang Q, Raoof M, Chen Y, Sumi Y, Sursal T, Junger W, Brohi K, Itagaki K, Hauser CJ (2010) Circulating mitochondrial DAMPs cause inflammatory responses to injury. Nature 464(7285):104–107. doi:10.1038/nature08780

Zhou R, Yazdi AS, Menu P, Tschopp J (2011) A role for mitochondria in NLRP3 inflammasome activation. Nature 469(7329):221–225. doi:10.1038/nature09663

Chapter 24
Oxidative Stress Regulation by Reactive Cysteine Persulfides in Inflammation

Tomohiro Sawa

Abstract Reactive oxygen species (ROS) such as superoxide anion radical and hydrogen peroxide are ubiquitously generated during metabolisms of aerobic organisms. Excess production of ROS due to imbalance between formation and removal of ROS by the antioxidant system causes oxidative stress-related tissue damage. Therefore reinforcement of antioxidant capacity has been considered as a beneficial approach for treatment and prevention of chronic inflammatory disorders where ROS production is persistently activated. Cysteine persulfide was recently identified regarding its endogenous formation in mammalian cells. Biochemical analyses revealed that cysteine persulfide and its derivatives such as glutathione persulfide act as a strong antioxidant in cells. Better understanding of the antioxidant actions of cysteine persulfides in chronic inflammation-associated diseases is a necessary basis to develop new strategies for disease treatment and prevention by modulating the process of oxidative stress.

Keywords Reactive oxygen species • Oxidative stress • Inflammatory mediators • Antioxidants • Cysteine persulfide • Cellular senescence

24.1 Introduction

Reactive oxygen species (ROS) such as superoxide anion radical and hydrogen peroxide are ubiquitously formed by reduction of molecular oxygen during a variety of aerobic metabolisms including aerobic respiration (Starkov 2008). Eukaryotic immune systems also utilise ROS as host defense mechanisms to combat invading microorganisms by activating ROS-producing enzyme NADPH oxidase (Sumimoto 2008). Because of the toxic nature of ROS, cellular levels of ROS should be maintained low by antioxidant molecules such as glutathione and by ROS scavenging enzymes such as superoxide dismutase and catalase (Niki 2010).

T. Sawa (✉)
Department of Microbiology, Graduate School of Medical Sciences, Kumamoto University, 1-1-1 Honjo, Chuo-ku, Kumamoto 860-8556, Japan

PRESTO, Japan Science and Technology Agency, Kawaguchi 332-0012, Japan
e-mail: sawat@kumamoto-u.ac.jp

M. Miyasaka, K. Takatsu (eds.), *Chronic Inflammation*,
DOI 10.1007/978-4-431-56068-5_24

However, under certain conditions, imbalance between formation and removal of ROS by an antioxidant system occurs resulting in oxidative stress-related tissue damage (Halliwell 2007). Therefore reinforcement of antioxidant capacity has been considered as a beneficial approach for treatment and prevention of chronic inflammatory disorders where ROS production is persistently activated. The endogenous formation of cysteine persulfide in mammalian cells was recently identified (Ida et al. 2014; Miranda and Wink 2014; Ono et al. 2014). Biochemical analyses revealed that cysteine persulfide and its derivatives such as glutathione persulfide act as strong antioxidants in cells (Ida et al. 2014; Ono et al. 2014). This chapter describes pathological aspects of oxidative stress in inflammatory diseases, with particular focus on ROS-induced cellular senescence. Newly identified endogenous antioxidant cysteine persulfides in the regulation of oxidative stress is also discussed.

24.2 Nitrated Nucleotide as a Second Messenger of ROS Effects

According to their chemical reactivities, ROS can cause a diverse type of chemical modifications such as oxidation and nitration on biological molecules including nucleic acids, proteins, enzymes, and lipids. Oxidation of guanine nucleotide has been studied extensively because guanine-base oxidation potentially induces a mismatch base pair with adenine resulting in G:C to T:A mutations (Wu et al. 2004). Guanine nitration has been reported to occur endogenously in various cultured cells and tissues/organs, most of which were associated with excess production of ROS and nitric oxide (NO) such as inflammatory conditions. Examples include the lungs of influenza virus-infected mice (Akaike et al. 2003) and human patients with idiopathic pulmonary fibrosis and lung cancer (Terasaki et al. 2006), the liver of mice infected with *Salmonella thyphimurium* (Zaki et al. 2009), as well as the urine of humans particularly that obtained from smokers (Sawa et al. 2006). Detailed analyses of cell culture experiments revealed that 8-nitroguanosine 3′,5′-cyclic monophosphate (8-nitro-cGMP), a nitrated derivative of cGMP, was the major product formed in cells (Sawa et al. 2007). 8-Nitro-cGMP was first identified in the mouse macrophage cell line RAW 264.7 cells when cells were stimulated with interferon-γ and lipopolysaccharide (LPS) to produce NO via the inducible isoform of NO synthase (iNOS) (Sawa et al. 2007). Infection with *Salmonella* facilitated the formation of 8-nitro-cGMP in mouse macrophages with the wild-type iNOS gene but not in iNOS-deficient macrophages (Zaki et al. 2009). ROS- and NO-dependent 8-nitro-cGMP formation was also evident in rat glioma C6 cells treated with LPS plus proinflammatory cytokines (interferon-γ, interleukin (IL)-1β, tumour necrosis factor-α) (Fujii et al. 2010; Ahmed et al. 2012). By using a mouse myocardial infarction (MI) model, there was a remarkable increase of 8-nitro-cGMP formation in the heart at 6 weeks after MI (Nishida et al. 2012).

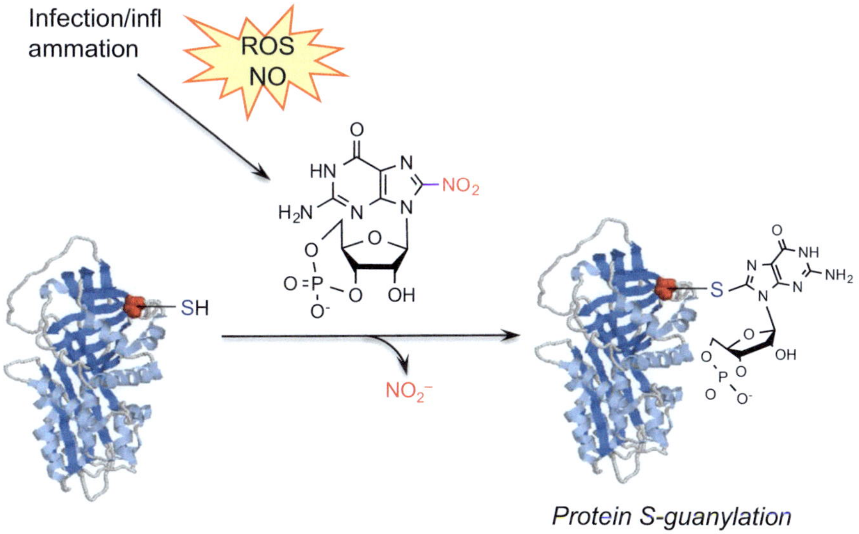

Fig. 24.1 Formation of 8-nitro-cGMP in infection and inflammation, and subsequent posttranslational cysteine modification via protein *S*-guanylation

This suggests the persistent generation of ROS and NO in the MI heart leading to 8-nitro-cGMP formation for a prolonged period of time.

It is noteworthy that 8-nitro-cGMP is the first endogenous cGMP derivative that functions as an electrophile and forms a stable conjugate with cysteine thiols at the purine base (Fig. 24.1; Sawa et al. 2007). This novel posttranslational modification is called 'protein *S*-guanylation' and proceeds without enzymatic stimulation in vitro. As shown in Fig. 24.1, in the reaction of *S*-guanylation, the nitro group of 8-nitro-cGMP is replaced by a thiol group, with the release of nitrite ion, resulting in the formation of irreversible cGMP adducts (8-thioalkoxy-cGMP adducts).

Keap1 is the first identified endogenous target for protein *S*-guanylation in cells (Sawa et al. 2007; Fujii et al. 2010). Keap1 binds to a transcriptional factor Nrf2 to act as a negative regulator for Nrf2 (Uruno and Motohashi 2011). We identified by means of mass spectrometry that cysteine residue at 434 of Keap1 was preferentially *S*-guanylated (Fujii et al. 2010). This Cys434 is localised near the interface for Nrf2 binding (Holland et al. 2008), and *S*-guanylation occurring at Cys434 may facilitate the dissociation of Keap1 from Nrf2, leading to activation of Nrf2-dependent gene expression (Uruno and Motohashi 2011). Other targets for protein *S*-guanylation thus far identified include mitochondrial heat shock protein (HSP-60; Rahaman et al. 2014) and small GTPase H-Ras (Nishida et al. 2012). Roles of H-Ras *S*-guanylation in induction of cellular senescence are described below. These observations suggest that 8-nitro-cGMP can act as a messenger molecule to transduce ROS formation via induction of protein *S*-guanylation.

24.3 Activation of Ras-p53 Pathway by a Nitrated Nucleotide in ROS-Dependent Cellular Senescence

Cellular senescence is defined as an irreversible arrest of the cell cycle (Rodier and Campisi 2011). Multiple mechanisms have been implicated in the induction of cellular senescence, including cellular stresses such as DNA damage and limited replicative capacity, also known as Hayflick's limit (Hayflick 1965). Excessive production of ROS may cause premature senescence via induction of DNA damage, telomere damage and shortening, and NF-kB activation (Jurk et al. 2013). Furthermore, senescent cells activate hyperproduction of ROS and secrete bioactive proinflammatory cytokines (Passos et al. 2010). This condition is called 'senescence-associated secretary phenotype (SASP)'. Hence, both senescence-associated ROS and NF-kB-driven proinflammatory cytokines, especially IL-6 and IL-8, contribute to positive feedback loops that stabilise stress-induced senescence. Sustained activation of H-Ras also can induce cellular senescence, called 'oncogene-induced premature senescence' (Kuilman et al. 2008). As mentioned above, we found that 8-nitro-cGMP caused protein *S*-guanylation of H-Ras in mouse heart during MI or in rat cardiac fibroblasts in culture (Nishida et al. 2012). In vitro analyses clearly demonstrated that 8-nitro-cGMP treatment remarkably activated the H-Ras-dependent signalling pathway including activation of ERK, p38 MAPK, and p53 in rat cardiomyocytes (Nishida et al. 2012). By using a rat cardiac fibroblast culture model, *S*-guanylation of H-Ras at cysteine 186 promoted alteration of H-Ras localisation from raft to nonraft region for full enzyme activation, by inhibiting palmitoylation at same cysteine residue. Taken together, *S*-guanylation-mediated H-Ras activation participated in ERK and p38 MAPK activation after MI, which led to induction of p53/Rb-mediated cardiac cellular senescence and resulted in the transition from hypertrophy to heart failure.

24.4 Regulation of Cellular Senescence by Reactive Sulfur Species: Implication of Cysteine Persulfides

Recent progress in understanding of sulfur metabolisms identified endogenous formation of highly reactive cysteine-based persulfide and polysulfide species in biological systems (Ida et al. 2014). We identified, by utilising RNA interference (RNAi) screening, that cystathionine β-synthase (CBS) and cystathionine γ-lyase (CSE) are critically involved in the metabolism of 8-nitro-cGMP (Nishida et al. 2012). Extensive chemical biology analyses revealed that these two enzymes are capable of producing cysteine persulfide from cystine as a substrate (5; Fig. 24.2). CBS and CSE are enzymes involved in transsulfuration from homocysteine to form cysteine (Fig. 24.2a). We demonstrated by means of mass spectrometry that both CBS and CSE catalyse CS lyase reaction to cleave CS bonding in cystine resulting in direct formation of cysteine persulfide (Fig. 24.2b).

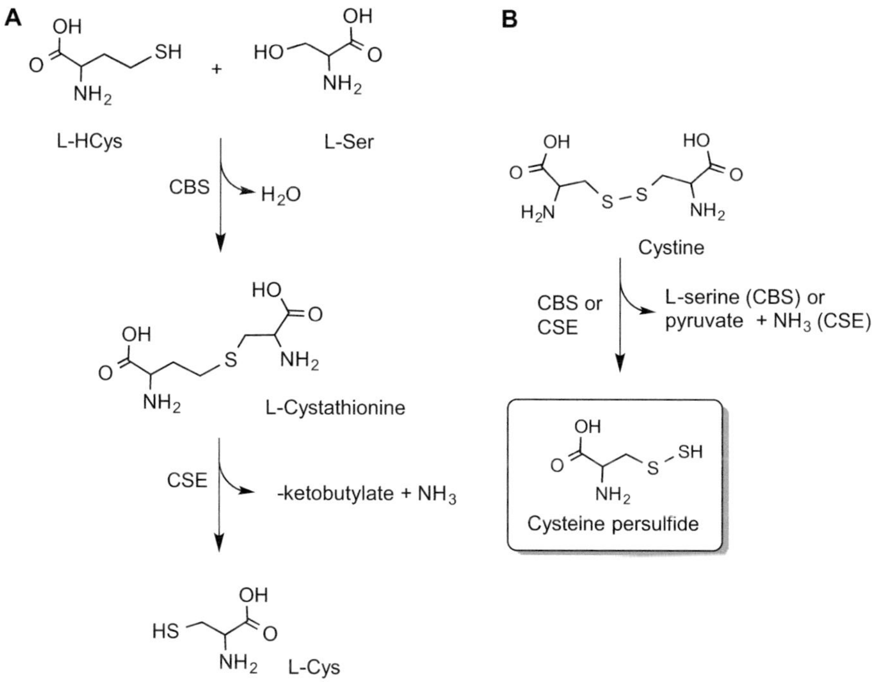

Fig. 24.2 Schematic representation for CBS and CSE-mediated transsulfuration (**a**) and CS lyase reaction (**b**). Abbreviations: *L-HCys* L-homocysteine, *L-Ser* L-serine, *L-Cys* L-cysteine, *CBS* cystathionine β-synthase, *CSE* cystathionine γ-lyase

Interestingly, cysteine persulfide thus formed is determined to be highly nucleophilic and a stronger antioxidant than cysteine. This unique nature of reactive cysteine persulfide is brought by adjacent electron pairs called the 'α-effect' (Ono et al. 2014; Edwards and Pearson 1962). According to the superior nucleophilicity, cysteine persulfide reacts very effectively with 8-nitro-cGMP to form 8-SH-cGMP with release of nitrite anion, and hence, is implicated in regulation ROS signalling. Endogenous 8-SH-cGMP formation was demonstrated to occur in cultured mammalian cells, in mice, as well as in plant cells, supporting cysteine-persulfide–dependent metabolism of 8-nitro-cGMP (Ida et al. 2014; Honda et al. 2015).

The importance of reactive cysteine persulfides in the regulation of oxidative stress is supported by the observation that these persulfide species exist in almost all organs with varying concentrations and molecular forms. In fact, we successfully identified endogenous formation of persulfides as cysteine, homocysteine, and glutathione forms in mouse heart, liver, brain, kidney, spleen, small intestine, lung, and eye (5 and our unpublished data). Another important finding is that cysteine persulfide residues are present in a variety of proteins, such as proteins involved in sulfur metabolism, protein folding, heat shock proteins, glycolysis, and so forth, which indicates an ubiquitous and possible regulatory function associated with protein persulfides and polysulfides (Ida et al. 2014).

Oxidant-dependent H-Ras activation can be a pharmacological target for prevention of heart failure after MI. This hypothesis was tested by MI-mice with sustained sodium hydrosulfide (NaHS) treatment. As shown in an in vitro experiment, cysteine and NaHS readily react each other to form cysteine persulfide in the presence of oxidants (Ida et al. 2014). Note that NaHS treatment almost completely nullified endogenous 8-nitro-cGMP formation in MI heart in mice. Furthermore, NaHS treatment markedly reduced H-Ras activation with simultaneous H-Ras *S*-guanylation in the same MI hearts (Nishida et al. 2012). Phosphorylation levels of p38 MAPK, ERK, and p53 were decreased by NaHS treatment.

24.5 Summary

NaHS treatment, possibly via formation of cysteine persulfide, improved heart functions after MI by suppressing *S*-guanylation–dependent activation of H-Ras in cardiac cells in mice (Fig. 24.3). However, application of NaHS for human treatment may be hampered because of toxicity and chemical instability of NaHS. Approaches that enable enhancing endogenous cysteine persulfide formation may thus become a potential target for prevention of ROS-dependent oxidative stress and cellular senescence. Better understanding of the antioxidant actions of cysteine persulfides in chronic inflammation-associated diseases is a necessary basis to developing new strategies for disease treatment and prevention by modulating the process of oxidative stress.

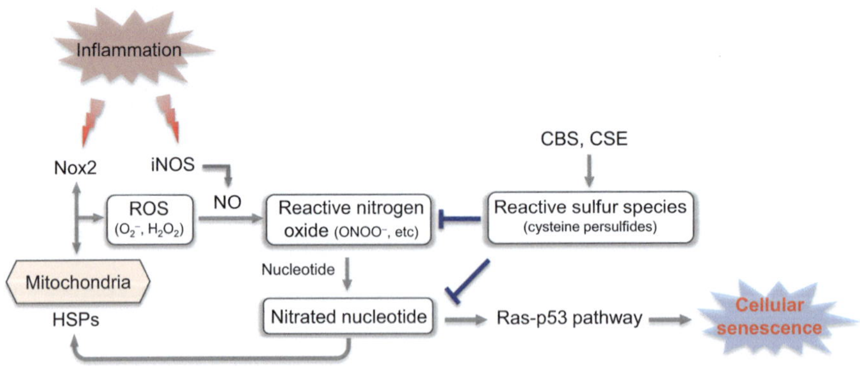

Fig. 24.3 Regulation by reactive cysteine persulfides of ROS-dependent activation of Ras-p53 pathway and cellular senescence. Abbreviations: *Nox2* NADPH oxidase, *iNOS* inducible NO synthase, *HSPs* heat shock proteins, *CBS* cystathionine β-synthase, *CSE* cystathionine γ-lyase

References

Ahmed KA, Sawa T, Ihara H et al (2012) Regulation by mitochondrial superoxide and NADPH oxidase of cellular formation of nitrated cyclic GMP: potential implications for ROS signalling. Biochem J 441:719–730

Akaike T, Okamoto S, Sawa T et al (2003) 8-Nitroguanosine formation in viral pneumonia and its implication for pathogenesis. Proc Natl Acad Sci U S A 100:685–690

Edwards JO, Pearson RG (1962) The factors determining nucleophilic reactivities. J Am Chem Soc 84:16–24

Fujii S, Sawa T, Ihara H et al (2010) The critical role of nitric oxide signaling, via protein *S*-guanylation and nitrated cyclic GMP, in the antioxidant adaptive response. J Biol Chem 285:23970–23984

Halliwell B (2007) Biochemistry of oxidative stress. Biochem Soc Trans 35:1147–1150

Hayflick L (1965) The limited in vitro lifetime of human diploid cell strains. Exp Cell Res 37:614–636

Holland R, Hawkins AE, Eggler AL et al (2008) Prospective type 1 and type 2 disulfides of Keap1 protein. Chem Res Toxicol 21:2051–2060

Honda K, Yamada N, Yoshida R et al (2015) 8-Mercapto-cyclic GMP mediates hydrogen sulfide-induced stomatal closure in Arabidopsis. Plant Cell Physiol 56:1481–1489

Ida T, Sawa T, Ihara H et al (2014) Reactive cysteine persulfides and S-polythiolation regulate oxidative stress and redox signaling. Proc Natl Acad Sci U S A 111:7606–7611

Jurk D, Wilson C, Passos JF et al (2013) Chronic inflammation induces telomere dysfunction and accelerates ageing in mice. Nat Commun 2:4172

Kuilman T, Michaloglou C, Vredeveld LC et al (2008) Oncogene-induced senescence relayed by an interleukin-dependent inflammatory network. Cell 133:1019–1031

Miranda KM, Wink DA (2014) Persulfides and the cellular thiol landscape. Proc Natl Acad Sci U S A 111:7505–7506

Niki E (2010) Assessment of antioxidant capacity in vitro and in vivo. Free Radic Biol Med 49:503–515

Nishida M, Sawa T, Kitajima N et al (2012) Hydrogen sulfide anion regulates redox signaling via electrophile sulfhydration. Nat Chem Biol 8:714–724

Ono K, Akaike T, Sawa T et al (2014) Redox chemistry and chemical biology of H_2S, hydropersulfides, and derived species: implications of their possible biological activity and utility. Free Radic Biol Med 77:82–94

Passos JF, Nelson G, Wang C et al (2010) Feedback between p21 and reactive oxygen production is necessary for cell senescence. Mol Syst Biol 6:347

Rahaman MM, Sawa T, Ahtesham AK et al (2014) *S*-Guanylation proteomics for redox-based mitochondrial signaling. Antioxid Redox Signal 20:295–307

Rodier F, Campisi J (2011) Four faces of cellular senescence. J Cell Biol 192:547–556

Sawa T, Tatemichi M, Akaike T et al (2006) Analysis of urinary 8-nitroguanine, a marker of nitrative nucleic acid damage, by high-performance liquid chromatography-electrochemical detection coupled with immunoaffinity purification: association with cigarette smoking. Free Radic Biol Med 40:711–720

Sawa T, Zaki MH, Okamoto T et al (2007) Protein *S*-guanylation by the biological signal 8-nitroguanosine 3′,5′-cyclic monophosphate. Nat Chem Biol 3:727–735

Starkov AA (2008) The role of mitochondria in reactive oxygen species metabolism and signaling. Ann N Y Acad Sci 1147:37–52

Sumimoto H (2008) Structure, regulation and evolution of Nox-family NADPH oxidases that produce reactive oxygen species. FEBS J 275:3249–3277

Terasaki Y, Akuta T, Terasaki M et al (2006) Guanine nitration in idiopathic pulmonary fibrosis and its implication for carcinogenesis. Am J Respir Crit Care Med 174:665–673

Uruno A, Motohashi H (2011) The Keap1-Nrf2 system as an in vivo sensor for electrophiles. Nitric Oxide 25:153–160

Wu LL, Chiou CC, Chang PY et al (2004) Urinary 8-OHdG: a marker of oxidative stress to DNA and a risk factor for cancer, atherosclerosis and diabetics. Clin Chim Acta 339:1–9
Zaki MH, Fujii S, Okamoto T et al (2009) Cytoprotective function of heme oxygenase 1 induced by a nitrated cyclic nucleotide formed during murine salmonellosis. J Immunol 182:3746–3756

Part V
Chronic Inflammation and Innate Immunity

Chapter 25
Posttranscriptional Regulation of Cytokine mRNA Controls the Initiation and Resolution of Inflammation

Osamu Takeuchi

Abstract Excess or sustained production of cytokines in innate immune cells leads to the development of chronic inflammatory diseases, such as autoimmune diseases. In physiological conditions, cytokine production undergoes transcriptional and posttranscriptional regulation. Posttranscriptional regulation modifies mRNA stability and translation, allowing for the rapid and flexible control of gene expression, which is important for coordinating the initiation and resolution of inflammation. Recent studies revealed that posttranscriptional cytokine mRNA regulation is critical not only for the resolution of inflammation but also the maintenance of immune homeostasis. This regulation is mediated by a set of RNA binding proteins (RBPs) such as Regnase-1, Roquin, tristetraprolin, and AUF1. These RBPs bind with RNA *cis*-elements present in 3' untranslated regions of cytokine mRNAs, and controls their fate. Lack of these RBPs leads to the development of autoimmunity in mice, demonstrating the importance of posttranscriptional regulation of inflammatory mRNAs in controlling chronic inflammation. In this chapter, we review a variety of posttranscriptional control mechanisms that regulate inflammation, and discuss how these mechanisms are integrated to coordinate expression of this important class of signalling proteins.

Keywords Innate immunity • Cytokine • mRNA stability • AU rich element • Regnase-1 • Roquin • Autoimmunity

---

O. Takeuchi (✉)
Laboratory of Infection and Prevention, Department of Biological Responses, Institute for Virus Research, Kyoto University, 53 Shogoin Kawara-cho, Sakyo-ku, Kyoto 606-8507, Japan

AMED-CREST, Kyoto 606-8507, Japan
e-mail: otake@virus.kyoto-u.ac.jp

M. Miyasaka, K. Takatsu (eds.), *Chronic Inflammation*,
DOI 10.1007/978-4-431-56068-5_25

25.1 Introduction

Inflammation is defined as the immune response to tissue injury or infection. Causes of inflammation include various factors such as microbial infection, tissue injury, and cellular stress (Chovatiya and Medzhitov 2014). Whereas appropriate inflammation is critical for the clearance of infected organisms and inflammatory insults, excess and chronic inflammation leads to various inflammatory diseases, such as septic shock, autoimmune diseases, atherosclerosis, metabolic diseases, and so on. Therefore, inflammation is tightly controlled under physiological conditions. It is well known that both innate and adaptive immune cells contribute to inflammation, and communication between immune cells by the secretion of cytokines is critical for the regulation of inflammation. Proinflammatory cytokines such as tumour necrosis factor (TNF), interleukin-1β (IL-1β), and IL-6 are produced predominantly in innate immune cells immediately after inflammatory stimuli and play a key role in subsequent immune responses.

The expression of cytokines is tightly suppressed in resting innate immune cells, and is rapidly induced in response to infection by pathogens via a set of pattern-recognition receptors (PRRs) such as Toll-like receptors (TLRs), RIG-I-like receptors (RLRs), Nod-like receptors (NLRs), C-type lectin receptors (CLRs), and a DNA sensor cyclic GMP–AMP synthase (cGAS; Beutler 2009; Takeuchi and Akira 2010; Lamkanfi and Dixit 2014; Cai et al. 2014). These receptors recognise molecular signatures that are broadly shared by microorganisms, called pathogen-associated molecular patterns (PAMPs). Among PRRs, TLRs were the first discovered and are the most well studied in terms of their ligands and signalling pathways. TLRs recognise various PAMPs such as lipopolysaccharide (LPS), bacterial peptidoglycan, CpG-DNA, and double- and single-stranded RNAs. TLRs initiate a series of signalling pathways via adaptor proteins, MyD88 and TRIF, that lead to the activation of transcription factors, including NF-kB, AP-1, interferon (IFN)-regulatory factors (IRFs), and CCAAT/enhancer-binding protein β (C/EBPβ). In addition to eliciting inflammation, PRR signalling simultaneously induces maturation of dendritic cells (DCs), which are responsible for stimulating the second line of host defence, adaptive immunity.

The expression of cytokines and proinflammatory factors is tightly regulated at multiple levels, including gene transcription, mRNA translation, and mRNA degradation (Mino and Takeuchi 2013; Carpenter et al. 2014; Anderson 2010; Kafasla et al. 2014). Although transcriptional regulation is highly important for generating new mRNA transcripts, it is a complicated process that cannot be rapidly stopped. Here, we focus on the emerging importance of posttranscriptional regulation, which is essential for the resolution of inflammation to prevent tissue damage brought about by transcriptional activation. Posttranscriptional regulation determines the fate of mRNA in association with RNA-binding proteins (RBPs) binding with characteristic *cis*-elements present in the 3' untranslated regions (UTR). Adenine- and uridine (AU)-rich elements (AREs) and stem-loop structures are well-studied *cis*-elements and a number of RBPs recognising them have been identified.

Impairment in mRNA decay results in excess and sustained inflammation, causing various inflammatory diseases.

In this chapter, we focus on posttranscriptional regulation mechanisms that regulate mRNA stability and modulate both the initiation and resolution of inflammation mediated by RBPs. We also discuss how different types of posttranscriptional control mechanisms are coordinated to regulate immune-mediated inflammation.

25.2 Roles of AU-Rich Element (ARE) Binding Proteins in Controlling mRNA Stability

Clusters of AREs found in the 3' UTRs of mRNAs encoding cytokines were the first to be identified (Caput et al. 1986) and subsequent studies showed that cytokines, chemokines, lymphocytes, proto-oncogenes, and proinflammatory genes are subject to ARE-mediated decay (Shaw and Kamen 1986). AREs provide binding sites for *trans*-acting RBPs, such as tristetraprolin (TTP), AUF1, and HuR (also known as ELAV1), which subsequently regulate the stability and translation of mRNA (Fig. 25.1). ARE basic motifs include pentamers of AUUUA (Wilusz et al. 2001). AREs are located at the 3′ UTRs of mRNA transcripts for cytokines such as granulocyte-macrophage colony-stimulating factor (GM-CSF), TNF, IL-2, IL-3, IL-6, and IL-8, as well as proinflammatory factors such as cyclooxygenase 2 (COX-2). mRNAs harboring AREs are prone to quick degradation in innate immune cells in general, although stimulation with LPS can extend their half-lives (Hao and Baltimore 2009). Thus, AREs are critical elements in controlling gene expression at the posttranscriptional level.

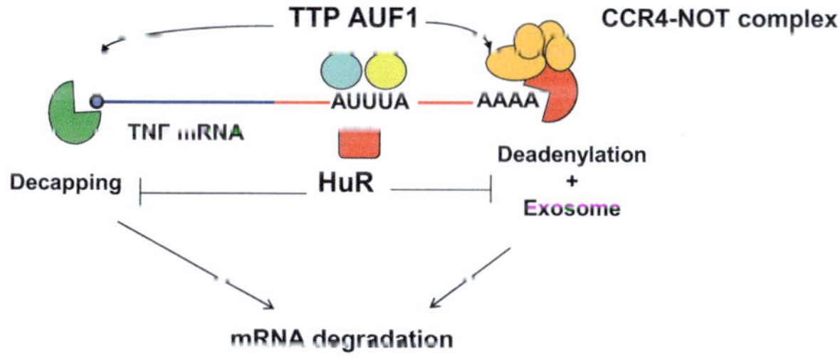

Fig. 25.1 ARE-mediated degradation of cytokine mRNA. AREs located in the 3' UTR of cytokine mRNAs are recognised by proteins including TTP, AUF1 and HuR. TTP, and AUF1 promote the degradation of mRNAs by recruiting a CCR4–NOT deadenylase complex and decapping enzymes. In contrast, HuR stabilises ARE containing mRNAs

25.2.1 *Tristetraprolin (TTP, Also Known as Zfp36)*

TTP harbors two CCCH-type zinc-finger domains and associates with mRNA via AREs present in the 3′ UTR (Carrick et al. 2004). TTP transcription is induced by various inflammatory stimuli such as TNF, LPS, glucocorticoids, and IFN-γ (Anderson 2010). TTP-deficient mice develop arthritis, dermatitis, and cachexia, a syndrome caused by the spontaneous overexpression of proinflammatory cytokines, particularly TNF (Taylor et al. 1996). Subsequent biochemical studies have shown that TTP decreased *TNF* mRNA stability by binding to its ARE (Carballo et al. 1998).

Recognition of target mRNAs by TTP leads to removal of the poly(A) tail by recruitment of a CCR4-NOT deadenylase complex, and interacting with decapping factors (Sandler et al. 2011; Fabian et al. 2013). Deadenylation promotes rapid mRNA degradation that is thought to occur in processing (P)-bodies, small cytoplasmic foci that contain many enzymes required for mRNA decay. Cells exposed to various stresses such as heat shock, oxidative stress, or virus infection promote the assembly of stress granules, small cytoplasmic foci that harbor translationally arrested mRNAs, stalled translation initiation factors, and ARE-binding proteins such as T-cell–restricted intracellular antigen-1 (TIA-1) and TIA-1-related protein (TIAR; Anderson and Kedersha 2008). TTP also destabilises mRNA transcripts encoding multiple inflammatory modulators, including GM-CSF, IL-2, IL-6, c-Fos, inducible nitric oxide synthase (iNOS), COX-2 (also known as PTGS2), IFN-γ, and IL-10 (Anderson 2010).

The activity of TTP was shown to be regulated in the course of inflammation. LPS stimulation leads to the phosphorylation of TTP by MAPKAP kinase 2 (MK2) downstream of p38 MAP kinase (Mahtani et al. 2001). Phosphorylated TTP interacts with 14-3-3 preventing stress granule localisation and recruitment of the CCR4–NOT1 complex, which results in the inhibition of TTP-mediated TNF-mRNA degradation (Stoecklin et al. 2004; Clement et al. 2011; Fig. 25.2a). Mice harboring point mutations in TTP phosphorylation sites showed impaired cytokine responses to LPS stimulation in vivo (Ross et al. 2015). Together, these studies demonstrate TTP is a key player in posttranscriptional gene regulation, particularly with regard to the regulation of ARE-mediated decay of cytokine mRNAs.

25.2.2 *ARE/Poly-(U) Binding Degradation Factor 1 (AUF1)*

AUF1 (also known as heterogeneous nuclear ribonucleoprotein D, hnRNP D) is another attenuator of inflammatory cytokine responses that acts by destabilising inflammatory cytokine mRNAs, such as *IL-2*, *TNF*, and *IL-1β*, and other mRNAs containing AREs in 3′-UTRs (Lu et al. 2006; Sadri and Schneider 2009). AUF1 also binds ARE and consists of four highly related protein isoforms generated by alternate RNA splicing from a single genetic locus (Moore et al. 2014). The four

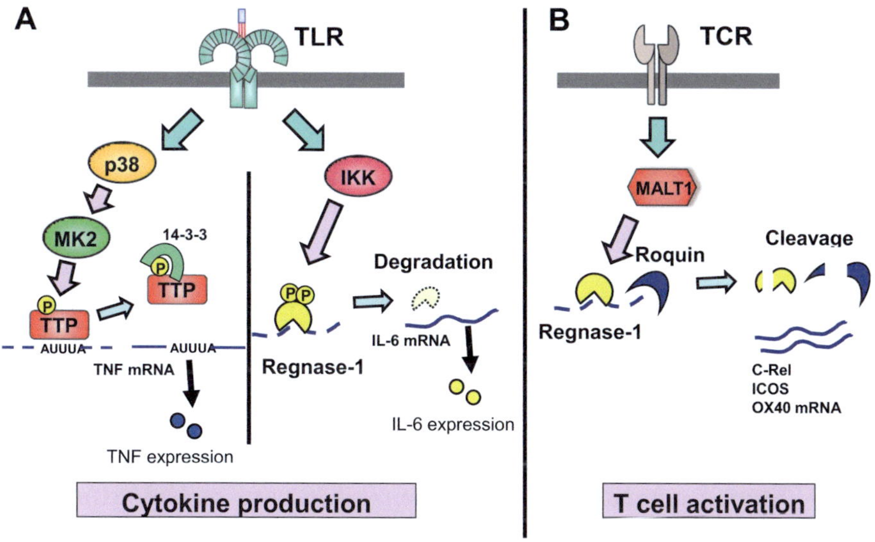

Fig. 25.2 Modification of RNA binding proteins in response to immune stimuli. (**a**) The TLR signalling activates MAP kinases and IKK in innate immune cells. The p38 MAP kinase activation leads to MK2-mediated phosphorylation of TTP, which sequester TTP from the target mRNAs by binding with 14-3-3 protein. IKK phosphorylates Regnase-1, leading to stabilization of *Il6* mRNAs by degrading Regnase-1. (**b**) The TCR signalling activates MALT1 protease, cleaving Regnase-1 and Roquin proteins. Decreased expression of Regnase-1 and Roquin stabilises c-Rel, Icos, and Ox40 mRNAs, facilitating T cell activation

AUF1 isoforms contain two RNA recognition motifs (RRMs) in the central portion of the protein (Moore et al. 2014). Mice lacking AUF1 develop chronic dermatitis with increasing age, characterised by pruritis and excoriations, which are associated with increased IL-2, TNF-α, and IL-1β (Sadri and Schneider 2009). Although AUF1 is primarily localised in the nucleus, it shuttles to the cytoplasm and attaches to ARE-containing mRNAs through MAPK phosphatase-1 (MKP-1)-induced translocation (Yu et al. 2011). MKP-1 functions as a negative regulator of the host inflammatory response to infection by specific dephosphorylation of activated mitogen-activated protein kinases (MAPKs), which, when activated, induce proinflammatory cytokines. Although the mechanisms underlying the regulation of cytokine mRNA stability by AUF1 have not been fully described, these findings show the complexity of the processes that control AUF1-regulated cytokine mRNA expression and cellular localisation.

25.2.3 Hu Antigen R (HuR)

HuR (also known as Embryonic Lethal, Abnormal Vision, Drosophila-Like 1; ELAVL1) is another protein-binding AREs found in the 3' UTR of cytokine mRNAs. In contrast to TTP and AUF1, HuR stabilises mRNAs harboring AREs (Fan and Steitz 1998; Herdy et al. 2015). However, HuR is reported to be involved in various regulatory processes such as guiding mRNA processing, export, translation, and miRNA-mediated mechanisms (Meisner and Filipowicz 2010). Indeed, PAR-CLIP analysis of HuR-binding mRNA sites has revealed that HuR- and microRNA-binding sites tend to reside in close proximity to each other in HEK293 cells, implying that dimerised HuR stabilises mRNAs by preventing microRNA-mediated degradation (Lebedeva et al. 2011). Another PAR-CLIP study identified HuR binding sites both in introns and 3' UTRs, suggesting coupling of pre-mRNA processing and mRNA stability (Mukherjee et al. 2011). Further studies are required for uncovering all functions of HuR in controlling cytokine mRNA expression.

25.3 Stem-Loop Structures as the *cis*-Element in Cytokine mRNA 3' UTR

In addition to AREs, stem-loop structures present in cytokine mRNAs 3' UTR, such as *TNF* and *IL6*, have been identified as a regulatory motif for posttranscriptional control (Stoecklin et al. 2003; Paschoud et al. 2006). In the *TNF* 3' UTR, this sequence is called the constitutive decay element (CDE) and contributes to the suppression of *TNF* mRNA by binding with an RNA-binding protein Roquin (Leppek et al. 2013). *IL6* mRNA 3' UTR also harbors an evolutionally conserved element potentially forming a stem-loop structure. Regnase-1 (also known as Zc3h12a, MCPIP1) was identified as an RNase responsible for the degradation of *IL6* mRNA via the stem-loop (Matsushita et al. 2009). Thus, stem-loop RNAs have emerged as important *cis*-elements for the regulation of cytokine mRNAs.

25.3.1 Roquin

The Roquin family of proteins Roquin-1 and Roquin-2 bind mRNAs involved in the activation of both innate and acquired immune systems, and induce their degradation. Roquin proteins harbour an N-terminal RING finger domain, followed by a RNA binding ROQ domain, and a CCCH-type zinc finger domain. Roquin was initially discovered as a protein controlling follicular helper T- (Tfh) cell development and the maintenance of self-tolerance (Vinuesa et al. 2005). The Roquin-1 Sanroque mutation in mice results in the increase in Tfh cells, and the development of lupus-like autoimmunity by controlling the stability of *ICOS* mRNA (Yu et al. 2007). In macrophages, Roquin binds to *TNF* mRNA 3' UTRs to destabilise their expression (Leppek et al. 2013). These notions indicate that Roquin is critical for the prevention of chronic inflammation and maintenance of immune homeostasis.

Although complete Roquin-1 deficiency in mice resulted in perinatal lethality without causing autoimmunity (Bertossi et al. 2011), mice lacking both Roquin-1 and Roquin-2 in T cells developed systemic inflammation accompanied by Tfh accumulation, as well as severe lung inflammation and enhanced Th17 differentiation (Jeltsch et al. 2014; Pratama et al. 2013; Vogel et al. 2013). Roquin-1 and Roquin-2 repress the expression of molecules involved in Tfh and Th17 differentiation, such as ICOS, Ox40, IL-6, IκBζ, and IκBNS. Although Roquin was first identified as a protein binding to the constitutive decay element in *TNF* mRNA, Roquin protein expression in T cells is tightly controlled under immune responses. T cell–receptor (TCR) signalling leads to the cleavage of Roquin via a protease MALT1 (Jeltsch et al. 2014). In this way, antigen-recognition by the TCR facilitates the induction of genes involved in Th17 differentiation by inducing degradation of Roquin.

Structural studies of Roquin-1 and Roquin-2 revealed that the ROQ domain forms an extended winged-helix–turn-helix motif and is responsible for the recognition of stem-loop structure with a pyrimidine–purine–pyrimidine triloop sequence present in TNF 3' UTR CDE (Codutti et al. 2015; Tan et al. 2014; Schuetz et al. 2014; Schlundt et al. 2014; Sakurai et al. 2015). Roquin-1 is localised in stress granules and P bodies, and the ROQ domain is essential for the localisation. Roquin binds to mRNAs such as ICOS, Ox40, TNF, and A20 via stem-loop structures present in their 3' UTR, and is reported to recruit components of the CCR4–NOT deadenylase complex and a decapping enhancer Edc4 for exonuclease-mediated mRNA decay (Leppek et al. 2013; Glasmacher et al. 2010; Fig. 25.3).

A PAR-CLIP analysis of human Roquin-1 identified about 3800 mRNAs binding to Roquin-1, which is much more than that expected by counting potential CDE-like stem-loop structures (Murakawa et al. 2015). In addition to mRNAs, Roquin was shown to be required for destabilisation of miR-146a by inducing its 3' end uridylation. Thus, further studies are required for clarifying the mechanisms of Roquin-mediated target mRNA recognition and degradation.

25.3.2 *Regnase-1*

Regnase-1 (also known as Zc3h12a or MCPIP1) harbours a PIN-like RNase domain and a CCCH-type zinc-finger domain (Matsushita et al. 2009), and is one of a family of four proteins; Zc3h12a, Zc3h12b, Zc3h12c, and Zc3h12d, with similar domain architectures (Liang et al. 2008). Regnase-1 is critical for preventing development of an autoimmune inflammatory disease in mice by controlling the stability of mRNA encoding IL-6, IL-12p40, and other inflammatory genes (Fig. 25.2a). The production of IL-6 and IL-12p40 in response to TLR ligands is greater in *Regnase-1*–deficient macrophages compared with normal macrophages.

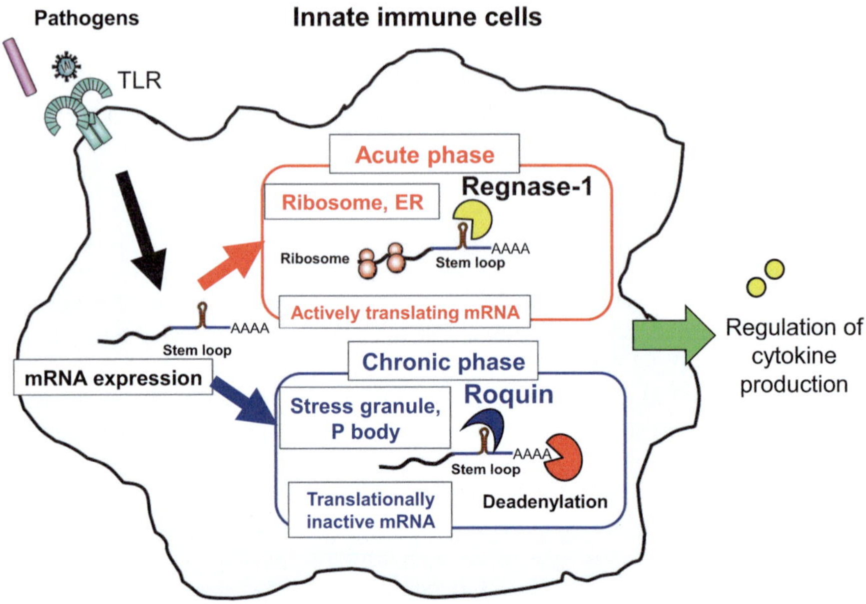

Fig. 25.3 Spatiotemporal regulation of inflammatory mRNAs by Regnase-1 and Roquin. A set of TLR-inducible inflammatory mRNAs harbour common stem-loop structures recognised by Regnase-1 and Roquin. Whereas Regnase-1 localised in the ER binding with ribosomes and degrades its target mRNAs in a translation-dependent faction, Roquin is present in the P bodies and stress granules, and degrades translationally inactive mRNAs. Regnase-1 and Roquin contribute to the control of inflammatory mRNA degradation in acute and late phases of inflammation, respectively

IL6 mRNA is stabilised in *Regnase-1*–deficient macrophages, and a conserved element present in the 3′ UTR of *IL6* mRNA is responsible for Regnase-1–mediated degradation (Matsushita et al. 2009). Regnase-1 also suppresses IL-17–mediated cellular responses, and *Regnase-1*$^{+/-}$ mice showed resistance to *Candida albicans* infection and exacerbated pathology in experimental autoimmune encephalomyelitis (Garg et al. 2015). Thus, Regnase-1–mediated posttranscriptional control of mRNA expression has an essential role in maintaining immune homeostasis.

Although Regnase-1 was identified as a TLR-inducible gene in macrophages, Regnase-1 protein is rapidly degraded in response to stimulation with IL-1β or TLR ligands, but not TNF (Iwasaki et al. 2011). The degradation of Regnase-1 is triggered by the phosphorylation of the DSGxxS sequence by inhibitors of transcription factor NF-κB (IκB) kinase (IKK) complex. Phosphorylated Regnase-1 undergoes ubiquitin-proteasome–mediated degradation via the E3 ligase β-TrCP complex (β-transducin repeat-containing protein). Degradation of Regnase-1 protein is found to be important for higher expression of *IL6* mRNA (Figure 25.2a). Following degradation, Regnase-1 is re-expressed in part due to Regnase-1–mediated suppression of its own *Regnase-1* mRNA via its 3′ UTR. These findings show that the IKK complex phosphorylates IκBα in order to activate transcription, and phosphorylates Regnase-1 in order to release the 'brake' on *IL-6* mRNA expression. Regnase-1–mediated mRNA degradation can be antagonised by another RNA binding protein Arid5a, which results in the stabilisation of *IL6* mRNA.

Regnase-1 is expressed not only in innate immune cells, but also in acquired immune cells such as T cells. Lack of Regnase-1 specifically in T cells resulted in the development of autoimmune inflammatory disease similar to the systemic deletion of Regnase-1 in mice (Uehata et al. 2013). Regnase-1 degrades mRNAs involved in T cell activation such as *c-Rel*, *ICOS*, *OX40*, and *IL2*. Among these, loss of c-Rel together with Regnase-1 partially rescued spontaneous activation of T cells, indicating that Regnase-1 controls T cell–activation in part by controlling c-Rel-mediated transcription. Regnase-1 deficiency augmented TCR-mediated effector functions, indicating that Regnase-1 is critical for balancing TCR sensitivity.

Furthermore, TCR signalling leads to the cleavage and degradation of Regnase-1 by MALT1 paracaspase activity (Uehata et al. 2013; Fig. 25.2b). Given that MALT1 cleaves both Regnsae-1 and Roquin, MALT1 protease activity can function to degrade negative regulators of T cell–activation for facilitating T-cell–effector responses. Like Roquin, Regnase-1 can suppress genes involved in Th17 differentiation including IL6, IκBζ, and IκBNS (Mino et al. 2015). Whereas Roquin-1/-2 deficiency resulted in augmentation of Th17 responses, Regnase-1 deficiency in T cells enhanced differentiation of Th1, Th2, and Th17 cells.

When Regnase-1– and Roquin-binding mRNAs were examined, they were found to overlap significantly (Mino et al. 2015). Furthermore, Regnase-1 and Roquin can bind and degrade a common set of mRNAs related with inflammation. Both Regnase-1 and Roquin bind common stem-loop structures harbouring a pyrimidine–purine–pyrimidine triloop, implicating the redundant roles of these proteins. However, Roquin and Regnase-1 were found to degrade their common target mRNAs by distinct mechanisms. Whereas Roquin localised in P bodies and stress granules and mainly target translationally inactive mRNAs for degradation, Regnase-1 binds the Ribosome, localises in the endoplasmic reticulum (ER), and degrades translationally active mRNAs (Fig. 25.3). In addition, Regnase-1 binds the ATP-dependent RNA helicase upframeshift 1 (UPF1) and depends on translation. Furthermore, UPF1 helicase activity is essential for Regnase-1–mediated mRNA decay. The molecular mechanism of Regnase-1–mediated mRNA degradation is similar to that of nonsense-mediated decay (NMD) in their requirement of active protein translation and a helicase UPF1.

In the control of inflammatory gene expression, Regnase-1 and Roquin function in distinct phases of inflammatory responses. Whereas Regnase-1 is more important for suppressing inflammatory mRNAs in acute phase, Roquin contributes to their control in later phases of inflammation (Mino et al. 2015). Although further studies are required to uncover the reason why these two RNA binding proteins act differently, it is tempting to speculate that translation can be activated in the acute phase of inflammation, when Regnase-1 contributes more in suppressing its target mRNA expression.

25.4 Relations Between RNA Binding Proteins in Human Inflammatory Diseases

As shown above, posttranscriptional control is critical for the regulation of chronic inflammation and autoimmunity in mice. Thus, it is easy to imagine that the same set of RNA binding proteins are involved in the control of autoimmunity in humans. Indeed, one single nucleotide polymorphism (SNP) in the promoter region of TTP has been implicated in the disease activity of rheumatoid arthritis in Japanese individuals (Suzuki et al. 2008). Although Roquin and Regnase-1 SNPs have not been linked to inflammatory diseases, Zc3h12c, a Regnase-1 family member, was found to be associated with a psoriasis susceptibility locus (Tsoi et al. 2012). Although the roles of Zc3h12c in psoriasis development have yet to be shown, further studies will uncover the relationship between mRNA degradation control and chronic inflammation in humans.

25.5 Conclusions

As demonstrated in this chapter, mRNA stability is tightly regulated in the course of inflammation via multiple mRNA 3' UTR elements and RNA binding proteins recognising them. In most cases, these RNA binding proteins dampen the expression of inflammatory mediators by promoting mRNA decay or by inhibiting protein translation. These effects are mediated by regulatory factors that bind to *cis*-acting elements in 3′ UTRs of mRNA transcripts. In addition to well-characterised AREs, stem-loop structures are emerging as critical *cis*-elements for the control of inflammation targeted by multiple RNA binding proteins. Loss of posttranscriptional regulation of cytokine mRNAs can dramatically increase cytokine production, leading to tissue destruction and increased mortality. Given the onset of cytokine production, several features of posttranscriptional control play a critical role in their maintenance. However, these mechanisms are complex and remain unclear. A greater understanding of how RBPs regulate mRNA stability could potentially lead to improved therapies for inflammatory diseases.

References

Anderson P (2010) Post-transcriptional regulons coordinate the initiation and resolution of inflammation. Nat Rev Immunol 10(1):24–35. doi:10.1038/nri2685, nri2685 [pii]

Anderson P, Kedersha N (2008) Stress granules: the Tao of RNA triage. Trends Biochem Sci 33 (3):141–150. doi:10.1016/j.tibs.2007.12.003

Bertossi A, Aichinger M, Sansonetti P, Lech M, Neff F, Pal M, Wunderlich FT, Anders HJ, Klein L, Schmidt-Supprian M (2011) Loss of Roquin induces early death and immune deregulation but not autoimmunity. J Exp Med 208(9):1749–1756. doi:10.1084/jem.20110578

Beutler B (2009) Microbe sensing, positive feedback loops, and the pathogenesis of inflammatory diseases. Immunol Rev 227(1):248–263. doi:10.1111/j.1600-065X.2008.00733.x

Cai X, Chiu YH, Chen ZJ (2014) The cGAS-cGAMP-STING pathway of cytosolic DNA sensing and signaling. Mol Cell 54(2):289–296. doi:10.1016/j.molcel.2014.03.040

Caput D, Beutler B, Hartog K, Thayer R, Brown-Shimer S, Cerami A (1986) Identification of a common nucleotide sequence in the 3'-untranslated region of mRNA molecules specifying inflammatory mediators. Proc Natl Acad Sci U S A 83(6):1670–1674

Carballo E, Lai WS, Blackshear PJ (1998) Feedback inhibition of macrophage tumor necrosis factor-alpha production by tristetraprolin. Science 281(5379):1001–1005

Carpenter S, Ricci EP, Mercier BC, Moore MJ, Fitzgerald KA (2014) Post-transcriptional regulation of gene expression in innate immunity. Nat Rev Immunol 14(6):361–376. doi:10.1038/nri3682

Carrick DM, Lai WS, Blackshear PJ (2004) The tandem CCCH zinc finger protein tristetraprolin and its relevance to cytokine mRNA turnover and arthritis. Arthritis Res Ther 6(6):248–264. doi:10.1186/ar1441

Chovatiya R, Medzhitov R (2014) Stress, inflammation, and defense of homeostasis. Mol Cell 54 (2):281–288. doi:10.1016/j.molcel.2014.03.030

Clement SL, Scheckel C, Stoecklin G, Lykke-Andersen J (2011) Phosphorylation of tristetraprolin by MK2 impairs AU-rich element mRNA decay by preventing deadenylase recruitment. Mol Cell Biol 31(2):256–266. doi:10.1128/MCB.00717-10

Codutti L, Leppek K, Zalesak J, Windeisen V, Masiewicz P, Stoecklin G, Carlomagno T (2015) A distinct, sequence-induced conformation is required for recognition of the constitutive decay element RNA by Roquin. Structure 23(8):1437–1447. doi:10.1016/j.str.2015.06.001

Fabian MR, Frank F, Rouya C, Siddiqui N, Lai WS, Karetnikov A, Blackshear PJ, Nagar B, Sonenberg N (2013) Structural basis for the recruitment of the human CCR4-NOT deadenylase complex by tristetraprolin. Nat Struct Mol Biol 20(6):735–739. doi:10.1038/nsmb.2572

Fan XC, Steitz JA (1998) Overexpression of HuR, a nuclear-cytoplasmic shuttling protein, increases the in vivo stability of ARE-containing mRNAs. EMBO J 17(12):3448–3460. doi:10.1093/emboj/17.12.3448

Garg AV, Amatya N, Chen K, Cruz JA, Grover P, Whibley N, Conti HR, Hernandez Mir G, Sirakova T, Childs EC, Smithgall TE, Biswas PS, Kolls JK, McGeachy MJ, Kolattukudy PE, Gaffen SL (2015) MCPIP1 endoribonuclease activity negatively regulates Interleukin-17-mediated signaling and inflammation. Immunity. doi:10.1016/j.immuni.2015.07.021

Glasmacher E, Hoefig KP, Vogel KU, Rath N, Du L, Wolf C, Kremmer E, Wang X, Heissmeyer V (2010) Roquin binds inducible costimulator mRNA and effectors of mRNA decay to induce microRNA-independent post-transcriptional repression. Nat Immunol 11(8):725–733. doi:10.1038/ni.1902

Hao S, Baltimore D (2009) The stability of mRNA influences the temporal order of the induction of genes encoding inflammatory molecules. Nat Immunol 10(3):281–288. doi:10.1038/ni.1699, ni.1699 [pii]

Herdy B, Karonitsch T, Vladimer GI, Tan CS, Stukalov A, Trefzer C, Bigenzahn JW, Theil T, Holinka J, Kiener HP, Colinge J, Bennett KL, Superti-Furga G (2015) The RNA-binding protein HuR/ELAVL1 regulates IFN-beta mRNA abundance and the type I IFN response. Eur J Immunol 45(5):1500–1511. doi:10.1002/eji.201444979

Iwasaki H, Takeuchi O, Teraguchi S, Matsushita K, Uehata T, Kuniyoshi K, Satoh T, Saitoh T, Matsushita M, Standley DM, Akira S (2011) The IkappaB kinase complex regulates the stability of cytokine-encoding mRNA induced by TLR-IL-1R by controlling degradation of regnase-1. Nat Immunol 12(12):1167–1175. doi:10.1038/ni.2137, ni.2137 [pii]

Jeltsch KM, Hu D, Brenner S, Zoller J, Heinz GA, Nagel D, Vogel KU, Rehage N, Warth SC, Edelmann SL, Gloury R, Martin N, Lohs C, Lech M, Stehklein JE, Geerlof A, Kremmer E, Weber A, Anders HJ, Schmitz I, Schmidt-Supprian M, Fu M, Holtmann H, Krappmann D, Ruland J, Kallies A, Heikenwalder M, Heissmeyer V (2014) Cleavage of roquin and regnase-1 by the paracaspase MALT1 releases their cooperatively repressed targets to promote T(H)17 differentiation. Nat Immunol 15(11):1079–1089. doi:10.1038/ni.3008

Kafasla P, Skliris A, Kontoyiannis D (2014) Post-trascriptional coordination of immunological responses by RNA-binding proteins. Nat Immunol 15(6):492–502. doi:10.1038/ni.2884

Lamkanfi M, Dixit VM (2014) Mechanisms and functions of inflammasomes. Cell 157 (5):1013–1022. doi:10.1016/j.cell.2014.04.007

Lebedeva S, Jens M, Theil K, Schwanhausser B, Selbach M, Landthaler M, Rajewsky N (2011) Transcriptome-wide analysis of regulatory interactions of the RNA-binding protein HuR. Mol Cell 43(3):340–352. doi:10.1016/j.molcel.2011.06.008

Leppek K, Schott J, Reitter S, Poetz F, Hammond MC, Stoecklin G (2013) Roquin promotes constitutive mRNA decay via a conserved class of stem-loop recognition motifs. Cell 153 (4):869–881. doi:10.1016/j.cell.2013.04.016, S0092-8674(13)00457-1 [pii]

Liang J, Song W, Tromp G, Kolattukudy PE, Fu M (2008) Genome-wide survey and expression profiling of CCCH-zinc finger family reveals a functional module in macrophage activation. PLoS One 3(8), e2880. doi:10.1371/journal.pone.0002880

Lu JY, Sadri N, Schneider RJ (2006) Endotoxic shock in AUF1 knockout mice mediated by failure to degrade proinflammatory cytokine mRNAs. Genes Dev 20(22):3174–3184. doi:10.1101/gad.1467606

Mahtani KR, Brook M, Dean JL, Sully G, Saklatvala J, Clark AR (2001) Mitogen-activated protein kinase p38 controls the expression and posttranslational modification of tristetraprolin, a regulator of tumor necrosis factor alpha mRNA stability. Mol Cell Biol 21(19):6461–6469

Matsushita K, Takeuchi O, Standley DM, Kumagai Y, Kawagoe T, Miyake T, Satoh T, Kato H, Tsujimura T, Nakamura H, Akira S (2009) Zc3h12a is an RNase essential for controlling immune responses by regulating mRNA decay. Nature 458(7242):1185–1190. doi:10.1038/nature07924, nature07924 [pii]

Meisner NC, Filipowicz W (2010) Properties of the regulatory RNA-binding protein HuR and its role in controlling miRNA repression. Adv Exp Med Biol 700:106–123

Mino T, Takeuchi O (2013) Post-transcriptional regulation of cytokine mRNA controls the initiation and resolution of inflammation. Biotechnol Genet Eng Rev 29(1–2):49–60. doi:10.1080/02648725.2013.801236

Mino T, Murakawa Y, Fukao A, Vandenbon A, Wessels HH, Ori D, Uehata T, Tartey S, Akira S, Suzuki Y, Vinuesa CG, Ohler U, Standley DM, Landthaler M, Fujiwara T, Takeuchi O (2015) Regnase-1 and Roquin regulate a common element in inflammatory mRNAs by spatiotemporally distinct mechanisms. Cell 161(5):1058–1073. doi:10.1016/j.cell.2015.04.029

Moore AE, Chenette DM, Larkin LC, Schneider RJ (2014) Physiological networks and disease functions of RNA-binding protein AUF1. Wiley Interdiscip Rev RNA 5(4):549–564. doi:10.1002/wrna.1230

Mukherjee N, Corcoran DL, Nusbaum JD, Reid DW, Georgiev S, Hafner M, Ascano M Jr, Tuschl T, Ohler U, Keene JD (2011) Integrative regulatory mapping indicates that the RNA-binding protein HuR couples pre-mRNA processing and mRNA stability. Mol Cell 43 (3):327–339. doi:10.1016/j.molcel.2011.06.007

Murakawa Y, Hinz M, Mothes J, Schuetz A, Uhl M, Wyler E, Yasuda T, Mastrobuoni G, Friedel CC, Dolken L, Kempa S, Schmidt-Supprian M, Bluthgen N, Backofen R, Heinemann U, Wolf J, Scheidereit C, Landthaler M (2015) RC3H1 post-transcriptionally regulates A20 mRNA and modulates the activity of the IKK/NF-kappaB pathway. Nat Commun 6:7367. doi:10.1038/ncomms8367

Paschoud S, Dogar AM, Kuntz C, Grisoni-Neupert B, Richman L, Kuhn LC (2006) Destabilization of interleukin-6 mRNA requires a putative RNA stem-loop structure, an AU-rich element, and the RNA-binding protein AUF1. Mol Cell Biol 26(22):8228–8241. doi:10.1128/MCB.01155-06

Pratama A, Ramiscal RR, Silva DG, Das SK, Athanasopoulos V, Fitch J, Botelho NK, Chang PP, Hu X, Hogan JJ, Mana P, Bernal D, Korner H, Yu D, Goodnow CC, Cook MC, Vinuesa CG (2013) Roquin-2 shares functions with its paralog Roquin-1 in the repression of mRNAs

controlling T follicular helper cells and systemic inflammation. Immunity 38(4):669–680. doi:10.1016/j.immuni.2013.01.011
Ross EA, Smallie T, Ding Q, O'Neil JD, Cunliffe HE, Tang T, Rosner DR, Klevernic I, Morrice NA, Monaco C, Cunningham AF, Buckley CD, Saklatvala J, Dean JL, Clark AR (2015) Dominant suppression of inflammation via targeted mutation of the mRNA destabilizing protein tristetraprolin. J Immunol 195(1):265–276. doi:10.4049/jimmunol.1402826
Sadri N, Schneider RJ (2009) Auf1/Hnrnpd-deficient mice develop pruritic inflammatory skin disease. J Invest Dermatol 129(3):657–670. doi:10.1038/jid.2008.298
Sakurai S, Ohto U, Shimizu T (2015) Structure of human Roquin-2 and its complex with constitutive-decay element RNA. Acta Crystallogr F Struct Biol Commun 71 (Pt 8):1048–1054. doi:10.1107/S2053230X15011887
Sandler H, Kreth J, Timmers HT, Stoecklin G (2011) Not1 mediates recruitment of the deadenylase Caf1 to mRNAs targeted for degradation by tristetraprolin. Nucleic Acids Res 39(10):4373–4386. doi:10.1093/nar/gkr011
Schlundt A, Heinz GA, Janowski R, Geerlof A, Stehle R, Heissmeyer V, Niessing D, Sattler M (2014) Structural basis for RNA recognition in roquin-mediated post-transcriptional gene regulation. Nat Struct Mol Biol 21(8):671–678. doi:10.1038/nsmb.2855
Schuetz A, Murakawa Y, Rosenbaum E, Landthaler M, Heinemann U (2014) Roquin binding to target mRNAs involves a winged helix-turn-helix motif. Nat Commun 5:5701. doi:10.1038/ncomms6701
Shaw G, Kamen R (1986) A conserved AU sequence from the 3' untranslated region of GM-CSF mRNA mediates selective mRNA degradation. Cell 46(5):659–667
Stoecklin G, Lu M, Rattenbacher B, Moroni C (2003) A constitutive decay element promotes tumor necrosis factor alpha mRNA degradation via an AU-rich element-independent pathway. Mol Cell Biol 23(10):3506–3515
Stoecklin G, Stubbs T, Kedersha N, Wax S, Rigby WF, Blackwell TK, Anderson P (2004) MK2-induced tristetraprolin:14-3-3 complexes prevent stress granule association and ARE-mRNA decay. EMBO J 23(6):1313–1324. doi:10.1038/sj.emboj.7600163
Suzuki T, Tsutsumi A, Suzuki H, Suzuki E, Sugihara M, Muraki Y, Hayashi T, Chino Y, Goto D, Matsumoto I, Ito S, Miyazawa K, Sumida T (2008) Tristetraprolin (TTP) gene polymorphisms in patients with rheumatoid arthritis and healthy individuals. Mod Rheumatol 18(5):472–479. doi:10.1007/s10165-008-0085-5
Takeuchi O, Akira S (2010) Pattern recognition receptors and inflammation. Cell 140(6):805–820. doi:10.1016/j.cell.2010.01.022
Tan D, Zhou M, Kiledjian M, Tong L (2014) The ROQ domain of Roquin recognizes mRNA constitutive-decay element and double-stranded RNA. Nat Struct Mol Biol 21(8):679–685. doi:10.1038/nsmb.2857
Taylor GA, Carballo E, Lee DM, Lai WS, Thompson MJ, Patel DD, Schenkman DI, Gilkeson GS, Broxmeyer HE, Haynes BF, Blackshear PJ (1996) A pathogenetic role for TNF alpha in the syndrome of cachexia, arthritis, and autoimmunity resulting from tristetraprolin (TTP) deficiency. Immunity 4(5):445–454
Tsoi LC, Spain SL, Knight J, Ellinghaus E, Stuart PE, Capon F, Ding J, Li Y, Tejasvi T, Gudjonsson JE, Kang HM, Allen MH, McManus R, Novelli G, Samuelsson L, Schalkwijk J, Stahle M, Burden AD, Smith CH, Cork MJ, Estivill X, Bowcock AM, Krueger GG, Weger W, Worthington J, Tazi-Ahnini R, Nestle FO, Hayday A, Hoffmann P, Winkelmann J, Wijmenga C, Langford C, Edkins S, Andrews R, Blackburn H, Strange A, Band G, Pearson RD, Vukcevic D, Spencer CC, Deloukas P, Mrowietz U, Schreiber S, Weidinger S, Koks S, Kingo K, Esko T, Metspalu A, Lim HW, Voorhees JJ, Weichenthal M, Wichmann HE, Chandran V, Rosen CF, Rahman P, Gladman DD, Griffiths CE, Reis A, Kere J, Collaborative Association Study of P, Genetic Analysis of Psoriasis C, Psoriasis Association Genetics E, Wellcome Trust Case Control C, Nair RP, Franke A, Barker JN, Abecasis GR, Elder JT, Trembath RC (2012) Identification of 15 new psoriasis susceptibility loci highlights the role of innate immunity. Nat Genet 44(12):1341–1348. doi:10.1038/ng.2467

Uehata T, Iwasaki H, Vandenbon A, Matsushita K, Hernandez-Cuellar E, Kuniyoshi K, Satoh T, Mino T, Suzuki Y, Standley DM, Tsujimura T, Rakugi H, Isaka Y, Takeuchi O, Akira S (2013) Malt1-induced cleavage of regnase-1 in CD4(+) helper T cells regulates immune activation. Cell 153(5):1036–1049. doi:10.1016/j.cell.2013.04.034

Vinuesa CG, Cook MC, Angelucci C, Athanasopoulos V, Rui L, Hill KM, Yu D, Domaschenz H, Whittle B, Lambe T, Roberts IS, Copley RR, Bell JI, Cornall RJ, Goodnow CC (2005) A RING-type ubiquitin ligase family member required to repress follicular helper T cells and autoimmunity. Nature 435(7041):452–458. doi:10.1038/nature03555, doi:nature03555 [pii]

Vogel KU, Edelmann SL, Jeltsch KM, Bertossi A, Heger K, Heinz GA, Zoller J, Warth SC, Hoefig KP, Lohs C, Neff F, Kremmer E, Schick J, Repsilber D, Geerlof A, Blum H, Wurst W, Heikenwalder M, Schmidt-Supprian M, Heissmeyer V (2013) Roquin paralogs 1 and 2 redundantly repress the Icos and Ox40 costimulator mRNAs and control follicular helper T cell differentiation. Immunity 38(4):655–668. doi:10.1016/j.immuni.2012.12.004

Wilusz CJ, Wormington M, Peltz SW (2001) The cap-to-tail guide to mRNA turnover. Nat Rev Mol Cell Biol 2(4):237–246. doi:10.1038/35067025

Yu D, Tan AH, Hu X, Athanasopoulos V, Simpson N, Silva DG, Hutloff A, Giles KM, Leedman PJ, Lam KP, Goodnow CC, Vinuesa CG (2007) Roquin represses autoimmunity by limiting inducible T-cell co-stimulator messenger RNA. Nature 450(7167):299–303. doi:10.1038/nature06253

Yu H, Sun Y, Haycraft C, Palanisamy V, Kirkwood KL (2011) MKP-1 regulates cytokine mRNA stability through selectively modulation subcellular translocation of AUF1. Cytokine 56 (2):245–255. doi:10.1016/j.cyto.2011.06.006

Chapter 26
Roles of C-Type Lectin Receptors in Inflammatory Responses

Shinobu Saijo

Abstract Myeloid cells act central roles in inflammatory responses. C-type lectin receptors (CLRs) are key players that are expressed mainly in these cells to orchestrate immune responses for maintaining immune homeostasis. These receptors recognise carbohydrate structures in microbes, including fungi, bacteria, viruses, and parasites, as pathogen-associated molecular patterns (PAMPs); some of the receptors recognise self-ligands and some recognise both. Ligand ligations of these receptors initiate various biological reactions such as proinflammatory, anti-inflammatory, endocytic, or phagocytic responses. Owing to these activities, the pleiotropic roles of CLRs have an important effect on homeostasis in the body. In addition, because of these varied abilities, it is thought that inhibitors or activators of this receptor signalling are good therapeutic agents for inflammatory disorders. In this section, we provide a detailed summary of the current knowledge of these receptors, describing their expression, ligand recognition, signalling, and associated diseases, and how they function in innate and adaptive immunity.

Keywords C-type lectin • Myeloid cell • Cytokine • Inflammatory diseases • IL-17 • Carbohydrate

26.1 Introduction

C-type lectin receptors (CLRs) are a group of proteins that bind self and/or nonself carbohydrates with the conserved C-type lectin-like domain. This family of receptors is composed of more than 1,000 molecules classified into 17 subgroups according to their domain organisation and structure (Zelensky and Gready 2005). Because of structural and functional similarities and the formation of gene clusters, it is thought that some members of this family have arisen from gene duplication. In this section, we focus on CLRs that abundantly expressed in myeloid

S. Saijo (✉)
Division of Molecular Immunology, Medical Mycology Research Center, Chiba University, Chiba, Japan, 260-8673

PRESTO, Japan Science and Technology Agency, Kawaguchi 332-0012, Japan
e-mail: saijo@faculty.chiba-u.jp

M. Miyasaka, K. Takatsu (eds.), *Chronic Inflammation*,
DOI 10.1007/978-4-431-56068-5_26

cells and drive the production of cytokines that strongly regulate host immune systems, including processes such as T helper cell differentiation, inflammatory cell recruitment, and cell damage induction by recognition of self or nonself ligands.

One of the CLRs, DC-associated C-type lectin-1 (Dectin-1, gene symbol: *Clec7a*), contains an immunoreceptor tyrosine-based activation motif (ITAM-like motif) in the cytoplasmic N-terminus (Brown and Gordon 2001). However, most CLRs, such as DC-associated C-type lectin-2 (Dectin-2, gene symbol: *Clec4n*) and macrophage-inducible C-type lectin (Mincle, gene symbol: *Clec4e*), do not have known signalling motifs in their cytoplasmic tails. They recruit the ITAM-bearing adaptor molecule Fc receptor γ chain (FcRγ, gene symbol: *Fcer1g*) for their signal transductions (Robinson et al. 2009). *Clec4n*, *Clec4e*, and *Clec7a* are mapped on chromosome 6F3 in mice and 12q13 in humans where this group of CLRs forms a cluster (Balch et al. 2002; Fig. 26.1). These CLRs constitute dimers or tetramers and recruit phosphorylated spleen tyrosine kinase (Syk) with the ligand ligations (Fig. 26.2). Then phosphorylated Syk promotes a CARD9–BCL10–MALT1 (CBM) complex, resulting in the activation of the nuclear factor of the k light polypeptide gene enhancer in B cells (NF-kB). NF-kB activation induces expression of inflammatory cytokines such as tumour necrosis factor (TNF), interleukin (IL)-6, IL-23, and pro-IL-1β. Concurrently, Syk activation induces reactive oxygen species (ROS) production. ROS can directly damage invaded microbes and simultaneously activate inflammasomes that drive the processing of pro-IL-1β into mature IL-1β. The mitogen-activated protein kinase (MAPK) pathway is also activated simultaneously with the recruitment of Syk; however, the biological significance of this pathway in the host defence mechanism is not known (Saijo and Iwakura 2011; Fig. 26.2).

The dendritic cell immunoreceptor (DCIR, gene symbol: *Clec4a2*) family of CLRs includes CLRs that consists of a transmembrane domain with a stalk and a CRD capable of recognising carbohydrate structures on pathogens. This receptor is also mapped in the C-type lectin cluster on mouse chromosome 6 and human chromosome 12; however, unlike other CLRs, some of the DCIR family receptors have an immunoreceptor tyrosine-based inhibitory motif (ITIM) in their cytoplasmic region (Kanazawa et al. 2003). In this section, we update paradigms and

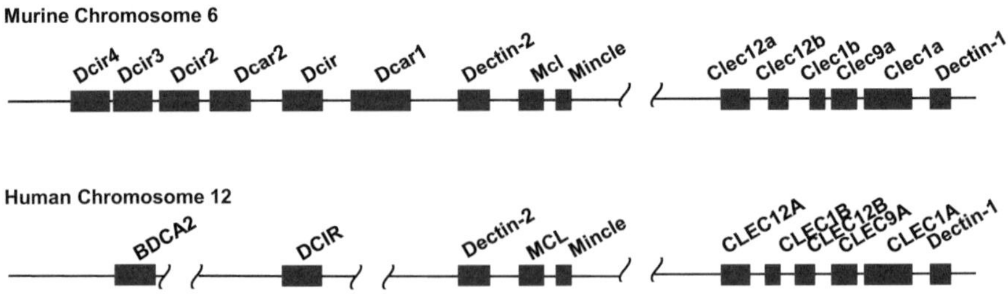

Fig. 26.1 Genes encoding myeloid CLRs form a cluster. The murine CLR gene cluster is placed in chromosome 6 F3, and the human counterpart is placed in chromosome 12q13

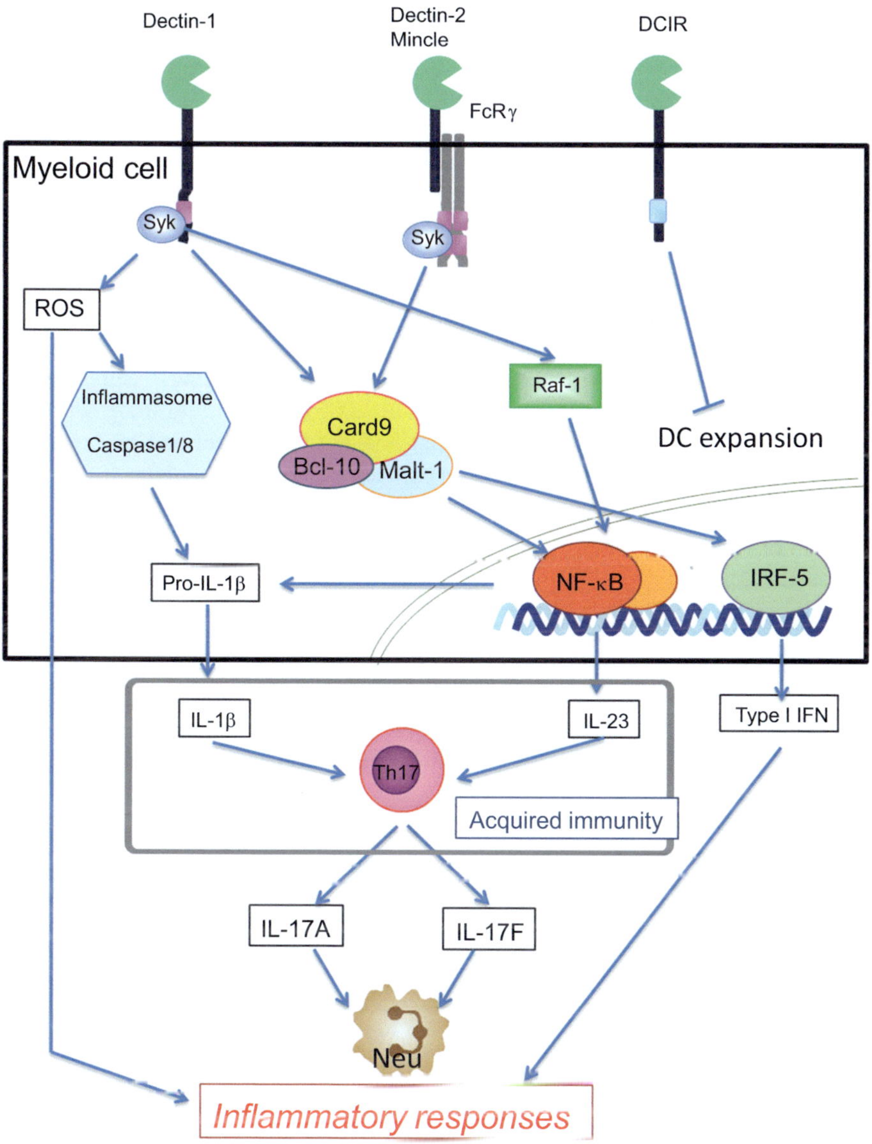

Fig. 26.2 Signalling pathway and outcome of myeloid CLRs. Dectin-2 and Mincle transduce their signalling via ITMA-bearing FcRγ. Dectin-1 signalling is initiated by its own hem-ITAM in the cytoplasmic region. These signallings all share the following steps: syk-recruitment leads to formation of the CARD9–Bcl10–Malt1 complex, which results in the activation of the transcriptional activator NF-kB. Then, Cytokine production is induced. These cytokines secreted from myeloid cells promote Th17 differentiation. DCIR maintains homeostasis by controlling the number of the myeloid cells, but the signalling pathway is still unclear

summarise findings that have obtained in recent years on these members of CLRs and their functions.

26.2 Dectin-1

Dectin-1 was originally cloned as a DC-specific CLR by subtracting mRNA purified from a DC line with the mRNAs from a macrophage line (Ariizumi et al. 2000a). However, in vivo, Dectin-1 expression is found in DCs, monocytes, macrophages, neutrophils, and γδ T cells in mice, and in humans, the expression is found in B cells, eosinophils, and mast cells, in addition to these cells (Taylor et al. 2002; Willment et al. 2005; Olynych et al. 2006). In the initial study, Dectin-1 was reported to have a unique ligand that had not been identified in T cells to stimulate their proliferation, but, using an expression cloning technique, it was soon found that Dectin-1 is a β-glucan receptor which has been sought for a long time (Brown and Gordon 2001).

β-glucans are polycarbohydrates which are found in the fungal cell wall; the backbone structure consists of polymerised β-1,3-linked β-D-glucopyranosyl units with β-1,6-linked side chains (Brown and Gordon 2003). This sugar chain potently drives the secretion of proinflammatory cytokines, including TNF, IL-6, and IL-1β as well as inhibitory cytokine IL-10 from the myeloid cells in a Dectin-1-CARD9 signalling-pathway–dependent fashion (Fig. 26.2; Saijo et al. 2007; Taylor et al. 2007). In vivo, Dectin-1–deficient mice are more susceptible to fungal infection, for example, with *Pneumocystis carinii* and *Candida albicans* (*C. albicans*). Because families with the inherited Dectin-1 gene mutations were found in patients with chronic mucocutaneous candidiasis, one of the opportunistic infections in humans, it is suggested that Dectin-1 is crucial to maintain the balance of mucocutaneous microflora (Ferwerda et al. 2009). Dectin-1 or Dectin-2 (see below) also induce interferon-β (IFN-β) production along with *C. albicans* recognition, which is crucial for fungal elimination (del Fresno et al. 2013). In this case, atypically, IFN-β production requires Syk and the transcriptional factor interferon regulatory factor 5 (IRF5), but not IRF3 and IRF7.

Dectin-1 also acts a central role in the sustenance of intestinal microflora (Iliev, Funari et al. 2012). A fertile fungal biome is detected in the mammalian digestive tracts and communicates with the host immune system through Dectin-1 in a certain circumstance. Consistent with these observations, polymorphisms in the gene encoding CARD9 in humans, an adaptor molecule downstream of Dectin-1, is highly linked with inflammatory bowel diseases such as Crohn's disease and ulcerative colitis, suggesting that CLRs are important for intestinal homeostasis in humans (de Vries et al. 2009). More recently, it was reported that inhibition of Dectin-1 signalling protects against experimental colitis by suppressing antimicrobial peptide secretion in order to allow growth of lactobacilli, which triggers T regulatory cell expansion in the gut (Tang et al. 2015).

Glycosylation is a well-known step in which a carbohydrate molecule is added to proteins in mammals. Abnormal glycosylation is often found in tumour cells, and it is thought that the altered glycosylation on the tumor cell surface facilitates migration and spreading to other organs. Recently, Chiba et al. found that Dectin-1 recognises not β-glucans but N-glycan structures, which are highly expressed in some tumour cell lines. Surprisingly, Dectin-1 has an affinity to the N-glycan structure on the tumour cell surface and initiates antitumour immunity though Natural Killer (NK) cell activation (Chiba et al. 2014). It would be very interesting to determine if this antitumour system by Dectin-1 plays a crucial role in preventing the evolution or the migration of cancer cells in the host.

26.3 Dectin-2

Dectin-2 was isolated as DC-specific CLR together with Dectin-1 by subtractive cloning, but is expressed in several DC subsets, macrophages, and monocytes (Ariizumi et al. 2000b). Dectin-2 has an EPN (Glu-Pro-Asn) motif, a Ca^{2+}-dependent mannose-binding amino acid sequence, in its C-terminus extracellular CRD (Ariizumi et al. 2000a, b). It lacks a known signalling motif in the short cytoplasmic domain, but transduces its signalling by associating with FcRγ, that carries ITAM (Taylor et al. 2005; Fig. 26.2). Dectin-2 binds to high-mannose structures that are found in a broad spectrum of species, including fungi, parasites, bacteria, and mammals (Table 26.1). *C. albicans* has a high-mannose–type N-glycan structure in the cell wall called a-mannans; a backbone consists of polycarbohydrate of α-1, 6-linked mannose unit with α-1,2-linked polymannose side chains. These *C. albicans* α-mannans are placed in the outermost part of the cell wall and are found abundantly, therefore, one may consider that mannose recognition should be the first step for the host to lead up the innate immune responses. As expected, Dectin-2–deficient mice show significantly higher mortality to *C. albicans* infection

Table 26.1 CLR sensing ligands

CLR	Ligands	Ligand locations
Dectin-1	β-glucans	fungi
	N-glycan	tumor cells
Dectin-2	High-mannose-type *N*-glycans	fungi
	α-mannans	fungi
	O-linked mannobioses	parasites
	Man-LAM	mycobacterium
Mincle	Highly mannosylated glycans	Fungi
	Mannosyl fatty acids	Fungi
	TDM	Mycobacterium
	SAP-130	Dead cells
DCIR	Not identified	HIV-1

than the wild-type mice (Robinson et al. 2009; Saijo et al. 2010). The importance of Dectin-2 in immunity against fungi can be seen not only for *C. albicans* but also for *C. glabrata* and *Aspergillus fumigatus*.

IL-17A is an essential cytokine for inflammatory responses. Mice with IL-17A, IL-17 receptor A (IL-17RA), or IL-23-p19 deficiency show higher susceptibility to *C. albicans* infection, indicating that IL-17A is crucial for host protection (Kagami et al. 2010; Saijo et al. 2010). In humans, mutations in IL-17F and IL-17 receptor A (IL-17RA) genes alter IL-17-based immunity, leading to the development of chronic mucoepithelial fungal infection (Puel et al. 2011). In this context, cytokines that are induced by both Dectin-1- and Dectin-2-signalling with fungal stimulation have been found to selectively promote Th17 cell differentiation in vitro, suggesting that these CLRs expressed in myeloid cells mobilise acquired immunity to ultimate antifungal immunity. However, IL-17A and IL-17F producing cells upon infection with fungi is still unclear. IL-17F is produced by epithelial cells and innate immune cells other than Th17 cells (Ishigame et al. 2009). Moreover, it was recently reported that, upon infection with *A. fumigatus*, Dectin-2-induced IL-6 and IL-23, which in turn drive IL-17A secretion in neutrophils that constitutively express the transcription factor retinoid-related orphan nuclear receptor-γ t (RORγt). The IL-6 and IL-23 also promote surface expression of IL-17RC and Dectin-2 in an autocrine fashion, leading to the ROS synthesis for the promotion of the fungal eliminations (Taylor et al. 2014).

Dectin-2 also recognises one of the major mycobacterial cell wall lipoglycans, mannose-capped lypoarabionomannan (Man-LAM; Yonekawa et al. 2014). Several receptors have been reported to have affinities for Man-LAM, but none of these molecules could explain the dual nature of Man-LAM, which induces both immune activation and suppuration in the host. The ligand recognition of Dectin-2 results in the production of proinflammatory cytokines as well as of the inhibitory cytokine IL-10 through the ITAM-bearing adaptor molecule, FcRγ. It is noteworthy that this IL-10 production in response to *Mycobacterium* is a unique aspect of Dectin-2. The strains of *Mycobacterium* lacking the mannose cap are sensed by another C-type lectin Mincle and Mcl, whose signalling induce only proinflammatory cytokines, and not IL-10, suggesting that the presence of the mannose cap contributes to the virulence of the bacterium. In vivo, Dectin-2 deficiency resulted in higher susceptibility to *Mycobacterium* infection, indicating that Dectin-2 also plays an important role in the host defence against *Mycobacterium* strains with a mannose cap.

Dectin-2 also triggers allergic responses with the recognition of mannose-bearing antigens in house dust mites or *Schistosoma mansoni*. These interactions induce cysteinyl leukotriene production which promotes Th2 and/or Th17 responses, that strongly mediate airway inflammation (Ritter et al. 2010; Norimoto et al. 2014; Parsons et al. 2014).

26.4 Mincle

Mincle is expressed in macrophages and DCs at a low level and can be induced by stimulation with inflammatory cytokines or Toll-like receptor (TLR) ligands such as lipopolysaccharide (Matsumoto et al. 1999). Similar to Dectin-2, Mincle has a short cytoplasmic tail without known signalling motifs, but associates with ITAM-containing adaptor molecule FcRγ (Yamasaki et al. 2008). Mincle also contains an EPN motif in its extracellular CRD, which could bind to polyvalent α-mannose in a Ca^{2+}-dependent manner (Fig 26.2). This receptor has affinities with manno-glycoconjugates in the fungal cell walls, including *C. albicans*, and *Malassezia* species (Bugarcic et al. 2008; Yamasaki et al. 2009; Table 26.1), driving proinflammatory cytokine production. Because Mincle-deficient mice were more vulnerable to systemic *C. albicans* infection than wild-type mice (Wells et al. 2008), and inflammatory responses such as cytokine production and neutrophil infiltration are suppressed upon infection with *Malassezia* (Yamasaki et al. 2009), it was indicated that Mincle plays an important role in the host defence against pathogenic fungi. As mentioned above, Mincle is found to be a crucial receptor for the antimycobacterial immunity. Mincle directly binds to cell wall components unique to *Mycobacterium*, such as trehalose-6,6'-dimycolate (TDM; also called cord factor), which is a potent adjuvant because of its high immune-stimulatory effect on the host immune system (Ishikawa et al. 2009). Binding of the TDM by Mincle results in the activation of NF-kB via the ITAM–Syk–CBM complex and induces robust proinflammatory cytokine production in macrophages and DCs. In vivo, intratracheal administration of TDM causes Mincle-dependent granuloma formation in the lungs, revealing that this interaction substantially contributes to the adjuvant activity of *Mycobacterium tuberculosis* (Ishikawa et al. 2009; Schoenen et al. 2010). However, the in vivo roles of Mincle in antimycobacterial immunity are still controversial; Mincle-deficient mice are fatally overwhelmed rapidly after *M. bovis* bacillus Calmette-Guerin (BCG) or *M. tuberculosis* Erdman infections, whereas, surprisingly, they show normal responses to much more virulent Mycobacterium tuberculosis (during *M. tuberculosis*) H37Rv infection (Schoenen et al. 2010; Lee et al. 2012; Heitmann et al. 2013). This is likely because different mycobacterial strains may contain differing exterior cell wall components, which are indispensable for their microbial survival and pathogenicity. Of note, TDM and its analogue, treharose-6,6-dibehenate (TDB) induced macrophage activation, as well as Th1/Th17 cell differentiations in a Mincle-dependent manner, suggesting that Mincle contributes to both innate and adaptive immunity for protection against mycobacterial infection (Schoenen et al. 2010).

Mincle also targets damaged cells in the body. This receptor has an affinity to the endogenous ribonucleoprotein SAP-130 that is released from dead cells as DAMPs (Yamasaki et al. 2008). The binding of SAP-130 by Mincle dose not require the EPN motif, but shares the signalling pathway, ITAM-syk-CARD9-NF-kB to promote cytokine secretions.

26.5 DCIR

DCIR is a type II C-type lectin with an intracellular inhibitory signalling motif, ITIM (Bates et al. 1999). There is one DCIR in humans, and four DCIR homologues have been found in mice (Dcir1 to 4); however, only human DCIR and murine Dcir1 and Dcir2 contain an ITIM (Kaden et al. 2009). Human DCIR and mouse Dcir1 are expressed in DCs, monocytes/macrophages, and B cells. The expression of Dcir2 is limited in $CD8^-$ DCs. Like Dectin-2 and Mincle, Murine Dcir1 and Dcir2, and human DCIR have a single CRD with a Ca^{2+}-dependent sugar binding amino acid sequence EPN that recognises mannose-type structures (Lee et al. 2011). Although the sugar chain structures of the DCIR ligand are not identified yet, human DCIR has been found to recognise HIV-1 (Lambert et al. 2008). With the recognition of HIV-1, forced expression of human DCIR with CD4 in a B-cell line induces the activation of phosphatase SHP-1 and SHP-2, as well as kinases Syk and Src. These events result in the internalisation of HIV-1, and then activate intracellular PKC-α, p38, and Erk1/2. However, generally, it is thought that ITIM signalling negatively regulates the cellular responses initiated by ITAM. In this context, the chimeric receptor, which is composed of extracellular FcγR-IIB and intracellular Dcir1, inhibits BCR signalling (Kanazawa et al. 2002).

The physiological role of DCIR was studied using Dcir1-deficient mice (Fujikado et al. 2008). DCIR-deficient bone marrow cells differentiate into DCs more expeditiously with augmentation of STAT5 phosphorylation in response to GM-CSF, suggesting that Dcir1 controls DC expansion in vivo. Indeed, Dcir1-deficient mice spontaneously develop autoimmune sialadenitis and enthesitis with a higher number of DCs. In addition, Dcir1-deficient mice demonstrate ankylosis, which is completely suppressed in Rag-2 deficiency (Maruhashi et al. 2015), suggesting that DCIR mediates immunological homeostasis in an acquired immunity-dependent manner. However, the endogenous ligand of DCIR is not isolated yet, therefore, further studies are required for understanding the molecular mechanisms of DCIR-dependent immunity.

26.6 Concluding Remarks

Myeloid CLRs show a variety of functions in the immune response, as described in this section. This family of CLRs transduces their signalling that directly leads to transcriptional activations; on the other hand, they involve transcription-independent activities, such as antigen uptake, endocytosis, or phagocytosis. However, the functions of many CLR genes, such as *Dcar1*, *Dcar2*, *Clec12b*, and *Clec1a*, of the same gene cluster, still need to be studied. These analyses should offer important indications for developing novel therapeutics for the treatment of inflammatory diseases. Several CLRs can recognise multiple ligands. For instance, Dectin-1 can bind to fungal β-glucans and tumour cell N-glycan, and Dectin-2 has a broader spectrum of ligands; it can bind to Man-LAM in the Mycobacterial cell walls, *Malassezia O*-linked mannobiose-rich glycoprotein, and house dust mite antigens other than α-mannans in the fungal cell walls. Mincle can recognise SAP-130 from dead cells and TDM in the Mycobacterial cell walls, and DCIR can bind to HIV-1 and might have endogenous ligand(s). These receptors seem to share their intracellular signalling pathway, however, the outcomes of the CLRs slightly differ; Both Dectin-2 and Mincle employ FcRγ to transduce their signalling, but only Dectin-2 induces the production of the inhibitory cytokine IL-10. This is probably because the outcome of CLR depends on the receptor itself, as well as on the nature of the ligand. Of note, some CLRs are reported to form a heterodimeric complex with other CLRs. For example, Mcl, a FcRγ-coupled CLR in the same cluster, couples with Dectin-2 through FcRγ ligation (Zhu et al. 2013), or directly contacts with Mincle (Miyake et al. 2015). This complexity and divergence is one of the distinguishing qualities of the CLR system, suggesting that the CLR system evaluated efficiently to utilise all the pathogens in the most powerful way by using a limited number of CLRs. Furthermore, some ligands of CLRs are known as opportunistic pathogens, which are harmless and are allowed to live along with the host. It is possible that these commensal microbes continually stimulate the host immune system via CLRs at a low level, resulting in development of inflammatory diseases. Indeed, Dectin-1 maintains intestinal homeostasis (Iliev et al. 2012; Tang et al. 2015). In addition, several linkage studies showed that the chromosomal region of the CLR gene cluster associates to the development of autoimmune diseases in both mice and humans (Fujikado et al. 2006). Thus, the continued study of the functions of myeloid CLRs to complete the whole picture of the CLR system will help us to understand the role of myeloid cells in pathology and homeostasis, and will provide breakthroughs in the development of therapy for inflammatory diseases.

References

Ariizumi K, Shen GL, Shikano S, Xu S, Ritter R 3rd, Kumamoto T, Edelbaum D, Morita A, Bergstresser PR, Takashima A (2000a) Identification of a novel, dendritic cell associated molecule, dectin-1, by subtractive cDNA cloning. J Biol Chem 275(26):20157–20167

Ariizumi K, Shen GL, Shikano S, Ritter R 3rd, Zukas P, Edelbaum D, Morita A, Takashima A (2000b) Cloning of a second dendritic cell-associated C-type lectin (dectin-2) and its alternatively spliced isoforms. J Biol Chem 275(16):11957–11963

Balch SG, Greaves DR, Gordon S, McKnight AJ (2002) Organization of the mouse macrophage C-type lectin (Mcl) gene and identification of a subgroup of related lectin molecules. Eur J Immunogenet 29(1):61–64

Bates EE, Fournier N, Garcia E, Valladeau J, Durand I, Pin JJ, Zurawski SM, Patel S, Abrams JS, Lebecque S, Garrone P, Saeland S (1999) APCs express DCIR, a novel C-type lectin surface receptor containing an immunoreceptor tyrosine-based inhibitory motif. J Immunol 163 (4):1973–1983

Brown GD, Gordon S (2001) Immune recognition. A new receptor for beta-glucans. Nature 413 (6851):36–37

Brown GD, Gordon S (2003) Fungal beta-glucans and mammalian immunity. Immunity 19 (3):311–315

Bugarcic A, Hitchens K, Beckhouse AG, Wells CA, Ashman RB, Blanchard H (2008) Human and mouse macrophage-inducible C-type lectin (Mincle) bind Candida albicans. Glycobiology 18 (9):679–685

Chiba S, Ikushima H, Ueki H, Yanai H, Kimura Y, Hangai S, Nishio J, Negishi H, Tamura T, Saijo S, Iwakura Y, Taniguchi T (2014) Recognition of tumor cells by Dectin-1 orchestrates innate immune cells for anti-tumor responses. Elife 3, e04177

de Vries HS, Plantinga TS, van Krieken JH, Stienstra R, van Bodegraven AA, Festen EA, Weersma RK, Crusius JB, Linskens RK, Joosten LA, Netea MG, de Jong DJ (2009) Genetic association analysis of the functional c.714T>G polymorphism and mucosal expression of dectin-1 in inflammatory bowel disease. PLoS One 4(11):e7818

del Fresno C, Soulat D, Roth S, Blazek K, Udalova I, Sancho D, Ruland J, Ardavin C (2013) Interferon-beta production via Dectin-1-Syk-IRF5 signaling in dendritic cells is crucial for immunity to C. albicans. Immunity 38(6):1176–1186

Ferwerda B, Ferwerda G, Plantinga TS, Willment JA, van Spriel AB, Venselaar H, Elbers CC, Johnson MD, Cambi A, Huysamen C, Jacobs L, Jansen T, Verheijen K, Masthoff L, Morre SA, Vriend G, Williams DL, Perfect JR, Joosten LA, Wijmenga C, van der Meer JW, Adema GJ, Kullberg BJ, Brown GD, Netea MG (2009) Human dectin-1 deficiency and mucocutaneous fungal infections. N Engl J Med 361(18):1760–1767

Fujikado N, Saijo S, Iwakura Y (2006) Identification of arthritis-related gene clusters by microarray analysis of two independent mouse models for rheumatoid arthritis. Arthritis Res Ther 8 (4):R100

Fujikado N, Saijo S, Yonezawa T, Shimamori K, Ishii A, Sugai S, Kotaki H, Sudo K, Nose M, Iwakura Y (2008) Dcir deficiency causes development of autoimmune diseases in mice due to excess expansion of dendritic cells. Nat Med 14(2):176–180

Heitmann L, Schoenen H, Ehlers S, Lang R, Holscher C (2013) Mincle is not essential for controlling Mycobacterium tuberculosis infection. Immunobiology 218(4):506–516

Iliev ID, Funari VA, Taylor KD, Nguyen Q, Reyes CN, Strom SP, Brown J, Becker CA, Fleshner PR, Dubinsky M, Rotter JI, Wang HL, McGovern DP, Brown GD, Underhill DM (2012) Interactions between commensal fungi and the C-type lectin receptor Dectin-1 influence colitis. Science 336(6086):1314–1317

Ishigame H, Kakuta S, Nagai T, Kadoki M, Nambu A, Komiyama Y, Fujikado N, Tanahashi Y, Akitsu A, Kotaki H, Sudo K, Nakae S, Sasakawa C, Iwakura Y (2009) Differential roles of interleukin-17A and -17F in host defense against mucoepithelial bacterial infection and allergic responses. Immunity 30(1):108–119

Ishikawa E, Ishikawa T, Morita YS, Toyonaga K, Yamada H, Takeuchi O, Kinoshita T, Akira S, Yoshikai Y, Yamasaki S (2009) Direct recognition of the mycobacterial glycolipid, trehalose dimycolate, by C-type lectin Mincle. J Exp Med 206(13):2879–2888

Kaden SA, Kurig S, Vasters K, Hofmann K, Zaenker KS, Schmitz J, Winkels G (2009) Enhanced dendritic cell-induced immune responses mediated by the novel C-type lectin receptor mDCAR1. J Immunol 183(8):5069–5078

Kagami S, Rizzo HL, Kurtz SE, Miller LS, Blauvelt A (2010) IL-23 and IL-17A, but not IL-12 and IL-22, are required for optimal skin host defense against Candida albicans. J Immunol 185 (9):5453–5462

Kanazawa N, Okazaki T, Nishimura H, Tashiro K, Inaba K, Miyachi Y (2002) DCIR acts as an inhibitory receptor depending on its immunoreceptor tyrosine-based inhibitory motif. J Invest Dermatol 118(2):261–266

Kanazawa N, Tashiro K, Inaba K, Miyachi Y (2003) Dendritic cell immunoactivating receptor, a novel C-type lectin immunoreceptor, acts as an activating receptor through association with Fc receptor gamma chain. J Biol Chem 278(35):32645–32652

Lambert AA, Gilbert C, Richard M, Beaulieu AD, Tremblay MJ (2008) The C-type lectin surface receptor DCIR acts as a new attachment factor for HIV-1 in dendritic cells and contributes to trans- and cis-infection pathways. Blood 112(4):1299–1307

Lee RT, Hsu TL, Huang SK, Hsieh SL, Wong CH, Lee YC (2011) Survey of immune-related, mannose/fucose-binding C-type lectin receptors reveals widely divergent sugar-binding specificities. Glycobiology 21(4):512–520

Lee HM, Yuk JM, Kim KH, Jang J, Kang G, Park JB, Son JW, Jo EK (2012) Mycobacterium abscessus activates the NLRP3 inflammasome via Dectin-1-Syk and p62/SQSTM1. Immunol Cell Biol 90(6):601–610

Maruhashi T, Kaifu T, Yabe R, Seno A, Chung SH, Fujikado N, Iwakura Y (2015) DCIR maintains bone homeostasis by regulating IFN-gamma production in T cells. J Immunol 194 (12):5681–5691

Matsumoto M, Tanaka T, Kaisho T, Sanjo H, Copeland NG, Gilbert DJ, Jenkins NA, Akira S (1999) A novel LPS-inducible C-type lectin is a transcriptional target of NF-IL6 in macrophages. J Immunol 163(9):5039–5048

Miyake Y, Masatsugu OH, Yamasaki S (2015) C-type lectin receptor MCL facilitates Mincle expression and signaling through complex formation. J Immunol 194(11):5366–5374

Norimoto A, Hirose K, Iwata A, Tamachi T, Yokota M, Takahashi K, Saijo S, Iwakura Y, Nakajima H (2014) Dectin-2 promotes house dust mite-induced T helper type 2 and type 17 cell differentiation and allergic airway inflammation in mice. Am J Respir Cell Mol Biol 51 (2):201–209

Olynych TJ, Jakeman DL, Marshall JS (2006) Fungal zymosan induces leukotriene production by human mast cells through a dectin-1-dependent mechanism. J Allergy Clin Immunol 118 (4):837–843

Parsons MW, Li L, Wallace AM, Lee MJ, Katz HR, Fernandez JM, Saijo S, Iwakura Y, Austen KF, Kanaoka Y, Barrett NA (2014) Dectin-2 regulates the effector phase of house dust mite-elicited pulmonary inflammation independently from its role in sensitization. J Immunol 192 (4):1361–1371

Puel A, Cypowyj S, Bustamante J, Wright JF, Liu L, Lim HK, Migaud M, Israel L, Chrabieh M, Audry M, Gumbleton M, Toulon A, Bodemer C, El-Baghdadi J, Whitters M, Paradis T, Brooks J, Collins M, Wolfman NM, Al-Muhsen S, Galicchio M, Abel L, Picard C, Casanova JL (2011) Chronic mucocutaneous candidiasis in humans with inborn errors of interleukin-17 immunity. Science 332(6025):65–68

Ritter M, Gross O, Kays S, Ruland J, Nimmerjahn F, Saijo S, Tschopp J, Layland LE, Prazeres da Costa C (2010) Schistosoma mansoni triggers Dectin-2, which activates the Nlrp3 inflammasome and alters adaptive immune responses. Proc Natl Acad Sci U S A 107 (47):20459–20464

Robinson MJ, Osorio F, Rosas M, Freitas RP, Schweighoffer E, Gross O, Verbeek JS, Ruland J, Tybulewicz V, Brown GD, Moita LF, Taylor PR, Reis e Sousa C (2009) Dectin-2 is a Syk-coupled pattern recognition receptor crucial for Th17 responses to fungal infection. J Exp Med 206(9):2037–2051

Saijo S, Iwakura Y (2011) Dectin-1 and Dectin-2 in innate immunity against fungi. Int Immunol 23 (8):467–472

Saijo S, Fujikado N, Furuta T, Chung SH, Kotaki H, Seki K, Sudo K, Akira S, Adachi Y, Ohno N, Kinjo T, Nakamura K, Kawakami K, Iwakura Y (2007) Dectin-1 is required for host defense against Pneumocystis carinii but not against Candida albicans. Nat Immunol 8(1):39–46

Saijo S, Ikeda S, Yamabe K, Kakuta S, Ishigame H, Akitsu A, Fujikado N, Kusaka T, Kubo S, Chung SH, Komatsu R, Miura N, Adachi Y, Ohno N, Shibuya K, Yamamoto N, Kawakami K, Yamasaki S, Saito T, Akira S, Iwakura Y (2010) Dectin-2 recognition of alpha-mannans and induction of Th17 cell differentiation is essential for host defense against Candida albicans. Immunity 32(5):681–691

Schoenen H, Bodendorfer B, Hitchens K, Manzanero S, Werninghaus K, Nimmerjahn F, Agger EM, Stenger S, Andersen P, Ruland J, Brown GD, Wells C, Lang R (2010) Cutting edge:

Mincle is essential for recognition and adjuvanticity of the mycobacterial cord factor and its synthetic analog trehalose-dibehenate. J Immunol 184(6):2756–2760

Tang C, Kamiya T, Liu Y, Kadoki M, Kakuta S, Oshima K, Hattori M, Takeshita K, Kanai T, Saijo S, Ohno N, Iwakura Y (2015) Inhibition of Dectin-1 signaling ameliorates colitis by inducing Lactobacillus-mediated regulatory T cell expansion in the intestine. Cell Host Microbe 18(2):183–197

Taylor PR, Brown GD, Reid DM, Willment JA, Martinez-Pomares L, Gordon S, Wong SY (2002) The beta-glucan receptor, dectin-1, is predominantly expressed on the surface of cells of the monocyte/macrophage and neutrophil lineages. J Immunol 169(7):3876–3882

Taylor PR, Reid DM, Heinsbroek SE, Brown GD, Gordon S, Wong SY (2005) Dectin-2 is predominantly myeloid restricted and exhibits unique activation-dependent expression on maturing inflammatory monocytes elicited in vivo. Eur J Immunol 35(7):2163–2174

Taylor PR, Tsoni SV, Willment JA, Dennehy KM, Rosas M, Findon H, Haynes K, Steele C, Botto M, Gordon S, Brown GD (2007) Dectin-1 is required for beta-glucan recognition and control of fungal infection. Nat Immunol 8(1):31–38

Taylor PR, Roy S, Leal SM Jr, Sun Y, Howell SJ, Cobb BA, Li X, Pearlman E (2014) Activation of neutrophils by autocrine IL-17A-IL-17RC interactions during fungal infection is regulated by IL-6, IL-23, RORgammat and dectin-2. Nat Immunol 15(2):143–151

Wells CA, Salvage-Jones JA, Li X, Hitchens K, Butcher S, Murray RZ, Beckhouse AG, Lo YL, Manzanero S, Cobbold C, Schroder K, Ma B, Orr S, Stewart L, Lebus D, Sobieszczuk P, Hume DA, Stow J, Blanchard H, Ashman RB (2008) The macrophage-inducible C-type lectin, mincle, is an essential component of the innate immune response to Candida albicans. J Immunol 180(11):7404–7413

Willment JA, Marshall AS, Reid DM, Williams DL, Wong SY, Gordon S, Brown GD (2005) The human beta-glucan receptor is widely expressed and functionally equivalent to murine Dectin-1 on primary cells. Eur J Immunol 35(5):1539–1547

Yamasaki S, Ishikawa E, Sakuma M, Hara H, Ogata K, Saito T (2008) Mincle is an ITAM-coupled activating receptor that senses damaged cells. Nat Immunol 9(10):1179–1188

Yamasaki S, Matsumoto M, Takeuchi O, Matsuzawa T, Ishikawa E, Sakuma M, Tateno H, Uno J, Hirabayashi J, Mikami Y, Takeda K, Akira S, Saito T (2009) C-type lectin Mincle is an activating receptor for pathogenic fungus, Malassezia. Proc Natl Acad Sci U S A 106 (6):1897–1902

Yonekawa A, Saijo S, Hoshino Y, Miyake Y, Ishikawa E, Suzukawa M, Inoue H, Tanaka M, Yoneyama M, Oh-Hora M, Akashi K, Yamasaki S (2014) Dectin-2 is a direct receptor for mannose-capped lipoarabinomannan of mycobacteria. Immunity 41(3):402–413

Zelensky AN, Gready JE (2005) The C-type lectin-like domain superfamily. FEBS J 272 (24):6179–6217

Zhu LL, Zhao XQ, Jiang C, You Y, Chen XP, Jiang YY, Jia XM, Lin X (2013) C-type lectin receptors Dectin-3 and Dectin-2 form a heterodimeric pattern-recognition receptor for host defense against fungal infection. Immunity 39(2):324–334

Chapter 27
Elucidation and Control of the Mechanisms Underlying Chronic Inflammation Mediated by Invariant Natural Killer T Cells

Hiroshi Watarai

Abstract Invariant natural killer T (iNKT) cells are a subpopulation of T lymphocytes with unique specificities against (glyco)lipids presented in the context of CD1d belonging to major histocompatibility complex (MHC) class Ib. They recognise microbially encoded or synthetic glycolipids directly through their CD1d-restricted T-cell receptors (TCRs) other than microbial components or lipids including pathogen-associated molecular patterns (PAMPs) and bacterial superantigens. Once activated, iNKT cells participate as early effectors and/or regulators of immune responses. Immunoregulatory cytokines produced in copious amounts by these cells target a wide range of downstream effector cells and help shape the ensuing immune responses. Mammals comprise various numbers and isoforms of CD1 molecules, suggesting that (glyco)lipid presentation is a rapidly evolving component of the immune system, which adapts to environmental threats. Recent progress in our understanding of CD1d-restricted iNKT cells contributes to their true potentials in immunotherapeutic applications for various diseases. Recent findings about iNKT cell subtypes (iNKT1, 2, 17, 10) and their roles in pathological inflammation are also introduced and discussed.

Keywords iNKT cells • Glycolipids • CD1d • TCR • Innate immunity • Adoptive immunity • Inflammation

27.1 Introduction

The immune system consists of separate but interacting components, the innate and acquired immune systems. Invariant natural killer T (iNKT) cells play unique intermediary roles, bridging between both immune systems.

H. Watarai (✉)
Division of Stem Cell Cellomics, Center for Stem Cells and Regenerative Medicine, Institute of Medical Science, The University of Tokyo, Tokyo, Japan

PRESTO, Japan Science and Technology Agency, Kawaguchi 332-0012, Japan
e-mail: hwatarai@ims.u-tokyo.ac.jp

M. Miyasaka, K. Takatsu (eds.), *Chronic Inflammation*,
DOI 10.1007/978-4-431-56068-5_27

NKT cells were originally characterised as having cell surface markers such as NK1.1 in mice (MacDonald 1995) and CD161 in humans (Schmidt et al. 1986), as well as a T-cell receptor (TCR) (Makino et al. 1995; Porcelli et al. 1993). However, recent studies have indicated that a certain but not all of the NKT cells express NK1.1 in mice or CD161 in humans, but all NKT cells are defined by a particular invariant TCR α chain from Trav11_Traj18 (Vα14-Jα18) mostly paired with Trbv13-2 (Vβ8.2), Trbv13-3 (Vβ8.1), Trbv29 (Vβ7), or Trbv1 (Vβ2) in mice and TRAV10_TRAJ18/TRBV25-1_TRBD2_TRBJ2-7 (Vα24-Jα18/Vβ11-Dβ2-Jβ2.7) in humans, giving rise to the moniker 'invariant' for NKT cells. Different from the conventional αβT cells which recognise the peptide/MHC complex, these stereotypic TCRs endow iNKT cells with some types of glycolipid antigens presented by monomorphic MHC class I-like molecule CD1d (Godfrey et al. 2004; Salio et al. 2014). The prototypic antigen is glycosphingolipid α-galactosylceramide (α-GalCer), originally isolated from the marine sponge *Agelas mauritianus* and identified from structure-activity relationship studies around Agelasphin 9b by the pharmaceutical division of Kirin Brewery Co. Ltd. in a screen for naturally occurring molecules that prevented tumour metastases in mice in vivo (Natori et al. 1994). The synthetic compound, also known as KRN7000, retains the activity of Agelasphin 9b while being much easier to synthesise (Morita et al. 1995). α-GalCer has been used in many different studies and is a highly potent iNKT cell agonist both in humans and in mice (Hayakawa et al. 2004).

When activated by α-GalCer, iNKT cells immediately produce Th1 (IFN-γ), Th2 (IL-13, IL-4), Th17 (IL-17A), and inhibitory (IL-10) cytokines, resulting in bystander immune modulating functions, activating and inhibiting various immune effector cells, including NK cells, macrophages, granulocytes, dendritic cells (DCs), basophils, and eosinophils in the innate system as well as $CD4^+$ T and $CD8^+$ T cells in the acquired system. Therefore, iNKT cells participate in the regulation of various disease states in a broad spectrum, including infection, autoimmunity, allergy, antitumour response, allograft rejection, and graft-versus-host disease.

The numerous studies investigating the role of iNKT cells were represented by mouse models of deficiency in iNKT cells. One makes use of mice deficient in CD1d (*Cd1d1*$^{-/-}$) (Mendiratta et al. 1997), which prevents the development of any CD1d-reactive T cells including iNKT cells. Another model directly targets Jα18 (*Traj18*$^{-/-}$) (Cui et al. 1997), which is required for iNKT-TCR formation. However, TCR repertoire diversity was impaired in *Traj18*$^{-/-}$ mice, in which the *PGK-Neo*^{r} cassette replaced from *Traj18* has inadvertent substantial effects on transcription and TCR rearrangements (Bedel et al. 2013). Consequently, any changes in immunological activity associated with *Traj18*$^{-/-}$ and *Cd1d1*$^{-/-}$ mice and for which a role has been ascribed to iNKT cells needs to be reassessed.

27.2 iNKT Cells and Host Defence

There is evidence that iNKT cells participate in protection of mice from a variety of bacterial, viral, protozoan, and fungal parasites, although some of these results are controversial because susceptibility is different between *Traj18*$^{-/-}$ and *Cd1d1*$^{-/-}$ mice or between C57BL/6 and BALB/c genetic backgrounds.

In some examples where the mechanism of iNKT cell action in host defence has been partially defined, IFN-γ secretion is important. iNKT cells act not only as effectors by themselves in these cases, but also as conductors of IFN-γ secretion from NK cells and CD8$^+$ T cells that have been stimulated as a result of iNKT cell activation (Hayakawa et al. 2001; Carnaud et al. 1999; Smyth et al. 2002a). However, iNKT cells can operate in either a stimulating or suppressive manner. For example, clearance of herpes simplex virus type 1 is reduced in *Traj18*$^{-/-}$ and *Cd1d1*$^{-/-}$ mice (Grubor-Bauk et al. 2003). By contrast, iNKT cells negatively regulate the response to lymphocytic choriomeningitis virus (Roberts et al. 2004).

A defect in iNKT cell development has been reported both in SAP (signalling lymphocytic activation molecule associated protein)-deficient mice and in humans with mutations in *SH2D1A*, the gene encoding SAP (Chung et al. 2005; Nichols et al. 2005; Pasquier et al. 2005). The *Sh2d1a*$^{-/-}$ mice are relatively normal in the development of T and B cells, whereas humans with SAP mutations are affected by a severe X-linked lymphoproliferative (XLP) syndrome and fail to control primary Epstein–Barr virus (EBV) infection, presenting with uncontrolled expansions of T and B cells. Although SAP-deficient XLP patients have a plethora of immune defects that could contribute to the lack of control of EBV infection, another group of XLP patients with defective iNKT cell numbers has been reported who bear mutations in the inhibitor of apoptosis XIAP gene (Rigaud et al. 2006). This raises questions of whether and how iNKT cells might be involved in the control of EBV and perhaps other viral infections.

Some microbes produce glycolipids different from viruses. iNKT cells are known to respond directly to some exogenous microbial glycolipids. *Sphingomonas* bacteria widely distributed in the environment are essentially universal antigens for iNKT cells and their glycosphingolipids containing either a galacturonic or a glucuronic acid have structures similar to α-GalCer and have an ability to activate iNKT cells directly (Kinjo et al. 2005; Mattner et al. 2005; Sriram et al. 2005).

Another example of the importance of iNKT cells is provided by a study of a Lyme disease model infected with *Borrelia burgadorferi*. After intradermal infection with *Borrelia burgadorferi*, *Cd1d1*$^{-/-}$ mice developed a greater thickening of the knee and tibiotarsal joints, indicative of arthritis compared to BALB/c mice (Kumar et al. 2000).

It has been suggested that some *Bacteroides* sphingolipids might influence host immune homeostasis. Species of *Bacteroides* and their relatives are a prevalent commensal organism of the human intestine, and many of them also have the capacity to produce sphingolipids. Recently, a glycosphingolipid antigen from *Bacteroides fragilis* has been purified and characterised and it also activates mouse and human iNKT cells (Wieland Brown et al. 2013).

The role of iNKT cells in infection and host defence is clearly an exciting area of investigation with clinical potential, which makes them important targets for future vaccine design.

27.3 iNKT Cells and Antitumour Responses

iNKT cells have been shown to be potentially important to tumour rejection (Smyth et al. 2002b). iNKT cells were found to be necessary for IL-12–mediated tumour therapy in mice (Cui et al. 1997), and α-GalCer/KRN7000 has an ability to promote iNKT cell-dependent rejection of a broad range of experimental tumour cell lines including melanoma, carcinoma, thymoma, and sarcoma (Hayakawa et al. 2004; Smyth et al. 2002b; Kawano et al. 1997). Recent evidence has also shown that treatment with α-GalCer/KRN7000 can protect against spontaneous oncogene- or carcinogen-induced primary tumour formation in mice (Hayakawa et al. 2003). Based on these findings in mice, a phase I clinical study of KRN7000 was carried out in patients with solid tumours (Giaccone et al. 2002). However, a detectable level of IFN-γ and IL-12 in serum was never observed after 3 times intravenous (i.v.) injection of KRN7000 (Giaccone et al. 2002). This may be partially due to the small population of iNKT cells in cancer patients compared to the healthy, although it is not clear whether this is a cause or effect of the cancer. Another possible reason is repeated injection of KRN7000 elicits changes in immunity as regulatory properties by IL-10 production from iNKT cells and dendritic cells (DCs) as shown in mice (Kojo et al. 2005).

Overcoming these problems has been demonstrated. Induced pluripotent stem cells (iPSCs) are known to hold tremendous potential for cell replacement therapy and functionally competent mouse iNKT cells from iPSCs were successfully generated that suppressed tumour growth in vivo (Watarai et al. 2010), which is an important advance in solving the problem of the limited number of iNKT cells in advanced cancer patients (Watarai et al. 2012a). Concerning the lesser effectiveness of KRN7000 in humans, i.v. injection of monocyte-derived DCs that were KRN7000-pulsed was evaluated and led to sustained expansion of iNKT cells in advanced cancer patients (Chang et al. 2005). iNKT-cell–mediated immunotherapy has clearly provided a proof of concept for future perspectives for cancer immunotherapy even though numerical tests and trials should be done and there is much to be learned before these cells can be safely and effectively manipulated in the clinic.

27.4 iNKT Cells and Autoimmunity

It has been suggested that iNKT cells naturally influence autoimmunity from some mouse models and the greatest number of studies was investigated in the type 1 diabetes model. iNKT cells are reduced in nonobese diabetic (NOD) mice (Ortaldo et al. 2004) and produce less IL-4 after stimulation (Hammond et al. 1998). Nonetheless, stimulation of iNKT cells with repeated administration of KRN7000 has a therapeutic effect and protection from diabetes by iNKT cells was associated with the induction of a Th2 response to islet autoantigens (Hong et al. 2001). Interestingly however, iNKT cells prevented differentiation of IFN-γ producing diabetogenic T cells in NOD mice in an iNKT-derived IL-4–independent manner but through activating NK cells and DCs (Beaudoin et al. 2002). Taken collectively, iNKT cells may act by inducing KRN7000-mediated anergy in islet-specific T cells and/or by altering the function of DCs with regulatory, in addition to Th2 polarisation.

Similar to the findings in NOD mice, KRN7000 was found to prevent experimental autoimmune encephalomyelitis (EAE; Jahng et al. 2001), an animal model of multiple sclerosis (MS), by shifting the balance from a pathogenic Th1 toward a Th2 response. A decrease in iNKT cells has also been found in the peripheral blood of MS patients, suggesting a regulatory role for iNKT cells in this human disease (Araki et al. 2003).

27.5 The Growing iNKT Cell Subsets

A major paradox is the ability of iNKT cells to produce various types of cytokines and to both promote and suppress immune responses as described above. This is probably due to the existence of functionally distinct subtypes of iNKT cells producing different cytokines. It was previously established that iNKT cells include both CD4-positive and CD4-negative subtypes (Godfrey et al. 2004; Bendelac et al. 2007), each of which produces different cytokines. Although a functional distinction was originally less apparent for mouse $CD4^+$ and $CD4^-$ iNKT cell subsets, the division of mouse iNKT cells on the basis of NK1.1 expression has revealed striking differences between NK1.1-positive and NK1.1-negative subsets. $NK1.1^+$ iNKT cells produce a large amount of IFN-γ and little IL-4, whereas $NK1.1^-$ iNKT cells produce less IFN-γ and more IL-4 (Benlagha et al. 2002; Pellicci et al. 2002).

More recently, there is further heterogeneity in $CD4^+$ iNKT cells, a subset of which bears IL-17 receptor B (also known as IL-25 receptor; Terashima et al. 2008). IL-25 is a key cytokine in Th2 immunity that eliminates helminth and promotes airway hyperreactivity (Fallon et al. 2006). $IL\text{-}17RB^+$ $CD4^+$ iNKT cells which can be designated iNKT2 cells produce large amounts of IL-13, IL-9, and IL-4 in response to IL-25 (Terashima et al. 2008; Angkasekwinai et al. 2010;

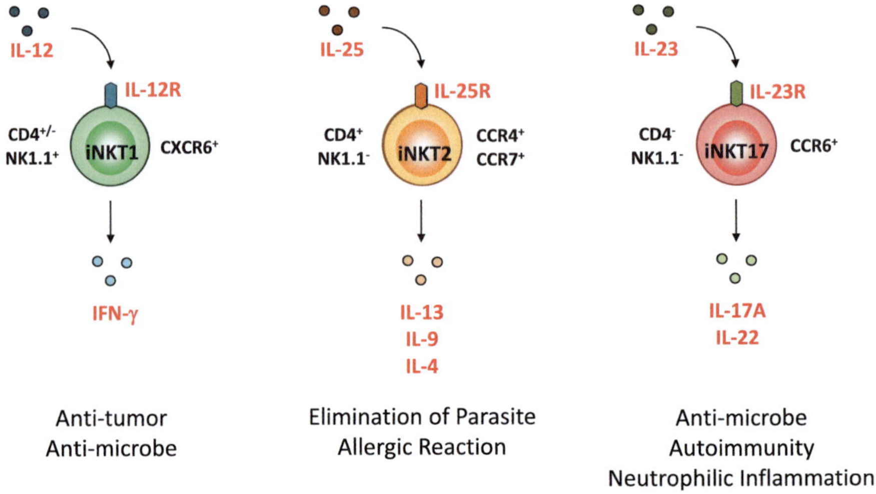

Fig. 27.1 Classification of iNKT cell subtypes based on the phenotypical and functional differences

Motomura et al. 2011) and are phenotypically NK1.1^{-} and present mainly in thymus, spleen, lung, and lymph nodes (Watarai et al. 2012b). Another newly identified subset of iNKT cells is producing predominant IL-17A (Michel et al. 2007). IL-17A is produced from Th17 cells and/or γδ T cells, which have a key role in autoimmune pathology in disease models such as EAE and collagen-induced arthritis (CIA; Weaver et al. 2007). IL-17A–producing iNKT cells (iNKT17 cells) express the transcription factor RORγt, IL-23 receptor, and CCR6, the same as Th17 cells (Michel et al. 2008) and are enriched in lymph nodes and the skin, resulting in rapid production of IL-17A in response to inflammation (Watarai et al. 2012b; Doisne et al. 2009, 2011; Fig. 27.1). Concerning the IL-10 producing iNKT cells with regulatory property, it is still unclear whether this type of cell exists in steady state. In regard to the α-GalCer/KRN7000-induced protection from EAE, iNKT cells have been reported to regulate immune responses by inducing IL-10 production by other cells or to produce IL-10 themselves (Kojo et al. 2005; Singh et al. 2001). However, in these cases, a distinct population of iNKT cells with a unique phenotype was not described. Recently, IL-10–producing iNKT cells were determined from mice pretreated with α-GalCer retaining cytotoxic activity and maintaining the ability to respond to TCR-dependent as well as -independent cytokine-mediated stimulation, phenotypically and functionally similar to inducible regulatory T cells (Tregs; Sag et al. 2014). Furthermore, it has been reported that early after α-GalCer injection, iNKT cells acquire a T follicular helper (Tfh) phenotype, characterised by the expression of the transcription factor BCL6 (Chang et al. 2011; King et al. 2011). Taken collectively, iNKT cells consist of functionally distinct subtypes rather than a multifunctional uniform population, similar to helper T cells or innate lymphoid cells (ILCs). It is important to elucidate the complexity, stability, plasticity, and the role of iNKT subsets in immunity and inflammation, including crosstalk with other immune cells and subtypes.

27.6 iNKT-Cell–Mediated Type 2 Inflammation

iNKT cells are also involved in the development of asthma, because *Traj18*$^{-/-}$ mice fail to develop or reduce ovalbumin (OVA) antigen-induced airway hyperreactivity (AHR) (Akbari et al. 2003). Th2 cells and ILC2 are not always essential for iNKT cell-mediated AHR development, because activation of iNKT cells induces AHR in the absence of CD4^{+} T cells in MHC class-II–deficient mice (Meyer et al. 2006). These findings suggest that iNKT cells are directly involved in the development of AHR, independent of Th2 response or ILC2 activation in some cases. IL-17RB expressing iNKT2 cells and IL-25, a ligand for IL-17RB, might be responsible for iNKT-cell–mediated OVA/IL-25–induced AHR (Terashima et al. 2008). IL-17RB is preferentially expressed in iNKT2 cells among cells in steady state. iNKT2 cells in the lung produce predominantly IL-13 upon stimulation with IL-25 in vitro. Depletion of iNKT2 cells but not of other subtypes by IL-17RB–specific monoclonal antibodies or *Traj18*$^{-/-}$ mice failed to develop OVA/IL-25–dependent AHR. Cell transfer of iNKT2 cells into *Traj18*$^{-/-}$ mice also successfully reconstituted AHR induction. These results strongly suggest that iNKT2 cells play a crucial role in the pathogenesis of allergen-induced asthma. Recently, IL-25–mediated mucosal inflammation appeared to be negatively regulated by IL-17B which antagonises shared receptor IL-17RB (Reynolds et al. 2015). It should be investigated whether iNKT2-mediated IL-25–dependent inflammation is also regulated by IL-17B (Fig. 27.2).

Not only allergens but also certain viruses, such as respiratory syncytial virus (RSV), Sendai virus (SeV), metapneumovirus, and parainfluenza virus, cause childhood asthma and chronic obstructive pulmonary disease (COPD)-like symptoms, which include AHR, airway inflammation, and mucus hypersecretion (Gern & Busse 2000; Sigurs et al. 2005; Hamelin et al. 2006). It has been reported that, in a mouse model of infection with a SeV, virus-induced chronic inflammation leads to asthma that resembles human asthma and COPD even long after the apparent clearance of viruses (Kim et al. 2008). The chronic pulmonary symptoms evolved independently of CD4^{+} T cells but required iNKT cells, leading to accumulation of M2 macrophage resulted in IL-13–dependent lung disease (Kim et al. 2008). The link between these events needs to be better defined as to whether iNKT2 cells are involved; it was found that TREM (triggering receptor expressed on myeloid cells)-2 promotes macrophage survival and lung disease after viral infection (Wu et al. 2015).

iNKT2 cells potentially trigger respiratory syncytial virus (RSV)-induced AHR (Watarai et al. 2012b). RSV is an unusual virus in that it can cause repeated reinfections throughout life and a major viral pathogen causing extensive outbreaks of respiratory tract infections both in the very young and in vulnerable adults.

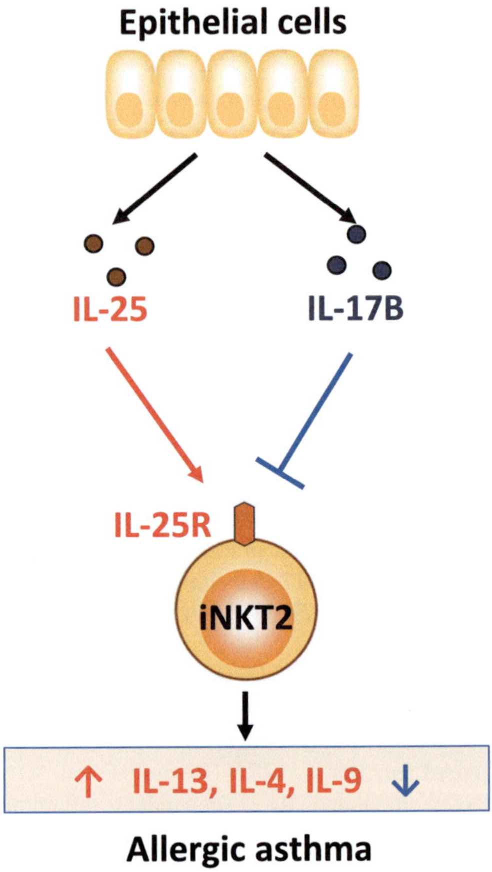

Fig. 27.2 Proposed model for IL-25R-dependent iNKT2 activation and suppression

iNKT2 cells produce Th2 cytokine after RSV infection, although triggering molecular mechanisms are still unclear.

27.7 Conclusion

iNKT cells have been characterised as a T-cell subtype with diverse and miscellaneous functions that recognise (glyco)lipids. However, recent studies revealed that iNKT cells consist of functionally and phenotypically distinct subtypes, each of which mediates different types of immunity and inflammation. iNKT cells can regulate the function of innate and adaptive immune cells in antimicrobial immunity, tumour rejection, and inflammatory diseases. Further studies are warranted to dissect the molecular mechanisms by which iNKT subtypes can modulate both pro- and anti-inflammatory responses. The challenge remains to translate the iNKT cell biology to the clinic.

References

Akbari O, Stock P, Meyer E, Kronenberg M, Sidobre S, Nakayama T, Taniguchi M, Grusby MJ, DeKruyff RH, Umetsu DT (2003) Essential role of NKT cells producing IL-4 and IL-13 in the development of allergen-induced airway hyperreactivity. Nat Med 9:582–588

Angkasekwinai P, Chang SH, Thapa M, Watarai H, Dong C (2010) Regulation of IL-9 expression by IL-25 signaling. Nat Immunol 11:250–256

Araki M, Kondo T, Gumperz JE, Brenner MB, Miyake S, Yamamura T (2003) Th2 bias of CD4+ NKT cells derived from multiple sclerosis in remission. Int Immunol 15:279–288

Beaudoin L, Laloux V, Novak J, Lucas B, Lehuen A (2002) NKT cells inhibit the onset of diabetes by impairing the development of pathogenic T cells specific for pancreatic beta cells. Immunity 17:725–736

Bedel R, Matsuda JL, Brigl M, White J, Kappler J, Marrack P, Gapin L (2013) Lower TCR repertoire diversity in Traj18-deficient mice. Nat Immunol 13:705–706

Bendelac A, Savage PB, Teyton L (2007) The biology of NKT cells. Annu Rev Immunol 25:297–336

Benlagha K, Kyin T, Beavis A, Teyton L, Bendelac A (2002) A thymic precursor to the NK T cell lineage. Science 296:553–555

Carnaud C, Lee D, Donnars O, Park SH, Beavis A, Koezuka Y, Bendelac A (1999) Cutting edge: Cross-talk between cells of the innate immune system: NKT cells rapidly activate NK cells. J Immunol 163:4647–4650

Chang DH, Osman K, Connolly J, Kukreja A, Krasovsky J, Pack M, Hutchinson A, Geller M, Liu N, Annable R, Shay J, Kirchhoff K, Nishi N, Ando Y, Hayashi K, Hassoun H, Steinman RM, Dhodapkar MV (2005) Sustained expansion of NKT cells and antigen-specific T cells after injection of alpha-galactosyl-ceramide loaded mature dendritic cells in cancer patients. J Exp Med 201:1503–1517

Chang P-P, Barral P, Fitch J, Pratama A, Ma CS, Kallies A, Hogan JJ, Cerundolo V, Tangye SG, Bittman R, Nutt SL, Brink R, Godfrey DI, Batista FD, Vinuesa CG (2011) Identification of Bcl-6-dependent follicular helper NKT cells that provide cognate help for B cell responses. Nat Immunol 13:35–43

Chung B, Aoukaty A, Dutz J, Terhorst C, Tan R (2005) Signaling lymphocytic activation moleculeassociated protein controls NKT cell functions. J Immunol 174:3153–3157

Cui J, Shin T, Kawano T, Sato H, Kondo E, Toura I, Kaneko Y, Koseki H, Kanno M, Taniguchi M (1997) Requirement for Valpha14 NKT cells in IL-12-mediated rejection of tumors. Science 278:1623–1626

Doisne JM, Becourt C, Amniai L, Duarte N, Le Luduec JB, Eberl G, Benlagha K (2009) Skin and peripheral lymph node invariant NKT cells are mainly retinoic acid receptor-related orphan receptor γt+and respond preferentially under inflammatory conditions. J Immunol 183:2142–2149

Doisne JM, Soulard V, Becourt C, Amniai L, Henrot P, Havenar-Daughton C, Blanchet C, Zitvogel L, Ryffel B, Cavaillon JM, Marie JC, Couillin I, Benlagha K (2011) Cutting edge: crucial role of IL-1 and IL-23 in the innate IL-17 response of peripheral lymph node NK1.1− invariant NKT cells to bacteria. J Immunol 186:662–666

Fallon PG, Ballantyne SJ, Mangan NE, Barlow JL, Dasvarma A, Hewett DR, McIlgorm A, Jolin HE, McKenzie AN (2006) Identification of an interleukin (IL)-25-dependent cell population that provides IL-4, IL-5, and IL-13 at the onset of helminth expulsion. J Exp Med 203:1105–1116

Gern JE, Busse WW (2000) The role of viral infections in the natural history of asthma. J Allergy Clin Immunol 106:201–212

Giaccone G, Punt CJ, Ando Y, Ruijter R, Nishi N, Peters M, von Blomberg BM, Scheper RJ, van der Vliet HJ, van den Eertwegh AJ, Roelvink M, Beijnen J, Zwierzina H, Pinedo HM (2002) A phase I study of the natural killer T-cell ligand alpha-galactosylceramide (KRN7000) in patients with solid tumors. Clin Cancer Res 8:3702–3709

Godfrey DI, MacDonald HR, Kronenberg M, Smyth MJ, Van Kaer L (2004) NKT cells: what's in a name? Nat Rev Immunol 4:231–237

Grubor-Bauk B, Simmons A, Mayrhofer G, Speck PG (2003) Impaired clearance of herpes simplex virus type 1 from mice lacking CD1d or NKT cells expressing the semivariant V alpha 14-J alpha 281 TCR. J Immunol 170:1430–1434

Hamelin ME, Prince GA, Gomez AM, Kinkead R, Boivin G (2006) Human metapneumovirus infection induces long-term pulmonary inflammation associated with airway obstruction and hyperresponsiveness in mice. J Infect Dis 193:1634–1642

Hammond KJ, Poulton LD, Palmisano LJ, Silveira PA, Godfrey DI, Baxter AG (1998) Alpha/beta-T cell receptor (TCR)+CD4-CD8- (NKT) thymocytes prevent insulin-dependent diabetes mellitus in nonobese diabetic (NOD/Lt) mice by the influence of interleukin (IL)-4 and/or IL-10. J Exp Med 187:1047–1056

Hayakawa Y, Takeda K, Yagita H, Kakuta S, Iwakura Y, Van Kaer L, Saiki I, Okumura K (2001) Critical contribution of IFN-gamma and NK cells, but not perforin-mediated cytotoxicity, to anti-metastatic effect of alpha-galactosylceramide. Eur J Immunol 31:1720–1727

Hayakawa Y, Rovero S, Forni G, Smyth MJ (2003) Alpha-galactosylceramide (KRN7000) suppression of chemical- and oncogene-dependent carcinogenesis. Proc Natl Acad Sci U S A 100:9464–9469

Hayakawa Y, Godfrey DI, Smyth MJ (2004) Alpha-galactosylceramide: potential immunomodulatory activity and future application. Curr Med Chem 11:241–252

Hong S, Wilson MT, Serizawa I, Wu L, Singh N, Naidenko OV, Miura T, Haba T, Scherer DC, Wei J, Kronenberg M, Koezuka Y, Van Kaer L (2001) The natural killer T-cell ligand alpha-galactosylceramide prevents autoimmune diabetes in non-obese diabetic mice. Nat Med 7:1052–1056

Jahng AW, Maricic I, Pedersen B, Burdin N, Naidenko O, Kronenberg M, Koezuka Y, Kumar V (2001) Activation of natural killer T cells potentiates or prevents experimental autoimmune encephalomyelitis. J Exp Med 194:1789–1799

Kawano T, Cui J, Koezuka Y, Toura I, Kaneko Y, Motoki K, Ueno H, Nakagawa R, Sato H, Kondo E, Koseki H, Taniguchi M (1997) CD1d-restricted and TCR-mediated activation of valpha14 NKT cells by glycosylceramides. Science 278:1626–1629

Kim EY, Battaile JT, Patel AC, You Y, Agapov E, Grayson MH, Benoit LA, Byers DE, Alevy Y, Tucker J, Swanson S, Tidwell R, Tyner JW, Morton JD, Castro M, Polineni D, Patterson GA, Schwendener RA, Allard JD, Peltz G, Holtzman MJ (2008) Persistent activation of an innate immune response translates respiratory viral infection into chronic lung disease. Nat Med 14:633–640

King IL, Amiel E, Tighe M, Mohrs K, Veerapen N, Besra G, Mohrs M, Leadbetter EA (2011) Invariant natural killer T cells direct B cell responses to cognate lipid antigen in an IL-21-dependent manner. Nat Immunol 13:44–50

Kinjo Y, Wu D, Kim G, Xing GW, Poles MA, Ho DD, Tsuji M, Kawahara K, Wong CH, Kronenberg M (2005) Recognition of bacterial glycosphingolipids by natural killer T cells. Nature 434:520–525

Kojo S, Seino K, Harada M, Watarai H, Wakao H, Uchida T, Nakayama T, Taniguchi M (2005) Induction of regulatory properties in dendritic cells by Valpha14 NKT cells. J Immunol 175:3648–3655

Kumar H, Belperron A, Barthold SW, Bockenstedt LK (2000) Cutting edge: CD1d deficiency impairs murine host defense against the spirochete, Borrelia burgdorferi. J Immunol 165:4797–4801

MacDonald HR (1995) $NK1.1^{+}$ T cell receptor-alpha/$beta^{+}$ cells: new clues to their origin, specificity, and function. J Exp Med 182:633–638

Makino Y, Kanno R, Ito T, Higashino K, Taniguchi M (1995) Predominant expression of invariant $V\alpha14^{+}$ TCR α chain in $NK1.1^{+}$ T cell populations. Int Immunol 7:1157–1161

Mattner J, Debord KL, Ismail N, Goff RD, Cantu C 3rd, Zhou D, Saint-Mezard P, Wang V, Gao Y, Yin N, Hoebe K, Schneewind O, Walker D, Beutler B, Teyton L, Savage PB, Bendelac A

(2005) Exogenous and endogenous glycolipid antigens activate NKT cells during microbial infections. Nature 434:525–529

Mendiratta SK, Martin WD, Hong S, Boesteanu A, Joyce S, Van Kaer L (1997) CD1d1 mutant mice are deficient in natural T cells that promptly produce IL-4. Immunity 6:469–477

Meyer EH, Goya S, Akbari O, Berry GJ, Savage PB, Kronenberg M, Nakayama T, DeKruyff RH, Umetsu DT (2006) Glycolipid activation of invariant T cell receptor + NKT cells is sufficient to induce airway hyperreactivity independent of conventional CD4 + T cells. Proc Natl Acad Sci U S A 103:2782–2787

Michel ML, Keller AC, Paget C, Fujio M, Trottein F, Savage PB, Wong CH, Schneider E, Dy M, Leite-de-Moraes MC (2007) Identification of an IL-17-producing NK1.1− iNKT cell population involved in airway neutrophilia. J Exp Med 204:995–1001

Michel ML, Mendes-da-Cruz D, Keller AC, Lochner M, Schneider E, Dy M, Eberl G, Leite-de-Moraes MC (2008) Critical role of ROR-γt in a new thymic pathway leading to IL-17-producing invariant NKT cell differentiation. Proc Natl Acad Sci U S A 105:19845–19850

Morita M, Motoki K, Akimoto K, Natori T, Sakai T, Sawa E, Yamaji K, Koezuka Y, Kobayashi E, Fukushima H (1995) Structure-activity relationship of alpha-galactosylceramides against B16-bearing mice. J Med Chem 38:2176–2187

Motomura Y, Kitamura H, Hijikata A, Matsunaga Y, Matsumoto K, Inoue H, Atarashi K, Hori S, Watarai H, Zhu J, Taniguchi M, Kubo M (2011) The transcription factor E4BP4 regulates the production of IL-10 and IL-13 in CD4+ T cells. Nat Immunol 12:450–459

Natori T, Morita M, Akimoto K, Koezuka Y (1994) Agelasphins, novel antitumor and immunostimulatory cerebrosides from the marine sponge Agelas mauritianus. Tetrahedron 50:2771–2784

Nichols KE, Hom J, Gong SY, Ganguly A, Ma CS, Gong SY, Ganguly A, Ma CS, Cannons JL, Tangye SG, Schwartzberg PL, Koretzky GA, Stein PL (2005) Regulation of NKT cell development by SAP, the protein defective in XLP. Nat Med 11:340–345

Ortaldo JR, Young HA, Winkler-Pickett RT, Bere EW Jr, Murphy WJ, Wiltrout RH (2004) Dissociation of NKT stimulation, cytokine induction, and NK activation in vivo by the use of distinct TCR-binding ceramides. J Immunol 172:943–953

Pasquier B, Yin L, Fondaneche MC, Relouzat F, Bloch-Queyrat C, Lambert N, Fischer A, de Saint-Basile G, Latour S (2005) Defective NKT cell development in mice and humans lacking the adapter SAP, the X-linked lymphoproliferative syndrome gene product. J Exp Med 201:695–701

Pellicci DG, Uldrich AP, Kyparissoudis K, Crowe NY, Brooks AG, Hammond KJ, Sidobre S, Kronenberg M, Smyth MJ, Godfrey DI (2002) A natural killer T (NKT) cell developmental pathway involving a thymus-dependent NK1.1− CD4+ CD1d-dependent precursor stage. J Exp Med 195:835–844

Porcelli S, Yockey CE, Brenner MB, Balk SP (1993) Analysis of T cell antigen receptor (TCR) expression by human peripheral blood CD4-8-alpha/beta T cells demonstrates preferential use of several V beta genes and an invariant TCR alpha chain. J Exp Med 178:1–16

Reynolds JM, Lee YH, Shi Y, Wang X, Angkasekwinai P, Nallaparaju KC, Flaherty S, Chang SH, Watarai H, Dong C (2015) Interleukin-17B antagonizes interleukin-25-mediated mucosal inflammation. Immunity 42:692–703

Rigaud S, Fondaneche MC, Lambert N, Pasquier B, Mateo V, Soulas P, Galicier L, Le Deist F, Rieux-Laucat F, Revy P, Fischer A, de Saint BG, Latour S (2006) XIAP deficiency in humans causes an X-linked lymphoproliferative syndrome. Nature 444:110–114

Roberts TJ, Lin Y, Spence PM, Van Kaer L, Brutkiewicz RR (2004) CD1d1-dependent control of the magnitude of an acute antiviral immune response. J Immunol 172:3454–3461

Sag D, Krause P, Hedrick CC, Kronenberg M, Wingender G (2014) IL-10-producing NKT10 cells are a distinct regulatory invariant NKT cell subset. J Clin Invest 124:3725–3740

Salio M, Silk JD, Jones EY, Cerundolo V (2014) Biology of CD1- and MR1-restricted T cells. Annu Rev Immunol 32:323–366

Schmidt RE, Murray C, Daley JF, Schlossman SF, Ritz J (1986) A subset of natural killer cells in peripheral blood displays a mature T cell phenotype. J Exp Med 164:351–356

Sigurs N, Gustafsson PM, Bjarnason R, Lundberg F, Schmidt S, Sigurbergsson F, Kjellman B (2005) Severe respiratory syncytial virus bronchiolitis in infancy and asthma and allergy at age 13. Am J Respir Crit Care Med 171:137–141

Singh AK, Wilson MT, Hong S, Olivares-Villagómez D, Du C, Stanic AK, Joyce S, Sriram S, Koezuka Y, Van Kaer L (2001) Natural killer T cell activation protects mice against experimental autoimmune encephalomyelitis. J Exp Med 194:1801–1811

Smyth MJ, Crowe NY, Pellicci DG, Kyparissoudis K, Kelly JM, Takeda K, Yagita H, Godfrey DI (2002a) Sequential production of interferon-gamma by NK1.1(+) T cells and natural killer cells is essential for the antimetastatic effect of alpha-galactosylceramide. Blood 99:1259–1266

Smyth MJ, Crowe NY, Takeda K, Yagita H, Godfrey DI (2002b) NKT cells – conductors of tumor immunity? Curr Opin Immunol 14:165–171

Sriram V, Du W, Gervay-Hague J, Brutkiewicz RR (2005) Cell wall glycosphingolipids of Sphingomonas paucimobilis are CD1d-specific ligands for NKT cells. Eur J Immunol 35:1692–1701

Terashima A, Watarai H, Inoue S, Sekine E, Nakagawa R, Hase K, Iwamura C, Nakajima H, Nakayama T, Taniguchi M (2008) A novel subset of mouse NKT cells bearing the IL-17 receptor B responds to IL-25 and contributes to airway hyperreactivity. J Exp Med 205:2727–2733

Watarai H, Fujii S, Yamada D, Rybouchkin A, Sakata S, Nagata Y, Iida-Kobayashi M, Sekine-Kondo E, Shimizu K, Shozaki Y, Sharif J, Matsuda M, Mochiduki S, Hasegawa T, Kitahara G, Endo TA, Toyoda T, Ohara O, Harigaya K, Koseki H, Taniguchi M (2010) Murine induced pluripotent stem cells can be derived from and differentiate into natural killer T cells. J Clin Invest 120:2610–2618

Watarai H, Yamada D, Fujii S, Taniguchi M, Koseki H (2012a) Induced pluripotency as a potential path towards iNKT cell-mediated cancer immunotherapy. Int J Hematol 95:624–631

Watarai H, Sekine-Kondo E, Shigeura T, Motomura Y, Yasuda T, Satoh R, Yoshida H, Kubo M, Kawamoto H, Koseki H, Taniguchi M (2012b) Development and function of invariant natural killer T cells producing Th2- and Th17-cytokines. PLoS Biol 10, e1001255

Weaver CT, Hatton RD, Mangan PR, Harrington LE (2007) IL-17 family cytokines and the expanding diversity of effector T cell lineages. Annu Rev Immunol 25:821–852

Wieland Brown LC, Penaranda C, Kashyap PC, Williams BB, Clardy J, Kronenberg M, Sonnenburg JL, Comstock LE, Bluestone JA, Fischbach MA (2013) Production of α-galactosylceramide by a prominent member of the human gut microbiota. PLoS Biol 11, e1001610

Wu K, Byers DE, Jin X, Agapov E, Alexander-Brett J, Patel AC, Cella M, Gilfilan S, Colonna M, Kober DL, Brett TJ, Holtzman MJ (2015) TREM-2 promotes macrophage survival and lung disease after respiratory viral infection. J Exp Med 212:681–697

Chapter 28
Understanding of the Role of Plasmacytoid Dendritic Cells in the Control of Inflammation and T-Cell Immunity

Katsuaki Sato

Abstract Dendritic cells (DCs) comprise several subsets that are critically involved in the initiation and regulation of innate and adaptive immunity. Plasmacytoid DCs (pDCs) are specialized cells in robust type I interferon (IFN-I) secretion by engagement of endosomal toll-like receptors (TLRs), and uniquely express sialic acid binding Ig-like lectin (Siglec)-H. However, how pDCs control immune responses in vivo remains unclear. Here we show a critical role of pDCs in the regulation of inflammation and T-cell immunity in vivo using knock-in mice with a conditional ablation of pDCs. pDC-ablated mice exhibited the reduced inflammatory responses triggered by TLR9 ligand. Upon antigenic immunization, pDC-ablated mice showed not only reduced antigen-specific $CD4^+$ T-cell responses but also impaired generation of cytotoxic T lymphocytes (CTLs) under the TLR9-mediated inflammatory conditions. Thus, our findings highlight previously unidentified roles of pDCs in the control of innate and adaptive immunity.

Keywords Dendritic cells • Innate immunity • Inflammation • Cytokine • Adaptive immunity • T cells

28.1 Introduction

Dendritic cells (DCs) are essential antigen (Ag)-presenting cells (APCs) that consist of heterogeneous subsets, mainly classified as conventional DCs (cDCs) and plasmacytoid DCs (pDCs) (Sato and Fujita 2007, Shortman and Naik 2007; Villadangos and Schnorrer 2007). DCs serve as sentinels, recognising the presence of invading pathogens or virus-infected cells through various pattern-recognition

K. Sato (✉)
Division of Immunology, Department of Infectious Diseases, Faculty of Medicine, University of Miyazaki, 5200 Kihara, Kiyotake, Miyazaki 889-1692, Japan

PRESTO, Japan Science and Technology Agency, Kawaguchi 332-0012, Japan
e-mail: katsuaki_sato@med.miyazaki-u.ac.jp

M. Miyasaka, K. Takatsu (eds.), *Chronic Inflammation*,
DOI 10.1007/978-4-431-56068-5_28

receptors (Sato and Fujita 2007; Shortman and Naik 2007; Villadangos and Schnorrer 2007). DCs process such exogenous Ags intracellularly and present them to $CD4^+$ T cells via major histocompatibility complex class II (MHC II) for induction of $CD4^+$ effector T ($CD4^+$ T_{eff}) cells (Dudziak et al. 2007; Villadangos and Schnorrer 2007; Hildner et al 2008). DCs also show an unusual specialisation in their MHC class I (MHC I) presentation pathway to prime $CD8^+$ T cells. Although most cells use their MHC I to present peptides derived from endogenously synthesised proteins, DCs have the capacity to deliver exogenous antigens to the (MHC I) pathway, a phenomenon known as cross-presentation, that underlies the generation of cytotoxic T lymphocyte (CTL) immunity (Dudziak et al. 2007; Villadangos and Schnorrer 2007; Hildner et al 2008). DCs thereby play a crucial role in the link between innate and adaptive immunity.

pDCs represent a specialised DC population that is capable of producing large amounts of type I interferon (IFN-I) by sensing single-stranded RNA or unmethylated CpG DNA through endosomal toll-like receptor (TLR) 7 and 9, respectively (Gilliet et al. 2008; Swiecki and Colonna 2010). Upon binding of their ligands, both TLRs recruit a cytoplasmic adaptor MyD88 and initiate downstream signalling pathways, in which the activation of nuclear factor kB (NF-kB) is involved in the production of proinflammatory cytokines, whereas the activation of inhibitor of NF-kB (IkB) kinase (IKK)-α leading to phosphorylation and nuclear translocation of IFN regulatory factor (IRF)-7 is critical for the production of IFN-I (Kaisho and Tanaka 2008).

pDCs are believed to provide an initial line of host defense against viral infection because of their capacity to produce IFN-I (Gilliet et al. 2008; Swiecki and Colonna 2010). In addition, pDCs may exert a pleiotropic activating or inhibitory role in the regulation of T-cell–mediated adaptive immune responses and immunopathogenesis (Salio et al. 2004; Ochando et al. 2006; Lande et al. 2007; Irla et al. 2010). Although pDCs are identified by the combination of multiple cell surface molecules such as Gr-1 (Asselin-Paturel et al. 2001; Yoneyama et al. 2005) or bone marrow stromal antigen 2 (BST-2; Krug et al. 2004; Yoneyama et al. 2005; Banchereau and Pascual 2006), sialic acid-binding immunoglobulin (Ig)-like lectin-H (Siglec-H) is predominantly found on the cell surface of pDCs in lymphoid organs (Blasius and Colonna 2006; Zhang et al. 2006; Puttur et al. 2013). To evaluate the contribution of pDCs to the immune system precisely, we have used a diphtheria toxin receptor (DTR)-based approach (Jung et al. 2002; Kissenpfennig et al. 2005) targeting Siglec-H to produce *Siglech*$^{dtr/dtr}$ knock-in (KI) mice that allows inducible in vivo selective ablation of pDCs (referred to as pDC-ablated mice; Takagi et al. 2011). Using these mice, we provided the evidence that pDCs play a crucial role in the regulation of inflammation and T-cell immunity in vivo.

28.2 Results

28.2.1 Inducible Ablation of pDCs in Mice

Siglech$^{dtr/dtr}$ mice were born at the expected frequencies, and homozygous mice were healthy. To validate the *Siglech*$^{dtr/dtr}$ mice in terms of DT-induced pDC elimination, *Siglech*$^{dtr/dtr}$ mice received a single injection of DT (pDC-ablated mice), and we monitored 2 days later the subsequent frequency of pDCs among leukocytes in spleen (Spl), peripheral lymph nodes (PLNs), and bone marrow (BM). Treatment with DT at 1 μg/mouse almost completely ablated pDCs among leukocytes in pDC-ablated mice within 24 h after initial DT treatment, whereas this treatment had no effect in wild-type (WT) mice (see Fig. 28.1a). Histological analysis of pDC-ablated mice confirmed that pDCs disappeared following the injection with DT, whereas it had no effect on the localisation of CD11c^{high} cDCs (Fig. 28.1b).

Therefore, pDC-ablated mice provide the possibility to analyse the effect of conditional ablation of pDCs on immune responses in vivo.

28.2.2 pDCs Exacerbates TLR9-Mediated Systemic Inflammatory Responses in vivo

We addressed the roles of pDCs in the TLR9-mediated inflammatory response in vivo (Fig. 28.2). After administration of CpG-A known as TLR9 ligand (Iwasaki

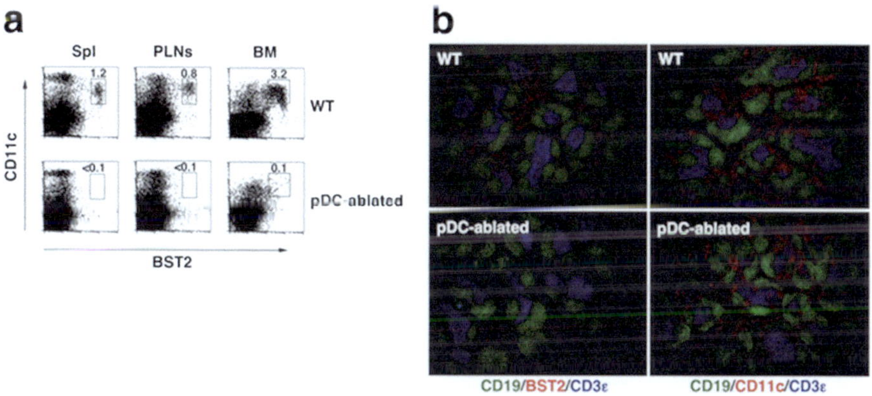

Fig. 28.1 Conditional ablation of pDCs in *Siglech*$^{dtr/dtr}$ mice. (**a**) The frequency of CD11c$^+$BST2$^+$ pDCs in Spl, PLNs, and BM obtained from WT mice and pDC-ablated mice was analysed by flow cytometry. Data are represented by a dot plot, and numbers represent the proportion of CD11c$^+$ BST2$^+$ cells among leukocytes in each quadrant. (**b**) Immunofluorescent microscopic analysis was performed on frozen horizontal sections of Spl obtained from WT mice and pDC-ablated mice. Sections were stained for CD19 (*green*), CD3ε (*blue*), and BST2 (*red*) or CD11c (*red*). All data are representative of at least three independent experiments

and Medzhitov 2004) plus D-galactosamine (D-GalN), WT mice died within 24 h after the administration of CpG-A plus D-GalN with a marked elevation in serum concentrations of IFN-α, IFN-β, interleukin (IL)-6, and IL-12p40. In contrast, pDC-ablated mice were more resistant to the TLR9-mediated lethality than WT mice, which was accompanied by a dramatic reduction in serum cytokine levels.

Taken together, these results indicate that pDCs are required for the initiation of the TLR9-mediated systemic inflammatory response in vivo.

28.2.3 pDCs Participate in the Induction of T-Cell Responses in vivo

We examined the roles of pDCs in the induction of Ag-specific response of $CD4^+$ T cells in vivo under TLR9-mediated inflammatory conditions (Fig. 28.3). Following immunisation with ovalbumin (OVA) protein plus CpG-A, $CD4^+$ T cells obtained from pDC-ablated mice showed lower proliferation and production of IFN-γ as well as frequency of T helper $1(T_H1)$ cells than did $CD4^+$ T cells from WT mice.

Collectively, these results indicate that pDCs are involved in the induction of Ag-specific $CD4^+$ T-cell responses in vivo.

We also examined the generation of CTLs after immunisation with OVA protein combined with CpG-A and anti-CD40 mAb (Fig. 28.4). Although WT mice showed efficient generation of MHC I-OVA $tetramer^+CD44^{high}CD8^+$ T cells and $CD8^+IFN$-γ^+ T cells as well as significant cytotoxic activity against targeted cells, pDC-ablated mice had a reduction in the generation of OVA-specific CTLs.

Collectively, these results indicate that pDCs contribute to the generation of CTLs throurgh the cross-presentation of soluble Ag.

28.3 Perspective and Conclusion

In this chapter, we describe a DTR-based gene targeting strategy that targets the *Siglech* gene and allows the selective elimination of pDCs in vivo. Using those mice, we have demonstrated the importance of the functions of pDCs for the control of innate and adaptive immune responses in vivo.

Although a variety of immune and nonhematopoietic cells reportedly express TLR9 and respond to its ligands (Iwasaki and Medzhitov 2004), the contribution of pDCs to the induction of the TLR9-mediated inflammation in vivo remains unclear. pDC-ablated mice manifested a near-complete abrogation of IFN-I secretion and modest reduction of proinflammatory cytokine production following injection with the TLR9 ligand. Thus, pDCs could be the primary and major source of IFN-I, but they also might contribute to the proinflammatory cytokine release in response to

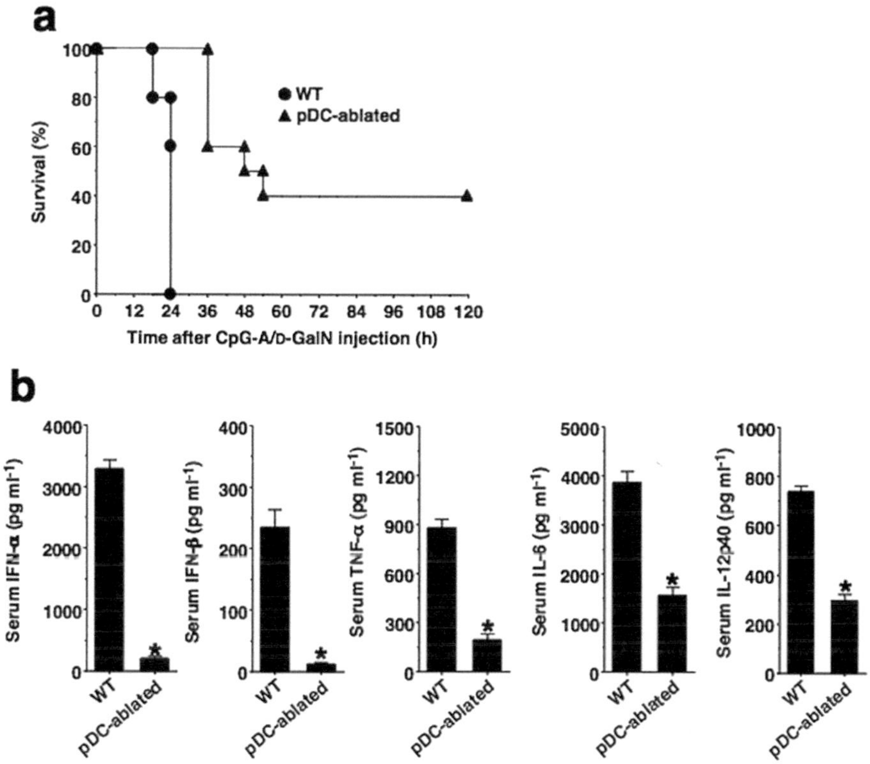

Fig. 28.2 pDCs induce TLR9-mediated systemic lethal inflammation. WT mice ($n = 6$) and pDC-ablated mice ($n = 6$) were injected with CpG-A complexed to in vivo-jetPEI™ transfection reagent plus d-GalN. (**a**) Survival was monitored for 5 days or (**b**) serum production of IFN-α, IFN β, TNF α, IL 6, and IL-12p40 was measured at 3 h after injection by enzyme-linked immunosorbent assay (ELISA). Data are the mean ± s.d. *$P < 0.01$ compared with WT mice. All data are representative of at least three independent experiments

TLR9 ligand in vivo. Collectively, these findings suggest that pDCs play a pivotal role in the initiation of TLR9-mediated inflammation.

There has been controversy regarding whether pDCs are capable of inducing T_{eff}-cell responses in vivo (Villadangos and Schnorrer 2007; Swiecki and Colonna 2010). Analysis of pDC-ablated mice revealed that pDCs enhanced the induction of the $CD4^+$ T_{eff}-cell responses when using antigenic protein plus TLR9 ligand for immunisation. Furthermore, pDCs appeared to act as APCs to generate CTLs through cross-presentation of antigenic protein under the TLR9-mediated inflammatory conditions in vivo. These phenomena suggest that the appropriate endosomal TLR ligands are needed for licensing of pDCs in terms of the activation status for the induction of the T_{eff}-cell responses. On the other hand, we observed that pDCs promoted the maturational changes of cDCs under TLR9-mediated inflammatory conditions (Takagi et al. 2011). Therefore, pDCs could provide help signals to cDCs for their optimal activation to initiate T_{eff}-cell responses

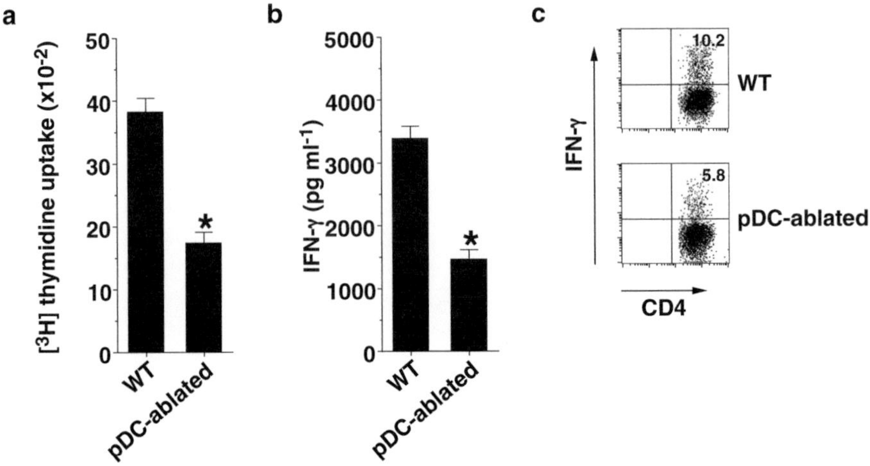

Fig. 28.3 pDCs potentiates Ag-specific $CD4^+$ T cell-responses in vivo. WT mice ($n = 6$) and pDC-ablated mice ($n = 6$) were immunised with CpG-A plus OVA protein. At 14 days after the immunisation, Spl $CD4^+$ T cells were cultured with WT $CD11c^+$ DCs in the presence or absence of OVA protein for the measurements of proliferative response by [^{3}H]thymidine incorporation (**a**) and production of IFN-γ by ELISA (**b**). Data are the mean ± s.d. $*P < 0.01$ compared with WT mice. (**c**) Intracellular production of IFN-γ in the cultured $CD4^+$ T cells was analysed by flow cytometry. Data are represented by a dot plot, and numbers represent the proportion of IFN-γ^+ cells among gated $CD4^+$ T cells in each quadrant. All data are representative of at least three independent experiments

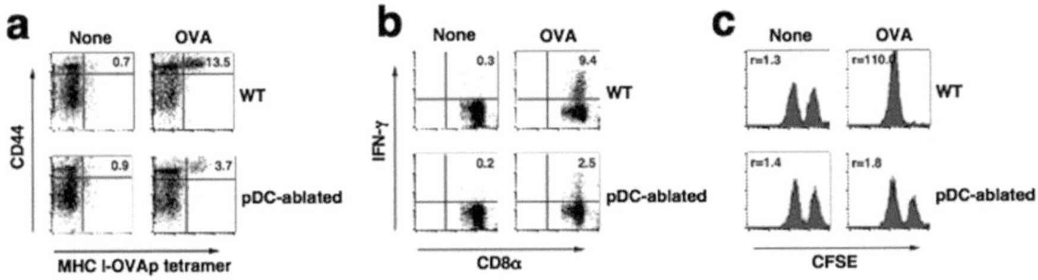

Fig. 28.4 pDCs promote Ag-specific $CD8^+$ T cell responses in vivo. WT mice ($n = 6$) and pDC-ablated mice ($n = 6$) were immunised with CpG-A complexed to in vivo-jetPEI™ transfection reagent, anti-CD40 mAb, and OVA protein, and then the mixture of unpulsed $CFSE^{low}$ cells plus Ag-pulsed $CFSE^{high}$ cells was injected 5 days after the immunisation. At 6 days after the immunisation, splenocytes were analysed for the generation of MHC I-OVA tetramer$^+$CD44^{high} $CD8^+$ T cells (**a**), for intracellular IFN-γ-producing $CD8^+$ T cells (**b**) and for in vivo cytotoxic activity (**c**) by flow cytometry. Data are represented by a dot plot (**a**, **b**), and numbers represent the proportion of MHC I-OVA tetramer$^+$CD44^{high} cells (**a**) and IFN-γ^+ cells (**b**) among gated $CD8^+$ T cells in each quadrant, or by a histogram (**c**), and numbers represent the ratio of unpulsed $CFSE^{low}$ cells to Ag-pulsed $CFSE^{high}$ cells. All data are representative of at least three independent experiments

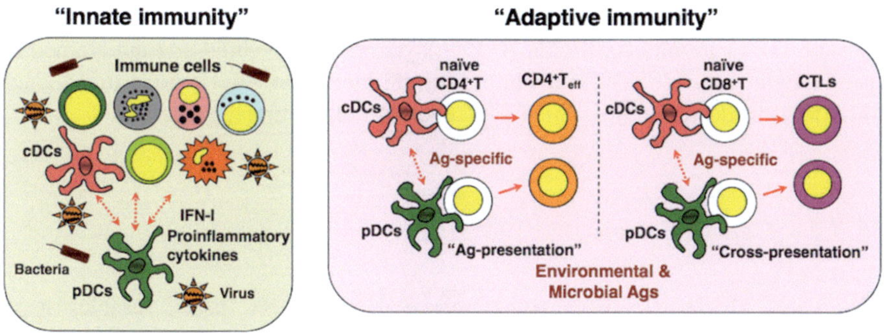

Fig. 28.5 Proposed model for the role of pDCs in the control of TLR9-mediated innate and adaptive immune responses

through the production of IFN-I as well as the interaction of CD40 and CD154 (Takagi et al. 2011).

In conclusion, we demonstrate that pDCs play a crucial role in initiating the TLR9-mediated inflammation and T-cell immunity in vivo. The observations led us to hypothesise that pDCs act as primary cells to induce inflammation upon the recognition of TLR9 ligands released from tissue cells under sterile pathophysiological conditions and/or microbes during the infection. Subsequently, pDCs could activate cDCs and other leukocytes through the secretion of IFN-I as well as costimulation for the amplification of inflammation. Concomitantly, pDCs and cDCs cooperatively activate Ag-specific naive T cells for the differentiation of T_{eff} cells under the TLR9-mediated inflammatory conditions. Thus, these scenarios might account for the establishment of TLR9-mediated innate and adaptive immune responses (Fig. 28.5). On the other hand, accumulating evidence proposes that pDCs act as a key source of aberrant IFN-I production in chronic autoimmune diseases such as psoriasis and systemic lupus erythematosus (SLE), in which patients often present with elevated levels of serum IFN-I (Gilliet et al. 2008) Thus, further study will be needed to analyse how pDCs control innate and adaptive immune responses leading to the initiation of psoriasis and SLE, and to test the possibility whether pDCs represent an attractive potential therapeutic target for the amelioration of these IFN-I-associated autoimmune diseases.

28.4 Methods

28.4.1 Mice

The following mice at 8- to 12-weeks-old were used in this study: *Siglech*$^{dtr/dtr}$ mice were cross-mated for more than nine generations with B6.mice (Charles River Laboratories), and *Siglech*$^{+/+}$ littermates were used as WT mice (Takagi et al. 2011). For systemic ablation of pDCs, *Siglech*$^{dtr/dtr}$ KI mice were injected i.p. with DT (1 μg/mouse; Sigma-Aldrich). All mice were bred and maintained in specific pathogen-free conditions in the animal facility at RIKEN Research Center for Allergy and Immunology in accordance with institutional guidelines.

28.4.2 Cell Isolation

To prepare single-cell suspensions from Spl and PLNs, tissue samples were digested with collagenase type III (Worthington Biochemical) at 37 °C for 20 min, and were ground between glass slides. BM cells were flushed from the femurs and tibias. Splenocytes and BM cells were treated with RBC lysis buffer (Sigma-Aldrich) before suspension. Single-cell suspensions were obtained by forcing through a 100-μm cell strainer (BD Bioscience). $CD4^+$ T cells and $CD8^+$ T cells were purified from spleen with mouse CD4 T lymphocyte Enrichment Set-DM and mouse CD8 T lymphocyte Enrichment Set-DM (both from BD Bioscience).

28.4.3 Flow Cytometry

Cells were stained with fluorescein-conjugated mAbs to mouse CD4 (RM4-5), CD8α (53–6.7), CD11c (HL3), CD44 (IM7), H-$2K^b$ (AF6-88.5), IFN-γ (XMG1.2), isotype-matched control mAb (BD Bioscience), mPDCA-1/BST2 (JF05-1C2.4.1), (Miltenyi Biotec), and H-$2K^b$ OVA tetramer (MBL). For the intracellular expression of cytokines, cells were incubated for 4 h with phorbol 12-myristate 13-acetate (PMA, 50 ng/ml; Sigma-Aldrich) plus ionomycin (500 ng/ml; Sigma-Aldrich) or $OVA_{257–264}$ peptide (10 μM), plus GolgiPlug (BD Biosciences) during the final 2 h. Subsequently, the cells were resuspended in Fixation-Permeabilisation solution (BD Cytofix/Cytoperm kit; BD Biosciences) and intracellular cytokine staining was done according to the manufacturer's directions. Fluorescence staining was analysed with a FACSCalibur flow cytometer and CELLQuest Software (both from BD Biosciences).

28.4.4 Immunohistochemical Analysis

Spl was embedded in OCT compound (Sakura Finetechnical) and frozen in liquid N_2. The tissue segments were sectioned with a cryostat at 8 μm. Frozen sections were fixed in cold acetone and blocked in TNB buffer (PerkinElmer Life Science) containing 5 % normal rat serum. To block endogenous biotin, the sections were further treated with the Streptavidin/Biotin Blocking Kit (Vector Laboratories), and endogenous peroxidase activity was quenched with 1 % H_2O_2. The primary Abs were anti-CD19 mAb-FITC (1D3, BD Bioscience), anti-mPDCA-1/BST2 mAb-biotin (JF05-1C2.4.1), anti-CD11c mAb-biotin (HL3), and anti-CD3ε mAb-APC (145-2C11). The CD19 signal was then amplified by incubation with Alexa Fluor 488-conjugated anti-rat IgG (Invitrogen). The mPDCA1 and CD11c staining were revealed with a TSA signal amplification kit (PerkinElmer Life Science) according to the manufacturer's instructions. The sections were then incubated with Streptavidine-HRP (PerkinElmer Life Science) followed by Tyramide-Cy3 conjugate. At the end of the staining, slides were washed and mounted with Vectashield (Vector laboratories). The stained slides were examined with a BIOREVO fluorescence microscope (BZ-9000; KEYENCE).

28.4.5 *In Vivo TLR Stimulation*

Mice were intravenously (i.v.) injected with CpG-A (10 μg/mouse; D19, ggTGCATCGATGCAgggggG, Sigma Aldrich) complexed to in vivo-jetPEI™ transfection reagent (10 μl/mouse; Polyplus transfection) with a peritoneal (i.p.) injection of D-GalN (20 mg/mouse; Sigma-Aldrich). Survival was then monitored for 5 days or sera were collected at 3 h later.

28.4.6 *Detection of Cytokines*

Sera were assayed for IFN-α, IFN-β (PBL), TNF-α, and IL-6 and IL-12p40 (Life Technologies) by ELISA kits according to the manufacturer's instructions.

28.4.7 *Immunisation*

For the analysis of Ag-specific $CD4^+$ T cell responses, mice were immunised i.v. with OVA protein (100 μg/mouse; Sigma-Aldrich) plus CpG-A (10 μg/mouse) complexed to in vivo-jetPEI™ transfection reagent (10 μl/mouse), and the Spl was obtained 14 days after the immunisation. For the generation of Ag-specific CTLs, mice received an i.v. injection of CpG-A (10 μg/mouse) complexed to in vivo-jetPEI™ transfection reagent (10 μl/mouse) in combination with an i.p. injection of OVA protein (500 μg/mouse) plus anti-CD40 mAb (10 μg/mouse; clone 1C10, eBioscience), and Spl was obtained from the mice 6 days later.

28.4.8 Antigen Presentation Assay

$CD4^+$ T cells (2×10^5) were cultured with irradiated (15 Gy) $CD11c^+$ DCs (2×10^4) in the presence or absence of OVA protein (1 mg/ml) for 3 days in 96-well flat-bottomed plates (BD Biosciences). [^{3}H]thymidine (GE Healthcare) incorporation was measured on day 3 for the last 18 h. In another experiment, the cells and the culture supernatants were collected to detect the production of cytokines.

28.4.9 In Vivo Cytotoxicity Assay

Syngeneic splenocytes were labeled with either 0.25 or 2.5 μM carboxyfluorescein diacetate-succinimidyl ester (CFSE; Molecular Probes) at 37 °C for 10 min, and washed twice with cold PBS. $CFSE^{high}$ cells were subsequently pulsed with 10 μM $OVA_{257–264}$ peptide (SIINFEKL) at 37 °C for 60 min. $CFSE^{low}$ cells served as an internal control and therefore were not pulsed with peptide. Cells were mixed at a 1:1 ratio, and then 1×10^7 total cells were injected i.v. into mice 5 days after immunisation as described above. 24 h later, splenocytes from each mouse were analysed by flow cytometry to detect the presence of CFSE-labelled cells. The ratio of unpulsed $CFSE^{low}$ cells to pulsed $CFSE^{high}$ cells was calculated as an indication of Ag-specific lytic activity.

28.4.10 Statistical Analysis

Data are expressed as the mean ±s.d. The statistical significance of the differences between the values obtained was evaluated by ANOVA and the Kaplan–Meier log-rank test. A p-value of <0.01 was considered significant.

Acknowledgments I thank all members of Laboratory for Dendritic Cell Immunobiology and the Central Facility at the RIKEN Research Center for Allergy and Immunology; Division of Immunology, Department of Infectious Diseases, Faculty of Medicine, and the animal facility at University of Miyazaki. This work was supported by Grants-in-aid for Scientific Research (B) (K.S.) from the Ministry of Education, Science and Culture of Japan, Japan Science and Technology Agency, Precursory Research for Embryonic Science and Technology (PRESTO) (K.S.), the Uehara Memorial Foundation (K.S.), and the Mochida Memorial Foundation for Medical and Pharmaceutical Research (K.S.),

References

Asselin-Paturel C, Boonstra A, Dalod M, Durand I, Yessaad N, Dezutter-Dambuyan C, Vicari A, O'Garra A, Biron C, Brière F, Trinchieri G (2001) Mouse type I IFN-producing cells are immature APCs with plasmacytoid morphology. Nat Immunol 2:1144–1150

Banchereau J, Pascual V (2006) Type I interferon in systemic lupus erythematosus and other autoimmune diseases. Immunity 25:383–392

Blasius AL, Colonna M (2006) Sampling and signaling in plasmacytoid dendritic cells: the potential roles of Siglec-H. Trends Immunol 27:255–260

Dudziak D, Kamphorst AO, Heidkamp GF, Buchholz VR, Trumpfheller C, Yamazaki S, Cheong C, Liu K, Lee HW, Park CG, Steinman RM, Nussenzweig NC (2007) Differential antigen processing by dendritic cell subsets in vivo. Science 315:107–111

Gilliet M, Cao W, Liu YJ (2008) Plasmacytoid dendritic cells: sensing nucleic acids in viral infection and autoimmune diseases. Nat Rev Immunol 8:594–606

Hildner K, Edelson BT, Purtha WE, Diamond M, Matsushita H, Kohyama M, Calderon B, Schraml BU, Unanue ER, Diamond MS, Schreiber RD, Murphy TL, Murphy KM (2008) Batf3 deficiency reveals a critical role for CD8α^+ dendritic cells in cytotoxic T cell immunity. Science 322:1097–1100

Irla M, Küpfer N, Suter T, Lissilaa R, Benkhoucha M, Skupsky J, Lalive PH, Fontana A, Reith W, Hugues S (2010) MHC class II-restricted antigen presentation by plasmacytoid dendritic cells inhibits T cell-mediated autoimmunity. J Exp Med 207:1891–1905

Iwasaki A, Medzhitov R (2004) Toll-like receptor control of the adaptive immune responses. Nat Immunol 10:987–995

Jung S, Unutmaz D, Wong P, Sano G, De los Santos K, Sparwasser T, Wu S, Vuthoori S, Ko K, Zavala F, Pamer EG, Littman DR, Lang RA (2002) In vivo depletion of CD11c$^+$ dendritic cells abrogates priming of CD8$^+$ T cells by exogenous cell-associated antigens. Immunity 17:211–220

Kaisho T, Tanaka T (2008) Turning NF-kB and IRFs on and off in DC. Trends Immunol 7:329–336

Kissenpfennig A, Henri S, Dubois B, Laplace-Builhé C, Perrin P, Romani N, Tripp CH, Douillard P, Leserman L, Kaiserlian D, Saeland S, Davoust J, Malissen B (2005) Dynamics and function of Langerhans cells in vivo: dermal dendritic cells colonise lymph node areas distinct from slower migrating Langerhans cells. Immunity 22:643–654

Krug A, Luker GD, Barchet W, Leib DA, Akira S, Colonna M (2004) Herpes simplex virus type 1 activates murine natural interferon-producing cells through toll-like receptor 9. Blood 103:1433–1437

Lande R, Gregorio J, Facchinetti V, Chatterjee B, Wang YH, Homey B, Cao W, Wang YH, Su B, Nestle FO, Zal T, Mellman I, Schröder JM, Liu YJ, Gilliet M (2007) Plasmacytoid dendritic cells sense self-DNA coupled with antimicrobial peptide. Nature 449:564–569

Ochando JC, Homma C, Yang Y, Hidalgo A, Garin A, Tacke F, Angeli V, Li Y, Boros P, Ding Y, Jessberger R, Trinchieri G, Lira SA, Randolph GJ, Bromberg JS (2006) Alloantigen-presenting plasmacytoid dendritic cells mediate tolerance to vascularised grafts. Nat Immunol 6:652–662

Puttur F, Arnold-Schrauf C, Lahl K, Solmaz G, Lindenberg M, Mayer CT, Gohmert M, Swallow M, van Helt C, Schmitt H, Nitschke L, Lambrecht BN, Lang R, Messerle M, Sparwasser T (2013) Absence of Siglec-H in MCMV infection elevates interferon alpha production but does not enhance viral clearance. PLoS Pathog 9, e1003648

Salio M, Palmowski MJ, Atzberger A, Hermans IF, Cerundolo V (2004) CpG-matured murine plasmacytoid dendritic cells are capable of in vivo priming of functional CD8 T cell responses to endogenous but not exogenous antigens. J Exp Med 199:567–579

Sato K, Fujita S (2007) Dendritic cells-nature and classification. Allergol Int 56:183–191

Shortman K, Naik SH (2007) Steady-state and inflammatory dendritic-cell development. Nat Rev Immunol 7:19–30

Swiecki M, Colonna M (2010) Unraveling the functions of plasmacytoid dendritic cells during viral infections, autoimmunity, and tolerance. Immunol Rev 234:142–162

Takagi H, Fukaya T, Eizumi K, Sato Y, Sato K, Shibazaki A, Otsuka H, Hijikata A, Watanabe T, Ohara O, Kaisho T, Malissen B, Sato K (2011) Plasmacytoid dendritic cells are crucial for the initiation of inflammation and T cell immunity in vivo. Immunity 35:958–971

Villadangos JA, Schnorrer P (2007) Intrinsic and cooperative antigen-presenting functions of dendritic-cell subsets in vivo. Nat Rev Immunol 7:543–555

Yoneyama H, Matsuno K, Toda E, Nishiwaki T, Matsuo N, Nakano A, Narumi S, Lu B, Gerard C, Ishikawa S, Matsushima K (2005) Plasmacytoid DCs help lymph node DCs to induce anti-HSV CTLs. J Exp Med 202:425–435

Zhang J, Raper A, Sugita N, Hingorani R, Salio M, Palmowski MJ, Cerundolo V, Crocker PR (2006) Characterization of Siglec-H as a novel endocytic receptor expressed on murine plasmacytoid dendritic cell precursors. Blood 107:3600–3608

Chapter 29
Mechanisms of Lysosomal Exocytosis by Immune Cells

Ji-hoon Song and Rikinari Hanayama

Abstract Lysosomal exocytosis is an essential process to regulate various immune responses. Many cells of immune systems have cell-specific secretory lysosomes, which are secreted in response to external stimuli, including neutrophil azurophil granules, platelet dense granules, eosinophil granules, basophil and mast cell histamine granules, and cytotoxic T lymphocyte (CTL) lytic granules. On the other hand, phagocytes such as macrophages, neutrophils, and dendritic cells contain many conventional lysosomes, which fuse with phagosomes to degrade the engulfed particles, and then the waste materials are expelled by lysosomal exocytosis. A failure of this process can lead to accumulation of waste materials, which in turn may aberrantly activate the phagocytes. Recent studies have identified various proteins that regulate the lysosomal exocytosis, and their dysfunctions were shown to cause several genetic immune disorders. This chapter highlights the current understandings of the molecular mechanisms of lysosomal exocytosis by immune cells and their relevance to the development of chronic inflammation.

Keywords Lysosomes • Exocytosis • Lysosomal digestive enzymes • Membrane trafficking • Membrane fusion • C2 domain-containing proteins • Lysosome-related disease • Chronic inflammation

J.-h. Song
Laboratory of Immune Network, WPI Immunology Frontier Research Center (IFReC), Osaka University, 3-1 Yamada-oka, Suita, Osaka 565-0871, Japan

R. Hanayama (✉)
Laboratory of Immune Network, WPI Immunology Frontier Research Center (IFReC), Osaka University, 3-1 Yamada-oka, Suita, Osaka 565-0871, Japan

Department of Immunology, Kanazawa University Graduate School of Medicine, 13-1 Takara, Kanazawa, Ishikawa 920-8640, Japan

PRESTO, Japan Science and Technology Agency, Kawaguchi 332-0012, Japan
e-mail: rikinari-hanayama@umin.ac.jp

M. Miyasaka, K. Takatsu (eds.), *Chronic Inflammation*,
DOI 10.1007/978-4-431-56068-5_29

29.1 Introduction

Exocytosis, in which subcellular vesicles fuse with plasma membrane and the contents within the vesicles are subsequently released to the extracellular environment, is crucial for various fundamental physiological processes (Wu et al. 2014). These include release of neurotransmitters within the synaptic vesicles from neurons (Jahn and Fasshauer 2012), secretion of insulin from pancreatic beta cells in response to the elevation of blood sugar level (Leibiger et al. 2008), and antigen presentation through which macrophages and dendritic cells display foreign antigens in complex with major histocompatibility complexes (MHCs) on their surface (Vyas et al. 2008). Most cells store their secretory products in specialised secretory granules, however, some cells have modified their lysosomal compartment so that it can function as secretory lysosomes (Blott and Griffiths 2002). Lysosomes are membrane-enclosed cytoplasmic organelles that hold approximately 60 types of acid-dependent digestive hydrolases including proteases, lipases, phospholipases, phosphatases, glycosidases, sulfatases, and nucleases for degradation of lysosomal membrane-enclosed particles (Luzio et al. 2007). Only a few cell types, mostly immune cells, contain secretory lysosomes, whose effector functions vary depending on the secretory cells (Blott and Griffiths 2002). For instance, cytotoxic T lymphocyte (CTL) and natural killer (NK) cells secrete membrane-bound Fas ligand, the pore-forming protein perforin, and granzyme serine protease from secretory lysosomes to induce the killing of cancer cells and cells infected by viruses (van Dommelen et al. 2006). Stimulation of mast cells and basophils through Immunoglobulin E (IgE) causes the secretion of histamine and serotonin from secretory lysosomes which elicits allergic reactions (Stone et al. 2010). This evidence strongly suggests that lysosomal exocytosis is a crucial process for proper immune responses. It is therefore considered that an impairment of lysosomal exocytosis can lead to the development of immune disorders. Actually, several lines of evidence have demonstrated that the dysfunctions of lysosomal exocytosis are closely related to several genetic disorders, such as Chediak–Higashi syndrome, Griscelli syndrome, and Hermansky–Pudlak syndrome, which are characterised by severe immunological deficiency (Stinchcombe et al. 2004).

In several cell types, conventional lysosomes are also secreted by exocytosis. Phagocytes such as macrophages, neutrophils, and dendritic cells take up a variety of extracellular particles and internalise them into a subcellular compartment, known as the phagosome, which in turn fuse with lysosomes (forming phagolysosomes), to digest these enclosed particles (Fairn and Grinstein 2012). Proteins in the phagolysosomes are digested into small peptides, loaded onto MHC class II molecules, and then transported to the surface of the phagocytes for antigen presentation by the fusion of the phagolysosomes with plasma membrane (Vyas et al. 2008). At the same time, the indigestible waste materials generated within these phagolysosomes are released into the extracellular environment by exocytosis (Medina et al. 2011; Xu et al. 2012). A failure of this exocytosis can lead to the accumulation of waste materials inside the phagocytes, which in turn aberrantly activate innate immune responses leading to the development of chronic inflammation (Okabe et al. 2005; Kawane et al. 2006). Thus, the exocytosis of both secretory lysosomes and conventional lysosomes have to be properly regulated in the physiological settings, and we describe the current understanding of its molecular mechanisms.

29.2 Biogenesis, Maturation, and Transport of Secretory Lysosomes

Secretory proteins destined for secretory lysosomes are synthesised in the rough endoplasmic reticulum, then surrounded by the vesicles coated with COP II protein to bud towards Golgi apparatus (Ramalho-Santos et al. 2001). During their passage, the sugar chains of the secretory proteins are modified with mannose 6-phosphate residue, thereby binding to mannose 6-phosphate receptor in *trans*-Golgi network (Le Borgne and Hoflack 1997). Then these proteins are aggregated and are packaged in clathrin-coated vesicles. The clathrin coat is subsequently disassembled, letting the vesicles fuse with late endosomes or phagosomes, where the secretory proteins are dissociated from mannose 6-phosphate receptor due to the acid environment. As the endosomes and phagosomes become more mature by fusing with one another, the protein aggregates become more condensed and their lumen becomes more acidic. This acidification within the vesicle lumen is driven by V-ATPases, which locate within the membrane of the vesicles and translate the energy from hydrolysis of ATP to transport protons (H^+) across intracellular and plasma membrane (Marshansky and Futai 2008). Within these acidified vesicular lumen, the protein aggregates are degraded by lysosomal hydrolase (Swanson 2006). These vesicles further fuse with lysosomes forming endolysosomes or phagolysosomes. For the release of these degraded products into the extracellular space, intracellular Ca^{2+} level should be elevated. The activation of signalling receptors and/or ion channels, which exist on the cell surface and the intracellular vesicles, cause the elevation of intracellular Ca^{2+} level (Cheng et al. 2010). For example, it is well known that signals from cell surface receptors such as Fc receptors activate the phospholipase C γ and inositol 1,4,5-triphosphate signalling pathways, thereby causing the release of Ca^{2+} from the endoplasmic reticulum (Sanchez-Mejorada and Rosales 1998). Ca^{2+} binding proteins on the lysosomal vesicle detect this Ca^{2+} elevation as a signal to drive the vesicles along microtubules towards the microtubule organising center (MTOC) by kinesin family members, from which the vesicles are transported to the cell periphery (Blott and Griffiths 2002). Once they reach the cellular periphery, the vesicles are detached from the kinesin, followed by actin-based movement towards the fusion site of the plasma membrane and the secretory proteins are released by fusion of endolysosomes or phagolysosomes with plasma membrane.

29.3 Mechanisms of Membrane Fusion

29.3.1 Rabs

The Rab family of small GTPases, which consists of more than 70 mammalian members, are localised to intracellular membranes and plays a central role in membrane vesicle trafficking, ensuring that vesicles are transported to their correct target destinations (Stenmark 2009). They play a central role in the process of membrane trafficking and fusion by associating with various components such as SNAREs and motor proteins. As do other small GTPases, Rab switches between the GDP-bound inactive form and the GTP-bound active form. A GDP/GTP exchange factor (GEF) catalyses the conversion from the GDP-bound to GTP-bound form, thereby facilitating the association of Rab with the intracellular organelle structures including plasma membrane and trafficking vesicles. Vesicles associating with GTP-bound Rab directly bind motor proteins or indirectly via motor adaptor proteins, and move along the actin filaments or microtubules. Subsequently, Rab proteins interact with tethering factors that are located at the target sites of the plasma membrane where exocytosis occurs (Blott and Griffiths 2002). After the fusion event, the active form of Rab is converted to the inactive form by a GTPase-activating protein (GAP), then promoting the conjugation of Rab to GDP dissociation inhibitor (GDI), which stabilises the inactive form (Seabra and Wasmeier 2004).

29.3.2 SNAREs

SNARE proteins mediate fusion of vesicles with their target membrane when the vesicles are recruited to the fusion site (Chen and Scheller 2001). An animal genome encodes at least 35 SNAREs, each of which is associated with the membrane of a particular organelle, such as plasma membrane and intracellular compartments that are destined to fuse. Several immune cells have SNARE proteins associated with their secretory lysosomes: For instance, mast cells have VAMP2, VAMP7, and SNAP23, and neutrophils have syntaxin3, syntaxin4, VAMP2, and SCAMP (Lorentz et al. 2012). There are several classifications for SNAREs based on their location, structural features, and sequence homology. Based on their intracellular location, SNAREs can be categorised into two types, v-SNAREs and t-SNAREs, which are located on the membrane of transport vesicles and the membrane of the target site, respectively. Transport vesicles can fuse with the fusion site of the target membrane through the distinct interaction of v-SNARE and t-SNARE. The process is promoted by Rab proteins and Rab-interacting tethering molecules, which associate with SNAREs and facilitate appropriate combination of v-SNARE and t-SNARE in two membranes which can find each other. During the process of membrane fusion, v-SNARE and t-SNARE are assembled to form a *trans*-SNARE complex, bridging them more closely and promoting the fusion. In the regulated exocytosis, however, fusion cannot be completed without the elevation of intracellular Ca^{2+} level which is sensed by Ca^{2+} sensor proteins. The representative Ca^{2+} sensor proteins are C2 domain-containing proteins such as synaptotagmin, Munc13, and ferlin protein family (Martens and McMahon 2008). These molecules bind to lipid bilayers of fusing membranes in a Ca^{2+}-dependent manner, triggering membrane fusion through the interaction with SNARE proteins via complexin, which is known as a priming molecule required for the assembly of SNARE proteins (Brose 2008).

29.3.3 Synaptotagmin and Munc13 Family

The timing of membrane fusion is regulated by Ca^{2+} sensor proteins. For example, synaptic vesicle fusion essential for neurotransmitter release from neurons is regulated by synaptotagmins which facilitate Ca^{2+}-dependent fusion of synaptic vesicle with the plasma membrane. Synaptotagmins are located on the synaptic vesicles containing two linked C2 domains that mediate Ca^{2+}-dependent binding to anionic phospholipids such as phosphatidylserine (PS) in the inner leaflet of the plasma membrane, which is a critical process for membrane fusion (Martens and McMahon 2011). The pocket structures that are shaped by loops of the C2 domains facilitate the binding of Ca^{2+} to synaptotagmins. Although this region is negatively charged due to the abundance of aspartic acid in the pockets, Ca^{2+} binding alters the net charge from negative to positive charges, thus allowing the C2 domain of synaptotagmins to interact with the anionic phospholipids in the plasma membrane (Martens 2010). The human synaptotagmin family consists of 17 members, and most of them are highly expressed in the nervous system. However, synaptotagmin VII (Syt VII) is also found in various tissues including immune cells (Martinez et al. 2000). Syt VII is present on the lysosomal membrane and regulates Ca^{2+}-triggered exocytosis by binding to the plasma membrane. In macrophages, Syt VII is required for efficient uptake of large particle loads, by promoting the delivery of lysosomal membrane to phagocytic cups (Czibener et al. 2006). The expression of Syt VII is upregulated in dendritic cells during LPS-induced maturation, and it mediates translocation of MHC class II to the cell surface for antigen presentation by the fusion of phagolysosomes with plasma membrane (Becker et al. 2009).

The Munc13 protein family is another C2 domain-containing protein family which works as a Ca^{2+} sensor for lysosomal exocytosis. The roles of Munc13 have been extensively studied in CTLs in terms of the exocytic machinery of the secretory lysosomes, which causes the release of membrane-bound Fas ligand, the pore forming protein perforin, and granzyme serine protease to induce the killing of target cells (Trapani and Smyth 2002). In CTLs, lysosomal exocytosis is mediated by a complex formed of small GTPase Rab27A and its effector Munc 13–4 that tethers the lytic granules to the plasma membrane to allow fusion (Elstak et al. 2011). Mutations in the *Munc 13–4* gene cause familial hemophagocytic lymphohistiocytosis type 3, which causes immunodeficiency due to defects in the lysosomal exocytosis of cytolytic enzymes by CTLs.

29.3.4 Ferlin Family

The ferlin family members consist of six mammalian members (Posey et al. 2011). These proteins carry at least four tandem cytosolic C2 domains and a single C-terminal transmembrane domain required for anchoring to the plasma membrane. The common function of ferlin family proteins is to mediate the fusion of cytoplasmic vesicles with the plasma membrane in exocytosis. Dysferlin is the most thoroughly studied ferlin, and its mutations cause limb-girdle muscular dystrophy 2B and Miyoshi myopathy (Han and Campbell 2007). Dysferlin localises to the cytoplasmic vesicles and plays a critical role in muscle membrane repair by mediating the fusion of vesicles with the plasma membrane, thereby serving as a membrane patch in the disrupted regions (Han and Campbell 2007). Myoferlin was identified as the protein that shares the highest homology with dysferlin (Davis et al. 2000). Both dysferlin and myoferlin possess seven C2 domains in the cytoplasmic region, and the first C2 domain was shown to bind PS in a calcium-dependent manner (Fig. 29.1; Posey et al. 2011). Myoferlin was identified as a protein expressed in the plasma membrane of myoblasts undergoing fusion and has been implicated in the repair of injured plasma membranes (Doherty et al. 2005). However, we have recently found that myoferlin is highly expressed in various phagocytes (particularly inflammatory macrophages), raising a possibility that myoferlin might play a role in the fusion process of lysosomal exocytosis in these phagocytes (Song and Hanayama, unpublished results).

29.3.5 Myoferlin

Myoferlin was specifically localised to lysosomes in macrophages. The myoferlin-deficient macrophages had three times more lysosomes, compared to the wild-type

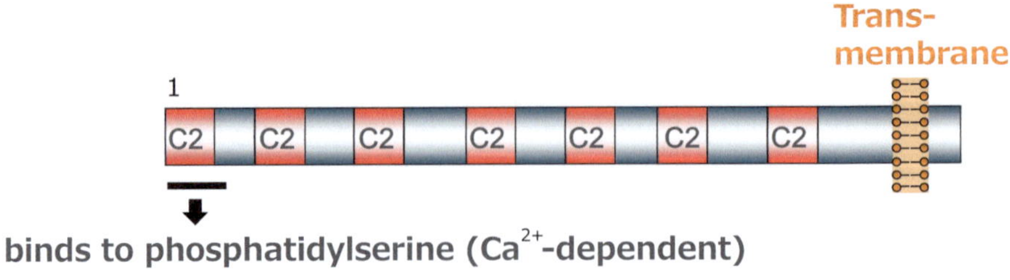

Fig. 29.1 Structure of myoferlin. Myoferlin is a novel regulator of lysosomal exocytosis by phagocytes which binds to phosphatidylserine via the first C2 domain in a Ca^{2+}-dependent manner

cells, suggesting that myoferlin deficiency caused accumulation of lysosomes. Electron microscopic analysis revealed that myoferlin-deficient macrophages carried many lysosomal vesicles filled with undigested debris and some membranous materials inside them. Myoferlin was also localised to the phagolysosomes that carry bacteria after phagocytosis, playing a role in the release of waste materials of the digested bacteria by mediating the fusion with cell plasma membrane. Due to impaired release of waste materials after bacterial phagocytosis, myoferlin-deficient macrophages carried much indigestible debris inside the phagolysosomes, which caused augmented and prolonged expression of various inflammatory cytokines, such as IL-6, IL-1β, and TNF-α. These findings indicate that proper release of waste materials after phagocytosis is essential for the resolution of inflammatory responses by macrophages. When we injected bacteria into the mouse peritoneal cavity to promote the phagocytosis by the peritoneal exudate cells, we found both gene and protein levels of inflammatory cytokines were strongly augmented in the myoferlin-deficient mice. However, myoferlin-deficient mice displayed more resistance to the bacterial injection, compared to the WT mice, all of which died within 3 days after injection.

To clarify the cause of the death, we considered the following model (Fig. 29.2): myoferlin is expressed at the lysosomal membrane (including phago-lysosomes). After completion of bacterial phagocytosis, myoferlin interacts with PS on the plasma membrane in a calcium-dependent manner, thus promoting a fusion for

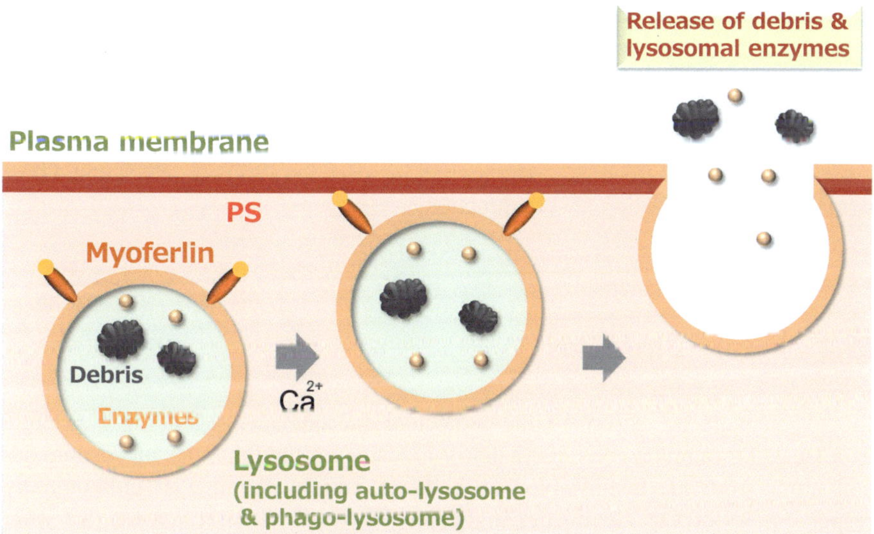

Fig. 29.2 Model of myoferlin-mediated lysosomal exocytosis. Myoferlin localises to the lysosomal membrane of various phagocytes. Upon calcium stimulation, myoferlin binds to phosphatidylserine in the inner leaflet of the plasma membrane, promoting the fusion with lysosomes for the release of indigestible waste materials and lysosomal enzymes inside the lysosomes

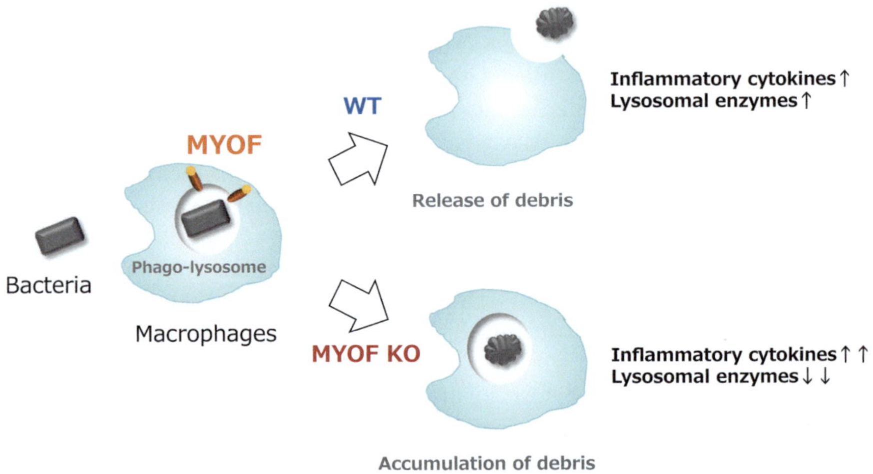

Fig. 29.3 Myoferlin is a double-edged sword in inflammation. After phagocytosis and digestion of bacteria, macrophages release the debris in a myoferlin-dependent manner. During these processes, inflammatory cytokines are upregulated and lysosome enzymes are secreted. Myoferlin deficiency causes the accumulation of debris in the macrophages, which leads to augmented inflammatory cytokine expressions but reduced lysosomal enzyme secretion

the release of waste materials. At the same time, myoferlin also promotes the release of cytolytic lysosomal enzymes from phagolysosomes, which increased the likelihood of mortality in severe infections. Indeed, we detected very strong cytolytic activity of lysosomal enzymes in the ascites of WT mice, but it was significantly reduced in the myoferlin-deficient mice. These findings define myoferlin's novel roles in lysosomal exocytosis by macrophages, which increases the mortality in severe infection (Fig. 29.3). Inhibiting myoferlin will prevent the release of cytolytic lysosomal enzymes and increase the survival, which could be a novel strategy for treating severe infection.

29.4 Future Directions

In this chapter, we described the molecular mechanisms regulating lysosomal exocytosis by immune cells, particularly focusing on the functions of C2 domain-containing proteins. In physiological settings, various immune or inflammatory responses elevate Ca^{2+} levels in phagocytes. However, the mechanisms that yield Ca^{2+} to allow the release of indigestible waste materials from phagolysosomes have not been elucidated completely. If the fusion between phagolysosomes and plasma membrane occurs during the early stage after phagocytosis, this may result in the escape of bacteria that have not been entirely killed in the phagolysosomes. Therefore, the timing of fusion must be strictly regulated, suggesting there may be a mechanism that senses the completion of bacterial digestion. We assume that the Ca^{2+} released from the digested bacteria inside phagolysosomes can be transferred to the cytoplasm via calcium ion channels that are localised to the phagolysosomes, such as mucolipin-1 (also known as TRPML1). Therefore, a local increase in Ca^{2+} may promote the binding of myoferlin in the phagolysosomes to the plasma membrane. It will be interesting to determine whether mucolipin-1 mediates this process. Clarifying the regulatory mechanisms will provide new insights into chronic inflammatory mechanisms and may facilitate the development of novel therapeutic approaches based on the regulating lysosomal exocytosis.

References

Becker SM, Delamarre L, Mellman I, Andrews NW (2009) Differential role of the Ca(2+) sensor synaptotagmin VII in macrophages and dendritic cells. Immunobiology 214:495–505

Blott EJ, Griffiths GM (2002) Secretory lysosomes. Nat Rev Mol Cell Biol 3:122–131

Brose N (2008) For better or for worse: complexins regulate SNARE function and vesicle fusion. Traffic 9:1403–1413

Chen YA, Scheller RH (2001) SNARE-mediated membrane fusion. Nat Rev Mol Cell Biol 2:98–106

Cheng X, Shen D, Samie M, Xu H (2010) Mucolipins: intracellular TRPML1-3 channels. FEBS Lett 584:2013–2021

Czibener C, Sherer NM, Becker SM, Pypaert M, Hui E, Chapman ER, Mothes W, Andrews NW (2006) Ca2+ and synaptotagmin VII-dependent delivery of lysosomal membrane to nascent phagosomes. J Cell Biol 174:997–1007

Davis DB, Delmonte AJ, Ly CT, Mcnally EM (2000) Myoferlin, a candidate gene and potential modifier of muscular dystrophy. Hum Mol Genet 9:217–226

Doherty KR, Cave A, Davis DB, Delmonte AJ, Posey A, Earley JU, Hadhazy M, Mcnally EM (2005) Normal myoblast fusion requires myoferlin. Development 132:5565–5575

Elstak ED, Neeft M, Nehme NT, Voortman J, Cheung M, Goodarzifard M, Gerritsen HC, Van Bergen en Henegouwen PM, Callebaut I, De Saint Basile G, van der Sluijs P (2011) The munc13-4-rab27 complex is specifically required for tethering secretory lysosomes at the plasma membrane. Blood 118:1570–1578

Fairn GD, Grinstein S (2012) How nascent phagosomes mature to become phagolysosomes. Trends Immunol 33:397–405

Han R, Campbell KP (2007) Dysferlin and muscle membrane repair. Curr Opin Cell Biol 19:409–416

Jahn R, Fasshauer D (2012) Molecular machines governing exocytosis of synaptic vesicles. Nature 490:201–207

Kawane K, Ohtani M, Miwa K, Kizawa T, Kanbara Y, Yoshioka Y, Yoshikawa H, Nagata S (2006) Chronic polyarthritis caused by mammalian DNA that escapes from degradation in macrophages. Nature 443:998–1002

le Borgne R, Hoflack B (1997) Mannose 6-phosphate receptors regulate the formation of clathrin coated vesicles in the TGN. J Cell Biol 137:335–345

Leibiger IB, Leibiger B, Berggren PO (2008) Insulin signaling in the pancreatic beta-cell. Annu Rev Nutr 28:233–251

Lorentz A, Baumann A, Vitte J, Blank U (2012) The SNARE machinery in mast cell secretion. Front Immunol 3:143

Luzio JP, Pryor PR, Bright NA (2007) Lysosomes: fusion and function. Nat Rev Mol Cell Biol 8:622–632

Marshansky V, Futai M (2008) The V-type H+ −ATPase in vesicular trafficking: targeting, regulation and function. Curr Opin Cell Biol 20:415–426

Martens S (2010) Role of C2 domain proteins during synaptic vesicle exocytosis. Biochem Soc Trans 38:213–216

Martens S, Mcmahon HT (2008) Mechanisms of membrane fusion: disparate players and common principles. Nat Rev Mol Cell Biol 9:543–556

Martens S, Mcmahon HT (2011) C2 domains and membrane fusion. Curr Top Membr 68:141–159

Martinez I, Chakrabarti S, Hellevik T, Morehead J, Fowler K, Andrews NW (2000) Synaptotagmin VII regulates Ca(2+)-dependent exocytosis of lysosomes in fibroblasts. J Cell Biol 148:1141–1149

Medina DL, Fraldi A, Bouche V, Annunziata F, Mansueto G, Spampanato C, Puri C, Pignata A, Martina JA, Sardiello M, Palmieri M, Polishchuk R, Puertollano R, Ballabio A (2011) Transcriptional activation of lysosomal exocytosis promotes cellular clearance. Dev Cell 21:421–430

Okabe Y, Kawane K, Akira S, Taniguchi T, Nagata S (2005) Toll-like receptor-independent gene induction program activated by mammalian DNA escaped from apoptotic DNA degradation. J Exp Med 202:1333–1339

Posey AD Jr, Demonbreun A, Mcnally EM (2011) Ferlin proteins in myoblast fusion and muscle growth. Curr Top Dev Biol 96:203–230

Ramalho-Santos J, Moreno RD, Wessel GM, chan EK, Schatten G (2001) Membrane trafficking machinery components associated with the mammalian acrosome during spermiogenesis. Exp Cell Res 267:45–60

Sanchez-Mejorada G, Rosales C (1998) Signal transduction by immunoglobulin Fc receptors. J Leukoc Biol 63:521–533

Seabra MC, Wasmeier C (2004) Controlling the location and activation of Rab GTPases. Curr Opin Cell Biol 16:451–457

Stenmark H (2009) Rab GTPases as coordinators of vesicle traffic. Nat Rev Mol Cell Biol 10:513–525

stinchcombe J, Bossi G, Griffiths GM (2004) Linking albinism and immunity: the secrets of secretory lysosomes. Science 305:55–59

Stone KD, Prussin C, Metcalfe DD (2010) IgE, mast cells, basophils, and eosinophils. J Allergy Clin Immunol 125:S73–S80

Swanson J (2006) CFTR: helping to acidify macrophage lysosomes. Nat Cell Biol 8:908–909

Trapani JA, Smyth MJ (2002) Functional significance of the perforin/granzyme cell death pathway. Nat Rev Immunol 2:735–747

van Dommelen SL, Sumaria N, Schreiber RD, Scalzo AA, Smyth MJ, Degli-Esposti MA (2006) Perforin and granzymes have distinct roles in defensive immunity and immunopathology. Immunity 25:835–848

Vyas JM, van der Veen AG, PLOEGH HL (2008) The known unknowns of antigen processing and presentation. Nat Rev Immunol 8:607–618

Wu LG, Hamid E, Shin W, Chiang HC (2014) Exocytosis and endocytosis: modes, functions, and coupling mechanisms. Annu Rev Physiol 76:301–331

Xu J, Toops KA, Diaz F, Carvajal-Gonzalez JM, Gravotta D, Mazzoni F, Schreiner R, Rodriguez-Boulan E, Lakkaraju A (2012) Mechanism of polarized lysosome exocytosis in epithelial cells. J Cell Sci 125:5937–43

Chapter 30
Potential Therapeutic Natural Products for the Treatment of Obesity-Associated Chronic Inflammation by Targeting TLRs and Inflammasomes

Yoshinori Nagai, Hiroe Honda, Yasuharu Watanabe, and Kiyoshi Takatsu

Abstract Obesity-associated chronic inflammation is a key event to type 2 diabetes mellitus. Recent advances in deciphering the various immune cells and signalling networks that link the immune system and metabolic system have contributed to our understanding of the pathogenesis of obesity-associated chronic inflammation. Recent studies have suggested that pattern recognition receptors in the innate immune system recognise various kinds of endogenous ligands, and play a crucial role in initiating or promoting obesity-associated chronic inflammation. These findings have also informed new therapeutic strategies based on immunomodulation. Chinese herbal medicines have been used to treat type 2 diabetes mellitus in Asian countries. With the rapid advancement of novel technologies and the increased research on natural products, many new plant-derived extracts and active compounds have been identified to exhibit anti-inflammatory effects. Here we overview natural products that inhibit activation of pattern recognition receptors, particularly Toll-like receptors and inflammasomes, as potential

Y. Nagai (✉)
Department of Immunobiology and Pharmacological Genetics, Graduate School of Medicine and Pharmaceutical Science for Research, University of Toyama, 2630 Sugitani, Toyama-shi, Toyama 930-0194, Japan

PRESTO, Japan Science and Technology Agency, Kawaguchi 332-0012, Japan
e-mail: ynagai@med.u-toyama.ac.jp

H. Honda • K. Takatsu (✉)
Department of Immunobiology and Pharmacological Genetics, Graduate School of Medicine and Pharmaceutical Science for Research, University of Toyama, 2630 Sugitani, Toyama-shi, Toyama 930-0194, Japan

Toyama Prefectural Institute for Pharmaceutical Research, 17-1 Nakataikouyama, Imizu City, Toyama 939-0363, Japan
e-mail: takatsuk@med.u-toyama.ac.jp

Y. Watanabe
Department of Immunobiology and Pharmacological Genetics, Graduate School of Medicine and Pharmaceutical Science for Research, University of Toyama, 2630 Sugitani, Toyama-shi, Toyama 930-0194, Japan

M. Miyasaka, K. Takatsu (eds.), *Chronic Inflammation*,
DOI 10.1007/978-4-431-56068-5_30

therapeutic agents for the treatment of obesity-associated chronic inflammation. We also discuss molecular mechanisms by which the natural products regulate pattern recognition receptors, with a particular focus on our findings regarding unique compounds of *Glycyrrhiza uralensis*.

Keywords Chronic inflammation • *Glycyrrhiza uralensis* • Glycyrrhizin • Inflammasome • Isoliquiritigenin • Metabolic disorder • Natural product • Toll-like receptor

30.1 Introduction

Obesity has become a worldwide health problem because it is strongly associated with metabolic syndromes including type 2 mellitus (T2DM), atherosclerosis, and ischemic heart diseases (Hotamisligil 2006; Schenk et al. 2008). Accumulating evidence indicates that chronic-low grade inflammation plays a crucial role in the pathogenesis of obesity-related metabolic dysfunction (Kalupahana et al. 2012; Osborn and Olefsky 2012). The chronic inflammatory alternations are associated with dynamic changes in the composition and function of immune cells in various tissues such as adipose tissue, pancreatic islet, liver, muscle, and hypothalamus (Ouchi et al. 2011; Ryan et al. 2012; Lumeng and Saltiel 2011). A large number of inflammatory immune cells infiltrate into adipose tissue during the course of obesity. M1-like macrophages accumulate in adipose tissue and are major sources of inflammatory mediators such as tumour necrosis factor (TNF)-α and IL-6 (Zeyda and Stulnig 2007). It has been reported that the infiltration of $CD8^+$ T cells (Nishimura et al. 2009) and neutrophils precedes macrophage accumulation and promotes the recruitment and activation of macrophages (Elgazar-Carmon et al. 2008).

An obese state results in an elevation of circulating levels of fatty acids (FAs) and, subsequently, an increase in inflammation of adipose tissue (Fielding et al. 1996). Adipose macrophages play critical roles in the immune responses through several FA-sensing mechanisms, such as pattern recognition receptors (PRRs). PRRs have provided great insights into immune responses against pathogens (Kawai and Akira 2009). Several families of PRRs such as Toll-like receptors (TLRs) and nucleotide oligomerisation domain (NOD)-like receptors (NLRs) have been discovered. It is now becoming ever more apparent that these PRRs are not only able to recognise microbial components but also mediate immune responses to endogenous molecules, including those arising in metabolic disorders, such as FAs. These endogenous molecules have been termed danger-associated molecular patterns (DAMPs) and have similar functions as microbial components to activate immune responses (Chen and Nunez 2010). For example, saturated FAs activate TLR4 and drive cytokine production in macrophages by activating the nuclear factor-kB (NF-kB) pathway, leading to increased expression of proinflammatory target genes (Shi et al. 2006; Senn 2006).

Inflammation induces insulin resistance via a two-hit process. Firstly, activated immune cells accumulate in tissues, such as adipose tissue, and release proinflammatory cytokines, such as TNF-α. Secondly, these cytokines act on insulin target cells, causing insulin resistance. In the target cells, activation of TLRs and/or TNF receptor leads to activation of serine kinases, including IkB kinase (IKK) and c-Jun N-terminal kinase (JNK), which inhibit insulin signalling (Osborn and Olefsky 2012). In addition, IL-1β is another key cytokine that impairs insulin sensitivity and insulin secretion by pancreatic β cells (Tack et al. 2012).

Even today, plant-derived natural products and their derivatives and synthetic mimics make up a considerable portion of current drugs. These products have played an important role in treating T2DM especially in Asian countries. With the rapid advancement of novel technologies and the increased research on antidiabetic natural products, many new plants, their extracts, and their active compounds have been identified to exhibit anti-inflammatory and antidiabetic effects. We reported that glycyrrhizin (GL) and isoliquiritigenin (ILG), components of *Glycyrrhiza uralensis* (*G. uralensis*), inhibit TLR4 signalling at the receptor level, resulting in inhibition of NF-kB and mitogen-activated protein kinases (MAPKs) activation (Honda et al. 2012). Furthermore, ILG potently inhibits NLRP3 inflammasome activation independent of its inhibitory action on TLR4 (Honda et al. 2014). Our in vivo study revealed that ILG attenuated high-fat diet (HFD)-induced adipose tissue inflammation and insulin resistance by inhibiting IL-1β and caspase-1 production in adipose tissue (Honda et al. 2014). On the basis of these findings, natural products derived from *G. uralensis* may serve as lead compounds for the development of new antidiabetic drugs.

This review article highlights the recent discoveries of the anti-inflammatory effects and mechanisms of action of natural products that target PRRs. We overview the roles of innate immunity and discuss potential mechanisms by which it participates in chronic inflammation of obesity-associated inflammation. We also overview our findings and other natural products that target TLRs and inflammasomes.

30.2 Innate Immune Receptors That Induce Obesity-Associated Inflammation and Insulin Resistance

30.2.1 Toll-Like Receptors

Exogenous signals can modify inflammatory responses in obesity. TLRs are transmembrane receptors that are important for sensing conserved structural moieties of microorganisms and for the subsequent induction of proinflammatory responses (Kawai and Akira 2009). Following ligand recognition, they activate the NF-kB and MAPK pathways to induce the production of proinflammatory cytokines that are important for evading pathogens. TLRs also sense nonmicrobial endogenous ligands, such as dietary FAs (Chen and Nunez 2010). Recent advances in deciphering TLR signalling networks reveal that ligation of TLRs by the endogenous ligands similarly activates proinflammatory pathways as microbial ligands in various organs, including adipose tissue, liver, and central nervous system.

TLR4 is the most important member of the TLR family proteins for LPS recognition and LPS-mediated inflammatory responses (Hoshino et al. 1999). There is a body of evidence suggesting that TLR4 is an attractive candidate for linking innate immune responses to obesity and insulin resistance. Firstly, TLR4 expression is increased in adipose tissue inflammatory macrophages in obesity (Shi et al. 2006; Nguyen et al. 2007). Secondly, TLR4 KO mice or mice with a loss-of-function mutation in the TLR4 gene are protected from obesity-induced insulin resistance (Shi et al. 2006; Tsukumo et al. 2007). Thirdly, hematopoietic cell-specific deletion of TLR4 ameliorates HFD-induced hepatic and adipose tissue insulin resistance (Saberi et al. 2009). Fourthly, saturated FAs released by adipocyte lipolysis activate the NF-kB pathway in vitro through TLR4 on macrophages (Suganami et al. 2007). Additionally, LPS derived from the gut microflora can leak into the circulation in obesity, resulting in higher LPS concentration in serum, and LPS signals through TLR4 stimulate production of proinflammatory cytokines in adipose tissue (Cani et al. 2008). Another report describes that an adipokine resistin derived from adipose tissue directly binds to TLR4 in the hypothalamus and leads to the activation of MAPKs signalling and promoting insulin resistance through MyD88 (Benomar et al. 2013), suggesting that resistin is an endogenous TLR4 ligand, which links hypothalamic inflammation with insulin resistance.

Radioprotective 105 (RP105) was identified as a TLR family protein expressed on B cells (Miyake et al. 1995). Both RP105 and TLR4 contain 22 LRRs in their extracellular portions, suggesting the possible involvement of RP105 in the LPS-induced responses. In fact, RP105-deficient mice show reduced LPS-dependent proliferation and CD86 upregulation in B cells, albeit to a lesser extent than TLR4-deficient mice (Ogata et al. 2000; Nagai et al. 2002). We have found that murine epididymal white adipose tissue (eWAT) expresses RP105 mRNA and this expression was markedly increased by HFD supplementation (Watanabe et al. 2012). The cell surface expression of RP105 is also increased on inflammatory M1 ATMs by HFD. Using a coculture system composed of 3 T3-L1 adipocytes and macrophage cell lines, we have shown that RP105 mRNA expression is increased in macrophages in parallel with the upregulation of TNF-α mRNA expression, although to a lesser extent expression of TLR4 mRNA is observed in the contact coculture system. Furthermore, HFD-induced obesity, adipose tissue inflammation, and insulin resistance are severely attenuated in RP105 KO mice compared with wild-type (WT) and TLR4 KO mice. These results clearly suggest that the induction of obesity-related inflammation and metabolic disorders by HFD may require or be dependent on the RP105 pathway rather than the TLR4 pathway. However, RP105 may recognise lipids other than palmitic acid, inasmuch as RP105-deficient macrophages normally respond to this FA as well as WT macrophages (Watanabe et al. 2012).

30.2.2 Nod-Like Receptors and Inflammasomes

NLRs and inflammasomes are cytoplasmic receptors and play an important role in the host defence against microbial infection and the homeostasis of the gut. Genetic evidence shows that the impaired recognition and killing of commensal bacteria in NLR-deficient mice can contribute to the development of metabolic disorders (De Nardo and Latz 2011). NOD proteins propagate inflammatory signals in response to peptidoglycan (PGN) (Carneiro et al. 2008). NOD1 recognises D-glutamyl-meso-diaminopimelic acid-containing PGN found in Gram-positive bacteria, whereas NOD2 recognises muramyl dipeptide present in all bacteria (Carneiro et al. 2004, 2008). NOD1/2 double KO mice are protected from HFD-induced inflammation, lipid accumulation, and insulin resistance in adipose tissue and liver (Schertzer et al. 2011). Conversely, NOD1 ligands induce profound acute insulin resistance in mice without major systemic inflammation, potentially mediated by direct action on hepatocytes and adipocytes and indirect action on skeletal muscle.

Activation of NLRP3 inflammasome is regulated by various sterile stimuli, including cholesterol crystals, β-amyloid, palmitic acid, and ceramides. It is generally accepted that two signals are required for NLRP3 inflammasome activation. One is a NF-kB-dependent priming step that induces the transcription of pro-IL-1β and NLRP3. Another is an activation step that induces the activation of caspase-1. Normal activation of NLRP3 inflammasome contributes to host defence, but several studies suggest that excessive activation leads to the development of obesity-associated inflammation and insulin resistance.

Islet amyloid polypeptide (IAPP) is deposited in the pancreas and associated with the loss of β cell function in T2DM. The observation of NLRP3-dependent IL-1β production by macrophages in response to IAPP implied a potential role for NLRP3 in promoting IL-1β secretion in T2DM (Masters et al. 2010). Interestingly, an antidiabetic drug glyburide blocks IL-1β production by macrophages in response to IAPP. Direct involvement of NLRP3 in obesity has been confirmed in studies that NLRP3 KO mice fed HFD display reduced caspase-1 activation and pro-IL-1β expression in adipose tissue compared with WT mice (Stienstra et al. 2011). Moreover, NLRP3 KO and caspase-1 KO mice are more protected from HFD-induced insulin resistance. The decrease in insulin sensitivity is a consequence of NLRP3 inflammasome-mediated activation of effector T cells that, through the release of interferon-γ, mediate downstream pathways resulting in insulin resistance.

30.3 Natural Products That target TLR or Inflammasome-Associated Inflammation

30.3.1 Herbal Medicines with Anti-Inflammatory Activity

Chinese herbal formulas with antidiabetic effects are well developed such that a number of these formulas have commonly been used in diabetic patients since ancient times. The 10 most frequently used Chinese herbs in the period from 2004 to 2009, for the treatment of T2DM, include *Radix Astragali*, *Rhizome Dioscoreae*, *Radix Rehnabbiae*, *Radix Salviae Miltiorrhizae*, *Radix Puerariae*, *Rhizoma Coptidis*, *Fructys Lycii*, *Poria*, *Rhizoma Alismatis*, and *Fructus Corni* (Qi et al. 2010). However, it is unclear whether these Chinese herbs have as potent anti-inflammatory effects as those of Western antidiabetic drugs. The antidiabetic Chinese drugs are mostly heat-clearing and detoxifying drugs (Zhang 2007). It has been reported that heat-clearing and detoxifying drugs usually have anti-inflammatory effects (Ren et al. 1994). The main active compounds of antidiabetic Chinese herbs, including polysaccharides, terpenoids, flavonoids, and alkaloids, usually have wide anti-inflammatory activities. Therefore, these activities may exert beneficial effects on the inflammatory process of T2DM.

30.3.2 Potent Inhibitors of TLR4 and NLRP3 Inflammasome Activation Derived from G. uralensis

Glycyrrhiza plants (licorice) have been used as herbal medicine worldwide for over 4000 years (Shibata 2000). Several studies have reported that Chinese and Japanese herbal medicines or their components regulate innate immunity. Roasted licorice extracted by ethanol demonstrates potent anti-inflammatory action by reducing NO, PGE2, and proinflammatory cytokines. Among *Glycyrrhiza* plants, *G. uralensis* is one of the most used herbal medicines in Asian countries. Various components have been isolated from licorice, for example, triterpene saponins, flavonoids, isoflavonoids, and chalcones. Therefore, we have focused on the anti-inflammatory effects of *G. uralensis* and its components on innate immune responses.

30.3.2.1 Glycyrrhizin

Glycyrrhizin (GL) is a triterpene saponin and considered to be the major biological active ingredient of *G. uralensis* (Fig. 30.1a). It has been reported that GL inhibits LPS-induced TLR4 internalisation (Schrofelbauer et al. 2009). We demonstrated that GL suppresses lipid A moiety of LPS-induced IL-6 production in mouse macrophages (Honda et al. 2012). Furthermore, administration of GL effectively suppresses the production of inflammatory cytokines, including TNF-α and IL-6, in

LPS-injected mice (Honda et al. 2012). We further demonstrated that GL attenuates lipid A-mediated activation of NF-kB and MAPKs, including JNK, p38, and ERK. It was suggested that GL might be incorporated into lipid bilayers and suppress the plasma membrane integrity (Schrofelbauer et al. 2009). LPS binds to MD-2 and this triggers homodimerisation of the TLR4/MD-2 complex, resulting in the induction of signal transduction (Saitoh et al. 2004). Then, we first examined whether GL affected the binding of LPS to the TLR4/MD-2 complex. GL stimulation inhibited its binding to the complex in a dose-dependent manner (Fig. 30.2a). Accordingly, LPS-induced TLR4 homodimerisation was not observed by GL stimulation. Moreover, GL inhibited not only LPS- but also TLR9 ligand CpG-DNA–induced inflammatory responses (our unpublished data), as suggested previously (Schrofelbauer et al. 2009). These results suggest that the inhibitory effect of GL is not specific to TLR4 and GL may not interact directly with TLRs to interfere with their signalling. GL may inhibit the activation of multiple TLRs at the plasma membrane by altering membrane integrity.

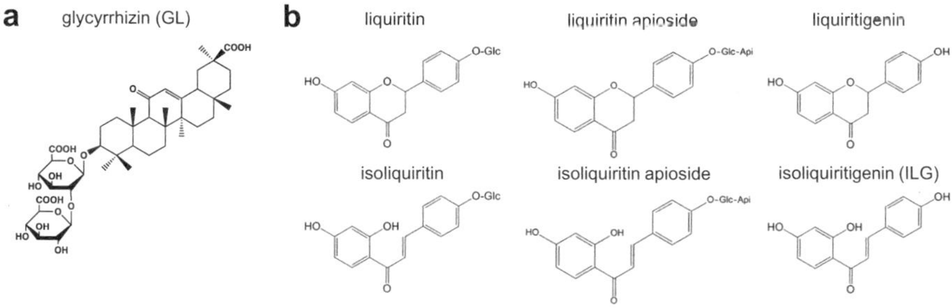

Fig. 30.1 Chemical structures of a saponin and flavanones/chalcones from extracts of *G. uralensis*. (**a**) Chemical structure of GL. (**b**) Chemical structures of ILG and its related flavanines/chalcones

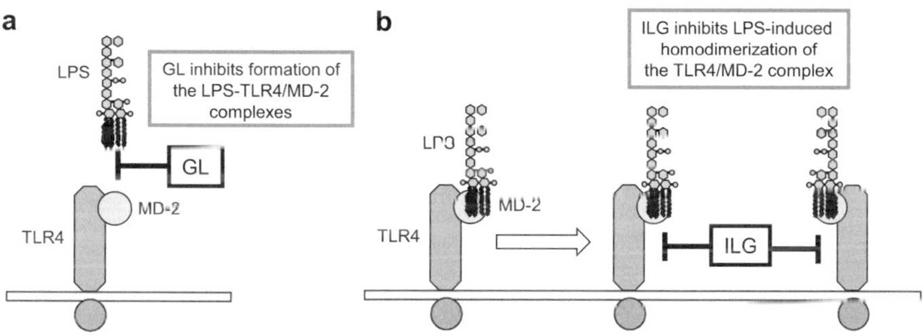

Fig. 30.2 Schematic diagrams of GL- and ILG-mediated suppression of the TLR4/MD-2 complex activation. (**a**) GL inhibits LPS binding to the TLR4/MD-2 complex. Accordingly, LPS-induced homodimerisation of the complex cannot be induced in the presence of GL. (**b**) LPS can bind to TLR4/MD-2 in the presence of ILG. However, LPS-induced homodimerisation of TLR4/MD-2 is inhibited by ILG

Interestingly, GL inhibits the activation of NLRP3 inflammasome induced by various stimuli, including adenosine triphosphate (ATP), monosodium urate (MSU), and nigericin (Fig. 30.3; Honda et al. 2014). It is noted that this inhibitory activity of GL is independent of its effect on TLR4 activation. LPS is a potent inducer of the priming of the NLRP3 inflammasome. LPS-induced pro-IL-1β protein expression and IL-1β mRNA expression were remarkably decreased by GL stimulation when added at the priming step (Honda et al. 2014). In contrast, these expressions were not affected when GL was added at the activation step. These findings clearly show that GL inhibits NLRP3 inflammasome activation without affecting TLR4 activation. Because absent in melanoma 2 (AIM2) inflammasome-dependent IL-1β and caspase-1 productions were decreased by GL stimulation, GL inhibits the activity of multiple inflammasomes (Fig. 30.3). However, a much higher concentration of GL is required for these inhibitory effects on inflammasomes than that of ILG. Therefore, GL may not be suggested to be a potential therapeutic agent for the treatment of obesity-associated inflammation.

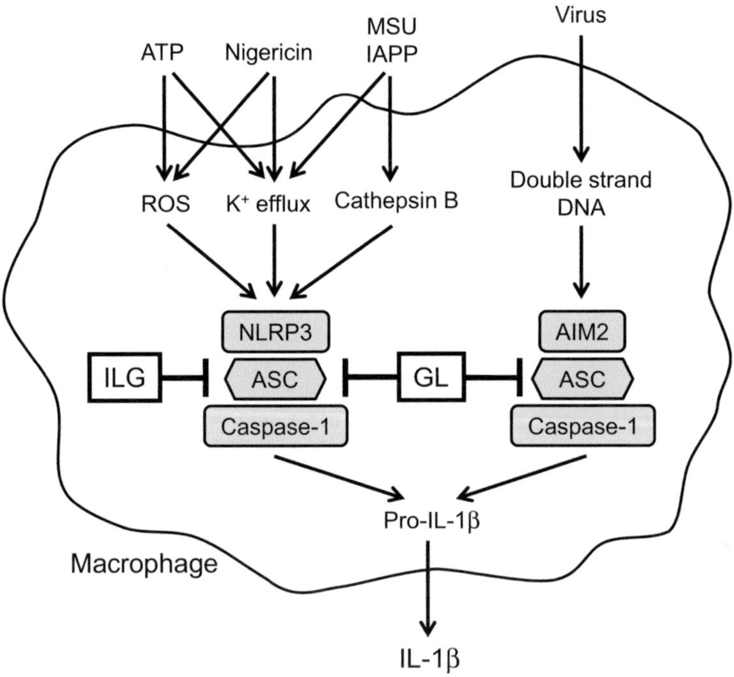

Fig. 30.3 Schematic diagram of GL- and ILG-mediated suppression of the inflammasome activation in macrophage. NLRP3 inflammasome is activated by ATP, nigericin, MSU, and IAPP. AIM2 inflammasome is activated viral ds-DNA. The inflammasomes assemble into an oligomeric complex with ASC and activate caspase-1. Activated caspase-1 processes pro-IL-1β into mature IL-1β. ILG potently inhibits NLRP3 but not AIM2 inflammasome activation. GL inhibits both NLRP3 and AIM2 inflammasome activation, but a high concentration of GL is needed for the inhibitory effects compared with ILG. ILG and GL suppress ligand-induced formation of ASC pyroptosome

30.3.2.2 Isoliquiritigenin

ILG is another component of *G. uralensis* and a flavonoid with a chalcone structure (Fig. 30.1b). It has a wide variety of biological activities, including antiallergic, antitumour growth, and antiplatelet aggregation activity. We recently demonstrated that ILG and the glycoside isoliquiritin potently suppresses lipid A-induced IL-6 production in mouse macrophages (Honda et al. 2012, Fig. 30.1).

It has been reported that ILG inhibits LPS-induced homodimerisation of TLR4 (Park and Youn 2010). We confirmed that LPS-induced homotypic interactin of TLR4 is potently inhibited by ILG stimulation (Honda et al. 2012, Fig. 30.2b). Then, we considered whether ILG affects the formation of the LPS-TLR4/MD-2 complexes on the cell surface. Whereas GL stimulation affects the binding of LPS to TLR4/MD-2, the amount of bound LPS is not decreased by ILG stimulation. It has been reported that an isothiocyanate sulforaphane suppresses TLR4 oligomerisation by forming adducts with cysteine residues in the extracellular domain of TLR4 (Youn et al. 2010). Because ILG has an α,β-unsaturated carbonyl group to react with the thiol group of cysteine (Kamei et al. 2003), it may interact with the extracellular domain of TLR4. It has been reported that ILG interacts with IKK directly and inhibits its kinase activity (Kumar et al. 2007). This is consistent with our finding that ILG strongly inhibits lipid A-induced IKKα/β phosphorylation (Honda et al. 2012). Thus, ILG suppresses TLR4/MD-2–mediated immune responses in multiple steps, at the receptor level and the downstream signalling level.

ILG also potently inhibits the activation of NLRP3 inflammasome independent of its inhibitory effects on TLR4 signalling (Fig. 30.3; Honda et al. 2014). Moreover, ILG is highly effective in inhibiting NLRP3 inflammasome activation by various stimuli including ATP, MSU, and nigericin, compared with GL and the NLRP3 inhibitor parthenolide. Among selected flavanones/chalcones of *G. uralensis* (Fig. 30.1b), an aglycone liquiritigenin as well as ILG markedly attenuates ATP-induced NLRP3 inflammasome activation. Glycosides (liquiritin, liquiritin apioside, isoliquiritin, and isoliquiritin apioside) have no inhibitory effects on the NLRP3 inflammasome. Contrary to GL, ILG does not inhibit poly (dA:dT)-induced activation of the AIM2 inflammasome. As ILG suppresses NLRP3- but not AIM2-induced formation of ASC pyroptosome, it is unlikely that ASC is a molecular target of ILG. NLRP3 itself or upstream of it may be a target of ILG.

An important observation is that ILG is highly effective in inhibiting islet amyloid polypeptide (IAPP)-induced NLRP3 inflammasome activation compared with GL and parthenolide (Honda et al. 2014). IAPP is a unique polypeptide constituent of amyloid deposited in pancreatic islets (Cooper et al. 1987; Westermark et al. 1987). This deposition is associated with disease progression of T2DM and triggers NLRP3 inflammasome activation in islet macrophages (Masters et al. 2010). The sulfonylurea drug glyburide inhibits IAPP-induced activation of NLRP3 inflammasome (Masters et al. 2010). It is noteworthy that a low concentration of ILG is more effective in inhibiting IAPP-induced activation of NLRP3 inflammasome than that of glyburide.

The above observations clearly demonstrate that ILG would be useful to modulate TLR4- and NLRP3-inflammasome–mediated inflammation (Fig. 30.4). Thus, we evaluated the inhibitory effect of ILG on obesity-associated inflammation. ILG supplementation remarkably improves obesity, hyperlipidemia, hepatic steatosis, and insulin resistance in HFD-fed wild-type mice (Honda et al. 2014). Furthermore, ILG supplementation inhibits IL-1β and caspase-1 production in eWAT from wild-type mice fed with HFD for 4 weeks. At this time point, TNF-α production is not increased in eWAT from HFD-fed mice compared with that from normal diet-fed mice. These results clearly suggest that inflammasome activation occurs in eWAT at an early time point during obesity before TNF-α–associated inflammation. Moreover, ILG is suggested to be a potential therapeutic agent that targets NLRP3-associated adipose tissue inflammation in obesity.

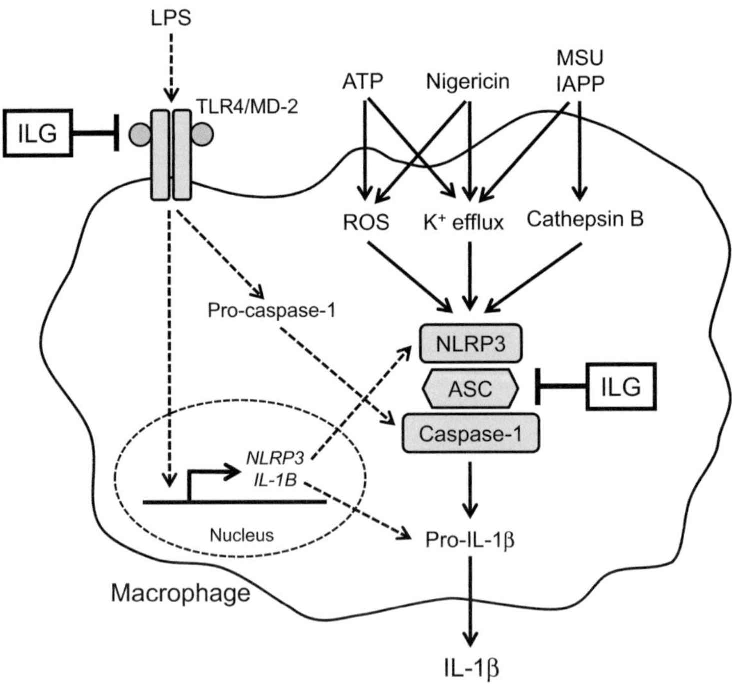

Fig. 30.4 Schematic diagram of ILG-mediated suppression of TLR4 and NLRP3 inflammasome activation in macrophage. ILG inhibits LPS-induced homodimerisation of TLR4/MD-2 at the cell surface, thereby suppressing procaspase-1, NLRP3, IL-1β expression. ILG inhibits activation of NLRP3 inflammasome via a pathway that is independent of its inhibitory effect on the TLR4/MD-2 pathway. *Solid arrows* indicate the pathways that are related to NLRP3 activation. *Dashed arrows* indicate the pathways that related to TLR4/MD-2 activation

30.3.3 Other Natural Products That Target Innate Immune Receptors

Activation of TLR4 and NLRP3 inflammasome has been suggested to be linked to the development of not only obesity-associated diseases but also other inflammatory diseases, including bacterial/viral infection, sepsis, brain infarction, and cancers. Recent studies have reported that many natural products have an antagonistic activity against TLRs and inflammasomes. Among them, flavones or flavonoids are frequently reported to have inhibitory effects on the receptors. We overview some natural products, including flavone, flavonoid, polyphenol, and terpenoid, which have a therapeutic potential to treat TLR- or inflammasome-related inflammatory diseases.

30.3.3.1 Sparstolonin B

Sparstolonin B (SsnB) is a novel polyphenol with structural features of xanthone and isocoumarin (Liang et al. 2011). SsnB is derived from a Chinese herb *Sparganium stoloniferum* (*S. stoloniferum*), which is a perennial aquatic plant grown in North and East China. Its tubers have been used in traditional Chinese medicine for the treatment of several inflammatory diseases. Liang at al. isolated SsnB from the tubers of *S. stoloniferum* and its biological activity (Liang et al. 2011). SsnB selectively inhibits TLR2- and TLR4-induced inflammatory cytokine expression in mouse macrophages, with attenuating phosphorylation of IkBα and MAPKs. Furthermore, SsnB may be associated with the intracellular domain of TLR4 and inhibits the recruitment of MyD88 to TLR4. Although further study is required to evaluate the biological properties of SsnB, it may be a promising lead for drug development to treat TLR2- or TLR4-associated inflammatory diseases.

30.3.3.2 Baicalein

Bovine mastitis is considered to be a bacterial infection of the mammary gland and causes the greatest economic loss to dairy farming around the world. *Scutellaria baicalensis Georgi* (*S. baicalensis Georgi*) is a traditional Chinese medicine and has been used to treat mastitis. The active components of the root of *S. baicalensis Georgi* are flavonoids. It contains four major flavonoids, baicalin and its aglycone baicalein, and wogonoside and its aglycone wogonin. It has been reported that baicalein has a therapeutic effect on endotoxin-induced tissue injury including acute liver failure, glomerulonephritis, and septic shock in mice. Then, He et al. investigated the anti-inflammatory effect and mechanism of baicalein in LPS-induced mastitis in mice (He et al. 2015). Baicalein treatment significantly attenuated LPS-induced pathological changes of mammary tissues with decreasing cytokine expression myeloperoxidase activity. Western blot analysis showed that baicalein inhibited LPS-induced phosphorylation of IkBα and MAPKs. Moreover, LPS-induced upregulation of TLR4 and MyD88 expression in mammary tissue was decreased by baicalein stimulation in a dose-dependent manner. This decrease may be responsible for the inhibitory effects of baicalein against TLR4-mediated responses. Interestingly, baicalin decreases TLR2 expression and suppresses TLR2-mediated responses in mammary gland cells in *Staphylococcus aureus*-induced mastitis. Therefore, the major flavonoids of *S. baicalensis Georgi* have inhibitory effects on multiple TLRs and their related mastitis. Other natural products including bergenin (Gao et al. 2015), salidroside (Li et al. 2013a), and curcumin (Fu et al. 2014) are suggested to inhibit LPS-induced mastitis in mice.

30.3.3.3 Resveratrol

A previous study identified the important roles of microtubules in the activation of NLRP3 inflammasome (Misawa et al. 2013). The authors demonstrated that inducers of NLRP3 inflammasome accumulated acetylated α-tubulin in the perinuclear region of mouse macrophages. These accumulated α-tubulins stimulate the spatial arrangement of the mitochondria and promote the assembly of NLRP3 and ASC. They also demonstrated that the NAD^+-dependent deacetylase SIRT2 regulates the acetylation of α-tubulin. Resveratrol is a phytoalexin found in the skin and seeds of grapes and has been reported to have beneficial effects on the treatment of a variety of diseases such as cancer, atherosclerosis, T2DM, and obesity (Park and Pezzuto 2015). Misawa et al. reported that resveratrol suppresses NLRP3 inflammasome activation (Misawa et al. 2015). Resveratrol treatment inhibits the activation of NLRP3 inflammasome induced by various stimuli, including ATP, nigericin, silica, and MSU in mouse macrophages. Furthermore, resveratrol treatment inhibits IL-1β production and neutrophil accumulation in the MSU-induced peritonitis mouse model. In contrast, flagellin- or ds-DNA–induced AIM2 inflammasome activation was not inhibited by resveratrol treatment. It has been reported that resveratrol regulates the enzymatic activity of SIRT family members including SIRT2 (Suzuki and Koike 2007). The authors speculate that the mechanism by which resveratrol suppresses NLRP3 inflammasome activation is that resveratrol inhibits acetylated α-tubulin accumulation by promoting SIRT2 activation.

30.3.3.4 Arglabin

Terpenoids are the largest class of secondary metabolites in plants and exhibit a broad range of biological activities (Toyang et al. 2013; Syrovets et al. 2005; Morad et al. 2013) and potential therapeutic effects on atherosclerosis, psoriasis, and liver cirrhosis (Cuaz-Perolin et al. 2008; Liu et al. 2013). Arglabin is an guaianolide sesquiterpene lactone and is mainly synthesised in *Artemisia glabella*, which grows in Kazakhstan. Abderrazak et al. reported that arglabin inhibited cholesterol crystal-

or ATP-induced activation of NLRP3 inflammasome in mouse macrophages (Abderrazak et al. 2015). LPS-induced cytokine production is not inhibited by arglabin stimulation, indicating that arglabin has no inhibitory effects on the NF-kB–dependent priming step of NLRP3 inflammasome. Furthermore, arglabin activates autophagy, which leads to increased degradation of NLRP3. In vivo analyses have revealed that arglabin supplementation decreases the levels of plasma IL-1β, total cholesterol, and triglycerides in an atherosclerosis mouse model $ApoE_2$.Ki mice fed HFD (Abderrazak et al. 2015). Furthermore, arglabin supplementation markedly reduces the atherosclerosis lesion size in the aortic sinus and in whole aorta of HFD-fed $ApoE_2$.Ki mice. Additionally, arglabin may convert M1 macrophages into M2 macrophages in the spleen and arterial lesions. Thus, arglabin is a promising natural compound that treats atherosclerosis through NLRP3 inflammasome inhibition.

30.3.3.5 Chrysophanol

Chrysophanol (CHR) is a member of the anthraquinone family and is extracted from plants of the *Rheum genus*. CHR has been shown to have anti-inflammatory activity. A previous study has demonstrated that CHR inhibits caspase-1 and its downstream cytokine expression in colitis (Kim et al. 2010). It has been reported that IL-1β plays a crucial role in the development of inflammation in cerebral infarction (Li et al. 2013b). Zhang et al. examined the role of NLRP3 inflammasome in inflammatory processes of cerebral infarction by using middle cerebral artery occlusion (MCAO) and investigated whether CHR has any protective effect on the disease process (Zhang et al. 2014). The result shows that expression levels of NLRP3, ASC, caspase-1, and IL-1β in the brain are upregulated after MCAO. Furthermore, systemic administration of CHR is effective in preventing the infarct volume, brain oedema, and neurological deficits during MCAO with reducing the elevated expressions of NLRP3, caspase-1, and IL-1β. Although the underlying mechanisms mediating the observed neuroprotective effects of CHR are still unclear, CHR is a potential therapeutic agent for the treatment of ischemic stroke and other NLRP3-associated diseases.

30.3.3.6 Luteoloside

Luteoloside is a flavone subclass of flavonoids and has potential anti-inflammatory effects (Hu and Kitts 2004). Furthermore, it has been reported that luteoloside inhibits the proliferation of colon cancer cells (Baskar et al. 2011). NLRP3 inflammasome plays an important role in the development of several cancers, including, colorectal cancer (Ungerback et al. 2012), intestinal cancer (Chen and Nunez 2011), and melanoma (Ahmad et al. 2013). Then, Fan et al. investigated whether NLRP3 inflammasome is involved in the process of hepatocellular carcinoma (HCC) and whether luteoloside is a potential agent against human hepatoma cells both in vitro and in vivo (Fan et al. 2014). The results have shown that luteoloside inhibits the proliferation, migration, and the invasion of several HCC cell lines in vitro. Furthermore, in vivo treatment of luteoloside significantly inhibits the development and metastasis of HCC cells without any adverse effects. Interestingly, the expression levels of NLRP3, caspase-1, and IL-1β proteins are elevated in two HCC cell lines and these expressions are decreased by luteoloside treatment *in vitro*. ROS has been associated with cancer development. Fan et al. found that intracellular ROS is accumulated in the HCC cell lines and this is inhibited by luteoloside in a dose-dependent manner (Fan et al. 2014). These results indicate that luteoloside suppresses the development of HCC cells by inhibiting the expression of NLRP3 components and the accumulation of intracellular ROS. Although it is still unclear whether luteoloside directly inhibits NLRP3 inflammasome activation in HCC cells, its effects may be beneficial for the prevention of NLRP3-associated tumour proliferation, invasion, and metastasis.

30.3.3.7 Korean Red Ginseng

Korean red ginseng (*Panax Ginseng C.A. Meyer*) is one of the most studied traditional herbal medicines. Ginsenosides are saponins and the major active component of ginseng. Ginsenosides have anti-inflammatory effects by inhibiting NF-kB activation (Lu et al. 2009). Kim et al. reported that Korean red ginseng extract inhibits NLRP3 inflammasome activation induced by various stimuli, including ATP, nigericin, and aluminium potassium sulfate (Kim et al. 2014). Interestingly, Korean red ginseng extract inhibits double-stranded DNA (ds-DNA)- or *Listeria*-mediated AIM2 inflammasome activation. Among several compounds of red ginseng, ginsenoside Rh1 and ginsenoside Rg3 are responsible for the inhibitory effects on both NLRP3 and AIM2 inflammasome activation. Because ginseng extract has been shown to decrease the elevation of ROS production and intracellular Ca^{2+} in various cell types, these events are suggested to be mediators of the inhibitory effect of ginseng extract on NLRP3 inflammasome.

30.4 Concluding Remarks

There is accumulating evidence suggesting that traditional herbal medicines contain various kinds of natural compounds that suppress chronic inflammation through the inhibition of PRRs activation. There is as yet insufficient data to demonstrate the interaction between innate and adaptive immune systems contributing to adipose tissue inflammation and further studies are needed. Meanwhile, we have identified critical pharmacological activities and molecular mechanisms of two components of *G. uralensis* in suppressing TLR4/MD-2 signalling and NLRP3 inflammasome activation. ILG may be one of the most useful components of *G. uralensis* for the treatment of TLR4- and NLRP3-associated inflammatory disease. Our findings could contribute to the development of therapeutic approaches for the treatment of obesity-associated inflammation. Additionally, such development will require a much greater understanding of PRR/endogenous ligand interactions.

Acknowledgment We thank all members of our laboratories in University of Toyama and Toyama Prefectural Institute for Pharmaceutical Research for their helpful discussions. We also sincerely thank Toyama Prefecture for supporting our laboratory.

This work was supported by grants from Grant-in-Aid for Scientific Research from the Japan Society for the Promotion of Science (JSPS) (JSPS KAKENHI Grant Numbers 25870257 to Y.W., 24590577 to Y.N., 15K07960 to H.H. and 24390119 to K.T.), JST, PRESTO (Y.N.), Hokuriku Innovation Cluster for Health Science, MEXT Regional Innovation Cluster Program, Toyama/Ishikawa Region (K.T.), Hokuriku Life Science Cluster, MEXT Regional Innovation Strategy Support Program (K.T.), and First Bank of Toyama Scholarship Foundation (Y.W.).

References

Abderrazak A, Couchie D, Mahmood DF, Elhage R, Vindis C, Laffargue M, Mateo V, Buchele B, Ayala MR, El Gaafary M, Syrovets T, Slimane MN, Friguet B, Fulop T, Simmet T, El Hadri K, Rouis M (2015) Anti-inflammatory and antiatherogenic effects of the NLRP3 inflammasome inhibitor arglabin in ApoE2.Ki mice fed a high-fat diet. Circulation 131(12):1061–1070. doi:10.1161/CIRCULATIONAHA.114.013730

Ahmad I, Muneer KM, Tamimi IA, Chang ME, Ata MO, Yusuf N (2013) Thymoquinone suppresses metastasis of melanoma cells by inhibition of NLRP3 inflammasome. Toxicol Appl Pharmacol 270(1):70–76. doi:10.1016/j.taap.2013.03.027

Baskar AA, Ignacimuthu S, Michael GP, Al Numair KS (2011) Cancer chemopreventive potential of luteolin-7-O-glucoside isolated from Ophiorrhiza mungos Linn. Nutr Cancer 63(1): 130–138. doi:10.1080/01635581.2010.516869

Benomar Y, Gertler A, De Lacy P, Crepin D, Ould Hamouda H, Riffault L, Taouis M (2013) Central resistin overexposure induces insulin resistance through Toll-like receptor 4. Diabetes 62(1): 102–114. doi:10.2337/db12-0237

Cani PD, Bibiloni R, Knauf C, Waget A, Neyrinck AM, Delzenne NM, Burcelin R (2008) Changes in gut microbiota control metabolic endotoxemia-induced inflammation in high-fat diet-induced obesity and diabetes in mice. Diabetes 57(6):1470–1481. doi:10.2337/db07-1403

Carneiro LA, Travassos LH, Philpott DJ (2004) Innate immune recognition of microbes through Nod1 and Nod2: implications for disease. Microbes Infect 6(6):609–616. doi:10.1016/j.micinf.2004.01.012

Carneiro LA, Magalhaes JG, Tattoli I, Philpott DJ, Travassos LH (2008) Nod-like proteins in inflammation and disease. J Pathol 214(2):136–148. doi:10.1002/path.2271

Chen GY, Nunez G (2010) Sterile inflammation: sensing and reacting to damage. Nat Rev Immunol 10(12):826–837. doi:10.1038/nri2873

Chen GY, Nunez G (2011) Inflammasomes in intestinal inflammation and cancer. Gastroenterology 141(6):1986–1999. doi:10.1053/j.gastro.2011.10.002

Cooper GJ, Willis AC, Clark A, Turner RC, Sim RB, Reid KB (1987) Purification and characterization of a peptide from amyloid-rich pancreases of type 2 diabetic patients. Proc Natl Acad Sci U S A 84(23):8628–8632

Cuaz-Perolin C, Billiet L, Bauge E, Copin C, Scott-Algara D, Genze F, Buchele B, Syrovets T, Simmet T, Rouis M (2008) Antiinflammatory and antiatherogenic effects of the NF-kappaB inhibitor acetyl-11-keto-beta-boswellic acid in LPS-challenged ApoE−/− mice. Arterioscler Thromb Vasc Biol 28(2):272–277. doi:10.1161/ATVBAHA.107.155606

De Nardo D, Latz E (2011) NLRP3 inflammasomes link inflammation and metabolic disease. Trends Immunol 32(8):373–379. doi:10.1016/j.it.2011.05.004

Elgazar-Carmon V, Rudich A, Hadad N, Levy R (2008) Neutrophils transiently infiltrate intra-abdominal fat early in the course of high-fat feeding. J Lipid Res 49(9):1894–1903. doi:10.1194/jlr.M800132-JLR200

Fan SH, Wang YY, Lu J, Zheng YL, Wu DM, Li MQ, Hu B, Zhang ZF, Cheng W, Shan Q (2014) Luteoloside suppresses proliferation and metastasis of hepatocellular carcinoma cells by inhibition of NLRP3 inflammasome. PLoS One 9(2), e89961. doi:10.1371/journal.pone.0089961

Fielding BA, Callow J, Owen RM, Samra JS, Matthews DR, Frayn KN (1996) Postprandial lipemia: the origin of an early peak studied by specific dietary fatty acid intake during sequential meals. Am J Clin Nutr 63(1):36–41

Fu Y, Gao R, Cao Y, Guo M, Wei Z, Zhou E, Li Y, Yao M, Yang Z, Zhang N (2014) Curcumin attenuates inflammatory responses by suppressing TLR4-mediated NF-kappaB signaling pathway in lipopolysaccharide-induced mastitis in mice. Int Immunopharmacol 20(1):54–58. doi:10.1016/j.intimp.2014.01.024

Gao XJ, Guo MY, Zhang ZC, Wang TC, Cao YG, Zhang NS (2015) Bergenin Plays an Anti-Inflammatory Role via the Modulation of MAPK and NF-kappaB Signaling Pathways in a Mouse Model of LPS-Induced Mastitis. Inflammation 38(3):1142–1150. doi:10.1007/s10753-014-0079-8

He X, Wei Z, Zhou E, Chen L, Kou J, Wang J, Yang Z (2015) Baicalein attenuates inflammatory responses by suppressing TLR4 mediated NF-kappaB and MAPK signaling pathways in LPS-induced mastitis in mice. Int Immunopharmacol 28(1):470–476. doi:10.1016/j.intimp.2015.07.012

Honda H, Nagai Y, Matsunaga T, Saitoh S, Akashi-Takamura S, Hayashi H, Fujii I, Miyake K, Muraguchi A, Takatsu K (2012) Glycyrrhizin and isoliquiritigenin suppress the LPS sensor toll-like receptor 4/MD-2 complex signaling in a different manner. J Leukoc Biol 91(6): 967–976. doi:10.1189/jlb.0112038

Honda H, Nagai Y, Matsunaga T, Okamoto N, Watanabe Y, Tsuneyama K, Hayashi H, Fujii I, Ikutani M, Hirai Y, Muraguchi A, Takatsu K (2014) Isoliquiritigenin is a potent inhibitor of NLRP3 inflammasome activation and diet-induced adipose tissue inflammation. J Leukoc Biol 96(6):1087–1100. doi:10.1189/jlb.3A0114-005RR

Hoshino K, Takeuchi O, Kawai T, Sanjo H, Ogawa T, Takeda Y, Takeda K, Akira S (1999) Cutting edge: Toll-like receptor 4 (TLR4)-deficient mice are hyporesponsive to lipopolysaccharide: evidence for TLR4 as the Lps gene product. J Immunol 162(7):3749–3752

Hotamisligil GS (2006) Inflammation and metabolic disorders. Nature 444(7121):860–867. doi:10.1038/nature05485

Hu C, Kitts DD (2004) Luteolin and luteolin-7-O-glucoside from dandelion flower suppress iNOS and COX-2 in RAW264.7 cells. Mol Cell Biochem 265(1–2):107–113

Kalupahana NS, Moustaid-Moussa N, Claycombe KJ (2012) Immunity as a link between obesity and insulin resistance. Mol Aspects Med 33(1):26–34. doi:10.1016/j.mam.2011.10.011

Kamei J, Nakamura R, Ichiki H, Kubo M (2003) Antitussive principles of Glycyrrhizae radix, a main component of the Kampo preparations Bakumondo-to (Mai-men-dong-tang). Eur J Pharmacol 469(1–3):159–163

Kawai T, Akira S (2009) The roles of TLRs, RLRs and NLRs in pathogen recognition. Int Immunol 21(4):317–337. doi:10.1093/intimm/dxp017

Kim SJ, Kim MC, Lee BJ, Park DH, Hong SH, Um JY (2010) Anti-Inflammatory activity of chrysophanol through the suppression of NF-kappaB/caspase-1 activation in vitro and in vivo. Molecules 15(9):6436–6451. doi:10.3390/molecules15096436

Kim J, Ahn H, Han BC, Lee SH, Cho YW, Kim CH, Hong EJ, An BS, Jeung EB, Lee GS (2014) Korean red ginseng extracts inhibit NLRP3 and AIM2 inflammasome activation. Immunol Lett 158(1–2):143–150. doi:10.1016/j.imlet.2013.12.017

Kumar S, Sharma A, Madan B, Singhal V, Ghosh B (2007) Isoliquiritigenin inhibits IkappaB kinase activity and ROS generation to block TNF-alpha induced expression of cell adhesion molecules on human endothelial cells. Biochem Pharmacol 73(10):1602–1612. doi:10.1016/j.bcp.2007.01.015

Li D, Fu Y, Zhang W, Su G, Liu B, Guo M, Li F, Liang D, Liu Z, Zhang X, Cao Y, Zhang N, Yang Z (2013a) Salidroside attenuates inflammatory responses by suppressing nuclear factor-kappaB and mitogen activated protein kinases activation in lipopolysaccharide-induced mastitis in mice. Inflamm Res 62(1):9–15. doi:10.1007/s00011-012-0545-4

Li H, Yin J, Li L, Deng J, Feng C, Zuo Z (2013b) Isoflurane postconditioning reduces ischemia-induced nuclear factor-kappaB activation and interleukin 1beta production to provide neuroprotection in rats and mice. Neurobiol Dis 54:216–224. doi:10.1016/j.nbd.2012.12.014

Liang Q, Wu Q, Jiang J, Duan J, Wang C, Smith MD, Lu H, Wang Q, Nagarkatti P, Fan D (2011) Characterization of sparstolonin B, a Chinese herb-derived compound, as a selective Toll-like receptor antagonist with potent anti-inflammatory properties. J Biol Chem 286(30): 26470–26479. doi:10.1074/jbc.M111.227934

Liu M, Chen P, Buchele B, Dong S, Huang D, Ren C, Zhang Y, Hou X, Simmet T, Shen J (2013) A boswellic acid-containing extract attenuates hepatic granuloma in C57BL/6 mice infected with Schistosoma japonicum. Parasitol Res 112(3):1105–1111. doi:10.1007/s00436-012-3237-7

Lu JM, Yao Q, Chen C (2009) Ginseng compounds: an update on their molecular mechanisms and medical applications. Curr Vasc Pharmacol 7(3):293–302

Lumeng CN, Saltiel AR (2011) Inflammatory links between obesity and metabolic disease. J Clin Invest 121(6):2111–2117. doi:10.1172/JCI57132

Masters SL, Dunne A, Subramanian SL, Hull RL, Tannahill GM, Sharp FA, Becker C, Franchi L, Yoshihara E, Chen Z, Mullooly N, Mielke LA, Harris J, Coll RC, Mills KH, Mok KH, Newsholme P, Nunez G, Yodoi J, Kahn SE, Lavelle EC, O'Neill LA (2010) Activation of the NLRP3 inflammasome by islet amyloid polypeptide provides a mechanism for enhanced IL-1beta in type 2 diabetes. Nat Immunol 11(10):897–904. doi:10.1038/ni.1935

Misawa T, Takahama M, Kozaki T, Lee H, Zou J, Saitoh T, Akira S (2013) Microtubule-driven spatial arrangement of mitochondria promotes activation of the NLRP3 inflammasome. Nat Immunol 14(5):454–460. doi:10.1038/ni.2550

Misawa T, Saitoh T, Kozaki T, Park S, Takahama M, Akira S (2015) Resveratrol inhibits the acetylated alpha-tubulin-mediated assembly of the NLRP3-inflammasome. Int Immunol 27(9): 425–434. doi:10.1093/intimm/dxv018

Miyake K, Yamashita Y, Ogata M, Sudo T, Kimoto M (1995) RP105, a novel B cell surface molecule implicated in B cell activation, is a member of the leucine-rich repeat protein family. J Immunol 154(7):3333–3340

Morad SA, Schmid M, Buchele B, Siehl HU, El Gafaary M, Lunov O, Syrovets T, Simmet T (2013) A novel semisynthetic inhibitor of the FRB domain of mammalian target of rapamycin blocks proliferation and triggers apoptosis in chemoresistant prostate cancer cells. Mol Pharmacol 83(2):531–541. doi:10.1124/mol.112.081349

Nagai Y, Shimazu R, Ogata H, Akashi S, Sudo K, Yamasaki H, Hayashi S, Iwakura Y, Kimoto M, Miyake K (2002) Requirement for MD-1 in cell surface expression of RP105/CD180 and B-cell responsiveness to lipopolysaccharide. Blood 99(5):1699–1705

Nguyen MT, Favelyukis S, Nguyen AK, Reichart D, Scott PA, Jenn A, Liu-Bryan R, Glass CK, Neels JG, Olefsky JM (2007) A subpopulation of macrophages infiltrates hypertrophic adipose tissue and is activated by free fatty acids via Toll-like receptors 2 and 4 and JNK-dependent pathways. J Biol Chem 282(48):35279–35292. doi:10.1074/jbc.M706762200

Nishimura S, Manabe I, Nagasaki M, Eto K, Yamashita H, Ohsugi M, Otsu M, Hara K, Ueki K, Sugiura S, Yoshimura K, Kadowaki T, Nagai R (2009) CD8+ effector T cells contribute to macrophage recruitment and adipose tissue inflammation in obesity. Nat Med 15(8):914–920. doi:10.1038/nm.1964

Ogata H, Su I, Miyake K, Nagai Y, Akashi S, Mecklenbrauker I, Rajewsky K, Kimoto M, Tarakhovsky A (2000) The toll-like receptor protein RP105 regulates lipopolysaccharide signaling in B cells. J Exp Med 192(1):23–29

Osborn O, Olefsky JM (2012) The cellular and signaling networks linking the immune system and metabolism in disease. Nat Med 18(3):363–374. doi:10.1038/nm.2627

Ouchi N, Parker JL, Lugus JJ, Walsh K (2011) Adipokines in inflammation and metabolic disease. Nat Rev Immunol 11(2):85–97. doi:10.1038/nri2921

Park EJ, Pezzuto JM (2015) The pharmacology of resveratrol in animals and humans. Biochim Biophys Acta 1852(6):1071–1113. doi:10.1016/j.bbadis.2015.01.014

Park SJ, Youn HS (2010) Suppression of homodimerization of toll-like receptor 4 by isoliquiritigenin. Phytochemistry 71(14–15):1736–1740. doi:10.1016/j.phytochem.2010.07.008

Qi LW, Liu EH, Chu C, Peng YB, Cai HX, Li P (2010) Anti-diabetic agents from natural products–an update from 2004 to 2009. Curr Top Med Chem 10(4):434–457

Ren Y, Song CS, Liu XH, Shi Y, Gao JF, He XD (1994) Experimental study on compatible application of heat-clearing and detoxifying drugs with blood circulation improving drugs. Zhongguo Zhong Yao Za Zhi 19(10):626–628, 640

Ryan KK, Woods SC, Seeley RJ (2012) Central nervous system mechanisms linking the consumption of palatable high-fat diets to the defense of greater adiposity. Cell Metab 15(2): 137–149. doi:10.1016/j.cmet.2011.12.013

Saberi M, Woods NB, de Luca C, Schenk S, Lu JC, Bandyopadhyay G, Verma IM, Olefsky JM (2009) Hematopoietic cell-specific deletion of toll-like receptor 4 ameliorates hepatic and adipose tissue insulin resistance in high-fat-fed mice. Cell Metab 10(5):419–429. doi:10.1016/j.cmet.2009.09.006

Saitoh S, Akashi S, Yamada T, Tanimura N, Kobayashi M, Konno K, Matsumoto F, Fukase K, Kusumoto S, Nagai Y, Kusumoto Y, Kosugi A, Miyake K (2004) Lipid A antagonist, lipid IVa, is distinct from lipid A in interaction with Toll-like receptor 4 (TLR4)-MD-2 and ligand-induced TLR4 oligomerization. Int Immunol 16(7):961–969. doi:10.1093/intimm/dxh097

Schenk S, Saberi M, Olefsky JM (2008) Insulin sensitivity: modulation by nutrients and inflammation. J Clin Invest 118(9):2992–3002. doi:10.1172/JCI34260

Schertzer JD, Tamrakar AK, Magalhaes JG, Pereira S, Bilan PJ, Fullerton MD, Liu Z, Steinberg GR, Giacca A, Philpott DJ, Klip A (2011) NOD1 activators link innate immunity to insulin resistance. Diabetes 60(9):2206–2215. doi:10.2337/db11-0004

Schrofelbauer B, Raffetseder J, Hauner M, Wolkerstorfer A, Ernst W, Szolar OH (2009) Glycyrrhizin, the main active compound in liquorice, attenuates pro-inflammatory responses by interfering with membrane-dependent receptor signalling. Biochem J 421(3):473–482. doi:10.1042/BJ20082416

Senn JJ (2006) Toll-like receptor-2 is essential for the development of palmitate-induced insulin resistance in myotubes. J Biol Chem 281(37):26865–26875. doi:10.1074/jbc.M513304200

Shi H, Kokoeva MV, Inouye K, Tzameli I, Yin H, Flier JS (2006) TLR4 links innate immunity and fatty acid-induced insulin resistance. J Clin Invest 116(11):3015–3025. doi:10.1172/JCI28898

Shibata S (2000) A drug over the millennia: pharmacognosy, chemistry, and pharmacology of licorice. Yakugaku Zasshi 120(10):849–862

Stienstra R, van Diepen JA, Tack CJ, Zaki MH, van de Veerdonk FL, Perera D, Neale GA, Hooiveld GJ, Hijmans A, Vroegrijk I, van den Berg S, Romijn J, Rensen PC, Joosten LA, Netea MG, Kanneganti TD (2011) Inflammasome is a central player in the induction of obesity and insulin resistance. Proc Natl Acad Sci U S A 108(37):15324–15329. doi:10.1073/pnas.1100255108

Suganami T, Tanimoto-Koyama K, Nishida J, Itoh M, Yuan X, Mizuarai S, Kotani H, Yamaoka S, Miyake K, Aoe S, Kamei Y, Ogawa Y (2007) Role of the Toll-like receptor 4/NF-kappaB pathway in saturated fatty acid-induced inflammatory changes in the interaction between adipocytes and macrophages. Arterioscler Thromb Vasc Biol 27(1):84–91. doi:10.1161/01.ATV.0000251608.09329.9a

Suzuki K, Koike T (2007) Resveratrol abolishes resistance to axonal degeneration in slow Wallerian degeneration (WldS) mice: activation of SIRT2, an NAD-dependent tubulin deacetylase. Biochem Biophys Res Commun 359(3):665–671. doi:10.1016/j.bbrc.2007.05.164

Syrovets T, Gschwend JE, Buchele B, Laumonnier Y, Zugmaier W, Genze F, Simmet T (2005) Inhibition of IkappaB kinase activity by acetyl-boswellic acids promotes apoptosis in androgen-independent PC-3 prostate cancer cells in vitro and in vivo. J Biol Chem 280(7): 6170–6180. doi:10.1074/jbc.M409477200

Tack CJ, Stienstra R, Joosten LA, Netea MG (2012) Inflammation links excess fat to insulin resistance: the role of the interleukin-1 family. Immunol Rev 249(1):239–252. doi:10.1111/j.1600-065X.2012.01145.x

Toyang NJ, Wabo HK, Ateh EN, Davis H, Tane P, Sondengam LB, Bryant J, Verpoorte R (2013) Cytotoxic sesquiterpene lactones from the leaves of Vernonia guineensis Benth. (Asteraceae). J Ethnopharmacol 146(2):552–556. doi:10.1016/j.jep.2013.01.022

Tsukumo DM, Carvalho-Filho MA, Carvalheira JB, Prada PO, Hirabara SM, Schenka AA, Araujo EP, Vassallo J, Curi R, Velloso LA, Saad MJ (2007) Loss-of-function mutation in Toll-like receptor 4 prevents diet-induced obesity and insulin resistance. Diabetes 56(8):1986–1998. doi:10.2337/db06-1595

Ungerback J, Belenki D, Jawad Ul-Hassan A, Fredrikson M, Fransen K, Elander N, Verma D, Soderkvist P (2012) Genetic variation and alterations of genes involved in NFkappaB/TNFAIP3- and NLRP3-inflammasome signaling affect susceptibility and outcome of colorectal cancer. Carcinogenesis 33(11):2126–2134. doi:10.1093/carcin/bgs256

Watanabe Y, Nakamura T, Ishikawa S, Fujisaka S, Usui I, Tsuneyama K, Ichihara Y, Wada T, Hirata Y, Suganami T, Izaki H, Akira S, Miyake K, Kanayama HO, Shimabukuro M, Sata M, Sasaoka T, Ogawa Y, Tobe K, Takatsu K, Nagai Y (2012) The radioprotective 105/MD-1 complex contributes to diet-induced obesity and adipose tissue inflammation. Diabetes 61(5): 1199–1209. doi:10.2337/db11-1182

Westermark P, Wernstedt C, Wilander E, Hayden DW, O'Brien TD, Johnson KH (1987) Amyloid fibrils in human insulinoma and islets of Langerhans of the diabetic cat are derived from a neuropeptide-like protein also present in normal islet cells. Proc Natl Acad Sci U S A 84(11): 3881–3885

Youn HS, Kim YS, Park ZY, Kim SY, Choi NY, Joung SM, Seo JA, Lim KM, Kwak MK, Hwang DH, Lee JY (2010) Sulforaphane suppresses oligomerization of TLR4 in a thiol-dependent manner. J Immunol 184(1):411–419. doi:10.4049/jimmunol.0803988

Zeyda M, Stulnig TM (2007) Adipose tissue macrophages. Immunol Lett 112(2):61–67. doi:10.1016/j.imlet.2007.07.003

Zhang JQ (2007) Progress of diabetes research in traditional Chinese medicine in recent years. Zhong Xi Yi Jie He Xue Bao 5(4):373–377

Zhang N, Zhang X, Liu X, Wang H, Xue J, Yu J, Kang N, Wang X (2014) Chrysophanol inhibits NALP3 inflammasome activation and ameliorates cerebral ischemia/reperfusion in mice. Mediators Inflamm 2014:370530. doi:10.1155/2014/370530

Part VI
Chronic Inflammation and Adaptive Immunity

Chapter 31
Human and Mouse Memory-Type Pathogenic Th2 (Tpath2) Cells in Airway Inflammation

Yusuke Endo, Kiyoshi Hirahara, Kenta Shinoda, Tomohisa Iinuma, Heizaburo Yamamoto, Shinichiro Motohashi, Yoshitaka Okamoto, and Toshinori Nakayama

Abstract Immunological memory has a central role in the adaptive immune systems that can efficiently eliminate pathogens such as virus and bacteria. Particularly, memory CD4 T cells function as a control tower of the adaptive immune systems. However, memory helper T (Th) cells are also involved in the pathogenesis of various chronic inflammatory diseases, including asthma. Recently, we have found IL-5-producing pathogenic population in memory Th2 cells in allergic asthma. We term this population memory-type 'pathogenic Th2 cells (Tpath2 cells)' in the airway. Several groups also reported distinct memory type Th2 populations, which produce a large amount of IL-5 or multiple cytokines in addition to IL-4 and IL-13. These populations seem to be responsible for the pathology of chronic inflammatory disorders. Our recent study has revealed that ST2 (a component of IL-33 receptor) expression on Tpath2 cells plays an important role for inducing pathogenicity in memory Th2 cells. Here, we highlight the

Y. Endo • K. Shinoda
Department of Immunology, Graduate School of Medicine, Chiba University, 1-8-1 Inohana, Chuo-ku, Chiba 260-8670, Japan

K. Hirahara
Department of Advanced Allergology of the Airway, Graduate School of Medicine, Chiba University, 1-8-1 Inohana, Chuo-ku, Chiba 260-8670, Japan

T. Iinuma • H. Yamamoto • Y. Okamoto
Department of Otorhinolaryngology, Head and Neck Surgery, Graduate School of Medicine, Chiba University, 1-8-1 Inohana, Chuo-ku, Chiba 260-8670, Japan

S. Motohashi
Department of Medical Immunology, Graduate School of Medicine, Chiba University, 1-8-1 Inohana, Chuo-ku, Chiba 260-8670, Japan

T. Nakayama (✉)
Department of Immunology, Graduate School of Medicine, Chiba University, 1-8-1 Inohana, Chuo-ku, Chiba 260-8670, Japan

AMED-CREST, Chiba 260-8670, Japan
e-mail: tnakayama@faculty.chiba-u.jp

M. Miyasaka, K. Takatsu (eds.), *Chronic Inflammation*,
DOI 10.1007/978-4-431-56068-5_31

regulation and function of IL-33 and ST2 on memory Th2 cells, and review their roles in the induction and progression of chronic airway inflammation in both mice and human systems.

Keywords Pathogenic Th2 cells (Tpath2 cells) • Asthma • IL-33 • IL-5 • ST2 • p38 signal • Chronic rhinosinusitis (CRS) • Eosinophilic CRS (ECRS)

31.1 Memory-Type Pathogenic Th2 Cells (Tpath2 Cells) in Allergic Airway Inflammation

Chronic inflammation acts as a driver of both systemic disorders (including autoimmune diseases, cancer, and obesity) and organ-specific inflammatory disorders. In the lung, chronic airway inflammation causes various treatment-resistant diseases such as bronchial asthma, COPD, and lung fibrosis. Among these chronic inflammatory diseases, bronchial asthma affects 300 million people worldwide (Cohn et al. 2004). The development of asthma is provoked by a combination of risk factors. Recent studies emphasise the importance of genetic susceptibility and exposure to particular environmental factors and a history of infection. Continuous exposure to inhaled substances including pollutants and environmental allergens such as pollen, animal dander, and house dust mites is also thought to be critically involved in the development of chronic airway inflammation (Gregory and Lloyd 2011; Kim et al. 2010; Gelfand and Kraft 2009).

CD4 T cells are crucial for immune cell homeostasis and host defence. In their key role as helper cells they also drive immune-related diseases, such as autoimmune diseases and allergies. Animal models of allergic airway inflammation, as well as human bronchial asthma, have revealed the importance of cytokines produced by Th2 cells, namely IL-4, IL-5, and IL-13 (Lloyd and Hessel 2010). The contribution of these various cytokines to the pathophysiology of airway inflammation, airway hyperresponsiveness (AHR), eosinophilia, fibrosis, increased serum IgE, and other responses is well-recognised (Cohn et al. 2004; Rosenberg et al. 2013; Porter et al. 2011). A series of studies regarding human asthma patients and animal models revealed that the memory Th2 cells, which reside in the lung during disease remission, contribute to the persistence and progression of asthma (Christianson and Alam 2013). However, the factors that induce and sustain chronic inflammation through memory CD4 T cells remain unclear. The purpose of our research is to elucidate the mechanisms through which airway inflammation becomes chronic, and to establish basic technologies for the early detection, control, resolution, and repair of chronic inflammation.

It is well established that upon antigen stimulation, effector Th2 cells produce large amounts of IL-4, IL-5, and IL-13, that Th1 cells produce IFN-γ, and that Th17 produce IL-17, IL-21, and IL-22 (Fig. 31.1, left). It remains unclear whether these polarised cytokine production profiles are preserved for an extended period of time in vivo during the memory phase. Recent studies have reported the identification of several distinct memory type Th2 cell subsets, which produce a large amount of

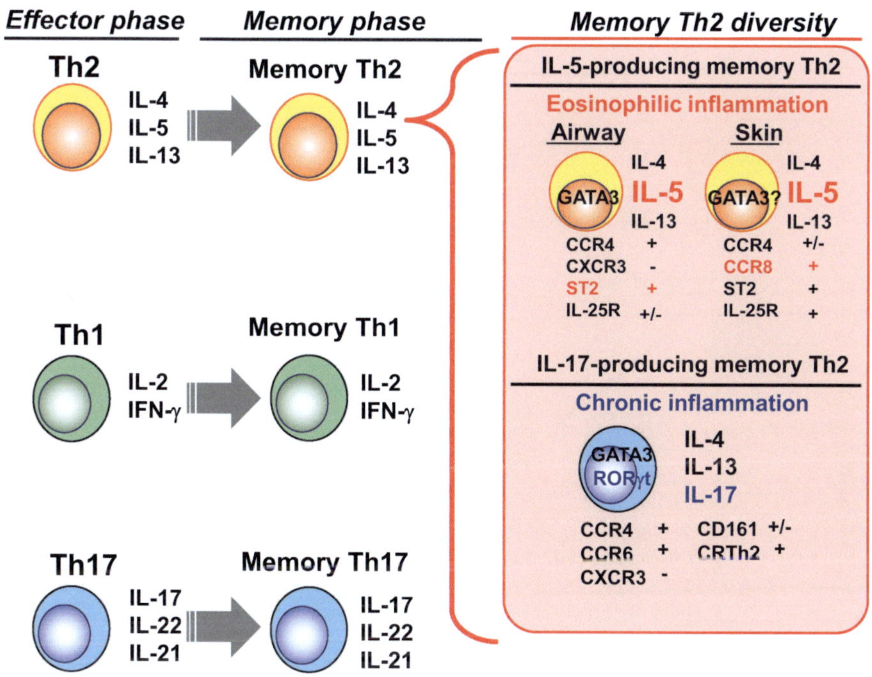

Fig. 31.1 The diversity of memory Th2 cells. Memory Th1/Th2/Th17 cells are generated from effector Th1/Th2/Th17 cells. Memory Th2 cells can be subdivided into several populations based on the expression of cell surface molecules and functional properties. In the airway and skin, IL-5-producing memory Th2 cells are characterised by the expression of ST2 and CCR8, respectively. IL-17-producing Th2 cells are also identified as CCR6^{+} and CRTH2^{+} populations in the airway. *Abbreviation*: CRTH2, Chemoattractant receptor–homologous molecule expressed on Th2 lymphocytes

IL-5 or IL-17 in addition to IL-4 and IL-13 (Fig. 31.1, right). These distinct memory Th2 cell populations appear to be important drivers of the pathology of chronic allergic diseases (Endo et al. 2011; Holt et al. 2009; Wang et al. 2010). We recently identified that a memory-type IL-5–producing highly pathogenic Th2 subpopulation induces eosinophilic airway inflammation, and named this population 'pathogenic Th2 (Tpath2) cells' (Figs. 31.1, right and 31.2). Moreover, we very recently demonstrated that the IL-33-ST2-p38 MAPK axis is crucial for the induction of the pathogenicity of memory Th2 cells in allergic airway inflammation in both mice and humans (Endo et al. 2015). Our findings revealed that IL-33 is one of the crucial factors in the induction and exacerbation of chronic airway inflammation through memory Th2 cellsk. In this chapter, we highlight recent findings in relation to the functions and characteristics of Tpath2 cells, focusing on the role of IL-33 in allergic airway inflammation.

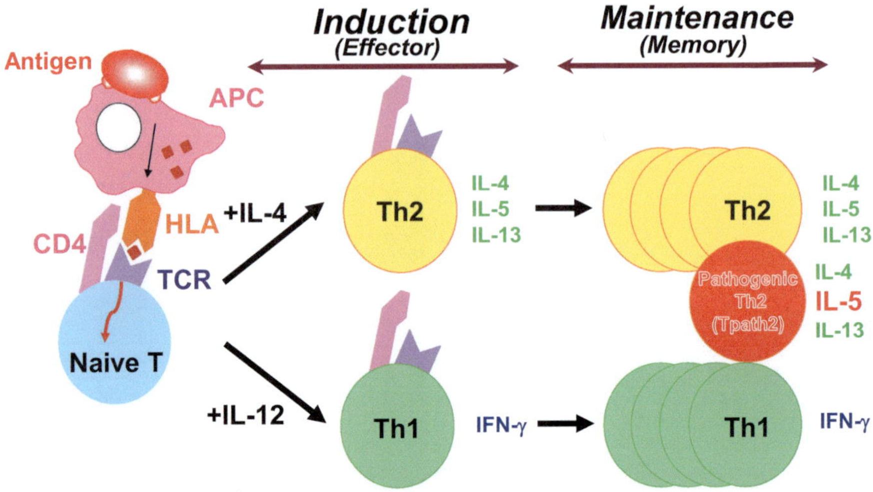

Fig. 31.2 Th1/Th2 cell differentiation and maintenance. After antigenic stimulation with APC, naïve CD4 T cells differentiate into effector Th1 or Th2 cells. Some of the effector Th1 or Th2 cells are maintained as memory Th1 or Th2 cells for long periods in vivo. Distinct IL-5–producing memory-type pathogenic Th2 cells (Tpath2 cells) are identified in allergic airway inflammation (Endo et al. 2011; Endo et al. 2015; Endo et al. 2014)

31.2 Crucial Roles for IL-33 and Its Receptor ST2 in the Induction of Pathogenicity in Memory Th2 Cells

IL-33 is a newly recognised member of the IL-1 superfamily and is present in the nucleus of various types of cells (Liew et al. 2010). IL-33 was originally identified as a ligand for the ST2 receptor (also known as IL-1RL1). Recent large-scale genome-wide association studies (GWAS) have shown that the *IL33* and *IL1RL1* genes are susceptible to the onset of asthma (Schmitz et al. 2005; Grotenboer et al. 2013). Furthermore, an increasing number of studies revealed the association between IL-33 and chronic airway inflammation (Table 31.1). Indeed, the levels of *Il33* mRNA and protein are increased in the serum and tissues of asthma patients and in mouse models of airway inflammation (Lloyd 2010). In asthma patients, epithelial cells and airway smooth muscle cells are two major sources of IL-33, especially in the event of tissue damage (Prefontaine et al. 2010). The administration of neutralising anti-IL-33 antibodies or anti-ST2 antibodies to allergic mice inhibits eosinophilic recruitment, Th2 cytokine production, serum IgE, and mucus hyperproduction (Liu et al. 2009; Coyle et al. 1999). Consistent with these findings, *Il33*$^{-/-}$ mice or *Il1rl1*$^{-/-}$ mice show attenuated eosinophil infiltration into the bronchoalveolar lavage (BAL) fluid and pulmonary inflammation during airway inflammation (Besnard et al. 2011; Kurowska-Stolarska et al. 2008; Oboki et al. 2010). IL-33 induces the differentiation of Th2 effector cells from naïve CD4 T cells (Schmitz et al. 2005; Kurowska-Stolarska et al. 2008). Additionally, it

Table 31.1 Summary of reviews and articles associated with IL-33 and chronic airway inflammation such as asthma

IL-33 and chronic airway inflammation
1. Prefontaine, D et al. Increased expression of IL-33 in severe asthma: evidence of expression by airway smooth muscle cells. *J. Immunol.* 183:5094, 2009 **ASMCs/ Human (Asthma)**
2. Liew, FY et al. Disease-associated functions of IL-33: the new kid in IL-1 family. (Review) Nat. Rev. Immunol. 10:103, 2010 **Th2, DC, Macrophage/ Human, mouse**
3. Ananda, SM et al. Interleukin-33 and the function of innate lymphoid cells. (Review) *Trends in Immunol.* 33:389, 2012 **Th2, DC, Macrophage/ Human, mouse**
4. Lloyd CM. IL-33 family members and asthma – bridging innate and adaptive immune responses. (Review) *Curr. Opin. Immunol.* 22:800, 2010 **Th2, DC, Macrophage/ Human, mouse**
5. IL-33: an alarmin cytokine with crucial roles in innate immunity, inflammation and allergy. *Curr. Opin. Immunol.* 31:31, 2014 **IL-33/ST2 Various types of asthma**
IL-33 and ST2/ILRL1 are the genes reproducibly identified in all major genome-wide association studies
1. Gudbjartsson, DF et al. Sequence variants affecting eosinophil numbers associate with asthma and myocardial infarction. *Nat Genet* 41:342, 2009.
2. Moffatt MF et al. A large-scale, consortium-based genomewide association study of asthma. *N Engl J Med* 363:1211, 2010.
3. Torgerson DG et al. Meta-analysis of genome-wide association studies of asthma in ethnically diverse North American populations *Nat Genet* 43:887, 2011
4. Ramassamy A, Genome-wide association studies of asthma in population-based cohorts confirm known and suggested loci and identify an additional association near HLA. *PLoS ONE* 7:e44008, 2012.
5. BonnelykkeK et al. A genome-wide association study identifies CDHR3 as a susceptibility locus for early childhood astham with severe exacerbations. *Nat Genet* 46:51 2014.

is now known that IL-33 functions on other cell types in an antigen-independent fashion, including mast cells, macrophages, eosinophils, basophils, and type 2 innate lymphoid (ILC2) cells (Fig. 31.3) (Oboki et al. 2010; Scanlon and McKenzie 2012). ILC2 cells are a novel population of lineage marker-negative cells which produce large amounts of Th2 cytokines IL-5 and IL-13 in response to IL-33 (Scanlon and McKenzie 2012). The better understanding of the mechanisms through which IL-33 enacts both antigen-specific and antigen-independent effects in a variety of immune cells will lead to the design of effective therapeutic agents for patients with allergic inflammatory disorders such as chronic asthma.

IL-33 has the ability to induce strong Th2-type immune responses and eosinophilic inflammation in the lung and intestine (Lloyd 2010). However, the type of cells that IL-33 affects in these settings is still being determined. We found that IL-5-producing 'pathogenic Th2 cells (Tpath2 cells)' showed high ST2 expression (Endo et al. 2011, 2015). Thus, we considered the possibility that IL 33 induces the pathogenicity in memory Th2 cells (Fig. 31.4a). In fact, stimulation with IL-33 for 5 days selectively induced IL-5 production by memory Th2 cells but not effector Th2 cells (Fig. 31.4b). In contrast, IL-33 could not induce the upregulation of IL-4. Thus, IL-33 selectively enhances IL-5 production by memory Th2 cells, but not by

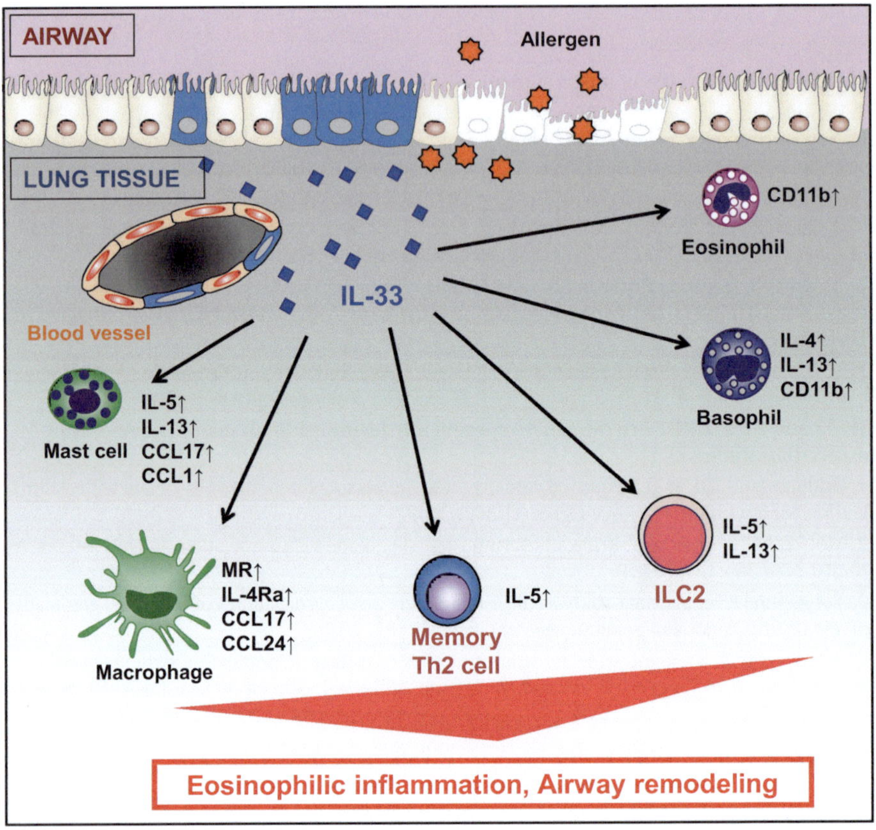

Fig. 31.3 The target cells of IL-33 during chronic airway inflammation. Exposure to allergens and infectious microorganisms causes damage to the pulmonary epithelial and/or endothelial cells leading to the release of IL-33. IL-33 likely drives type-2 allergic responses via effects on type-2 innate lymphoid cells (ILC2) and Th2 cells via secretion of IL-4, IL-5, and IL-13. IL-33 leads to the generation of macrophages with an alternatively activated phenotype and to the recruitment, maturation, and survival of eosinophils. IL-33 is also able to induce airway hyperresponsiveness via the activation of mast cells and basophils

effector Th2 cells. Regarding the signal cascades activated by IL-33, the activation of MAPKs and NF-kB in mast cells has been reported (Fig. 31.5a) (Schmitz et al. 2005). In memory Th2 cells, we found that SB203580, a p38 MAPK inhibitor inhibited the IL-33–induced production of IL-5 (Fig. 31.5b). IL-33 selectively activates p38 MAPK, increases the expression of ST2, and augments the production of IL-5 in memory Th2 cells.

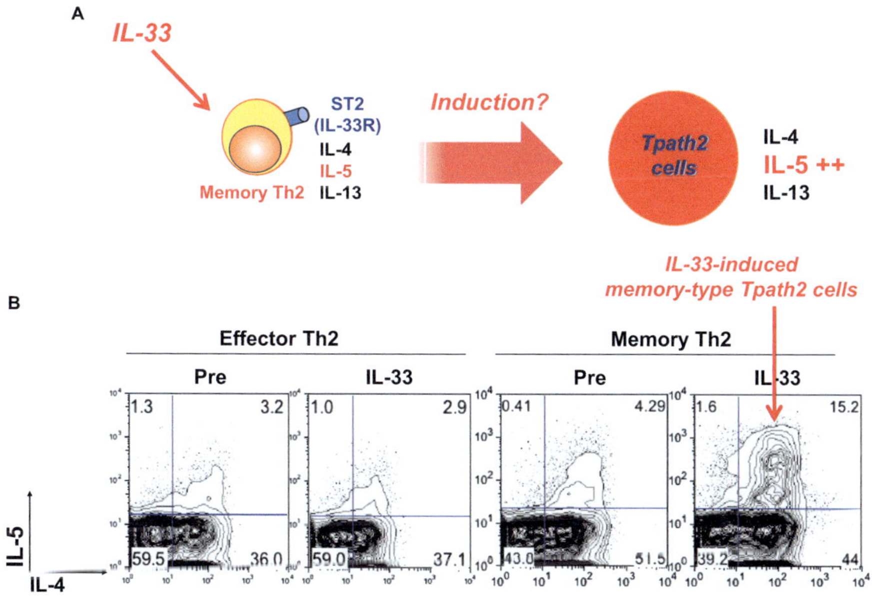

Fig. 31.4 IL-33 selectively induces IL-5 production by memory Th2 cells. (**a**) Schematic representation of the effects of IL-33 on memory Th2 cells, in which pathogenic Th2 cells are induced. (**b**) Antigen-specific memory and effector Th2 cells were cultured with IL-33 (1μg/ml) for 5 days. The cultured cells were stimulated with immobilised anti-TCRβ for 6 h. The intracellular-staining profiles of IL-5 and IL-4 are shown

31.3 IL-33-ST2 Pathway Drives Airway Inflammation Through Pathogenic Th2 Cells

The pathological role of the IL-33-ST2 pathway in memory Th2 cells was demonstrated in a memory Th2 cell-dependent airway inflammation model (Fig. 31.6a). In this model, ST2-expressing memory Th2 cells were increased after an OVA challenge in the lung. The genetic deletion of ST2 resulted in a significant decrease in mononuclear cell infiltration into the peribronchiolar regions of the lungs and in the number of inflammatory eosinophils in BAL fluid (Fig. 31.6b and c). Consistent with the results of the genetic deletion of ST2 from donor memory Th2 cells, the deletion of IL-33 in recipient mice abrogates inflammatory responses, including the recruitment of eosinophils, Th2 cytokine production, and mucus hyperproduction. Interestingly, there was no difference in the level of eosinophilic infiltration in the BAL fluid from mice that received *Il1rl1*$^{+/+}$ and *Il1rl1*$^{-/-}$ effector Th2 cells. This suggests that effector Th2 cells possess little reactivity to IL-33 and that this pathway is less important for the induction of eosinophilic inflammation in effector Th2 cells than it is in memory Th2 cells. Whether ILC2 cells can respond to IL-33 by producing IL-5 in the lung has previously been investigated (Halim et al. 2012). Thus, we assessed the relative contribution of memory Th2 cells to eosinophilic

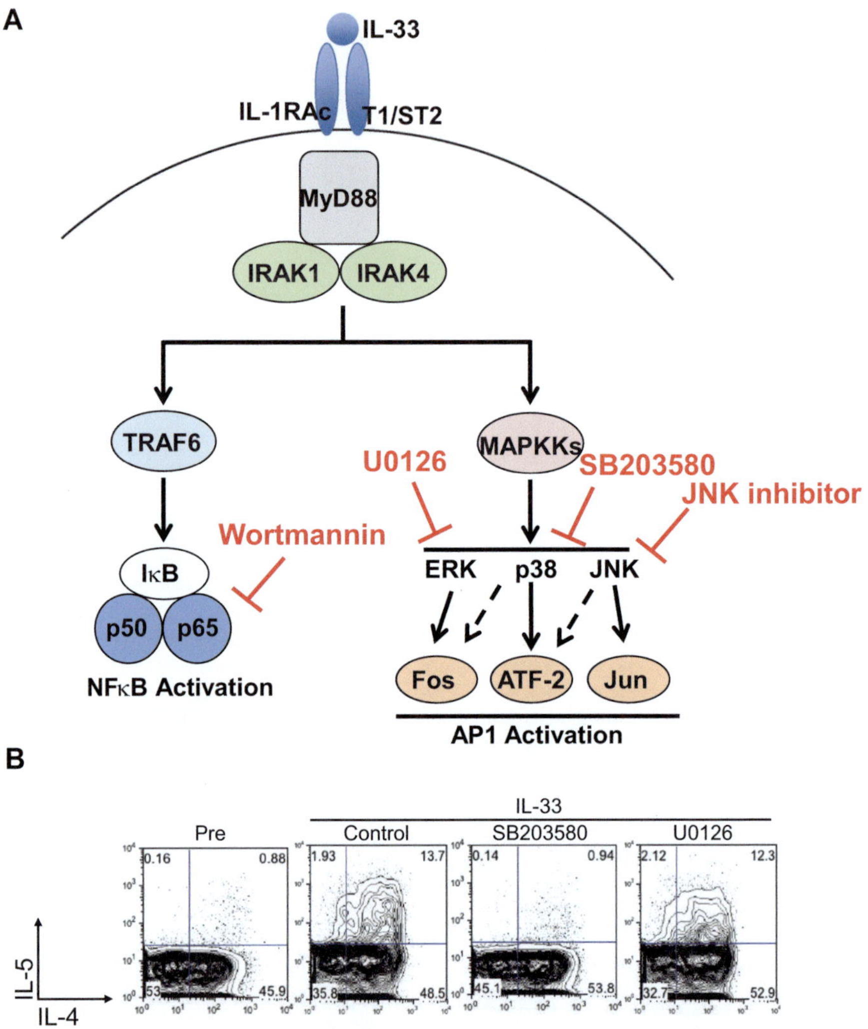

Fig. 31.5 IL-33 activates the p38-MAPK signalling pathway in memory Th2 cells. (**a**) IL-33 mediates its function by binding to a receptor complex comprising the ST2 and IL-1 receptor accessory protein (IL-1RAP), leading to the recruitment of the myeloid differentiation primary-response protein 88 (MYD88) complex. Although the precise mechanism is unclear, the complex is thought to activate at least two independent pathways involving the mitogen-activated protein kinase (MAPK) pathway as well as the NFkB pathway. (**b**) The intracellular-staining profiles of IL-5 and IL-4 in IL-33–treated memory Th2 cells in the presence of indicated signalling inhibitors. SB203580; p38 inhibitor, U0126; MEK inhibitor, SP600125; JNK inhibitor, Wortmanin; PI3K inhibitor

inflammation. The depletion of ILC2 cells in memory Th2 cell-induced airway inflammation in the lung using CD90.2 antibody did not affect the degree of airway inflammation. Therefore, at least in this model, the contribution of ILC2 cells to

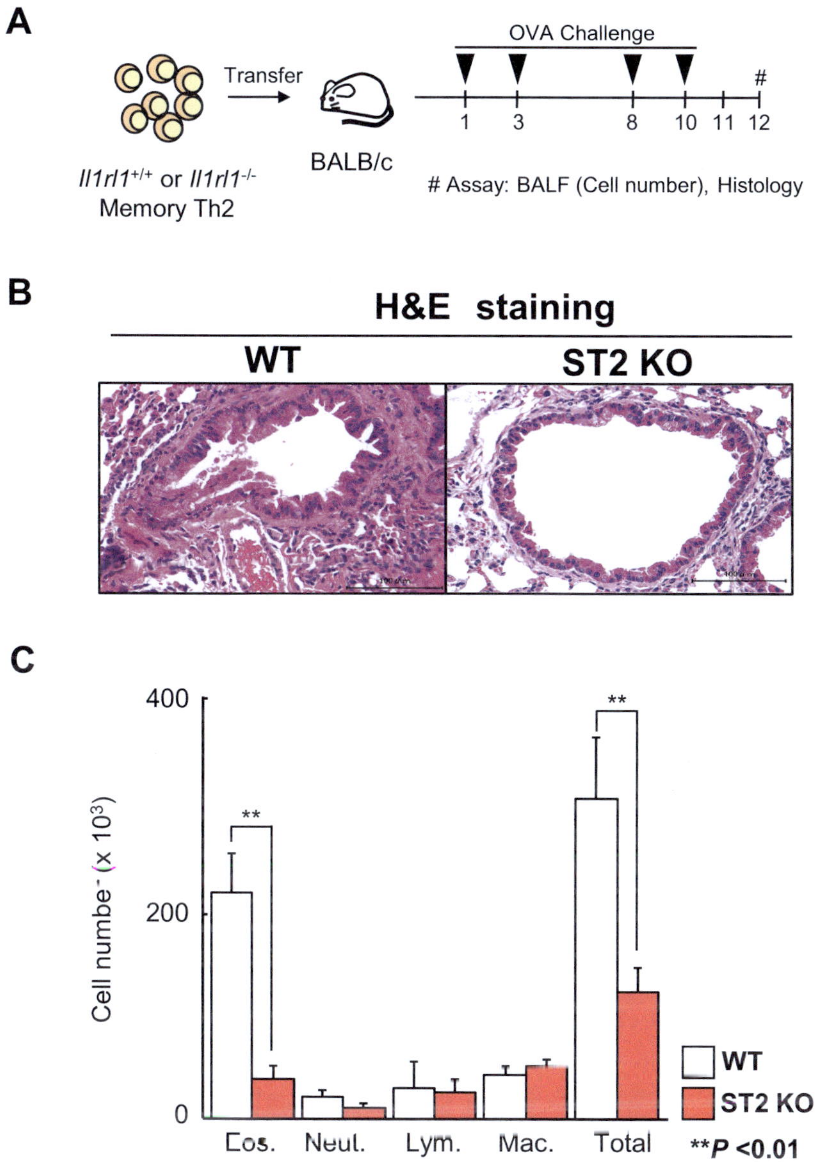

Fig. 31.6 The pathogenicity of memory Th2 cells is ameliorated by the loss of IL-33-ST2 signalling. (**a**) The experimental protocols for memory-Th2-cell–dependent allergic inflammation. (**b**) Lung tissue specimens were fixed and stained with H&E. A representative staining pattern is shown. Scale bars: 100 mm. (**c**) The absolute cell numbers of eosinophils (Eos), neutrophils (Neu), lymphocytes (Lym), and macrophages (Mac) in the BAL fluid are shown. The values are the mean ± SD of five mice per group

eosinophilic inflammation seems to be relatively smaller. Interestingly, endogenous IL-33 maintains the expression of ST2 and IL-5 production in memory-type Tpath 2 cells at steady state. The proportion of ST2-expressing memory CD4 T cells was increased even 42 days after OVA-Alum immunisation. Thus, IL-33 may have a crucial role in the progression and persistence of airway inflammation via the maintenance of Tpath 2 cells. Altogether, the IL-33/ST2 pathway appears to be critical for the induction and also the progression and persistence of memory-type Th2 cell-dependent eosinophilic airway inflammation.

31.4 IL-33–ST2–p38MAPK Axis Is Critical for the Induction of IL-5–Producing Tpath2 Cells in Human ECRS Patients

To gain more insight into the possible pathophysiological role of the effect of IL-33 on memory Th2 cells in human disease, we intensively investigated patients who suffered from chronic inflammation of the upper airway. Chronic rhinosinusitis (CRS) is a heterogeneous disease which is characterised by distinct cytokine production profiles and tissue-remodelling patterns (Hamilos 2011; Van Bruaene et al. 2008; Zhang et al. 2008). CRS can be divided into two phenotypes according to the presence of nasal polyps. CRS with nasal polyps (CRSwNP) is often accompanied by Th2-cell–skewed eosinophilic inflammation, whereas CRS without nasal polyps (CRSsNP) is characterised by a predominantly Th1-cell–skewed response (Hamilos 2011). IL-5 is more abundant in the nasal mucosal tissues of patients with CRSwNP than in patients with CRSsNP (Van Bruaene et al. 2008; Gevaert et al. 2006). CRSwNP is further subdivided into two types of diseases based on the extent of eosinophilic inflammation (particularly in people of East Asian descent) (Zhang et al. 2008): eosinophilic CRS (ECRS) and noneosinophilic rhinosinusitis (NECRS). We have found that the numbers of IL-33$^+$PECAM1$^+$ endothelial cells were significantly elevated in nasal polyps from ECRS patients in comparison to nasal polyps from NECRS patients (Fig. 31.7a) in addition to substantial numbers of epithelial cells in the airway (Fig. 31.7b). The baseline expression levels of Th2 cytokines (*IL4* and *IL5*), *IL1RL1* and *GATA3* are also higher in ECRS polyps (Fig. 31.8). In contrast, the expression levels of *TBX21* are lower in ECRS polyps. Interestingly, IL-33 dramatically enhanced *IL5* expression in CD45RO$^+$ memory CD4 T cells from the nasal polyps of ECRS patients (Fig. 31.8). Furthermore, the expression of *IL1RL1* was elevated after treatment with IL-33, which supports the results obtained from the murine memory Th2 cell experiments. In contrast, IL-33 was not observed to affect the expression of *IL5* or *IL13* in CD45RO$^+$ memory CD4 T cells from the nasal polyps of NECRS patients. The IL-33-enhanced induction of IL-5 and *IL1RL1* in CD45RO$^+$ memory CD4 T cells from the nasal polyps of ECRS patients was selectively inhibited by treatment with SB203580.

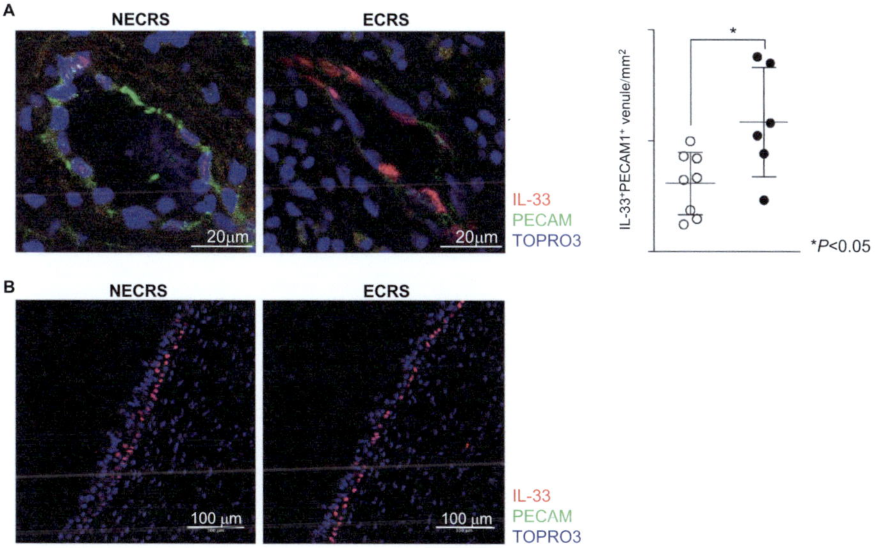

Fig. 31.7 IL-33^{+}PECAM1^{+} endothelial cells were detected in the nasal polyps of ECRS patients. (**a**) An immunofluorescence analysis (with staining for IL-33 [*red*], PECAM [*green*], and TOPRO3 [*blue*]) shows a representative section of nasal polyps from NECRS and ECRS patients. The frequency of IL-33^{+}PECAM1^{+} cells among the PECAM^{+} cells of the nasal polyps from patients is shown (mean ± SD; ECRS, $n = 6$; NECRS, $n = 8$) (*right panel*). (**b**) An immunofluorescence analysis (with staining for IL-33 [*red*] and TOPRO3 [*blue*]) of epithelial cells shows a representative section of nasal polyps from NECRS and ECRS patients

IL-25 is another epithelium-derived cytokine which causes eosinophilic inflammation via its receptor, IL-17RB (Fort et al. 2001; Petersen et al. 2012). IL-25 is produced by airway epithelial cells in response to various antigens and pathogens. It causes Th2-skewed inflammation in a specific but antigen-independent manner. Interestingly, IL-25 levels were increased in the nasal polyps obtained from patients with ECRS (Iinuma et al. 2015). Moreover, the enhanced production of Th2-cytokines from memory CD4 T cells was observed to be accompanied by increased IL-17RB expression in the nasal polyps obtained from patients with ECRS. Note that IL-17RB expression in tissue CD4 T cells was significantly correlated with the number of eosinophils in nasal polyps and with disease severity as measured by the CT score. Overall, IL-33/ST2-p38 and IL-25 signalling may play an important role in the induction of pathogenicity in tissue infiltrating memory CD4 T cells in human chronic allergic inflammatory diseases such as ECRS.

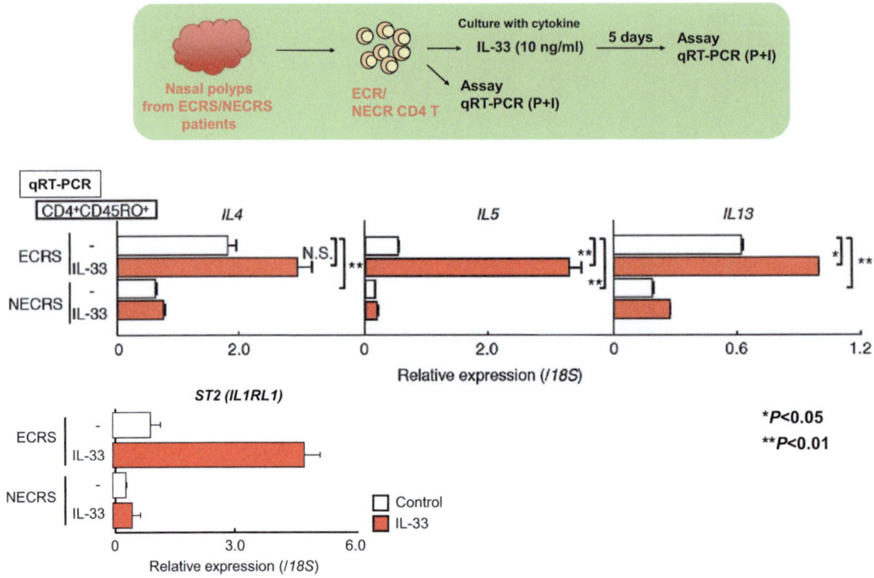

Fig. 31.8 IL-33 enhances IL-5 production of memory phenotype CD4 T cells in the polyps of ECRS patients. Schema of experimental procedures are shown (*top*). A quantitative RT-PCR analysis of *IL4*, *IL5*, *IL13*, and *ST2 (IL1RL1)* in CD45RO$^+$ memory CD4 T cells from ECRS and NECRS patients was performed, and representative results in the culture group with or without IL-33 are shown

31.5 Conclusions

In summary, our study identifies memory Th2 cells as an important target of chronic inflammatory cues, including IL-33 and IL-25, in the pathogenesis of airway inflammation (Endo et al. 2015; Iinuma et al. 2015). The p38-mediated signalling pathway is critical for the TCR-independent IL-33-induced expression of IL-5 in both murine and human memory Th2 cells (Endo et al. 2015). For more than two decades, the 'Th1/Th2 paradigm' has been proposed as the cornerstone of immune responses ('Th1/Th2 balance disease induction model', Fig. 31.9, upper panel). Based on this 'classical' model, allergic diseases are considered to be the results of an imbalance in the favor of Th2 responses, and to be negatively regulated by Th1 cells. Conversely, autoimmune diseases are considered to be induced by the favoring of Th1 responses, and to be negatively regulated by Th2 cells (Szab et al. 2003). However, attempts to link common allergic diseases or autoimmune diseases to a simple 'Th1/Th2 paradigm' have been problematic (Gor et al. 2003). A number of lines of evidence suggest that autoimmune mechanisms cannot be reduced to the action of Th1 cells alone. We propose another model regarding the pathogenesis of so-called Th1, Th2, or Th17 diseases ('Pathogenic Th population disease induction model,' Fig. 31.9, lower panel). In this model, a pathogenic subpopulation of Th cells which possesses a distinct effector function, with the

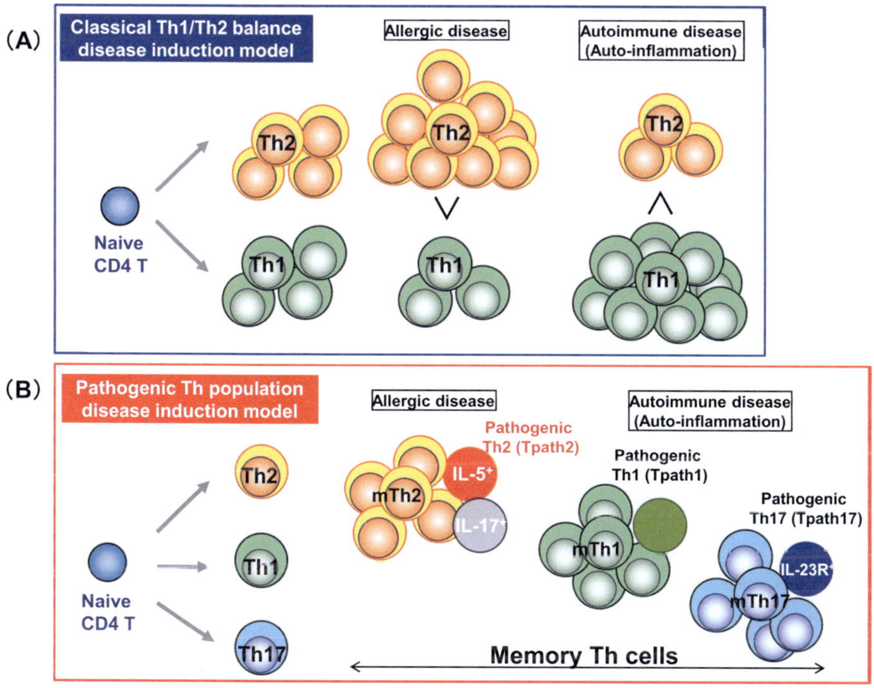

Fig. 31.9 The pathogenic Th population disease induction model. (**a**) The 'classical Th1/Th2 balance disease induction model'. One theory of immune regulation involves homeostasis between Th1 and Th2 activity. The overactivation of either pattern can cause disease, and either pathway can downregulate the other. Accordingly, the unbalanced generation of Th1 and Th2 cells is associated with the pathogenesis of Th1 diseases (autoimmune diseases) or Th2 diseases (allergic diseases), respectively. (**b**) The 'pathogenic Th population disease induction model'. In this model, the pathogenic subpopulations of Th cells that possess a distinct effector function are generated in vivo and are crucial for the pathogenesis of Th1, Th2, or Th17 diseases, regardless of the balance of the Th subsets. In Th2-mediated pathologies such as asthma or chronic dermatitis, the interleukin-5 (IL-5)-producing or IL-17–producing memory Th2 cell subpopulations are considered to be pathogenic Th2 cells (Tpath2). In Th17-related pathologies such as autoimmune diseases, the IL-23R$^+$ and T-bet$^+$ populations are considered to be pathogenic (Tpath17). A pathogenic Th1 (Tpath1) cell subpopulation has not been reported at this time. *Abbreviation*: T-bet, T-box transcription factor Tbx21

unique expression of cytokines and chemokine receptors, are generated in vivo and are crucial for the pathogenesis of Th1, Th2, or Th17 diseases regardless of the balance of these Th subsets (Endo et al. 2014). Further detailed studies focusing on memory-type pathogenic Th2 cells (Tpath2 cells) may lead to the better understanding of the 'Pathogenic Th population disease induction model' and the discovery of novel therapeutic targets for the curative treatment of chronic allergic inflammatory diseases.

References

Besnard AG, Togbe D, Guillou N, Erard F, Quesniaux V, Ryffel B (2011) IL-33-activated dendritic cells are critical for allergic airway inflammation. Eur J Immunol 41:1675–1686

Christianson CA, Alam R (2013) Mechanisms of sustained signaling in asthma. Curr Opin Allergy Clin Immunol 1:100–105

Cohn L et al (2004) Asthma: mechanisms of disease persistence and progression. Annu Rev Immunol 22:789–815

Coyle AJ, Lloyd C, Tian J, Nguyen T, Erikkson C, Wang L, Ottoson P, Persson P, Delaney T, Lehar S et al (1999) Crucial role of the interleukin 1 receptor family member T1/ST2 in T helper cell type 2-mediated lung mucosal immune responses. J Exp Med 190:895–902

Endo Y, Iwamura C, Kuwahara M, Suzuki A, Sugaya K, Tumes DJ, Tokoyoda K, Hosokawa H, Yamashita M, Nakayama T (2011) Eomesodermin controls interleukin-5 production in memory T helper 2 cells through inhibition of activity of the transcription factor GATA3. Immunity 35:733–745

Endo Y, Hirahara K, Yagi R, Nakayama T (2014) Pathogenic memory type Th2 cells in allergic inflammation. Trends Immunol 35:69–78

Endo Y, Hirahara K, Iinuma T, Shinoda K, Tumes DJ, Asou HK, Matsugae N, Obata-Ninomiya K, Yamamoto H, Motohashi S, Oboki K, Nakae S, Saito H, Okamoto Y, Nakayama T (2015) The Interleukin-33-p38 kinase axis confers memory T helper 2 cell pathogenicity in the airway. Immunity 42:294–308

Fort MM et al (2001) IL-25 induces IL-4, IL-5, and IL-13 and Th2-Associated pathologies in vivo. Immunity 15:985–995

Gelfand EW, Kraft M (2009) The importance and features of the distal airways in children and adults. J Allergy Clin Immunol 124:S84–S87

Gevaert P, Lang-Loidolt D, Lackner A, Stammberger H, Staudinger H, Van Zele T, Holtappels G, Tavernier J, van Cauwenberge P, Bachert C (2006) Nasal IL-5 levels determine the response to anti-IL-5 treatment in patients with nasal polyps. J Allergy Clin Immunol 118:1133–1141

Gor DO, Rose NR, Greenspan NS (2003) TH1-TH2: a procrustean paradigm. Nat Immunol 4:503–505

Gregory LG, Lloyd CM (2011) Orchestrating house dust mite-associated allergy in the lung. Trends Immunol 32:402–411

Grotenboer NS, Ketelaar ME, Koppelman GH, Nawijn MC (2013) Decoding asthma: translating genetic variation in IL33 and IL1RL1 into disease pathophysiology. J Allergy Clin Immunol 131:856–865

Halim TY, Krauss RH, Sun AC, Takei F (2012) Lung natural helper cells are a critical source of Th2 cell-type cytokines in protease allergen-induced airway inflammation. Immunity 36:451–463

Hamilos DL (2011) Chronic rhinosinusitis: epidemiology and medical management. J Allergy Clin Immunol 128:693–707, quiz 708–699

Holt PG et al (2009) Pathogenic mechanisms of allergic inflammation: atopic asthma as a paradigm. Adv Immunol 104:51–113

Iinuma T, Okamoto Y, Yamamoto H, Inamine-Sasaki A, Ohki Y, Sakurai T, Funakoshi U, Yonekura S, Sakurai D, Hirahara K, Nakayama T (2015) Interleukin-25 and mucosal T cells in noneosinophilic and eosinophilic chronic rhinosinusitis. Ann Allergy Asthma Immunol 114:289–298

Kim HY et al (2010) The many paths to asthma: phenotype shaped by innate and adaptive immunity. Nat Immunol 11:577–584

Kurowska-Stolarska M, Kewin P, Murphy G, Russo RC, Stolarski B, Garcia CC, Komai-Koma M, Pitman N, Li Y, Niedbala W et al (2008) IL-33 induces antigen-specific IL-5+ T cells and promotes allergic-induced airway inflammation independent of IL-4. J Immunol 181:4780–4790

Liew FY, Pitman NI, McInnes IB (2010) Disease-associated functions of IL-33: the new kid in the IL-1 family. Nat Rev Immunol 10:103–110

Liu X, Li M, Wu Y, Zhou Y, Zeng L, Huang T (2009) Anti-IL-33 antibody treatment inhibits airway inflammation in a murine model of allergic asthma. Biochem Biophys Res Commun 386:181–185

Lloyd CM (2010) IL-33 family members and asthma – bridging innate and adaptive immune responses. Curr Opin Immunol 22:800–806

Lloyd CM, Hessel EM (2010) Functions of T cells in asthma: more than just T(H)2 cells. Nat Rev Immunol 10:838–848

Oboki K, Ohno T, Kajiwara N, Arae K, Morita H, Ishii A, Nambu A, Abe T, Kiyonari H, Matsumoto K et al (2010) IL-33 is a crucial amplifier of innate rather than acquired immunity. Proc Natl Acad Sci U S A 107:18581–18586

Petersen BC, Budelsky AL, Baptist AP, Schaller MA, Lukacs NW (2012) IL-25 induces type 2 cytokine production in a novel, steroid resistant IL-17RB+ myeloid population that exacerbates asthmatic pathology. Nat Med 18:751–758

Porter PC et al (2011) Seeking common pathophysiology in asthma, atopy and sinusitis. Trends Immunol 32:43–49

Prefontaine D, Nadigel J, Chouiali F, Audusseau S, Semlali A, Chakir J, Martin JG, Hamid Q (2010) Increased IL-33 expression by epithelial cells in bronchial asthma. J Allergy Clin Immunol 125:752–754

Rosenberg HF et al (2013) Eosinophils: changing perspectives in health and disease. Nat Rev Immunol 13:9–22

Scanlon ST, McKenzie AN (2012) Type 2 innate lymphoid cells: new players in asthma and allergy. Curr Opin Immunol 24:707–712

Schmitz J, Owyang A, Oldham E, Song Y, Murphy E, McClanahan TK, Zurawski G, Moshrefi M, Qin J, Li X et al (2005) IL-33, an interleukin-1-like cytokine that signals via the IL-1 receptor-related protein ST2 and induces T helper type 2-associated cytokines. Immunity 23:479–490

Szab OSH, Sullivan BM, Peng SL, Glimcher LH (2003) Molecular mechanisms regulating Th1 immune responses. Annu Rev Immunol 21:713–758

Van Bruaene N, Perez-Novo CA, Basinski TM, Van Zele T, Holtappels G, De Ruyck N, Schmidt-Weber C, Akdis C, Van Cauwenberge P, Bachert C, Gevaert P (2008) T-cell regulation in chronic paranasal sinus disease. J Allergy Clin Immunol 121:1435–1441, 1441 e1431–1433

Wang YH et al (2010) A novel subset of CD4(+) T(H)2 memory/effector cells that produce inflammatory IL-17 cytokine and promote the exacerbation of chronic allergic asthma. J Exp Med 207:2479–2491

Zhang N, Van Zele T, Perez-Novo C, Van Bruaene N, Holtappels G, DeRuyck N, Van Cauwenberge P, Bachert C (2008) Different types of T-effector cells orchestrate mucosal inflammation in chronic sinus disease. J Allergy Clin Immunol 122:961–968

Chapter 32
Controlling the Mechanism Underlying Chronic Inflammation Through the Epigenetic Modulation of CD4 T Cell Senescence

Masakatsu Yamashita, Makoto Kuwahara, Junpei Suzuki, and Takeshi Yamada

Abstract Senescent cells acquire a senescence-associated secretory phenotype (SASP), which is characterised by the increased production of pro-inflammatory factors.

Senescent CD4 T cells also exhibit the SASP and may contribute to a component of age-associated inflammatory responses called inflammaging. However, the

M. Yamashita (✉)
Department of Immunology, Graduate School of Medicine, Ehime University, Shitsukawa, Toon, Ehime 791-0295, Japan

Translational Research Center, Ehime University Hospital, Shitsukawa, Toon, Ehime 791-0295, Japan

Division of Immune Regulation, Department of Proteo-Inovation, Proteo-Science Center, Ehime University, Matsuyama, Ehime 790-8577, Japan

PRESTO, Japan Science and Technology Agency, Kawaguchi 332-0012, Japan
e-mail: yamamasa@m.ehime-u.ac.jp

M. Kuwahara
Department of Immunology, Graduate School of Medicine, Ehime University, Shitsukawa, Toon, Ehime 791-0295, Japan

Translational Research Center, Ehime University Hospital, Shitsukawa, Toon, Ehime 791-0295, Japan

Division of Immune Regulation, Department of Proteo-Inovation, Proteo-Science Center, Ehime University, Matsuyama, Ehime 790-8577, Japan

J. Suzuki
Translational Research Center, Ehime University Hospital, Shitsukawa, Toon, Ehime 791-0295, Japan

Department of Hematology, Clinical Immunology and Infectious Diseases, Graduate School of Medicine, Ehime University, Shitsukawa, Toon, Ehime 791-0295, Japan

T. Yamada
Department of Infection and Host Defenses, Graduate School of Medicine, Ehime University, Shitsukawa, Toon, Ehime 791-0295, Japan

M. Miyasaka, K. Takatsu (eds.), *Chronic Inflammation*,
DOI 10.1007/978-4-431-56068-5_32

mechanism behind the CD4 T-cell SASP remains unclear. We recently found that a T-cell–specific *Menin* deficiency results in the premature senescence of T cells after antigenic stimulation, which is accompanied by the SASP. Menin, in part, acts by targeting *Bach2*, which is known to regulate immune homeostasis. Menin binds to the *Bach2* gene locus and controls its expression through the maintenance of active histone modification. Menin-dependent recruitment of PCAF and subsequent histone acetylation appears to be important for the maintenance of *Bach2* expression. Menin binding at the *Bach2* locus leads to a reduced *Bach2* expression in the senescent CD4 T cells. The forced expression of *Bach2* in *menin*-deficient CD4 T cells, as well as in senescent CD4 T cells, normalises the SASP. In addition, *Bach2*-deficient CD4 T cells exhibit SASP. These findings indicate that the Menin–Bach2 axis therefore plays a critical role in regulating the SASP in CD4 T cells.

Keywords Menin • Bach2 • Immunosenescence • Senescence-associated secretory phenotype (SASP) • Inflammaging • Epigenetics

32.1 Introduction

Immunosenescence is defined by age-induced alterations of innate and adaptive immunity, and is involved in the age-related decline of immune functions. Immunosenescence causes an increase in the susceptibility of elderly individuals to infectious diseases and certain types of cancers (Finkel et al. 2007; Gavazzi and Krause 2002). In addition, immunosenescence induces a proinflammatory state and increases the susceptibility of autoimmunity such as rheumatoid arthritis (Lindstrom and Robinson 2010; Cavanagh et al. 2012). Immunosenescence is accompanied by alterations of CD4 T-cell–mediated hormonal immunity. Although the CD4 T-cell functions are more susceptible to ageing than CD8 T cells, the molecular mechanism of different sensitivity between CD4 and CD8 T cells has not been fully elucidated.

One of the characteristic features of age-related alterations of immune responses is an increased inflammatory trait. Senescent cells can significantly harm the tissue microenvironment through the acquisition of a senescent-associated secretory phenotype (SASP), which is characterised by a striking increase in the secretion of proinflammatory factors including cytokines, chemokines, matrix remodelling factors, and proangiogenic factors (Rodier and Campisi 2011; Tchkonia et al. 2013). The proinflammatory status induced by SASP is referred to as inflammaging. It is possible that senescent CD4 T cells exhibit the SASP and may thus contribute to the induction of inflammaging (Franceschi et al. 2000).

The number of naïve CD4 T cells decreases, whereas memory phenotype (MP) CD4 T cells increase with ageing. The age-dependent decline of the thymus is a major cause of this phenomenon. MP CD4 T cells increased with aging produce a large amount of inflammatory cytokines as compared with memory CD4 T cells that are induced by normal immunisation (Maue et al. 2009; Haynes and Lefebvre 2011). Furthermore, the characteristic features of naïve CD4 T cells change drastically with ageing. Reduced IL-2 production and decreased clonal expansion were reported in naïve CD4 T cells from aged mice and humans. The formation of an immunological synapse, helper activity for B lymphocytes, and the differentiation of memory CD4 T cells are also impaired in naïve CD4 T cells from aged mice. These data suggest that naïve CD4 T cells senesce with ageing. Therefore, elucidating immunosenescence requires knowledge of CD4 T cell senescence.

Epigenetics is the study of heritable changes that affect gene expression without causing any changes in the DNA sequence. It was reported that one phenomenon associated with senescence is a change in the patterns of epigenetic modifications including histone modifications and DNA methylation (Jones et al. 2015; Salminen et al. 2015). Epigenetic change in CD4 T cells induced by environmental stimuli such as foreign antigens and pathogens can alter the CD4 T cell functions (Agarwal et al. 2000). Furthermore, global changes in the epigenetic landscape are sometimes observed in chronic-inflammation–driven diseases such as rheumatoid arthritis (Klein and Gay 2013). In addition, it was recently reported that inhibitors for histone deacetylase have been shown to have an anti-inflammatory effect (Cantley and Haynes 2013). Therefore, we focused on the epigenetic regulation of CD4 T-cell senescence and the subsequent induction of a proinflammatory status, which leads to the development of chronic inflammation in the future.

32.2 *Menin* Deficiency Induces CD4 T-Cell Senescence

We have been studying the regulatory mechanisms of peripheral CD4 T cell differentiation and function and previously demonstrated that the epigenetic regulation of gene expressions plays a critical role in the acquisition and maintenance of the CD4 T-cell function (Nakayama and Yamashita 2009). For instance, we found that mixed-lineage leukemia 1 (MLL1), a histone methyltransferase, is required for the maintenance of the Th2 cell functions (Yamashita et al. 2006). MLL1 heterozygote Th2-cells gradually lose their ability to produce Th2 cytokines.

Next, we focused on the role of Menin, which is a component of the MLL1 complex (Yokoyama et al. 2004, 2005). A schematic representation of Menin is displayed in Fig. 32.1. MENIN, a tumour suppressor protein, is encoded by the *MEN1* gene in humans. Certain germinal mutations of MEN1 induce an autosomal-dominant syndrome characterised by concurrent parathyroid adenomas, gastro-enteropancreatic tumours and several other tumour types. Although Menin contains nuclear localisation signals within its C terminus, it has no other significant similarity with other proteins. The interactions between menin and MLL1, JunD, nuclear factor of kappa light polypeptide gene enhancer in B cells (NF-kB), peroxisome proliferator-activated receptor gamma (PPAR-γ), SMAD family member 3 (SMAD3), β-catenin, and RNA polymerase II have been previously reported (Balogh et al. 2006; Wu and Hua 2011). Therefore, menin is involved in both transcriptional activation and repression.

We crossed *Menin*$^{flox/flox}$ mice with *CD4-Cre* transgenic mice to generate T-cell–specific *Menin*-deficient (menin KO) mice (Kuwahara et al. 2014). The

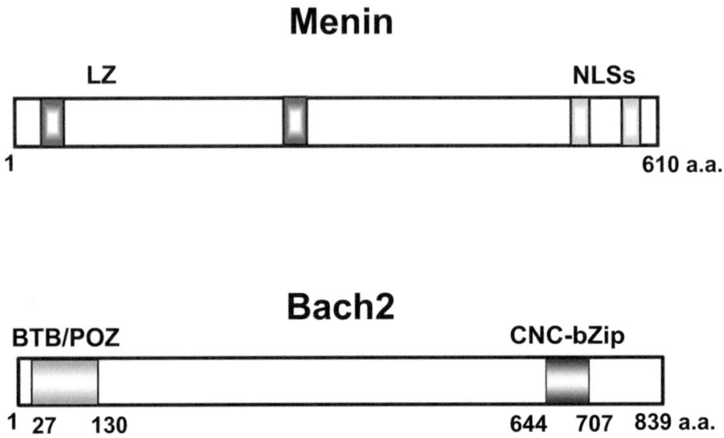

Fig. 32.1 Schematic representation of Menin (*upper panel*) and Bach2 (*lower panel*). *LZ* leucine zipper, *NLS* nuclear localisation signal, *BTB* for BR-C, ttk and bab, *POZ domain* for Pox virus and the zinc finger domain, *CNC* Cap'n'collar, *bZip* basic leucine zipper

number of naïve CD4 T cells decreased, whereas MP CD4 T cells increased in Menin KO mice. The survival of the antigen-experienced CD4 T cells significantly decreased in the *Menin*-deficient mice, and the recall responses against antigens were impaired in a murine model of allergic airway inflammation and *Listeria monocytogenes* infection model.

Early cell-cycle arrest accompanied by the increased expression of cycle-dependent kinase (CDK) inhibitors is a hallmark of cellular senescence (Kuilman et al. 2010). An increased expression of CDK inhibitors *Cdkn1a*, *Cdkn2a*, and *Cdkn2b* was detected in in vitro activated *Menin* KO CD4 T cells. *Menin* KO CD4 T cells normally proliferate after TCR-stimulation in vitro, however, the growth rate was reduced after 4 days and did not further proliferate after 9 days. These results suggest that early cell arrest is induced in the *Menin* KO-activated CD4 T cells.

The decreased cell surface expression of CD27 and CD62L and increased expression of inhibitory receptors such as programmed cell death-1 (PD-1) and NK cell markers are characteristics of senescent and/or exhausted T cells (van Gisbergen et al. 2011; Fearon 2007; Moro-Garcia et al. 2012). The number of CD27^{low}/CD62L^{low} cells markedly increased in the *Menin* KO CD4 T cells. The upregulation of PD-1 was also detected in the *Menin* KO CD4 T cells.

The high-level production of IL-6 and osteopontin (OPN) was detected in the *Menin* KO CD4 T cells. Proinflammatory chemokines, CCL-3, CCL-4, and CCL-5 were also secreted from the activated *Menin* KO CD4 T cells, and proangiogenic factors *Esm1*, *Pdgfα*, *S100a4*, and *Vegfc* were detected. Furthermore, the expression of genes encoding cytotoxic granules (*Gzm a*, *b*, *c*, *d*, *e*, *f*, and *Prf1*), which are involved in age-related cardiovascular and pulmonary diseases, strikingly increased in *Menin* KO CD4 T cells. These results suggest that the SASP is induced in the *Menin* KO CD4 T cells by TCR-stimulation.

Increased senescence-associated β-galactosidase (SA βGal) activity is another maker of cellular senescence (Kuilman et al. 2010). Although the *Menin* KO CD4 T cells did not express a SA βGal activity, a portion of the SA-βGal-positive cells rapidly increased in the *Menin* KO CD4 T-cell cultures after TCR stimulation, and a strong SA-βGal activity was detected in the *Menin* KO cells on day 7.

These experimental results indicate that *Menin* deficiency induces premature cellular senescence of the activated CD4 T cells.

32.3 The Prolonged Activation of NF-kB Is Involved in the Induction of the SASP in the *menin* KO-Activated CD4 T Cells

NF-kB signaling induces cellular senescence, and the transcription factor NF-kB subunit p65 (RelA) is on the chromatin of senescent cells (Chien et al. 2011). Enhanced phosphorylation of RelA (pSer534) was detected, and its DNA-binding activity increased in the *Menin* KO effector CD4 T cells. In contrast, the DNA binding activity of RelB did not increase. To assess the role of NF-kB signalling in the induction of the SASP in the *Menin* KO-activated CD4 T cells, the *Menin* KO CD4 T cells were treated with an NF-kB inhibitor, SM-7368, during the TCR-mediated activation phase. The treatment of *Menin* KO CD4 T cells with SM-7368 normalised the enhanced generation of IL-6-, IL-10, and OPN-producing cells in *Menin* KO CD4 T cells (Fig. 32.2). In addition, the expression of cytotoxic granules decreased in the SM-7368-treated *Menin* KO CD4 T cells. In contrast, the expressions of proinflammatory chemokines, angiogenic factors, and Cdk inhibitors did not normalise after SM-7368 treatment. These results indicate that the induction of the SASP in the *Menin* KO-activated CD4 T cells is in part regulated by the activation of the NF-kB signalling pathway.

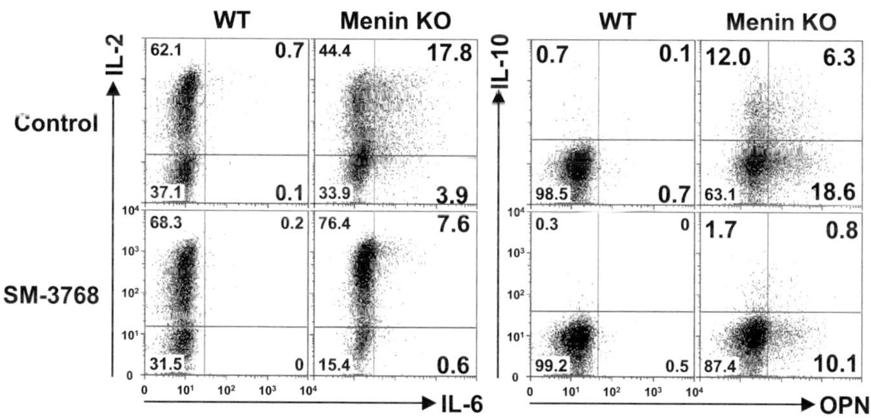

Fig. 32.2 Effect of pharmacological inhibition of the NF-kB activation on increased generation of IL-6-, IL-10, and OPN-producing CD4 T cells in the *Menin*-deficient activated CD4 T cells

32.4 Menin Inhibits the SASP by Maintaining Bach2 Expression

We identified Bach2 (BTB and Cap'n'collar (CNC) homology 1; basic leucine zipper transcription factor 2) as a downstream target of menin according to the results of a DNA microarray analysis. A schematic representation of Bach2 is displayed in Fig. 32.1. Bach2 belongs to the CNC gene family (Sykiotis and Bohmann 2010). The CNC gene family consists of two transcriptional repressors (Bach1 and Bach2) and four activators (NF-E2, Nrf1, Nrf2, and Nrf3) (Motohashi et al. 2002; Sykiotis and Bohmann 2010). Bach2 interacts with small Maf proteins and acts as a genetic inhibitor of the gene expression directed by the TPA response element, Maf recognition element, and antioxidant response element (Blank 2008; Oyake et al. 1996). Bach2 regulates Treg-mediated immune homeostasis and suppresses multiple CD4 T-cell effector programmes. Polymorphisms in *BACH2* are associated with multiple inflammatory and autoimmune diseases including rheumatoid arthritis and celiac disease. The involvement of Bach2 in memory CD8 T-cell formation has been recently reported. Therefore, Bach2 acts as a key regulator of T-cell–mediated immune homeostasis.

The expression of Bach2 (mRNA and protein) was reduced in the *Menin* KO effector CD4 T cells compared with those in the WT cells, whereas its expression in the *Menin* KO naïve CD4 T cells showed only marginal reduction. The augmented production of IL-6, IL-10, OPN, and proinflammatory chemokines (CCL-3, CCL-4, and CCL-5) in the *Menin* KO effector CD4 T cells was normalised by the transduction of *Bach2*. The transduction of the *Menin* KO effector CD4 T cells with *Bach2* also restored the enhanced expressions of cytotoxic granules and the genes that encode proinflammatory enzymes and proangiogenic factors. Furthermore, the augmented production of IL-6, OPN, and proinflammatory chemokines was detected in the *Bach2* KO CD4 T cells. The expression of cytotoxic granules was also upregulated in the activated *Bach2* KO CD4 T cells. However, the early induction of cell death and cell-cycle arrest, which were observed in the *Menin*-deficient activated CD4 T cells, were not induced by Bach2 deficiency. These data indicate that a SASP of *Menin* KO CD4 T cells is induced, in part, through a reduction of Bach2 expression.

32.5 Epigenetic Regulation of Bach2 Expression by Menin

Menin is a component of the KMT (lysine methyl transferase) 2A/2B complex, which is involved in the trimethylation of histone H3K4 (H3K4me3), an active histone marker. Menin binds the transcriptional start site (TSS) of the *Bach2* gene. The level of H3K4me3 at *Bach2* peaked at the TSS and showed only a slight reduction in the *Menin* KO effector CD4 T cells compared with the WT effector CD4 T cells. In contrast, the level of H3K27me3, a repressive histone marker, at the

Bach2 TSS was increased in the *Menin* KO effector CD4 T cells. Furthermore, the wide range of acetylated (ac) histone H3K27 level was significantly reduced in the *Menin* KO effector CD4 T cells. In *Menin* KO naïve CD4 T cells, only a small reduction of H3K27ac was detected, and the level was rapidly decreased in the *Menin* KO CD4 T cells after TCR stimulation. The reduced expression of Bach2 in the *Menin* KO effector CD4 T cells was restored by treatment with trichostatin A (TSA), an HDAC inhibitor. The menin-dependent recruitment of lysine acetyltransferase 2B (PCAF) appears to be important for the maintenance of H3K27ac at the *Bach2* gene locus. The increased recruitment of polycomb repressive complex 2 (PRC2) components such as Ezh2 and Suz12 was detected in the *menin* KO effector CD4 T cells. These results suggest that Menin maintains *Bach2* expression by regulating histone acetylation (Fig. 32.3).

Recently, O'Shea's group has reported that the upstream region of the *Bach2* gene locus is the most prominent T-cell superenhancer (Vahedi et al. 2015). Superenhancers, also known as stretch-enhancers, are a type of enhancer that is

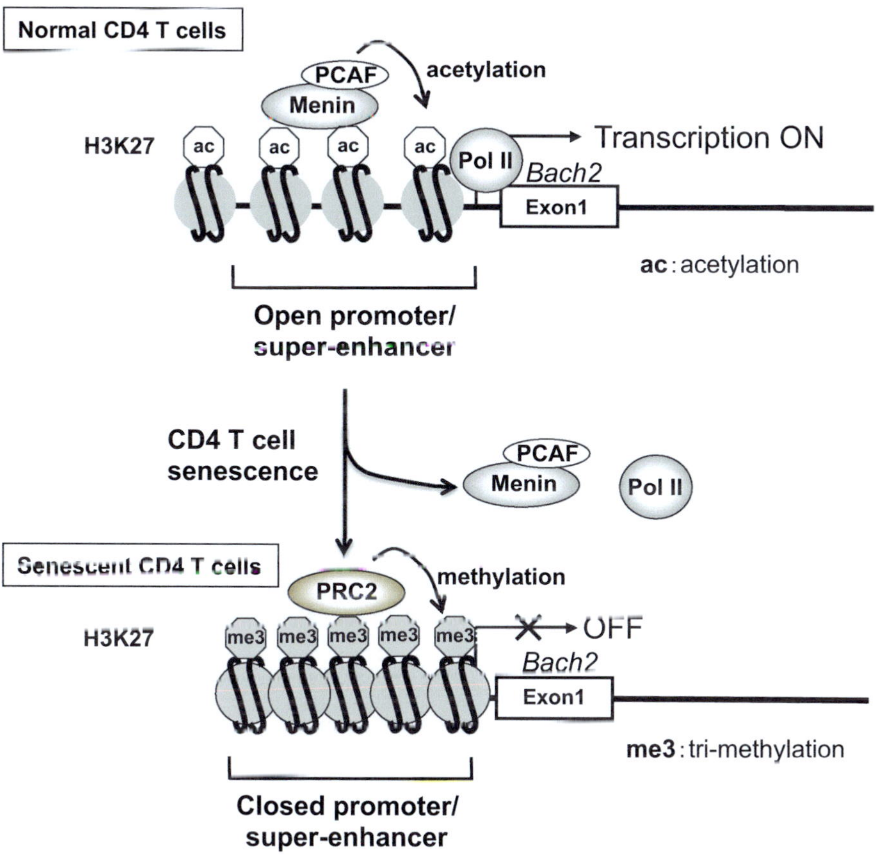

Fig. 32.3 Epigenetic regulation of *Bach2* transcription by menin (Hypothesis)

particularly important for genes associated with cell identity. The superenhancer regions are marked by highly acetylated histone H3K27, and the binding of RNA polymerase II, mediator, cohesion, p300, CBP, and Lsd1-NuRD complexes are enriched (Pott and Lieb 2015; Hnisz et al. 2013). PCAF (a CBP/p300 associated factor) recruits to the superenhancer region of the *Bach2* gene locus in a Menin-dependent manner. In addition, Menin is required for the maintenance of H3K27ac at the 5' upstream region of the *Bach2* gene locus. It is possible that Menin maintains the *Bach2* expression by regulating the superenhancer activity. Therefore, Menin may inhibit CD4 T-cell senescence in part by controlling superenhancer activity.

32.6 Reduced Expression of *Bach2* in Senescent CD4 T Cells

We established a senescent Th (sTh)-cell induction method to determine whether the expression of Menin and/or Bach2 is reduced in senescent CD4 T cells (Kuwahara et al. 2012). Increased CD62L^{low}/CD27^{high} and CD62L^{low}/CD27^{low} populations were observed in the sTh cells in comparison to the primary effector CD4 T cells (pTh). The number of IL-6-, IL-10, and OPN-producing cells was also increased in the sTh cell cultures.

The mRNA and protein expressions of Bach2 in the sTh cells were lower than those observed in the pTh cells. However, the expression of Menin in the sTh cells was comparable to that observed in the pTh cells. A ChIP sequencing analysis using an anti-Menin antibody revealed that the level of Menin binding at the *Bach2* gene locus was decreased in the sTh cells, and a reduction in the H3K4me3 and H3K27ac levels was also detected at the *Bach2* gene locus in the sTh cells. A reduction in the binding of RNA polymerase II (PNAPII), PCAF, MLL1, Rbbp5, and Wdr5, as well as an increase in the binding of Suz12 at the *Bach2* gene locus was also detected in the sTh cells. Although the reduced expression of *Bach2* mRNA and the level of H3K27ac in the sTh cells were completely restored by TSA treatment, the recruitment of Menin at the *Bach2* locus was not increased in the TSA-treated sTh cells. These results indicate that the Menin-dependent maintenance of histone acetylation is required for the proper expression of *Bach2*, whereas Menin is not essential for *Bach2* transcription.

32.7 Conclusion

Our experimental results demonstrate that Menin inhibits CD4 T-cell senescence. Menin regulates CD4 T-cell senescence via multiple pathways (Fig. 32.4). Recently, we found that Menin also regulates CD8 T cell senescence. The *Menin-*

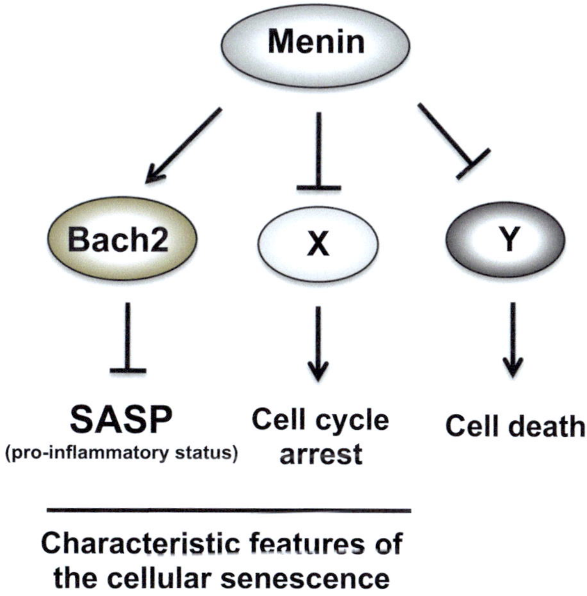

Fig. 32.4 Menin regulates the SASP (formation of the proinflammatory status), cell cycle, and cell death in T cells

deficient activated CD8 T-cells exhibit the characteristic features of premature cellular senescence (unpublished data). Among the Menin-regulated pathways, the Menin–Bach2 axis plays a crucial role in repressing the SASP. However, the downstream targets of Menin that mediate cell survival and cell cycle remain to be elucidated. The molecular mechanism for regulating Menin recruitment at the *Bach2* gene locus also should be addressed. Taken together, further analysis of the function of Menin in T cells may help to elucidate the age-related dysregulation of the T-cell–mediated immune responses and subsequent development of chronic inflammation.

References

Agarwal S, Avni O, Rao A (2000) Cell-type-restricted binding of the transcription factor NFAT to a distal IL-4 enhancer in vivo. Immunity 12:643–652

Balogh K, Racz K, Patocs A, Hunyady L (2006) Menin and its interacting proteins: elucidation of menin function. Trends Endocrinol Metab 17:357–364

Blank V (2008) Small Maf proteins in mammalian gene control: mere dimerization partners or dynamic transcriptional regulators? J Mol Biol 376:913–925

Cantley MD, Haynes DR (2013) Epigenetic regulation of inflammation: progressing from broad acting histone deacetylase (HDAC) inhibitors to targeting specific HDACs. Inflammopharmacology 21:301–307

Cavanagh MM, Weyand CM, Goronzy JJ (2012) Chronic inflammation and aging: DNA damage tips the balance. Curr Opin Immunol 24:488–493

Chien Y, Scuoppo C, Wang X, Fang X, Balgley B, Bolden JE, Premsrirut P, Luo W, Chicas A, Lee CS et al (2011) Control of the senescence-associated secretory phenotype by NF-kappaB promotes senescence and enhances chemosensitivity. Genes Dev 25:2125–2136

Fearon DT (2007) The expansion and maintenance of antigen-selected CD8(+) T cell clones. Adv Immunol 96:103–139

Finkel T, Serrano M, Blasco MA (2007) The common biology of cancer and ageing. Nature 448: 767–774

Franceschi C, Bonafe M, Valensin S, Olivieri F, De Luca M, Ottaviani E, De Benedictis G (2000) Inflamm-aging. An evolutionary perspective on immunosenescence. Ann N Y Acad Sci 908: 244–254

Gavazzi G, Krause KH (2002) Ageing and infection. Lancet Infect Dis 2:659–666

Haynes L, Lefebvre JS (2011) Age-related deficiencies in antigen-specific CD4 T cell responses: lessons from mouse models. Aging Dis 2:374–381

Hnisz D, Abraham BJ, Lee TI, Lau A, Saint-Andre V, Sigova AA, Hoke HA, Young RA (2013) Super-enhancers in the control of cell identity and disease. Cell 155:934–947

Jones MJ, Goodman SJ, Kobor MS (2015) DNA methylation and healthy human aging. Aging Cell 14:924–932

Klein K, Gay S (2013) Epigenetic modifications in rheumatoid arthritis, a review. Curr Opin Pharmacol 13:420–425

Kuilman T, Michaloglou C, Mooi WJ, Peeper DS (2010) The essence of senescence. Genes Dev 24:2463–2479

Kuwahara M, Yamashita M, Shinoda K, Tofukuji S, Onodera A, Shinnakasu R, Motohashi S, Hosokawa H, Tumes D, Iwamura C et al (2012) The transcription factor Sox4 is a downstream target of signaling by the cytokine TGF-beta and suppresses T(H)2 differentiation. Nat Immunol 13:778–786

Kuwahara M, Suzuki J, Tofukuji S, Yamada T, Kanoh M, Matsumoto A, Maruyama S, Kometani K, Kurosaki T, Ohara O et al (2014) The Menin-Bach2 axis is critical for regulating CD4 T-cell senescence and cytokine homeostasis. Nat Commun 5:3555

Lindstrom TM, Robinson WH (2010) Rheumatoid arthritis: a role for immunosenescence? J Am Geriatr Soc 58:1565–1575

Maue AC, Yager EJ, Swain SL, Woodland DL, Blackman MA, Haynes L (2009) T-cell immunosenescence: lessons learned from mouse models of aging. Trends Immunol 30: 301–305

Moro-Garcia MA, Alonso-Arias R, Lopez-Larrea C (2012) Molecular mechanisms involved in the aging of the T-cell immune response. Curr Genomics 13:589–602

Motohashi H, O'Connor T, Katsuoka F, Engel JD, Yamamoto M (2002) Integration and diversity of the regulatory network composed of Maf and CNC families of transcription factors. Gene 294:1–12

Nakayama T, Yamashita M (2009) Critical role of the Polycomb and Trithorax complexes in the maintenance of CD4 T cell memory. Semin Immunol 21:78–83

Oyake T, Itoh K, Motohashi H, Hayashi N, Hoshino H, Nishizawa M, Yamamoto M, Igarashi K (1996) Bach proteins belong to a novel family of BTB-basic leucine zipper transcription factors that interact with MafK and regulate transcription through the NF-E2 site. Mol Cell Biol 16: 6083–6095

Pott S, Lieb JD (2015) What are super-enhancers? Nat Genet 47:8–12

Rodier F, Campisi J (2011) Four faces of cellular senescence. J Cell Biol 192:547–556

Salminen A, Kauppinen A, Kaarniranta K (2015) 2-Oxoglutarate-dependent dioxygenases are sensors of energy metabolism, oxygen availability, and iron homeostasis: potential role in the regulation of aging process. Cell Mol Life Sci 72:3897–914

Sykiotis GP, Bohmann D (2010) Stress-activated cap'n'collar transcription factors in aging and human disease. Sci Signal 3:re3

Tchkonia T, Zhu Y, van Deursen J, Campisi J, Kirkland JL (2013) Cellular senescence and the senescent secretory phenotype: therapeutic opportunities. J Clin Invest 123:966–972

Vahedi G, Kanno Y, Furumoto Y, Jiang K, Parker SC, Erdos MR, Davis SR, Roychoudhuri R, Restifo NP, Gadina M et al (2015) Super-enhancers delineate disease-associated regulatory nodes in T cells. Nature 520:558–562

van Gisbergen KP, Klarenbeek PL, Kragten NA, Unger PP, Nieuwenhuis MB, Wensveen FM, ten Brinke A, Tak PP, Eldering E, Nolte MA, van Lier RA (2011) The costimulatory molecule CD27 maintains clonally diverse CD8(+) T cell responses of low antigen affinity to protect against viral variants. Immunity 35:97–108

Wu T, Hua X (2011) Menin represses tumorigenesis via repressing cell proliferation. Am J Cancer Res 1:726–739

Yamashita M, Hirahara K, Shinnakasu R, Hosokawa H, Norikane S, Kimura MY, Hasegawa A, Nakayama T (2006) Crucial role of MLL for the maintenance of memory T helper type 2 cell responses. Immunity 24:611–622

Yokoyama A, Wang Z, Wysocka J, Sanyal M, Aufiero DJ, Kitabayashi I, Herr W, Cleary ML (2004) Leukemia proto-oncoprotein MLL forms a SET1-like histone methyltransferase complex with menin to regulate Hox gene expression. Mol Cell Biol 24:5639–5649

Yokoyama A, Somervaille TC, Smith KS, Rozenblatt-Rosen O, Meyerson M, Cleary ML (2005) The menin tumor suppressor protein is an essential oncogenic cofactor for MLL-associated leukemogenesis. Cell 123:207–218

Chapter 33
Adrenergic Control of Lymphocyte Dynamics and Inflammation

Kazuhiro Suzuki

Abstract Relationships between the nervous and immune systems have been documented in the literature since the beginning of the last century. However, the cellular and molecular basis for their interconnections has emerged only in the past decade. Recent studies on the control of immune cell dynamics by adrenergic nerves have provided evidence for cell-type–specific mechanisms. In lymphocytes, activation of β_2-adrenergic receptors (β_2ARs) augments the responsiveness of a particular set of chemoattractant receptors. The crosstalk between these receptors expressed on lymphocytes restricts their exit from lymph nodes (LNs) and contributes to heamostasis of their distribution among lymphoid organs. Moreover, this mechanism is implicated in limiting the progression of inflammatory diseases and could lead to a new therapeutic approach. This chapter summarises the current knowledge of adrenergic control of lymphocyte dynamics and locates it in the historical context of the field.

Keywords Adrenergic nerves • β_2-adrenergic receptors • Chemokine receptors • Inflammation • Lymph-node egress • Lymphocyte trafficking • Multiple sclerosis • Stress

33.1 Introduction

More than a hundred years ago, immediately after the initial characterisation of adrenaline (Aldrich 1901; Takamine 1901), it was reported that subcutaneous injection of adrenaline rapidly induced leukocytosis in humans (Loeper and Crouzon 1904), which was the first demonstration that the immune system could respond to adrenergic stimulation (Fig. 33.1). A Japanese bacteriologist Tohru Ishigami (1919) showed that phagocytic activity of leukocytes in tuberculosis patients was decreased during the periods of adverse mental conditions, suggesting for the first time that psychological stress could alter immune cell functions.

K. Suzuki (✉)
WPI Immunology Frontier Research Center, Osaka University, 3-1 Yamada oka, Suita, Osaka 565-0871, Japan

PRESTO, Japan Science and Technology Agency, Kawaguchi 332-0012, Japan
e-mail: ksuzuki@ifrec.osaka-u.ac.jp

M. Miyasaka, K. Takatsu (eds.), *Chronic Inflammation*,
DOI 10.1007/978-4-431-56068-5_33

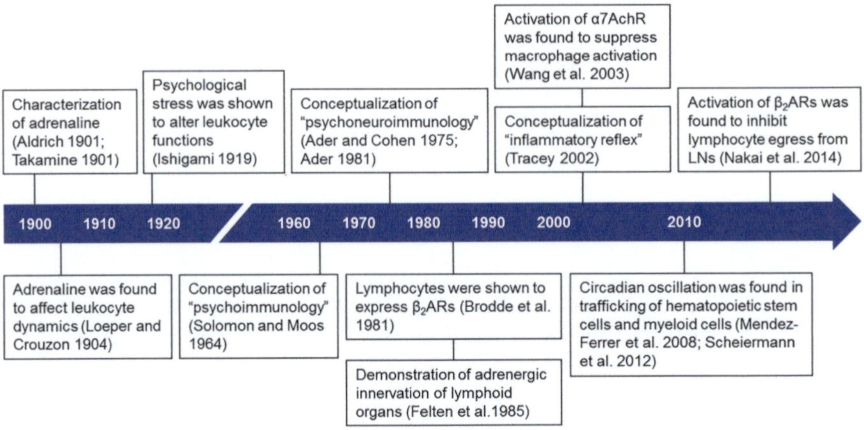

Fig. 33.1 A timeline of discoveries about neural control of immunity. The history of this research field is comprised of three phases. Studies during the first half of the twentieth century described phenomenological connections between the nervous and immune systems, which culminated in the concept of 'psychoneuroimmunology'. Studies during the 1980s and 1990s provided the first indications of how neuroimmune interactions occur. Studies since the beginning of the twenty-first century have clarified molecular mechanisms for immune regulation by the nervous system

Nevertheless, the immune system had been recognised as essentially autonomous until the middle of the twentieth century when 'psychoimmunology' was conceptualised by Solomon and Moos (1964) based on the observation that stress increased the susceptibility to autoimmune diseases. A decade later, the field was further formulated as 'psychoneuroimmunology'" by Ader and Cohen, who discovered that Pavlovian behavioral conditioning affected immune responses in rodents (Ader 1981; Ader and Cohen 1975; Ader et al. 1995).

Studies during the 1980s and 1990s provided the first mechanistic insights into immune regulation by the nervous system (Fig. 33.1). Comprehensive morphological studies demonstrated that both primary and secondary lymphoid organs are innervated by adrenergic nerves (Felten et al. 1985), establishing the anatomical connection of the nervous system with the immune system. Notably, in secondary lymphoid organs, including spleen and LNs, the adrenergic nerve fibres exhibit a uniform pattern of distribution, where they mainly supply T-cell zones but are absent from B-cell follicles. The neurotransmitter noradrenaline released from adrenergic nerve terminals was found to bind the surface of lymphocytes, which was almost exclusively mediated by β_2ARs (Brodde et al. 1981). Additionally, stimulation of β_2ARs was shown to inhibit differentiation and functions of T helper type 1 (Th1) cells in vitro (Ramer-Quinn et al. 1997; Sanders et al. 1997). These findings indicated that neuronal inputs could directly act on immune cells to alter their behaviors.

It was not until the beginning of the twenty-first century that a solid molecular basis for the neural regulation of immunity was uncovered (Fig, 33.1). Pioneering studies by Tracey and colleagues led to establishing a concept of 'inflammatory reflex' as a neural circuit that contributes to homeostasis of the immune system (Andersson and Tracey 2012; Tracey 2002). Molecular products of infection or tissue damage activate sensory neurons in the vagus nerve that travel to the brainstem nucleus. These signals in turn generate action potentials that travel from the brainstem to the spleen. This results in the release of acetylcholine from a small population of T cells, which binds to the nicotinic acetylcholine receptor subunit α7 (α7nAchR) expressed on macrophages to inhibit their production of proinflammatory cytokines (Rosas-Ballina et al. 2011; Wang et al. 2003). Recently, roles of adrenergic nerves in the regulation of haematopoietic cell dynamics were revisited by Frenette and colleagues (Scheiermann et al. 2013). Adrenergic nerves were shown to control the bone marrow homing of haematopoietic stem cells and tissue recruitment of granulocytes by establishing circadian oscillations of adhesion molecule and chemoattractant expression by bone marrow stromal cells and tissue vascular endothelial cells, respectively (Mendez-Ferrer et al. 2008; Scheiermann et al. 2012). However, it had been unclear how the inputs from adrenergic nerves affect the trafficking of B and T lymphocytes, major immune cell populations that mediate adaptive immunity. Our recent study demonstrated that adrenergic inputs control lymphocyte recirculation through LNs in a cell-intrinsic manner (Nakai et al. 2014), which marks a sharp contrast to the adrenergic control of granulocyte dynamics that largely depends on the alternation of environmental cues. In this chapter, I discuss cellular and molecular mechanisms for adrenergic control of lymphocyte trafficking and their relevance to the pathology of inflammatory diseases.

33.2 Adrenergic Control of Lymphocyte Egress from LNs

The observations from early studies on adrenergic alternation of immune cell dynamics led to the general agreement that in an acute phase after adrenergic stimulation, circulating granulocyte numbers are increased, but in contrast, lymphocyte numbers are decreased (Benschop et al. 1996). Given the fact that lymphocytes almost exclusively express β_2ARs among subtypes of adrenergic receptors (Sanders 2012), we investigated the role of β_2ARs in the regulation of lymphocyte dynamics. As expected, intravenous administration of selective β_2AR agonists into mice resulted in a rapid reduction of lymphocyte numbers in blood (Nakai et al. 2014). Notably, the reduction of blood lymphocytes was accompanied by a sharp decline of lymphocyte numbers in lymph, which had not been noted in the previous studies. These effects were prominent in B cells compared with T cells, which was consistent with the highest expression of β_2ARs in B cells among lymphocyte populations in mice. The lymphopenic effect of β_2AR agonists was diminished in irradiated wild-type mice reconstituted with β_2AR-deficient bone marrow cells, which lack β_2AR expression in the haematopoietic compartment. In contrast, bone marrow chimeras generated with β_2AR-deficient recipients and wild-type donors responded normally to β_2AR agonists. These observations indicated that lymphopoenia induced by β_2AR stimulation is mainly mediated by cell-intrinsic mechanisms.

Lymphocytes continuously recirculate through lymphoid organs, allowing immunosurveillance of the whole body. After entering a LN from blood, lymphocytes travel to subcompartments designated for each cell type, where they survey for antigen. After spending several hours to a day in the LN, lymphocytes exit into lymph and eventually return to the bloodstream through the thoracic duct (Cyster and Schwab 2012). Given this fact, we speculated that reduction of lymphocyte numbers in lymph as well as blood upon agonist stimulation of β_2ARs might be caused by inhibition of lymphocyte egress from LNs. To test this hypothesis, we treated mice with neutralising antibodies against α4 and αL integrins to block lymphocyte entry to LNs during continuous administration of a β_2AR agonist. After entry blockade, the numbers of lymphocyte in LNs are decreased over time because of their egress into lymph without supply from blood. Thus, the frequency of lymphocyte egress can be determined by measuring the extent of reduction of lymphocyte numbers in LNs (Lo et al. 2005). We found that treatment with a β_2AR agonist increased the numbers of lymphocytes that were retained in LNs after entry blockade (Nakai et al. 2014), indicating that β_2AR stimulation inhibits lymphocyte egress from LNs.

To test whether physiological levels of adrenergic inputs are involved in the control of lymphocyte egress from LNs, we examined the frequency of LN egress of transferred β_2AR-deficient lymphocytes in wild-type mice. Our result showed that β_2AR-deficient lymphocytes are more prone to exit LNs than β_2AR-sufficient lymphocytes (Nakai et al. 2014). Additionally, chemical depletion of adrenergic nerve fibres promoted lymphocyte egress from LNs. These findings indicated that physiological adrenergic inputs through lymphocyte β_2ARs, which are at least in part derived from adrenergic nerves, limit LN egress or help LN retention of lymphocytes.

33.3 Crosstalk Between β_2ARs and Chemokine Receptors

Lymphocyte egress from LNs is strongly dependent on sphingosine-1-phosphate receptor type 1 (S1PR1), by which lymphocytes sense a high concentration of S1P in lymph (~100 nM) compared with LN parenchyma (~1 nM) to exit LNs. S1PR1 acts to overcome retention signals mediated by CCR7, CXCR4, and possibly additional chemoattractant receptors (Pham et al. 2008; Schmidt et al. 2013). Thus, the rate of lymphocyte egress from LNs appears to be determined by the relative strength of egress-promoting signals versus retention-promoting signals. All of these chemoattractant receptors as well as β_2ARs are G protein-coupled receptors (GPCRs). Studies showed that different types of GPCRs form heteromeric complexes on the cell surface and mutually regulate their signals (Fribourg et al. 2011; Gonzalez-Maeso et al. 2008). Given these facts, we hypothesised that β_2AR activation in lymphocytes might alter responsiveness of the chemoattractant receptors involved in LN egress. Treatment of lymphocytes with a β_2AR agonist enhanced activation of a small GTPase Rac1 in response to the ligand for CCR7 or CXCR4, whereas the effect was unremarkable in response to S1P (Nakai et al. 2014). Consistent with this observation, β_2AR stimulation promoted lymphocyte chemotaxis mediated by CCR7 or CXCR4, but not S1PR1. These findings indicated that activation of lymphocyte β_2ARs selectively enhances signals through the retention-promoting receptors CCR7 and CXCR4. Notably, β_2ARs were immunoprecipitated together with CCR7 or CXCR4, but not S1PR1, suggesting that β_2ARs form heteromeric complexes with CCR7 or CXCR4. We speculate that the physical association with β_2ARs might enable the selective signal enhancement through these chemokine receptors.

The effect of β_2AR stimulation on LN egress (or retention) of lymphocytes was diminished by CCR7 deficiency in lymphocytes or pharmacological blockade of CXCR4 (Nakai et al. 2014). This result confirmed that β_2AR-mediated control of lymphocyte recirculation through LNs is largely dependent on CCR7 or CXCR4 although it is possible that other undefined receptors might be involved. Collectively, the above findings established a model for adrenergic control of lymphocyte dynamics, where activation of β_2ARs expressed on lymphocytes enhances responsiveness of the retention-promoting chemokine receptors and consequently inhibits their egress from LNs (Fig. 33.2).

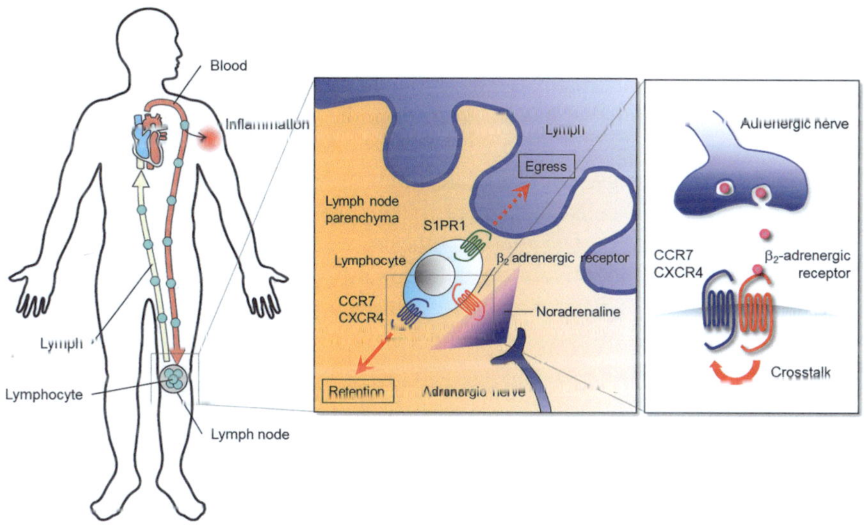

Fig. 33.2 β_2AR-mediated control of lymphocyte trafficking. Activation of β_2ARs expressed on lymphocytes enhances signals through chemokine receptors CCR7 and CXCR4, leading to inhibition of lymphocyte egress from LNs. In the context of T-cell–mediated inflammation, β_2AR-mediated signals inhibit trafficking of pathogenic T cells, reducing their numbers in inflamed tissues

33.4 Adrenergic Control of T-Cell–Mediated Inflammation

Although influence of adrenergic inputs on inflammation is complex, it appears that signals through β_2ARs generally act to suppress T-cell–mediated inflammatory diseases, including rheumatoid arthritis and multiple sclerosis (MS) (Bellinger et al. 2008). An early study showed that by increasing noradrenaline levels through administration of tri- or tetracyclic antidepressants or L-dopa, the clinical symptoms of MS were ameliorated (Berne-Fromell et al. 1987). This observation was supported by animal studies showing that treatment with β_2AR agonists suppressed experimental autoimmune encephalomyelitis (EAE), an animal model of MS (Chelmicka-Schorr et al. 1989; Wiegmann et al. 1995). However, mechanisms by which β_2AR-mediated signals affect the pathogenesis had been poorly understood. To confirm the role of β_2ARs in T-cell–mediated inflammation, we employed mouse EAE and another representative model, delayed-type hypersensitivity (DTH) in the skin. Administration of a β_2AR agonist at the onset of EAE symptoms, or at the elicitation of DTH responses reduced the magnitude of inflammation (Nakai et al. 2014), recapitulating the previous findings. Moreover, in both models, β_2AR-deficient mice developed more severe symptoms compared with wild-type mice, reinforcing the notion that β_2AR-mediated signals attenuate T-cell–mediated inflammation.

It has been established that modulation of lymphocyte trafficking is effective for the treatment of inflammatory diseases (Steinman 2014). The functional S1PR1 antagonist FTY720 (Fingolimod/Gilenya^{TM}) was proved to be effective for the treatment of MS and approved for clinical use. A major proposed action of FTY720 is to inhibit LN egress of autoreactive T cells and consequently their invasion into inflammatory sites (Brinkmann et al. 2010). Thus, lymphocyte egress from LNs represents an important therapeutic target for the treatment of T-cell–mediated inflammatory diseases.

Based on our findings, we speculated that β_2AR-mediated inhibition of LN egress of pathogenic T cells might contribute to the suppression of inflammation at the peripheral sites. To test this hypothesis, we used the skin DTH model in which trafficking of adoptively transferred antigen-specific T cells could be tracked. We found that agonist-induced and even physiological signals through β_2ARs limit LN egress of antigen-primed T cells of a central memory phenotype, but not those of an effector memory phenotype (Nakai et al. 2014). This might be attributable to higher expression of β_2ARs in central memory T cells compared with effector memory T cells. Accordingly, inputs through β_2ARs reduced the numbers of antigen-primed T cells found in the circulation and site of inflammation. It was suggested that central memory T cells are reactivated by dendritic cells at the sites of inflammation, and undergo proliferation and local differentiation to effector cells to induce inflammation (Kivisakk et al. 2004). Therefore, our findings suggested that signals through β_2ARs inhibit LN egress of pathogenic T cells and reduce their recruitment to inflamed peripheral tissues, which might prevent the initiation and propagation of inflammation (Fig. 33.2).

At this moment, however, the direct causal connection between the β_2AR-mediated effects on lymphocyte dynamics and the attenuated inflammation has not been established. Recent studies showed that stimulation of β_2ARs expressed on dendritic cells reduced their capacities for antigen presentation (Nijhuis et al. 2014) and production of proinflammatory cytokines, including interleukin-6, −12, and −23 (Herve et al. 2013). Therefore, the alteration of dendritic cell functions might impair reactivation of central memory T cells at sites of inflammation. We speculate that combination of multiple β_2AR-mediated effects on immune functions might contribute to the suppression of inflammatory responses. Although it is possible that the aforementioned effects on Th cell differentiation might affect the pathology of T-cell–mediated inflammation, we found that antigen-induced differentiation of Th cells were not altered by agonist stimulation or deficiency of β_2ARs (Nakai et al. 2014). Thus, it is unlikely that alternation of Th cell differentiation could contribute to the suppressed inflammation at least in our experimental settings.

33.5 Conclusion

Studies on the roles of adrenergic nerves in immune regulation during the last three decades revealed close connections between the nervous and immune systems at cellular and molecular levels. In this context, our study provided an answer to a long-standing question of how adrenergic nerves regulate lymphocyte dynamics (Fig. 33.1). Activation of β_2ARs in lymphocytes inhibits their egress from LNs through augmentation of retention-promoting signals mediated by chemokine receptors that might be physically associated with β_2ARs. Physiological inputs provided by adrenergic nerves contribute to homeostasis of lymphocyte recirculation through β_2ARs. Moreover, the β_2AR-mediated control of lymphocyte dynamics could be an additional mechanism of immunosuppression by adrenergic nerves (Fig. 33.2). However, some questions still remain open. The molecular basis for the crosstalk between β_2ARs and chemokine receptors is unclear. Additionally, the real picture of interactions between adrenergic nerves and lymphocytes in LNs remains to be visualised. A More important question is what is the physiological significance of the adrenergic control of lymphocyte trafficking. These questions should be addressed in future studies for comprehensive understanding of immune regulation by adrenergic nerves.

Stress response is a mechanism that helps an organism adapt physiologically to deal with physical and psychological threats. Peripheral adrenergic nerves constitute the efferent arc of the sympathetic nervous system, one of the major pathways that mediate stress response. Given the immunosuppressive effects of β_2AR-mediated signals, one might assume that stress would act to dampen inflammation. However, reality is not that simple. Consistent with a prevailing belief that stress is harmful for human health, substantial evidence indicates that stress precipitates symptoms of inflammatory diseases, which has been clearly demonstrated in MS

(Karagkouni et al. 2013). In addition to the sympathetic nervous system, stress activates the hypothalamic–pituitary–adrenal axis through the hypothalamic secretion of corticotropin-releasing hormone (CRH), which is supposed to suppress immune functions through the release of glucocorticoids from adrenal glands. Why then does stress worsen inflammation? Part of the reason might be attributable to dysfunction of these stress-adaptation systems that is observed under inflammatory conditions. MS patients with active lesions in the hypothalamus exhibited impaired activation of CRH neurons (Huitinga et al. 2004). Additionally, less glucocorticoid production in response to CRH stimulation was observed in MS patients (Wei and Lightman 1997). A recent study demonstrated that response to sympathetic activation is also altered by inflammation (Arima et al. 2012). In the mouse EAE, inputs from adrenergic nerves were shown to induce production of a chemokine CCL20 from vascular endothelial cells at the sites of inflammation, but not in intact tissues, promoting invasion of pathogenic T cells. This effect appeared to be mediated by β_1ARs because treatment with a selective β_1AR antagonist diminished chemokine production by endothelial cells and ameliorated the disease. Thus, stress response under inflammatory conditions is highly complicated and has been hindering medical interventions. However, our study established that specific activation of β_2ARs could inhibit progression of T-cell–mediated inflammation. This principle would lead to a novel therapeutic strategy that exploits the beneficial aspect inherent in stress response.

During the last century, extensive studies were performed on individual organ systems and yielded huge progress of biomedical science. However, because the organ systems in our body mutually communicate to cross-regulate their functions, it is nearly impossible to predict responses of one organ system without knowing the states of other organ systems. Thus, the importance of studies that clarify interconnections among multiple organ systems has been increasingly recognised. In this regard, the recent studies that have revealed the cellular and molecular basis for the connection between the nervous and immune systems represent the new trend in the field. In addition to the prototypical autonomic neurotransmitters, noradrenaline and acetylcholine, substantial evidence has been accumulating on the immunoregulatory roles of other neurotransmitters, including dopamine (Torres-Rosas et al. 2014; Yan et al. 2015), calcitonin gene-related peptide (Granstein et al. 2015), and vasoactive intestinal peptide (Smalley et al. 2009; Yadav and Goetzl 2008). Further elucidation of the multimodal information transfer at the interface between the two organ systems would help visualise the whole picture of neuro-immune communications.

Acknowledgments I would like to thank my lab members for critical reading and helpful comments on the manuscript. Our studies described in this chapter were supported by grants from the Japan Science and Technology Agency (PRESTO), the Japan Society for the Promotion of Science, the Naito Foundation, the Senri Life Science Foundation, the Daiichi-Sankyo Foundation of Life Science, the Sumitomo Foundation, and the Kowa Life Science Foundation.

The author has no conflicting financial interests.

References

Ader R (1981) Psychoneuroimmunology. Academic, New York

Ader R, Cohen N (1975) Behaviorally conditioned immunosuppression. Psychosom Med 37(4): 333–340

Ader R, Cohen N, Felten D (1995) Psychoneuroimmunology: interactions between the nervous system and the immune system. Lancet 345(8942):99–103

Aldrich T (1901) A preliminary report on the active principle of the suprarenal gland. Am J Physiol–Legacy Content 5(7):457–461

Andersson U, Tracey KJ (2012) Reflex principles of immunological homeostasis. Annu Rev Immunol 30:313–335. doi:10.1146/annurev-immunol-020711-075015

Arima Y, Harada M, Kamimura D, Park JH, Kawano F, Yull FE, Kawamoto T, Iwakura Y, Betz UA, Marquez G, Blackwell TS, Ohira Y, Hirano T, Murakami M (2012) Regional neural activation defines a gateway for autoreactive T cells to cross the blood–brain barrier. Cell 148(3):447–457. doi:10.1016/j.cell.2012.01.022

Bellinger DL, Millar BA, Perez S, Carter J, Wood C, ThyagaRajan S, Molinaro C, Lubahn C, Lorton D (2008) Sympathetic modulation of immunity: relevance to disease. Cell Immunol 252(1–2):27–56. doi:10.1016/j.cellimm.2007.09.005

Benschop RJ, Rodriguez-Feuerhahn M, Schedlowski M (1996) Catecholamine-induced leukocytosis: early observations, current research, and future directions. Brain Behav Immun 10(2) 77–91. doi:10.1006/brbi.1996.0009

Berne-Fromell K, Fromell H, Lundkvist S, Lundkvist P (1987) Is multiple sclerosis the equivalent of Parkinson's disease for noradrenaline? Med Hypotheses 23(4):409–415

Brinkmann V, Billich A, Baumruker T, Heining P, Schmouder R, Francis G, Aradhye S, Burtin P (2010) Fingolimod (FTY720): discovery and development of an oral drug to treat multiple sclerosis. Nat Rev Drug Discov 9(11):883–897. doi:10.1038/nrd3248

Brodde O-E, Engel G, Hoyer D, Bock K, Weber F (1981) The β-adrenergic receptor in human lymphocytes: Subclassification by the use of a new radio-ligand,(±) – 125 iodocyanopindolol. Life Sci 29(21):2189–2198

Chelmicka-Schorr E, Kwasniewski MN, Thomas BE, Arnason BG (1989) The beta-adrenergic agonist isoproterenol suppresses experimental allergic encephalomyelitis in Lewis rats. J Neuroimmunol 25(2–3):203–207

Cyster JG, Schwab SR (2012) Sphingosine-1-phosphate and lymphocyte egress from lymphoid organs. Annu Rev Immunol 30:69–94. doi:10.1146/annurev-immunol-020711-075011

Felten DL, Felten SY, Carlson SL, Olschowka JA, Livnat S (1985) Noradrenergic and peptidergic innervation of lymphoid tissue. J Immunol 135(2 Suppl):755s–765s

Fribourg M, Moreno JL, Holloway T, Provasi D, Baki L, Mahajan R, Park G, Adney SK, Hatcher C, Eltit JM, Ruta JD, Albizu L, Li Z, Umali A, Shim J, Fabiato A, MacKerell AD Jr, Brezina V, Sealfon SC, Filizola M, Gonzalez-Maeso J, Logothetis DE (2011) Decoding the signaling of a GPCR heteromeric complex reveals a unifying mechanism of action of antipsychotic drugs. Cell 147(5):1011–1023. doi:10.1016/j.cell.2011.09.055

Gonzalez-Maeso J, Ang RL, Yuen T, Chan P, Weisstaub NV, Lopez Gimenez JF, Zhou M, Okawa Y, Callado LF, Milligan G, Gingrich JA, Filizola M, Meana JJ, Sealfon SC (2008) Identification of a serotonin/glutamate receptor complex implicated in psychosis. Nature 452(7183):93–97. doi:10.1038/nature06612

Granstein RD, Wagner JA, Stohl LL, Ding W (2015) Calcitonin gene-related peptide: key regulator of cutaneous immunity. Acta Physiol (Oxf) 213(3):586–594. doi:10.1111/apha.12442

Herve J, Dubreil L, Tardif V, Terme M, Pogu S, Anegon I, Rozec B, Gauthier C, Bach JM, Blancou P (2013) Beta2-Adrenoreceptor agonist inhibits antigen cross-presentation by dendritic cells. J Immunol 190(7):3163–3171. doi:10.4049/jimmunol.1201391

Huitinga I, Erkut ZA, van Beurden D, Swaab DF (2004) Impaired hypothalamus-pituitary-adrenal axis activity and more severe multiple sclerosis with hypothalamic lesions. Ann Neurol 55(1): 37–45. doi:10.1002/ana.10766

Ishigami T (1919) The influence of psychic acts on the progress of pulmonary tuberculosis. Am Rev Tuberculosis 2:470–484
Karagkouni A, Alevizos M, Theoharides TC (2013) Effect of stress on brain inflammation and multiple sclerosis. Autoimmun Rev 12(10):947–953. doi:10.1016/j.autrev.2013.02.006
Kivisakk P, Mahad DJ, Callahan MK, Sikora K, Trebst C, Tucky B, Wujek J, Ravid R, Staugaitis SM, Lassmann H, Ransohoff RM (2004) Expression of CCR7 in multiple sclerosis: implications for CNS immunity. Ann Neurol 55(5):627–638. doi:10.1002/ana.20049
Lo CG, Xu Y, Proia RL, Cyster JG (2005) Cyclical modulation of sphingosine-1-phosphate receptor 1 surface expression during lymphocyte recirculation and relationship to lymphoid organ transit. J Exp Med 201(2):291–301. doi:10.1084/jem.20041509
Loeper M, Crouzon O (1904) L'action de l'adrenaline sur le sang. Arch Med Exp Anat Pathol 16: 83–108
Mendez-Ferrer S, Lucas D, Battista M, Frenette PS (2008) Haematopoietic stem cell release is regulated by circadian oscillations. Nature 452(7186):442–447. doi:10.1038/nature06685
Nakai A, Hayano Y, Furuta F, Noda M, Suzuki K (2014) Control of lymphocyte egress from lymph nodes through beta2-adrenergic receptors. J Exp Med 211(13):2583–2598. doi:10.1084/jem.20141132
Nijhuis LE, Olivier BJ, Dhawan S, Hilbers FW, Boon L, Wolkers MC, Samsom JN, de Jonge WJ (2014) Adrenergic beta2 receptor activation stimulates anti-inflammatory properties of dendritic cells in vitro. PLoS One 9(1), e85086. doi:10.1371/journal.pone.0085086
Pham TH, Okada T, Matloubian M, Lo CG, Cyster JG (2008) S1P1 receptor signaling overrides retention mediated by G alpha i-coupled receptors to promote T cell egress. Immunity 28(1): 122–133. doi:10.1016/j.immuni.2007.11.017
Ramer-Quinn DS, Baker RA, Sanders VM (1997) Activated T helper 1 and T helper 2 cells differentially express the beta-2-adrenergic receptor: a mechanism for selective modulation of T helper 1 cell cytokine production. J Immunol 159(10):4857–4867
Rosas-Ballina M, Olofsson PS, Ochani M, Valdes-Ferrer SI, Levine YA, Reardon C, Tusche MW, Pavlov VA, Andersson U, Chavan S, Mak TW, Tracey KJ (2011) Acetylcholine-synthesizing T cells relay neural signals in a vagus nerve circuit. Science 334(6052):98–101. doi:10.1126/science.1209985
Sanders VM (2012) The beta2-adrenergic receptor on T and B lymphocytes: do we understand it yet? Brain Behav Immun 26(2):195–200. doi:10.1016/j.bbi.2011.08.001
Sanders VM, Baker RA, Ramer-Quinn DS, Kasprowicz DJ, Fuchs BA, Street NE (1997) Differential expression of the beta2-adrenergic receptor by Th1 and Th2 clones: implications for cytokine production and B cell help. J Immunol 158(9):4200–4210
Scheiermann C, Kunisaki Y, Lucas D, Chow A, Jang JE, Zhang D, Hashimoto D, Merad M, Frenette PS (2012) Adrenergic nerves govern circadian leukocyte recruitment to tissues. Immunity 37(2):290–301. doi:10.1016/j.immuni.2012.05.021
Scheiermann C, Kunisaki Y, Frenette PS (2013) Circadian control of the immune system. Nat Rev Immunol 13(3):190–198. doi:10.1038/nri3386
Schmidt TH, Bannard O, Gray EE, Cyster JG (2013) CXCR4 promotes B cell egress from Peyer's patches. J Exp Med 210(6):1099–1107. doi:10.1084/jem.20122574
Smalley SG, Barrow PA, Foster N (2009) Immunomodulation of innate immune responses by vasoactive intestinal peptide (VIP): its therapeutic potential in inflammatory disease. Clin Exp Immunol 157(2):225–234. doi:10.1111/j.1365-2249.2009.03956.x
Solomon GF, Moos RH (1964) Emotions, immunity, and disease: a speculative theoretical integration. Arch Gen Psychiatry 11(6):657–674
Steinman L (2014) Immunology of relapse and remission in multiple sclerosis. Annu Rev Immunol 32:257–281. doi:10.1146/annurev-immunol-032713-120227
Takamine J (1901) The isolation of the active principle of the suprarenal gland. J Physiol 27:29–30
Torres-Rosas R, Yehia G, Pena G, Mishra P, del Rocio T-BM, Moreno-Eutimio MA, Arriaga-Pizano LA, Isibasi A, Ulloa L (2014) Dopamine mediates vagal modulation of the immune system by electroacupuncture. Nat Med 20(3):291–295. doi:10.1038/nm.3479

Tracey KJ (2002) The inflammatory reflex. Nature 420(6917):853–859. doi:10.1038/nature01321
Wang H, Yu M, Ochani M, Amella CA, Tanovic M, Susarla S, Li JH, Wang H, Yang H, Ulloa L, Al-Abed Y, Czura CJ, Tracey KJ (2003) Nicotinic acetylcholine receptor alpha7 subunit is an essential regulator of inflammation. Nature 421(6921):384–388. doi:10.1038/nature01339
Wei T, Lightman SL (1997) The neuroendocrine axis in patients with multiple sclerosis. Brain 120(Pt 6):1067–1076
Wiegmann K, Muthyala S, Kim DH, Arnason BG, Chelmicka-Schorr E (1995) Beta-adrenergic agonists suppress chronic/relapsing experimental allergic encephalomyelitis (CREAE) in Lewis rats. J Neuroimmunol 56(2):201–206
Yadav M, Goetzl EJ (2008) Vasoactive intestinal peptide-mediated Th17 differentiation: an expanding spectrum of vasoactive intestinal peptide effects in immunity and autoimmunity. Ann N Y Acad Sci 1144:83–89. doi:10.1196/annals.1418.020
Yan Y, Jiang W, Liu L, Wang X, Ding C, Tian Z, Zhou R (2015) Dopamine controls systemic inflammation through inhibition of NLRP3 inflammasome. Cell 160(1–2):62–73. doi:10.1016/j.cell.2014.11.047

Chapter 34
The Multifaceted Role of PD-1 in Health and Disease

Mohamed El Sherif Gadelhaq Badr, Kikumi Hata, Masae Furuhata, Hiroko Toyota, and Tadashi Yokosuka

Abstract Properly functioning T cells should clear invading pathogens and tumors without causing damage to the host. Programmed cell death-1 (PD-1) is a negative costimulatory receptor that is expressed on an effector T cell to regulate its functions; otherwise, uncontrolled activation of T cells might cause collateral tissue damage and autoimmune diseases. However, such a beneficial regulatory tool has been hijacked by several pathogens and tumors to promote chronic infection and evade antitumor immunity, respectively. PD-1 ligands are expressed by a variety of haematopoietic and non-haematopoietic cells all over the body under steady state and their expression is further modulated within different pathological conditions, suggesting that PD-1 is an immune checkpoint that is utilised by the body to regulate T-cell functions in different contexts. Furthermore, blocking PD-1 has recently received attention as a promising therapeutic approach for the treatment of cancer and chronic viral infections. Herein, we shed some light on the current knowledge of PD-1 biology and role in its health and disease.

Keywords Programmed cell death 1 • PD-1 • Immune checkpoint • Tumor immunotherapy • T-cell signaling • T-cell receptor • Immunological synapse • Microcluster • Imaging • Regulatory T cell

---

M.E.S.G. Badr
Laboratory for Cell Signaling, Center for Integrative Medical Sciences, RIKEN Yokohama Institute, Kanagawa 230-0045, Japan

WPI Immunology Frontier Research Center, Osaka University, Suita, Osaka 565-0871, Japan

K. Hata • M. Furuhata • H. Toyota
Department of Immunology, Tokyo Medical University, 6-1-1 Shinjuku, Shinjuku-ku, Tokyo 160-8402, Japan

T. Yokosuka (✉)
Department of Immunology, Tokyo Medical University, 6-1-1 Shinjuku, Shinjuku-ku, Tokyo 160-8402, Japan

PRESTO, Japan Science and Technology Agency, Kawaguchi 332-0012, Japan
e-mail: yokosuka-ths@umin.ac.jp

M. Miyasaka, K. Takatsu (eds.), *Chronic Inflammation*,
DOI 10.1007/978-4-431-56068-5_34

34.1 Introduction

Inflammation is one of the biological responses to restore body homeostasis that was disrupted by insult from the outer world such as infection with pathogenic invaders such as bacteria, fungi, and viruses and physical injuries. However, uncontrolled or exaggerated inflammatory responses can backfire on the body causing allergic or devastating autoimmune diseases. An inflammatory response is launched mainly by innate leukocytes, including neutrophils, which sense infecting pathogens and tissue disruption and then secrete various chemical mediators in response. On the contrary, a prolonged inflammatory response, called ‘chronic inflammation’, is mediated mainly by acquired immunity through the antibody production by B cells and various helper functions by T cells. Because T cells are playing the role of the maestro in coordinating the orchestra of the immune response symphony, it should be quite important to unveil the mechanisms of T-cell activation regulation to reach the ultimate goal of bringing chronic inflammation under control.

T cells are activated upon recognition of their cognate antigen peptide loaded on a major histocompatibility complex (MHC) and the functional characteristics of an activated T cell are further modified by various signals through costimulatory receptors and cytokines within the surrounding environment. T cells express several costimulatory receptors, either with positive or negative functions, the expression of which is changing according to the activation status of a T cell. On the other hand, the ligands for those costimulatory receptors are expressed on various antigen presenting cell (APC) subsets and nonhaematopoietic cells all over the body; and the types and the expression levels of the ligands also change according to the inflammation status of the surrounding microenvironment within the tissue. Therefore, T cells, with help from costimulatory receptors, sense the changes within their environment to respond with the proper functional outputs. In this review, we discuss the regulatory functions of the negative costimulatory receptor, programmed cell death-1 (PD-1; CD279), which has become recently appreciated as an important ‘immune checkpoint’ molecule.

34.2 Characteristics of PD-1 and Its Ligands

T cell costimulatory receptors contain two receptor superfamilies (SFs), immunoglobulin (Ig) SF and tumor necrosis factor (TNF) SF. The IgSF is further divided into two functionally opposing groups: activating IgSF receptors, for example, CD28 and inducible T-cell costimulator (ICOS, CD278) and inhibitory IgSF receptors, for example, cytotoxic T-lymphocyte associated protein 4 (CTLA-4 or CD152), PD-1, and B and T lymphocyte attenuator (BTLA, CD272). PD-1 was originally discovered as an apoptosis-related molecule and characterised as a type I membrane protein of 288 amino acids; composed of an extracellular Ig-like variable domain, transmembrane domain (27 amino acids), and a cytoplasmic (94 amino acids; Ishida et al. 1992). Similar to CTLA-4, PD-1 is expressed in a T cell after activation; but, because PD-1 possesses neither a membrane-proximal cysteine residue as in CD28 nor a clathrin–adaptor protein complex 2 (AP-2) binding domain as in CTLA-4, therefore, PD-1 exists on the plasma membrane as a monomeric receptor (Fig. 34.1, left). There are two known ligands for PD-1: PD-1 ligand-1 (PD-L1; B7-H1; CD274; Freeman et al. 2000) and PD-L2 (B7-DC; CD273; Latchman et al. 2001; Tseng et al. 2001). Both ligands belong to the B7 family, which is characterised by bearing one IgV and one IgC domains in the

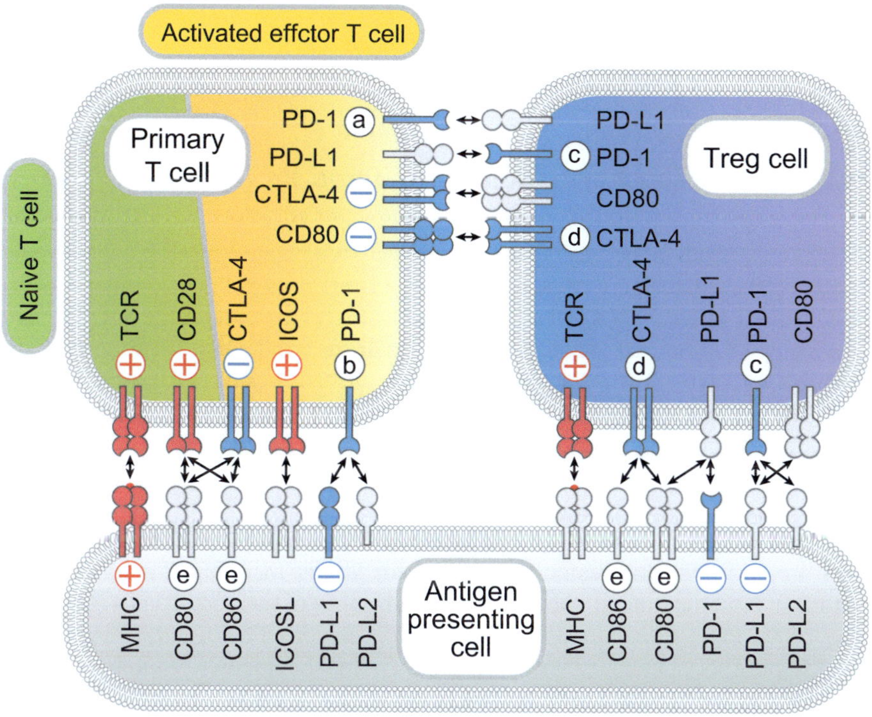

Fig. 34.1 The B7-CD28 family-dependent crosstalk between APCs and T cells. The diagram shows the B7-CD28 costimulatory or coinhibitory receptors and their ligands' dependent interaction between a naïve T cell (*green*), an activated effector T cell (*orange*), a regulatory T cell (*blue*), and an antigen-presenting cell. TCR, costimulatory (*red*) and coinhibitory receptors (*blue*), occasionally, through binding to their ligands convey positive and negative signaling, respectively. Treg cells suppress effector T-cell functions and also promote conversion of effector T cells into iTreg cells through the interaction between PD-1 and PD-L1 (**a**). The PD-1–PD-L1 interaction suppresses T-cell functions and promotes iTreg cell differentiation (**b**). The sustained PD-1 signaling maintains the Foxp3 expression, suppressive functions, and survival of Treg cells (**c**). Signaling through CTLA-4 enhances the Treg-cell–suppressive functions (**d**). Crosslinking of CD80/CD86 by CD28 or CTLA-4 induces either immune enhancing or suppressive outcomes in antigen-presenting cells (APCs), respectively. Treg cells strip off CD80/CD86, the ligands for CD28, from the surface of APCs via CTLA-4, a phenomenon known as trogocytosis (**e**)

extracellular region. PD-L1 consists of 290 amino acids, with a 30 amino acid transmembrane region, and is expressed on a variety of cells of both haematopoietic and nonhaematopoietic origins. PD-L1 was also reported to bind the CD28 ligand, CD80, with introducing bidirectional negative signals (Butte et al. 2007). In contrast, PD-L2 is inducibly expressed on professional APCs after their activation and is known to have triple higher affinity for PD-1 than PD-L1 (Fig. 34.1, bottom). Although human PD-L2 consist of 273 amino acids with a 30 amino acid cytoplasmic tail and the murine counterpart consists of 274 amino acids with only a 4 amino acid cytoplasmic tail, PD-L2 is speculated to induce functional signaling to dendritic cells (DCs), because the activation or survival signals in dendritic cells were augmented by treatment with the anti-PD-L2 monoclonal antibody, sHIgM12, originally isolated from the primary (Waldenstorm's) macroglobulinemia patients (Nguyen et al. 2002).

34.3 Expression Regulation of PD-1 and Its Ligands

PD-1 is expressed on various types of cells, such as activated T and B cells, human naïve B cells, exhausted T cells, regulatory T (Treg) cells, follicular helper T (Tfh) cells, natural killer T (NKT) cells, activated $CD11c^+$ myeloid DCs (mDCs), and monocytes, but not on macrophages and plasmacytoid DCs (pDCs). In vitro experiments have shown that anti-CD3 plus anti-CD28 and anti-IgM plus anti-CD40 induce the expression of PD-1 on T and B cells, respectively; and the expression level was proportional to the strength and duration of the stimulation. PD-1 is expressed in T cells within 2–6 h following TCR stimulation; earlier than CTLA-4 and BTLA or than the first cell division. Because PD-1 expression on T cells is sustained for more than one week, starting from the initial activation of naïve T cells till the differentiation into effector T cells, therefore, it is considerable that PD-1 is modulating all the developmental stages of memory T cells (Chikuma et al. 2009). PD-1 expression in T cells is further enhanced by stimulation with type I interferon (IFN). In fact, the promoter region of PD-1 contains a binding region for the transcription factor interferon regulatory factor 9 (IRF9) which is activated downstream of type I IFN receptor (Terawaki et al. 2011) as well as a binding region for the transcription factor nuclear factor of activated T cells, cytoplasmic 1 (NFATc1) which is activated downstream of TCR (Oestreich et al. 2008). Moreover, chronic viral infection has been shown to play a role in regulating the methylation status of the CpG islands within the promoter region of PD-1 (Youngblood et al. 2011). Independent of TCR activation, the common gamma chain (cγ) cytokines; IL-2, IL-7, IL-15, and IL-21 have been reported to induce the expression of PD-1 in T cells (Kinter et al. 2008). On the other hand, Listeria infection induced the expression of PD-1 in mDCs through the activation of Toll-like receptor 2 (TLR2), TLR3, TLR4, and NOD-like receptors. It is reported that estrogen enhances the expression of PD-1 on Treg cells, macrophages, B cells, and mDCs and this inhibitory machinery contributes to the immune suppression during pregnancy (Polanczyk et al. 2007).

PD-L1 is constitutively expressed not only by immune cells, such as T cells, B cells, macrophages, m/pDCs, and bone marrow mast cells, but also by tissue-constructing cells, such as endothelial cells, fibroblastic reticular cells, gut epithelial cells, pancreatic islet cells, astrocytes, central neurons, placental trophoblasts, retinal pigmented epithelial cells, and optic nerve cells. The expression of PD-L1 is enhanced upon stimulation with type I IFNs, IFN-γ, and TNF-α consistent with the presence of the IRF1 and signal transducer and activator of transcription 3 (STAT3) binding domains in its promoter region. T and B cells also express more PD-L1 upon stimulation with cγ cytokines whereas DCs and macrophages upregulate PD-L1 upon stimulation with granulocyte/macrophage-colony stimulating factor (GM-CSF) and IL-4. PD-L1 is also expressed by many cancer cells; furthermore, PD-L1 is known to be a target for the transcription factor Hypoxia-inducible factor-1α (Hif-1α), therefore an hypoxic tumor microenvironment can further augment the expression of PD-L1 (Noman et al. 2014). It was also reported that multiple myeloma cells upregulate the expression of PD-L1 upon stimulation with IFN-γ and TLR ligands via myeloid differentiation primary response (MyD88)–TNF receptor associated factor 6 (TRAF6)–mitogen-activated protein kinase (MAPK)/extracellular signal-regulated kinase (Erk) kinase (MEK) pathway (Liu et al. 2007).

As mentioned previously, PD-L2 is expressed by m/pDCs, macrophages, B-1 cells, and bone marrow mast cells and its expression is further enhanced by stimulation with several cytokines, for example, GM-CSF, IFN-γ, IL-4, cγ cytokines, and TNFα. Therefore, it is plausible that the increased expression of IFNs and other cytokines, responsible for the maintenance of activated and memory T cells especially in the case of chronic infection and inflammation, induce the expression of PD-1 and its ligands PD-L1/2; the interaction between them transduces bidirectional suppressive signals that induce tolerance and suppress that excessive activation of immune system and ultimately lead to chronic inflammation

34.4 PD-1 Signaling Machinery

The cytoplasmic domain of PD-1 contains two tyrosine-based signaling motifs: an immunoreceptor tyrosine-based switch motif (ITSM, TxYxxL/I) and an immunoreceptor tyrosine-based inhibitory motif (ITIM, S/I/LxYxxI/V/L). Upon crosslinking of PD-1, the tyrosine residues get phosphorylated and consequently they act as docking sites to recruit the Src homology 2 (SH2) domain-containing tyrosine phosphatase 2 (SHP-2) and SHP-1 which subsequently dephosphorylate the signalling molecules downstream of TCR resulting in the diminished activation of the Erk and Akt signaling pathways. In fact, it was shown that PD-1 suppressed Erk activation in human T cells which led to decreased transcription of the ubiquitin ligase SKP2 that is responsible for the degradation of the inhibitors of cycling-dependent kinases (Cdks), ending up blocking cell-cycle progression at G1 the phase (Patsoukis et al. 2012).

Although the kinases important for the phosphorylation of the cytoplasmic domain of PD-1 have not been definitively shown, the pull-down experiments, using the ITSM/ITIM motifs as bait followed by mass spectrometry, have suggested that Lck and C-terminal Src kinase (Csk) could be the responsible kinases (Sheppard et al. 2004). The phosphorylated ITSM motif recruits both SHP-1 and SHP-2 whereas the phosphorylated ITIM motif recruits only SHP-2 and mutation experiments of the tyrosine residues have suggested that the suppressive functions of PD-1 are mainly attributed to the ITSM motif (Chemnitz et al. 2004). In T cells, SHP-2 and possibly SHP-1 associate with both the ITSM and ITIM motifs and dephosphorylate the CD3ζ-chain, zeta-associated protein of 70kD (Zap70), and protein kinase C θ (PKCθ) (Sheppard et al. 2004). In B cells, only SHP-2 is recruited to the PD-1 cytoplasmic tail and dephosphorylate Ig-β-chain, spleen tyrosine kinase (Syk), Erk, and phospholipase Cγ2 (PLCγ2) to suppress BCR signaling (Okazaki et al. 2001).

Formation of an immunological synapse is required for the antigen recognition and activation of T cells, when T cells contact with APCs bearing their cognate antigens. The immunological synapse is a concentric, bull's eye-like, structure generated at the interface between a T cell and an APC, where various immune receptors including TCRs, coreceptors CD4 and CD8, costimulatory receptors, and their signaling transducers are gathering together for the efficient signal transduction and are surrounded by adhesion molecules such as lymphocyte function-associated antigen-1 (LFA-1), intercellular adhesion molecule-1 (ICAM-1), very late antigen-4 (VLA-4), and vascular cell adhesion molecule-1 (VCAM-1) to keep a stable conjugation between a T cell and an APC. We discovered that the immunological synapse is constructed by a number of more tiny signaling units that we named 'TCR microclusters' (Yokosuka et al. 2005; Fig. 34.2, left). Our imaging analysis has shown that a TCR microcluster is composed of a few decades of TCRs and that PD-1 is recruited to a TCR microcluster upon binding to PD-L1, where its ITSM/ITIM motifs get phosphorylated and subsequently recruit SHP-2 (Yokosuka et al. 2012). In fact, the SHP-2 substrates, such as CD3ζ, Zap70, and PKCθ, are also localised within the TCR microclusters suggesting that PD-1-mediated dephosphorylation reactions and regulation of T-cell activation primarily take place within the microclusters. Interestingly, SHP-2 which is recruited to the phosphorylated ITSM/ITIM in the PD-1 cytoplasmic tail could dephosphorylate not only the SHP-2 substrates but also PD-1 itself and subsequently SHP-2 dissociates from the PD-1-TCR microclusters (Fig. 34.2, right). Although the negative costimulatory receptor CTLA-4 also has a cytoplasmic ITIM-like motif that is believed to recruit SHP-2 and SHP-1, its suppressive mechanism is different from PD-1. Whereas PD-1 dephosphorylates phosphatidylinositol-3-kinase (PI3K) which is an upstream regulator of Akt, CTLA-4 recruits the protein phosphatase 2A (PP2A) which directly dephosphorylates Akt (Parry et al. 2005). However, our imaging studies demonstrate that the mechanism of CTLA-4–mediated suppression is mainly accomplished by the competition with CD28 for binding to the ligand CD80/CD86

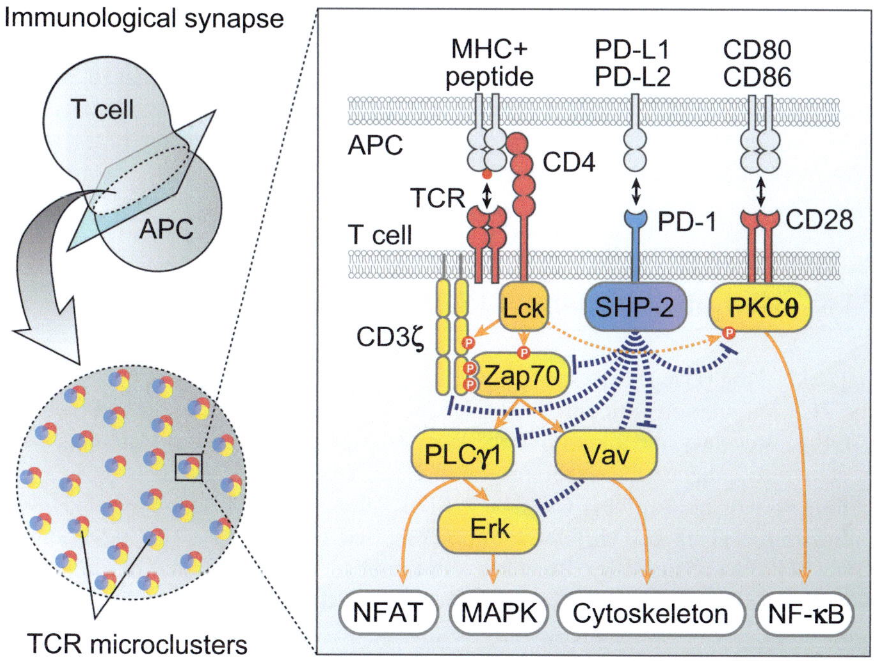

Fig. 34.2 PD-1-mediated inhibition of TCR signaling at microclusters. A T cell and an APC form the immunological synapse at the interface between the two cells, which is initially constructed by tiny clusters of TCRs with their downstream signaling molecules 'TCR microclusters'. At TCR microclusters, the engagement of TCRs and CD28 triggers several downstream kinases (*orange arrow*) to activate distinct but synergistic signaling cascades that modify the T cells' cytoskeleton, transcriptional, and functional program. PD-1 also translocates at TCR microclusters via binding to PD-L1/2 and specifically recruits SHP-2, which dephosphorylates various signaling molecules downstream of TCRs (*blue dotted line*)

which is critical for NF-kB signaling, whereas PD-1 is suppressing the proximal TCR downstream signaling via recruiting SHP-2.

34.5 PD-1 and Thymic T-Cell Development

The expression of PD-1 is detected on $CD4^-$ $CD8^-$ double negative (DN) αβ T cells and γδ T cells in the thymus. PD-1 expression starts from $CD44^+$ $CD25^-$ DN II stage, peaks at the $CD44^-$ $CD25^-$ DN IV stage with the termination of the β-chain rearrangement, and then ceases at the double positive (DP) stage (Nishimura et al. 1996). PD-L1 is expressed in both thymic cortical epithelial cells and thymocytes and PD-L2 is in the thymic medulla. TCR transgenic mice, 2C and H-Y, on PD-1 deficient background showed an increased number of DP thymocytes and decreased number of mature $CD8^+$ single positive T cells indicating that PD-1 in the thymocytes adjust the TCR signaling threshold to regulate β-selection negatively, modify positive selection, and ultimately select a TCR repertoire with reduced autoreactivity (Nishimura et al. 2000). Indeed, microarray analysis of NOD mice, that spontaneously develop type I diabetes, suggested that PD-1 expression in the thymus could be involved in the induction of tolerance (Zucchelli et al. 2005).

34.6 Multiple Functions of PD-1 in Regulatory T Cells

Inducible Treg (iTreg) cells could be developed from naïve T cells in vitro through the TCR stimulation with anti-TCR/CD3 in the presence of tumor growth factor-β (TGF-β). Recently, the addition of the PD-L1-immunoglobulin (Ig) fusion proteins is shown to enhance the development of iTreg cells further in the same in vitro culture system and similarly *Cd274*$^{-/-}$ (PD-L1-deficient) APCs are less efficient in inducing the iTreg cell differentiation (Francisco et al. 2009). The induction of iTreg cells is enhanced by a mechanistic target of rapamycin (mTOR) inhibitor, Rapamycin; similarly, PD-L1-Ig could suppress the Akt-mTOR activation and induce the PTEN enzymatic activity to upregulate Foxp3 expression and sustain the suppressor functions of iTreg cells (Francisco et al. 2009). In contrast, because the expression of PD-L1 is dependent on PTEN, the PD-1–mediated suppression of PTEN itself interferes with the iTreg cell development. The expression of PD-L1 on APCs or tissue stromal cells is also enhanced under inflammatory conditions or within the tumor microenvironment. This alteration of PD-L1 expression induces the differentiation of naïve or effector T cells into iTreg cells to sustain chronic infection and preserve peripheral immune tolerance (Fig. 34.1, right). Not only iTreg cells but also natural Treg (nTreg) cells express both PD-1 and PD-L1. Treg cells expressing PD-L1 suppress other effector T cells expressing PD-1 by the PD-L1–PD-1 interaction and furthermore promote conversion of effector T cells into iTreg cells. Treg cells also suppress the expression of CD80/CD86 on DCs expressing PD-1 by the PD-L1–PD-1 interaction (Yao et al. 2009). One of the splice variants of PD-1, the soluble PD-1, is reported to induce $CD4^+$ T cells and DCs to produce the inhibitory cytokine, IL-10, without indoleamine 2,3 dioxygenase (IDO) production (Kuipers et al. 2006).

34.7 PD-1-Deficient Mice Develop Spontaneous Autoimmune Diseases

Pdcd1$^{-/-}$ mice were originally generated on the 129 genetic background and these mice have shown augmented antibody production upon immunisation with foreign antigens. However, by backcrossing *Pdcd1*$^{-/-}$ mice to the B6 genetic background, they started to develop symptoms of lupus-like autoimmune disease, for example, splenomegaly, autoantibody production, glomerulonephritis, and lupus-like proliferative arthritis as they exceeded 6 months of age (Nishimura et al. 1999). In contrast, *Pdcd1*$^{-/-}$ mice on the BALB/c genetic background developed dilated cardiomyopathy (DCM) and autoimmune gastritis (Nishimura et al. 2001). In fact, that was the first experimental model to show that DCM could have an autoimmune basis; later on, it was found that deletion of the negative costimulatory receptor lymphocyte activation gene 3 (LAG-3) further exacerbated the pathology of the disease in *Pdcd1*$^{-/-}$ mice (Okazaki et al. 2011). Diabetes with early onset and severity is shown by the NOD mice crossed with *Pdcd1*$^{-/-}$ (Wang et al. 2005) or inoculation with anti-PD-1 or PD-L1. Furthermore, fatal autoimmune myocarditis was developed by the autoimmune-prone mice, MRL, crossed with *Pdcd1*$^{-/-}$ mice (Wang et al. 2010). Focusing on the differences in function among negative costimulatory receptors, *Ctla4*$^{-/-}$ mice, irrespective of the genetic background, showed infiltration of the inflammatory cells, mainly T cells, into the entire body without a clear specific tissue tropism and died within a few weeks after birth; in contrast, *Pdcd1*$^{-/-}$ mice demonstrate the tissue-specific autoimmune diseases in a much later phase. Generally, CTLA-4 suppresses T-cell activation via both cell intrinsic and extrinsic mechanisms, whereas PD-1 inhibitory effect is mediated mainly by extrinsic factors. The study with bone marrow chimera from wild-type and receptor-deficient mice demonstrates that autoimmune diseases are developing in the wild type–*Ctla4*$^{-/-}$ chimeras, not in the wild type–*Pdcd1*$^{-/-}$ chimeras.

34.8 PD-1 Polymorphism and Diseases

The first report to correlate between single-nucleotide polymorphisms (SNPs) in PD-1 and autoimmune diseases was about a group of systemic erythematous lupus (SLE) patients in Europe and Mexico. These SLE patients demonstrated SNPs in the runt-related transcription factor 1 (Runx1)-binding region in the enhancer element of PD-1 (Prokunina et al. 2002). The same SNPs were also identified in type I diabetes patients in Denmark, multiple sclerosis in Germany, arthritis in Australia, Basedow disease in Germany, Kawasaki disease in Korea, and ankylosing spondylitis in China. In addition, SNPs in PD-L1 and PD-L2 are reported in Basedow disease patients in Japan and SLE patients in Taiwan, respectively.

34.9 Control of Inflammation During Viral Infection by PD-1

Inhibition of clearance of viral infection by PD-1 was first demonstrated in a murine adenovirus-induced hepatitis model (Iwai et al. 2003). *Pdcd1*$^{-/-}$ mice showed faster clearance of adenovirus but more severe destruction of liver tissues. Therefore, PD-1, rather than promoting viral infection by its immunosuppressive functions, prevents tissue damage by infiltrating immune cells and cytokines during the acute phase of infection and consequently promotes chronic viral infections. In fact, the lymphocytic choriomeningitis virus (LCMV) Armstrong strain, an acute infectious strain, was similarly cleared from WT and *Pdcd1*$^{-/-}$ mice. However, upon infection with clone 13, a chronic infectious strain, the virus antigen-specific PD-1^{hi} CD8^{+} T cells were rendered unresponsive or so-called anaergic, a phenomenon that was called T-cell exhaustion (Barber et al. 2006). Clone 13-induced exhausted T cells did not terminally lose cytotoxicity, because these cells recover cytotoxicity upon treatment with anti-PD-1 or anti-PD-L1. In contrast, *Cd274*$^{-/-}$ (PD-L1–deficient) mice that were infected by Clone 13 died of liver failure caused by excessive T-cell responses and resistance to exhaustion. Virus-specific PD-1^{hi} CD8^{+} T cells in HIV-infected individuals, as an example of chronic viral infection in humans, showed exhausted phenotypes and blocking PD-L1 could restore the function of those exhausted cells (Day et al. 2006) (Trautmann et al. 2006). In chronic hepatitis B or C virus infections, virus-specific CD8^{+} T cells express PD-1 and both CD14^{+} monocytes and mDCs are shown to upregulate PD-L1. Other than viral infections, the PD-1–PD-L1 axis can promote for the induction of chronic inflammation during bacterial or helminthic infections; Helicobacter pylori infection enhanced the expression of PD-L1 in the gastric mucosal epithelium; Taenia crassiceps infection induced upregulation of PD-L1/PD-L2 and PD-1 in macrophages and CD4^{+} T cells, respectively; Schistosoma mansoni infection induced anaergy in T cells via upregulation of PD-L1 in macrophages (Smith et al. 2004). Therefore, breaking the PD-1–PD-L1 axis, to restore the functions of exhausted pathogen-specific T cells, is a promising approach for the treatment of chronic infections.

34.10 Blocking PD-1 Signaling as a Cancer Therapy

The first evidence for a role of PD-1–PD-L interaction in tumor immune evasion came from the study of a murine mastocytoma cell line, P815, transgenically overexpressing PD-L1, which showed resistance to the cytolytic activity of cytotoxic CD8^{+} T cells (Iwai et al. 2002). Moreover, the administration of anti-PD-L1 not only decreased the tumor growth but also suppressed its metastasis resulting in a higher survival rate. In fact, PD-L1 is expressed in the human lung, colon, and ovarian cancers and in malignant melanoma. The expression of PD-L1 on tumor cells is further up-regulated by IFN-γ and PD-L1 could itself suppress tumor-specific $CD8^+$ T cells and induce their apoptosis through FasL and IL-10 (Dong et al. 2002). Recently, other types of tumor cells bearing PD-L1 were also reported, such as breast cancer, bladder cancer, hepatocellular carcinoma, salivary gland carcinoma, gastric cancer, glioma, and thyroid cancer. From a clinical viewpoint, PD-L1 expression is correlated with the poor prognosis of kidney, ovarian, bladder, breast, gastric, esophageal, and pancreatic cancers (Thompson et al. 2004). Simultaneous expression of PD-1 and PD-L1 was also observed in lymphoid tumors such as multiple myeloma and Hodgkin's lymphoma. Furthermore, PD-L1 is expressed by monocyte-derived mDCs residing within the tumor draining lymph nodes and blocking the PD-1–PD-L1 interaction enhances DC-mediated autologous tumor-specific T-cell activation and antitumor immunity (Curiel et al. 2003).

In appreciation for the role of the PD-1–PD-L1 axis in tumor immune evasion, the humanised anti-PD-1 monoclonal antibody (Nivolumab) was approved in last year in Japan for the treatment of malignant melanoma. In a phase I clinical trial in the United States, anti-PD-1 blocking antibody has shown efficacy in the treatment of recurrent solid tumors (Brahmer et al. 2010). In another large-scale phase I clinical trial, the cumulative response rates were 28 % in melanoma, 18 % in non-small cell lung cancer, and 27 % in renal cell cancer patients (Topalian et al. 2012). On the other hand, a phase I clinical trial for anti-PD-L1 conducted on 207 patients with advanced cancers, the cumulative response rates were 17 % in malignant melanoma, 10 % in non-small cell lung cancer, 12 % in renal cell cancer, and 6 % in ovarian cancer patients, which is an historically remarkable outcome (Brahmer et al. 2012). The combination therapy of anti-CTLA-4 (Ipilimumab) and anti-PD-1 was reported to double the response rate, compared to monotherapy, in a phase I clinical trial for the treatment of advanced melanoma (Wolchok et al. 2013). Therefore, targeting the PD-1–PD-L1 axis is expected to introduce a breakthrough in cancer treatment when applied in combination with chemotherapy, radiotherapy, and surgical interventions.

34.11 PD-1 and Autoimmune Diseases

As mentioned earlier, the PD-1–PD-L1/2 binding not only controls the functions of T cells and APCs but also contributes to the differentiation of iTreg cells that are induced by the endothelial and epithelial cells in the peripheral tissues. Therefore, the dysfunction of PD-1 both directly or indirectly, through induction of iTreg cells, leads to uncontrolled functions of effector T cells which end up in autoimmune diseases. On the other side, when the autoimmune diseases develop from other causes, the chronic inflammation within the disease microenvironment enhances the expression of PD-L1 and iTreg cell differentiation.

It was experimentally shown that the deficiency of PD-1 increases the severity of type I diabetes in NOD mice. Although the relationship between PD-1 and the destruction of β cells of the pancreatic islets was not definitively shown in humans, the number of Treg cells and their suppressive functions in pancreatic islets was markedly reduced in type I diabetic patients (Kukreja et al. 2002). The inhibitory function of Treg cells in the peripheral blood of type I diabetic patients appears to be normal but it is unsure that the same is true for Treg cells residing within the microenvironment of the pancreatic islets. In mice, pancreatic island β cells are expressing PD-L1, thereby directly suppressing the cytolytic activity of effector $CD8^+$ T cells and possibly inducing the differentiation of $CD4^+$ T cells into iTreg cells. Moreover, it seems that PD-L1 expressed by the pancreatic epithelial and vascular endothelial cells has a similar suppressive function. Therefore, PD-1 exerts its inhibitory activity on T cells from different angles. Using two-photon microscopy, it was shown that antibody blockade of PD-L1 resulted in decreased motility of tolerant islet-antigen–specific $CD4^+$ T cells within the pancreatic island concomitant with breaking the tolerance and induction of autoimmune diabetes (Fife et al. 2009). Functions and differentiation of effector T cells could be cooperatively determined by not only the direct signaling through the PD-1–PD-L1 binding but also the duration of TCR signaling and the cytokine concentration within the microenvironment.

Immunosuppressive mechanisms by PD-1 are also believed to control the development of multiple sclerosis. PD-1 and PD-L1 are widely expressed by various types of cells within the central nervous system (CNS); furthermore, retinal ganglion cells not only express PD-1 but also upregulate the PD-L1/2 expression under inflammatory conditions. Astrocytes and brain vascular endothelial cells express PD-L1. Microglia, the antigen-presenting cells within the CNS, also express PD-L1 upon stimulation with IFN-γ. In experimental autoimmune encephalomyelitis (EAE), the mouse model for multiple sclerosis, PD-1 and PD-L1/2 are expressed by the inflammatory cells infiltrating within the cerebrospinal fluid. Furthermore, the deficiency of PD-1 or the antibody blockade of the PD-1–PD-L1 interaction leads to more severe EAE with an earlier onset, suggesting that PD-1 plays a critical role in limiting inflammation within CNS. Bone marrow chimera experiments have further shown that the expression of PD-1 on macrophages and other myeloid cells is as important as its expression on lymphocytes for the control of EAE (Rui et al. 2013). Under inflammatory conditions, nerve cells themselves express PD-1 and TGF-β; those induce the differentiation of the infiltrating encephalitogenic effector T cells into iTreg cells, promoting the relief of inflammation (Liu et al. 2006).

In inflammatory bowel disease (IBD) patients, PD-L1 is highly expressed by intestinal epithelial cells suggesting that intestinal mucosal epithelia could contribute to the tolerance induction via PD-1 (Nakazawa et al. 2004). $CD103^+$ DCs within the gut-associated lymphoid tissue (GALT) could induce oral tolerance via the induction of iTreg cell differentiation. Additionally, iTreg cells could also be induced by PD-L1–expressing intestinal epithelial cells. Inflammation-related cytokines such as IL-12, IL-17, IL-21, IL-23, and IFN-γ were locally concentrated in the microenvironments under the IBD conditions and induce T-cells differentiation into Th17 cells, the major T-helper subset responsible for IBD. Although the number of Foxp3^{+} cells is increased in the gut, Th17 cells drastically increase in number compared to Treg cells and consequently the decreasing Treg/Th17 ratio results in IBD. The murine IBD model, in which CD4^{+} CD45RB^{hi} T cells were transferred into severe combined immunodeficient (SCID) mice, demonstrated that not only iTreg cells but also CD4^{+} CD25^{-} PD-1^{+} Foxp3^{+} effector T cells contribute to the control of inflammation; indeed, the pathogenic CD4^{+} CD45RB^{hi} T cells that expressed PD-1 became nonpathogenic after several rounds of sequential transfer into SCID mice and finally could suppress colitis upon cotransfer with colitogenic CD4^{+} T cells (Totsuka et al. 2008)

34.12 Closing Remarks

Although PD-1 was originally cloned as an apoptosis-related molecule, now its negative costimulatory function has become very clear. Biological characteristics of PD-1 are unique compared to the other IgSF costimulatory molecules; first, its ligand expression pattern is complicated as it is not only restricted to immune cells but rather expressed by several cell types all over the body. The expression of PD-1 and its ligands are controlled by various environmental cues and also both the receptor and their ligands transduce inhibitory signaling as counterreceptors. Nonhaematopoietic cells expressing PD-1 ligands can regulate T-cell activation through an MHC Class II-independent fashion. Second, the receptor PD-1 and the ligands PD-L1/2 are simultaneously expressed on the same cell forming a cell intrinsic inhibitory machinery that might also regulate the developmental or functional outputs in concert with the extrinsic machinery. Collectively, PD-1 was unequivocally proved to be a superior target molecule for cancer and chronic inflammatory disease therapy, probably because its immune-suppressive functions are mediated at multiple stages of T-cell life and via crosstalk between T cells, haematopoietic cells, and even almost all types of cells within the body.

References

Barber DL, Wherry EJ, Masopust D, Zhu B, Allison JP, Sharpe AH, Freeman GJ, Ahmed R (2006) Restoring function in exhausted CD8 T cells during chronic viral infection. Nature 439 (7077):682–687. doi:10.1038/nature04444

Brahmer JR, Drake CG, Wollner I, Powderly JD, Picus J, Sharfman WH, Stankevich E, Pons A, Salay TM, McMiller TL, Gilson MM, Wang C, Selby M, Taube JM, Anders R, Chen L, Korman AJ, Pardoll DM, Lowy I, Topalian SL (2010) Phase I study of single-agent anti-programmed death-1 (MDX-1106) in refractory solid tumors: safety, clinical activity, pharmacodynamics, and immunologic correlates. J Clin Oncol 28(19):3167–3175. doi:10.1200/JCO.2009.26.7609

Brahmer JR, Tykodi SS, Chow LQ, Hwu WJ, Topalian SL, Hwu P, Drake CG, Camacho LH, Kauh J, Odunsi K, Pitot HC, Hamid O, Bhatia S, Martins R, Eaton K, Chen S, Salay TM, Alaparthy S, Grosso JF, Korman AJ, Parker SM, Agrawal S, Goldberg SM, Pardoll DM,

Gupta A, Wigginton JM (2012) Safety and activity of anti-PD-L1 antibody in patients with advanced cancer. N Engl J Med 366(26):2455–2465. doi:10.1056/NEJMoa1200694

Butte MJ, Keir ME, Phamduy TB, Sharpe AH, Freeman GJ (2007) Programmed death-1 ligand 1 interacts specifically with the B7-1 costimulatory molecule to inhibit T cell responses. Immunity 27(1):111–122. doi:10.1016/j.immuni.2007.05.016

Chemnitz JM, Parry RV, Nichols KE, June CH, Riley JL (2004) SHP-1 and SHP-2 associate with immunoreceptor tyrosine-based switch motif of programmed death 1 upon primary human T cell stimulation, but only receptor ligation prevents T cell activation. J Immunol 173 (2):945–954

Chikuma S, Terawaki S, Hayashi T, Nabeshima R, Yoshida T, Shibayama S, Okazaki T, Honjo T (2009) PD-1-mediated suppression of IL-2 production induces CD8+ T cell anergy in vivo. J Immunol 182(11):6682–6689. doi:10.4049/jimmunol.0900080

Curiel TJ, Wei S, Dong H, Alvarez X, Cheng P, Mottram P, Krzysiek R, Knutson KL, Daniel B, Zimmermann MC, David O, Burow M, Gordon A, Dhurandhar N, Myers L, Berggren R, Hemminki A, Alvarez RD, Emilie D, Curiel DT, Chen L, Zou W (2003) Blockade of B7-H1 improves myeloid dendritic cell-mediated antitumor immunity. Nat Med 9(5):562–567. doi:10.1038/nm863

Day CL, Kaufmann DE, Kiepiela P, Brown JA, Moodley ES, Reddy S, Mackey EW, Miller JD, Leslie AJ, DePierres C, Mncube Z, Duraiswamy J, Zhu B, Eichbaum Q, Altfeld M, Wherry EJ, Coovadia HM, Goulder PJ, Klenerman P, Ahmed R, Freeman GJ, Walker BD (2006) PD-1 expression on HIV-specific T cells is associated with T-cell exhaustion and disease progression. Nature 443(7109):350–354. doi:10.1038/nature05115

Dong H, Strome SE, Salomao DR, Tamura H, Hirano F, Flies DB, Roche PC, Lu J, Zhu G, Tamada K, Lennon VA, Celis E, Chen L (2002) Tumor-associated B7-H1 promotes T-cell apoptosis: a potential mechanism of immune evasion. Nat Med 8(8):793–800. doi:10.1038/nm730

Fife BT, Pauken KE, Eagar TN, Obu T, Wu J, Tang Q, Azuma M, Krummel MF, Bluestone JA (2009) Interactions between PD-1 and PD-L1 promote tolerance by blocking the TCR-induced stop signal. Nat Immunol 10(11):1185–1192. doi:10.1038/ni.1790

Francisco LM, Salinas VH, Brown KE, Vanguri VK, Freeman GJ, Kuchroo VK, Sharpe AH (2009) PD-L1 regulates the development, maintenance, and function of induced regulatory T cells. J Exp Med 206(13):3015–3029. doi:10.1084/jem.20090847

Freeman GJ, Long AJ, Iwai Y, Bourque K, Chernova T, Nishimura H, Fitz LJ, Malenkovich N, Okazaki T, Byrne MC, Horton HF, Fouser L, Carter L, Ling V, Bowman MR, Carreno BM, Collins M, Wood CR, Honjo T (2000) Engagement of the PD-1 immunoinhibitory receptor by a novel B7 family member leads to negative regulation of lymphocyte activation. J Exp Med 192(7):1027–1034

Ishida Y, Agata Y, Shibahara K, Honjo T (1992) Induced expression of PD-1, a novel member of the immunoglobulin gene superfamily, upon programmed cell death. EMBO J 11 (11):3887–3895

Iwai Y, Ishida M, Tanaka Y, Okazaki T, Honjo T, Minato N (2002) Involvement of PD-L1 on tumor cells in the escape from host immune system and tumor immunotherapy by PD-L1 blockade. Proc Natl Acad Sci U S A 99(19):12293–12297. doi:10.1073/pnas.192461099

Iwai Y, Terawaki S, Ikegawa M, Okazaki T, Honjo T (2003) PD-1 inhibits antiviral immunity at the effector phase in the liver. J Exp Med 198(1):39–50. doi:10.1084/jem.20022235

Kinter AL, Godbout EJ, McNally JP, Sereti I, Roby GA, O'Shea MA, Fauci AS (2008) The common gamma-chain cytokines IL-2, IL-7, IL-15, and IL-21 induce the expression of programmed death-1 and its ligands. J Immunol 181(10):6738–6746

Kuipers H, Muskens F, Willart M, Hijdra D, van Assema FB, Coyle AJ, Hoogsteden HC, Lambrecht BN (2006) Contribution of the PD-1 ligands/PD-1 signaling pathway to dendritic cell-mediated CD4+ T cell activation. Eur J Immunol 36(9):2472–2482. doi:10.1002/eji.200635978

Kukreja A, Cost G, Marker J, Zhang C, Sun Z, Lin-Su K, Ten S, Sanz M, Exley M, Wilson B, Porcelli S, Maclaren N (2002) Multiple immuno-regulatory defects in type-1 diabetes. J Clin Invest 109(1):131–140. doi:10.1172/JCI13605

Latchman Y, Wood CR, Chernova T, Chaudhary D, Borde M, Chernova I, Iwai Y, Long AJ, Brown JA, Nunes R, Greenfield EA, Bourque K, Boussiotis VA, Carter LL, Carreno BM, Malenkovich N, Nishimura H, Okazaki T, Honjo T, Sharpe AH, Freeman GJ (2001) PD-L2 is a second ligand for PD-1 and inhibits T cell activation. Nat Immunol 2(3):261–268. doi:10.1038/85330

Liu Y, Teige I, Birnir B, Issazadeh-Navikas S (2006) Neuron-mediated generation of regulatory T cells from encephalitogenic T cells suppresses EAE. Nat Med 12(5):518–525. doi:10.1038/nm1402

Liu J, Hamrouni A, Wolowiec D, Coiteux V, Kuliczkowski K, Hetuin D, Saudemont A, Quesnel B (2007) Plasma cells from multiple myeloma patients express B7-H1 (PD-L1) and increase expression after stimulation with IFN-{gamma} and TLR ligands via a MyD88-, TRAF6-, and MEK-dependent pathway. Blood 110(1):296–304. doi:10.1182/blood-2006-10-051482

Nakazawa A, Dotan I, Brimnes J, Allez M, Shao L, Tsushima F, Azuma M, Mayer L (2004) The expression and function of costimulatory molecules B7H and B7-H1 on colonic epithelial cells. Gastroenterology 126(5):1347–1357

Nguyen LT, Radhakrishnan S, Ciric B, Tamada K, Shin T, Pardoll DM, Chen L, Rodriguez M, Pease LR (2002) Cross-linking the B7 family molecule B7-DC directly activates immune functions of dendritic cells. J Exp Med 196(10):1393–1398

Nishimura H, Agata Y, Kawasaki A, Sato M, Imamura S, Minato N, Yagita H, Nakano T, Honjo T (1996) Developmentally regulated expression of the PD-1 protein on the surface of double-negative (CD4-CD8-) thymocytes. Int Immunol 8(5):773–780

Nishimura H, Nose M, Hiai H, Minato N, Honjo T (1999) Development of lupus-like autoimmune diseases by disruption of the PD-1 gene encoding an ITIM motif-carrying immunoreceptor. Immunity 11(2):141–151

Nishimura H, Honjo T, Minato N (2000) Facilitation of beta selection and modification of positive selection in the thymus of PD-1-deficient mice. J Exp Med 191(5):891–898

Nishimura H, Okazaki T, Tanaka Y, Nakatani K, Hara M, Matsumori A, Sasayama S, Mizoguchi A, Hiai H, Minato N, Honjo T (2001) Autoimmune dilated cardiomyopathy in PD-1 receptor-deficient mice. Science 291(5502):319–322. doi:10.1126/science.291.5502.319

Noman MZ, Desantis G, Janji B, Hasmim M, Karray S, Dessen P, Bronte V, Chouaib S (2014) PD-L1 is a novel direct target of HIF-1alpha, and its blockade under hypoxia enhanced MDSC mediated T cell activation. J Exp Med 211(5):781–790. doi:10.1084/jem.20131916

Oestreich KJ, Yoon H, Ahmed R, Boss JM (2008) NFATc1 regulates PD-1 expression upon T cell activation. J Immunol 181(7):4832–4839

Okazaki T, Maeda A, Nishimura H, Kurosaki T, Honjo T (2001) PD-1 immunoreceptor inhibits B cell receptor-mediated signaling by recruiting src homology 2-domain-containing tyrosine phosphatase 2 to phosphotyrosine. Proc Natl Acad Sci U S A 98(24):13866–13871. doi:10.1073/pnas.231486598

Okazaki T, Okazaki IM, Wang J, Sugiura D, Nakaki F, Yoshida T, Kato Y, Fagarasan S, Muramatsu M, Eto T, Hioki K, Honjo T (2011) PD-1 and LAG-3 inhibitory co-receptors act synergistically to prevent autoimmunity in mice. J Exp Med 208(2):395–407. doi:10.1084/jem.20100466

Parry RV, Chemnitz JM, Frauwirth KA, Lanfranco AR, Braunstein I, Kobayashi SV, Linsley PS, Thompson CB, Riley JL (2005) CTLA-4 and PD-1 receptors inhibit T-cell activation by distinct mechanisms. Mol Cell Biol 25(21):9543–9553. doi:10.1128/MCB.25.21.9543-9553.2005

Patsoukis N, Brown J, Petkova V, Liu F, Li L, Boussiotis VA (2012) Selective effects of PD-1 on Akt and Ras pathways regulate molecular components of the cell cycle and inhibit T cell proliferation. Sci Signal 5(230):ra46. doi:10.1126/scisignal.2002796

Polanczyk MJ, Hopke C, Vandenbark AA, Offner H (2007) Treg suppressive activity involves estrogen-dependent expression of programmed death-1 (PD-1). Int Immunol 19(3):337–343. doi:10.1093/intimm/dxl151

Prokunina L, Castillejo-Lopez C, Oberg F, Gunnarsson I, Berg L, Magnusson V, Brookes AJ, Tentler D, Kristjansdottir H, Grondal G, Bolstad AI, Svenungsson E, Lundberg I, Sturfelt G, Jonssen A, Truedsson L, Lima G, Alcocer-Varela J, Jonsson R, Gyllensten UB, Harley JB, Alarcon-Segovia D, Steinsson K, Alarcon-Riquelme ME (2002) A regulatory polymorphism in PDCD1 is associated with susceptibility to systemic lupus erythematosus in humans. Nat Genet 32(4):666–669. doi:10.1038/ng1020

Rui Y, Honjo T, Chikuma S (2013) Programmed cell death 1 inhibits inflammatory helper T-cell development through controlling the innate immune response. Proc Natl Acad Sci U S A 110 (40):16073–16078. doi:10.1073/pnas.1315828110

Sheppard KA, Fitz LJ, Lee JM, Benander C, George JA, Wooters J, Qiu Y, Jussif JM, Carter LL, Wood CR, Chaudhary D (2004) PD-1 inhibits T-cell receptor induced phosphorylation of the ZAP70/CD3zeta signalosome and downstream signaling to PKCtheta. FEBS Lett 574 (1–3):37–41. doi:10.1016/j.febslet.2004.07.083

Smith P, Walsh CM, Mangan NE, Fallon RE, Sayers JR, McKenzie AN, Fallon PG (2004) Schistosoma mansoni worms induce anergy of T cells via selective up-regulation of programmed death ligand 1 on macrophages. J Immunol 173(2):1240–1248

Terawaki S, Chikuma S, Shibayama S, Hayashi T, Yoshida T, Okazaki T, Honjo T (2011) IFN-alpha directly promotes programmed cell death-1 transcription and limits the duration of T cell-mediated immunity. J Immunol 186(5):2772–2779. doi:10.4049/jimmunol.1003208

Thompson RH, Gillett MD, Cheville JC, Lohse CM, Dong H, Webster WS, Krejci KG, Lobo JR, Sengupta S, Chen L, Zincke H, Blute ML, Strome SE, Leibovich BC, Kwon ED (2004) Costimulatory B7-H1 in renal cell carcinoma patients: Indicator of tumor aggressiveness and potential therapeutic target. Proc Natl Acad Sci U S A 101(49):17174–17179. doi:10.1073/pnas.0406351101

Topalian SL, Hodi FS, Brahmer JR, Gettinger SN, Smith DC, McDermott DF, Powderly JD, Carvajal RD, Sosman JA, Atkins MB, Leming PD, Spigel DR, Antonia SJ, Horn L, Drake CG, Pardoll DM, Chen L, Sharfman WH, Anders RA, Taube JM, McMiller TL, Xu H, Korman AJ, Jure-Kunkel M, Agrawal S, McDonald D, Kollia GD, Gupta A, Wigginton JM, Sznol M (2012) Safety, activity, and immune correlates of anti-PD-1 antibody in cancer. N Engl J Med 366 (26):2443–2454. doi:10.1056/NEJMoa1200690

Totsuka T, Kanai T, Nemoto Y, Tomita T, Tsuchiya K, Sakamoto N, Okamoto R, Watanabe M (2008) Immunosenescent colitogenic CD4(+) T cells convert to regulatory cells and suppress colitis. Eur J Immunol 38(5):1275–1286. doi:10.1002/eji.200737914

Trautmann L, Janbazian L, Chomont N, Said EA, Gimmig S, Bessette B, Boulassel MR, Delwart E, Sepulveda H, Balderas RS, Routy JP, Haddad EK, Sekaly RP (2006) Upregulation of PD-1 expression on HIV-specific CD8+ T cells leads to reversible immune dysfunction. Nat Med 12(10):1198–1202. doi:10.1038/nm1482

Tseng SY, Otsuji M, Gorski K, Huang X, Slansky JE, Pai SI, Shalabi A, Shin T, Pardoll DM, Tsuchiya H (2001) B7-DC, a new dendritic cell molecule with potent costimulatory properties for T cells. J Exp Med 193(7):839–846

Wang J, Yoshida T, Nakaki F, Hiai H, Okazaki T, Honjo T (2005) Establishment of NOD-Pdcd1−/− mice as an efficient animal model of type I diabetes. Proc Natl Acad Sci U S A 102(33):11823–11828. doi:10.1073/pnas.0505497102

Wang J, Okazaki IM, Yoshida T, Chikuma S, Kato Y, Nakaki F, Hiai H, Honjo T, Okazaki T (2010) PD-1 deficiency results in the development of fatal myocarditis in MRL mice. Int Immunol 22(6):443–452. doi:10.1093/intimm/dxq026

Wolchok JD, Kluger H, Callahan MK, Postow MA, Rizvi NA, Lesokhin AM, Segal NH, Ariyan CE, Gordon RA, Reed K, Burke MM, Caldwell A, Kronenberg SA, Agunwamba BU, Zhang X, Lowy I, Inzunza HD, Feely W, Horak CE, Hong Q, Korman AJ, Wigginton JM, Gupta A,

Sznol M (2013) Nivolumab plus ipilimumab in advanced melanoma. N Engl J Med 369 (2):122–133. doi:10.1056/NEJMoa1302369
Yao S, Wang S, Zhu Y, Luo L, Zhu G, Flies S, Xu H, Ruff W, Broadwater M, Choi IH, Tamada K, Chen L (2009) PD-1 on dendritic cells impedes innate immunity against bacterial infection. Blood 113(23):5811–5818. doi:10.1182/blood-2009-02-203141
Yokosuka T, Sakata-Sogawa K, Kobayashi W, Hiroshima M, Hashimoto-Tane A, Tokunaga M, Dustin ML, Saito T (2005) Newly generated T cell receptor microclusters initiate and sustain T cell activation by recruitment of Zap70 and SLP-76. Nat Immunol 6(12):1253–1262
Yokosuka T, Takamatsu M, Kobayashi-Imanishi W, Hashimoto-Tane A, Azuma M, Saito T (2012) Programmed cell death 1 forms negative costimulatory microclusters that directly inhibit T cell receptor signaling by recruiting phosphatase SHP2. J Exp Med 209 (6):1201–1217. doi:10.1084/jem.20112741
Youngblood B, Oestreich KJ, Ha SJ, Duraiswamy J, Akondy RS, West EE, Wei Z, Lu P, Austin JW, Riley JL, Boss JM, Ahmed R (2011) Chronic virus infection enforces demethylation of the locus that encodes PD-1 in antigen-specific CD8(+) T cells. Immunity 35(3):400–412. doi:10.1016/j.immuni.2011.06.015
Zucchelli S, Holler P, Yamagata T, Roy M, Benoist C, Mathis D (2005) Defective central tolerance induction in NOD mice: genomics and genetics. Immunity 22(3):385–396. doi:10.1016/j.immuni.2005.01.015

Chapter 35
The Role of Lysophospholipids in Immune Cell Trafficking and Inflammation

Masayuki Miyasaka, Akira Takeda, Erina Hata, Naoko Sasaki, Eiji Umemoto, and Sirpa Jalkanen

Abstract Lysophospholipids are phospholipids that lack an acyl chain. Thus, they are less hydrophobic than diacyl phospholipids and can act as intercellular signaling molecules. Like cytokines, they are locally acting, short-lived molecules, which signal through specific cell-surface receptors. Accumulating evidence indicates that at least two of the lysophospholipids, namely sphingosine-1-phosphate (S1P) and lysophosphatidic acid (LPA), play critical roles in determining the spatial distribution of immune cells in lymphoid tissues. Both of these molecules act on lymphocytes, dendritic cells, and lymphoid tissue stromal cells, via specific G-protein–coupled receptors. Under physiological conditions, S1P regulates lymphocyte egress from lymphoid tissues, whereas LPA regulates lymphocyte ingress into and migration within lymph nodes. The aberrant production and/or metabolism of these lysophospholipids results in the dysregulated distribution of immune cells and the induction of various types of inflammatory responses in vivo. Here we discuss the specific roles of these lysophospholipids in immune cell trafficking and inflammation.

M. Miyasaka (✉)
Interdisciplinary Program for Biomedical Sciences, Institute for Academic Initiatives, Osaka University, 2-2, Yamada-oka, Suita 565-0871, Japan

WPI Immunology Frontier Research Center, 2-2, Yamada-oka, Suita 565-0871, Japan

MediCity Research Laboratory, University of Turku, Tykistökatu 6A, Turku 20520, Finland
e-mail: mmiyasak@orgctl.med.osaka-u.ac.jp

A. Takeda • S. Jalkanen
MediCity Research Laboratory, University of Turku, Tykistökatu 6, Turku 20520, Finland
e-mail: akitak@utu.fi; sirjal@utu.fi

E. Hata • N. Sasaki
Laboratory of Immunodynamics, Graduate School of Medicine, Osaka University, 2-2, Yamada-oka, Suita 565-0871, Japan
e-mail: erinaht@orgctl.med.osaka-u.ac.jp; sasaki@orgctl.med.osaka-u.ac.jp

E. Umemoto
Laboratory of Immune Regulation, Graduate School of Medicine, Osaka University, 2-2, Yamada-oka, Suita 565-0871, Japan
e-mail: umemoto@ongene.med.osaka-u.ac.jp

M. Miyasaka, K. Takatsu (eds.), *Chronic Inflammation*,
DOI 10.1007/978-4-431-56068-5_35

Keywords Lysophospholipids • Sphingosine-1-phosphate (S1P) • S1P receptors • Lysophosphatidic acid (LPA) • LPA receptors • Lymphocytes • Endothelial cells • Immune cell trafficking • Inflammation

35.1 Sphingosine-1-Phosphate (S1P) and Immune Cell Trafficking

35.1.1 S1P Is a Bioactive Lysophospholipid with Pleiotropic Cellular Functions

S1P is an amphipathic lipid with a single acyl chain and a phosphate-containing polar head group that is structurally similar to lysophosphatidic acid (LPA; Fig. 35.1). S1P regulates many processes, including cell proliferation, survival, migration, and differentiation, as well as immune cell trafficking and angiogenesis (reviewed by Spiegel and Milstien 2011; Mirendil et al. 2013; Blaho and Hla 2014).

S1P, a metabolite of cell membrane sphingolipids, is produced intracellularly and subsequently released to the extracellular environment as follows (Fig. 35.2). A plasma-membrane component, ceramide, which is produced from sphingomyelins by sphingomyelinases, is first converted to sphingosine by ceramidase. Sphingosine is subsequently converted into S1P by two sphingosine kinase isoforms (SphK1 and SphK2), which are localised to different intracellular compartments; SphK1 localises to the plasma membrane, whereas SphK2 localises to the mitochondria and nucleus. The SphKs are ubiquitously expressed in most tissues and have both distinct and overlapping functions. Mice deficient in both kinases are embryonic lethal and exhibit severe neural and vascular defects with no detectable phosphorylation of sphingosine (Mizugishi et al. 2005), confirming that S1P is generated by these enzymes and is essential for embryogenesis. S1P is released into the extracellular compartment by specific S1P transporter proteins, such as spinster homologue 2 (Spns2), and functions as a soluble lipid mediator, acting on specific G-protein–coupled receptors (GPCRs) termed $S1P_1$-$S1P_5$ (reviewed by Jung and Hla 2013). As described later, S1P may also act within the cell, independently of the S1P receptors expressed on the cell surface.

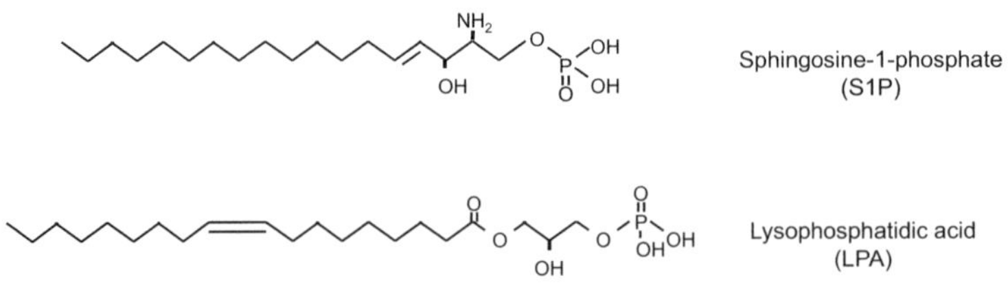

Fig. 35.1 Structures of sphingosine-1-phosphate (S1P) and lysophosphatidic acid (LPA). Each is a phospholipid bearing a single acyl chain. LPA does not reflect a single molecular species, but represents several related species with fatty acids of different length and degree of saturation

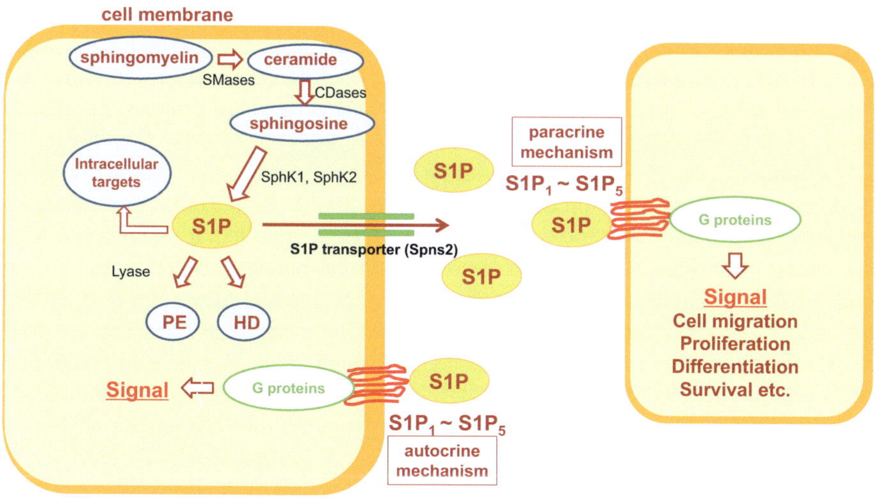

Fig. 35.2 S1P metabolism and the S1P/S1P receptor signalling axis. S1P is produced intracellularly and subsequently released to the extracellular environment as shown here. Its synthesis and degradation are tightly regulated

In the peripheral blood, ~65 % of the S1P is associated with high-density lipoprotein (HDL) and ~35 % is associated with albumin. In HDL, HDL-associated apolipoprotein M serves as an S1P carrier and delivers S1P to S1P receptors on the cell surface for signaling (Christoffersen et al. 2011). S1P is then rapidly degraded by S1P phosphatases, which dephosphorylate S1P to yield sphingosine, and by S1P lyase, which cleaves S1P into phosphoethanolamine and hexadecenal (reviewed by Jung and Hla 2013).

35.1.2 *S1P's Role in Immune Cell Trafficking*

The prevailing hypothesis is that S1P regulates immune cell trafficking by serving as a chemotactic ligand that controls lymphocyte egress from lymphoid tissues (reviewed by Cyster and Schwab 2012; Mendoza et al. 2012). According to this postulation, S1P is stored in, and released from, erythrocytes (Hla 2004) and vascular endothelial cells (Venkataraman et al. 2008), resulting in high S1P concentrations (up to several μM) in the peripheral blood. High S1P levels are also produced by the lymphatic endothelium (Pham et al. 2010), and eventually drain into the blood. In contrast, the S1P concentration is relatively low in lymphoid tissues, due to the presence of the S1P-degrading activity of S1P lyase. This differential concentration of S1P in the blood and lymphoid tissue results in a concentration gradient between the two compartments. Lymphocytes within the lymph nodes express high levels of $S1P_1$ and undergo chemotaxis in response to the S1P gradient (Matloubian et al. 2004). Thus, S1P/$S1P_1$-mediated signals are sufficient to drive lymphocyte egress from the lymphoid tissue and to overcome the retention signals mediated by CCR7 and other chemokine receptors (Pham et al. 2008). In contrast, peripheral blood lymphocytes express low levels of $S1P_1$, probably due to its downregulation by internalisation in response to high S1P concentrations in the blood (Schwab et al. 2005). Blood-borne lymphocytes enter the lymph nodes by transmigrating across high endothelial venules (HEVs), and begin to express $S1P_1$ at high levels due to the paucity of S1P in the lymph nodes. The lymphocytes then migrate to the cortical sinuses (Grigorova et al. 2010), the medullary sinus, and finally to the efferent lymphatics, by sensing the S1P concentration gradient in an $S1P_1$-dependent manner. Thus, it has been postulated that cyclical changes in lymphocyte $S1P_1$ expression and differential S1P concentrations in the blood and specific tissue compartments direct lymphocyte egress from the lymph nodes (Cyster and Schwab 2012). A similar scenario is envisioned for the regulation of lymphocyte egress from the thymus.

The postulation that the S1P/$S1P_1$ axis regulates lymphocyte egress is supported by the results of numerous animal studies. The genetic inactivation of *S1PR1* in mice results in a marked inhibition of lymphocyte egress from various lymphoid tissues, including the lymph nodes, spleen, and thymus (Matloubian et al. 2004). Notably, the disruption of S1P gradients by S1P lyase inhibition (Schwab et al. 2005) or by the conditional ablation of sphingosine kinases (Pappu et al. 2007; Pham et al. 2010) also inhibits lymphocyte egress from the lymph nodes into the lymphatic system. Furthermore, mice deficient in the S1P transporter protein Spns2, which is expressed by endothelial cells but not by erythrocytes, have substantially reduced lymph S1P with only minor changes in plasma S1P levels (Mendoza et al. 2012), and exhibit a block in lymphocyte egress from the lymph nodes (Mendoza et al. 2012; Fukuhara et al. 2012). Lipid phosphate phosphatase 3 (LPP3), a phosphatase expressed in vascular endothelial cells, has been shown to promote the efficient export of mature T cells from the thymus into the circulation by dephosphorylating thymic S1P, thus contributing to the generation of an S1P gradient between the thymus and circulation (Bréart et al. 2011). S1P lyase is also expressed by vascular endothelial cells and stromal cells in the perivascular spaces in the thymic medulla, and S1P lyase inhibition results in prominent lymphocyte accumulation in the perivascular spaces (Maeda et al. 2014). Thus, thymic blood vessels supporting thymocyte emigration appear to be equipped with a mechanism that mediates S1P degradation.

However, this widely accepted S1P concentration gradient hypothesis is in apparent conflict with several other reported findings. First, although $S1P_1$ is also expressed at high levels on macrophages, dendritic cells, and natural killer cells, only lymphocytes exit the lymph nodes in response to physiological concentrations of S1P. Second, *S1P1* transcripts are also abundant in the endothelial cells, pericytes, and vascular smooth muscle cells surrounding blood vessels. Proia's group recently characterised the activity of $S1P_1$ at the cellular level in vivo using $S1P_1$ GFP signaling mice, in which $S1P_1$ activation can be observed by the expression of a GFP reporter gene (Kono et al. 2014). They detected prominent GFP signals in both the lymphatic and vascular endothelial cells in lymphoid tissues, indicating strong $S1P_1$ activation in these cell types, whereas most lymphocytes showed minimal signs of $S1P_1$ activation under homeostatic conditions, raising the possibility that locally produced, but not plasma-derived S1P, induces these $S1P_1$ activation events. Although we cannot exclude the possibility that the methodology used in this study was not sensitive enough to detect low levels of lymphocyte $S1P_1$ activation, the apparent absence of $S1P_1$ activation in most lymphocytes is surprising, given that cyclical changes in $S1P_1$ expression are postulated to be involved in lymphocyte egress from lymphoid tissues (Cyster and Schwab 2012).

The potent immunosuppressant FTY720, which is structurally similar to S1P, is rapidly phosphorylated in vivo and binds to all of the S1P receptors except $S1P_2$. In addition, FTY720 induces the internalisation and degradation of $S1P_1$ and is thus thought to act as a functional antagonist of this receptor. At least two studies have reported contradictory findings on the effect of FTY720 on lymphocyte trafficking across HEVs. Yopp et al. (2005) reported that FTY720 does not enhance lymphocyte migration across HEVs, whereas Pabst et al. (2006) reported that FTY720 facilitates lymphocyte migration across HEVs, in an integrin-dependent manner. Thus, although the involvement of the S1P/$S1P_1$ axis in lymphocyte egress is well established, its effect on lymphocyte topology in lymphoid tissues remains unclear.

35.1.3 S1P and Inflammation

S1P biosynthetic pathways are regulated by proinflammatory mediators, and this regulatory mechanism underlies the potential contribution of S1P signalling to inflammation (reviewed by Rosen and Goetzl 2005; Kunkel et al. 2013). For instance, TNFα and IL-1 can induce the conversion of sphingomyelin to ceramide (Spiegel and Merrill 1996). TNFα can also induce SphK1 activation and its translocation from the cytosol to the plasma membrane, resulting in the production of S1P. Similarly, the crosslinking of high affinity IgE receptors on mast cells activates SphK1, leading to S1P production. In both cases, S1P is secreted and acts in an autocrine or paracrine manner to activate S1P-responsive cells. This process is known as S1P 'inside-out signalling' and appears to play an important role in inflammation (Spiegel and Milstien 2011). Indeed, in endothelial cells, TNFα enhances SphK activities and generates S1P, which in turn enhances proinflammatory adhesion molecule expression and the activation of ERK kinases and NF-kB, all of which can be inhibited by an SphK inhibitor. Thus, TNFα-induced endothelial activation appears to be dependent on SphK activities (Xia et al. 1998). In addition, in certain situations, S1P appears to induce NF-kB activation independently of S1P receptor activation (Alvare et al. 2010). In these cases, intracellular S1P acts directly on target molecules, such as TRAF2, a key component of TNFα-induced NF-kB activation.

In tumour cells, S1P's binding to $S1P_1$ also activates STAT3, a downstream target of the proinflammatory cytokine, IL-6, and induces the expression of both IL-6 and $S1P_1$. These signals activate a positive feedback loop leading to persistent STAT3 activation (Lee et al. 2010). Activation of the SphK1/S1P/$S1P_1$ axis was recently shown to induce chronic intestinal inflammation, suggesting that this pathway could be linked to colitis-associated cancer (Liang et al. 2013). Using an experimental model of colitis-induced tumourigenesis, in which mice are treated with a chemical carcinogen and a colitis-inducing chemical, dextran sulfate, Liang et al. (2013) reported that SphK2-deficient mice develop increased intestinal inflammation and markedly increased tumour burden, compared with wild-type mice. Notably, these mice also exhibit a prominent increase in the expression of SphK1. Furthermore, in SphK2-deficient mice, the S1P-induced activation of NF-kB and STAT3, and downstream production of TNFα and IL-6, are all enhanced, and exhibit further enhancement during colitis development. SphK1 or $S1P_1$ inhibition reduces the activation of NF-kB and STAT3, providing additional support for the role of the SphK1/S1P pathway in inflammatory signalling. Using bone marrow transfer experiments, the authors demonstrated that haematopoietic cells drive the NF-kB–induced IL-6 production and STAT3 activation, and the induction of colitis. Treating mice with FTY720, which antagonises both $S1P_1$ and SphK1, was found to block the persistent activation of NF-kB and STAT3, IL-6 production, and the development of colitis-associated tumours. Taken together, this study showed that the SphK1/S1P axis drives a feedforward signal amplification loop involving the $S1P_1$/NF-kB/IL-6/STAT3 pathway that exacerbates inflammation and tumourigenesis in the large intestine and is critical for the progression from chronic inflammation to colitis-associated tumour cancer.

35.2 Lysophosphatidic Acid (LPA) and Immune Cell Trafficking

35.2.1 LPA Is a Pleiotropic Lipid Mediator Detected in Various Tissues

Similar to S1P, LPA is a phospholipid bearing a single acyl chain (Fig. 35.1). LPA regulates multiple biological processes, including cell proliferation, migration, and differentiation (Mirendil et al. 2013). The term LPA does not reflect a single molecular species, but represents several related species with fatty acids of different lengths and degrees of saturation.

LPA can be generated both intra- and extracellularly in multiple tissues (Fig. 35.3). Under steady-state conditions, LPA is produced extracellularly primarily by autotaxin (ATX), a secreted lysophospholipase D that converts lysophosphatidylcholine (LPC) in the circulation to LPA, as described later. Under certain conditions, LPA is generated by phospholipases on membrane

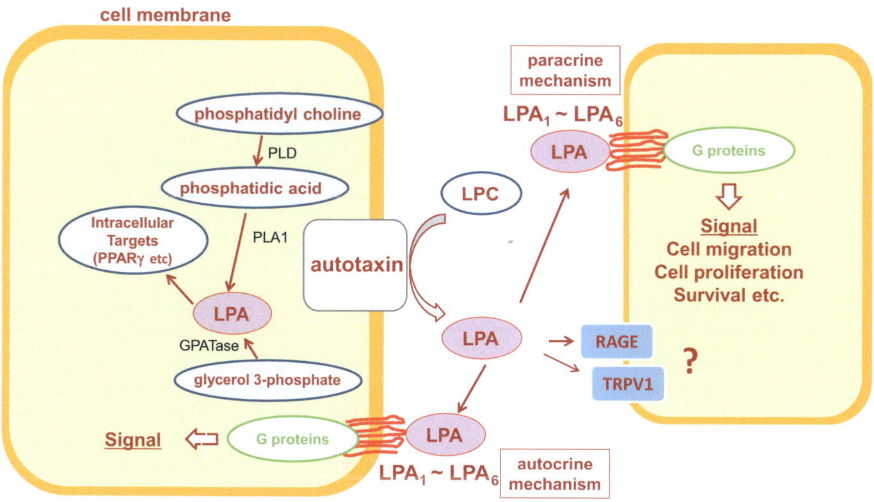

Fig. 35.3 LPA metabolism and the LPA/LPA receptor signalling axis. LPA can be generated both intra- and extracellularly, however, the latter appears to be mainly involved in the regulation of immune cell trafficking

vesicles shed by activated platelets and red blood cells. In addition, activated platelets produce various lysophospholipids, including LPC, which are converted to LPA by ATX in the systemic circulation (Aoki et al. 2002).

Intracellular production of LPA involves the enzymatic activity of phospholipase D (PLD), which hydrolyses phosphatidylcholine to produce phosphatidic acid, which is in turn metabolised into LPA via phospholipase A_1 (PLA_1). Acylation of glycerol-3-phosphate by glycerophosphate acyltransferase also results in LPA production (Nakanaga et al. 2010).

LPA is found in plasma at concentrations of ~0.1 μM, where it is primarily derived from lysophosphatidylcholine (LPC) by the enzymatic action of the lysophospholipase D, ATX, which is also present in plasma (~1.5 μM in mouse). Once produced in vivo, LPA is rapidly degraded by a group of integral membrane exophosphatases, termed lipid phosphate phosphatases (LPP1-3), and cleared from the circulation.

LPA binds specific cell-surface receptors, LPA_1-LPA_6, which are all GPCRs that couple to one or more of four heterotrimeric Gα proteins ($G_{12/13}$, $G_{q/11}$, $G_{i/o}$, and G_s). Ligand binding to these receptors results in the activation of multiple signalling pathways with various downstream physiological and pathological effects (Table 35.1). LPA also binds to noncanonical, non-GPCR–type receptors, such as the receptor for advanced glycation end-product (RAGE; Rai et al. 2012), the transient receptor potential vanilloid 1 (TRPV1) ion channel (Nieto-Posadas et al. 2012), and the peroxisome proliferator-activated receptor gamma (PPARγ; Zhang et al. 2004); however, the biological significance of these interactions remains unclear.

Table 35.1 LPA receptors

Canonical receptors	Expressing cells/ tissues	G proteins involved	Phenotype of KO mice	Remarks
LPA_1	Ubiquitous	$G_{i/o}$, $G_{q/11}$, $G_{12/13}$	Neurodevelopmental defects and reduced inflammation-induced fibrosis	Implicated in neuropathic pain, abundantly expressed in activated T cells
LPA_2	Highly expressed in kidney, testis, and leukocytes	$G_{i/o}$, $G_{q/11}$, $G_{12/13}$	Phenotypically normal	Protumourigenic in cancer cells, abundantly expressed in naïve T cells
LPA_3	Many tissues including testis, kidney, lung, heart, ovary, and brain	$G_{i/o}$, $G_{q/11}$	Delayed embryo implantation, and reduced litter size	Triple genetic deletion of LPA_1, LPA_2, and LPA_3 results in defective germ cell survival
LPA_4	Many tissues including heart, skin, thymus, and brain	$G_{12/13}$, $G_{i/o}$, $G_{q/11}$, Gs	Decreased prenatal survival, enlarged lymphatic vessels, and increased bone trabecular thickness	
LPA_5	Highly expressed in spleen, heart, platelets, lymphocytes, and dorsal root ganglia	$G_{12/13}$, $G_{q/11}$	Protected from partial sciatic nerve ligation-induced neuropathic pain, and reduced incidence of melanoma-derived lung metastasis in LPA_5 KO mice	
LPA_6	Inner root sheaths of hair follicles, blood vessels	$G_{12/13}$, $G_{i/o}$, $G_{q/11}$	Apparently normal	A mutation results in autosomal recessive wooly hair in humans

Noncanonical receptors	Expressing cells/tissues	G proteins involved	Phenotype of KO mice	Remarks
RAGE	Ubiquitous	None	Protected against bleomycin-induced pulmonary fibrosis	
PPARγ	Ubiquitous	None	Embryonic lethal, adipose-tissue–specific KO mice are protected from diet-induced obesity and insulin resistance	Only activated by unsaturated LPA species, involved in mast cell proliferation and differentiation?
TRPV1	Nociceptive neurons of the peripheral nervous system	None	Impaired thermal sensitivity to pain triggered by heat or capsaicin	LPA produces acute pain-like behaviors, which are substantially reduced in *Trpv1*-null mice

35.2.2 The ATXLPA Axis Regulates Lymphocyte Extravasation in Lymph Nodes

ATX is a member of the ectonucleotide pyrophosphatase and phosphodiesterase (ENPP) family and is also known as ENPP2. Initially, ATX was isolated as a tumour-motility–stimulating protein and was later found to be identical to lysophospholipase D. ATX was subsequently shown to catalyse the generation of LPA from serum-derived LPC, indicating that its ability to stimulate cell mobility is based on its lysophospholipase D activity (reviewed by Nakanaga et al. 2010).

In the peripheral blood, ATX is the primary LPA-producing enzyme. It is exported from various types of cells through the classical secretory pathway, and may act as an ectoenzyme on the cell surface by binding to proteoglycans and integrins. ATX is transcribed abundantly in HEV endothelial cells (Nakasaki et al. 2008; Kanda et al. 2008). Accumulating evidence indicates that the ATX on the HEV luminal surface regulates lymphocyte trafficking by generating LPA locally. Kanda et al. (2008) reported that the intravenous injection of enzymatically inactive ATX inhibits lymphocyte migration into the lymph node parenchyma, although the specific step affected (rolling, adhesion, or transmigration) was not examined in that study. Bai et al. (2013) demonstrated that ATX is selectively expressed in HEVs and showed that local inhibition of the ATX/LPA axis substantially inhibits lymphocyte transmigration across HEVs. Furthermore, the local administration of LPA was found to abrogate this effect. In agreement with these findings, Zhang et al. (2012) showed that the systemic inhibition of ATX by a specific small-molecule inhibitor, HA-130, attenuates lymphocyte transmigration across HEVs.

Two hypotheses, which are not mutually exclusive, could explain the mechanism by which ATX affects lymphocyte transmigration. One is that ATX acts primarily on lymphocytes, and the other is that ATX acts on HEV endothelial cells. Zhang et al. (2012) showed that treating mouse T cells with LPA alone, or with LPC plus ATX, enhances lymphocyte shape changes and transendothelial migration in vitro. They detected substantial ATX binding to the lymphocyte surface via Mn^{2+}-activatable receptors, implying that ATX generates LPA on the lymphocyte surface. However, the binding of ATX to undisturbed naïve lymphocytes was found to be minimal, and hence, its role in regulating lymphocyte trafficking across HEVs remains unclear. Recently, Knowlden et al. (2014) reported that LPA_2-deficient CD4 T cells move more slowly than their wild-type counterparts in the lymph node parenchyma, a finding that was also confirmed by our group. ATX is in fact expressed strongly by the fibroblastic reticular cells in the parenchyma, and ATX expressed by these cells produces LPA locally, which in turn stimulates lymphocyte motility in the parenchyma via LPA_2 (Takeda et al. 2016).

The evidence that LPA acts on endothelial cells is also compelling. Bai et al. (2013) reported that treatment with both LPA and LPC increases the motility of ATX-expressing HEV endothelial cells; LPC's effect can be abrogated by ATX inhibition, whereas LPA's effect is abrogated by ATX/LPA receptor inhibition, in agreement with the previous finding that ATX is required to convert LPC to LPA. In an in vitro transmigration assay, ATX inhibition was found to impair the release of lymphocytes that had migrated underneath HEV endothelial cells, and this defect could be abrogated by the addition of LPA. This effect of LPA was found to depend on myosin II activity expressed in the HEV endothelial cells. These findings indicate that LPA acts at least in part on endothelial cells to regulate lymphocyte transmigration. We have also recently shown that LPA_4 expressed on HEV endothelial cells plays an important role in regulating the lymphocyte transmigration across the HEV endothelial cell layer (Hata et al. in press).

Finally, it should be mentioned that LPA may play a role in stimulating S1P receptor-mediated signalling. LPA binds to one of the S1P receptors, $S1P_1$ with an apparent Kd of 2.3 μM, suggesting that it may function as a low-affinity agonist for this receptor (Lee et al. 1998). Given that HEV endothelial cells produce high levels of ATX and LPA, the promiscuous activation of $S1P_1$ by LPA may occur in HEVs. These observations are compatible with a recent report by Proia's group showing that strong $S1P_1$ activation can be observed in HEV endothelial cells (Kono et al. 2014).

35.2.3 LPA and Inflammation

ATX expression is induced at inflammatory sites, which leads to the local production of LPA. Multiple lines of evidence indicate that the locally produced LPA promotes inflammation by acting on both leukocytes and stromal cells, which leads to further local LPA production (reviewed by Sevastou et al. 2013). In vitro, LPA stimulates multiple proinflammatory activities, including the polarisation and degranulation of neutrophils, the migration of eosinophils, immature dendritic cells, Th1 cells, Th2 cells, and activated NK cells, and the generation of various chemokines in mast cells, including CCL4/MIP-1β, CXCL8/IL-8, and CCL2/MCP-1, in an IL-4–dependent manner. In addition, treating cells with LPA causes a loss of barrier function in endothelial cells and increased CXCL8/IL-8 production in epithelial cells. The in vivo injection of LPA into the murine air pouch induces robust leukocyte infiltration, together with the local release of CXCL1/KC and CXCL10/IP-10, which is strongly enhanced by prior treatment with TNFα (Zhao et al. 2011). Furthermore, elevated levels of ATX (Zhao et al. 2008) and LPA (Nochi et al. 2008) are found in the synovial fluids of rheumatoid arthritis (RA) patients. Synoviocytes from RA patients express enhanced levels of LPA_1 compared with those from osteoarthritis patients and show strong LPA_1-induced proliferation and the production of proinflammatory molecules, including IL-6, VEGF, and MMP-3 (Miyabe et al. 2014). LPA_1-deficient mice do not develop collagen-induced experimental arthritis (Miyabe et al. 2013). Taken together, these results support the importance of LPA/LPA_1 signalling in the development of arthritis.

LPA has also been implicated in the tissue fibrosis induced by chronic inflammation. LPA is elevated in the bronchoalveolar lavage fluids of patients with idiopathic pulmonary fibrosis (Tager et al. 2008), and LPA_1 knockout mice are protected from the development of bleomycin-induced pulmonary fibrosis (Tager et al. 2008). Renal fibrosis experimentally induced by ureteral obstruction is accompanied by upregulated LPA_1 expression and enhanced LPA production by the obstructed kidney, and the fibrotic changes are significantly reduced by treatment with an LPA_1 antagonist and are attenuated in LPA_1-deficient mice (Pradère et al. 2007).

In addition to the considerable amount of evidence indicating a proinflammatory role for LPA, there is also evidence that LPA plays a role in controlling inflammation. For example, LPA has been shown to inhibit endotoxin-induced inflammatory responses in vivo (Fan et al. 2008) and to promote wound healing when applied topically (Balazs et al. 2001). Therefore, depending on the stage and/or type of inflammation, LPA may exhibit either pro- or antiinflammatory effects.

References

Alvare SE, Harikuma KB, Hait NC et al (2010) Sphingosine-1-phosphate is a missing cofactor for the E3 ubiquitin ligase TRAF2. Nature 465:1084–1088

Aoki J, Taira A, Takanezawa Y et al (2002) Serum lysophosphatidic acid is produced through diverse phospholipase pathways. J Biol Chem 277:48737–48744

Bai Z, Cai L, Umemoto E et al (2013) Constitutive lymphocyte transmigration across the basal lamina of high endothelial venules is regulated by the autotaxin/lysophosphatidic acid axis. J Immunol 190:2036–2048

Balazs L, Okolicany J, Ferrebee M et al (2001) Topical application of the phospholipid growth factor lysophosphatidic acid promotes wound healing *in vivo*. Am J Physiol Regul Integr Comp Physiol 280:R466–R472

Blaho VA, Hla T (2014) An update on the biology of sphingosine 1-phosphate receptors. J Lipid Res 55:1596–1608

Bréart B, Ramos-Perez WD, Mendoza A et al (2011) Lipid phosphate phosphatase 3 enables efficient thymic egress. J Exp Med 208:1267–1278

Christoffersen C, Obinata H, Kumaraswamy SB et al (2011) Endothelium-protective sphingosine-1-phosphate provided by HDL-associated apolipoprotein M. Proc Natl Acad Sci U S A 108:9613–9618

Cyster JG, Schwab SR (2012) Sphingosine-1-phosphate and lymphocyte egress from lymphoid organs. Annu Rev Immunol 30:69–94

Fan H, Zingarelli B, Harris V et al (2008) Lysophosphatidic acid inhibits bacterial endotoxin-induced pro-inflammatory response: potential anti-inflammatory signaling pathways. Mol Med 14:422–428

Fukuhara S, Simmons S, Kawamura S et al (2012) The sphingosine-1-phosphate transporter Spns2 expressed on endothelial cells regulates lymphocyte trafficking in mice. J Clin Invest 122:1416–1426

Grigorova IL, Panteleev M, Cyster JG et al (2010) Lymph node cortical sinus organization and relationship to lymphocyte egress dynamics and antigen exposure. Proc Natl Acad Sci U S A 107:20447–20452

Hata E, Sasaki N, Takeda A, et al (2016) Lysophosphatidic acid receptors LPA4 and LPA6 differentially promote lymphocyte transmigration across high endothelial venules in lymph nodes. Int Immunol (in press)

Hla T (2004) Physiological and pathological actions of sphingosine 1-phosphate. Semin Cell Dev Biol 15:513–520

Jung B, Hla T (2013) Sphingosine 1-phosphate receptors. In: Chun J, Hla T, Moolenaar W, Spiegel S (eds) Lysophospholipid Receptors. Wiley, New York, pp 41–60

Kanda H, Newton R, Klein R et al (2008) Autotaxin, an ectoenzyme that produces lysophosphatidic acid, promotes the entry of lymphocytes into secondary lymphoid organs. Nat Immunol 9:415–423

Knowlden SA, Capece T, Popovic M et al (2014) Regulation of T cell motility *in vitro* and *in vivo* by LPA and LPA_2. PLoS One 9, e101655

Kono M, Tucker AE, Tran J et al (2014) Sphingosine-1-phosphate receptor 1 reporter mice reveal receptor activation sites in vivo. J Clin Invest 124:2076–2086

Kunkel GT, Maceyka M, Milstien S et al (2013) Targeting the sphingosine-1-phosphate axis in cancer, inflammation and beyond. Nat Rev Drug Discov 12:688–702

Lee MJ, Jr VB, Thangada S et al (1998) Sphingosine-1-phosphate as a ligand for the G protein-coupled receptor EDG-1. Science 279:1552–1555

Lee H, Deng J, Kujawski M et al (2010) STAT3-induced $S1PR_1$ expression is crucial for persistent STAT3 activation in tumors. Nat Med 16:1421–1428

Liang J, Nagahashi M, Kim EY et al (2013) Sphingosine-1-phosphate links persistent STAT3 activation, chronic intestinal inflammation, and development of colitis-associated cancer. Cancer Cell 23:107–120

Maeda Y, Yagi H, Takemoto K et al (2014) S1P lyase in thymic perivascular spaces promotes egress of mature thymocytes via up-regulation of S1P receptor 1. Int Immunol 26:245–255

Matloubian M, Lo CG, Cinamon G et al (2004) Lymphocyte egress from thymus and peripheral lymphoid organs is dependent on S1P receptor 1. Nature 427:355–360

Mendoza A, Bréart B, Ramos-Perez WD et al (2012) The transporter Spns2 is required for secretion of lymph but not plasma sphingosine-1-phosphate. Cell Rep 2:1104–1110

Mirendil H, Lin M-E, Chun J (2013) Lysophosphatidic acid (LPA) receptor signaling. In: Chun J, Hla T, Moolenaar W, Spiegel S (eds) Lysophospholipid Receptors. Wiley, New York, pp p1–p39

Miyabe Y, Miyabe C, Iwai Y et al (2013) Necessity of lysophosphatidic acid receptor 1 for development of arthritis. Arthritis Rheum 65:2037–2047

Miyabe Y, Miyabe C, Iwai Y et al (2014) Activation of fibroblast-like synoviocytes derived from rheumatoid arthritis via lysophosphatidic acid-lysophosphatidic acid receptor 1 cascade. Arthritis Res Ther 16:461–469

Mizugishi K, Yamashita T, Olivera A et al (2005) Essential role for sphingosine kinases in neural and vascular development. Mol Cell Biol 25:11113–11121

Nakanaga K, Hama K, Aoki J (2010) Autotaxin–an LPA producing enzyme with diverse functions. J Biochem 148:13–24

Nakasaki T, Tanaka T, Okudaira S et al (2008) Involvement of the lysophosphatidic acid-generating enzyme autotaxin in lymphocyte-endothelial cell interactions. Am J Pathol 173:1566–1576

Nieto-Posadas A, Picazo-Juarez G, Llorente I et al (2012) Lysophosphatidic acid directly activates TRPV1 through a C-terminal binding site. Nat Chem Biol 8:78–85

Nochi H, Tomura H, Tobo M et al (2008) Stimulatory role of lysophosphatidic acid in cyclooxygenase-2 induction by synovial fluid of patients with rheumatoid arthritis in fibroblast-like synovial cells. J Immunol 181:5111–5119

Pabst O, Herbrand H, Willenzon S et al (2006) Enhanced FTY720-mediated lymphocyte homing requires $G\alpha_i$ signaling and depends on beta 2 and beta 7 integrin. J Immunol 176:1474–1480

Pappu R, Schwab SR, Cornelissen I et al (2007) Promotion of lymphocyte egress into blood and lymph by distinct sources of sphingosine-1-phosphate. Science 316:295–298

Pham TH, Okada T, Matloubian M et al (2008) $S1P_1$ receptor signaling overrides retention mediated by Gαi-coupled receptors to promote T cell egress. Immunity 28:122–133
Pham TH, Baluk P, Xu Y et al (2010) Lymphatic endothelial cell sphingosine kinase activity is required for lymphocyte egress and lymphatic patterning. J Exp Med 207:17–27
Pradère JP, Klein J, Gres S et al (2007) LPA_1 receptor activation promotes renal interstitial fibrosis. J Am Soc Nephrol 18:3110–3118
Rai V, Toure F, Chitayat S et al (2012) Lysophosphatidic acid targets vascular and oncogenic pathways via RAGE signaling. J Exp Med 209:2339–2350
Rosen H, Goetzl EJ (2005) Sphingosine 1-phosphate and its receptors: an autocrine and paracrine network. Nat Rev Immunol 5:560–570
Schwab SR, Pereira JP, Matloubian M et al (2005) Lymphocyte sequestration through S1P lyase inhibition and disruption of S1P gradients. Science 309:1735–1739
Sevastou I, Kaffe E, Mouratis MA et al (2013) Lysoglycerophospholipids in chronic inflammatory disorders: the PLA_2/LPC and ATX/LPA axes. Biochim Biophys Acta 1831:42–60
Spiegel S, Merrill AH Jr (1996) Sphingolipid metabolism and cell growth regulation. FASEB J 10:1388–1397
Spiegel S, Milstien S (2011) The outs and the ins of sphingosine-1-phosphate in immunity. Nat Rev Immunol 11:403–415
Tager AM, LaCamera P, Shea BS et al (2008) The lysophosphatidic acid receptor LPA_1 links pulmonary fibrosis to lung injury by mediating fibroblast recruitment and vascular leak. Nat Med 14:45–54
Takeda A, Kobayashi D, Aoi K et al (2016) Fibroblastic reticular cell derived lysophosphatidic acid regulates confined intranodal T-cell motility. Elife 5, e10561
Venkataraman K, Lee YM, Michau J et al (2008) Vascular endothelium as a contributor of plasma sphingosine 1-phosphate. Circ Res 102:669–676
Xia P, Gamble JR, Rye KA (1998) Tumor necrosis factor-α induces adhesion molecule expression through the sphingosine kinase pathway. Proc Natl Acad Sci U S A 95:14196–14201
Yopp AC, Ochando JC, Mao M et al (2005) Sphingosine 1-phosphate receptors regulate chemokine-driven transendothelial migration of lymph node but not splenic T cells. J Immunol 175:2913–2924
Zhang C, Baker DL, Yasuda S et al (2004) Lysophosphatidic acid induces neointima formation through PPARγ activation. J Exp Med 199:763–774
Zhang Y, Chen YC, Krummel MF et al (2012) Autotaxin through lysophosphatidic acid stimulates polarization, motility, and transendothelial migration of naive T cells. J Immunol 189:3914–3924
Zhao C, Fernandes MJ, Prestwich GD et al (2008) Regulation of lysophosphatidic acid receptor expression and function in human synoviocytes: implications for rheumatoid arthritis? Mol Pharmacol 73:587–600
Zhao C, Sardella A, Chun J et al (2011) TNF-alpha promotes LPA_1- and LPA_3-mediated recruitment of leukocytes *in vivo* through CXCR2 ligand chemokines. J Lipid Res 52:1307–1318

Part VII
Chronic Inflammation and Autoimmune Diseases

Chapter 36
Devising Novel Methods to Control Chronic Inflammation Via Regulatory T Cells

James B. Wing, Atsushi Tanaka, and Shimon Sakaguchi

Abstract Regulatory T (Treg) cells expressing the transcription factor Foxp3 make up approximately 10 % of CD4 T-cells and are essential for the maintenance of immune self-tolerance and homeostasis. Tregs have a critical role in the control of a variety of chronic inflammatory diseases, for example, autoimmune diseases such as Type 1 diabetes (T1D), Rheumatoid arthritis (RA), and Systemic lupus erythematosus (SLE). In addition, Treg control of chronic inflammation is relevant to the control of metabolic disorders, such as obesity, that accompany chronic inflammation. Increasing Treg number and augmenting suppressive function by various strategies such as transfer of ex vivo expanded autologous Treg cells and in vivo IL-2 treatments for Treg cell expansion are a promising avenue of investigation for the control of chronic inflammation. In this chapter, we discuss our current understanding of the roles of Tregs in chronic inflammation and recent advances in strategies to quantitatively or qualitatively enhance their function.

Keywords Regulatory T-cells • Chronic inflammation • Arthritis • Colitis • Diabetes • SLE

36.1 Phenotype and Function of Tregs

Many self-reactive T cells are deleted in the thymus in the course of negative selection; however, this process is imperfect, allowing some self-reactive T cells to emigrate from the thymus to the periphery where they may mediate autoimmunity. This creates a need for peripheral suppression. Regulatory T cells (Tregs) make up approximately 10 % of $CD4^+$ T cells and are critically required for prevention of

J.B. Wing • A. Tanaka
Laboratory of Experimental Immunology, Immunology Frontier Research Center, Osaka University, Suita 565-0871, Japan

S. Sakaguchi (✉)
Laboratory of Experimental Immunology, Immunology Frontier Research Center, Osaka University, Suita 565-0871, Japan

AMED-CREST, Suita 565-0871, Japan
e-mail: shimon@ifrec.osaka-u.ac.jp

M. Miyasaka, K. Takatsu (eds.), *Chronic Inflammation*,
DOI 10.1007/978-4-431-56068-5_36

autoimmunity. Abrogation of Treg cell development in mice by neonatal thymectomy around day 3 after birth or specific deletion of Tregs in adult mice results in the development of acute and severe autoimmune diseases (reviewed (Sakaguchi 2011)). More subtle loss of Treg function or reduction of their number is likely to contribute to the development of chronic inflammation.

Tregs were initially identified as CD25-positive $CD4^+$ T cells. CD25 is the alpha chain of the IL-2 receptor and IL-2 signalling via the IL-2 receptor is essential for the maintenance of regulatory T cells (reviewed (Bayer et al. 2013)). Tregs are controlled by the lineage-defining transcription factor Foxp3. Ectopic expression of Foxp3 confers suppressive function on Foxp3-negative conventional T cells, and mutations in the Foxp3 gene either in the scurfy mouse strain or human subjects suffering from IPEX syndrome leads to severe autoimmunity (Bennett et al. 2001; Sakaguchi 2011; Hori et al. 2003). Foxp3 expression is essential, but not alone sufficient, to maintain the Treg identity as it must form complexes with a wide range of binding partners and transcription factors such as Eos, GATA-1, GATA-3, Satb1, Lef1, and IRF4 to exert Treg cell functions. Treg cells are also maintained at the epigenetic level in which demethylation and permissive histone modifications at key genes such as *Foxp3* and *Ctla4* allow their constitutive expression (reviewed (Ohkura et al. 2013)).

Tregs have a number of suppressive mechanisms that may act in different circumstances (reviewed (Yamaguchi et al. 2011)). The primary mechanism that underpins immune homeostasis is CTLA-4–dependent control of CD80 and CD86 expression by APCs. Treg specific loss of CTLA-4 produces fatal autoimmunity similar to that seen in Foxp3- or Treg-deficient mice (Wing et al. 2008). In humans, heterozygous mutations in CTLA-4 lead to severe autoimmunity and chronic inflammation due to a partial loss of Treg function (Schubert et al. 2014; Kuehn et al. 2014). Treg-induced CD80/CD86 downregulation on APCs can indeed determine the cell fate of responder T cells, rendering them anergic, apoptotic, or dormant depending on their TCR affinity (Maeda et al. 2014). In addition to CTLA-4 expression, other Treg functions such as the production of IL-10 and TGF-β seem to have a critical role in controlling active immune responses. For example, loss of either total or Treg-specific IL-10 expression results in the development of chronic inflammation at the mucosa of the lung and the gut where the host is constantly exposed to invading microbes.

In this chapter, we discuss our current understanding of the roles of Treg cells in chronic inflammation and how they can be targeted to control chronic inflammation.

36.2 Tregs in Chronic Inflammation

36.2.1 SLE

Systemic lupus erythematosus (SLE) is a chronic autoimmune disease characterised by development of autoantibodies targeting a range of systemic self-antigens including double-stranded DNA (dsDNA). Such autoantibodies and immune complexes thus formed with relevant self-antigens cause tissue damage such as glomerulonephritis and vasculitis. SLE is mediated by T-dependent B-cell antibody production; in particular T-follicular helper (Tfh) cells are likely to have a critical role due to their importance in driving the germinal center reaction (reviewed (Ueno et al. 2015)). Circulating Tfh and activated PD-1^{hi} CCR7^{lo} Tfh cells have been demonstrated to increase in SLE patients (Simpson et al. 2010; He et al. 2013). Additionally, mice with dysregulated Tfh formation develop lupus-like symptoms (Vinuesa et al. 2005), whereas in another model of mouse lupus, the deficiency of the IL-21 receptor, which is critical for Tfh function, ameliorates lupus symptoms (Bubier et al. 2009).

Treg cells and their follicular resident subset T-follicular regulatory cells (Tfr) play a critical role in suppressing Tfh development and function (reviewed (Wing and Sakaguchi 2014; Vanderleyden et al. 2014)). Both Tregs and Tfr control Tfh formation via CTLA-4 (Wing et al. 2014; Sage et al. 2014). Scurfy mice and IPEX patients lacking functional Treg cells develop autoantibodies (Huter et al. 2010) and have enhanced Tfh formation, exhibiting lupus-like pathologies such as presence of antinuclear antibody (ANA), anti-dsDNA antibodies, nephritis, pneumonitis, and arthritis (Hadaschik et al. 2015). Correspondingly, dysregulation of Treg function by blocking IL-2, the cytokine critical for their maintenance, or Treg depletion via anti-CD25 antibodies worsens lupus-like disease in mouse models (Hsu et al. 2006; Humrich et al. 2010). Further to this, specific loss of Tfr function due to reduced CXCR5 expression by NFAT2-deficient Tregs exacerbates lupus-like disease (Vaeth et al. 2014). In humans the importance of changes to circulating Tregs in SLE patients is unclear with studies reporting both decreased or no change to circulating Tregs (reviewed (Chavele and Ehrenstein 2011)). This highlights a persistent problem with studies examining the phenotype of circulating Tregs; that is, there is little consensus about which markers to use, partly because it is clear that in humans both Foxp3 and CD25 expression are not entirely restricted to Tregs, as effector T cells also express low levels of Foxp3 on activation (Miyara et al. 2009).

36.2.2 Tregs in the Control of Diabetes

Type 1 diabetes (T1D) is a chronic autoimmune syndrome in which insulin-producing pancreatic beta cells are targeted and destroyed by both humoral and cellular components of the immune system. This commonly occurs in children and adolescents although more rarely adults can also develop the disease. The nonobese diabetic (NOD) mouse model is widely used as a model for T1D (Anderson and Bluestone 2005). Loss of Treg function and defective IL-2 signalling is reportedly associated with T1D in both humans and NOD mice (Lindley et al. 2005; Long et al. 2010; Tang et al. 2008). In particular, mutations of the Foxp3 gene in IPEX patients cause a range of endocrinopathies, especially T1D (Barzaghi et al. 2012). Further to this, effector T cells from T1D patients have been shown to be refractory to Treg suppression, whereas loss of Treg suppressive function may also inversely correlate with an increase in Th17 cell formation in T1D patients (Ferraro et al. 2011).

In contrast to T1D, Type 2 diabetes (T2D) is primarily a disease of later life and associated with obesity and metabolic syndrome. Although not classically considered a chronic inflammatory disorder, it has become clear recently that inflammation in obese adipose tissue may be a key factor in the development of metabolic disorder. Tregs are found in visceral adipose tissues (VAT) at high frequencies of the CD4 subset (40–50 %). These cells play a critical role in the downregulation of ongoing inflammation that otherwise may lead to the development of metabolic syndrome, as demonstrated by a loss of insulin-induced insulin receptor signalling upon Treg depletion (Feuerer et al. 2009).

36.2.3 Arthritis

Rheumatoid arthritis (RA) is a systemic autoimmune disease characterised by a debilitating, chronic inflammatory condition with progressive articular cartilage destruction and bone resorption. The pathogenesis of RA is multifaceted; however, it is well accepted that RA arises from a breakdown of immunological self-tolerance. In humans, an accumulating number of studies suggested increased levels of Tregs at the local site of inflammation (i.e. synovial fluid) in RA patients (van Amelsfort et al. 2004; Cao et al. 2004; Mottonen et al. 2005). However, no clear evidence of Treg anomaly in function or number has been identified in RA patients; and whether Treg anomaly can be causative of RA is unclear. In mice, there are several models of RA, including the induced arthritis models such as collagen or antigen-induced arthritis (Trentham et al. 1977; Courtenay et al. 1980; Brackertz et al. 1977), and genetically manipulated spontaneous arthritis models such as TNF-α transgenic mice, K/BxN mice, TS1 x HACII mice, and SKG mice (Keffer et al. 1991; Kouskoff et al. 1996; Rankin et al. 2008; Sakaguchi et al. 2003). In the SKG arthritis model, a hypomorphic mutation of the gene encoding ZAP-70, an essential TCR signalling molecule, produces chronic and progressing arthritis by altering thymic selection of self-reactive T cells as well as Tregs and their respective functions (Sakaguchi et al. 2011, 2003; Tanaka et al. 2010). Depletion of Tregs from normal BALB/c mice, for example, produces a variety of autoimmune diseases including autoimmune gastritis and oophoritis; however, incidences of arthritis remain low, suggesting that Treg deficiency alone is insufficient to cause arthritis in BALB/c mice (Sakaguchi et al. 1995; Tanaka et al. 2010) . In contrast, when SKG $CD4^+$ T cells devoid of Tregs were transferred to nude mice, the onset and severity of arthritis increased whereas incidences of other autoimmune diseases including oophoritis and gastritis decreased significantly, indicating an altered TCR repertoire by the ZAP-70 mutation. Functionally, SKG Tregs had reduced suppressive function compared to wild-type Tregs. When self-reactive SKG CD4+ T cells together with wild-type Tregs were transferred to nude mice, arthritis induction was prevented. These results suggest that SKG $CD4^+$ T cells are susceptible to suppression by normal Tregs. Furthermore, SKG mice with knock-in of chimeric eGFP-Foxp3 protein spontaneously develop arthritis in SPF conditions (Ito et al. 2014). The chimeric fusion of eGFP and Foxp3 was suggested to have subtly altered and partial loss of Foxp3 function which seems potent enough to change Treg function and shift the susceptibility of SKG mice in SPF conditions from 'inducible' to 'spontaneous'. Therefore, in the SKG model, self-reactive $CD4^+$ T cells together with defective Tregs are necessary for the induction and of arthritis. By using these eGFP-Foxp3 KI mice in SKG background, an arthritis-inducing TCR and its target self-antigen were recently identified (Ito et al. 2014). One of the isolated antigens turned out to be ubiquitously expressed 60S ribosomal protein L23a (RPL23A). Furthermore, in RA patients, autoantibodies and T cells reacting against RPL23A were detected. It remains to be determined how a ubiquitously expressed self-antigen is involved in the etiology and/or maintenance of RA.

36.2.4 Inflammatory Bowel Diseases

Inflammatory bowel diseases (IBD) are a group of chronic inflammatory disorders of the gastrointestinal tract encompassing Crohn's disease and ulcerative colitis. Although the etiology of these diseases is still unclear, studies from humans and mouse models suggest that IBD results from the induction of aberrant immune responses to innocuous antigens of the intestinal microflora, dietary proteins, or host cells. It has been suggested that an imbalance of effector T cells, such as Th17, and Treg cells, may trigger and perpetuate intestinal inflammation. One key piece of evidence indicating the role of Tregs in IBD comes from a T-cell transfer model of colitis. In this model, the transfer of $CD45RB^{hi}$ T cells (i.e. naïve T cells) into SCID mice induces Th1-mediated colitis (Powrie et al. 1993), however, when $CD45RB^{lo}$ T cells containing Tregs were transferred together, colitis development was prevented (Mottet et al. 2003). In support of Treg's role in IBD, a deficiency of Tregs, by a mutation of Foxp3, leads to IBD in both humans and mice (Barzaghi et al. 2012) . Furthermore, GWAS study on Crohn's disease patients showed SNPs in genes important for Tregs (i.e. IL2RA, SMAD3, IL10) and Th17 (i.e. IL23R) are associated with Crohn's disease (Franke et al. 2010). Recently, the importance of Tregs in mucosal tolerance as well as the induction of oral tolerance to dietary antigens by mucosal route has been focused on extensively (Hadis et al. 2011). The mechanistic understanding of Tregs in mucosal tolerance may be applicable for therapies targeting Tregs in IBD as well as generation of oral tolerance to suppress immune responses against antigens challenged by parenteral routes for prevention of autoimmune diseases and allergy.

36.2.5 Others

Although we have focused on SLE, Diabetes, arthritis, and colitis, Tregs also have a critical role in a wide range of other chronic autoimmune inflammatory disorders including endocrinopathies, dermatitis such as psoriasis, and hepatic disorders such as primary biliary cirrhosis (reviewed (Grant et al. 2015).

36.3 Strategies to Enhance Treg Numbers/Function

The critical role of Tregs in a variety of chronic autoimmune diseases has made therapies targeting Treg function a critical issue for the development of better treatments for these autoimmune syndromes (reviewed (Bluestone et al. 2015; Miyara et al. 2014).

36.3.1 Treg Transfer Therapies

One approach to treating chronic autoimmune diseases is to transfer large numbers of purified Tregs into patients to suppress autoimmunity, whether a loss of Treg number or function is involved in their etiologies (Fig. 36.1a). Mouse models have demonstrated that transfer of Treg cells may be a useful strategy in a number of settings. Transfer of purified Tregs decreases pathologies in models of SLE (Scalapino et al. 2006), T1D (Mukherjee et al. 2003), and RA (Morgan et al. 2005), whereas the ability of Tregs to prevent murine colitis is well established (Izcue et al. 2009). In addition to transfer of the normal pool of Tregs, in some cases transfer of specific Treg subsets targeted to the disease in question may prove effective. In a murine model of T2D, transfer of a specific Treg subset can be effective, as transfer of GATA3 expressing VAT Tregs into RAG-deficient mice was sufficient to reverse weight gain and insulin resistance from diet-induced obesity (Winer et al. 2009).

Isolation and transfer of large numbers of Tregs between genetically identical inbred mice is relatively simple, however, Treg numbers are more limiting in humans, making in vitro expansion of Tregs before transfer a critical issue. In order to do this, Tregs are first purified and then expanded in the presence of IL-2, by two rounds of either polyclonal stimulation with anti-CD3 and anti-CD28 coated beads or CD40L expressing alloreactive B cells to expand polyreactive or alloreactive Tregs, respectively, which allows between 100 and 1600-fold expansion of the original Treg population (Putnam et al. 2013). Similar protocols have already been used to expand Tregs from T1D patients and administer them to the patients, and to prevent graft-versus-host disease (GvHD) following allogenic haematopoietic stem cell transplantation (reviewed (Bluestone et al. 2015)).

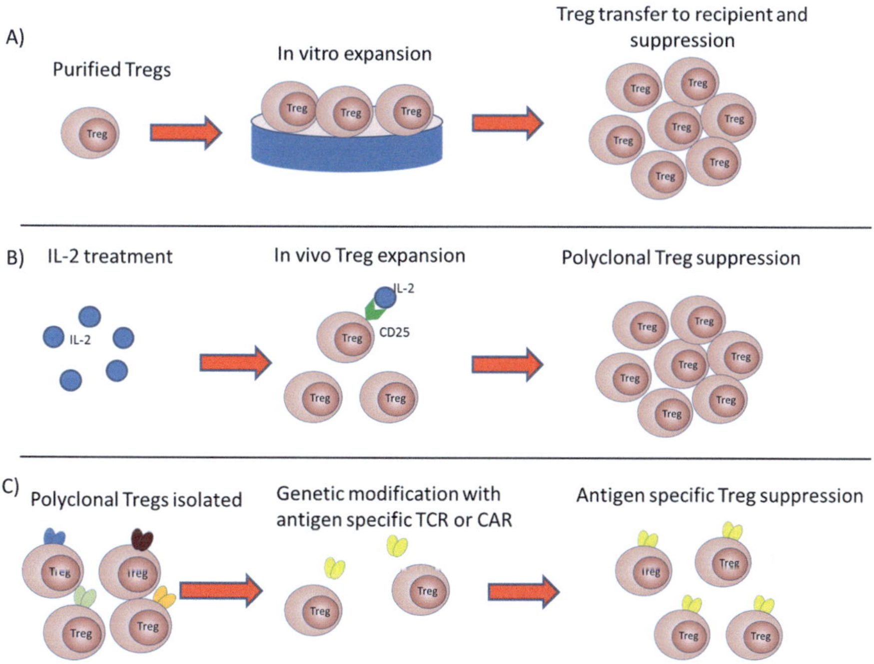

Fig. 36.1 Treg treatment strategies. (**a**) Treg transfer. Tregs are purified from donor blood and then expanded in vitro via polyclonal or antigen-specific stimulation in the presence of IL-2. The expanded Tregs are then transferred to the donor. (**b**) IL-2 treatment. IL-2 is directly injected into the recipient, and engagement with the high-affinity IL-2 receptor CD25 induces Treg proliferation in vivo. (**c**) Genetic manipulation of Tregs. Purified polyclonal Tregs are modified by addition or an antigen-specific TCR or CAR. These cells are then transferred to the recipient where they will suppress autoimmunity against the target antigen

36.3.2 Expansion/Enhancement of Tregs In Vivo

IL-2 is a critical factor for Treg survival (reviewed (Bayer et al. 2013) and as such it is possible to expand Tregs in vivo by addition of IL-2 (Fig. 36.1b). IL-2 may, however, both enhance Treg and effector T-cell function, making the dosage of IL-2 critical. Because Tregs have significantly higher levels of CD25 expression, they are able to outcompete effector T cells for limiting quantities of IL-2, and both expand and deny IL-2 to effector T cells. However, when large quantities of IL-2 are present, effector T cells, particularly Th1 cells, may expand causing further autoimmune pathology. As a result, low dose but not high dose IL-2 therapies have proven successful in reversing T1D in NOD mice (Grinberg-Bleyer et al. 2010). Thus far in humans, IL-2 has been used either as a monotherapy or in combination with the mTOR inhibitor rapamycin, added to suppress effector T-cell expansion, in a number of clinical trials for treatment of T1D therapies (reviewed (Long et al. 2013)) and SLE (Humrich et al. 2015). Currently, these treatments are at an early stage as determining the correct dose of IL-2 is a nontrivial issue (reviewed (Bayer et al. 2013)).

Another approach is the use of IL-2 complexes made up of IL-2 and anti-IL-2 antibodies. Careful selection of antibody clones allows preferential targeting of IL-2 to CD25 enriched on Tregs, rather than CD122 (IL-2 receptor beta) which is more uniformly expressed by effector cells. Also, IL-2 complexes are more stable in vivo allowing the persistence of low doses of IL-2 over a longer period of time (Boyman et al. 2006). This has proven successful in mice in the treatment of several mouse models such the NOD T1D model (Grinberg-Bleyer et al. 2010) and collagen-induced arthritis (Lee et al. 2012) and T2D (Feuerer et al. 2009), but has not yet made its way to clinical use in humans. IL-2–based therapies may also have a second benefit; in addition to inducing the expansion of Tregs, IL-2 also suppresses Tfh and Th17 cell formation due to IL-2–induced STAT5 signalling upregulating BLIMP-1, which in turn downregulates the Tfh specific transcription factor BCL6 (Johnston et al. 2012).

Although IL-2 therapies polyclonally expand Tregs, another approach is selectively to selectively enhance the formation of antigen-specific Tregs by vaccination with self-antigens. For example, either insulin or the Mr 65,000 isoform of glutamic acid decarboxylase (GAD65), autoantibodies against which are strongly correlated with T1D in both NOD mice and humans. Both induction of oral tolerance by intranasally administered insulin and vaccination have proven successful in the treatment of NOD mice (Petersen et al. 1994). To date, a number of clinical trials have not been able to replicate this success in humans, possibly because this form of vaccination may not be sufficient to reverse established disease (reviewed (Harrison et al. 2013)). A better understanding of how these treatments induce the formation of antigen-specific Tregs and possible combinations with agents (such as IL-2) to expand Tregs specifically at the time of vaccination may hold the key to enhancing the potential of these treatments.

The Treg epigenetic signature is important for the expression of key genes (reviewed (Ohkura et al. 2013)). Promotor methylation of *Ctla4* leading to loss of expression by Tregs has been demonstrated to be a factor in RA (Cribbs et al. 2014), suggesting that methylation may be a therapeutic target. The drug methotrexate appears able to restore demethylation of the Foxp3 gene in RA patients, leading to increased Foxp3 and CTLA-4 expression and lessening of RA symptoms (Cribbs et al. 2015).

In addition to therapies that aim to expand or enhance Treg function, another approach is to replace or supplement Treg function with immunosuppressive molecules. The clearest example of this is CTLA-4Ig, in which CTLA-4 has been fused with an immunoglobulin Fc fragment that allows it to be used as a soluble form of CTLA-4. CTLA-4Ig directly binds to CD80 and CD86, rendering them unavailable for CD28 ligation. CTLA-4Ig (Abatacept) is already used as a therapeutic agent for rheumatoid arthritis (Emery et al. 2015).

36.3.3 Genetic Manipulation of Tregs

Because Foxp3 is critical for Treg function and stability, it is possible to induce ectopic overexpression of Foxp3 genetically in conventional T cells in order to convert them to suppressive Treg cells (Hori et al. 2003). This has the benefit of maintaining the antigen specificity of the cells, for example, it should be noted, that although Foxp3 induction in non-Tregs may induce some suppressive capacity or repression of key genes such as IL-2, the cells may still lack the transcriptional landscape and epigenetic profile that would allow them to become bona fide Tregs.

Another approach that allows bypassing the difficulty in generating large numbers of antigen-specific Tregs, which are more effective than polyclonal Tregs, is to either transfer new T-cell receptors (TCRs) into existing Tregs (Brusko et al. 2010), or to make use of chimeric antigen receptors (CARs) in which the TCR is replaced/bypassed by an engineered receptor consisting of an antibody-derived variable region with the specificity of interest, and conjugated to a signalling and costimulatory domain (Fig. 36.1C). This approach has been used to generate functional antigen-specific Tregs which are able to suppress the development of colitis or EAE in mouse models (reviewed (Jethwa et al. 2014)).

36.4 Summary

It is clear that regulatory T cells play a critical role in the development and control of chronic inflammation. Already a number of strategies to enhance Treg numbers and/or function have been developed and are yielding clinical benefits. However, our understanding of both Treg homeostasis and suppressive function in certain disease contexts remains incomplete. We are still in the infancy of our ability to control human autoimmune disease via manipulation of Tregs. Further advances in accurate Treg purification, in vitro and in vivo generation of antigen-specific Tregs, and genetic manipulation of Tregs are required to translate success in mouse models and improve on the already promising results from early clinical trials.

Acknowledgments We would like to thank Hannah Hume for proofreading the manuscript. S.S. is the recipient of Grants-in-Aid for Specially Promoted Research (20002007) and for Scientific Research (A) (26253030) from the Ministry of Education, Culture, Sports, Science and Technology of Japan (MEXT); Core Research for Evolutional Science and Technology (CREST) from Japan Science and Technology Agency. J.B.W. is the recipient of Grant-in-Aid for Young Scientists (B) (15 K19129) from JSPS. A.T. is the recipient of Grant-in-Aid for Young Scientists (B) (26860331) from JSPS.

References

Anderson MS, Bluestone JA (2005) The NOD mouse: a model of immune dysregulation. Annu Rev Immunol 23:447–485. doi:10.1146/annurev.immunol.23.021704.115643

Barzaghi F, Passerini L, Bacchetta R (2012) Immune dysregulation, polyendocrinopathy, enteropathy, x-linked syndrome: a paradigm of immunodeficiency with autoimmunity. Front Immunol 3:211. doi:10.3389/fimmu.2012.00211

Bayer AL, Pugliese A, Malek TR (2013) The IL-2/IL-2R system: from basic science to therapeutic applications to enhance immune regulation. Immunol Res 57(1–3):197–209. doi:10.1007/s12026-013-8452-5

Bennett CL, Christie J, Ramsdell F, Brunkow ME, Ferguson PJ, Whitesell L, Kelly TE, Saulsbury FT, Chance PF, Ochs HD (2001) The immune dysregulation, polyendocrinopathy, enteropathy, X-linked syndrome (IPEX) is caused by mutations of FOXP3. Nat Genet 27(1):20–21. doi:10.1038/83713

Bluestone JA, Trotta E, Xu D (2015) The therapeutic potential of regulatory T cells for the treatment of autoimmune disease. Expert Opin Ther Targets 19:1–13. doi:10.1517/14728222.2015.1037282

Boyman O, Kovar M, Rubinstein MP, Surh CD, Sprent J (2006) Selective stimulation of T cell subsets with antibody-cytokine immune complexes. Science (New York) 311 (5769):1924–1927. doi:10.1126/science.1122927

Brackertz D, Mitchell GF, Mackay IR (1977) Antigen-induced arthritis in mice. I Induction of arthritis in various strains of mice. Arthritis Rheum 20(3):841–850

Brusko TM, Koya RC, Zhu S, Lee MR, Putnam AL, McClymont SA, Nishimura MI, Han S, Chang LJ, Atkinson MA, Ribas A, Bluestone JA (2010) Human antigen-specific regulatory T cells generated by T cell receptor gene transfer. PLoS One 5(7), e11726. doi:10.1371/journal.pone.0011726

Bubier JA, Sproule TJ, Foreman O, Spolski R, Shaffer DJ, Morse HC 3rd, Leonard WJ, Roopenian DC (2009) A critical role for IL-21 receptor signaling in the pathogenesis of systemic lupus erythematosus in BXSB-Yaa mice. Proc Natl Acad Sci U S A 106(5):1518–1523. doi:10.1073/pnas.0807309106

Cao D, van Vollenhoven R, Klareskog L, Trollmo C, Malmstrom V (2004) CD25brightCD4+ regulatory T cells are enriched in inflamed joints of patients with chronic rheumatic disease. Arthritis Res Ther 6(4):R335–R346. doi:10.1186/ar1192

Chavele K-M, Ehrenstein MR (2011) Regulatory T-cells in systemic lupus erythematosus and rheumatoid arthritis. FEBS Lett 585(23):3603–3610. doi:10.1016/j.febslet.2011.07.043

Courtenay JS, Dallman MJ, Dayan AD, Martin A, Mosedale B (1980) Immunisation against heterologous type II collagen induces arthritis in mice. Nature 283(5748):666–668

Cribbs AP, Kennedy A, Penn H, Read JE, Amjadi P, Green P, Syed K, Manka SW, Brennan FM, Gregory B, Williams RO (2014) Treg cell function in rheumatoid arthritis is compromised by ctla-4 promoter methylation resulting in a failure to activate the indoleamine 2,3-dioxygenase pathway. Arthritis Rheumatol 66(9):2344–2354. doi:10.1002/art.38715

Cribbs AP, Kennedy A, Penn H, Amjadi P, Green P, Read JE, Brennan F, Gregory B, Williams RO (2015) Methotrexate restores regulatory T cell function through demethylation of the FoxP3 upstream enhancer in patients with rheumatoid arthritis. Arthritis Rheumatol 67(5):1182–1192. doi:10.1002/art.39031

Emery P, Burmester GR, Bykerk VP, Combe BG, Furst DE, Barre E, Karyekar CS, Wong DA, Huizinga TW (2015) Evaluating drug-free remission with abatacept in early rheumatoid arthritis: results from the phase 3b, multicentre, randomised, active-controlled AVERT study of 24 months, with a 12-month, double-blind treatment period. Ann Rheum Dis 74(1):19–26. doi:10.1136/annrheumdis-2014-206106

Ferraro A, Socci C, Stabilini A, Valle A, Monti P, Piemonti L, Nano R, Olek S, Maffi P, Scavini M, Secchi A, Staudacher C, Bonifacio E, Battaglia M (2011) Expansion of Th17 cells and functional defects in T regulatory cells are key features of the pancreatic lymph nodes in patients with type 1 diabetes. Diabetes 60(11):2903–2913. doi:10.2337/db11-0090

Feuerer M, Herrero L, Cipolletta D, Naaz A, Wong J, Nayer A, Lee J, Goldfine AB, Benoist C, Shoelson S, Mathis D (2009) Lean, but not obese, fat is enriched for a unique population of regulatory T cells that affect metabolic parameters. Nat Med 15(8):930–939. doi:10.1038/nm.2002

Franke A, McGovern DP, Barrett JC, Wang K, Radford-Smith GL, Ahmad T, Lees CW, Balschun T, Lee J, Roberts R, Anderson CA, Bis JC, Bumpstead S, Ellinghaus D, Festen EM, Georges M, Green T, Haritunians T, Jostins L, Latiano A, Mathew CG, Montgomery GW, Prescott NJ, Raychaudhuri S, Rotter JI, Schumm P, Sharma Y, Simms LA, Taylor KD, Whiteman D, Wijmenga C, Baldassano RN, Barclay M, Bayless TM, Brand S, Buning C, Cohen A, Colombel JF, Cottone M, Stronati L, Denson T, De Vos M, D'Inca R, Dubinsky M, Edwards C, Florin T, Franchimont D, Gearry R, Glas J, Van Gossum A, Guthery SL, Halfvarson J, Verspaget HW, Hugot JP, Karban A, Laukens D, Lawrance I, Lemann M, Levine A, Libioulle C, Louis E, Mowat C, Newman W, Panes J, Phillips A, Proctor DD, Regueiro M, Russell R, Rutgeerts P, Sanderson J, Sans M, Seibold F, Steinhart AH, Stokkers PC, Torkvist L, Kullak-Ublick G, Wilson D, Walters T, Targan SR, Brant SR, Rioux JD, D'Amato M, Weersma RK, Kugathasan S, Griffiths AM, Mansfield JC, Vermeire S, Duerr RH, Silverberg MS, Satsangi J, Schreiber S, Cho JH, Annese V, Hakonarson H, Daly MJ, Parkes M (2010) Genome-wide meta-analysis increases to 71 the number of confirmed Crohn's disease susceptibility loci. Nat Genet 42(12):1118–1125. doi:10.1038/ng.717

Grant CR, Liberal R, Mieli-Vergani G, Vergani D, Longhi MS (2015) Regulatory T-cells in autoimmune diseases: challenges, controversies and--yet--unanswered questions. Autoimmun Rev 14(2):105–116. doi:10.1016/j.autrev.2014.10.012

Grinberg-Bleyer Y, Baeyens A, You S, Elhage R, Fourcade G, Gregoire S, Cagnard N, Carpentier W, Tang Q, Bluestone J, Chatenoud L, Klatzmann D, Salomon BL, Piaggio E (2010) IL-2 reverses established type 1 diabetes in NOD mice by a local effect on pancreatic regulatory T cells. J Exp Med 207(9):1871–1878. doi:10.1084/jem.20100209

Hadaschik EN, Wei X, Leiss H, Heckmann B, Niederreiter B, Steiner G, Ulrich W, Enk AH, Smolen JS, Stummvoll GH (2015) Regulatory T cell-deficient scurfy mice develop systemic autoimmune features resembling lupus-like disease. Arthritis Res Ther 17(1):35. doi:10.1186/s13075-015-0538-0

Hadis U, Wahl B, Schulz O, Hardtke-Wolenski M, Schippers A, Wagner N, Muller W, Sparwasser T, Forster R, Pabst O (2011) Intestinal tolerance requires gut homing and expansion of FoxP3+ regulatory T cells in the lamina propria. Immunity 34(2):237–246. doi:10.1016/j.immuni.2011.01.016

Harrison LC, Wentworth JM, Zhang Y, Bandala-Sanchez E, Bohmer RM, Neale AM, Stone NL, Naselli G, Bosco JJ, Auyeung P, Rashidi M, Augstein P, Morahan G (2013) Antigen-based vaccination and prevention of type 1 diabetes. Curr Diab Rep 13(5):616–623. doi:10.1007/s11892-013-0415-7

He J, Tsai LM, Leong YA, Hu X, Ma CS, Chevalier N, Sun X, Vandenberg K, Rockman S, Ding Y, Zhu L, Wei W, Wang C, Karnowski A, Belz GT, Ghali JR, Cook MC, Riminton DS, Veillette A, Schwartzberg PL, Mackay F, Brink R, Tangye SG, Vinuesa CG, Mackay CR, Li Z, Yu D (2013) Circulating precursor CCR7(lo)PD-1(hi) CXCR5(+) CD4(+) T cells indicate Tfh cell activity and promote antibody responses upon antigen reexposure. Immunity 39(4):770–781. doi:10.1016/j.immuni.2013.09.007

Hori S, Nomura T, Sakaguchi S (2003) Control of regulatory T cell development by the transcription factor Foxp3. Science 299(5609):1057–1061. doi:10.1126/science.1079490

Hsu WT, Suen JL, Chiang BL (2006) The role of CD4CD25 T cells in autoantibody production in murine lupus. Clin Exp Immunol 145(3):513–519. doi:10.1111/j.1365-2249.2006.03173.x

Humrich JY, Morbach H, Undeutsch R, Enghard P, Rosenberger S, Weigert O, Kloke L, Heimann J, Gaber T, Brandenburg S, Scheffold A, Huehn J, Radbruch A, Burmester GR, Riemekasten G (2010) Homeostatic imbalance of regulatory and effector T cells due to IL-2 deprivation amplifies murine lupus. Proc Natl Acad Sci U S A 107(1):204–209. doi:10.1073/pnas.0903158107

Humrich JY, von Spee-Mayer C, Siegert E, Alexander T, Hiepe F, Radbruch A, Burmester GR, Riemekasten G (2015) Rapid induction of clinical remission by low-dose interleukin-2 in a patient with refractory SLE. Ann Rheum Dis 74(4):791–792. doi:10.1136/annrheumdis-2014-206506

Huter EN, Natarajan K, Torgerson TR, Glass DD, Shevach EM (2010) Autoantibodies in scurfy mice and IPEX patients recognize keratin 14. J Invest Dermatol 130(5):1391–1399. doi:10.1038/jid.2010.16

Ito Y, Hashimoto M, Hirota K, Ohkura N, Morikawa H, Nishikawa H, Tanaka A, Furu M, Ito H, Fujii T, Nomura T, Yamazaki S, Morita A, Vignali DA, Kappler JW, Matsuda S, Mimori T, Sakaguchi N, Sakaguchi S (2014) Detection of T cell responses to a ubiquitous cellular protein in autoimmune disease. Science 346(6207):363–368. doi:10.1126/science.1259077

Izcue A, Coombes JL, Powrie F (2009) Regulatory lymphocytes and intestinal inflammation. Annu Rev Immunol 27:313–338. doi:10.1146/annurev.immunol.021908.132657

Jethwa H, Adami AA, Maher J (2014) Use of gene-modified regulatory T-cells to control autoimmune and alloimmune pathology: is now the right time? Clin Immunol 150(1):51–63. doi:10.1016/j.clim.2013.11.004

Johnston RJ, Choi YS, Diamond JA, Yang JA, Crotty S (2012) STAT5 is a potent negative regulator of TFH cell differentiation. J Exp Med 209(2):243–250. doi:10.1084/jem.20111174

Keffer J, Probert L, Cazlaris H, Georgopoulos S, Kaslaris E, Kioussis D, Kollias G (1991) Transgenic mice expressing human tumour necrosis factor: a predictive genetic model of arthritis. Embo J 10(13):4025–4031

Kouskoff V, Korganow AS, Duchatelle V, Degott C, Benoist C, Mathis D (1996) Organ-specific disease provoked by systemic autoimmunity. Cell 87(5):811–822

Kuehn HS, Ouyang W, Lo B, Deenick EK, Niemela JE, Avery DT, Schickel JN, Tran DQ, Stoddard J, Zhang Y, Frucht DM, Dumitriu B, Scheinberg P, Folio LR, Frein CA, Price S, Koh C, Heller T, Seroogy CM, Huttenlocher A, Rao VK, Su HC, Kleiner D, Notarangelo LD, Rampertaap Y, Olivier KN, McElwee J, Hughes J, Pittaluga S, Oliveira JB, Meffre E, Fleisher TA, Holland SM, Lenardo MJ, Tangye SG, Uzel G (2014) Immune dysregulation in human subjects with heterozygous germline mutations in CTLA4. Science 345(6204):1623–1627. doi:10.1126/science.1255904

Lee SY, Cho ML, Oh HJ, Ryu JG, Park MJ, Jhun JY, Park MK, Stone JC, Ju JH, Hwang SY, Park SH, Surh CD, Kim HY (2012) Interleukin-2/anti-interleukin-2 monoclonal antibody immune complex suppresses collagen-induced arthritis in mice by fortifying interleukin-2/STAT5 signalling pathways. Immunology 137(4):305–316. doi:10.1111/imm.12008

Lindley S, Dayan CM, Bishop A, Roep BO, Peakman M, Tree TI (2005) Defective suppressor function in CD4(+)CD25(+) T-cells from patients with type 1 diabetes. Diabetes 54(1):92–99

Long SA, Cerosaletti K, Bollyky PL, Tatum M, Shilling H, Zhang S, Zhang ZY, Pihoker C, Sanda S, Greenbaum C, Buckner JH (2010) Defects in IL-2R signaling contribute to diminished maintenance of FOXP3 expression in CD4(+)CD25(+) regulatory T-cells of type 1 diabetic subjects. Diabetes 59(2):407–415. doi:10.2337/db09-0694

Long SA, Buckner JH, Greenbaum CJ (2013) IL-2 therapy in type 1 diabetes: "Trials" and tribulations. Clin Immunol 149(3):324–331. doi:10.1016/j.clim.2013.02.005

Maeda Y, Nishikawa H, Sugiyama D, Ha D, Hamaguchi M, Saito T, Nishioka M, Wing JB, Adeegbe D, Katayama I, Sakaguchi S (2014) Detection of self-reactive CD8(+) T cells with an anergic phenotype in healthy individuals. Science 346(6216):1536–1540. doi:10.1126/science.aaa1292

Miyara M, Yoshioka Y, Kitoh A, Shima T, Wing K, Niwa A, Parizot C, Taflin C, Heike T, Valeyre D, Mathian A, Nakahata T, Yamaguchi T, Nomura T, Ono M, Amoura Z, Gorochov G, Sakaguchi S (2009) Functional delineation and differentiation dynamics of human CD4+ T cells expressing the FoxP3 transcription factor. Immunity 30(6):899–911. doi:10.1016/j.immuni.2009.03.019

Miyara M, Ito Y, Sakaguchi S (2014) TREG-cell therapies for autoimmune rheumatic diseases. Nat Rev Rheumatol 10(9):543–551. doi:10.1038/nrrheum.2014.105

Morgan ME, Flierman R, van Duivenvoorde LM, Witteveen HJ, van Ewijk W, van Laar JM, de Vries RR, Toes RE (2005) Effective treatment of collagen-induced arthritis by adoptive transfer of CD25+ regulatory T cells. Arthritis Rheum 52(7):2212–2221. doi:10.1002/art.21195

Mottet C, Uhlig HH, Powrie F (2003) Cutting edge: cure of colitis by CD4 + CD25+ regulatory T cells. J Immunol 170(8):3939–3943

Mottonen M, Heikkinen J, Mustonen L, Isomaki P, Luukkainen R, Lassila O (2005) CD4+ CD25+ T cells with the phenotypic and functional characteristics of regulatory T cells are enriched in the synovial fluid of patients with rheumatoid arthritis. Clin Exp Immunol 140(2):360–367. doi:10.1111/j.1365-2249.2005.02754.x

Mukherjee R, Chaturvedi P, Qin HY, Singh B (2003) CD4 + CD25+ regulatory T cells generated in response to insulin B:9–23 peptide prevent adoptive transfer of diabetes by diabetogenic T cells. J Autoimmun 21(3):221–237

Ohkura N, Kitagawa Y, Sakaguchi S (2013) Development and maintenance of regulatory T cells. Immunity 38(3):414–423. doi:10.1016/j.immuni.2013.03.002

Petersen JS, Karlsen AE, Markholst H, Worsaae A, Dyrberg T, Michelsen B (1994) Neonatal tolerization with glutamic acid decarboxylase but not with bovine serum albumin delays the onset of diabetes in NOD mice. Diabetes 43(12):1478–1484

Powrie F, Leach MW, Mauze S, Caddle LB, Coffman RL (1993) Phenotypically distinct subsets of CD4+ T cells induce or protect from chronic intestinal inflammation in C. B-17 scid mice. Int Immunol 5(11):1461–1471

Putnam AL, Safinia N, Medvec A, Laszkowska M, Wray M, Mintz MA, Trotta E, Szot GL, Liu W, Lares A, Lee K, Laing A, Lechler RI, Riley JL, Bluestone JA, Lombardi G, Tang Q (2013) Clinical grade manufacturing of human alloantigen-reactive regulatory T cells for use in transplantation. Am J Transplant 13(11):3010–3020. doi:10.1111/ajt.12433

Rankin AL, Reed AJ, Oh S, Cozzo Picca C, Guay HM, Larkin J 3rd, Panarey L, Aitken MK, Koeberlein B, Lipsky PE, Tomaszewski JE, Naji A, Caton AJ (2008) CD4+ T cells recognizing a single self-peptide expressed by APCs induce spontaneous autoimmune arthritis. J Immunol 180(2):833–841

Sage PT, Paterson AM, Lovitch SB, Sharpe AH (2014) The coinhibitory receptor ctla-4 controls B cell responses by modulating T follicular helper, T follicular regulatory, and T regulatory cells. Immunity 41(6):1026–1039. doi:10.1016/j.immuni.2014.12.005

Sakaguchi S (2011) Regulatory T cells: history and perspective. Methods Mol Biol 707:3–17. doi:10.1007/978-1-61737-979-6_1

Sakaguchi S, Sakaguchi N, Asano M, Itoh M, Toda M (1995) Immunologic self-tolerance maintained by activated T cells expressing IL-2 receptor alpha-chains (CD25). Breakdown of a single mechanism of self-tolerance causes various autoimmune diseases. J Immunol 155 (3):1151–1164

Sakaguchi N, Takahashi T, Hata H, Nomura T, Tagami T, Yamazaki S, Sakihama T, Matsutani T, Negishi I, Nakatsuru S, Sakaguchi S (2003) Altered thymic T-cell selection due to a mutation of the ZAP-70 gene causes autoimmune arthritis in mice. Nature 426(6965):454–460. doi:10.1038/nature02119

Sakaguchi S, Tanaka S, Tanaka A, Ito Y, Maeda S, Sakaguchi N, Hashimoto M (2011) Thymus, innate immunity and autoimmune arthritis: interplay of gene and environment. FEBS Lett 585 (23):3633–3639. doi:10.1016/j.febslet.2011.10.026

Scalapino KJ, Tang Q, Bluestone JA, Bonyhadi ML, Daikh DI (2006) Suppression of disease in New Zealand Black/New Zealand White lupus-prone mice by adoptive transfer of ex vivo expanded regulatory T cells. J Immunol 177(3):1451–1459

Schubert D, Bode C, Kenefeck R, Hou TZ, Wing JB, Kennedy A, Bulashevska A, Petersen BS, Schaffer AA, Gruning BA, Unger S, Frede N, Baumann U, Witte T, Schmidt RE, Dueckers G, Niehues T, Seneviratne S, Kanariou M, Speckmann C, Ehl S, Rensing-Ehl A, Warnatz K,

Rakhmanov M, Thimme R, Hasselblatt P, Emmerich F, Cathomen T, Backofen R, Fisch P, Seidl M, May A, Schmitt-Graeff A, Ikemizu S, Salzer U, Franke A, Sakaguchi S, Walker LS, Sansom DM, Grimbacher B (2014) Autosomal dominant immune dysregulation syndrome in humans with CTLA4 mutations. Nat Med. doi:10.1038/nm.3746

Simpson N, Gatenby PA, Wilson A, Malik S, Fulcher DA, Tangye SG, Manku H, Vyse TJ, Roncador G, Huttley GA, Goodnow CC, Vinuesa CG, Cook MC (2010) Expansion of circulating T cells resembling follicular helper T cells is a fixed phenotype that identifies a subset of severe systemic lupus erythematosus. Arthritis Rheum 62(1):234–244. doi:10.1002/art.25032

Tanaka S, Maeda S, Hashimoto M, Fujimori C, Ito Y, Teradaira S, Hirota K, Yoshitomi H, Katakai T, Shimizu A, Nomura T, Sakaguchi N, Sakaguchi S (2010) Graded attenuation of TCR signaling elicits distinct autoimmune diseases by altering thymic T cell selection and regulatory T cell function. J Immunol 185(4):2295–2305. doi:10.4049/jimmunol.1000848

Tang Q, Adams JY, Penaranda C, Melli K, Piaggio E, Sgouroudis E, Piccirillo CA, Salomon BL, Bluestone JA (2008) Central role of defective interleukin-2 production in the triggering of islet autoimmune destruction. Immunity 28(5):687–697. doi:10.1016/j.immuni.2008.03.016

Trentham DE, Townes AS, Kang AH (1977) Autoimmunity to type II collagen an experimental model of arthritis. J Exp Med 146(3):857–868

Ueno H, Banchereau J, Vinuesa CG (2015) Pathophysiology of T follicular helper cells in humans and mice. Nat Immunol 16(2):142–152. doi:10.1038/ni.3054

Vaeth M, Muller G, Stauss D, Dietz L, Klein-Hessling S, Serfling E, Lipp M, Berberich I, Berberich-Siebelt F (2014) Follicular regulatory T cells control humoral autoimmunity via NFAT2-regulated CXCR5 expression. J Exp Med 211(3):545–561. doi:10.1084/jem.20130604

van Amelsfort JM, Jacobs KM, Bijlsma JW, Lafeber FP, Taams LS (2004) CD4(+)CD25(+) regulatory T cells in rheumatoid arthritis: differences in the presence, phenotype, and function between peripheral blood and synovial fluid. Arthritis Rheum 50(9):2775–2785. doi:10.1002/art.20499

Vanderleyden I, Linterman MA, Smith KG (2014) Regulatory T cells and control of the germinal centre response. Arthritis Res Ther 16(5):471

Vinuesa CG, Cook MC, Angelucci C, Athanasopoulos V, Rui L, Hill KM, Yu D, Domaschenz H, Whittle B, Lambe T, Roberts IS, Copley RR, Bell JI, Cornall RJ, Goodnow CC (2005) A RING-type ubiquitin ligase family member required to repress follicular helper T cells and autoimmunity. Nature 435(7041):452–458. doi:10.1038/nature03555

Winer S, Chan Y, Paltser G, Truong D, Tsui H, Bahrami J, Dorfman R, Wang Y, Zielenski J, Mastronardi F, Maezawa Y, Drucker DJ, Engleman E, Winer D, Dosch HM (2009) Normalization of obesity-associated insulin resistance through immunotherapy. Nat Med 15 (8):921–929. doi:10.1038/nm.2001

Wing JB, Sakaguchi S (2014) Foxp3(+) T(reg) cells in humoral immunity. Int Immunol 26 (2):61–69. doi:10.1093/intimm/dxt060

Wing K, Onishi Y, Prieto-Martin P, Yamaguchi T, Miyara M, Fehervari Z, Nomura T, Sakaguchi S (2008) CTLA-4 control over Foxp3(+) regulatory T cell function. Science 322 (5899):271–275. doi:10.1126/Science.1160062

Wing JB, Ise W, Kurosaki T, Sakaguchi S (2014) Regulatory T cells control antigen-specific expansion of Tfh cell number and humoral immune responses via the coreceptor CTLA-4. Immunity 41(6):1013–1025. doi:10.1016/j.immuni.2014.12.006

Yamaguchi T, Wing JB, Sakaguchi S (2011) Two modes of immune suppression by Foxp3(+) regulatory T cells under inflammatory or non-inflammatory conditions. Semin Immunol 23 (6):424–430. doi:10.1016/j.smim.2011.10.002

Chapter 37
Control of Chronic Inflammation Through Elucidation of Organ-Specific Autoimmune Disease Mechanisms

Mitsuru Matsumoto

Abstract Our body is equipped with an immune system that normally distinguishes between microorganisms (non-self) and components of our body (self), thereby protecting us from the invasion by many pathogens. However, intractable autoimmune disease, in which somehow our immune system attacks our own body, could develop by unknown mechanisms. With aiming toward novel therapeutic approaches to chronic inflammation caused by autoimmunity, we are studying the mechanisms underlying the development of organ-specific autoimmune diseases caused by the abnormal function of antigen-presenting cells in the thymus. We are particularly interested in how the immunological self is presented to developing thymocytes to establish the self-tolerance within the thymic microenvironment. NF-κB activation pathway in medullary thymic epithelial cells (mTECs) and Aire in mTECs, a transcription factor which is responsible for the development of rather rare hereditary type of autoimmune disease, are the main focus of our research, hoping that manipulations of those factors could control the chronic inflammation observed in many organ-specific autoimmune diseases.

Keywords Organ-specific autoimmunity • Thymus • Tolerance • NF-κB • Aire • Thymic epithelial cell • Treg • Immune network

37.1 Introduction

Autoimmunity against particular self-antigens in our body elicits a prototypic chronic inflammation in the corresponding organs. Breakdown of central tolerance plays a critical role for developing organ-specific autoimmune diseases typically

M. Matsumoto (✉)
Division of Molecular Immunology, Institute for Enzyme Research, Tokushima University, Tokushima 770-8503, Japan

AMED-CREST, Tokushima 770-8503, Japan
e-mail: mitsuru@tokushima-u.ac.jp

M. Miyasaka, K. Takatsu (eds.), *Chronic Inflammation*,
DOI 10.1007/978-4-431-56068-5_37

seen in Aire-deficient mice and, most likely, in Aire-deficient humans as well (Mathis and Benoist 2009; Matsumoto 2009). Acute phase reaction of inflammation might be caused by the autoimmune attack by the autoreactive T cells against particular self-antigens. Although the mechanisms underlying the persistence of inflammation have not been clearly demonstrated, it is possible that continuous breakdown of the central tolerance is responsible for the long-acting diseased conditions. In other words, if we could repair the breakdown of central tolerance, organ-specific autoimmune diseases can be therapeutically terminated. Based on this premise, it would be essential to understand the molecular mechanisms for the establishment of self-tolerance both in the thymus and periphery. Elucidating the factors that are responsible for the breakdown of self-tolerance and studying the physiological mechanisms for the establishment of self-tolerance are the opposite sides of the same coin. Our approaches to investigate the chronic inflammation caused by the continuous breakdown of central tolerance mainly focus on the understanding of the immunological characteristics of the stromal element within the thymic microenvironment. Among them, thymic epithelial cells (TECs) play a pivotal role for expressing the immunological self to developing thymocytes. Thymic dendritic cells (DCs) are also important players because they can pick up self-antigens expressed by medullary thymic epithelial cells (mTECs) and present them to the developing thymocytes, a process called cross-presentation (Kyewski and Klein 2006).

This chapter focusses on two topics, the NF-κB activation pathway and a transcription factor Aire both acting in mTECs, hoping that these approaches could manipulate the chronic inflammation observed in many other organ-specific autoimmune diseases.

37.2 Factors That Affect the Development of Chronic Inflammation: Stromal Element That Controls the Expression of Immunological Self Within the Thymus

37.2.1 NF-κB Activation Pathway in mTECs

Physical contact between thymocytes and the thymic stroma is essential for T-cell maturation and shapes the T-cell repertoire in the periphery (Klein et al. 2014; van Ewijk et al. 1994). Stromal elements that control these processes still remain elusive. Fortunately, in the last decade, several strains of mice that have the defect in the NF-κB activation pathway have been reported that resulted in the development of organ-specific autoimmunity (Matsumoto 2007). Among them, the alymphoplasia (*aly*) strain of mice merits much attention.

NF-κB–inducing kinase (NIK) is structurally related to mitogen-activated protein kinase kinase kinase (Elewaut et al. 2003; Malinin et al. 1997) and has been shown to phosphorylate both IκB kinase (IKK)-α and IKK-β, which sequentially activate the downstream IκB proteins (Ling et al. 1998). The *aly* mice carry a natural mutation of the *NIK* gene (Miyawaki et al. 1994; Shinkura et al. 1999) in which a G855R substitution in the C terminus of the protein results in inability to bind to downstream IKK-α (Matsushima et al. 2001). *aly* mice lack all lymph nodes and Peyer's patches, and spleen architecture such as development of germinal centers and follicular dendritic cell clusters is disturbed (Koike et al. 1996; Miyawaki et al. 1994; Shinkura et al. 1996). They have provided a unique model for the abnormal development of secondary lymphoid organs. We have previously demonstrated that this is due to defective NF-κB activation through the lymphotoxin (LT)-β receptor (LTβR; Matsumoto et al. 1999; Matsushima et al. 2001), a receptor essential for the development of secondary lymphoid organs (Matsumoto et al. 1997). In addition to secondary lymphoid organs, however, thymic structure is also disorganized in *aly* mice (Miyawaki et al. 1994); the medulla in *aly* mice is smaller than that in control mice, and the boundary of the cortex and medulla is unclear. We have investigated *aly* mice as a model of organ-specific autoimmune disease due to the structural abnormality of the thymus because histopathological analysis of *aly* mice has revealed chronic inflammation in many organs including liver, pancreas, lung, salivary gland, and lacrimal gland (Kajiura et al. 2004); we reasoned that the autoimmune disease phenotype seen in *aly* mice might be associated with the altered thymic microenvironment. Indeed, this turned out to be the case because embryonic thymi taken from *aly* mice induced organ-specific autoimmune diseases when grafted onto thymus-deficient nude mice (Kajiura et al. 2004).

Both impaired production of regulatory T cells (Tregs) and possibly an impaired negative selection process contribute to the development of autoimmunity in *aly* mice. Interestingly, the exogenous supply of sufficient numbers of Tregs was able to rescue the autoimmune disease phenotype of *aly* mice (Tamura et al. 2006). Thus, therapeutic benefits of Tregs for the control of chronic inflammation caused by these types of organ-specific autoimmunity are promising.

37.2.2 Novel Thymic Crosstalk Organized by NF-κB Activities in mTECs

mTECs particularly play a pivotal role in eliminating pathogenic autoreactive T cells by negative selection through expressing a large set of tissue-restricted self-antigens (TRAs) representing many parenchymal organs within the body (Kyewski and Klein 2006). In addition to mTECs, thymic DCs constitute an important component that presents self-antigens for the induction of negative selection as well as for the production of Tregs (Dresch et al. 2012). Recently, cross-presentation of mTEC-derived self-antigens by thymic DCs has been demonstrated to be an important interplay between mTECs and DCs (Gallegos and Bevan 2004;

Kyewski and Klein 2006). However, little was known about how the thymic stroma affects the development and function of thymic DCs. *aly* mice again provided a unique opportunity to study this issue. The issue was not so simple though, because NIK has also been implicated in the intrinsic function of T cells (Matsumoto et al. 2002; Yamada et al. 2000) as well as DCs (Hofmann et al. 2011). Given that thymic DCs play an important role in the production of Tregs in a cell-intrinsic or cell-extrinsic manner, we examined how the function of DCs may be affected by the interaction with thymic stroma that also expresses NIK. In other words, we hypothesised that thymic crosstalk is not confined to the interplay between developing thymocytes and mTECs. Instead, crosstalk between DCs and mTECs should also be taken into account (Mouri et al. 2014). To investigate the characteristics of thymic crosstalk required for the establishment of self-tolerance, we investigated the thymic stroma-dependent development of thymocytes, including Tregs, together with the development and function of thymic DCs in *aly* mice. The phenotypes of thymic DCs were strongly influenced by their interaction with mTECs expressing NIK, as is the case for developing thymocytes (Mouri et al. 2014). In light of the fact that, in turn, thymic DCs can affect the function of thymocytes in an NIK-dependent manner (Hofmann et al. 2011), we suggest that the thymus can now be viewed as a platform where two essential types of interplay (i.e. between thymocytes and mTECs and between DCs and mTECs) further cross.

37.2.3 Aire Functions in mTECs: Role of Aire in mTEC Differentiation Programme

One of the best characterised molecules expressed in mTECs that contributes to the expression of wide varieties of TRAs thereby contributing to the central tolerance is Aire, a gene responsible for the hereditary type of organ-specific autoimmune disease in humans (Mathis and Benoist 2009; Matsumoto et al. 2013). Loss of Aire has been demonstrated to result in the defective expression of many TRAs, and consequently in allowing the escape of autoreactive T cells (Anderson et al. 2002, 2005; Liston et al. 2003). In contrast to the remarkable changes in the expression profiles of TRA genes in Aire-deficient mTECs, morphological alterations in the medullary components from Aire-deficient mice were not initially appreciated. However, subsequent studies of Aire-deficient thymi in more detail have revealed several important aspects of the Aire-dependent differentiation programmes of mTECs (Matsumoto 2011), such as increased numbers of mTECs with a globular cell shape (Gillard et al. 2007; Yano et al. 2008) and, in contrast, reduced numbers of terminally differentiated mTECs expressing involucrin, the latter being associated with reduced numbers of Hassall's corpuscles (Yano et al. 2008). Increased percentages of mTECs expressing CD80 at high levels (CD80^{high}) were another suggested aspect of the Aire-dependent mTEC differentiation programme (Gray et al. 2007; Kuroda et al. 2005). All of these findings together illuminated the importance of a full understanding of the Aire-dependent maturation process of the Aire-expressing lineage of mTECs (AEL-mTECs). It is noteworthy that the Aire-dependent mTEC differentiation programme can be linked with the control of TRA gene expression, in which Aire may play a role (Matsumoto 2011; Matsumoto et al. 2013): any defect in the Aire-dependent maturation programme described above could also account for the defects of TRA gene expression in Aire-deficient mTECs from a different viewpoint.

Another enigmatic aspect of Aire function was whether Aire exerts any proapoptotic activity within cells. It is believed that mTECs contribute to self-antigen expression by being phagocytosed by professional APCs at the expense of their death (Gray et al. 2007). In line with this notion, we have observed that many AEL-mTECs are in close contact with thymic DCs, suggesting efficient cross-presentation of TRAs from AEL-mTECs (Nishikawa et al. 2010). Nevertheless, the issue of whether Aire itself exerts any proapoptotic activities has not been directly addressed especially with in vivo experimental settings.

We therefore decided to monitor the maturation programme of AEL-mTECs by temporal lineage tracing, in which bacterial artificial chromosome transgenic mice express tamoxifen-inducible Cre recombinase under control of the Aire regulatory element: mice were further crossed with reporter strains in order to allow permanent marking of AEL-mTECs with fluorescent proteins. With this in vivo experimental system, we estimated that the half-life of AEL-mTECs subsequent to Aire expression was ~7–8 days, which was much longer than that reported previously, owing to the existence of a 'post-Aire' stage we described for the first time (Nishikawa et al. 2010). We also found that loss of Aire did not alter the overall lifespan of AEL-mTECs, inconsistent with the previous notion that Aire expression in mTECs might result in their apoptosis for efficient cross-presentation of self-antigens expressed by AEL-mTECs. In contrast, Aire was required for the full maturation programme of AEL-mTECs, as exemplified by the lack of physiological downregulation of CD80 during the post-Aire stage in Aire-deficient mice, thus accounting for the abnormally increased $CD80^{high}$ mTECs seen in such mice (Nishikawa et al. 2010).

Finally, a conflicting idea has come to light that Aire can either inhibit or promote the differentiation programme of AEL-mTECs (Matsumoto 2011; Matsumoto et al. 2013). The inhibition model assumes that only an absence of Aire would reveal the full programme of terminal differentiation of AEL-mTECs (Dooley et al. 2008): it is probably important to mention that this model was constructed based partly on the concept of the proposed proapoptotic activity of Aire. In contrast, we have suggested that Aire helps to promote the differentiation programme in AEL-mTECs, and this model assumes defective accomplishment of the differentiation programme in the absence of Aire, associated with impaired expression of the TRA gene in AEL-mTECs through the mechanisms described above (Matsumoto 2011; Yano et al. 2008). These approaches for investigating the cellular dynamics of AEL-mTECs with temporal lineage tracing have revealed many fundamental and previously unknown characteristics of AEL-mTECs.

Table 37.1 Mechanisms responsible for the development of organ-specific autoimmunity due to the abnormal function of mTECs

	Defect in NF-κB activation pathway: NIK mutation	Defect in Aire expression: Aire mutation
Defective development of mTECs	$mTEC^{high}$ population is missing	Increased $mTEC^{high}$ population; Hassall's bodies are missing
TRA expression	Impaired expression of both Aire-dependent and Aire-independent TRAs	Impaired expression of Aire-dependent TRAs
Negative selection	Impaired	Impaired
Production of Tregs	Reduced, but functionally competent	Reduced at neonatal stage (Yang et al. 2015)
Mechanisms underlying the breakdown of self-tolerance	Defective TRA expression due to defective development of mTECs	Defective TRA expression due to incomplete terminally differentiation of mTECs

There must be a group of genes that together control promiscuous gene expression and establishment of self-tolerance in the thymus through their unique actions, as exemplified by the NF-κB activation pathway and Aire as summarised in Table 37.1 (Gardner et al. 2009).

37.3 A Novel Animal Model for the Study of Chronic Inflammation Caused by Organ-Specific Autoimmunity

37.3.1 Paradoxical Development of Muscle-Specific Autoimmunity by the Additive Expression of Aire

Development of harmful autoimmunity is prevented by two distinct but functionally cooperative mechanisms: central tolerance established in the thymus and peripheral tolerance operational outside the thymus. In the thymus, autoreactive T cells are eliminated by the clonal deletion by the interaction with self-antigens presented by mTECs and/or thymic DCs (Klein et al. 2014). This process of negative selection, however, is not perfect (Yu et al. 2015), and there exist T cells that are capable of reacting with self-antigens in the periphery. The major function of Tregs is considered to maintain these autoreactive T cells in an aenergic state in periphery (Maeda et al. 2014). Thus, cooperation between central tolerance and peripheral tolerance is essential to generate and maintain a self-tolerant T-cell repertoire.

In addition to mTECs, Aire expression has been demonstrated in BM-derived nonconventional antigen-presenting cells (APCs) that show MHC^{high} expression as a distinct tolerogenic cell population in secondary lymphoid organs (Gardner et al. 2008, 2013). Although it is not clear whether conventional DCs express

Aire, impaired DC function as evidenced by altered cytokine expression profiles and DC maturation/activation has been implicated for the higher susceptibility to candida infection in patients with Aire mutation (i.e. APECED; Brannstrom et al. 2006; Ryan et al. 2008). Thus, it is possible that Aire exerts its tolerogenic function not only within the thymic microenvironment but also in the periphery (Metzger and Anderson 2011).

NOD is a well-studied animal model of type I diabetes, in which insulin is a primary autoantigen (Nakayama et al. 2005) although it is controversial whether NOD mice have a defect in central tolerance (Kishimoto and Sprent 2001; Mingueneau et al. 2012). Interestingly, insulin is one of the typical Aire-dependent TRAs in mTECs whose expression level is strongly affected by the loss of Aire in mice (Anderson et al. 2002). Because of the action of Aire for activating TRA expression in mTECs and in the peripheral lymphoid organs (Gardner et al. 2008), we decided to examine whether additive expression of Aire in APCs of both mTECs and peripheral APCs by the use of MHC class II (MHC-II) promoter activity would ameliorate the autoimmune pathology possibly through the augmentation of TRAs such as insulin in NOD. We have established several lines of transgenic animals that express human Aire in MHC-II$^+$ APCs both in the thymus and periphery. Quite unexpectedly, however, we found that mice expressing human Aire at high levels paradoxically developed muscle-specific autoimmunity resembling the human disease of polymyositis (Venalis and Lundberg 2014): mice harboring muscle-specific autoimmunity produced autoantibodies against myosin heavy chain (MyHC). Consistent with the autoimmune nature of mice, supplementing prednisolone in drinking water ameliorated the disease. Our results suggest that expression of Aire needs to be tightly controlled within appropriate levels to exert its authentic tolerogenic function. The results also emphasise the importance of coordinated action between central tolerance and peripheral tolerance controlled by Aire in common.

37.3.2 Pathogenesis of Polymyositis-Like Autoimmunity by the Additive Expression of Aire

Because Aire's primary action is considered to promote TRA expression, thereby contributing to the clonal deletion and/or Treg generation both in the thymus and periphery (Gardner et al. 2009), it was striking that additive expression of Aire in thymic and peripheral APCs resulted in the development of muscle-specific autoimmunity. In the thymus, it is possible that altered development of mTECs by the additive expression of Aire within mTEC^{high} was associated with defective expression of wide varieties of TRAs, although why the breakdown of self-tolerance was confined to muscle tissues remains unknown. In this regard, there are some other reports describing the spontaneous development of muscle-specific autoimmunity similar to our animal model. Two independently generated transgenic mice expressing HLA-DQ8 (DQA1*0301/DQB1*0302), a human MHC-II subtype

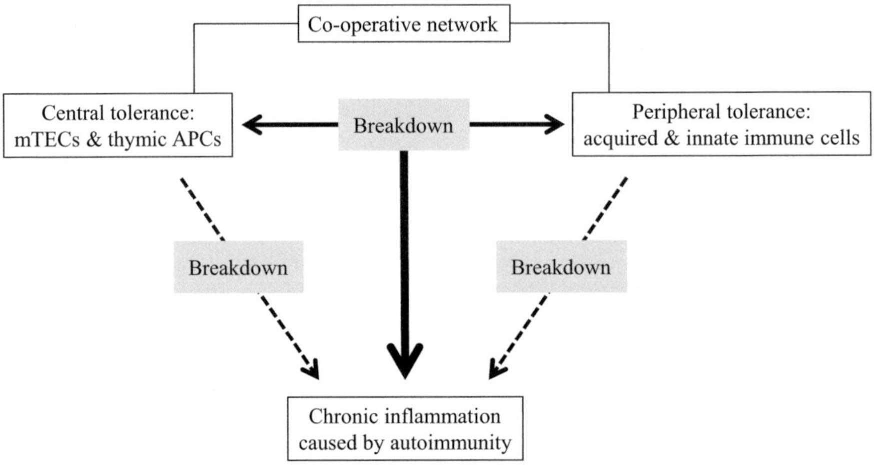

Fig. 37.1 Cooperative tolerogenic network and its breakdown leading to the development of chronic inflammation caused by autoimmunity. Breakdown of either central or peripheral tolerance has been shown to result in the development of autoimmune reaction. The polymyositis-like autoimmune model in our human Aire transgenic model has unveiled the existence of a cooperative immune network between central and peripheral tolerance. Elucidation of the specific markers that characterise chronic inflammation caused by the breakdown of this network is an important step toward overcoming organ-specific autoimmunity

associated with type 1 diabetes, simultaneously lacking endogenous mouse MHC (I-Ab) developed autoimmune myocarditis but not polymyositis, although autoantibodies produced in mice reacted with both types of MyHC, cardiac type (α-MyHC) and skeletal muscle type (β-MyHC), suggesting an epitope spreading (Elliott et al. 2003; Hayward et al. 2006; Taylor et al. 2004). Similar to our polymyositis-like mouse model, HLA-DQ8 transgenic mice developed the disease only on NOD background.

Our animal model has resulted in highlighting the cooperative role of central and peripheral tolerance in preventing autoimmunity through the action of Aire in common (Fig. 37.1). We assume that finding the specific markers expressed on immune cells and/or target tissues that define chronic inflammation, but not acute inflammation, must be an important task to control this type of organ-specific autoimmunity.

37.4 Concluding Remarks

Even the transcription factor necessary for establishing self-tolerance can become harmful once its expression level becomes too high as demonstrated in our novel model of muscle-specific autoimmunity. Thus, inappropriate Aire expression induced by some extrinsic factors such as infection and inflammatory stimuli might be feasible mechanisms for the development of autoimmunity and persistence of inflammation (chronic inflammation). With the advent of thymic organogenesis using thymic precursor cells (Parent et al. 2013; Sun et al. 2013), it may be feasible to manipulate the thymic microenvironment, thereby controlling the processes for the establishment of self-tolerance and ameliorating the undesired condition of chronic inflammation.

References

Anderson MS, Venanzi ES, Klein L, Chen Z, Berzins SP, Turley SJ, von Boehmer H, Bronson R, Dierich A, Benoist C, Mathis D (2002) Projection of an immunological self shadow within the thymus by the Aire protein. Science 298:1395–1401

Anderson MS, Venanzi ES, Chen Z, Berzins SP, Benoist C, Mathis D (2005) The cellular mechanism of Aire control of T cell tolerance. Immunity 23:227–239

Brannstrom J, Hassler S, Peltonen L, Herrmann B, Winqvist O (2006) Defect internalization and tyrosine kinase activation in Aire deficient antigen presenting cells exposed to Candida albicans antigens. Clin Immunol 121:265–273

Dooley J, Erickson M, Farr AG (2008) Alterations of the medullary epithelial compartment in the Aire-deficient thymus: implications for programs of thymic epithelial differentiation. J Immunol 181:5225–5232

Dresch C, Leverrier Y, Marvel J, Shortman K (2012) Development of antigen cross-presentation capacity in dendritic cells. Trends Immunol 33:381–388

Elewaut D, Shaikh RB, Hammond KJ, De Winter H, Leishman AJ, Sidobre S, Turovskaya O, Prigozy TI, Ma L, Banks TA, Lo D, Ware CF, Cheroutre H, Kronenberg M (2003) NIK-dependent RelB activation defines a unique signaling pathway for the development of V alpha 14*i* NKT cells. J Exp Med 197:1623–1633

Elliott JF, Liu J, Yuan ZN, Bautista-Lopez N, Wallbank SL, Suzuki K, Rayner D, Nation P, Robertson MA, Liu G, Kavanagh KM (2003) Autoimmune cardiomyopathy and heart block develop spontaneously in HLA-DQ8 transgenic IAβ knockout NOD mice. Proc Natl Acad Sci U S A 100:13447–13452

Gallegos AM, Bevan MJ (2004) Central tolerance to tissue-specific antigens mediated by direct and indirect antigen presentation. J Exp Med 200:1039–1049

Gardner JM, Devoss JJ, Friedman RS, Wong DJ, Tan YX, Zhou X, Johannes KP, Su MA, Chang HY, Krummel MF, Anderson MS (2008) Deletional tolerance mediated by extrathymic Aire-expressing cells. Science 321:843–847

Gardner JM, Fletcher AL, Anderson MS, Turley SJ (2009) AIRE in the thymus and beyond. Curr Opin Immunol 21:582–589

Gardner JM, Metzger TC, McMahon EJ, Au-Yeung BB, Krawisz AK, Lu W, Price JD, Johannes KP, Satpathy AT, Murphy KM, Tarbell KV, Weiss A, Anderson MS (2013) Extrathymic Aire-expressing cells are a distinct bone marrow-derived population that induce functional inactivation of CD4(+) T cells. Immunity 39:560–572

Gillard GO, Dooley J, Erickson M, Peltonen L, Farr AG (2007) Aire-dependent alterations in medullary thymic epithelium indicate a role for Aire in thymic epithelial differentiation. J Immunol 178:3007–3015

Gray D, Abramson J, Benoist C, Mathis D (2007) Proliferative arrest and rapid turnover of thymic epithelial cells expressing Aire. J Exp Med 204:2521–2528

Hayward SL, Bautista-Lopez N, Suzuki K, Atrazhev A, Dickie P, Elliott JF (2006) CD4 T cells play major effector role and CD8 T cells initiating role in spontaneous autoimmune

myocarditis of HLA-DQ8 transgenic IAb knockout nonobese diabetic mice. J Immunol 176:7715–7725
Hofmann J, Mair F, Greter M, Schmidt-Supprian M, Becher B (2011) NIK signaling in dendritic cells but not in T cells is required for the development of effector T cells and cell-mediated immune responses. J Exp Med 208:1917–1929
Kajiura F, Sun S, Nomura T, Izumi K, Ueno T, Bando Y, Kuroda N, Han H, Li Y, Matsushima A, Takahama Y, Sakaguchi S, Mitani T, Matsumoto M (2004) NF-κB-inducing kinase establishes self-tolerance in a thymic stroma-dependent manner. J Immunol 172:2067–2075
Kishimoto H, Sprent J (2001) A defect in central tolerance in NOD mice. Nat Immunol 2:1025–1031
Klein L, Kyewski B, Allen PM, Hogquist KA (2014) Positive and negative selection of the T cell repertoire: what thymocytes see (and don't see). Nat Rev Immunol 14:377–391
Koike R, Nishimura T, Yasumizu R, Tanaka H, Hataba Y, Hataba Y, Watanabe T, Miyawaki S, Miyasaka M (1996) The splenic marginal zone is absent in alymphoplastic *aly* mutant mice. Eur J Immunol 26:669–675
Kuroda N, Mitani T, Takeda N, Ishimaru N, Arakaki R, Hayashi Y, Bando Y, Izumi K, Takahashi T, Nomura T, Sakaguchi S, Ueno T, Takahama Y, Uchida D, Sun S, Kajiura F, Mouri Y, Han H, Matsushima A, Yamada G, Matsumoto M (2005) Development of autoimmunity against transcriptionally unrepressed target antigen in the thymus of Aire-deficient mice. J Immunol 174:1862–1870
Kyewski B, Klein L (2006) A central role for central tolerance. Annu Rev Immunol 24:571–606
Ling L, Cao Z, Goeddel DV (1998) NF-κB-inducing kinase activates IKK-α by phosphorylation of Ser-176. Proc Natl Acad Sci U S A 95:3792–3797
Liston A, Lesage S, Wilson J, Peltonen L, Goodnow CC (2003) Aire regulates negative selection of organ-specific T cells. Nat Immunol 4:350–354
Maeda Y, Nishikawa H, Sugiyama D, Ha D, Hamaguchi M, Saito T, Nishioka M, Wing JB, Adeegbe D, Katayama I, Sakaguchi S (2014) Detection of self-reactive CD8(+) T cells with an anergic phenotype in healthy individuals. Science 346:1536–1540
Malinin NL, Boldin MP, Kovalenko AV, Wallach D (1997) MAP3K-related kinase involved in NF-κB induction by TNF, CD95 and IL-1. Nature 385:540–544
Mathis D, Benoist C (2009) Aire. Annu Rev Immunol 27:287–312
Matsumoto M (2007) Transcriptional regulation in thymic epithelial cells for the establishment of self tolerance. Arch Immunol Ther Exp (Warsz) 55:27–34
Matsumoto M (2009) The role of autoimmune regulator (Aire) in the development of the immune system. Microbes Infect 11:928–934
Matsumoto M (2011) Contrasting models for the roles of Aire in the differentiation program of epithelial cells in the thymic medulla. Eur J Immunol 41:12–17
Matsumoto M, Fu YX, Molina H, Chaplin DD (1997) Lymphotoxin-α-deficient and TNF receptor-I-deficient mice define developmental and functional characteristics of germinal centers. Immunol Rev 156:137–144
Matsumoto M, Iwamasa K, Rennert PD, Yamada T, Suzuki R, Matsushima A, Okabe M, Fujita S, Yokoyama M (1999) Involvement of distinct cellular compartments in the abnormal lymphoid organogenesis in lymphotoxin-α-deficient mice and alymphoplasia (*aly*) mice defined by the chimeric analysis. J Immunol 163:1584–1591
Matsumoto M, Yamada T, Yoshinaga SK, Boone T, Horan T, Fujita S, Li Y, Mitani T (2002) Essential role of NF-κB-inducing kinase in T cell activation through the TCR/CD3 pathway. J Immunol 169:1151–1158
Matsumoto M, Nishikawa Y, Nishijima H, Morimoto J, Matsumoto M, Mouri Y (2013) Which model better fits the role of aire in the establishment of self-tolerance: the transcription model or the maturation model? Front Immunol 4:210

Matsushima A, Kaisho T, Rennert PD, Nakano H, Kurosawa K, Uchida D, Takeda K, Akira S, Matsumoto M (2001) Essential role of nuclear factor (NF)-κB-inducing kinase and inhibitor of κB (IκB) kinase alpha in NF-κB activation through lymphotoxin beta receptor, but not through tumor necrosis factor receptor I. J Exp Med 193:631–636

Metzger TC, Anderson MS (2011) Control of central and peripheral tolerance by Aire. Immunol Rev 241:89–103

Mingueneau M, Jiang W, Feuerer M, Mathis D, Benoist C (2012) Thymic negative selection is functional in NOD mice. J Exp Med 209:623–637

Miyawaki S, Nakamura Y, Suzuka H, Koba M, Yasumizu R, Ikehara S, Shibata Y (1994) A new mutation, *aly*, that induces a generalized lack of lymph nodes accompanied by immunodeficiency in mice. Eur J Immunol 24:429–434

Mouri Y, Nishijima H, Kawano H, Hirota F, Sakaguchi N, Morimoto J, Matsumoto M (2014) NF-κB-inducing kinase in thymic stroma establishes central tolerance by orchestrating cross-talk with not only thymocytes but also dendritic cells. J Immunol 193:4356–4367

Nakayama M, Abiru N, Moriyama H, Babaya N, Liu E, Miao D, Yu L, Wegmann DR, Hutton JC, Elliott JF, Eisenbarth GS (2005) Prime role for an insulin epitope in the development of type 1 diabetes in NOD mice. Nature 435:220–223

Nishikawa Y, Hirota F, Yano M, Kitajima H, Miyazaki J, Kawamoto H, Mouri Y, Matsumoto M (2010) Biphasic Aire expression in early embryos and in medullary thymic epithelial cells before end-stage terminal differentiation. J Exp Med 207:963–971

Parent AV, Russ HA, Khan IS, LaFlam TN, Metzger TC, Anderson MS, Hebrok M (2013) Generation of functional thymic epithelium from human embryonic stem cells that supports host T cell development. Cell Stem Cell 13:219–229

Ryan KR, Hong M, Arkwright PD, Gennery AR, Costigan C, Dominguez M, Denning D, McConnell V, Cant AJ, Abinun M, Spickett GP, Lilic D (2008) Impaired dendritic cell maturation and cytokine production in patients with chronic mucocutanous candidiasis with or without APECED. Clin Exp Immunol 154:406–414

Shinkura R, Matsuda F, Sakiyama T, Tsubata T, Hiai H, Paumen M, Miyawaki S, Honjo T (1996) Defects of somatic hypermutation and class switching in alymphoplasia (*aly*) mutant mice. Int Immunol 8:1067–1075

Shinkura R, Kitada K, Matsuda F, Tashiro K, Ikuta K, Suzuki M, Kogishi K, Serikawa T, Honjo T (1999) Alymphoplasia is caused by a point mutation in the mouse gene encoding Nf-κ b-inducing kinase. Nat Genet 22:74–77

Sun X, Xu J, Lu H, Liu W, Miao Z, Sui X, Liu H, Su L, Du W, He Q, Chen F, Shi Y, Deng H (2013) Directed differentiation of human embryonic stem cells into thymic epithelial progenitor-like cells reconstitutes the thymic microenvironment in vivo. Cell Stem Cell 13:230–236

Tamura C, Nakazawa M, Kasahara M, Hotta C, Yoshinari M, Sato F, Minami M (2006) Impaired function of dendritic cells in alymphoplasia (*aly/aly*) mice for expansion of $CD25^+CD4^+$ regulatory T cells. Autoimmunity 39:445–453

Taylor JA, Havari E, McInerney MF, Bronson R, Wucherpfennig KW, Lipes MA (2004) A spontaneous model for autoimmune myocarditis using the human MHC molecule HLA-DQ8. J Immunol 172:2651–2658

van Ewijk W, Shores EW, Singer A (1994) Crosstalk in the mouse thymus. Immunol Today 15:214–217

Venalis P, Lundberg IE (2014) Immune mechanisms in polymyositis and dermatomyositis and potential targets for therapy. Rheumatology 53:397–405

Yamada T, Mitani T, Yorita K, Uchida D, Matsushima A, Iwamasa K, Fujita S, Matsumoto M (2000) Abnormal immune function of hemopoietic cells from alymphoplasia (*aly*) mice, a natural strain with mutant NF-κB-inducing kinase. J Immunol 165:804–812

Yang S, Fujikado N, Kolodin D, Benoist C, Mathis D (2015) Immune tolerance. Regulatory T cells generated early in life play a distinct role in maintaining self-tolerance. Science 348:589–594

Yano M, Kuroda N, Han H, Meguro-Horike M, Nishikawa Y, Kiyonari H, Maemura K, Yanagawa Y, Obata K, Takahashi S, Ikawa T, Satoh R, Kawamoto H, Mouri Y, Matsumoto M (2008) Aire controls the differentiation program of thymic epithelial cells in the medulla for the establishment of self-tolerance. J Exp Med 205:2827–2838

Yu W, Jiang N, Ebert PJ, Kidd BA, Muller S, Lund PJ, Juang J, Adachi K, Tse T, Birnbaum ME, Newell EW, Wilson DM, Grotenbreg GM, Valitutti S, Quake SR, Davis MM (2015) Clonal deletion prunes but does not eliminate self-specific αβ CD8(+) T lymphocytes. Immunity 42:929–941

Chapter 38
Lysophosphatidylserine as an Inflammatory Mediator

Kumiko Makide, Asuka Inoue, and Junken Aoki

Abstract Lysophosphatidylserine (LysoPS), a deacylated form of phosphatidylserine (PS), has been assumed to serve as a bioactive lysophospholipid mediator. However, LysoPS has received little attention because its mode of actions as well as its synthetic pathways have been obscure. Recently G protein-coupled receptors (GPCRs) specific for LysoPS, and enzymes responsible for the generation of LysoPS have been identified. These LysoPS-related molecules are mainly expressed on immune cells, which led us to assume that LysoPS may have some roles in inflammation. Here we summarised the newly discovered LysoPS receptors and enzymes including our recent works.

Keywords Lysophospholipid mediator • GPCR • Lysophosphatidylserine • Phospholipase

38.1 Introduction

Lysophospholipid (LPL, 1-acyl-2-LPLs or 2-acyl-1-LPLs) is a deacylated form of phospholipids with a single fatty acid chain. Various LPLs, including lysophosphatidylcholine (LPC), lysophosphatidic acid (LPA), lysophosphatidylethanolamine (LPE), lysophosphatidylserine (LysoPS), lysophosphatidylinositol (LPI), lysophosphatidylglycerol (LPG), sphingosine-1-phosphate (S1P), and sphingosylphosphorylcholine (SPC) are present. They serve as precursors for diacyl phospholipids, and at least some of them also have roles as lipid mediators. Among

K. Makide (✉) • A. Inoue
Graduate School of Pharmaceutical Sciences, Tohoku University, 6-3, Aoba, Aramaki, Aoba-ku, Sendai 980-8578, Japan

PRESTO, Japan Science and Technology Corporation, Kawaguchi 332-0012, Japan
e-mail: makide@m.tohoku.ac.jp

J. Aoki
Graduate School of Pharmaceutical Sciences, Tohoku University, 6-3, Aoba, Aramaki, Aoba-ku, Sendai 980-8578, Japan

AMED-CREST, Sendai 980-8578, Japan

M. Miyasaka, K. Takatsu (eds.), *Chronic Inflammation*,
DOI 10.1007/978-4-431-56068-5_38

them, LPA and S1P have been studied intensively on their receptors, producing enzymes, and catabolic enzymes (Mendelson et al. 2014; Yung et al. 2014). Studies using knockout mice and human hereditary diseases of receptors and synthetic enzymes of LPA and S1P have revealed their in vivo significance. Furthermore some of them were found to be drug targets and have received a lot of attention in terms of drug discovery. Notably FTY720 (Fingolimod), an $S1P_1$ functional antagonist, has been approved for treatment of multiple sclerosis (Groves et al. 2013).

On the other hand, lysophosphatidylserine (LysoPS) is known to induce various cellular responses when applied in vivo or in vitro, but has received little attention because its receptors, producing pathways, and its in vivo occurrence have remained unknown. Recently LysoPS receptors and producing enzymes have been identified, which enables us to examine the role of LysoPS in vivo using knockout mice. In addition, the new methods using LC-MS/MS can sensitively detect LysoPS in vivo. In this review we summarise the recent advance in the study of LysoPS in terms of receptors, synthetic enzymes, and detection, with particular focus on the relevance of LysoPS to both physiological and pathological states.

38.2 LysoPS Receptors

Lysophosphatidylserine is known to induce several cellular responses both in vitro and in vivo. The most characterised response has been the stimulatory response of mast cell degranulation. In vitro, LysoPS enhances histamine release from peritoneal rodent mast cells triggered by the crosslinking of high-affinity IgE receptors (FcεRI) (Martin and Lagunoff 1979). It also induces rapid degranulation of mast cells and a consequent anaphylactic shock and hypothermia when administered i.v. in rodents. The mast cell degranulation-stimulating activity is not induced by other LPLs and strictly requires serine residue of LysoPS. This suggests the presence of a receptor for LysoPS on mast cells, although it remains unidentified.

In the course of a ligand fishing study of orphan GPCRs, Sugo et al. found that LysoPS is a ligand for GPR34, which is thought to be a member of P2Y family (Sugo et al. 2006). The P2Y family includes receptors for nucleotide, UDP-glucose, LPA, and the orphan GPCRs. They showed that LysoPS caused a dose-dependent inhibition of forskolin-stimulated cAMP accumulation in human GPR34-expressing Chinese hamster ovary (CHO) cells. They also showed that LysoPS induced phosphorylation of ERK in GPR34-expressing CHO cells. These responses were completely abolished by treatment of pertussis toxin (PTX), which indicates that GPR34 couples to a $G_{i/o}$-type G-protein. Later, Liebscher et al. and Makide et al. demonstrated that GPR34 are conserved in a wide range of vertebrates from fish to mammals by showing that GPR34 from carp and zebrafish react with LysoPS (Liebscher et al. 2011; Makide et al. 2014). Kitamura et al. used two assays for measuring GPR34 activation, a Ca^{2+} mobilisation assay and a newly developed transforming growth factor-α (TGFα) shedding assay, and confirmed that mammalian GPR34s from human, rat, and mouse origins reacted specifically with LysoPS but not with other LPLs (Kitamura et al. 2012). Notably, GPR34 reacted strongly with LysoPS species with an unsaturated fatty acid at the *sn*-2 position, showing that *sn*-2 unsaturated LysoPS is a preferable ligand for GPR34. GPR34 mRNA is expressed in many tissues but is most highly expressed in mast cells. Thus, it was once proposed that LysoPS enhances mast cell degranulation through GPR34. However, peritoneal mast cells from GPR34-deficient mice still responded to LysoPS, which suggested that GPR34 is not involved in the mast cell degranulation response induced by LysoPS. Until now the LysoPS receptor on mast cells has remained unknown.

Frasch et al. showed that LysoPS was generated in neutrophils by an oxidation-dependent mechanism and that is served as an endogenous anti-inflammatory mediator by stimulating the clearance of recruited neutrophils by macrophages, contributing to the resolution of inflammation (Frasch et al. 2008; Frasch and Bratton 2012). In addition, their results suggest that G2A (GPR132) on macrophages is responsible for the clearance of neutrophils by macrophages, raising the possibility that LysoPS is an endogenous ligand for G2A. G2A was once proposed as a receptor for LPC but the proposal was later retracted. Now many reports have confirmed that G2A is a receptor for protons (Murakami et al. 2004) and for 9-hydroxyoctadecadienoic acid (9-HODE), a kind of oxidised fatty acid (linoleic acid) (Obinata et al. 2005). We observed that G2A responded to 9-HODE, but not to LysoPS using a TGFα shedding assay (unpublished data). Thus, it is currently not clear if G2A directly recognises LysoPS.

Recently, we identified three additional GPCRs (P2Y10, A630033H20, and GPR174) as novel LysoPS receptors through a ligand screening of orphan GPCRs using a TGFα shedding assay (Inoue et al. 2012). These receptors are members of the P2Y family, and are located in tandem in a clustered locus on the X chromosome. We propose that, according to the nomenclature of lysophospholipid receptors, GPR34, P2Y10, A630033H20, and GPR174 be designated as LPS_1, LPS_2, LPS_2-like (LPS_{2L}), and LPS_3, respectively (Fig. 38.1).

P2Y10 is coupled with $G\alpha_{12/13}$ but not with other G proteins. Expression of P2Y10 is restricted to lymphoid organs such as spleen, thymus, and lymph node. The expression of P2Y10 is dependent on PU.1 and Spi-B, which are closely related Ets transcription factors (Rao et al. 1999). In *PU.1*$^{+/-}$Spi-B$^{-/-}$ mice, the expression of P2Y10 is dramatically reduced. These Ets transcription factors have a role in the signal transduction of B-cell receptors (BCRs), which suggests that P2Y10 has a role in regulating BCR signalling. In addition, $G_{12/13}$ signalling in B cells has a role in regulating the marginal zone B cells and germinal center B cells (Muppidi et al. 2014; Rieken et al. 2006), which raises the possibility that LysoPS functions in B cells through P2Y10.

Although human A630033H20 is truncated in the open reading frame and becomes a pseudogene, mouse and rat A630033H20 function as LysoPS receptors. A630033H20 is the closest homologue of P2Y10, with a 75 % homology to P2Y10 at the amino acid level, coupled with $G_{12/13}$, and its expression pattern (it is highly expressed in lymphoid tissue) is similar to that of P2Y10.

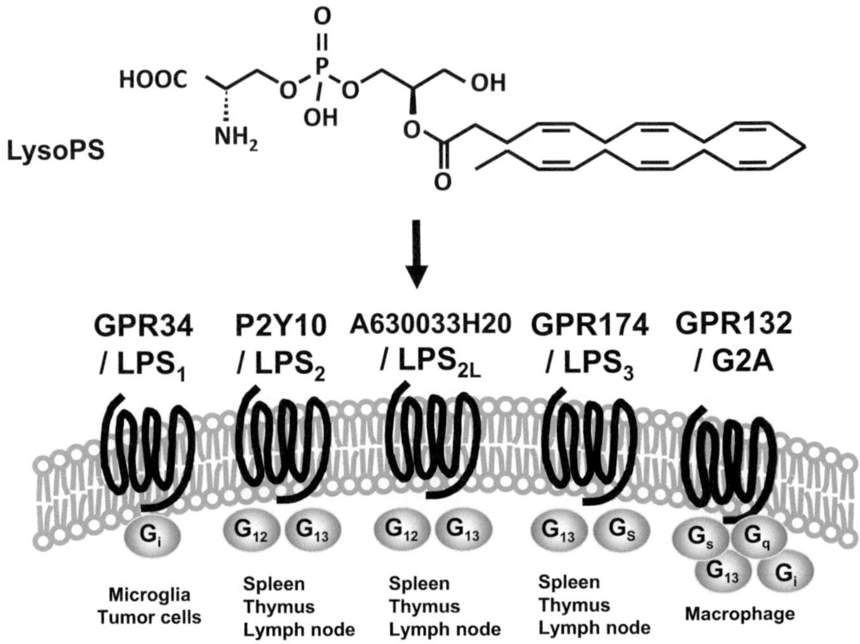

Fig. 38.1 LysoPS receptors

GPR174 shows the highest homology to P2Y10 and A630033H20 with ~50 % identity at the amino acid level. Like P2Y10 and A630033H20, GPR174 is activated by LysoPS. The expression pattern of GPR174 is similar to the expression pattern of P2Y10 and A630033H20 with high expression in lymphoid tissues. However, GPR174 is also strongly expressed in some melanoma cells (Qin et al. 2011). GPR174 mainly coupled with both $G\alpha_s$ and $G\alpha_{13}$. Given that $G\alpha_{13}$ signalling is induced by the three LysoPS receptors (P2Y10, A630033H20, and GPR174), which have similar expression patterns, it is likely that these three LysoPS receptors share redundant functions in activating the $G\alpha_{13}$ pathway. On the other hand, inasmuch as only one LysoPS receptor (GPR174) is coupled with $G\alpha_s$, GPR174 may have a unique role in regulating $G\alpha_s$ signalling. Recent genome-wide association studies show that single nucleotide polymorphisms (SNPs) of GPR174 are associated with the risk for Graves' disease (Chu et al. 2013; Szymanski et al. 2014) or Addison's disease (Napier et al. 2015), autoimmune diseases. The expression of GPR174 is elevated in blood cells in vasovagal syncope patients (Huang et al. 2015). Thus, LysoPS may serve as an immunomodulator through GPR174.

Although P2Y10 was reported to be a receptor for LPA and S1P (Murakami et al. 2008), we and other researchers were unable to confirm it. In recent studies, we throughly examined the ligand specificity of cloned LysoPS receptors, that is, LPS_1/GPR34, LPS_2/P2Y10, and LPS_3/GPR174, using chemically synthesised

LysoPS analogues (Ikubo et al. 2015; Iwashita et al. 2009; Kitamura et al. 2012; Uwamizu et al. 2015). These studies revealed that modifications of the serine residue resulted in the loss of agonistic activity for all three LysoPS receptors, demonstrating that LysoPS receptors strictly recognise the head group of LysoPS, that is, serine. They also revealed that some modifications conferred certain selectivity toward each LysoPS receptor to the LysoPS analogues. For example, deoxy LysoPS, in which an *sn*-2 hydroxy group (−OH) was removed, was found to be selective for LPS_2/P2Y10 and lysophosphatidylallothreonine, in which a methyl group ($-CH_3$) was introduced in the serine residue, was found to be selective for LPS_3/GPR174. These ligand specificities suggest that LysoPS is the real ligand for the LysoPS receptors. This is supported by the fact that all four LysoPS receptors belong to the P2Y family, to which the LPA receptors (LPA_{4-6}) belong.

38.3 Generation of LysoPS

The main pathway of LysoPS production is probably the hydrolysis of PS by phospholipase (Figure 38.2). We recently established a method to detect LysoPS with high sensitivity (Okudaira et al. 2014). In this method, acyl migration reaction, in which an acyl chain of 2-acyl-1-LPLs is quickly moved to the *sn*-1 position, generating 1-acyl-2-LPLs, was completely suppressed by lowering the pH. With this method, we detected two types of LysoPS (1-acyl-2-LysoPS and 2-acyl-1-LysoPS) in various tissues and cells. The results, together with previous knowledge, suggest that both phospholipase A_1 (PLA_1) and A_2 (PLA_2) are involved in the production of LysoPS. In fact, rat platelets express two extracellular PLA enzymes, secretory PLA_2 group IIA ($sPLA_2$IIA) and PS-specific PLA_1 (PS-PLA_1), and secrete them upon activation. In the course of activation of rat platelets, $sPLA_2$-IIA and PS-PLA_1 produce 1-acyl LysoPS and 2-acyl LysoPS, respectively.

PS-PLA_1 is stored in α granules of rat platelets and is secreted into the medium when activated. Although PS-PLA_1 is structurally homologous to triglyceride (TG) lipase, it selectively hydrolyses PS and doesn't have lipase activity for TG. PS-PLA_1 expression is dramatically induced at the mRNA and protein levels by various inflammatory stimuli. Under some inflammatory conditions, we detected an increase of LysoPS in parallel with the induction of PS-PLA_1 and found that the increase was partially abolished in PS-PLA_1 knockout mice.

Because secretory PLA_2 and PS-PLA_1 are secreted proteins, they should act on PS extracellularly. On the contrary, PS localises exclusively to the inner leaflet of the plasma membrane. Recently, TMEM16F and Xkr8 were identified as scramblases that trigger exposure of PS in activated platelets and apoptotic cells, respectively (Suzuki et al. 2010; 2013). It is likely that secretory PLAs such as PS-PLA_1 and $sPLA_2$ deacylate PS exposed by TMEM16F or Xkr8 during platelet activation or apoptosis.

Recently ABHD16A was identified as a PS lipase that generates LysoPS (Kamat et al. 2015). ABHD16A, also known as lymphocyte antigen B-associated transcript

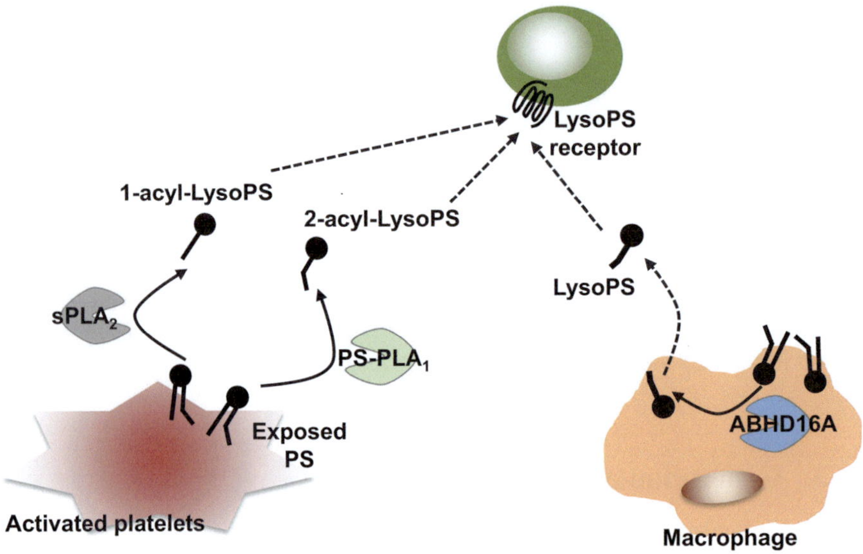

Fig. 38.2 Producing pathways of LysoPS

5 (BAT5), is a member of the alpha beta hydrolase domain (ABHD) enzyme family and is predicted to be a multipass membrane protein. In mice, ABHD16A mRNA is abundantly and ubiquitously expressed with highest expression in skeletal muscle and brain. The number of newborn ABHD16A$^{-/-}$ mice was much less than the value expected from Mendelian ratio. The body weight of ABHD16A$^{-/-}$ mice is smaller than wild-type mice, however, their behavior and survival rate appeared to be normal. The amount of various species of LysoPS in ABHD16A$^{-/-}$ brain were reduced, which suggests that ABHD16A does not discriminate between saturated fatty acids (mainly in the *sn*-1 position) and polyunsaturated fatty acids (mainly in the *sn*-2 position). Thioglycollate-elicited peritoneal macrophages derived from ABHD16A$^{-/-}$ mice have less LysoPS and release fewer inflammatory cytokines following stimulation with lipopolysaccharide. It is not known whether ABHD16A produces LysoPS intracellularly or extracellularly. If LysoPS is produced intracellularly, LysoPS must be released through some transporter on the plasma membrane such as Spns2 for S1P.

The degradation pathways of LysoPS are also important, because degradation is involved in the termination of LysoPS signaling. ABHD12 is reported to have lysoPS lipase activity in the mammalian brain (Blankman et al. 2013). In addition to having elevated brain LysoPS and microglial activation, ABHD12$^{-/-}$ mice show the phenotype of the human neurogenerative disorder PHARC (polyneuropathy, hearing loss, ataxia, retinosis pigmentosa, and cataract). Taken together, these results suggest that the interplay of ABHD12 and ABHD16A regulates the LysoPS level in neuronal diseases.

38.4 Conclusion

The LysoPS field has advanced with an expanding repertoire of receptors and metabolic enzymes. Recent studies strongly suggest that LysoPS plays important roles in processes related to inflammation. Further studies are needed to identify LysoPS-generating cells and enzymes, and LysoPS signaling pathways. The results of these studies will help to develop drugs that target LysoPS enzymes and receptors.

References

Blankman JL, Long JZ, Trauger SA, Siuzdak G, Cravatt BF (2013) ABHD12 controls brain lysophosphatidylserine pathways that are deregulated in a murine model of the neurodegenerative disease PHARC. Proc Natl Acad Sci U S A 110(4):1500–1505. doi:10.1073/pnas.1217121110

Chu X, Shen M, Xie F, Miao XJ, Shou WH, Liu L, Yang PP, Bai YN, Zhang KY, Yang L, Hua Q, Liu WD, Dong Y, Wang HF, Shi JX, Wang Y, Song HD, Chen SJ, Chen Z, Huang W (2013) An X chromosome-wide association analysis identifies variants in GPR174 as a risk factor for Graves' disease. J Med Genet 50(7):479–485. doi:10.1136/jmedgenet-2013-101595

Frasch SC, Bratton DL (2012) Emerging roles for lysophosphatidylserine in resolution of inflammation. Prog Lipid Res 51(3):199–207. doi:10.1016/j.plipres.2012.03.001

Frasch SC, Berry KZ, Fernandez-Boyanapalli R, Jin HS, Leslie C, Henson PM, Murphy RC, Bratton DL (2008) NADPH oxidase-dependent generation of lysophosphatidylserine enhances clearance of activated and dying neutrophils via G2A. J Biol Chem 283(48):33736–33749. doi:10.1074/jbc.M807047200

Groves A, Kihara Y, Chun J (2013) Fingolimod: direct CNS effects of sphingosine 1-phosphate (S1P) receptor modulation and implications in multiple sclerosis therapy. J Neurol Sci 328(1–2):9–18. doi:10.1016/j.jns.2013.02.011

Huang YJ, Zhou ZW, Xu M, Ma QW, Yan JB, Wang JY, Zhang QQ, Huang M, Bao L (2015) Alteration of gene expression profiling including GPR174 and GNG2 is associated with vasovagal syncope. Pediatr Cardiol 36(3):475–480. doi:10.1007/s00246-014-1036-x

Ikubo M, Inoue A, Nakamura S, Jung S, Sayama M, Otani Y, Uwamizu A, Suzuki K, Kishi T, Shuto A, Ishiguro J, Okudaira M, Kano K, Makide K, Aoki J, Ohwada T (2015) Structure-activity relationships of lysophosphatidylserine analogs as agonists of G-protein-coupled receptors GPR34, P2Y10, and GPR174. J Med Chem 58(10):4204–4219. doi:10.1021/jm5020082

Inoue A, Ishiguro J, Kitamura H, Arima N, Okutani M, Shuto A, Higashiyama S, Ohwada T, Arai H, Makide K, Aoki J (2012) TGFalpha shedding assay: an accurate and versatile method for detecting GPCR activation. Nat Methods 9(10):1021–1029. doi:10.1038/nmeth.2172

Iwashita M, Makide K, Nonomura T, Misumi Y, Otani Y, Ishida M, Taguchi R, Tsujimoto M, Aoki J, Arai H, Ohwada T (2009) Synthesis and evaluation of lysophosphatidylserine analogues as inducers of mast cell degranulation. Potent activities of lysophosphatidylthreonine and its 2-deoxy derivative. J Med Chem 52(19):5837–5863. doi:10.1021/jm900598m

Kamat SS, Camara K, Parsons WH, Chen DH, Dix MM, Bird TD, Howell AR, Cravatt BF (2015) Immunomodulatory lysophosphatidylserines are regulated by ABHD16A and ABHD12 interplay. Nat Chem Biol 11(2):164–171. doi:10.1038/nchembio.1721

Kitamura H, Makide K, Shuto A, Ikubo M, Inoue A, Suzuki K, Sato Y, Nakamura S, Otani Y, Ohwada T, Aoki J (2012) GPR34 is a receptor for lysophosphatidylserine with a fatty acid at the sn-2 position. J Biochem 151(5):511–518. doi:10.1093/jb/mvs011

Liebscher I, Muller U, Teupser D, Engemaier E, Engel KM, Ritscher L, Thor D, Sangkuhl K, Ricken A, Wurm A, Piehler D, Schmutzler S, Fuhrmann H, Albert FW, Reichenbach A, Thiery J, Schoneberg T, Schulz A (2011) Altered immune response in mice deficient for the G protein-coupled receptor GPR34. J Biol Chem 286(3):2101–2110. doi:10.1074/jbc.M110.196659

Makide K, Uwamizu A, Shinjo Y, Ishiguro J, Okutani M, Inoue A, Aoki J (2014) Novel lysophosphoplipid receptors: their structure and function. J Lipid Res 55(10):1986–1995. doi:10.1194/jlr.R046920

Martin TW, Lagunoff D (1979) Interactions of lysophospholipids and mast cells. Nature 279 (5710):250–252

Mendelson K, Evans T, Hla T (2014) Sphingosine 1-phosphate signalling. Development 141 (1):5–9. doi:10.1242/dev.094805

Muppidi JR, Schmitz R, Green JA, Xiao W, Larsen AB, Braun SE, An J, Xu Y, Rosenwald A, Ott G, Gascoyne RD, Rimsza LM, Campo E, Jaffe ES, Delabie J, Smeland EB, Braziel RM, Tubbs RR, Cook JR, Weisenburger DD, Chan WC, Vaidehi N, Staudt LM, Cyster JG (2014) Loss of signalling via Galpha13 in germinal centre B-cell-derived lymphoma. Nature 516 (7530):254–258. doi:10.1038/nature13765

Murakami N, Yokomizo T, Okuno T, Shimizu T (2004) G2A is a proton-sensing G-protein-coupled receptor antagonized by lysophosphatidylcholine. J Biol Chem 279(41):42484–42491. doi:10.1074/jbc.M406561200

Murakami M, Shiraishi A, Tabata K, Fujita N (2008) Identification of the orphan GPCR, P2Y (10) receptor as the sphingosine-1-phosphate and lysophosphatidic acid receptor. Biochem Biophys Res Commun 371(4):707–712. doi:10.1016/j.bbrc.2008.04.145

Napier C, Mitchell AL, Gan E, Wilson I, Pearce SH (2015) Role of the X-linked gene GPR174 in autoimmune Addison's disease. J Clin Endocrinol Metab 100(1):E187–E190. doi:10.1210/jc.2014-2694

Obinata H, Hattori T, Nakane S, Tatei K, Izumi T (2005) Identification of 9-hydroxyoctadecadienoic acid and other oxidized free fatty acids as ligands of the G protein-coupled receptor G2A. J Biol Chem 280(49):40676–40683. doi:10.1074/jbc.M507787200

Okudaira M, Inoue A, Shuto A, Nakanaga K, Kano K, Makide K, Saigusa D, Tomioka Y, Aoki J (2014) Separation and quantification of 2-acyl-1-lysophospholipids and 1-acyl-2-lysophospholipids in biological samples by LC-MS/MS. J Lipid Res 55(10):2178–2192. doi:10.1194/jlr.D048439

Qin Y, Verdegaal EM, Siderius M, Bebelman JP, Smit MJ, Leurs R, Willemze R, Tensen CP, Osanto S (2011) Quantitative expression profiling of G-protein-coupled receptors (GPCRs) in metastatic melanoma: the constitutively active orphan GPCR GPR18 as novel drug target. Pigment Cell Melanoma Res 24(1):207–218. doi:10.1111/j.1755-148X.2010.00781.x

Rao S, Garrett-Sinha LA, Yoon J, Simon MC (1999) The Ets factors PU.1 and Spi-B regulate the transcription in vivo of P2Y10, a lymphoid restricted heptahelical receptor. J Biol Chem 274 (48):34245–34252

Rieken S, Sassmann A, Herroeder S, Wallenwein B, Moers A, Offermanns S, Wettschureck N (2006) G12/G13 family G proteins regulate marginal zone B cell maturation, migration, and polarization. J Immunol 177(5):2985–2993

Sugo T, Tachimoto H, Chikatsu T, Murakami Y, Kikukawa Y, Sato S, Kikuchi K, Nagi T, Harada M, Ogi K, Ebisawa M, Mori M (2006) Identification of a lysophosphatidylserine receptor on mast cells. Biochem Biophys Res Commun 341(4):1078–1087

Suzuki J, Umeda M, Sims PJ, Nagata S (2010) Calcium-dependent phospholipid scrambling by TMEM16F. Nature 468(7325):834–838. doi:10.1038/nature09583

Suzuki J, Denning DP, Imanishi E, Horvitz HR, Nagata S (2013) Xk-related protein 8 and CED-8 promote phosphatidylserine exposure in apoptotic cells. Science 341(6144):403–406. doi:10.1126/science.1236758

Szymanski K, Miskiewicz P, Pirko K, Jurecka-Lubieniecka B, Kula D, Hasse-Lazar K, Krajewski P, Bednarczuk T, Ploski R (2014) rs3827440, a nonsynonymous single nucleotide polymorphism within GPR174 gene in X chromosome, is associated with Graves' disease in Polish Caucasian population. Tissue Antigens 83(1):41–44. doi:10.1111/tan.12259

Uwamizu A, Inoue A, Suzuki K, Okudaira M, Shuto A, Shinjo Y, Ishiguro J, Makide K, Ikubo M, Nakamura S, Jung S, Sayama M, Otani Y, Ohwada T, Aoki J (2015) Lysophosphatidylserine analogues differentially activate three LysoPS receptors. J Biochem 157(3):151–160. doi:10.1093/jb/mvu060

Yung YC, Stoddard NC, Chun J (2014) LPA receptor signaling: pharmacology, physiology, and pathophysiology. J Lipid Res 55(7):1192–1214. doi:10.1194/jlr.R046458

Chapter 39
Aberrant Activation of RIG-I–Like Receptors and Autoimmune Diseases

Hiroki Kato and Takashi Fujita

Abstract RIG-I–like receptors (RLRs) are known as viral RNA sensors that trigger the antiviral interferon (IFN) response by the recognition of the nonself signatures in viral RNAs, including 5' triphosphate structure and double-strand. Self-RNAs generally escape their recognition by several modifications such as 5' cap; however, it has recently been shown that endogenous RNAs without an adenosine deaminase ADAR1-dependent modification aberrantly activate RLR signalling and lead to severe IFN signature, resulting in autoimmune disorders. Also gain-of-function mutations of RLRs have been found in patients with autoimmune diseases. We herein provide a recent overview of the RLR-mediated antiviral IFN response and autoimmunity and discuss how atypical activation of RLRs triggers autoimmune diseases.

Keywords RIG-I-like receptors (RLRs) • Interferon (IFN) • Virus • Autoimmune disease • SLE • AGS • SMS

39.1 Introduction

For the elimination of intracellularly invading viruses, host cells evoke antiviral interferon (IFN) responses by the recognition of viral nucleic acids. Three major sensors, membrane-bound Toll-like receptors (TLRs), cytoplasmic RNA helicases RIG-I–like receptors (RLRs), and cytosolic DNA sensors including cGAS have been identified as key sensors that recognise viral DNAs and RNAs for the type I and III IFN production (IFN-α, IFN-β, and IFN-λ)(Ablasser et al. 2014; Berke et al. 2012; Bursztejn et al. 2015). These IFNs secreted from virus-infected cells

H. Kato
Laboratory of Molecular Genetics, Institute for Virus Research, Kyoto University, Kyoto, Japan

PRESTO, Japan Science and Technology Agency, Kawaguchi 332-0012, Japan

T. Fujita (✉)
Laboratory of Molecular Genetics, Institute for Virus Research, Kyoto University, Kyoto, Japan
e-mail: tfujita@virus.kyoto-u.ac.jp

M. Miyasaka, K. Takatsu (eds.), *Chronic Inflammation*,
DOI 10.1007/978-4-431-56068-5_39

activate themselves and surrounding noninfected cells by binding with cell surface IFN receptors, which transduces signalling and promotes the expression of hundreds of antiviral IFN-stimulated genes (ISGs) including 2'-5'-oligoadenylate synthetases, RNase L, PKR, and IFITs. Generally the IFN production is tightly regulated: IFN production is undetectable in uninfected cells and is rapidly induced and augmented upon viral infections through a positive feedback regulation. However, this IFN response is programmed to be transient by the actions of multiple negative regulators. When this tight regulation gets out of order for genetic reasons, a defect in IFN production or its overproduction occurs, resulting in high susceptibility to viruses or autoimmune disorders, respectively. For instance, in the case of dysfunction of nucleases such as DNase I, DNase II, or TREX1, aberrant antiviral signalling caused by failure in the clearance of endogenous DNA has been shown to lead to overproduction of IFN and result in autoimmune disorders such as systemic lupus erythematosus (SLE) (Cai et al. 2014; Crampton et al. 2012; Crow and Rehwinkel 2009; Crow et al. 2006). Endogenous RNAs without an adenosine deaminase ADAR1-dependent modification have also been reported to activate RLRs abnormally and lead to severe IFN signature, consequently causing autoimmune disorders such as Aicardi–Goutières syndrome (AGS)(Cunninghame Graham et al. 2011; Funabiki et al. 2014). Moreover, it has recently been reported that mutations of RLRs leading to constitutive activation have been found in patients with type I interferonopathies and autoimmune diseases such as SLE, AGS, and Singleton–Merten Syndrome (SMS). In this review, we focused on the involvement of RLRs in different autoimmune disorders and their possible causative mechanisms.

39.2 RLR-Mediated Recognition of Viral RNA and Signalling

Cytoplasmic RNA helicases RIG-I–like receptors (RLRs) consist of three family molecules: retinoic-acid inducible gene-I (RIG-I), melanoma differentiation-associated gene 5 (MDA5), and laboratory of genetics and physiology 2 (LGP2; Gack et al. 2007). RLRs play a critical role in triggering antiviral responses, including the production of type I interferon (IFN) and inflammatory cytokines including IL-6, IL-12, and TNFα, to eliminate invading viruses from the host. Domain and structural analyses of these helicases revealed that RLRs were composed of three structural domains: the caspase activation and recruitment domain (CARD) at the N-terminus for signal transduction, central DExD/H box RNA helicase domain with RNA-dependent ATPase activity, and C-terminal domain (CTD; Fig. 39.1). Both the helicase domain and CTD of RLRs are cooperatively involved in the recognition of viral RNA. Although LGP2 lacks CARD (Fig. 39.1), a loss-of-function analysis revealed that it acts as a positive regulator for both RIG-I- and MDA5-mediated signalling (Gao et al. 2015). The determination of

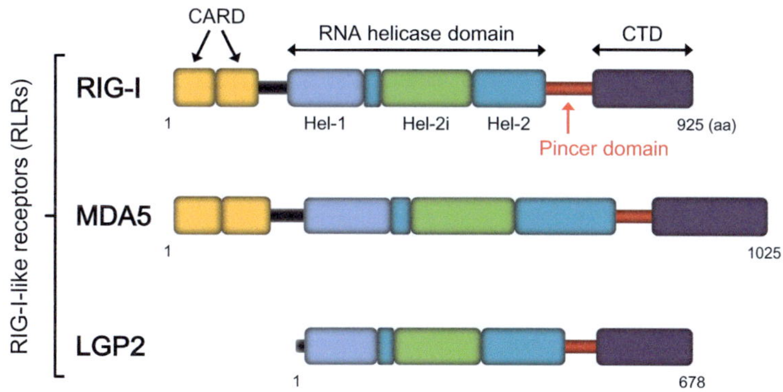

Fig. 39.1 Domain structure of RLRs. RIG-I and MDA5, respectively encoded by DDX58 and IFIH1, are composed of three main structural domains: the tandem caspase activation and recruitment domains (CARDs) at the N-terminus for signal transduction, central DExD/H box RNA helicase domain (Hel-1, Hel-2i, and Hel-2) with RNA-dependent ATP hydrolysis activities, and C-terminal domain (CTD) for RNA binding. LGP2 consists of a helicase domain and CTD. The recently identified Pincer domain (*red bar*) is composed of two alpha helices physically tethering the CTD to the helicase core

crystal structures revealed that the linker region (also known as the pincer) of RIG-I, which is located between the helicase domain and CTD (Fig. 39.1), was critical for maintaining RIG-I in a repressed conformation, in which CARD is masked by an interaction with the helicase domain (Gray et al. 2015; Hornung et al. 2006). Mutagenesis of the linker conferred the constitutive activity of RIG-I and loss of responsiveness to viral RNA (Hou et al. 2011).

Several steps have been proposed for the activation of RLRs, leading to antiviral signal transduction. The first step involves a physical association between viral RNA and RLRs (Fig. 39.2). Initial binding studies revealed that RLRs bound to double-stranded RNA and binding affinities to single-stranded RNA and dsDNA were very low (Gack et al. 2007; Jang et al. 2015). RIG-I was previously reported to sense the 5'-ppp structure of the primary viral transcript (Kageyama et al. 2011; Kato et al. 2006); however, this finding overlooked the possibility of the in vitro transcript having a copy back structure; partial transcription using product RNA as a template occurs after template transcription, producing a partial dsRNA structure. Subsequent studies revealed that the dsRNA structure was a prerequisite and the 5'-ppp structure enhanced signalling by RIG-I (Kato et al. 2008). Influenza A virus and Sendai virus produce panhandle and defective interfering RNA (with copy back), respectively, and these activate RIG-I. Although several early studies showed that RIG-I specifically bound to the 5'-ppp structure, the involvement of a copy back dsRNA structure was not considered. Therefore, there is no evidence to suggest that RIG-I solely recognises 5'-ppp in the absence of a dsRNA structure. A previous study clearly demonstrated that RIG-I and MDA5 recognised short and long dsRNA molecules and also that the virus specificities of RIG-I and MDA5

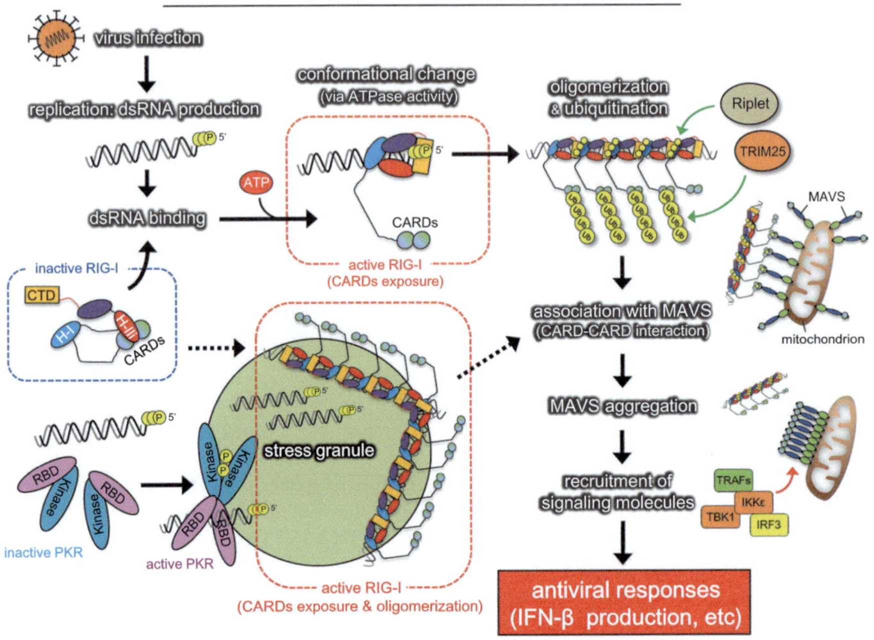

Fig. 39.2 Activation steps of RIG-I in the cytoplasm. The first step involves a physical association between viral RNA and RIG-I. RIG-I undergoes a prominent conformational change: the CTD binds tightly to 5'ppp or the blunt ends of 5' OH dsRNA and the helicase domains (Hel I, Hel II, Hel IIi) wrap around dsRNA, leading to a more compact composition. These conformational alterations via ATPase activity in RLRs result in the exposure of CARD and lead an oligomerisation of RIG-I. K63-linked polyubiquitination by TRIM25 and Riplet promotes and stabilises the oligomerisation. The exposure of oligomerised CARDs leads to a signal relay to another CARD-containing protein, MAVS via CARD–CARD interactions. The aggregate formation of MAVS recruits signalling molecules and finally induces the production of type I IFNs. In addition, viral dsRNA also activates PKR, leading to the phosphorylation of eIF2α, which ultimately induces the formation of stress granules (SGs) containing SG components, RLRs, and viral RNAs, to promote the activation of RLR

could be roughly correlated with the sizes of dsRNA produced by the respective viruses (Kawai et al. 2005; Kawane et al. 2006).

The next step is conformational alterations in RLRs to expose CARD (Fig. 39.2). The ATP hydrolysis activities (ATPase) of RIG-I and MDA5 may be involved in these conformational changes because a mutation at the ATP binding site (Walker's A motif) was shown to inactivate these sensors. Although ATPase activity was required for dsRNA unwinding (helicase activity), the helicase-resistant substrate (dsRNA with a 5'-overhang), but not the helicase-sensitive substrate (dsRNA with a 3'-overhang) induced signalling, which suggested that helicase activity may be irrelevant for antiviral signalling (Kowalinski et al. 2011). Once the CARD of RIG-I or MDA5 is exposed, the signal is relayed to another CARD-containing protein, MAVS (also termed IPS-1, VISA, or Cardif) via CARD–CARD interactions (Liddicoat et al. 2015; Marques et al. 2006; Meylan et al. 2005; Napirei et al. 2000) (Fig. 39.2). Recent in vitro reconstitution studies revealed that MDA5 formed a filament-like complex (also RIG-I-oligomer formation) on its ligand dsRNA and also that these MDA5 filaments promoted the prion-like aggregation of MAVS (Fig. 39.2) (Narita et al. 2014; Nejentsev et al. 2009; Ng et al. 2013; Oda et al. 2014). However, ATP hydrolysis induced filament disassembly, suggesting that these filaments may be unstable under physiological conditions. The importance of ubiquitination by TRIM25 and Riplet in RLR-oligomelisation and signaling has been reported (Onoguchi et al. 2010; Onomoto et al. 2012). Mitochondrial fusion and fission are known to be essential for the aggregation of MAVS (Oshiumi et al. 2010). In addition to mitochondria, several groups reported that virus-induced stress granules (SGs) functioned as critical loci for the activation of RLR. In several viral infections, viral dsRNA activates PKR and induces the formation of SGs, including SG components, RLRs, and viral RNAs, to promote the activation of RLR (Fig. 39.2) (Peisley et al. 2013; 2015; Pichlmair et al. 2006; Rice et al. 2014). Signalling molecules including ubiquitin ligases (TRAFs) and kinase complexes (TBK1/IKKe and IKKa/b/g) are recruited after the formation of MAVS aggregates and eventually activate the transcription factors IRF3/7 and NF-kB (Robinson et al. 2011), leading to the production of type I IFNs and inflammatory cytokines, such as IL-6.

39.3 The Relationship Between Aberrant RLR Signalling and Autoimmune Diseases

Recently accumulating evidence has indicated the strong correlation between autoimmune diseases and RLRs especially MDA5. Previous GWAS revealed the association of SNPs of the *IFIH1* gene, encoding MDA5, with the risk of many kinds of autoimmune diseases including Type 1 Diabetes (T1D), Multiple Sclerosis (MS), psoriasis, selective IgA deficiency, dilated cardiomyopathy, and SLE. For instance, Nejentsev et al. reported that two canonical splice site variants in *IFIH1*, a nonsense variant and missense substitution, were protective against T1D (Rutsch et al. 2015). These loss-of-function mutations in *IFIH1* may attenuate innate immune responses against viruses especially coxsackievirus B4, which has been strongly implicated in the triggering of T1D. However, the contribution of *IFIH1* with these SNPs to the pathogenesis of T1D and other autoimmune diseases remains unclear. It is also important to clarify the involvement of RNA virus infection in the development of these diseases.

In addition to SNPs, mutations in RLRs leading to their aberrant activation have been found in patients with autoimmune diseases. We summarise each case of three autoimmune diseases, SLE, AGS, and SMS below and discuss how atypical activation of RLRs triggers these autoimmune diseases.

39.3.1 Systemic Lupus Erythematosus (SLE)

Systemic lupus erythematosus (SLE) is the prevalent systemic autoimmune disease which has a large spectrum of clinical presentations. SLE can affect multiple organs, including kidneys, skin, joints, lungs and nervous system. Especially nephritis has been reported in severe cases and called lupus-nephritis.

Elevated levels of type I IFN, termed the IFN signature, have been reported in patients with SLE and the central role of type I IFN in disease pathogenesis has been suggested. SNPs in several molecules including nucleases such as DNase I as well as TLRs have been strongly correlated with the onset of SLE.

A previous study reported that human SNPs in *IFIH1* encoding MDA5 such as rs1990760 (A946T) were correlated with an increase in the susceptibility to SLE, suggesting that the atypical activation of RLR signalling may lead to SLE (Sato et al. 2000; Satoh et al. 2010). Fubabiki et al. recently reported that mice with the *Ifih1* missense mutation spontaneously developed lupus-like symptoms including nephritis and skin rash (Schlee et al. 2009). This is the first study to have directly demonstrated that mutations in RLRs directly led to an autoimmune disease. The upregulation of type I IFNs, IFN-inducible genes, and inflammatory cytokines including IL-6 and TNF-a was detected in multiple organs in this mutant mouse, reflecting the ubiquitous expression of MDA5. This missense mutation enhanced the basal activation level of IFN by MDA5, but abrogated responsiveness to viral infection as well as the ATPase activity induced by dsRNA. More important, the autoimmune phenotype was not observed in the mouse background of *Mavs−/−*. These findings suggested that the mutant MDA5 may confer constitutive activity rather than being hypersensitive to endogenous or viral RNA. In contrast, a simple increase in the wild-type *Ifih1* gene dosage was insufficient to cause spontaneous nephritis (Seth et al. 2005), suggesting that a dysregulation in MDA5 by a mutation may be essential for triggering autoimmunity. Note that de novo p.R779H IFIH1 gain-of-function mutation has recently been identified in a patient with severe early-onset SLE, selective IgA deficiency (Takeuchi and Akira 2010; Fig. 39.3). Together with SLE-like phenotypes observed in MDA5 G821S mutant mice. theis evidence indicates the strong correlation between aberrant activation of MDA5 and SLE.

39.3.2 Aicardi–Goutieres Syndrome (AGS)

Aicardi–Goutieres syndrome (AGS) is an inflammatory disease that particularly affects the brain and skin. AGS patients exhibit profound intellectual disabilities and dystonia, and lethality by the age of 17 years has been reported in approximately 25 % of patients. Yanick J Crow's group showed that mutations in the genes functioning in nucleic acid metabolism, including *TREX1*, *SAMHD1*, *ADAR1*, *RNASEH2A*, *RNASEH2B*, and *RNASEH2C*, were strongly linked to AGS (Van

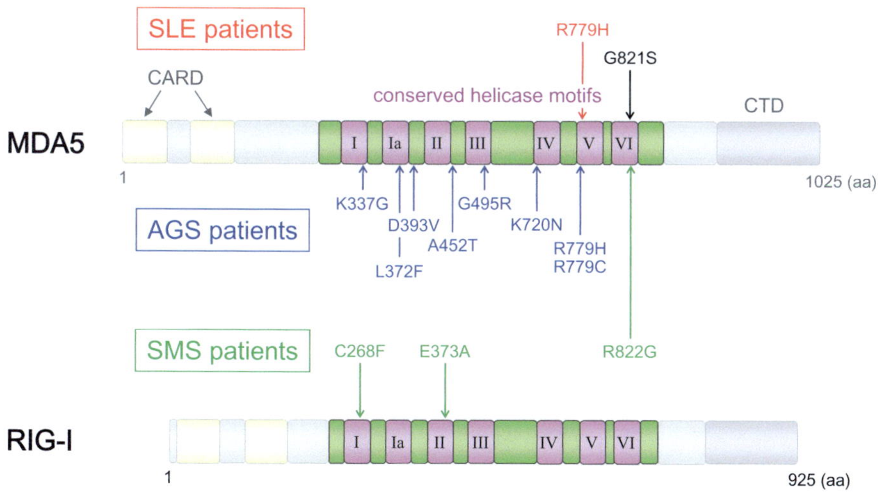

Fig. 39.3 Mutations of RLR in autoimmune diseases. Mutations of RLR identified in AGS, SMS, and SLE patients are indicated in *blue*, *green*, and *orange*, respectively. All currently reported mutations are located in the conserved helicase motifs. G821S (*red*) is a missense mutation found in mice

Eyck et al. 2015; Crow and Rehwinkel 2009) (Fig. 39.4). For instance, the exodeoxyribonuclease TREX1 is known to eliminate viral and aberrant cellular DNAs in the cytoplasm, which potentially activate the DNA sensor cGAS to transduce IFN signalling via the essential adaptor molecule, STING; thus, TREX1 has been identified as a negative regulator of the cGAS-STING-dependent IFN signalling pathway (Wang et al. 2010; Wu et al. 2013; Xu et al. 2005) (Fig. 39.4). Recently *Trex1*-deficient mice have been shown to develop lethal, IFN-driven autoimmune disorders such as SLE-like pehenotypes including nephritis, which is in fact rescued by deficiency in the cGAS-STING pathway. These findings indicate cGAS as a key driver of autoimmune disease in the case of TREX1 deficiency.

In 2014, Crow's group identified mutations in *IFIH1* in AGS patients(Yang et al. 2007). These mutations conferred the constitutive activity of MDA5, but occurred at different positions from the mouse *Ifih1* mutation reported by Fubabiki et al. (Schlee et al. 2009) (Fig. 39.3). Human MDA5 mutants exhibited hyperresponsiveness to ligand stimulation, thereby suggesting the possible involvement of endogenous and/or viral RNA in the onset of AGS.

Another group more recently identified *IFIH1* heterozygous missense mutations in AGS patients(Yoneyama et al. 2004). The encoded MDA5 mutants exhibited constitutive activity, but failed to respond to a viral stimulus similar to the mouse MDA5 mutation. These findings indicated a link between constitutive MDA5 activity and AGS; however, responsiveness to viral infections remains controversial, particularly the same mutation (G2336A:R779H) included in these studies. This discrepancy needs to be re-examined using a common assay. Furthermore, the

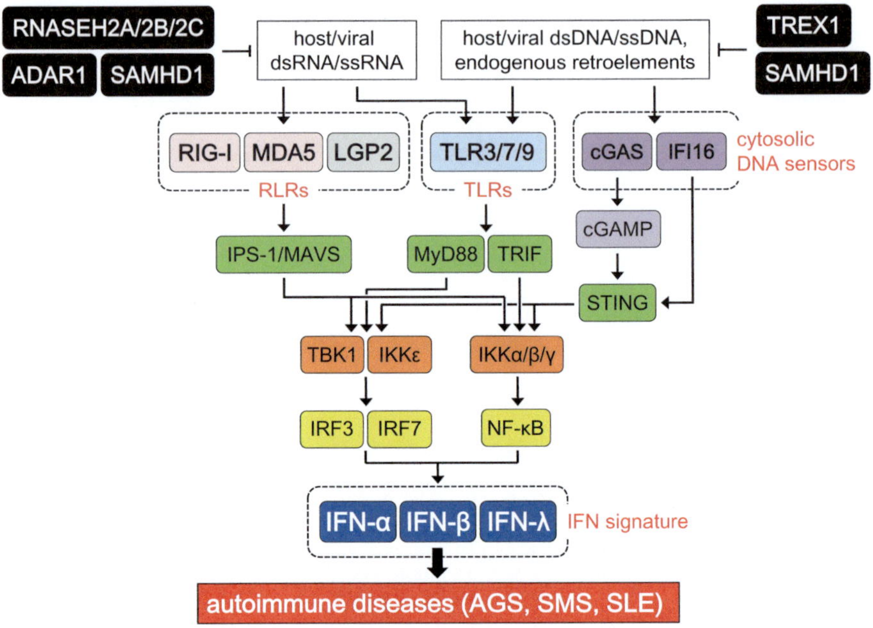

Fig. 39.4 Abnormal activation of DNA/RNA sensing pathway leads to IFN signature. Three major nucleic-acid–sensing pathways (RLR, TLR, and cytoplasmic DNA sensor) that trigger type I and III IFNs are indicated. Defect in the deoxynucleoside triphosphate triphosphohydrolase SAMHD1, adenosine deaminase ADAR1, and nucleases including TREX1 and RNASEH 2A, 2B, and 2C will cause the accumulation of immunostimulatory RNA and/or DNA, which abnormally activate these pathways. Also autoactivation of sensors MDA5 and RIG-I has been shown to lead aberrant signaling. Finally IFN signature induced by atypical activation of nucleic-acid–sensing pathways will trigger autoimmune diseases such as AGS, SMS, and SLE

R779H mutation was identified in an SLE patient as described above (Fig. 39.3). Thus, it is intriguing to investigate how the same mutation causes different diseases, SLE and AGS.

The early onset of autoimmunity in mutant mice and AGS suggests the autonomous cause of the activation of MDA5 rather than a strong association with a particular viral infection. Considering the high expression of IFN-inducible genes detectable in AGS patients, the so-called interferon signature, mutations in the deoxynucleoside triphosphate triphosphohydrolase SMAHD1, the dsRNA editing enzyme ADAR1, and the degradation enzyme of RNA:DNA heteroduplexes RNASEHs are likely to activate DNA and/or RNA sensors (including cGAS and RLRs) directly, leading to the production of type I IFN (Fig. 39.4).

Recently, several groups clearly demonstrated that RNA editing of endogenous RNAs by ADAR1 prevents the activation of MDA5. *Adar1*-deficient mice and mice with editing-deficient knock-in mutations (E861A/ E861A) have been shown to be embryonic lethal but their lethality was rescued by concurrent depletion of MDA5 (Cunninghame Graham et al. 2011). It will be interesting to explore the responsible

RNAs for MDA5 activation, which are usually edited by ADAR1 but not in the AGS patients with ADAR1 mutations.

39.3.3 Singleton–Merten Syndrome (SMS)

Singleton–Merten syndrome is an extremely rare, multisystem disorder that is characterised by dental dysplasia, calcifications of the aorta, glaucoma, and osteoporosis. SMS patients also exhibit IFN signature. Jang et al. recently described a family affected by these symptoms and identified a missense mutation (E373A) in *DDX58*, encoding RIG-I (Yoneyama et al. 2005; Fig. 39.3). This is the first report to show mutations of RIG-I in patients with autoimmune disease. RIG-I E373A leads to the constitutive production of type I IFN. Moreover, a further analysis of *DDX58* in 100 individuals with congenital glaucoma resulted in the identification of another mutation (C268F; Fig. 39.3). Glu373 and Cys268 of RIG-I reside in the ATP-binding motifs I and II in the Hel-1 domain, respectively (Fig. 39.3). The constitutive activities and responsiveness to ligand RNA of these mutants have not yet been characterised.

Another group reported a novel missense mutation (R822G) in *IFIH1* in a SMS patient (Yoneyama et al. 2015) (Fig. 39.3). MDA5 R822G exhibits constitutive activity. Considering that some AGS patients exhibit glaucoma or premature tooth loss, further clinical examination in AGS patients, especially those with an *IFIH1* mutation may help us to understand the pathogenesis of SMS inasmuch as only one SMS patient with an *IFIH1* mutation has been reported thus far. It is worth noting that this mutation is next to the G821S mutation in *Ifih1*. Therefore it would be interesting to explore the SMS-like phenotypes in MDA5 G821S mutant mice, which were reported to exhibit lupus-like nephritis.

Recently a family segregating a heterozygous pathogenic mutation A489T in IFIH1 was found (Yoo et al. 2014). All affected individuals exhibited an increased level of ISGs in sera and showed dermatological manifestations as a prominent feature, variably associated with neurological disturbance and premature tooth loss. These symptoms could be diagnosed as AGS and/or SMS, which highlight phenotypic overlap with AGS and SMS in the case of a heterozygous pathogenic mutation in IFIH1.

39.4 Discussion and Future Perspectives

It has clearly been demonstrated in a mutant mouse model that a single missense mutation in *Ifih1* caused spontaneous SLE-like nephritis, whereas *Ifih1* transgenic mice were reported to accelerate SLE-like nephritis, but did not spontaneously develop the nephritis in spite of chronically elevated levels of type I IFN. These findings suggested that the chronic activation of IFN may be insufficient to trigger autoimmune diseases. Although quantitative comparisons of IFN levels between these mice are critical, qualitative differences between mutations and gene multiplication may account for the triggering of these diseases. Also it is interesting to explore what is the factor causing autoimmunity in addition to augmented IFN level, comparing these two mouse models.

The *IFIH1* and *DDX58* mutations detected in humans and mice cause amino acid substitutions within the helicase domain of MDA5 and RIG-I as indicated in Fig. 39.3, however, no obvious hot spot has been identified to date. These mutations may commonly induce conformational changes, resulting in the unmasking of CARD for constitutive activity. Some mutations may also enhance the detection of endogenous RNA derived from host cells to which wild-type MDA5 and RIG-I do not respond and/or viral RNAs derived from chronic infections. In this regard, other causative mutations in AGS (*TREX, RNASEH2A, B, C*, and *ADAR*) may be implicated in the loss-of-function of nucleic acid clearance. Especially ADAR1 has recently been shown to activate MDA5. However, the endogenous ligand RNAs that activate MDA5 and how this is prevented by homeostatic mechanisms remain unknown. Therefore, the detailed molecular phenotypes of these mutations, especially requirement of RNA ligands for activation, need to be characterised.

The activation and suppression mechanisms of RLRs in virus-infected cells have been extensively investigated. Thus it appears that RLR research is advancing to the next stage, focusing on the correlation between RLR and autoimmunity and endogenous RLR ligands. However, important questions in the study of RLR-mediated virus recognition still remain unclear, including detailed mechanisms of how RLRs recognise each virus and how each virus evades the activation of RLR. Exact characterisation of RLR ligands in individual virus infection will also help to understand what is recognised by RLRs in autoimmune conditions. Finally, further comprehensive studies will be required for understanding how excess antiviral innate immunity causes autoimmunity.

References

Ablasser A et al (2014) TREX1 deficiency triggers cell-autonomous immunity in a cGAS-dependent manner. J Immunol 192:5993–5997. doi:10.4049/jimmunol.1400737

Berke IC, Yu X, Modis Y, Egelman EH (2012) MDA5 assembles into a polar helical filament on dsRNA. Proc Natl Acad Sci U S A 109:18437–18441. doi:10.1073/pnas.1212186109

Bursztejn AC et al (2015) Unusual cutaneous features associated with a heterozygous gain-of-function mutation in IFIH1: overlap between Aicardi-Goutieres and Singleton-Merten syndromes. Br J Dermatol 173:1505–1513. doi:10.1111/bjd.14073

Cai X, Chiu YH, Chen ZJ (2014) The cGAS-cGAMP-STING pathway of cytosolic DNA sensing and signaling. Mol Cell 54:289–296. doi:10.1016/j.molcel.2014.03.040

Crampton SP, Deane JA, Feigenbaum L, Bolland S (2012) Ifih1 gene dose effect reveals MDA5-mediated chronic type I IFN gene signature, viral resistance, and accelerated autoimmunity. J Immunol 188:1451–1459. doi:10.4049/jimmunol.1102705

Crow YJ, Rehwinkel J (2009) Aicardi-Goutieres syndrome and related phenotypes: linking nucleic acid metabolism with autoimmunity. Hum Mol Genet 18:R130–R136. doi:10.1093/hmg/ddp293

Crow YJ et al (2006) Mutations in the gene encoding the 3'-5' DNA exonuclease TREX1 cause Aicardi-Goutieres syndrome at the AGS1 locus. Nat Genet 38:917–920. doi:10.1038/ng1845

Cunninghame Graham DS et al (2011) Association of NCF2, IKZF1, IRF8, IFIH1, and TYK2 with systemic lupus erythematosus. PLoS Genet 7, e1002341. doi:10.1371/journal.pgen.1002341

Funabiki M et al (2014) Autoimmune disorders associated with gain of function of the intracellular sensor MDA5. Immunity 40:199–212. doi:10.1016/j.immuni.2013.12.014

Gack MU et al (2007) TRIM25 RING-finger E3 ubiquitin ligase is essential for RIG-I-mediated antiviral activity. Nature 446:916–920. doi:10.1038/nature05732

Gao D et al (2015) Activation of cyclic GMP-AMP synthase by self-DNA causes autoimmune diseases. Proc Natl Acad Sci U S A 112:E5699–E5705. doi:10.1073/pnas.1516465112

Gray EE, Treuting PM, Woodward JJ, Stetson DB (2015) Cutting Edge: cGAS Is Required for Lethal Autoimmune Disease in the Trex1-Deficient Mouse Model of Aicardi-Goutieres Syndrome. J Immunol 195:1939–1943. doi:10.4049/jimmunol.1500969

Hornung V et al (2006) 5'-Triphosphate RNA is the ligand for RIG-I. Science 314:994–997. doi:10.1126/science.1132505

Hou F et al (2011) MAVS forms functional prion-like aggregates to activate and propagate antiviral innate immune response. Cell 146:448–461. doi:10.1016/j.cell.2011.06.041

Jang MA et al (2015) Mutations in DDX58, which encodes RIG-I. Cause atypical Singleton-Merten syndrome. Am J Human Genet 96:266–274. doi:10.1016/j.ajhg.2014.11.019

Kageyama M et al (2011) 55 Amino acid linker between helicase and carboxyl terminal domains of RIG-I functions as a critical repression domain and determines inter-domain conformation. Biochem Biophys Res Commun 415:75–81. doi:10.1016/j.bbrc.2011.10.015

Kato H et al (2006) Differential roles of MDA5 and RIG-I helicases in the recognition of RNA viruses. Nature 441:101–105. doi:10.1038/nature04734

Kato H et al (2008) Length-dependent recognition of double-stranded ribonucleic acids by retinoic acid-inducible gene-I and melanoma differentiation-associated gene 5. J Exp Med 205:1601–1610. doi:10.1084/jem.20080091

Kawai T et al (2005) IPS-1, an adaptor triggering RIG-I- and Mda5-mediated type I interferon induction. Nat Immunol 6:981–988. doi:10.1038/ni1243

Kawane K et al (2006) Chronic polyarthritis caused by mammalian DNA that escapes from degradation in macrophages. Nature 443:998–1002. doi:10.1038/nature05245

Kowalinski E et al (2011) Structural basis for the activation of innate immune pattern-recognition receptor RIG-I by viral RNA. Cell 147:423–435. doi:10.1016/j.cell.2011.09.039

Liddicoat BJ et al (2015) RNA editing by ADAR1 prevents MDA5 sensing of endogenous dsRNA as nonself. Science 349:1115–1120. doi:10.1126/science.aac7049

Marques JT et al (2006) A structural basis for discriminating between self and nonself double-stranded RNAs in mammalian cells. Nat Biotechnol 24:559–565. doi:10.1038/nbt1205

Meylan E et al (2005) Cardif is an adaptor protein in the RIG-I antiviral pathway and is targeted by hepatitis C virus. Nature 437:1167–1172. doi:10.1038/nature04193

Napirei M et al (2000) Features of systemic lupus erythematosus in Dnase1-deficient mice. Nat Genet 25:177–181. doi:10.1038/76032

Narita R et al (2014) A novel function of human Pumilio proteins in cytoplasmic sensing of viral infection. PLoS Pathog 10, e1004417. doi:10.1371/journal.ppat.1004417

Nejentsev S, Walker N, Riches D, Egholm M, Todd JA (2009) Rare variants of IFIH1, a gene implicated in antiviral responses, protect against type 1 diabetes. Science 324:387–389. doi:10.1126/science.1167728

Ng CS et al (2013) Encephalomyocarditis virus disrupts stress granules, the critical platform for triggering antiviral innate immune responses. J Virol 87:9511–9522. doi:10.1128/JVI.03248-12

Oda H et al (2014) Aicardi-Goutieres syndrome is caused by IFIH1 mutations. Am J Hum Genet 95:121–125. doi:10.1016/j.ajhg.2014.06.007
Onoguchi K et al (2010) Virus-infection or 5'ppp-RNA activates antiviral signal through redistribution of IPS-1 mediated by MFN1. PLoS Pathog 6, e1001012. doi:10.1371/journal.ppat.1001012
Onomoto K et al (2012) Critical role of an antiviral stress granule containing RIG-I and PKR in viral detection and innate immunity. PLoS One 7, e43031. doi:10.1371/journal.pone.0043031
Oshiumi H et al (2010) The ubiquitin ligase Riplet is essential for RIG-I-dependent innate immune responses to RNA virus infection. Cell Host Microbe 8:496–509. doi:10.1016/j.chom.2010.11.008
Peisley A, Wu B, Yao H, Walz T, Hur S (2013) RIG-I forms signaling-competent filaments in an ATP-dependent, ubiquitin-independent manner. Mol Cell 51:573–583. doi:10.1016/j.molcel.2013.07.024
Pestal K et al (2015) Isoforms of RNA-editing enzyme ADAR1 independently control nucleic acid sensor MDA5-driven autoimmunity and multi-organ development. Immunity 43:933–944. doi:10.1016/j.immuni.2015.11.001
Pichlmair A et al (2006) RIG-I-mediated antiviral responses to single-stranded RNA bearing 5'-phosphates. Science 314:997–1001. doi:10.1126/science.1132998
Rice GI et al (2014) Gain-of-function mutations in IFIH1 cause a spectrum of human disease phenotypes associated with upregulated type I interferon signaling. Nat Genet 46:503–509. doi:10.1038/ng.2933
Robinson T et al (2011) Autoimmune disease risk variant of IFIH1 is associated with increased sensitivity to IFN-alpha and serologic autoimmunity in lupus patients. J Immunol 187:1298–1303. doi:10.4049/jimmunol.1100857
Rutsch F et al (2015) A specific IFIH1 gain-of-function mutation causes Singleton-Merten syndrome. Am J Hum Genet 96:275–282. doi:10.1016/j.ajhg.2014.12.014
Sato M et al (2000) Distinct and essential roles of transcription factors IRF-3 and IRF-7 in response to viruses for IFN-alpha/beta gene induction. Immunity 13:539–548
Satoh T et al (2010) LGP2 is a positive regulator of RIG-I- and MDA5-mediated antiviral responses. Proc Natl Acad Sci U S A 107:1512–1517. doi:10.1073/pnas.0912986107
Schlee M et al (2009) Recognition of 5' triphosphate by RIG-I helicase requires short blunt double-stranded RNA as contained in panhandle of negative-strand virus. Immunity 31:25–34. doi:10.1016/j.immuni.2009.05.008
Seth RB, Sun L, Ea CK, Chen ZJ (2005) Identification and characterization of MAVS, a mitochondrial antiviral signaling protein that activates NF-kappaB and IRF 3. Cell 122:669–682. doi:10.1016/j.cell.2005.08.012
Takeuchi O, Akira S (2010) Pattern recognition receptors and inflammation. Cell 140:805–820. doi:10.1016/j.cell.2010.01.022
Van Eyck L et al (2015) Brief report: IFIH1 mutation causes systemic Lupus Erythematosus with selective IgA deficiency. Arthritis Rheumatol 67:1592–1597. doi:10.1002/art.39110
Wang Y et al (2010) Structural and functional insights into 5'-ppp RNA pattern recognition by the innate immune receptor RIG-I. Nat Struct Mol Biol 17:781–787. doi:10.1038/nsmb.1863
Wu B et al (2013) Structural basis for dsRNA recognition, filament formation, and antiviral signal activation by MDA5. Cell 152:276–289. doi:10.1016/j.cell.2012.11.048
Xu LG et al (2005) VISA is an adapter protein required for virus-triggered IFN-beta signaling. Mol Cell 19:727–740. doi:10.1016/j.molcel.2005.08.014
Yang YG, Lindahl T, Barnes DE (2007) Trex1 exonuclease degrades ssDNA to prevent chronic checkpoint activation and autoimmune disease. Cell 131:873–886. doi:10.1016/j.cell.2007.10.017
Yoneyama M et al (2004) The RNA helicase RIG-I has an essential function in double-stranded RNA-induced innate antiviral responses. Nat Immunol 5:730–737. doi:10.1038/ni1087
Yoneyama M et al (2005) Shared and unique functions of the DExD/H-box helicases RIG-I, MDA5, and LGP2 in antiviral innate immunity. J Immunol 175:2851–2858

Yoneyama M, Onomoto K, Jogi M, Akaboshi T, Fujita T (2015) Viral RNA detection by RIG-I-like receptors. Curr Opin Immunol 32C:48–53. doi:10.1016/j.coi.2014.12.012
Yoo JS et al (2014) DHX36 enhances RIG-I signaling by facilitating PKR-mediated antiviral stress granule formation. PLoS Pathog 10, e1004012. doi:10.1371/journal.ppat.1004012

Chapter 40
Elucidation of the Exacerbation Mechanism of Autoimmune Diseases Caused by Disruption of the Ion Homeostasis

Masatsugu Oh-hora

Abstract (150–250 words) Ions, such as calcium, magnesium, potassium and sodium ions, are critically involved in the development and function of various tissues. Among these ions, calcium is known to function as second messenger to regulate a variety of immune cell functions including cytokine production, cytotoxicity, chemotaxis, and differentiation to effector cells. Intracellular concentration of calcium is increased after the engagement of immune receptors such as T-cell receptor and Fc receptor and strictly controlled by a network of calcium-permeable channels, transporters, and pumps. Among various mechanisms of extracellular calcium entry in lymphocytes, store-operated calcium (SOC) entry through calcium–release-activated calcium (CRAC) channels is known to be the predominant mechanism of antigen receptor-activated increase in intracellular calcium levels. Induction of calcium entry via CRAC channels is mediated by two molecular components, ORAI1, a subunit of the CRAC channels expressed in the plasma membrane, and stromal interaction molecule (STIM)1, a calcium sensor in the endoplasmic reticulum. Dysregulation of SOC entry has been reported to induce immunological and nonimmunological diseases. In this chapter, we first introduce the molecular mechanism of SOC entry, and then describe the physiological consequences of impaired SOC entry in human and mouse. Finally, the mechanism of the development of autoimmune diseases by the loss of SOC entry is described.

Keywords Calcium • Immunodeficiency • Tolerance • Regulatory T cells • Sjörgren syndrome • Foxp3 • NFAT

M. Oh-hora (✉)
Division of Molecular Immunology, Research Center for Infectious Diseases, Medical Institute of Bioregulation, Kyusyu University, 3-1-1 Maidashi, Higashi-ku, Fukuoka 812-8582, Japan

PRESTO, Japan Science and Technology Agency, Kawaguchi 332-0012, Japan
e-mail: ohora@bioreg.kyushu-u.ac.jp

M. Miyasaka, K. Takatsu (eds.), *Chronic Inflammation*,
DOI 10.1007/978-4-431-56068-5_40

40.1 Introduction

Intracellular concentration of various ions, such as calcium (Ca^{2+}), magnesium (Mg^{2+}), zinc (Zn^{2+}), potassium (K^{+}), sodium (Na^{+}), and chloride (Cl^{-}) are regulated by selective or nonselective ion channels and ion transporters expressed in the plasma membrane. Recent studies have shown that dysregulation of these ions by the impairment of their channels or transporters influences immune cell functions and causes severe immune diseases, such as immunodeficiency (Feske et al. 2012).

Among these ions, Ca^{2+} is one of the most intensively studied. Ca^{2+} is involved in signal transduction as a second messenger in almost all cell types and activates various Ca^{2+} signalling pathways, such as a calcineurin – nuclear factor of activated T-cell (NFAT) pathway (Hogan et al. 2003; Lewis 2001). Ca^{2+} signals are known crucially to regulate immune responses by immune cells including T cells, B cells, NK cells, mast cells, dendritic cells, and macrophages (Hogan et al. 2003; Feske 2007). The concentration of free intracellular Ca^{2+} is maintained around 100–200 nM under resting conditions. Upon stimulation of immune receptors including the T-cell receptor (TCR), intracellular Ca^{2+} concentration is rapidly increased by Ca^{2+} influx from extracellular space and reaches over 1 μM, which in turn activates Ca^{2+} signals (Hogan et al. 2003; Lewis 2001). Although T cells express multiple Ca^{2+}-permeable channels, electrophysiological and pharmacological studies demonstrated that T cells utilise store-operated Ca^{2+} (SOC) entry through Ca^{2+}-release–activated Ca^{2+} (CRAC) channels as the chief mechanism for entry of extracellular Ca^{2+} across the plasma membrane (Lewis 2001; Oh-hora 2009). Physiological importance of SOC entry is highlighted by the fact that patients who naturally lack the SOC entry were reported to suffer from primary immunodeficiency and nonimmunological diseases (Feske et al. 1996). Intriguingly, autoimmune complications have been reported in some newly identified patients with immunodeficiency (Fuchs et al. 2012; McCarl et al. 2009; Picard et al. 2009). However, its mechanism has been unclear.

In this chapter, we describe the background of SOC entry including its molecular mechanism and clinical phenotypes observed in SOC entry-deficient human and mouse. Then, the mechanism of autoimmune diseases caused by impaired SOC entry is explained.

40.2 Store-Operated Calcium Entry Through CRAC Channels

In T cells, an influx of extracellular Ca^{2+} is induced after the engagement of TCR with complexes of peptide and the major histocompatibility complex (pMHC). This Ca^{2+} entry across the plasma membrane is triggered by depletion of Ca^{2+} store in the endoplasmic reticulum (ER). Therefore, this is called "store-operated" Ca^{2+} (SOC) entry, formerly known as "capacitative Ca^{2+} entry" (Putney 1986). SOC entry has been observed in a variety of nonexcitable and excitable cells including lymphocytes (Oh-hora and Rao 2008), pancreatic acinar (Mogami et al. 1997), and skeletal muscle cells (Vazquez et al. 1997). Note that, lack of SOC entry results in fatal immunodeficiency in humans (Feske et al. 1996). The key event for triggering SOC entry is the activation of phospholipase C (PLC)γ or PLCβ by the stimulation with agonists to different types of cell surface receptors, such as TCR or chemokine receptors. PLC hydrolyses phosphatidylinositol-3,4-bisphosphate (PIP_2) to inositol-1,4,5-trisphosphate ($InsP_3$) and diacylglycerol. $InsP_3$ binds to the $InsP_3$ receptors in the ER membrane, then induces the release of Ca^{2+} from the ER to the cytoplasm, which in turn activates SOC channels in the plasma membrane (Fig. 40.1). The prototypical SOC channel is the Ca^{2+}-release–activated Ca^{2+} (CRAC) channel originally characterised in T cells and mast cells (Hoth and Penner 1992; Zweifach and Lewis 1993). Although the electrophysiological properties of the activity of CRAC channels were well characterised, their molecular identity

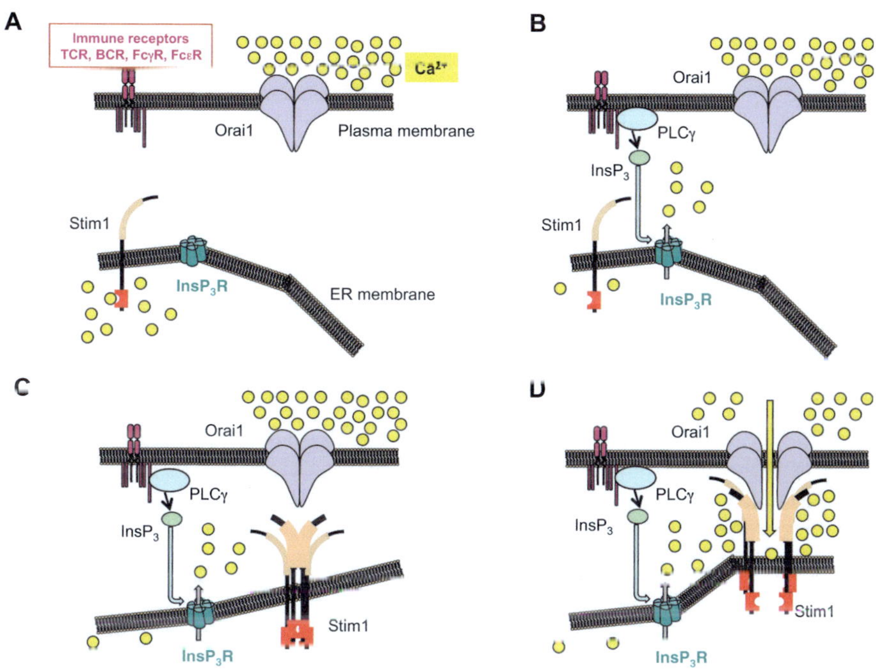

Fig. 40.1 Mechanism of store-operated Ca^{2+} entry. (**a**) In the resting cells with replete ER Ca^{2+} stores, Stim1 binds to Ca^{2+} with its EF hand motif in the ER lumen portion and is localised as an inactive monomer in the ER membrane. (**b**). PLCγ produces inositol-1,4,5-trisphosphate ($InsP_3$) after the stimulation with antigens to immune receptors (such as TCR, BCR or FcR), which in turn induces depletion of ER Ca^{2+} stores through $InsP_3R$. As a result of decrease of ER Ca^{2+} concentration, Stim1 is unbound to Ca^{2+}. (**c**). Stim1 forms oligomers by a conformational change. (**d**). Oligomerised Stim1 translocates to junctional ER sites in which the ER membrane is juxtaposed to the plasma membrane, then directly interacts with binding sites in the C- and N-termini of Orai1 through its CAD/SOAR domain. As a result, Orai1 CRAC channels are opened

remained a mystery for over 20 years. In 2005 and 2006, by high-throughput RNA interference (RNAi) screens, several groups identified two key proteins of the SOC entry, STIM1 (stromal interaction molecule 1) as the ER Ca^{2+} sensor (Liou et al. 2005; Roos et al. 2005) and ORAI1 (also known as CRACM1 or TMEM142A) as a pore-forming subunit of the CRAC channels (Feske et al. 2006; Vig et al. 2006; Zhang et al. 2006).

40.2.1 Orai1

In 2006, Orai1 (also known as CRACM1) was identified as an essential regulator of SOC entry and NFAT activation by RNAi screen in *Drosophila* S2 cells by three groups (Zhang et al. 2006; Vig et al. 2006; Feske et al. 2006). By positional cloning, one of these groups also demonstrated that ORAI1 is the causative gene in immunodeficient patients lacking SOC entry (Feske et al. 2006). Orai1 is a small and widely expressed tetraspanning membrane glycoprotein, whose N- and C-termini are both located in the cytoplasm. Orai1 is a novel class of channel protein with no sequence homology to other known ion channels except its paralogues Orai2 and Orai3 (also known as CRACM2 and CRACM3, respectively). Although Orai2 and Orai3 also form Ca^{2+} channels when they are ectopically overexpressed together with Stim1 (DeHaven et al. 2007; Lis et al. 2007), it remains elusive whether they are functional in immune cells. Orai1 predominantly exists as a dimer in the plasma membrane under resting conditions (Penna et al. 2008). After the stimulation, Orai1 forms a tetramer as the functional CRAC channels (Penna et al. 2008; Ji et al. 2008).

40.2.2 Stromal Interaction Molecule (Stim)1 and Stim2

Stim1 was originally identified as a tumor suppressor GOK and as SIM, a protein that promotes the survival and proliferation of pre-B cells (Parker et al. 1996; Sabbioni et al. 1997; Oritani and Kincade 1996). At the time, Stim1 seemed to be unrelated with SOC entry (Manji et al. 2000). As well as the identification of Orai1, RNAi screen revealed that Stim1 is an essential regulator of the SOC entry. Stim1 is a type I transmembrane protein predominantly localised in the ER membrane and expressed in broad tissues (Williams et al. 2001; Baba et al. 2008; Oh-Hora et al. 2008). Stim1 has several functional domains including a Ca^{2+}-binding EF hand motif in the ER lumen portion and the CRAC activation domain (CAD)/the STIM Orai activating region (SOAR) domain in the cytoplasm (Williams et al. 2001; Park et al. 2009; Yuan et al. 2009). C-terminal cytosolic portion of Stim1 contains a polybasic lysine-rich region, which is required for attachment to the plasma membrane (Yuan et al. 2009; Jardin et al. 2013). Stim2, a paralogue of Stim1, shares its overall protein domain architecture with Stim1, however, Stim2 has longer N- and C-termini than those of STIM1 (Williams et al. 2001). As well as

Stim1, Stim2 is capable of initiating SOC entry via Orai1 in some cell types (Oh-Hora et al. 2008). Interestingly, Stim2 regulates sustained Ca^{2+} influx and activation of NFAT in differentiated T cells (Oh-Hora et al. 2008). Furthermore, Stim2 is known to regulate basal cytosolic Ca^{2+} concentration in the Hela cell line (Brandman et al. 2007).

The mechanism by which Stim proteins regulate ORAI proteins has been extensively investigated in the past 10 years. Stim1 senses the decrease of ER Ca^{2+} concentration through its EF-hand motif. After Stim1 is unbound to Ca^{2+}, Stim1 forms the tetrameric complexes (Stathopulos et al. 2009; Stathopulos et al. 2008). Oligomerised Stim1 then accumulates in unique regions within 10–25 nm underneath of plasma membrane, commonly referred to as "puncta" (Liou et al. 2005; Liou et al. 2007; Wu et al. 2006), where Stim1 directly binds to a coiled-coil domain at the C-terminus or an additional domain at the N-terminus of Orai1 through its CAD/SOAR domain (Yuan et al. 2009; Park et al. 2009). This direct interaction allosterically activates the Orai1 channel (Fig. 40.1).

40.3 CRAC Channelopathies by Dysfunction of Orai1 and Stim1 in Human and Mouse

The in vivo importance of SOC entry through CRAC channels is highlighted by the phenotypes of human patients who have nonfunctional mutations in the *ORAI1* or the *STIM1* gene and by a series of studies using mice lacking the *Orai1*, *Stim1*, or *Stim2* genes.

40.3.1 Immunodeficiency

As mentioned in sect. 40.2.1, ORAI1 was identified as the gene mutated in immunodeficient patients born to consanguineous Turkish parents (Feske et al. 2006). The clinical phenotypes including the severity of infections, the time of onset, and the spectrum of pathogens in these patients are similar to those of severe combined immunodeficiency (SCID) patients. So far, 5 and 2 loss-of-function mutations in the *ORAI1* gene (R91W, A103E, L194P, H165P, and A88SfsX25) (Chou et al. 2015; Feske et al. 2006; McCarl et al. 2009) and the *STIM1* gene (E128RfsX9 and R429C; Picard et al. 2009; Fuchs et al. 2012), respectively, have been identified from individual immunodeficient patients. Although these patients have normal numbers of immune cells in the periphery, their T cells were not able to express some cytokine genes including IL-2 by a partial NFAT activation (Feske et al. 2000). We have shown that, using mice lacking the *Orai1*, *Stim1*, and/or *Stim2* genes, lack of SOC entry results in impaired proliferation and differentiation into effector T cells, such as type I helper T (T_H1)

cells, T_H2 cells, IL-17-producing helper T (T_H17) cells, and cytotoxic $CD8^+$ T cells (Oh-Hora et al. 2008; Gwack et al. 2008; McCarl et al. 2010), whereas positive selection in the thymus seems to be normal (Oh-Hora et al. 2013). In addition, another study demonstrated that cytolytic activity of natural killer (NK) cells is also ablated (Fuchs et al. 2012). Likely due to the same defects, patients suffer from recurrent severe infections with viral, bacterial, and fungal pathogens, for example, cytomegalovirus, Epstain–Barr virus, *Streptococcus pneumonia,* and *Candida albicans* (Feske et al. 1996; Le Deist et al. 1995; McCarl et al. 2009; Partiseti et al. 1994; Picard et al. 2009; Chou et al. 2015).

40.3.2 Inflammation and Autoimmune Diseases

Unlike immunodeficiency, severe chronic inflammation was originally found in mice lacking both Stim1 and Stim2 specifically in T cells (Stim-deficient mice, hereafter; Oh-Hora et al. 2008). These mice developed lymphadenopathy, hepatosplenomegaly, and severe inflammation including dermatitis with skin thickening and lichenification, alveolar destruction by lymphocyte infiltration in the lung, and blepharitis (Oh-Hora et al. 2008; unpublished observations). The most likely cause of inflammation is the reduced number of regulatory T (Treg) cells, defined as $CD4^+$ $CD25^+$ Forkhead box P3 $(Foxp3)^+$ T cells, in the thymus and peripheral lymphoid tissues, because transplantation of Treg cells dramatically ameliorated these symptoms (Oh-Hora et al. 2008). However, the mechanism of impaired development of $CD4^+$ $CD25^+$ $Foxp3^+$ Treg cells by lacking the SOC entry was unclear at the time.

Thereafter, STIM1-deficient patients with immunodeficiency were reported to develop lymphoproliferative disorders by reduced Treg cells in the periphery (Picard et al. 2009; Fuchs et al. 2012). Note that, these patients also developed autoimmune hemolytic anemia and thrombocytopenia. However, we did not observe these diseases in Stim-deficient mice. This may be due to the genetic background, but its mechanism of their difference remains unclear. Instead, we found that Stim-deficient mice develop severe primary Sjögren's syndrome such as autoimmune diseases (Cheng et al. 2012). We detected autoantibodies, such as antinuclear, SSA/Ro, and SSB/La antibodies, in their sera. In addition, the structure of their salivary and submandibular glands was progressively destroyed by T-cell infiltration, which resulted in a significant decrease of saliva secretion. We further found that peripheral blood mononuclear cells (PBMCs) from patients of primary Sjögren's syndrome displayed reduced STIM1 and STIM2 protein expression with the severity of diseases. SOC entry was significantly decreased in their PBMCs. This finding provides a noble insight that a postnatal loss of SOC entry may cause autoimmune diseases. However, it remains unclear how expressions of STIM1 and STIM2 are downregulated in patients.

In contrast to Stim deficiency, autoimmunity and lymphoproliferative disorders have been observed in only one out of eight ORAI1-deficient patients (McCarl et al. 2009; Feske et al. 2006; Chou et al. 2015), whereas Orai1-deficient mice did not show any inflammatory signs (Gwack et al. 2008; McCarl et al. 2010).

40.4 Control of T-Cell Development by Store-Operated Calcium Entry

Ca^{2+} signalling has long been thought to be crucial for T-cell development in the thymus. Pharmacological or genetic inactivation of calcineurin activity has been reported to be completely blocked T-cell development at CD4 and CD8 double-positive (DP) thymocytes by impairment of positive selection (Jenkins et al. 1988; Gao et al. 1988; Neilson et al. 2004; Bueno et al. 2002). Given the predominant mechanism of increase of intracellular Ca^{2+} concentration in T cells, the SOC entry has been predicted to be a critical Ca^{2+} entry pathway for T-cell development. Although this assumption has not been proved, we have found that lack of SOC entry results in the decrease of $CD4^+$ $CD25^+$ $Foxp3^+$ Treg cells and generation of deleterious T cells, suggesting that SOC entry is involved in the regulation of thymic selection. Therefore we speculate that abnormal T-cell development may explain the cause of a severe chronic inflammation in both mice and humans deficient for the SOC entry.

40.4.1 Conventional T Cells

As mentioned above, patients lacking the *ORAI1* gene or the *STIM1* gene have a normal number of T cells in the periphery, suggesting that SOC entry is dispensable for positive selection or their paralogues may compensate for their function. To clarify the role of SOC entry in T-cell development in the thymus, we analysed mice in which both Stim1 and Stim2 are deleted systemically or specifically in T cells or haematopoietic cells using CMV-Cre Tg, Lck-Cre Tg, or Vav-iCre Tg mice, respectively (Oh-Hora et al. 2013). Despite lack of SOC entry in DP thymocytes, number of thymocytes, frequencies of various thymic populations, and TCR repertoires were comparable between control and each Stim-deficient strain, suggesting that SOC entry is not essential for positive selection of conventional $TCR\alpha\beta^+$ T cells, consistent with the fact that ORAI1- or STIM1-deficient patients have normal number of T cells in the periphery. We have observed that Ca^{2+} entry form extracellular Ca^{2+} is required for T-cell development using an in vitro T-cell development system (unpublished observation), however, a specific Ca^{2+} entry pathway(s) essential for positive selection remains unidentified.

Ca^{2+} influx is induced by peptides promoting negative selection (Daniels et al. 2006). We then examined negative selection. Using TCR Tg mice and superantigens, we found that negative selection is partially impaired by the lack of SOC entry, suggesting that some autoreactive T cells may exist in the periphery and become pathogenic T cells. Indeed, despite ablated T-cell effector functions in Stim-deficient mice, these symptoms are still dependent on T cells because Stim-deficient mice having a monoclonal TCR did not develop severe chronic inflammation. Furthermore, we have found that one particular population may be involved in the development or exacerbation of inflammation in Stim-deficient mice (unpublished observation). However, further studies are needed to evaluate this population.

40.4.2 Regulatory T Cells and Other Agonist-Selected T Cells

$Foxp3^+$ Treg cells are crucial for maintaining immunological tolerance and are categorised in agonist-selected T cells, which are defined as self–antigen-experienced T cells during thymic selection and are considered to play immunoregulatory roles (Sakaguchi et al. 1995; Baldwin et al. 2004; Hsieh et al. 2012; Stritesky et al. 2012). Invariant natural killer T (iNKT) cells and $CD8\alpha\alpha^+$ $TCR\alpha\beta^+$ intestinal intraepithelial lymphocytes (IELs) are also classified into agonist-selected T cells (Stritesky et al. 2012; Baldwin et al. 2004). In contrast to minor phenotypes in conventional $TCR\alpha\beta^+$ T cells, mixed bone marrow reconstitution assays revealed that the numbers of $Foxp3^+$ Treg cells, iNKT cells, and $CD8\alpha\alpha^+$ $TCR\alpha\beta^+$ IELs are significantly reduced by Stim deficiency in both the thymus and peripheral tissues, indicating that SOC entry is cell-intrinsically required for the development of these populations. Then we intensively analysed the development and function of $Foxp3^+$ Treg cells. Mature $Foxp3^+$ Treg ($CD4^+$ $CD25^+$ $Foxp3^+$) cells have been demonstrated to be derived from the specific precursor cells defined as CD4-single positive CD25-high thymocytes (Lio and Hsieh 2008). We found that there are almost no mature $Foxp3^+$ Treg cells with impaired proliferation, whereas the precursor cells normally exist in the thymus of Stim-deficient mice. Stim-deficient precursor cells were capable of differentiating into mature $Foxp3^+$ Treg cells by the treatment with IL-2 both in vitro and in vivo. However, these recovered Treg cells by the treatment with IL-2 lost a suppressive activity due to decreased expressions of inhibitory surface molecules regulated by NFAT, such as CTLA-4 (Takahashi et al. 2000) and TIGIT (Yu et al. 2009). The development of iNKT cells and $CD8\alpha\alpha^+$ $TCR\alpha\beta^+$ IELs was also blocked at each precursor cell in the thymus due to impaired upregulation of genes downstream of NFAT, such as PLZF and Egr2. Consistently, with decreasing NFAT expression, we found that their maturation was inhibited. Furthermore, precursor cells of agonist-selected T cells showed larger and more sustained Ca^{2+} entry than that of DP thymocytes. These findings demonstrate that SOC entry is essential for postselection maturation of agonist-selected T cells through efficient and prolonged activation of Ca^{2+} signals.

40.5 Concluding Remarks

Activation of Ca^{2+} signalling by SOC entry has long been thought to be essential for lymphocyte activation. Now we recognise that the SOC entry is also essential for immune tolerance through Foxp3^{+} Treg cells and other immunoregulatory agonist-selected T cells. An imbalance between immune activation and tolerance may be presumably caused due to unexpected activation of escapees from negative selection, which in turn induces autoimmune diseases in human and mouse (Fig. 40.2). In addition to our findings, the SOC entry has been reported to have immunoregulatory roles in B cells (Matsumoto et al. 2011). The SOC entry is required for the generation of B cells producing IL-10, an anti-inflammatory cytokine. Due to lack of these cells, experimental autoimmune encephalomyelitis, a mouse model for multiple sclerosis, was exacerbated in mice deficient for both Stim1 and Stim2 in B cells. Inasmuch as NFAT is known to control IL-10 expression directly (Lee et al. 2009) and other immune cells also use the SOC entry (Feske 2007), it is tempting to speculate that the SOC entry may be involved in the regulation of anti-inflammatory functions in other regulatory immune cells, such as regulatory dendritic cells (Schmidt et al. 2012) or M2 macrophages (also known as alternatively activated macrophages; Biswas and Mantovani 2010). Further genetic and functional studies will be necessary to elucidate the precise role of SOC entry in immune system.

A chronic inflammation in Stim-deficient mice seems to be caused by T cells, which may be autoreactive. Recent studies in mice lacking functional Orai1, Stim1, and Stim2 have demonstrated that Orai1- or Stim-deficient T cells are not capable of differentiating into effector T cells including T_H1, T_H2, and T_H17 cells in vitro

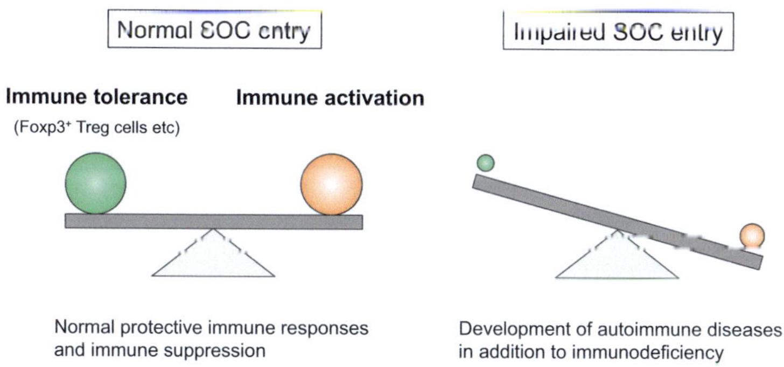

Fig. 40.2 Role of SOC entry in T-cell–mediated immune control. SOC entry is essential not only for protective immune responses against pathogens but also regulatory cell-mediated immune suppression (Such as Treg cells or CD8αα^{+} TCRαβ^{+} IELs). The balance between immune activation and tolerance is maintained under physiological conditions (*left*). In contrast, loss of SOC entry impairs both T-cell–mediated immune activation and tolerance, however, the effect of dysfunction of all immunoregulatory T cells (agonist-selected T cells) is more severe than that of immune activation. Due to this, autoimmune complications are also developed in Orai1- or Stim-deficient patients with immunodeficiency (*right*)

and in vivo. Nevertheless, these pathogenic T cells can still exert an effector function. One possibility is that these T cells may utilise other Ca^{2+}-permeable channels for their activation. Indeed, it has been reported that TRP channels, purinergic receptor X 7 (P2X7) and some voltage-gated Ca^{2+} channels, are involved in T-cell activation in the periphery (Oh-hora 2009; Feske et al. 2012). Alternatively, the Stim-mediated SOC entry may negatively regulate the function of a particular pathogenic T cell subset(s) in a cell-intrinsic manner. Identification and characterisation of these T cells will be helpful for the development of a novel therapeutic strategy for chronic inflammation and autoimmune diseases.

Accumulating evidence shows that K^+, Na^+, and Mg^{2+} control Ca^{2+} influx. Blockade of K^+ channels decreases Ca^{2+} influx by failing to protect against membrane depolarisation, which results in impaired T-cell activation (Cahalan and Chandy 2009; Di et al. 2010). Na^+ is also known to be involved in the regulation of membrane potential. Voltage-gated Na^+ channel (VGSC) has been recently reported to be essential for positive selection in the thymus through regulating sustained Ca^{2+} influx (Lo et al. 2012). Although Lo and colleagues did not provide mechanistic insights, Na^+ influx via VGSC might lead to an increase of Ca^{2+} through the activation of voltage-gated Ca^{2+} channels in T cells by regulating membrane depolarisation. Indeed, L-type voltage-gated Ca^{2+} channel $Ca_v1.4$ has been reported to be involved in T-cell development and peripheral T-cell maintenance (Omilusik et al. 2011). In addition to K^+ and Na^+, a recent study demonstrates that Mg^{2+} influx also influences Ca^{2+} influx. Mg^{2+} transporter protein 1 (MAGT1) is a highly selective Mg^{2+} transporter whose physiological function was not well understood. Li and colleagues have identified nonfunctional mutations of the *MAGT1* gene in patients with a novel type of immunodeficiency (Li et al. 2011). Intriguingly, MAGT1-deficient T cells lack activation of PLCγ-1 and SOC entry after the engagement of TCR, which results in the development of immunodeficiency. Other ions, such as Zn^{2+} and Cl^-, may also regulate Ca^{2+} influx directly or indirectly. These findings suggest that the balance of each ion influx/efflux is critical for maintaining immune homeostasis through the modulation of SOC entry. Thus, control of the influx of other ions may allow for finely tuned regulation of SOC entry, which leads to fine-tuned and optimal overall immune responses.

Finally, here we discussed the mechanism of immune diseases caused by dysfunction of Ca^{2+} entry in T cells. Results that SOC entry is required for T-cell effector functions suggest that CRAC channels may be a potential therapeutic drug target for immune modulation in the context of inflammation. However, complete deletion of CRAC channel function results in autoimmune diseases due to the blockade of Treg cell development and function. A therapeutic window for CRAC channel inhibition needs to be clarified that is effective for immune activation but not for Treg-mediated immune suppression.

References

Baba Y, Nishida K, Fujii Y, Hirano T, Hikida M, Kurosaki T (2008) Essential function for the calcium sensor STIM1 in mast cell activation and anaphylactic responses. Nat Immunol 9 (1):81–88. doi:10.1038/ni1546

Baldwin TA, Hogquist KA, Jameson SC (2004) The fourth way? Harnessing aggressive tendencies in the thymus. J Immunol 173(11):6515–6520

Biswas SK, Mantovani A (2010) Macrophage plasticity and interaction with lymphocyte subsets: cancer as a paradigm. Nat Immunol 11(10):889–896. doi:10.1038/ni.1937

Brandman O, Liou J, Park WS, Meyer T (2007) STIM2 is a feedback regulator that stabilizes basal cytosolic and endoplasmic reticulum Ca^{2+} levels. Cell 131(7):1327–1339. doi:10.1016/j.cell.2007.11.039

Bueno OF, Brandt EB, Rothenberg ME, Molkentin JD (2002) Defective T cell development and function in calcineurin Aβ -deficient mice. Proc Natl Acad Sci U S A 99(14):9398–9403. doi:10.1073/pnas.152665399

Cahalan MD, Chandy KG (2009) The functional network of ion channels in T lymphocytes. Immunol Rev 231(1):59–87. doi:10.1111/j.1600-065X.2009.00816.x

Cheng KT, Alevizos I, Liu X, Swaim WD, Yin H, Feske S, Oh-hora M, Ambudkar IS (2012) STIM1 and STIM2 protein deficiency in T lymphocytes underlies development of the exocrine gland autoimmune disease, sjögren's syndrome. Proc Natl Acad Sci U S A 109 (36):14544–14549. doi:10.1073/pnas.1207354109

Chou J, Badran YR, Yee CS, Bainter W, Ohsumi TK, Al-Hammadi S, Pai SY, Feske S, Geha RS (2015) A novel mutation in ORAI1 presenting with combined immunodeficiency and residual T-cell function. J Allergy Clin Immunol 136(2):479–482. doi:10.1016/j.jaci.2015.03.050, e471

Daniels MA, Teixeiro E, Gill J, Hausmann B, Roubaty D, Holmberg K, Werlen G, Hollander GA, Gascoigne NR, Palmer E (2006) Thymic selection threshold defined by compartmentalization of Ras/MAPK signalling. Nature 444(7120):724–729. doi:10.1038/nature05269

DeHaven WI, Smyth JT, Boyles RR, Putney JW Jr (2007) Calcium inhibition and calcium potentiation of orai1, orai2, and orai3 calcium release-activated calcium channels. J Biol Chem 282(24):17548–17556. doi:10.1074/jbc.M611374200

Di L, Srivastava S, Zhdanova O, Ding Y, Li Z, Wulff H, Lafaille M, Skolnik EY (2010) Inhibition of the K^+ channel KCa3.1 ameliorates T cell-mediated colitis. Proc Natl Acad Sci U S A 107 (4):1541–1546. doi:10.1073/pnas.0910133107

Feske S (2007) Calcium signalling in lymphocyte activation and disease. Nat Rev Immunol 7 (9):690–702. doi:10.1038/nri2152

Feske S, Muller JM, Graf D, Kroczek RA, Drager R, Niemeyer C, Baeuerle PA, Peter HH, Schlesier M (1996) Severe combined immunodeficiency due to defective binding of the nuclear factor of activated T cells in T lymphocytes of two male siblings. Eur J Immunol 26 (9):2119–2126. doi:10.1002/eji.1830260924

Feske S, Draeger R, Peter HH, Eichmann K, Rao A (2000) The duration of nuclear residence of NFAT determines the pattern of cytokine expression in human SCID T cells. J Immunol 165 (1):297–305

Feske S, Gwack Y, Prakriya M, Srikanth S, Puppel SH, Tanasa B, Hogan PG, Lewis RS, Daly M, Rao A (2006) A mutation in orai1 causes immune deficiency by abrogating CRAC channel function. Nature 441(7090):179–185. doi:10.1038/nature04702

Feske S, Skolnik EY, Prakriya M (2012) Ion channels and transporters in lymphocyte function and immunity. Nat Rev Immunol 12(7):532–547. doi:10.1038/nri3233

Fuchs S, Rensing-Ehl A, Speckmann C, Bengsch B, Schmitt-Graeff A, Bondzio I, Maul-Pavicic A, Bass T, Vraetz T, Strahm B, Ankermann T, Benson M, Caliebe A, Folster-Holst R, Kaiser P, Thimme R, Schamel WW, Schwarz K, Feske S, Ehl S (2012) Antiviral and regulatory T cell immunity in a patient with stromal interaction molecule 1 deficiency. J Immunol 188 (3):1523–1533. doi:10.4049/jimmunol.1102507

Gao EK, Lo D, Cheney R, Kanagawa O, Sprent J (1988) Abnormal differentiation of thymocytes in mice treated with cyclosporin A. Nature 336(6195):176–179. doi:10.1038/336176a0

Gwack Y, Srikanth S, Oh-Hora M, Hogan PG, Lamperti ED, Yamashita M, Gelinas C, Neems DS, Sasaki Y, Feske S, Prakriya M, Rajewsky K, Rao A (2008) Hair loss and defective T- and B-cell function in mice lacking ORAI1. Mol Cell Biol 28(17):5209–5222. doi:10.1128/MCB.00360-08

Hogan PG, Chen L, Nardone J, Rao A (2003) Transcriptional regulation by calcium, calcineurin, and NFAT. Genes Dev 17(18):2205–2232. doi:10.1101/gad.1102703

Hoth M, Penner R (1992) Depletion of intracellular calcium stores activates a calcium current in mast cells. Nature 355(6358):353–356. doi:10.1038/355353a0

Hsieh CS, Lee HM, Lio CW (2012) Selection of regulatory T cells in the thymus. Nat Rev Immunol 12(3):157–167. doi:10.1038/nri3155

Jardin I, Dionisio N, Frischauf I, Berna-Erro A, Woodard GE, Lopez JJ, Salido GM, Rosado JA (2013) The polybasic lysine-rich domain of plasma membrane-resident STIM1 is essential for the modulation of store-operated divalent cation entry by extracellular calcium. Cell Signal 25 (5):1328–1337. doi:10.1016/j.cellsig.2013.01.025

Jenkins MK, Schwartz RH, Pardoll DM (1988) Effects of cyclosporine A on T cell development and clonal deletion. Science 241(4873):1655–1658

Ji W, Xu P, Li Z, Lu J, Liu L, Zhan Y, Chen Y, Hille B, Xu T, Chen L (2008) Functional stoichiometry of the unitary calcium-release-activated calcium channel. Proc Natl Acad Sci U S A 105(36):13668–13673. doi:10.1073/pnas.0806499105

Le Deist F, Hivroz C, Partiseti M, Thomas C, Buc HA, Oleastro M, Belohradsky B, Choquet D, Fischer A (1995) A primary T-cell immunodeficiency associated with defective transmembrane calcium influx. Blood 85(4):1053–1062

Lee CG, Kang KH, So JS, Kwon HK, Son JS, Song MK, Sahoo A, Yi HJ, Hwang KC, Matsuyama T, Yui K, Im SH (2009) A distal cis-regulatory element, CNS-9, controls NFAT1 and IRF4-mediated IL-10 gene activation in T helper cells. Mol Immunol 46 (4):613–621. doi:10.1016/j.molimm.2008.07.037

Lewis RS (2001) Calcium signaling mechanisms in T lymphocytes. Annu Rev Immunol 19:497–521. doi:10.1146/annurev.immunol.19.1.497

Li FY, Chaigne-Delalande B, Kanellopoulou C, Davis JC, Matthews HF, Douek DC, Cohen JI, Uzel G, Su HC, Lenardo MJ (2011) Second messenger role for Mg^{2+} revealed by human T-cell immunodeficiency. Nature 475(7357):471–476. doi:10.1038/nature10246

Lio CW, Hsieh CS (2008) A two-step process for thymic regulatory T cell development. Immunity 28(1):100–111. doi:10.1016/j.immuni.2007.11.021

Liou J, Kim ML, Heo WD, Jones JT, Myers JW, Ferrell JE Jr, Meyer T (2005) STIM is a Ca^{2+} sensor essential for Ca^{2+}-store-depletion-triggered Ca^{2+} influx. Curr Biol 15(13):1235–1241. doi:10.1016/j.cub.2005.05.055

Liou J, Fivaz M, Inoue T, Meyer T (2007) Live-cell imaging reveals sequential oligomerization and local plasma membrane targeting of stromal interaction molecule 1 after Ca^{2+} store depletion. Proc Natl Acad Sci U S A 104(22):9301–9306. doi:10.1073/pnas.0702866104

Lis A, Peinelt C, Beck A, Parvez S, Monteilh-Zoller M, Fleig A, Penner R (2007) CRACM1, CRACM2, and CRACM3 are store-operated Ca^{2+} channels with distinct functional properties. Curr Biol 17(9):794–800. doi:10.1016/j.cub.2007.03.065

Lo WL, Donermeyer DL, Allen PM (2012) A voltage-gated sodium channel is essential for the positive selection of $CD4^+$ T cells. Nat Immunol 13(9):880–887. doi:10.1038/ni.2379

Manji SS, Parker NJ, Williams RT, van Stekelenburg L, Pearson RB, Dziadek M, Smith PJ (2000) STIM1: a novel phosphoprotein located at the cell surface. Biochim Biophys Acta 1481 (1):147–155

Matsumoto M, Fujii Y, Baba A, Hikida M, Kurosaki T, Baba Y (2011) The calcium sensors STIM1 and STIM2 control B cell regulatory function through interleukin-10 production. Immunity 34(5):703–714. doi:10.1016/j.immuni.2011.03.016

McCarl CA, Picard C, Khalil S, Kawasaki T, Rother J, Papolos A, Kutok J, Hivroz C, Ledeist F, Plogmann K, Ehl S, Notheis G, Albert MH, Belohradsky BH, Kirschner J, Rao A, Fischer A, Feske S (2009) ORAI1 deficiency and lack of store-operated Ca^{2+} entry cause immunodeficiency, myopathy, and ectodermal dysplasia. J Allergy Clin Immunol 124(6):1311–1318. doi:10.1016/j.jaci.2009.10.007, e1317

McCarl CA, Khalil S, Ma J, Oh-hora M, Yamashita M, Roether J, Kawasaki T, Jairaman A, Sasaki Y, Prakriya M, Feske S (2010) Store-operated Ca^{2+} entry through ORAI1 is critical for T cell-mediated autoimmunity and allograft rejection. J Immunol 185(10):5845–5858. doi:10.4049/jimmunol.1001796

Mogami H, Nakano K, Tepikin AV, Petersen OH (1997) Ca2+ flow via tunnels in polarized cells: recharging of apical Ca^{2+} stores by focal Ca^{2+} entry through basal membrane patch. Cell 88 (1):49–55

Neilson JR, Winslow MM, Hur EM, Crabtree GR (2004) Calcineurin B1 is essential for positive but not negative selection during thymocyte development. Immunity 20(3):255–266

Oh-hora M (2009) Calcium signaling in the development and function of T-lineage cells. Immunol Rev 231(1):210–224. doi:10.1111/j.1600-065X.2009.00819.x

Oh-hora M, Rao A (2008) Calcium signaling in lymphocytes. Curr Opin Immunol 20(3):250–258. doi:10.1016/j.coi.2008.04.004

Oh-Hora M, Yamashita M, Hogan PG, Sharma S, Lamperti E, Chung W, Prakriya M, Feske S, Rao A (2008) Dual functions for the endoplasmic reticulum calcium sensors STIM1 and STIM2 in T cell activation and tolerance. Nat Immunol 9(4):432–443. doi:10.1038/ni1574

Oh-Hora M, Komatsu N, Pishyareh M, Feske S, Hori S, Taniguchi M, Rao A, Takayanagi H (2013) Agonist-selected T cell development requires strong T cell receptor signaling and store-operated calcium entry. Immunity 38(5):881–895. doi:10.1016/j.immuni.2013.02.008

Omilusik K, Priatel JJ, Chen X, Wang YT, Xu H, Choi KB, Gopaul R, McIntyre-Smith A, Teh HS, Tan R, Bech-Hansen NT, Waterfield D, Fedida D, Hunt SV, Jefferies WA (2011) The $Ca_v1.4$ calcium channel is a critical regulator of T cell receptor signaling and naive T cell homeostasis. Immunity 35(3):349–360. doi:10.1016/j.immuni.2011.07.011

Oritani K, Kincade PW (1996) Identification of stromal cell products that interact with pre-B cells. J Cell Biol 134(3):771–782

Park CY, Hoover PJ, Mullins FM, Bachhawat P, Covington ED, Raunser S, Walz T, Garcia KC, Dolmetsch RE, Lewis RS (2009) STIM1 clusters and activates CRAC channels via direct binding of a cytosolic domain to orai1. Cell 136(5):876–890. doi:10.1016/j.cell.2009.02.014

Parker NJ, Begley CG, Smith PJ, Fox RM (1996) Molecular cloning of a novel human gene (D11S4896E) at chromosomal region 11p15.5. Genomics 37(2):253–256. doi:10.1006/geno.1996.0553

Partiseti M, Le Deist F, Hivroz C, Fischer A, Korn H, Choquet D (1994) The calcium current activated by T cell receptor and store depletion in human lymphocytes is absent in a primary immunodeficiency. J Biol Chem 269(51):32327–32335

Penna A, Demuro A, Yeromin AV, Zhang SL, Safrina O, Parker I, Cahalan MD (2008) The CRAC channel consists of a tetramer formed by stim-induced dimerization of orai dimers. Nature 456 (7218):116–120. doi:10.1038/nature07338

Picard C, McCarl CA, Papolos A, Khalil S, Luthy K, Hivroz C, LeDeist F, Rieux-Laucat F, Rechavi G, Rao A, Fischer A, Feske S (2009) STIM1 mutation associated with a syndrome of immunodeficiency and autoimmunity. N Engl J Med 360(19):1971–1980. doi:10.1056/NEJMoa0900082

Putney JW Jr (1986) A model for receptor-regulated calcium entry. Cell Calcium 7(1):1–12

Roos J, DiGregorio PJ, Yeromin AV, Ohlsen K, Lioudyno M, Zhang S, Safrina O, Kozak JA, Wagner SL, Cahalan MD, Velicelebi G, Stauderman KA (2005) STIM1, an essential and conserved component of store-operated Ca^{2+} channel function. J Cell Biol 169(3):435–445. doi:10.1083/jcb.200502019

Sabbioni S, Barbanti-Brodano G, Croce CM, Negrini M (1997) GOK: a gene at 11p15 involved in rhabdomyosarcoma and rhabdoid tumor development. Cancer Res 57(20):4493–4497

Sakaguchi S, Sakaguchi N, Asano M, Itoh M, Toda M (1995) Immunologic self-tolerance maintained by activated T cells expressing IL-2 receptor alpha-chains (CD25). Breakdown of a single mechanism of self-tolerance causes various autoimmune diseases. J Immunol 155 (3):1151–1164

Schmidt SV, Nino-Castro AC, Schultze JL (2012) Regulatory dendritic cells: there is more than just immune activation. Front Immunol 3:274. doi:10.3389/fimmu.2012.00274

Stathopulos PB, Zheng L, Li GY, Plevin MJ, Ikura M (2008) Structural and mechanistic insights into STIM1-mediated initiation of store-operated calcium entry. Cell 135(1):110–122. doi:10.1016/j.cell.2008.08.006

Stathopulos PB, Zheng L, Ikura M (2009) Stromal interaction molecule (STIM) 1 and STIM2 calcium sensing regions exhibit distinct unfolding and oligomerization kinetics. J Biol Chem 284(2):728–732. doi:10.1074/jbc.C800178200

Stritesky GL, Jameson SC, Hogquist KA (2012) Selection of self-reactive T cells in the thymus. Annu Rev Immunol 30:95–114. doi:10.1146/annurev-immunol-020711-075035

Takahashi T, Tagami T, Yamazaki S, Uede T, Shimizu J, Sakaguchi N, Mak TW, Sakaguchi S (2000) Immunologic self-tolerance maintained by $CD25^{+}CD4^{+}$ regulatory T cells constitutively expressing cytotoxic T lymphocyte-associated antigen 4. J Exp Med 192(2):303–310

Vazquez G, de Boland AR, Boland R (1997) Stimulation of Ca^{2+} release-activated Ca^{2+} channels as a potential mechanism involved in non-genomic 1,25$(OH)_2$-vitamin D_3-induced Ca^{2+} entry in skeletal muscle cells. Biochem Biophys Res Commun 239(2):562–565. doi:10.1006/bbrc.1997.7501

Vig M, Peinelt C, Beck A, Koomoa DL, Rabah D, Koblan-Huberson M, Kraft S, Turner H, Fleig A, Penner R, Kinet JP (2006) CRACM1 is a plasma membrane protein essential for store-operated Ca^{2+} entry. Science 312(5777):1220–1223. doi:10.1126/science.1127883

Williams RT, Manji SS, Parker NJ, Hancock MS, Van Stekelenburg L, Eid JP, Senior PV, Kazenwadel JS, Shandala T, Saint R, Smith PJ, Dziadek MA (2001) Identification and characterization of the STIM (stromal interaction molecule) gene family: coding for a novel class of transmembrane proteins. Biochem J 357(Pt 3):673–685

Wu MM, Buchanan J, Luik RM, Lewis RS (2006) Ca^{2+} store depletion causes STIM1 to accumulate in ER regions closely associated with the plasma membrane. J Cell Biol 174 (6):803–813. doi:10.1083/jcb.200604014

Yu X, Harden K, Gonzalez LC, Francesco M, Chiang E, Irving B, Tom I, Ivelja S, Refino CJ, Clark H, Eaton D, Grogan JL (2009) The surface protein TIGIT suppresses T cell activation by promoting the generation of mature immunoregulatory dendritic cells. Nat Immunol 10 (1):48–57. doi:10.1038/ni.1674

Yuan JP, Zeng W, Dorwart MR, Choi YJ, Worley PF, Muallem S (2009) SOAR and the polybasic STIM1 domains gate and regulate Orai channels. Nat Cell Biol 11(3):337–343. doi:10.1038/ncb1842

Zhang SL, Yeromin AV, Zhang XH, Yu Y, Safrina O, Penna A, Roos J, Stauderman KA, Cahalan MD (2006) Genome-wide RNAi screen of Ca^{2+} influx identifies genes that regulate Ca^{2+} release-activated Ca^{2+} channel activity. Proc Natl Acad Sci U S A 103(24):9357–9362. doi:10.1073/pnas.0603161103

Zweifach A, Lewis RS (1993) Mitogen-regulated Ca^{2+} current of T lymphocytes is activated by depletion of intracellular Ca^{2+} stores. Proc Natl Acad Sci U S A 90(13):6295–6299

Part VIII
Chronic Inflammation and Ageing

Chapter 41
Pathophysiological Role of Chronic Inflammation in Ageing-Associated Diseases

Yuichi Ikeda, Hiroshi Akazawa, and Issei Komuro

Abstract Chronic low-grade systemic inflammation in the absence of overt infection is a hallmark of aged tissues and age-related disorders. Although chronic inflammation is primarily mediated by the local production of inflammatory mediators, recent studies suggest that blood-borne factors may also directly regulate the ageing process. In this chapter, we describe an unexpected role of complement C1q in canonical Wnt signalling and age-related phenotypes. Consistent with a previous report demonstrating that canonical Wnt signalling is augmented in various aged tissues, serum and tissue levels of C1q increase with age. C1q directly binds to Frizzled with high affinity and recruits C1r and C1s to form the C1 complex. Activated C1s in the C1 complex cleave the extracellular domain of LRP6, resulting in constitutive activation of canonical Wnt signalling. This newly identified C1q-canonical Wnt pathway promotes ageing-associated impairment of skeletal muscle regeneration and hypertensive arterial remodeling. These findings demonstrate that C1q promotes age-related phenotypes in a complement pathway-independent manner. The results also provide a molecular link between age-related disorders and a critical component of the classical complement pathway involved in chronic inflammation.

Keywords Chronic inflammation • Age-related phenotypes • Complement pathway • C1q • Wnt signaling • Ectodomain shedding • Hypertensive arterial remodeling • Skeletal muscle regeneration

Y. Ikeda
Department of Cardiovascular Medicine, Graduate School of Medicine, The University of Tokyo, Tokyo, Japan

H. Akazawa • I. Komuro (✉)
Department of Cardiovascular Medicine, Graduate School of Medicine, The University of Tokyo, Tokyo, Japan

AMED-CREST, Tokyo, Japan
e-mail: komuro-tky@umin.ac.jp

M. Miyasaka, K. Takatsu (eds.), *Chronic Inflammation*,
DOI 10.1007/978-4-431-56068-5_41

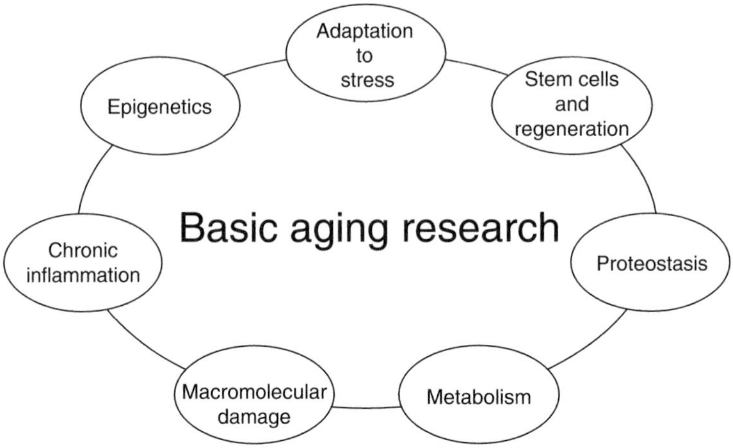

Fig. 41.1 The seven major themes in basic ageing research

41.1 Introduction

41.1.1 Chronic Inflammation Contributes to the Pathogenesis of Age-Associated Diseases

Ageing is a significant risk factor for chronic diseases, including neurodegenerative disorders, metabolic syndromes, cancers, and cardiovascular diseases. In order to develop preventative therapeutics for age-associated diseases, it is critical to understand the pathophysiological processes that promote ageing. Basic research into the ageing process is based on seven interrelated themes: (1) adaptation to stress, (2) epigenetics, (3) chronic inflammation, (4) macromolecular damage, (5) metabolism, (6) proteostasis and (7) stem cells and regeneration (Fig. 41.1; Kennedy et al. 2014).

Chronic low-grade systemic inflammation in the absence of overt infection is a hallmark of aged tissues and most, if not all, age-related disorders. This complex phenomenon termed "inflammaging" is one of the major themes of basic ageing research as noted above (Franceschi and Campisi 2014). However, the etiology of inflammaging and its roles in the pathogenesis of age-associated diseases remains elusive. Acute inflammation facilitates tissue repair, and therefore exerts beneficial effects in pathological conditions such as acute infections and traumatic tissue injury. In contrast, low-grade chronic inflammation usually results in tissue degeneration. Chronic inflammation in aged tissues can be driven by the production of endogenous host-derived cell debris and proinflammatory cytokines. Self-debris released from damaged cells accumulates due to impaired removal capacity during ageing, and thereby activates the innate immune response by mimicking bacterial products. Unlike cell debris, proinflammatory cytokines are produced from intact senescent cells residing in aged tissues. Senescent cells, which do not proliferate, are metabolically active and secrete numerous humoural factors that promote age-related phenotypes by altering the tissue microenvironment (Lasry and Ben-Neriah 2015).

Age-related alterations in the immune system also affect inflammaging. Adaptive immunity declines with age, whereas innate immunity becomes hyperactive. Notably, dysregulation of the complement pathway, one of the critical components of the innate immune system, has been reported to trigger age-related macular degeneration (AMD), the most common cause of blindness in the elderly (Anderson et al. 2010; Gallenga et al. 2014).

41.1.2 *Blood-Borne Factors Directly Regulate Age-Related Phenotypes*

Age-related phenotypes can be modulated by the balance between intrinsic tissue-derived factors and extrinsic factors from the systemic circulation. Although there is compelling evidence that the local production of proinflammatory cytokines promotes age-associated disorders, the contribution of systemic factors to the pathogenesis of age-associated disorders remains relatively uncertain. Increased levels of inflammatory mediators in the serum are therefore generally considered simply to reflect leakage from the local production. However, exciting findings indicating that systemic factors directly regulate the ageing processes have recently emerged (Conboy et al. 2005); (Brack et al. 2007); (Villeda et al. 2011); (Loffredo et al. 2013); (Villeda et al. 2014).

Some of these results were demonstrated by adopting an experimental paradigm known as heterochronic parabiosis, in which two mice of different ages are surgically joined to share a single systemic circulation (Fig. 41.2; Conboy et al. 2013). Isochronic parabiosis, in which either two young mice or two old mice are joined, serves as a control. In the model of heterochronic parabiosis, old mice are

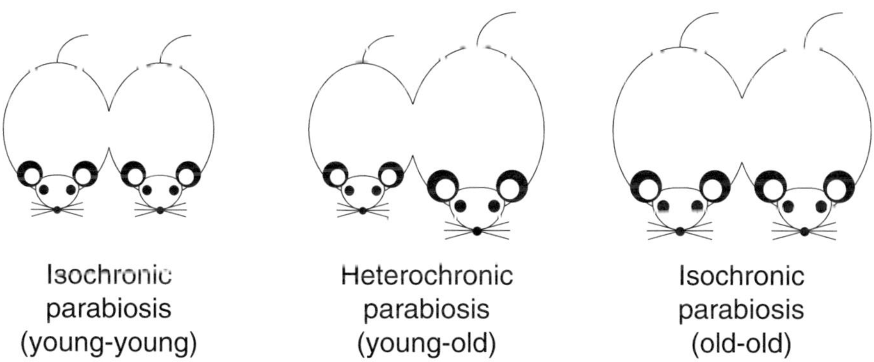

Fig. 41.2 Schematic of parabiotic pairings

continuously exposed to the systemic circulation of young mice, and freely exchange cells and humoural factors under physiological conditions.

Utilising heterochronic parabiosis, Loffredo et al. examined whether circulating factors specific to a young circulatory environment might reverse age-related cardiac hypertrophy (Loffredo et al. 2013). Surprisingly, antihypertrophic effects of the young circulation were immediately observed by blinded morphological analysis. After a 4-week exposure to the young circulation, cardiac hypertrophy in old mice dramatically regressed. Furthermore, a comprehensive proteomic analysis of plasma samples identified growth differentiation factor 11 (GDF11), a TGF-β superfamily member, as a potential age-regulated antihypertrophic factor because its plasma levels decrease with age. Indeed, age-related cardiac hypertrophy in old mice could be reversed by daily intraperitonial injection of GDF11 for 30 days.

In contrast to GDF11, the plasma levels of the chemokine CCL11 were found to increase with age and inversely correlate with adult neurogenesis and cognitive function (Villeda et al. 2011). Consistent with this, it has been reported that exposing young mice to an old systemic circulation resulted in decreased synaptic plasticity and impaired spatial learning and memory. Notably, elevating the plasma levels of CCL11 in young mice by injecting the recombinant protein led to impaired adult neurogenesis and cognitive function deficits.

An age-related decline in stem-cell activity results in impaired tissue regenerative potential, which has been proposed to drive, at least in part, age-related disorders. Blood-borne factors may also be involved in adult stem-cell function. Heterochronic parabiosis was performed to examine the influence of blood-borne factors on skeletal muscle stem cell (i.e. satellite cell) function. The exposure of old mice to a young circulation restored the impaired proliferation and regenerative potential of aged satellite cells, suggesting that systemic factors in the young circulation exert beneficial effects under these conditions (Conboy et al. 2005). Conversely, blood-borne factors in the old circulation were found to mediate the detrimental lineage conversion of satellite cells (Brack et al. 2007). Aged satellite cells in old mice convert from a myogenic to a fibrogenic lineage when they proliferate. This lineage conversion is associated with an increase in connective tissue deposition (i.e. fibrosis) in aged skeletal muscle. Brack et al. elegantly demonstrated that canonical Wnt signalling was activated in aged satellite cells, and that the age-related myogenic-to-fibrogenic conversion was suppressed by inhibition of the activated Wnt signalling. They further showed that fibrosis in regenerating muscle of old mice was improved by the systemic injection of a soluble Wnt inhibitor, such as DKK1 and sFRP3. Together, these findings indicate that increased Wnt signalling in myogenic progenitors in old mice may result from the increased levels of Wnt or Wnt-like molecules in aged serum.

These three important examples described above thus clearly demonstrate that age-related phenotypes in rodents can be reversed simply by changing the systemic milieu.

41.2 Unexpected Molecular Link Between Chronic Inflammation and Age-Related Phenotypes

41.2.1 Identification of Complement C1q as an Activator of Canonical Wnt Signalling in Aged Serum

Inspired by the work of Brack et al., we hypothesised that activators of canonical Wnt signalling exist in aged mouse serum. Because Wnt molecules are tightly trapped in the extracellular matrix, they usually function in an autocrine/paracrine manner. We thus assumed that the Wnt activators in aged serum are distinct from authentic Wnt ligands (Naito et al. 2012).

Consistent with the previous report, Wnt activity in aged mouse serum was found to be higher than that in young serum in a TOPFLASH assay (Fig. 41.3a). TOPFLASH assays were performed utilising HEK293 cells harbouring a luciferase reporter driven by eight TCF/LEF-binding sites for monitoring β-catenin–mediated transcription. Interestingly, even higher activity was observed in the serum obtained from mice with heart failure (Fig. 41.3a). In order to purify the Wnt activators, a chimeric protein (Fz8/Fc) was constructed by fusing a cysteine-rich domain (CRD) of Frizzled 8 (Fz8), a critical component of the Wnt receptor complex, to IgG/Fc. Using Fz8/Fc as a probe, binding partners of Fz8/Fc were precipitated from the mouse serum, and SDS-PAGE revealed that a 26 kDa protein was upregulated in the serum from mice with heart failure (Fig. 41.3b). Mass spectrometric analysis identified this protein as complement C1qa, a major component of complement C1q. C1q is composed of 18 polypeptides: 6 C1qa, 6 C1qb, and 6 C1qc, each encoded by three different genes.

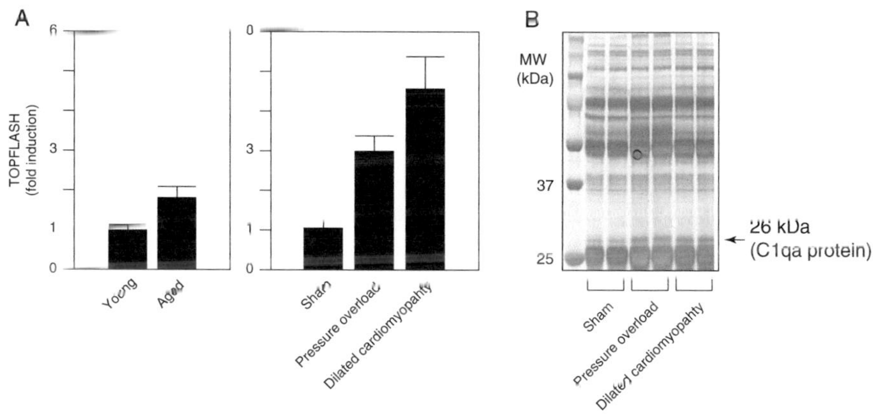

Fig. 41.3 Enhanced Wnt-like activity in aged mouse serum (Reprinted from *Cell*, 149, Naito AT et al., 'Complement C1q activates canonical Wnt signalling and promotes ageing-related phenotypes', 1298–1313, 2012, with permission from Elsevier). (**a**) Serum-induced Wnt signalling assessed in the TOPFLASH assay was higher in aged mouse serum and serum from mice with heart failure. (**b**) SDS-PAGE demonstrated that a 26 kDa protein was upregulated in the serum of mice with heart failure

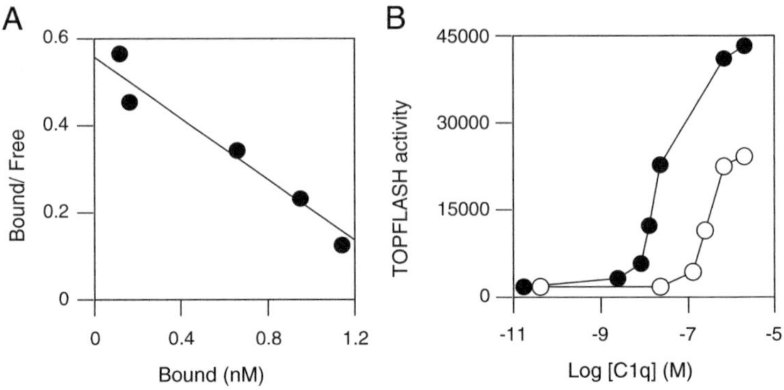

Fig. 41.4 In vitro pharmacology of C1q at the Frizzled receptor (Fz8; Reprinted from *Cell*, 149, Naito AT et al., 'Complement C1q activates canonical Wnt signalling and promotes ageing-related phenotypes', 1298–1313, 2012, with permission from Elsevier). (**a**) A Scatchard plot analysis of the binding of C1q to Fz8 CRD. (**b**) Dose–response analyses of C1q in the TOPFLASH assay. The activity of C1q was potentiated by Fz8 overexpression (*filled circles*)

Specific interactions between C1q and Fz receptors were then tested. A Scatchard plot analysis demonstrated a specific binding of C1q to Fz8 CRD with high-affinity (Kd = 2.8 nM) comparable to that of Wnt3a (Kd = 1.25 nM; Fig. 41.4a). Although the binding affinity was similar, the activity of C1q in the TOPFLASH assay was 200-fold weaker (EC_{50} = 259 nM) than that of Wnt3a (EC_{50} = 1.27 nM). Interestingly, however, overexpression of Fz8 potentiated the C1q activity by 13-fold (EC_{50} 259 nM to 22.8 nM), whereas it did not affect Wnt3a activity (Fig. 41.4b). These results suggest that C1q and authentic Wnt ligands activate canonical Wnt signalling in a distinct manner.

We then asked if C1q is required for serum-induced activation of Wnt signalling. Less TOPFLASH activity was observed in C1q-depleted serum or serum from C1qa-knockout mice, and Fz8/Fc treatment did not further decrease the TOPFLASH activity induced by these serums (Fig. 41.5a). Moreover, plasma levels of C1qa were confirmed to increase with age in wild-type (WT) mice (Fig. 41.5b), and there was no age-dependent increase in serum-induced TOPFLASH activity when serum from C1qa-knockout mice was tested (Fig. 41.5c). Based on these findings, we concluded that serum-induced activation of canonical Wnt signalling is mediated, at least in part, by C1q.

41.2.2 C1q Activates Canonical Wnt Signalling by Recruiting C1 Complex That Cleaves the Extracellular Domain of LRP6

The classical complement pathway is initiated by formation of the C1 complex composed of C1q, C1r, and C1s. Host cells utilised in the TOPFLASH assay secrete

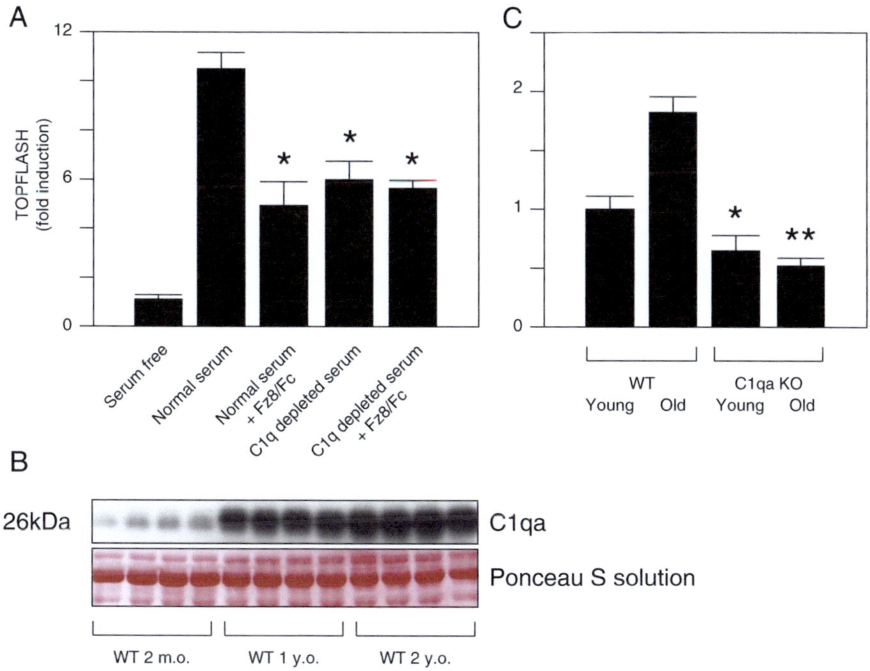

Fig. 41.5 C1q is required for serum-induced activation of canonical Wnt signalling (Reprinted from *Cell*, 149, Naito AT et al., 'Complement C1q activates canonical Wnt signalling and promotes aging-related phenotypes', 1298–1313, 2012, with permission from Elsevier). (**a**) Wnt signalling activation by mouse serum in the TOPFLASH assay was partially inhibited by Fz/Fc (10 μg/ml) or C1q depletion. (**b**) Serum levels of C1qa increase with age. (**c**) Age-dependent increase in serum-induced TOPFLASH activity was observed in wild-type mice, but not in C1qa knockout mice

both C1r, and C1s into the cell culture media. Inasmuch as purified C1q activated Wnt signalling even in a serum-free condition, we hypothesised that endogenous C1r and C1s from the host cells may be required for C1q-induced activation of Wnt signalling in the TOPFLASH assay. Consistent with this hypothesis, TOPFLASH activity was strongly inhibited either by knockdown of endogenous C1r/C1s in the host cells or by C1s inhibitor treatment. It is known that deletion of the extracellular domain of LRP6, a component of the Wnt receptor complex, leads to constitutive activation of canonical Wnt signalling (Mao et al. 2001); (Liu et al. 2003). C1s is a proenzyme, and once activated, it is capable of cleaving cell surface molecules such as MHC class I (Eriksson and Nissen 1990). We thus asked if LRP6 acts as a substrate for activated C1s protease. As expected, the extracellular domain of LRP6 was cleaved by activated C1s in vitro. Notably, the amino acid sequence of the cleavage site was well conserved among species. To test if this C1s-mediated cleavage is dependent on C1q, cells overexpressing LRP6 were treated with normal serum or C1q-depleted serum. The cleaved extracellular fragment of LRP6 was detected in the normal serum, but not in the C1q-depleted serum. Likewise, the

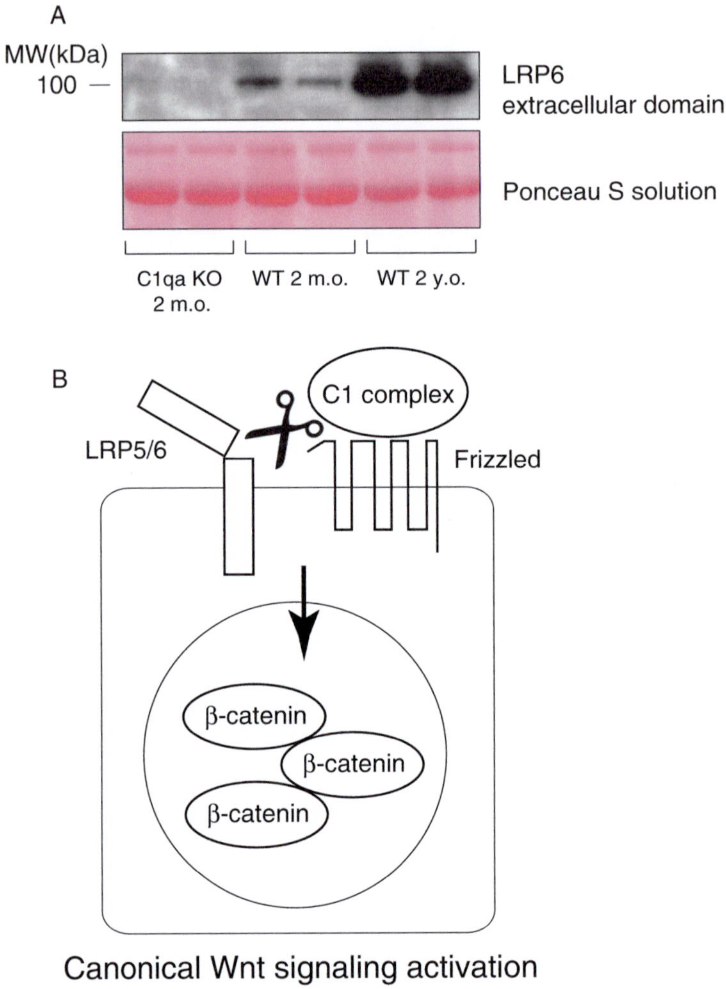

Fig. 41.6 C1q activates canonical Wnt signalling by recruiting the C1 complex that cleaves the extracellular domain of LRP6 (Reprinted from *Cell*, 149, Naito AT et al., 'Complement C1q activates canonical Wnt signalling and promotes aging-related phenotypes', 1298–1313, 2012, with permission from Elsevier). (**a**) Age-dependent increase in the amount of cleaved extracellular fragment of LRP6 was observed in serum from wild-type mice. In C1qa knockout mice, no cleaved fragment was detected. (**b**) Schematic of C1q-induced activation of canonical Wnt signalling. Cleavage of the extracellular domain of LRP6 by the C1 complex leads to constitutive activation of canonical Wnt signalling

cleaved fragment of LRP6 was detected in the serum from wild-type mice, but not in the serum from C1qa knockout mice (Fig. 41.6a).

C1s-mediated cleavage generates LRP6 lacking amino acids 21–792 (Del-LRP6). Transfection of Del-LRP6 alone stimulated TOPFLASH activity in the absence of authentic Wnt ligands, indicating that C1s-mediated cleavage of

LRP6 is sufficient to activate canonical Wnt signalling (Fig. 41.6b). Del-LRP6 was found to be phosphorylated in the transfected cells even without ligand stimulation. Furthermore, transfection of Del-LRP6 induced the phosphorylation of wild-type LRP6 (WT-LRP6) that was simultaneously transfected. These findings suggest that cleavage of a small fraction of LRP6 can amplify canonical Wnt signalling by phosphorylating endogenous LRP6 that remains uncleaved (Tamai et al. 2004); (Zeng et al. 2005).

41.2.3 Impaired Tissue Regeneration in Aged Skeletal Muscle is Mediated by Increased Levels of C1q

We next asked if increased levels of C1q during ageing result in augmented Wnt signalling in vivo. The expression levels of C1q in skeletal muscle were higher in aged mice than those in young mice (Fig. 41.7a). Consistent with this, an age-associated increase in the expression of Axin2, a well-established Wnt target gene that can track overall Wnt signalling activity, was observed in skeletal muscle from WT mice, but not in that from C1qa knockout mice (Fig. 41.7b).

We then cryoinjured the gastrocnemius muscle of aged WT and C1qa knockout mice followed by treatment with either a neutralising antibody against C1s (M241) or an anti-C5 antibody (BB5.1; Matsumoto and Nagaki 1986). A neutralising antibody against C1s inhibits both C1q-induced Wnt signalling and the complement pathway, whereas an anti-C5 antibody selectively inhibits the complement pathway. In aged mice, C1s inhibition or targeted disruption of C1qa attenuated canonical Wnt signalling and improved tissue fibrosis induced by cryoinjury (Fig. 41.7c). However, selective inhibition of the complement pathway by an anti C5 antibody had no effect (Fig. 41.7c). These results suggest that the impaired tissue regenerative potential in aged skeletal muscle is mediated by augmented C1q-induced canonical Wnt signalling.

41.2.4 C1q Secreted from M2 Macrophages Recruited to the Vasculature Regulates Arterial Remodelling Through the Activation of Canonical Wnt Signalling

Prolonged high blood pressure triggers the structural remodelling of arteries. This arterial remodelling is characterised by proliferation of vascular smooth muscle cells (VSMCs) and infiltration of immune cells. It has been reported that activation of canonical Wnt signalling is involved in VSMC proliferation during injury-induced intimal thickening, and that C1q is abundantly produced from the cells of monocyte/macrophage lineage (Petry et al. 2001). We thus hypothesised that C1q secreted from macrophages recruited to the arterial wall promotes VSMC

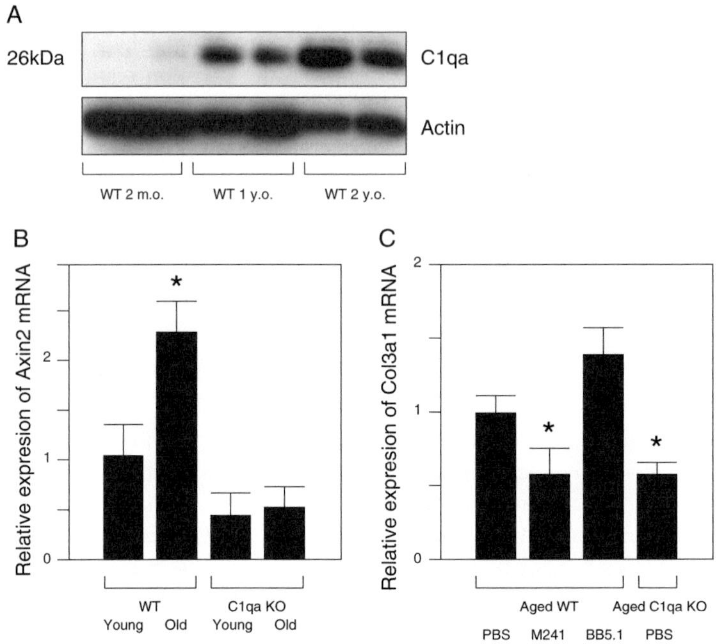

Fig. 41.7 Impaired tissue regeneration in aged skeletal muscle is mediated by increased levels of C1q (Reprinted from *Cell*, 149, Naito AT et al., 'Complement C1q activates canonical Wnt signalling and promotes ageing-related phenotypes', 1298–1313, 2012, with permission from Elsevier). (**a**) C1qa levels in skeletal muscle increase with age. (**b**) Expression levels of Axin2 mRNA in skeletal muscle increase with age in wild type mice, but not in C1qa knockout mice. (**c**) Expression levels of Col3a1 in skeletal muscle were decreased by M241 (a neutralising antibody against C1s) treatment or in C1qa knockout mice, but not by BB5.1 (an anti-C5 antibody) treatment, suggesting that impaired tissue regeneration in aged skeletal muscle is mediated by enhanced C1q-induced canonical Wnt signalling

proliferation and arterial remodelling through the activation of canonical Wnt signalling (Sumida et al. 2015).

It is well known that infusion of angiotensin II (AngII) in mice results in blood-pressure elevation followed by arterial remodelling. In this model, VSMC proliferation was observed as early as 1 week after peptide infusion began. Consistent with the previous report, canonical Wnt signalling was activated in these proliferating VSMCs. To test if this activation of canonical Wnt signalling is required for VSMC proliferation, in vivo administration of a small-molecule β-catenin inhibitor as well as VSMC-specific β-catenin disruption were performed. As expected, less VSMC proliferation was observed in both of these experiments.

Flow cytometric analysis revealed that most of the macrophages recruited to the arterial walls were M2-type. By depleting the cells of monocyte lineage, it was revealed that recruitment of these M2-type macrophages was essential for the activation of canonical Wnt signalling in VSMCs and VSMC proliferation. Furthermore, C1qa knockout mice showed attenuated canonical Wnt signalling in

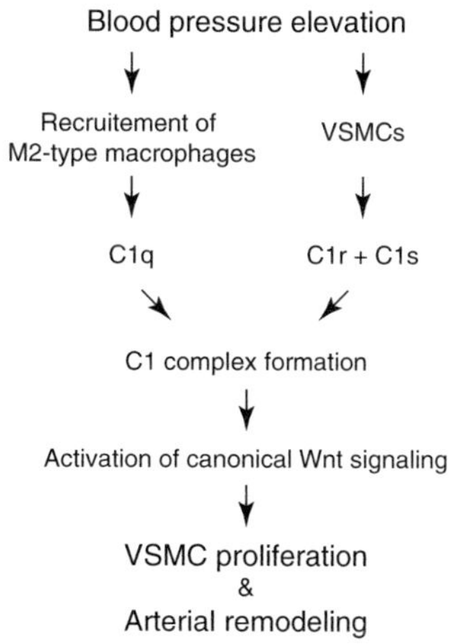

Fig. 41.8 Schematic of a molecular mechanism linking blood pressure elevation to arterial remodelling

VSMCs and less VSMC proliferation. Taken together, these results suggest that C1q secreted from arterial M2-type macrophages plays a critical role in hypertensive arterial remodelling through the activation of canonical Wnt signalling (Fig. 41.8).

41.3 Conclusions

Chronic inflammation is one of the seven major themes in basic ageing research. However, molecular mechanisms that link chronic inflammation with age-related phenotypes are not fully understood.

In this chapter, we described an unexpected role of C1q in the pathogenesis of age-related phenotypes. The complement system, an important component of innate immunity, consists of four activation pathways including the classical pathway that is initiated by the C1 complex composed of C1q, C1r, and C1s. Surprisingly, we found that C1q acts as an activator of canonical Wnt signalling. C1q directly binds with high affinity to Frizzled, one of the two components of the Wnt receptor, and subsequently recruits C1r and C1s to form the C1 complex. Activated C1s in the C1 complex cleaves LRP6, the other component of the Wnt receptor, and this cleavage leads to the activation of canonical Wnt signalling. Recent studies suggested that Wnt signalling is augmented in various aged tissues, and this augmented Wnt signalling promotes mammalian ageing. Consistent with this hypothesis, serum and tissue levels of C1q were found to increase with age. Furthermore, we demonstrated that increased levels of C1q mediate the ageing-associated impairment of skeletal muscle regeneration and hypertensive arterial remodelling through the activation of canonical Wnt signalling. We thus conclude that complement C1q links chronic inflammation to age-related disorders in a complement pathway-independent manner.

References

Anderson DH, Radeke MJ, Gallo NB, Chapin EA, Johnson PT, Curletti CR, Hancox LS, Hu J, Ebright JN, Malek G, Hauser MA, Rickman CB, Bok D, Hageman GS, Johnson LV (2010) The pivotal role of the complement system in aging and age-related macular degeneration: hypothesis re-visited. Prog Retin Eye Res 29(2):95–112. doi:10.1016/j.preteyeres.2009.11.003

Brack AS, Conboy MJ, Roy S, Lee M, Kuo CJ, Keller C, Rando TA (2007) Increased Wnt signaling during aging alters muscle stem cell fate and increases fibrosis. Science 317 (5839):807–810. doi:10.1126/science.1144090

Conboy IM, Conboy MJ, Wagers AJ, Girma ER, Weissman IL, Rando TA (2005) Rejuvenation of aged progenitor cells by exposure to a young systemic environment. Nature 433 (7027):760–764. doi:10.1038/nature03260

Conboy MJ, Conboy IM, Rando TA (2013) Heterochronic parabiosis: historical perspective and methodological considerations for studies of aging and longevity. Aging Cell 12(3):525–530. doi:10.1111/acel.12065

Eriksson H, Nissen MH (1990) Proteolysis of the heavy chain of major histocompatibility complex class I antigens by complement component C1s. Biochim Biophys Acta 1037(2):209–215

Franceschi C, Campisi J (2014) Chronic inflammation (inflammaging) and its potential contribution to age-associated diseases. J Gerontol Ser A Biol Med Sci 69(1):S4–S9. doi:10.1093/gerona/glu057

Gallenga CE, Parmeggiani F, Costagliola C, Sebastiani A, Gallenga PE (2014) Inflammaging: should this term be suitable for age related macular degeneration too? Inflamm Res 63 (2):105–107. doi:10.1007/s00011-013-0684-2

Kennedy BK, Berger SL, Brunet A, Campisi J, Cuervo AM, Epel ES, Franceschi C, Lithgow GJ, Morimoto RI, Pessin JE, Rando TA, Richardson A, Schadt EE, Wyss-Coray T, Sierra F (2014) Geroscience: linking aging to chronic disease. Cell 159(4):709–713. doi:10.1016/j.cell.2014.10.039

Lasry A, Ben-Neriah Y (2015) Senescence-associated inflammatory responses: aging and cancer perspectives. Trends Immunol 36(4):217–228. doi:10.1016/j.it.2015.02.009

Liu G, Bafico A, Harris VK, Aaronson SA (2003) A novel mechanism for Wnt activation of canonical signaling through the LRP6 receptor. Mol Cell Biol 23(16):5825–5835

Loffredo FS, Steinhauser ML, Jay SM, Gannon J, Pancoast JR, Yalamanchi P, Sinha M, Dall'Osso C, Khong D, Shadrach JL, Miller CM, Singer BS, Stewart A, Psychogios N, Gerszten RE, Hartigan AJ, Kim MJ, Serwold T, Wagers AJ, Lee RT (2013) Growth differentiation factor 11 is a circulating factor that reverses age-related cardiac hypertrophy. Cell 153 (4):828–839. doi:10.1016/j.cell.2013.04.015

Mao B, Wu W, Li Y, Hoppe D, Stannek P, Glinka A, Niehrs C (2001) LDL-receptor-related protein 6 is a receptor for dickkopf proteins. Nature 411(6835):321–325. doi:10.1038/35077108

Matsumoto M, Nagaki K (1986) Functional analysis of activated C1s, a subcomponent of the first component of human complement, by monoclonal antibodies. J Immunol 137(9):2907–2912

Naito AT, Sumida T, Nomura S, Liu M-L, Higo T, Nakagawa A, Okada K, Sakai T, Hashimoto A, Hara Y, Shimizu I, Zhu W, Toko H, Katada A, Akazawa H, Oka T, Lee J-K, Minamino T, Nagai T, Walsh K, Kikuchi A, Matsumoto M, Botto M, Shiojima I, Komuro I (2012) Complement C1q activates canonical Wnt signaling and promotes aging-related phenotypes. Cell 149(6):1298–1313. doi:10.1016/j.cell.2012.03.047

Petry F, Botto M, Holtappels R, Walport MJ, Loos M (2001) Reconstitution of the complement function in C1q-deficient (C1qa−/−) mice with wild-type bone marrow cells. J Immunol 167 (7):4033–4037

Sumida T, Naito AT, Nomura S, Nakagawa A, Higo T, Hashimoto A, Okada K, Sakai T, Ito M, Yamaguchi T, Oka T, Akazawa H, Lee JK, Minamino T, Offermanns S, Noda T, Botto M, Kobayashi Y, Morita H, Manabe I, Nagai T, Shiojima I, Komuro I (2015) Complement

C1q-induced activation of beta-catenin signalling causes hypertensive arterial remodelling. Nat Commun 6:6241. doi:10.1038/ncomms7241
Tamai K, Zeng X, Liu C, Zhang X, Harada Y, Chang Z, He X (2004) A mechanism for Wnt coreceptor activation. Mol Cell 13(1):149–156
Villeda SA, Luo J, Mosher KI, Zou B, Britschgi M, Bieri G, Stan TM, Fainberg N, Ding Z, Eggel A, Lucin KM, Czirr E, Park JS, Couillard-Despres S, Aigner L, Li G, Peskind ER, Kaye JA, Quinn JF, Galasko DR, Xie XS, Rando TA, Wyss-Coray T (2011) The ageing systemic milieu negatively regulates neurogenesis and cognitive function. Nature 477(7362):90–94. doi:10.1038/nature10357
Villeda SA, Plambeck KE, Middeldorp J, Castellano JM, Mosher KI, Luo J, Smith LK, Bieri G, Lin K, Berdnik D, Wabl R, Udeochu J, Wheatley EG, Zou B, Simmons DA, Xie XS, Longo FM, Wyss-Coray T (2014) Young blood reverses age-related impairments in cognitive function and synaptic plasticity in mice. Nat Med 20(6):659–663. doi:10.1038/nm.3569
Zeng X, Tamai K, Doble B, Li S, Huang H, Habas R, Okamura H, Woodgett J, He X (2005) A dual-kinase mechanism for Wnt co-receptor phosphorylation and activation. Nature 438 (7069):873–877. doi:10.1038/nature04185

Chapter 42
Uterine Cellular Senescence in the Mouse Model of Preterm Birth

Yasushi Hirota

Abstract Recent literature shows that ageing-related chronic inflammation called inflammaging is involved in the pathophysiology of several diseases such as cancers, atherosclerosis, diabetes mellitus, autoimmune diseases, and neurodegenerative diseases. Cellular senescence is one of the major contributors to inflammaging. As for the field of reproductive and perinatal medicine, preterm birth is a serious problem, and it is reported that maternal ageing is associated with the incidence of preterm birth. We recently established a mouse model of preterm delivery induced by uterine p53 deletion, in which uterine cellular senescence in early pregnancy via the activation of mTORC1-p21-Cox2-$PGF_{2\alpha}$ axis leads to both preterm birth and neonatal death and these phenotypes are reversed by the inhibition of mTORC1, p21, or Cox2. Inasmuch as it is reported that p53 activity declines according to ageing, maternal ageing may promote preterm delivery through enhanced cellular senescence in humans. The role of uterine cellular senescence in determining the timing of parturition in mice may promote a better understanding of the mechanism of labour and develop novel strategies against preterm birth in humans.

Keywords Inflammaging • Cellular senescence • Preterm birth • mTOR • p21 • Cox2

42.1 Inflammaging

Ageing is a process in which living animals and organisms irreversibly decline in the cellular and tissue functions. Cancers, arteriosclerosis, diabetes mellitus, autoimmune diseases such as rheumatoid arthritis, and neurodegenerative diseases such as Alzheimer disease are all known as typical age-related diseases, in which chronic inflammation and activated immune responses are lastingly observed. Circulating levels of proinflammatory cytokines such as IL-6 and TNF-α are elevated with age

Y. Hirota (✉)
Department of Obstetrics and Gynecology, Graduate School of Medicine, The University of Tokyo, 7-3-1 Hongo, Bunkyo-ku, Tokyo 113-8655, Japan

PRESTO, Japan Science and Technology Agency, Kawaguchi 332-0012, Japan
e-mail: yhirota-tky@umin.ac.jp

M. Miyasaka, K. Takatsu (eds.), *Chronic Inflammation*,
DOI 10.1007/978-4-431-56068-5_42

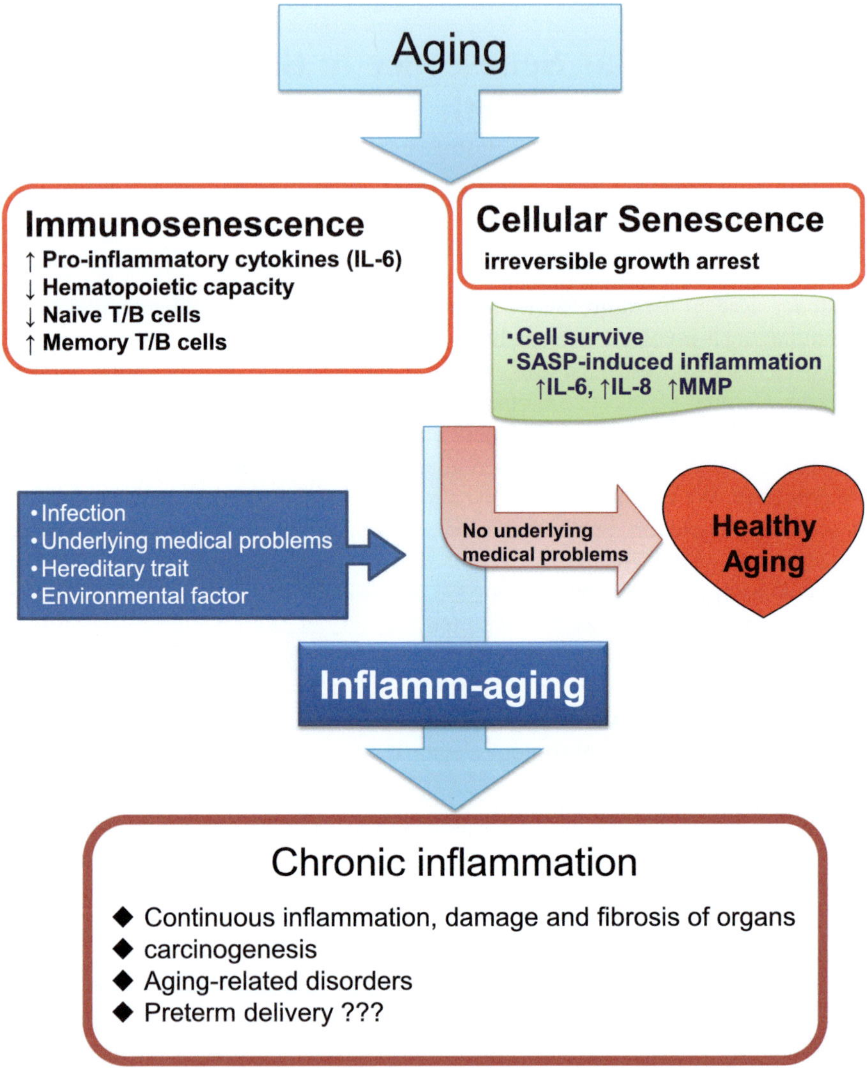

Fig. 42.1 Inflammaging induced by immunosenescence and cellular senescence

(Franceschi et al 2000). Upregulation of proinflammatory cytokines or downregulation of anti-inflammatory cytokines occur asymptomatically and systemically without chronic inflammation due to pathological conditions in the body of elderly people, and may be a potential inducer of various inflammatory diseases in the long run. Therefore, ageing-related inflammation has been sometimes called as 'inflamm-aging' (Franceschi et al 2000). The balance between proinflammatory and anti-inflammatory cytokines are considered to be important for the pathophysiology of inflammaging. Additional stimuli such as genetic and environmental factors are often required for the formation of inflamm-aging, and unless there are those factors, elderly persons can probably acquire healthy ageing (Fig. 42.1).

42.2 Immunosenescence

One possible cause of ageing-related inflammatory conditions is an irreversible reduction of immune function with age called immunosenescence. Immunosenescence occurs in all immune organs, and mainly eliminates functions of acquired immunity by lymphocytes, especially T lymphocytes (Ginaldi et al 2001). The organs in elderly people tend to have inflammatory conditions by their environment factors and underlying diseases. Thus, it is considered that immunosenescence may contribute to the formation of pathological condition in ageing-related disorders.

42.3 Cellular Senescence

Another possible cause of ageing-related inflammatory conditions is an irreversible arrest of cell growth, called cellular senescence. Although it was discovered in 1961 (Hayflick and Moorhead 1961), its molecular aspects have been unclear compared to apoptosis. Cellular senescence can be divided into two different forms: replicative and premature senescence. The former is the growth arrest by telomere shortening after the repeat of normal replication phase, and regulates cellular longevity (Fig. 42.2; (Blasco 2007; Wright and Shay 1995). The latter occurs prior to telomere shortening, and is alternatively stimulated by acute stress such as DNA damage, oxidative stress, and metabolic stress (Fig. 42.2; Campisi and d'Adda di Fagagna 2007; Collado and Serrano 2010; Kuilman et al 2010; Serrano et al 1997).

Senescent cells change cell shape into larger and flattened form with a vacuole-rich cytoplasm and display increased senescence-associated β-galactosidase (SA-β-gal) activity detected by SA-β-gal staining. The cells also show dynamic changes in chromatin texture that can be frequently visualised as so called senescence-associated heterochromatin foci (SAHF). Cyclin-dependent kinase inhibitors such as p21 and p16, the molecules involved in DNA damage response such as γH2AX, and those involved in the formation of SAHF such as the heterochromatin protein HP1γ are generally used as the surrogate markers of cellular senescence. In addition, senescent cells are characterised by global reprogramming of gene expression, including expression and secretion of proinflammatory cytokines and tissue remodelling enzymes referred to as the senescence-associated secretory phenotype (SASP), suggesting that senescent cells communicate with each other and with nonsenescent cells in the environment (Campisi and d'Adda di Fagagna 2007; Collado and Serrano 2010; Kuilman et al 2010). For instance, SASP

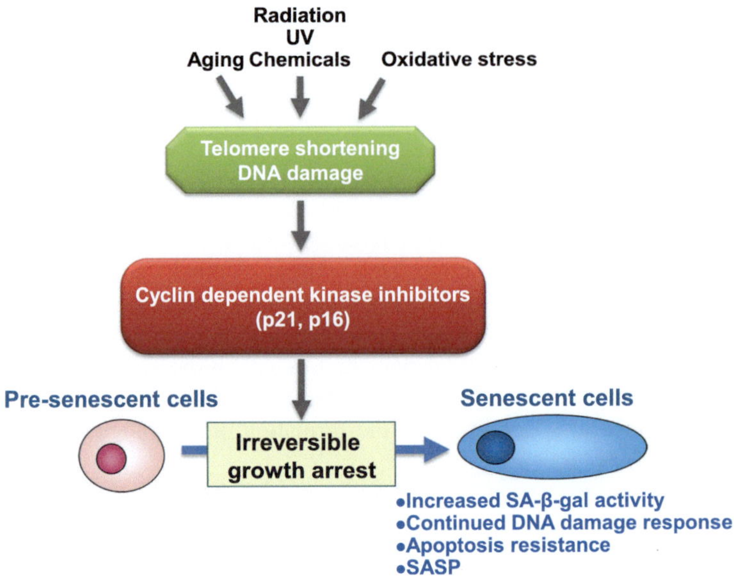

Fig. 42.2 Molecular pathway of cellular senescence

cytokines such as interleukin-6 (IL-6) and IL-8 promote the malignant phenotypes of neighbouring cells (Campisi and d'Adda di Fagagna 2007; Collado and Serrano 2010; Kuilman et al 2010). Moreover, these cytokines autonomously promote cellular senescence in a paracrine manner (Campisi and d'Adda di Fagagna 2007; Collado and Serrano 2010; Kuilman et al 2010). These findings indicate that cellular senescence may form the feedforward loop of chronic inflammation using SASP cytokines. Indeed, cellular senescence is observed in typical chronic inflammatory diseases such as cancers and metabolic syndrome (Campisi and d'Adda di Fagagna 2007; Collado and Serrano 2010; Kuilman et al 2010).

42.4 Preterm Birth

Parturition is an intricate process involved in both endocrine and mechanical stimulation. Nearly 15 million premature births occur worldwide each year (WHO 2012). Preterm birth in humans is defined by birth occurring earlier than 37 weeks of gestation. Premature delivery sometimes damages newborn babies severely. For examples, prematurity is one of the direct causes of neonatal deaths and often leads to developmental impairment and long-term disabilities in many survivors (WHO 2012). Although many factors, including infection, uterine overdistension, maternal age, progesterone (P_4) resistance, and cervical aberration are identified as contributors to preterm birth (Fig. 42.3) (Hirota et al 2010a), this disorder in pregnancy remains to be fully understood. The current therapies such as

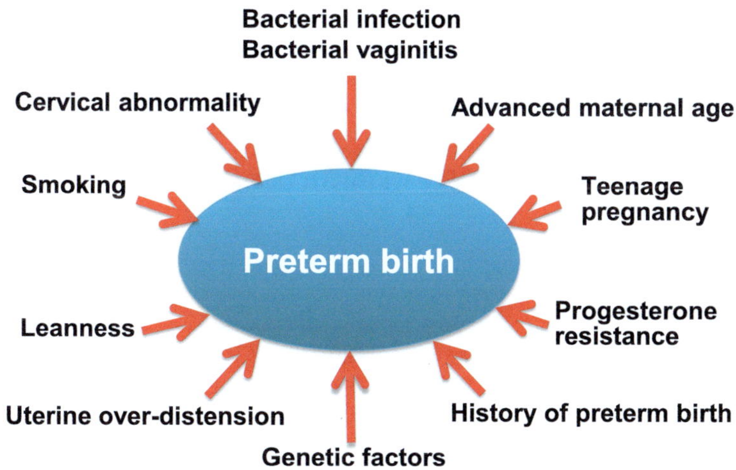

Fig. 42.3 Multiple factors are associated with the pathogenesis of preterm delivery

tocolytic drugs, antibiotics, surgical cerclage, and P_4 supplementation did not reduce the incidence of preterm birth sufficiently, and therefore, it is necessary to establish a new fundamental approach to target this disease (Hirota et al 2010a). Because advanced maternal age is epidemiologically associated with preterm birth (Carolan 2003; Cnattingius et al 1992; Roberts et al 1994), maternal ageing may irreversibly damage the uterus by inflammaging to increase the risk of preterm birth.

42.5 Animal Models of Preterm Birth

Animal models of spontaneous preterm delivery are good tools to understand the underlying mechanism, and to establish new strategies for its prevention and treatment. Although rodent models of preterm delivery induced by proinflammatory drugs such as lipopolysaccharide (LPS), are popularly used (Elovitz and Mrinalini 2004), these models are not ideal because the proinflammatory drugs directly affect ovaries to trigger labour by inducing luteolysis with P_4 withdrawal, which does not occur in human preterm delivery. In addition, LPS produces both uterine and systemic inflammation making the interpretation difficult as to the cause of preterm birth. Rodent models of preterm labour without luteolysis are required, although such models have scarcely existed. Remarkably, we have developed a novel mouse model of spontaneous preterm birth without P_4 (Cha et al 2013). The mice with conditional deletion of uterine *p53* ($p53^{d/d}$) generated by crossing p53-floxed ($p53^{f/f}$) and progesterone receptor-Cre ($Pgr^{Cre/+}$) mice showed increased levels of uterine cyclooxygenase-2 (Cox2) and

$PGF_{2\alpha}$ that induce premature delivery with neonatal death without early luteolysis, mimicking an aspect of human preterm delivery (Cha et al 2013).

42.6 A Mouse Model of Preterm Birth by Conditional Deletion of Uterine p53

Although *p53*$^{d/d}$ mice demonstrated normal implantation of the embryos, decidual cells, uterine stromal cells starting to be differentiated after the attachment of embryos to the uterus, proliferate limitedly in *p53*$^{d/d}$ mice. According to the defective decidual formation, SA-β-gal activity, the most popular marker of cellular senescence, was increased in *p53*$^{d/d}$ deciduae (Cha et al 2013). In addition, p21, a possible regulator of cellular senescence, as well as γH2AX, a DNA damage marker, was upregulated in *p53*$^{d/d}$ deciduae (Cha et al 2013). Because p21 and γH2AX are often used as surrogate markers of cellular senescence, these findings indicate that uterine p53 deletion promotes decidual cellular senescence. The proteomic analyses showed downregulation of a cluster of antioxidant enzymes and proteins regulating mitochondrial functions in p53 null deciduae, suggesting increased oxidative stress and mitochondrial dysfunction in *p53*$^{d/d}$ deciduae (Burnum et al 2012). Despite the restricted decidual growth after implantation, embryonic resorption did not increase. Interestingly, approximately half of pregnant *p53*$^{d/d}$ dams showed preterm delivery and neonatal death without a drop in circulating P_4 levels, mimicking characteristics of human parturition. We also found not only persistent signs of increased uterine cellular senescence but increased levels of Cox2 in later pregnancy before parturition (Cha et al 2013). It has been reported that Cox-derived prostaglandins (PGs) are crucial for the initiation of labor, therefore we presume that an early elevation of uterine Cox2 initiates preterm labor in *p53*$^{d/d}$ females. This hypothesis was clearly supported by the evidence of heightened uterine $PGF_{2\alpha}$ in the confrontation with unchanged levels of Cox1 and other PGs in these mice, and the prevention of their preterm births by oral administration of a Cox2 selective inhibitor celecoxib (Cha et al 2013). Taken together, these findings provide evidence that uterine deficiency of *p53* induces cellular senescence and elevates Cox2-derived $PGF_{2\alpha}$ levels.

42.7 Uterine Cellular Senescence and Preterm Birth

To investigate whether uterine cellular senescence triggers preterm birth, we performed further analyses using *p53*$^{d/d}$ mice. We found that uterine cellular senescence in *p53*$^{d/d}$ females is associated with heightened signalling of mammalian target of rapamycin complex 1 (mTORC1; (Hirota et al 2011). Heightened mTORC1 signalling is a significant contributor to preterm birth, because the phenotype of prematurity and the increase of uterine Cox2 in *p53*$^{d/d}$ mice was rescued by an oral gavage of a low dose of rapamycin, an mTORC1 inhibitor (Hirota et al 2010a). In the in vitro experiment using *Tsc1*$^{-/-}$ mouse embryonic fibroblasts where mTORC1 is activated, the inhibition of mTORC1 signalling negatively regulates the status of p21 and Cox2. We further established the mice with double deletion of p53 and p21, and these mice did not show any symptoms of preterm deliveries (Hirota et al 2011). Moreover, the superimposition of p21 deletion with p53 deletion suppressed cellular senescence and Cox2 expression. These findings suggest that the p53-mTORC1-p21-Cox2-PGF$_{2\alpha}$ axis is a critical component in the timing of birth and an intervention of any of these three targets is capable of reversing preterm delivery in *p53*$^{d/d}$ mice (Cha et al 2013; Hirota et al 2011; Fig. 42.4). This study also discloses that progressive uterine cellular senescence approaching term birth occurs in the normal course of mouse uterus. It is reported that p53 functions decline according to ageing in mice (Feng et al 2007), thus maternal ageing may promote preterm delivery through enhanced cellular

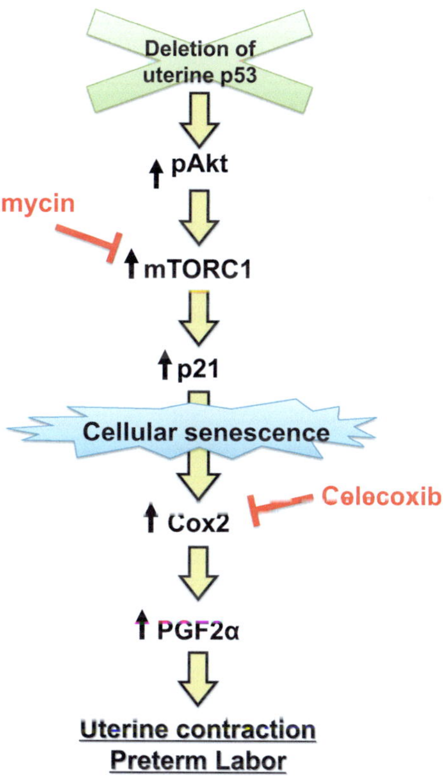

Fig. 42.4 Potential contribution of cellular senescence in the preterm birth of uterine p53-deleted mice

senescence in humans. These studies may help to clarify the mechanism of human parturition and establish novel approaches against preterm delivery.

This mouse model of uterine cellular-senescence–induced preterm labor can be used to explore interactions between genetic predisposition and infection/inflammation during pregnancy. We have shown that *p53*$^{d/d}$ dams subject to even low-grade immunological insults by LPS remarkably increase their predilection towards preterm birth, suggesting that a close relationship between genetic predisposition and environmental insults in exacerbating preterm delivery (Cha et al 2013; Fig. 42.5). These findings have raised many more questions. How and why is the pregnant uterus programmed to undergo cellular senescence during pregnancy? Is maternal ageing involved in the process of this uterine cellular senescence? How does uterine cellular senescence occurring in early pregnancy trigger preterm birth? With infection/inflammation known to amplify secretion of SASP cytokines IL-6 and IL-8 (Liang et al 1996; Sheldon and Roberts 2010), does the SASP contribute to premature birth? Can diets or endocrine disruptors influence cellular senescence to alter timing of birth? Answering these questions will require more intensive efforts which may enable the development of new preventive strategies by handling key regulators in the labour pathway.

42.8 Conclusions

Our results emphasise a new role of uterine cellular senescence involving the p53-mTORC1-p21-Cox2-$PGF_{2\alpha}$ signalling axis in determining the timing of birth (Fig. 42.5; Cha et al 2013; Hirota et al 2011; Hirota et al 2010b). Because ageing is a contributing factor to cellular senescence (Collado et al 2007) and advanced maternal age is epidemiologically associated with preterm birth (Carolan 2003; Cnattingius et al 1992; Roberts et al 1994), it is possible that uterine cellular senescence due to maternal ageing can increase the risk of preterm birth. *p53* polymorphisms have been implicated with ageing and life span in humans (de Keizer et al 2010; Rodier et al 2007). Although there is evidence that certain *p53* polymorphisms in women correlate with recurrent pregnancy failure (Kang et al 2009), there have not been any reports about the relationship between *p53* polymorphisms and the incidence of preterm birth. It also remains to be examined whether the patients with high risk of preterm birth retain genetic alterations of any members of this signalling axis. Further studies are needed to solve this issue.

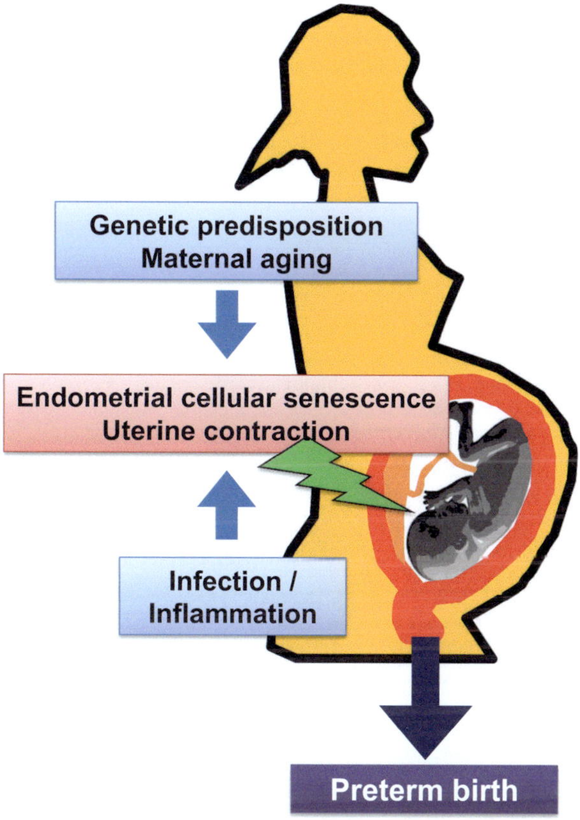

Fig. 42.5 Maternal ageing may promote preterm birth through uterine cellular senescence

Acknowledgments This work was supported by the Precursory Research for Embryonic Science and Technology and the Grant-in-Aid for Scientific Research from Japan Society for the Promotion of Science (Project Numbers: 24689062, 26670713, 26112506, 26112703, 40598653)

References

Blasco MA (2007) Telomere length, stem cells and aging. Nat Chem Biol 3(10):640–649

Burnum KE, Hirota Y, Baker ES, Yoshie M, Ibrahim YM, Monroe ME, Anderson GA, Smith RD, Daikoku T, Dey SK (2012) Uterine deletion of Trp53 compromises antioxidant responses in the mouse decidua. Endocrinology 153(9):4568–4579

Campisi J, d'Adda di Fagagna F (2007) Cellular senescence: when bad things happen to good cells. Nat Rev Mol Cell Biol 8(9):729–740

Carolan M (2003) The graying of the obstetric population: implications for the older mother. J Obstet Gynecol Neonatal Nurs 32(1):19–27

Cha J, Bartos A, Egashira M, Haraguchi H, Saito-Fujita T, Leishman E, Bradshaw H, Dey SK, Hirota Y (2013) Combinatory approaches prevent preterm birth profoundly exacerbated by gene-environment interactions. J Clin Invest 123(9):4063–4075

Cnattingius S, Forman MR, Berendes HW, Isotalo L (1992) Delayed childbearing and risk of adverse perinatal outcome. A population-based study. JAMA 268(7):886–890
Collado M, Blasco MA, Serrano M (2007) Cellular senescence in cancer and aging. Cell 130 (2):223–233
Collado M, Serrano M (2010) Senescence in tumours: evidence from mice and humans. Nat Rev Cancer 10(1):51–57
de Keizer PL, Laberge RM, Campisi J (2010) p53: pro-aging or pro-longevity? Aging (Albany NY) 2(7):377–379
Elovitz MA, Mrinalini C (2004) Animal models of preterm birth. Trends Endocrinol Metab 15 (10):479–487
Feng Z, Hu W, Teresky AK, Hernando E, Cordon-Cardo C, Levine AJ (2007) Declining p53 function in the aging process: a possible mechanism for the increased tumor incidence in older populations. Proc Natl Acad Sci U S A 104(42):16633–16638
Franceschi C, Bonafe M, Valensin S, Olivieri F, De Luca M, Ottaviani E, De Benedictis G (2000) Inflamm-aging. An evolutionary perspective on immunosenescence. Ann N Y Acad Sci 908:244–254
Ginaldi L, Loreto MF, Corsi MP, Modesti M, De Martinis M (2001) Immunosenescence and infectious diseases. Microbes Infect 3(10):851–857
Hayflick L, Moorhead PS (1961) The serial cultivation of human diploid cell strains. Exp Cell Res 25:585–621
Hirota Y, Cha J, Dey SK (2010a) Revisiting reproduction: prematurity and the puzzle of progesterone resistance. Nat Med 16(5):529–531
Hirota Y, Cha J, Yoshie M, Daikoku T, Dey SK (2011) Heightened uterine mammalian target of rapamycin complex 1 (mTORC1) signaling provokes preterm birth in mice. Proc Natl Acad Sci U S A 108(44):18073–18078
Hirota Y, Daikoku T, Tranguch S, Xie H, Bradshaw HB, Dey SK (2010b) Uterine-specific p53 deficiency confers premature uterine senescence and promotes preterm birth in mice. J Clin Invest 120(3):803–815
Kang HJ, Feng Z, Sun Y, Atwal G, Murphy ME, Rebbeck TR, Rosenwaks Z, Levine AJ, Hu W (2009) Single-nucleotide polymorphisms in the p53 pathway regulate fertility in humans. Proc Natl Acad Sci U S A 106(24):9761–9766
Kuilman T, Michaloglou C, Mooi WJ, Peeper DS (2010) The essence of senescence. Genes Dev 24(22):2463–2479
Liang L, Kover K, Dey SK, Andrews GK (1996) Regulation of interleukin-6 and interleukin-1 beta gene expression in the mouse deciduum. J Reprod Immunol 30(1):29–52
Roberts CL, Algert CS, March LM (1994) Delayed childbearing--are there any risks? Med J Aust 160(9):539–544
Rodier F, Campisi J, Bhaumik D (2007) Two faces of p53: aging and tumor suppression. Nucleic Acids Res 35(22):7475–7484
Serrano M, Lin AW, McCurrach ME, Beach D, Lowe SW (1997) Oncogenic ras provokes premature cell senescence associated with accumulation of p53 and p16INK4a. Cell 88 (5):593–602
Sheldon IM, Roberts MH (2010) Toll-like receptor 4 mediates the response of epithelial and stromal cells to lipopolysaccharide in the endometrium. PLoS One 5(9), e12906
WHO (2012) Born too soon: the global action report on preterm birth. World Health Organization, Geneva
Wright WE, Shay JW (1995) Time, telomeres and tumours: is cellular senescence more than an anticancer mechanism? Trends Cell Biol 5(8):293–297

Part IX
Chronic Inflammation and Bowel Diseases

Chapter 43
Physiological and Pathological Inflammation at the Mucosal Frontline

Yosuke Kurashima and Hiroshi Kiyono

Abstract The intestinal mucosa is covered by a single layer of epithelium, which comprises columnar epithelial cells, goblet cells, Paneth cells, and microfold (M) cells. This tightly associated physical and interactive barrier also provides bactericidal agents, mucins, and fucose to create a symbiotic and protective environment. The initial layers of the mucosal compartment constitutively regulate symbiosis with host and commensal microbiota. This regulation of symbiosis occurs through homeostatic or physiological inflammatory signalling, which is mediated by innate cell populations, including innate lymphoid cells and mast cells. However, damage to the epithelial barrier due to dysbiosis or chemical and physical stressors induces intestinal inflammation (e.g., inflammatory bowel disease [IBD]). The inflamed epithelium then releases damage-associated molecules, such as adenosine triphosphate (ATP), into the extracellular compartment. Extracellular ATP-mediated, purinergic signalling initiates the activation of innate cells (e.g., mast cells) located at the mucosal frontline, which typically function as sentinels against invasion from intestinal infections but also promote the inflammatory process upon its abnormal activation. In this chapter, we discuss these complex mucosal molecular–cellular connections and introduce advanced strategies to

Y. Kurashima
Division of Mucosal Immunology, Department of Microbiology and Immunology, The Institute of Medical Science, The University of Tokyo, 4-6-1 Shirokanedai, Minato-ku, Tokyo 108-8639, Japan

H. Kiyono (✉)
Division of Mucosal Immunology, Department of Microbiology and Immunology, The Institute of Medical Science, The University of Tokyo, 4-6-1 Shirokanedai, Minato-ku, Tokyo 108-8639, Japan

International Research and Development Center for Mucosal Vaccines, The Institute of Medical Science, The University of Tokyo, 4-6-1 Shirokanedai, Minato-ku, Tokyo 108-8639, Japan

AMED-CREST, Tokyo 108-8639, Japan

Department of Immunology, Graduate School of Medicine, Chiba University, 1-8-1 Inohana, Chuo-ku, Chiba-city, Chiba 260-8670, Japan
e-mail: kiyono@ims.u-tokyo.ac.jp

M. Miyasaka, K. Takatsu (eds.), *Chronic Inflammation*,
DOI 10.1007/978-4-431-56068-5_43

control chronic inflammation (e.g., IBD), especially focusing on (1) commensal mutualism, (2) the epithelial barrier, and (3) purinergic inflammatory signalling.

Keywords IBD • Dysbiosis • Mast cells • Innate lymphoid cells • Epithelial cells • Commensal mutualism • Probiotics • Fecal transplantation • Purinergic signaling

43.1 Introduction to Mucosal Immunity

The intestinal mucosa is constantly exposed to numerous and diverse substances, including foods, nutrients, and commensal bacteria, and is an entry site for pathogenic bacteria (e.g., *Vibrio cholerae* and *Salmonella typhi*) (Weinstein et al. 1998; Millet et al. 2014). Various physiological and pathological stimuli initiate well-organized immune responses orchestrated by both innate and acquired immune systems at mucosal surfaces to create the system known as mucosal immunity. The mucosal immune system consists of various cell populations and subsets, such as T cells (e.g., T helper 1, 2, 3, 9, 17, and T regulatory cells [T reg]), IgA-producing cells, myeloid cells (e.g., M1 and M2 macrophages and $CD103^{+}$ and CX_3CR1^{+} dendritic cells [DCs]), mast cells, eosinophils, and recently discovered innate lymphoid cells (ILC1, 2, 3); approximately half of the immune cells in our body are located in the gut. These cell populations dynamically interact and maintain both quiescence and active immune responses to create gut homeostasis. Immune-suppressive cell populations (e.g., Treg, Tr1, Th3, Breg, macrophage, and DC) play important roles in the maintenance of quiescent immunological homeostasis as previously reviewed by Sun M et al. (Sun et al. 2015). However, genetic factors or environmental stressors can disrupt those immunosuppressive properties and the balance between quiescent and active responses in the mucosal compartment, leading to intestinal inflammation, such as in the case of inflammatory bowel disease (IBD) as reviewed elsewhere (Kaser et al. 2010).

The appropriate development and physiological proliferation of immune-cell populations are directly and indirectly regulated through stimulation from non-pathogenic bacteria residing in the luminal compartment of the intestines, namely the commensal microbiota. More than 1000 species of bacteria reside within the lumen of the intestinal compartment; each gram of luminal content contains approximately 10^{12} commensal organisms, almost tenfold the number of human cells in the body. These commensal bacteria provide many beneficial metabolic contributions (e.g., vitamin K, short-chain fatty acids, and folate) (reviewed by Kau et al. 2011; Kunisawa and Kiyono 2015). Commensal bacterial stimulation leads to the production of physiologically suitable amounts of proinflammatory cytokines, such as IL-1β, IL-6, lymphotoxin, and tumour necrosis factor (TNF) α, which are required for the maintenance of mucosal homeostasis (Zaki et al. 2010; Goto et al. 2014); this process is likened to 'physiological inflammation' because proinflammatory cytokines facilitate the physiologic balancing act rather than contribute to pathology (Fig. 43.1). When the inflammatory cytokine signalling cascade (e.g., lymphotoxin and TNFα) is blocked, Peyer's patches (PP: an example

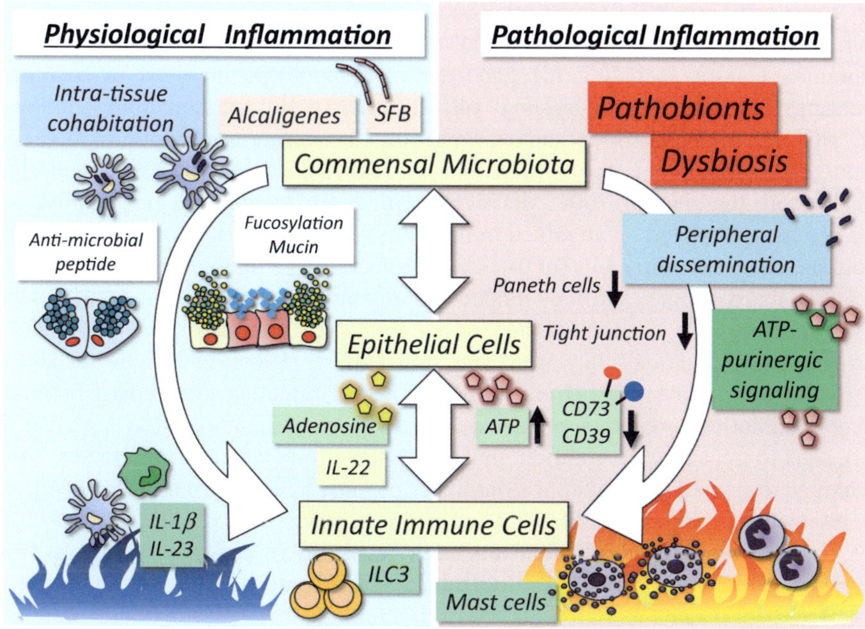

Fig. 43.1 Role of triangular interactions among commensal microbiota, epithelial cells, and innate immune cells for physiological and pathological inflammation. The luminal (e.g., segmented filamentous bacteria [SFB]) and intratissue (e.g., *Alcaligenes*) commensal bacteria regulate both quiescent and active immune responses by stimulating various immune cell populations. Even during steady-state conditions, innate cell populations (especially ILC3s) are constantly reacting to commensal bacteria via DC-derived IL-1β and IL-23 (physiological inflammation). Stimulated ILC3s release IL-22, leading to the induction of fucosylation in epithelial cells and the secretion of mucin and antimicrobial peptides (AMPs) from them. However, once tight junctions are disrupted or when Paneth cells become reduced in number, pathobionts are increased or dysbiosis with bacterial dissemination occurs. In addition, an increase in bacterial or cell-death–mediated ATP–purinergic signalling, accompanied by a reduction in CD73–CD39 pathways, initiates intestinal inflammation (pathological inflammation). Triangular interactions among commensal microbiota, epithelial cells, and innate immune cells contribute to both homeostatic and pathological inflammation. In the context of mucosal barrier biology, further understanding and control of IBD requires focusing on intestinal epithelium, commensal bacterial communication, commensal mutualism, damage-associated inflammatory mediators, and purinergic signalling

of gut-associated lymphoid tissues) shrink (Randall et al. 2008). Mice lacking commensal bacteria, such as germ-free or antibiotic-treated mice, have smaller PPs (Kiyono et al. 1980) and fewer intestinal immune cells, such as IgA-producing cells and intraepithelial lymphocytes (IELs; $CD8^+$ T cells that reside in the intraepithelium), than do their wild-type counterparts (Umesaki et al. 1993; Macpherson et al. 2000). These findings further emphasise the crucial well-being of the intestinal microbiota; its orchestrated production of cytokines contributes to the creation of a healthy intestinal immune system.

In contrast, abnormality of commensal bacteria, known as dysbiosis, induces inappropriate immunological signals and eventually causes intestinal inflammation (Petersen and Round 2014; Fig. 43.1). IBD is a chronic, immune-mediated disease that generally occurs in young adulthood and confers a debilitating life cycle of remission and relapse (Huttenhower et al. 2014). IBD results from numerous and complex factors, including the genotype of components involved in bacterial sensing (e.g., nucleotide-binding oligomerisation domain-containing protein 2 [NOD2]), endoplasmic reticulum stress (e.g., activating transcription factor 6), and glycosylation (e.g., fucosyltransferase 2 [FUT2]), expressed in intestinal epithelial cells (Ogura et al. 2001; McGovern et al. 2010). In addition, the composition of commensal bacteria is involved in the pathogenesis of IBD, and the invasion of luminal bacteria to intracells (or tissues) stimulates innate immune cell populations, such as macrophages, and thus induces an inflammatory milieu (Uo et al. 2013). Anti-inflammatory or immunomodulatory drugs (e.g., sulfasalazine, olsalazine, cyclosporine, 5-aminosalicylates, and anti-TNFα) (Klotz and Schwab 2005; Altwegg and Vincent 2014) as well as strategies targeting commensal bacteria (e.g., antibiotics, probiotics, prebiotics, and fecal transplantation) (Xu et al. 2015) are used to treat IBD. In addition, purinergic signalling, which is initiated by both immune responses and bacterial stimuli, has recently been shown to play important roles in the onset of intestinal inflammation as reviewed previously (Eltzschig et al. 2012; Idzko et al. 2014; Kurashima et al. 2015), and extracellular ATP–purinergic receptor pathways are potential, novel therapeutic targets for Crohn's disease (CD) (Ochoa-Cortes et al. 2014). In this chapter, we discuss these complex connections among the commensal flora, epithelial–mesenchymal cells, and immune cells in the intestinal environment, addressing recent progress and advanced strategies to control chronic IBD and especially focusing on (1) commensal mutualism, (2) the epithelial barrier, and (3) purinergic inflammatory signalling (Fig. 43.1).

43.2 Commensal Mutualism: Dysbiosis and Inflammation

Dysbiosis – the abnormal alteration of/ or disease-associated changes in the intestinal microbiota – is critically involved in the pathology of IBD (Petersen and Round 2014). As reviewed by Simpson HL et al., in general, *Bacteroidetes*, *Firmicutes*, *Proteobacteria*, and *Actinobacteria* are the most prevalent species in the gut microbiota (Simpson et al. 2014). Among these, *Actinobacteria* and *Proteobacteria* are increased in patients with IBD (Simpson et al. 2014). *Proteobacteria* are resident microbes with pathogenic potential (so-called pathobionts) (Chow and Mazmanian 2010; Petersen and Round 2014; Fig. 43.1). In addition to having pathogenic potential, bacterial species in the intestinal lumen must be able to access the intestinal epithelium (i.e., the lamina propria and inner mucosa) to induce intestinal inflammation. Two layers of mucosa segregate the microbiota in the intestinal lumen from the intestinal epithelium in colon; thus most commensal microbiota cannot directly or physically interact with the epithelium and the immune cells underneath (Johansson et al. 2008). In contrast, mice lacking mucus spontaneously develop intestinal inflammation due to penetration of the intestinal epithelium by commensal bacteria (Johansson et al. 2008). Furthermore, detergent-like emulsifiers (contained in processed foods) capable of destroying mucus may increase bacterial translocation across intestinal epithelium and the risk of intestinal inflammation (Chassaing et al. 2015).

To increase beneficial bacterial populations, various strategies have aimed to modulate imbalanced commensal biota or reduce pathobionts; these strategies include probiotics and prebiotics, reviewed by Simpson HL et al (Simpson et al. 2014). As clearly reviewed by Scaldaferri et al. and Verbeke et al., probiotic therapy is the application of live, beneficial bacteria (as a single strain or a mixture) that are considered to have properties of improving health (Scaldaferri et al. 2013; Verbeke et al. 2014). Prebiotic therapy is the application of compounds that induce growth and activity of beneficial bacteria in the gut (Scaldaferri et al. 2013; Verbeke et al. 2014). These strategies recently have emerged as beneficial and alternative options for patients with intractable diseases. In particular, increases in beneficial *Lactobacillus*, *Bifidobacterium*, and *Streptococcus* species are important for the suppression of intestinal inflammation (Scaldaferri et al. 2013). When provided as the probiotic compound VSL#3, a mixture of these bacteria effectively ameliorated the intestinal inflammation of patients with ulcerative colitis (UC) (Miele et al. 2009).

In detail, VSL#3, a mixture comprising live bacteria of three strains of *Bifidobacterium* (*B. longum*, *B. breve*, and *B. infantis*), *Streptococcus salivarius* subsp. *thermophilus*, and four strains of *Lactobacillus* (*L. casei, L. plantarum, L. acidophilus*, and *L. delbrueckii* subsp. *bulgaricus*) effectively suppressed chronic intestinal inflammation in adult human patients of UC (Bibiloni et al. 2005). The amelioration of dextran sulfate sodium-induced murine colitis by VSL#3 supplementation is associated primarily with changes in ileac microbiota composition, particularly *Enterobacteriaceae* enrichment (Mar et al. 2014; Verbeke et al. 2014). Also included in VSL#3, *Lactobacillus* strains promote an anti inflammatory milieu through their cell wall components (e.g., peptidoglycan [PGN]) (Sartor 2004; Macho Fernandez et al. 2011); in particular, the PGN from *Lactobacillus* strains reduces inflammatory responses by inducing a regulatory type of DC that produces IL-10, TGF-β, and indole amine 2, 3-dioxygenase and by promoting the generation of Treg cells (Kwon et al. 2010). In addition, lactoceptin, a protease produced by *Lactobacillus* strains, degrades various chemoattractants, such as the chemokines CXCL9, CXCL10, CXCL11, and CXCL12, thus inhibiting the infiltration of pathogenic lymphocytes (e.g., CXCR3-expressing, activated, pathologic T cells) into inflammatory sites and protecting against excess inflammation (Hormannsperger et al. 2013). Furthermore, *Bifidobacterium* species, such as the VSL#3 component *B. breve*, have the potential to stimulate regulatory-type $CD103^+$ DCs via the Toll-like receptor 2 (TLR2)–myeloid differential factor 88 (MyD88) pathway in the gut, leading to the production of IL-10 and IL-27 and subsequently generating Tr1 cells, a regulatory-type T-cell family involved in oral tolerance (Zhou et al. 2010; Jeon et al. 2012). Indeed, the oral administration of *B. breve* ameliorated colitis in murine models (Jeon et al. 2012). Another potential probiotic genus is *Clostridium* species. In patients with IBD, the amount of *Clostridium* species (clusters XIV, XVIII, IV) capable of generating Treg is reduced (Atarashi et al. 2011). Oral administration of these *Clostridium* species early in the lives of conventionally reared mice induced the production of Treg cells and inhibited intestinal inflammation and hypersensitivity (Atarashi et al. 2011).

Fecal transfer, or fecal microbiota transplantation [FMT], has recently been gathering attention as a potential therapy for patients with IBD. In FMT, feces are extracted from a healthy donor and transferred into the intestinal tract of the patient suffering from intestinal disorders. As reviewed by Brandt LJ, this treatment is effective, inexpensive, and rapid for intestinal inflammation (Brandt 2013); for example, FMT has cured cases of *Clostridium difficile* infection (Cammarota et al. 2014). Furthermore, as reviewed by Brandt LJ, the commensal microbiota affects not only intestinal disorders but potentially other manifestations of systemic pathological inflammation, including arthritis, asthma, systemic lupus erythematosus, and type 2 diabetes (Wu et al. 2010; Amar et al. 2011; Van Praet et al. 2015); neuronal disorders (Ochoa-Reparaz et al. 2011); and even behaviour and emotion (Bravo et al. 2011). For example, several case reports describe various beneficial effects of FMT in multiple sclerosis and Parkinson's disease as previously listed (Xu et al. 2015). Therefore, FMT might be considered an alternative strategy in controlling both mucosal and systemic diseases, further indicating the tremendous effect of luminal bacteria on both physiological and pathological conditions. The safety, efficacy, and potential disease applications of FMT merit continued careful investigation.

In addition to luminal commensal bacteria, we have recently identified the importance of intratissue-cohabiting commensal bacteria for the creation and maintenance of intestinal homeostasis (Obata et al. 2010). For example, *Alcaligenes*, a genus of opportunistic gram-negative bacteria (Sonnenberg et al. 2012), resides within the gut-associated lymphoid tissues (GALT), including PPs and mesenteric lymph nodes, of healthy humans, nonhuman primates, and mice (Obata et al. 2010). *Alcaligenes* enters the PPs via M cells (Sato et al. 2013), and the PPs of IBD patients contain fewer *Alcaligenes* than do those of healthy persons (Obata et al. 2010). Dysbiosis-induced trafficking of commensal bacteria to the mesenteric lymph nodes may depend on CX_3CR1^{high} mononuclear phagocytes and induce IgA (Diehl et al. 2013). Thus, *Alcaligenes*-specific secretory IgA (SIgA) but not serum IgG antibodies are generally found in healthy conditions. In contrast, *Alcaligenes* and its IgG antibodies are found in the systemic compartments of IBD patients, perhaps as a result of bacterial dissemination to the systemic compartments from PPs (Sonnenberg et al. 2012). Furthermore, this segregation of commensal bacteria from the systemic compartment may be mediated by IL-22, which is produced mainly by the group 3 subset of ILCs and potentially important in the containment of *Alcaligenes* in the GALT. Deletion of IL-22-producing ILCs results in systemic inflammation (e.g., splenomegaly) due to the diffusion of *Alcaligenes* to the periphery (Sonnenberg et al. 2012). Therefore, the mechanisms that restrict intratissue *Alcaligenes* from systemic dissemination would be a novel target for preventing both systemic and mucosal inflammation.

Multidirectional communication among commensal microorganisms is an important biological component in the creation of the intestinal ecosystem. The chemical interplay between commensal bacteria known as 'quorum sensing signals', which is mediated through proteins such as autoinducer-2, influences bacterial colonisation (Thompson et al. 2015). In addition, bacteriophages and viruses have recently been discovered to be involved in commensal mutualism (Reyes et al. 2010; Dutilh et al. 2014). For instance, a common bacteriophage, crAssphage, resides in *Bacteroides fragilis* and is involved in colonic inflammation and cancer (Reyes et al. 2010; Dutilh et al. 2014). Therefore, understanding commensal mutualism, especially microbial communication between bacteria and viruses, is important for the regulation of symbiosis versus dysbiosis and intestinal physiological versus pathological inflammation (Sears et al. 2014).

43.3 Epithelial Barrier

Composed of tightly adherent epithelial cells differentiated from crypt stem cells, the intestinal epithelium is a monolayer of intestinal epithelial cells (IEC), including columnar epithelial cells, goblet cells, Paneth cells, enteroendocrine cells, and M cells (Goto and Kiyono 2012; Fig. 43.2). As reviewed by Goto and Kiyono, IECs comprise the outermost layer of the intestinal compartment and are covered by a thick mucus barrier; this barrier contains complex mucin glycoproteins (GPs), antimicrobial peptides (AMPs), and SIgA and is required for both host defence and bacterial ecology (Goto and Kiyono 2012). Several loci for increased risk of IBD include genes involved in epithelial barrier and intestinal permeability (e.g., HNF4A, CDH1, and ITLN1; Barrett et al. 2008; Consortium et al. 2009; Muise et al. 2009). For example, CDH1 (E cadherin) is a calcium-dependent, cell–cell adhesion molecule required for the maintenance of epithelial integrity (Liu et al. 2009).

The mucin layers fulfill both lubrication and physical barrier functions. In patients with ulcerative colitis, these layers are reduced in thickness due to decreases in the number of goblet cells that specifically produce mucin 2 (MUC2) (Johansson et al. 2008; Fig. 43.2). The importance of goblet cells in protecting the intestinal mucosa is underscored by the roles of two goblet-cell products, MUC2 and the small peptide trefoil factor 3 (TFF3). For example, in mice, the lack of MUC2 allows commensal bacteria to contact the intestinal epithelium directly and induce spontaneous colitis (Velcich et al. 2002; Van der Sluis et al. 2006). TFF3, whose expression is regulated by TLR2-mediated signals (Podolsky et al. 2009), enhances the protective barrier properties of the mucus layer and promotes healing of the intestinal epithelium. Notably, mice lacking TFF3 showed high susceptibility to experimental colitis (Mashimo et al. 1996). In addition to mucin GPs, SIgA transcytosed by epithelial cells accumulated in the mucus layer limits bacterial access to the epithelial surface (Fig. 43.2). These mucus components (i.e., SIgA and mucins) are induced by bacterial stimulation via TLR-dependent pathways

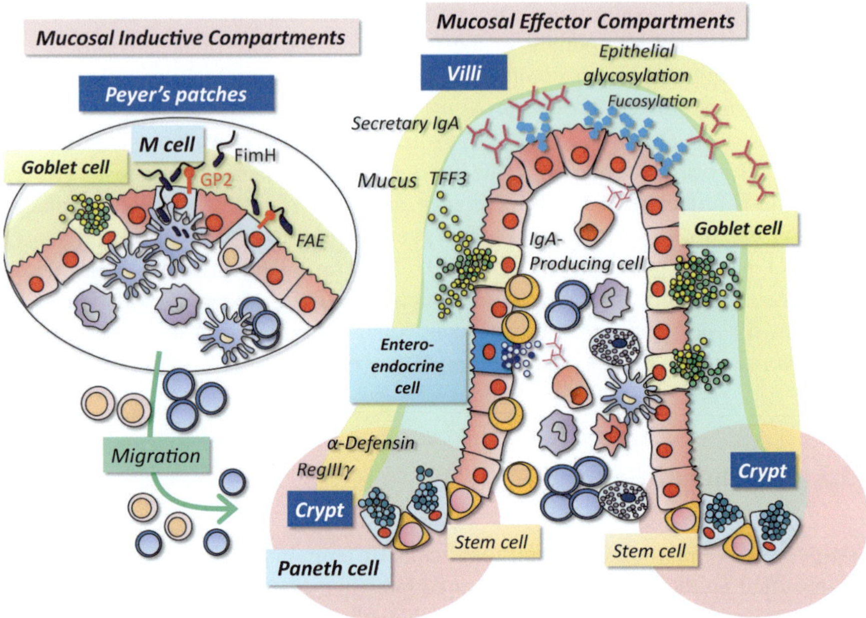

Fig. 43.2 Variety and uniqueness of epithelial cells associated with mucosal inductive and effector compartments. The intestinal immune system is functionally divided into inductive (e.g., PP) and effective sites (e.g., villi). At both sites, the epithelium has a unique variety of epithelial cell populations (e.g., M, goblet, Paneth, and columnar epithelial cells). M cells exist in the FAE and function as a sampling site of bacterial translocation via FimH recognition of GP2. Transcytosed bacteria are processed by antigen-presenting cells, which induce bacteria-specific immune responses. Goblet cells produce TFF3 and mucin, which is a major component of the mucus layer and prevents commensal bacteria from accessing the epithelium. In the mucus layer, SIgA is produced through harmonised interactions between poly IgR-expressing epithelial cells and IgA plasma cells in the intestinal lamina propria; this important antigen-specific humoural immune response occurs at the mucosal frontline. Paneth cells reside beneath the intestinal stem cells in the crypt region and produce antibacterial peptides (e.g., α-defensin and RegIIIγ) to protect against bacterial invasion and infection

(McGuckin et al. 2011). Indeed, deletion of the central adaptor molecule for nearly all TLRs, MyD88, decreases the expression of polymeric Ig receptor (poly IgR) (a major component in the formation and transportation of SIgA to the intestinal lumen) and MUC2 in the intestinal epithelium (Frantz et al. 2012).

As clearly summarised by Clevers HC et al., Paneth cells reside beneath the stem cells in the crypt in the small intestinal villi and are key effector cells in mucosal defence (Clevers and Bevins 2013; Fig. 43.2). Decreased Paneth cell functionality leads to intestinal inflammatory disorders in patients with CD, who often lack Paneth cells (Clevers and Bevins 2013). One important role of Paneth cells is the production of various AMPs, including lysozymes, secretory phospholipase A_2, and defensins, which together guard the intestinal crypt stem cells from bacterial invasion (Clevers and Bevins 2013). Defensin production is induced when bacterial antigens, such as lipopolysaccharide (LPS) and muramyl dipeptide (MDP) stimulate TLRs and invading bacteria trigger intercellular sensors, such as NOD and NOD-like receptors (NLRs) (Vora et al. 2004; Hirota et al. 2011). Furthermore, MyD88 in Paneth cells regulates the production of the antibacterial lectin RegIIIγ (Vaishnava et al. 2008). In addition, a mutation in NOD2 reduces production of AMPs, including α-defensin, in Paneth cells, and predisposes to the development of CD (Wehkamp et al. 2004).

M cells are mainly located in the follicle-associated epithelium (FAE) of PPs and make up 5% of the cells in PPs (Giannasca et al. 1999; Fig. 43.2); M cells can also reside in the villous epithelium (Jang et al. 2004). The unique morphological features of M cells include the short and irregular microvilli on their apical surfaces for readily contacting luminal antigens. In addition, a pocket structure on the basolateral side of M cells enfolds lymphocytes and antigen-presenting cells for the prompt transfer of antigens to DCs and rapid initiation of antigen-specific mucosal immunity (Nochi et al. 2007).

GP2 (the receptor for the type 1 bacterial fimbrial protein [FimH]) is specifically expressed on the surface of M cells in the intestinal compartment and is produced in the pancreas (Terahara et al. 2008; Hase et al. 2009; Fig. 43.2). Antipancreatic IgA and IgG autoantibodies are associated with the pathology of CD (Roggenbuck et al. 2013). The major target of humoural autoimmunity by autoantibodies may be the initial site of inflammation. For example, an autoantibody against GP2 is highly expressed in patients with CD and is considered a novel CD disease marker (Roggenbuck et al. 2009). This evidence suggests two theories by Roggenbuck et al., for the involvement of anti-GP2 antibodies or anti-M cell autoantibodies in the pathogenicity of IBD (Roggenbuck et al. 2013). For the first theory, anti-GP2 antibodies attack M cells, initiating the disruption of the surrounding PPs; consequently, inflammation occurs. In the second theory, because IgA is absorbed into M cells via IgA receptors expressed on their surfaces, the binding of anti GP2 IgA antibody to live bacteria facilitates their M-cell–mediated uptake or invasion into PPs; accelerated bacterial stimulation then initiates inflammation (Roggenbuck et al. 2013). In addition, adherent–invasive *Escherichia coli* with long polar fimbriae are enriched in IBD patients, and their excess translocation across M cells via GP2 may promote the initiation of aberrant inflammation processes (Dogan et al. 2014). However, the contribution of M cells to intestinal inflammation has not been clarified and warrants further investigation.

Columnar epithelial cells help to maintain the integrity of the physical and chemical barrier of the intestinal epithelium. Tight junctions between columnar epithelial cells seal the paracellular space and prevent passive movement of hydrophilic solutes and bacteria into the lamina proprial region as reviewed by Fries et al. (2013). As reported extensively elsewhere, tight junctions are controlled by various signalling molecules (e.g., myosin light chain kinase, protein kinase C, mitogen-activated protein kinases), and their integrity is regulated by various dietary components (e.g., vitamin D) and commensal bacterial stimulation (Du et al. 2015). In this regard, the lack of a bacterial sensor (e.g., MyD88) increases the severity of inflammatory symptoms, with weakened tight junctions and reduced epithelial regeneration, in murine colitis models (Rakoff-Nahoum et al. 2004; Cario et al. 2007). Epithelial–cell-specific deletion of one of the regulatory subunits of the IkB kinase complex (i.e., NEMO) spontaneously induced intestinal inflammation in mice (Nenci et al. 2007). NEMO deficiency may induce cell death and weaken epithelial integrity, consequently permitting commensal bacteria to penetrate into intestinal tissues (Nenci et al. 2007).

The glycosylation-mediating enzyme FUT2 is responsible for synthesis of the H antigen on the intestinal columnar epithelial cells (Rausch et al. 2011; Goto et al. 2014). Fut2 adds terminal α1, 2-fucose residues onto carbohydrate chains expressed on IECs (Fig. 43.2). Intestinal fucosylation is required for symbiosis between the host and commensal bacteria, such as *Bactereoides fragilis* and *B. thetaiotaomicron*, because they use epithelial fucose for their energy source (Bry et al. 1996; Coyne et al. 2005). Thus, dysbiosis caused by defects in epithelial fucosylation may create an inflammation-prone gut microbiota. A nonsense polymorphism of FUT2, or nonsecretors, which do not express active FUT2, is associated with CD. Recent genome-wide association studies have linked the FUT2 W143X mutation with the pathogenesis of CD, as well as vitamin B12 deficiency; a significant proportion of patients with CD suffer from vitamin B12 deficiency (Battat et al. 2014). In addition, mice lacking FUT2 are susceptible to inflammation induced by *Citrobacter rodentium* and *Salmonella typhimurium* (Goto et al. 2014).

The regulation of epithelial FUT2 expression requires ILC3s (Goto et al. 2014). Commensal and pathogenic bacteria stimulate ILC3s to produce IL-22 and induce the expression of FUT2 to IEC-expressing IL-22 receptors (Goto et al. 2014; Pham et al. 2014; Pickard et al. 2014). Furthermore, ILC3 can induce epithelial cell FUT2 expression by producing the inflammatory cytokine lymphotoxin (Goto et al. 2014). These data indicate the importance of epithelial fucosylation, which is induced and regulated by inflammatory signals mediated by innate immune cells, at the mucosal frontline of the intestinal homeostatic ecosystem.

In summary, despite being composed of only a single layer of cells, the epithelial barrier system – by fulfilling its roles of mucus secretion, the production of bactericidal molecules, and glycosylation – is crucial for maintaining intestinal homeostasis and preventing deleterious inflammation.

43.4 Extracellular ATP Purinergic Signalling for Host–Commensal Mutualism and Inflammation in the Gut

ATP is an essential energy source (the so-called 'energy currency') of most multicellular animals. In the 1980s, the involvement of ATP in neuronal transmission was recognised for its participation via its receptors (Ralevic and Burnstock 1998). Afterwards, the roles of extracellular ATP in various extracellular events, including cell activation, proliferation, migration, cell swelling, and apoptosis have been recognised as previously reviewed (Idzko et al. 2014; Kurashima et al. 2015).

Furthermore, extracellular ATP is a danger-/damage-associated molecular pattern molecule, as are high-mobility group box 1 (HMGB1) and uric acid crystals; DAMPs activate inflammasomes and lead to the production of IL-1β as reviewed by Di Virgilio F (Di Virgilio 2007). Upon stress and damage, including chemical, physical, and inflammatory signals, tissues release ATP; this extracellular ATP binds to various ATP receptors, namely members of the P2X ion-channel family (P2X1-7) and P2Y G protein-coupled receptor family (P2Y1,2,4,6,11,12,13; Abbracchio and Burnstock 1994). In addition to the damage-dependent extracellular release of ATP, activated immune cells (e.g., T cells, mast cells, and macrophages) can be sources of ATP (Idzko et al. 2014). In these activated immune cells, a haemichannel protein, pannexin-1, transports ATP from the intracellular to extracellular compartment (Iglesias et al. 2008; Chekeni et al. 2010). In the case of activated neutrophils, the gap junction molecule connexin 43 is required for the extracellular release of ATP (Eltzschig et al. 2006). In addition, the P2X receptor P2X7 reportedly has the potential to act as an ATP transporter (Idzko et al. 2014).

Several ATP-secreting bacteria (e.g., *Enterococcus gallinarum, E. mundtii,* and *E. faecalis*) have recently been identified in the gut lumen. Adding extracellular ATP to a bacterial culture of *Enterococcus* enhances cell survival via an unknown mechanism (Iwase et al. 2010; Hironaka et al. 2013). Bacterial ATP secretion requires glucose but not amino acids or vitamins; thus, glycolysis is an essential pathway (Hironaka et al. 2013). In the gut, bacterial ATP is important for the generation and regulation of various T-cell subpopulations. For instance, small intestinal Th17 cells are diminished in mice receiving an oral treatment of apyrase, which degrades extracellular ATP to adenosine diphosphate (ADP) and adenosine monophosphate (AMP) (Atarashi et al. 2008; Fig. 43.3). ATP from commensal bacteria and damaged intestinal epithelial cells stimulates DCs; promotes the production of IL-6, IL-12, IL-23, and TGF-β; and polarises inflammatory Th1 and Th17 cells (Yao et al. 2012; Fig. 43.3). Indeed, the rates of apoptosis are higher in the epithelium and lamina propria of CD patients than of healthy controls (Neves et al. 2014).

In addition, follicular helper T (Tfh) cells, a critical T-cell subset for IgA B-cell development in PPs, express high levels of the P2X7 receptor (Proietti et al. 2014). Tfh cells expressing C-X-C chemokine receptor type 5 and programmed cell death 1 help B cells to generate high-affinity Ig-producing plasma cells via IL-21 production (Proietti et al. 2014; Fig. 43.3). Commensal-derived ATP stimulates P2X7 and induces the cell death of Tfh cells and reduces their number; therefore, mice lacking P2X7 have increased numbers of Tfh cells in the PPs, thus upregulating the production of high-affinity IgA against luminal commensal bacteria (Proietti et al. 2014; Fig. 43.3). Consequently, P2X7-deficient mice have increased neutralising IgA-coated commensal bacteria, such as segmented filamentous bacteria; thus, their commensal bacterial load is decreased (Proietti et al. 2014). Due to the clearance mechanism of IgA-coated bacteria as part of mucosal immunity, the numbers of bacteria at luminal sites are significantly diminished in P2X7-deficient mice (Proietti et al. 2014; Fig. 43.3).

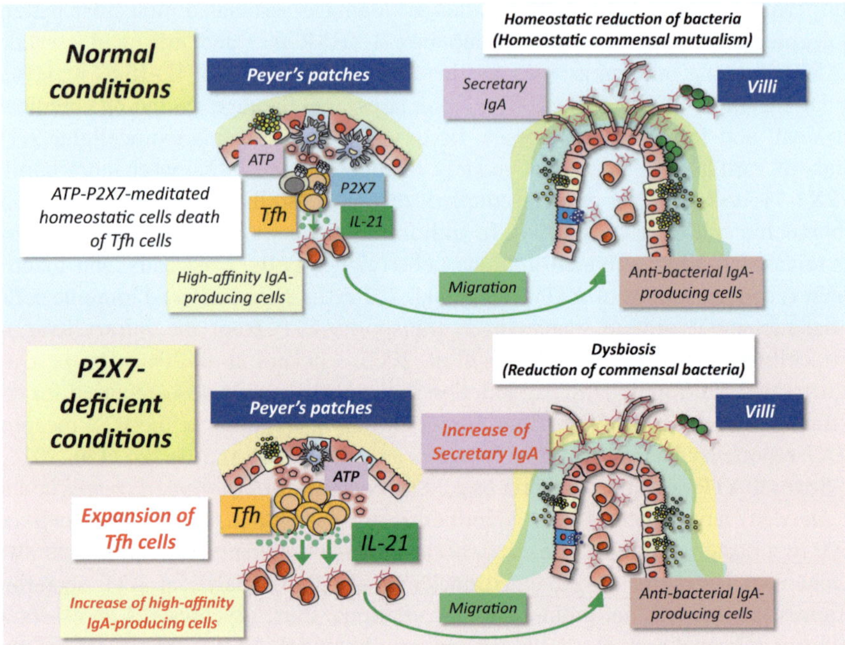

Fig. 43.3 Critical role of ATP–P2X7 cascade for physiological inflammation and homeostatic commensal mutualism. Extracellular ATP is released from commensal bacteria (e.g., *Enterococcus gallinarum, E. mundtii, E. faecalis*), damaged host cells, and activated immune cells. The ATP–P2X7 pathway induces the homeostatic cell death of P2X7-high–expressing follicular helper T cells (Tfh). Tfh cells induce high-affinity IgA-producing cells via IL-21 production. High-affinity IgA against commensal bacteria reduces the bacterial load by neutralising binding to epithelium. However, when those cascades are disrupted, Tfh cells can escape homeostatic cell death. Increased numbers of Tfh cells promote enhanced secretion of high-affinity IgA and the reduction or alteration of commensal bacteria, leading to dysbiosis

Components of the extracellular ATP–purinergic receptor pathway, especially P2X7, are novel and potential therapeutic targets for IBD (Fig. 43.4). Inhibition of P2X7 by A740003, brilliant blue G, or KN62 ameliorates experimental colitis by decreasing the recruitment of neutrophils, T cells, and macrophages and by reducing fibrosis (Marques et al. 2014; Hofman et al. 2015). Indeed, P2X7-deficient mice develop less severe experimental colitis than do their P2X7-bearing counterparts (Hofman et al. 2015). In patients with CD, P2X7 expression is upregulated during the active phase of inflammation, and AZD9056 (a selective, orally active inhibitor of P2X7) decreases the clinical symptoms of patients with moderate-to-severe CD (Neves et al. 2014; Eser et al. 2015).

Various types of immune cells accumulate at the site of inflammation in IBD. Histology of patients with CD reveals mast cells (Dvorak et al. 1978). Mast cells orchestrate immune responses against antigens, thus they are considered to be an

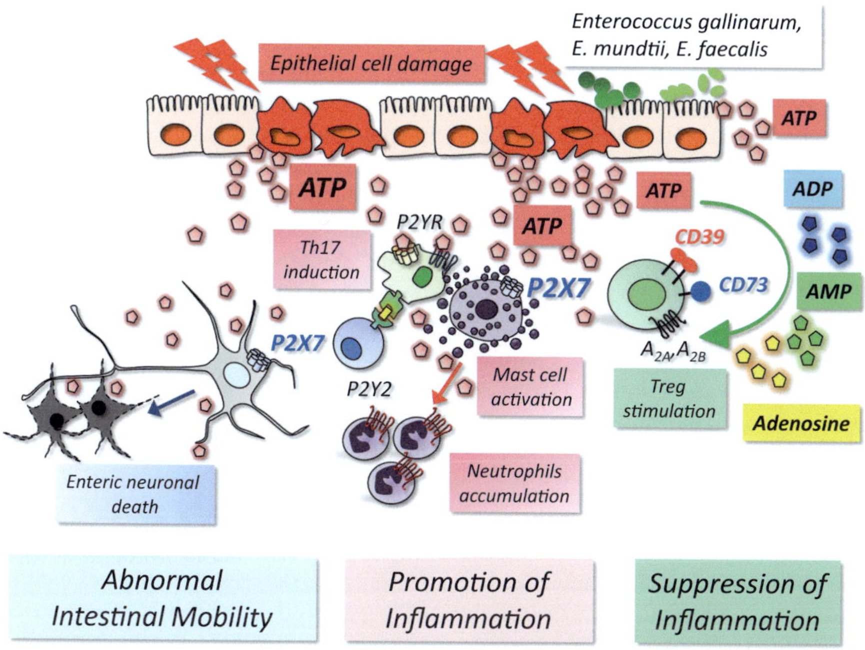

Fig. 43.4 Critical role of ATP–P2X7 cascade for pathological inflammation. The increase in extracellular ATP during inflammation induces neuronal cell death and causes abnormal intestinal motility via P2X7. In addition, the stimulation of mast cells or myeloid cells via P2X7 or P2Y2 receptors promotes inflammatory responses (e.g., release of inflammatory mediators, Th17 induction, and neutrophil accumulation). In contrast, extracellular ATP is degraded into adenosine by ectoenzymes (e.g., CD39, CD73) expressed by Treg. Adenosine suppresses excess intestinal inflammation and resolves inflammation

effective adjuvant target for vaccinations; however, at the site of inflammation, mast cells are major contributors to an inflammatory milieu in the gut as previously reviewed (Kurashima and Kiyono 2014). Indeed, degranulated mast cells have been identified at the site of intestinal inflammation (Kurashima et al. 2012). Degranulation occurs due to calcium influx into the cytoplasmic compartment in response to various signals, such as IgE–high-affinity IgE receptor pathways that depend on Th2 and B cells as precisely reviewed by Rivera et al. (2008). However, colitogenic activation of mast cells is independent of T and B cells, because the number of degranulated mast cells in the intestinal compartment is comparable between lymphocyte-deficient mice with colitis and wild-type mice. Furthermore, the treatment of mice with P2X7 inhibitory agents or an inhibitory antibody suppresses inflammatory symptoms in the gut and reduces the number of degranulated–activated mast cells (Kurashima et al. 2012); therefore, an extracellular ATP–P2X7 pathway is indispensable to mast cell activation (Kurashima et al. 2012). ATP–P2X7 pathways induce degranulation and promote the production of inflammatory cytokines (e.g., IL-1β, IL-6, and TNFα), chemokines (e.g., CCL1, 2, 4, 7, and

CXCL1, 2), and leukotrienes (LTB4 and LTC4; Fig. 43.4). These mediators accelerate inflammatory responses through the recruitment of inflammatory cells (e.g., neutrophils and macrophages).

Mast cell proteases, such as chymase and tryptase, also play important roles in the onset of IBD (Hamilton et al. 2011). For example, tryptase-deficient mice have reduced inflammation, and tryptase may promote capillary permeability via the degradation of cell–cell junctions in blood vessels and may enhance the accumulation of inflammatory cells at these sites. The concentrations of tryptase are increased in samples of peripheral blood and colonic tissue from CD patients (Hamilton et al. 2011). An inhibitory receptor on colonic mast cells, CD300f, is involved in the suppression of degranulation, likely by binding to ceramide (Matsukawa et al. 2015). Ceramide–CD300f pathways inhibit ATP-dependent mast cell activation; thus, CD300f-defcient mice exhibit aggravation of intestinal inflammation. The administration of a ceramide-containing liposome for stimulating CD300f-dependent suppression of mast cell activation reduces inflammatory reactions (Matsukawa et al. 2015).

In addition to mast cell activation, the ATP–P2X7 cascade is involved in the motility of the intestine via the regulation of enteric neurons. In patients with CD, physical movement of the intestine is limited due to a decreased number of enteric neurons as a result of extracellular ATP-mediated cell death (Gulbransen et al. 2012). The ATP–P2X7 cascade might also be involved in irritable bowel syndrome (IBS) (Ochoa-Cortes et al. 2014), a common gastrointestinal disorder characterised by discomfort, chronic abdominal pain, and altered bowel habits. Infectious stimulation in the intestine triggers the onset of so-called ‘post-infectious IBS’ (PI-IBS) (Keating et al. 2011). Indeed, after gastrointestinal infection, the risk of developing IBS increases sixfold (Thabane et al. 2007). Furthermore, P2X7 inhibition reduces IBS symptoms (e.g., visceral hypersensitivity) (Keating et al. 2011). In mice, transient intestinal infection with *Trichinella spiralis* leads to postinflammatory visceral hypersensitivity, which is associated with elevated IL-1β production via extracellular ATP–P2X7–IL-1β pathways (Keating et al. 2011). Mast cells are likely a central trigger for the onset of IBS. Indeed, numerous tryptase-positive mast cells are often present in the intestinal mucosa of patients with IBS. Thus, the extracellular ATP–P2X7 pathway in mast cells may be involved in IBD and IBS (Ohman and Simren 2010). Consequently, the extracellular ATP–P2X7 purinergic receptor-mediated pathways are attractive targets for the discovery of new drugs and innovative treatments of intestinal inflammation and its complications, including skin inflammation, abdominal pain, and arthritis (Brumfield et al. 2011).

Once extracellular ATP is released, ATP is hydrolysed to AMP and adenosine by ectonucleotidases, such as CD39 (nucleoside triphosphate diphosphohydrolase-1), CD73 (5′-nucleotidase), and CD203 (ectonucleotide pyrophosphatase phosphodiesterase). Adenosine is derived by dephosphorylation of extracellular ATP or directly diffuses from intracellular compartments via equilibrative nucleoside transporters as reviewed by Ye and Rajendran (2009). Extracellular adenosine has important roles in immunomodulation. Adenosine derived due to the metabolism of ATP via CD39 and CD73 binds to adenosine receptors, such as A2A and A3 receptors that are expressed on T cells and myeloid cells (Fig. 43.4). This signalling cascade inhibits intestinal inflammation (Ye and Rajendran 2009; Kurtz et al. 2014). Indeed, adenosine receptor agonists, such as those for the A2A and A3 adenosine receptors (ATP-146e and IB-MECA, respectively), are potent anti-inflammatory agents devoid of systemic side effects and effective against inflammatory diseases (Ye and Rajendran 2009; Kurtz et al. 2014). The suppressive effects of adenosine receptors on the recruitment of inflammatory cells and production of inflammatory cytokines subsequently ameliorate colitis (Odashima et al. 2005; Kurtz et al. 2014). Indeed, in a model of endothelial or epithelial–cell-specific, A2B receptor-deficient mice, A2B adenosine receptors are indispensable for epithelial barrier integrity in intestinal inflammation (Aherne et al. 2015).

Single-nucleotide polymorphisms that tag low levels of CD39 expression are associated with increased susceptibility to CD (Friedman et al. 2009). In contrast, IL-27 induces CD39 expression on DCs, limits the induction of Th1 and Th17 immune responses by downmodulating the activation of the NLRP3 inflammasome, and thus terminates inflammatory reactions (Mascanfroni et al. 2013). In addition, some therapeutic agents for IBD, such as cyclosporine, methotrexate, and sulfasalazine, reduce the concentrations of extracellular ATP and increase adenosine levels partly via the stimulation of CD73-dependent adenosine production. In mice, CD73 deficiency or the administration of a selective CD73 inhibitor (i.e., α, β-methylene ADP) increases susceptibility to colitis (Louis et al. 2008). These CD73-deficient mice exhibit unresolved inflammatory symptoms, with slow body-weight recovery and increased intestinal permeability owing to reduced expression of tight junction-related proteins (e.g., JAM-A, α-catenin, and claudin 2). Therefore, CD73–adenosine–IFN-αA pathways are important for the suppression of intestinal inflammation (Louis et al. 2008). In addition, CD203c (E-NPP3), which is specifically expressed on basophils (and increased on activated ones) and mast cells, is involved in the clearance of extracellular ATP (Odashima et al. 2005). A deficiency of CD203c increases extracellular ATP, leading to excess activation of mast cells (Tsai et al. 2015).

Elucidating the complexity underlying how extracellular ATP appropriately orchestrates inflammatory responses likely will facilitate our understanding and control of both acute and chronic inflammation.

43.5 Conclusion and Future Directions

Accumulating evidence has revealed the importance of crosstalk among the commensal bacterial community, epithelial barrier, and immune cell network in the development of pathological conditions and the maintenance of physiological homeostasis in mucosal compartments. Despite their complexity, understanding these tri-interactive mechanisms is crucial and necessary to establish new and efficient strategies to cure or control intestinal inflammation. Together with achieving immunological remission and developing new anti-inflammatory agents, targeting wound or mucosal healing has gained attention as a novel central approach for the treatment of IBD (Papi et al. 2013; Florholmen 2015). Several studies have demonstrated the requirement of ILC3 for the maintenance of epithelial integrity and intestinal stem cells, regulation of commensal-bacteria–specific T cells, and induction of mucosal healing (Longman et al. 2014; Aparicio-Domingo et al. 2015; Hepworth et al. 2015). In this context, the reduction of colonic ILC3s has led to intestinal inflammation in pediatric IBD patients (Hepworth et al. 2015). Therefore, strategies for activating and expanding ILC3s may be novel mechanisms for inducing mucosal healing.

In contrast, as clearly reviewed by Rieder and Fiocchi and Mifflin et al., inappropriate mucosal or wound healing leads to chronic inflammation with fibrosis (Rieder and Fiocchi 2009), which is mediated by the activation and proliferation of fibroblasts, myofibloblasts, and fibrocytes (a migratory type of collagen-expressing haematopoietic cell) (Mifflin et al. 2011). In addition, epithelial-to-mesenchymal transition, endothelial-to-mesenchymal transition, and pericyte differentiation are alternative sources of excessive collagen deposition (Rieder and Fiocchi 2009). ATP–P2X7 receptor pathways are considered to be involved in fibrosis in light of the expression of these signalling molecules in the fibroblasts of liver, kidney, and lung; however, the involvement of ATP–P2X7 signalling in colonic fibrosis has not been elucidated (Gentile et al. 2015). To devise a novel therapeutic approach against chronic intestinal inflammation, the appropriate targets among fibroblasts, myofibroblasts, fibrocytes, and pericytes at different stages of inflammation (e.g., acute and chronic) must be identified.

Acknowledgments We appreciate our former and current colleagues and collaborators who have been providing their expertise for our understanding of (1) commensal mutualism, (2) the epithelial barrier, and (3) purinergic inflammatory signalling.

Financial Support and Sponsorship
This work was supported by grants from the Core Research for Evolutional Science and Technology Program of the Japan Science and Technology Agency (to H.K.); the Ministry of Education, Science, Sports, and Technology of Japan (MEXT: Grant-in Aid for Scientific Research S [H.K.] and Research Activity Start-up [26893047 to Y.K].); the Young Researcher Overseas Visits Program for Vitalising Brain Circulation (Japan Society for the Promotion of Science, H.K.); and MEXT Translational Research Network Program (at the University of Tokyo) Seeds B and C.

Conflicts of Interest
The authors have no competing interests to report.

References

Abbracchio MP, Burnstock G (1994) Purinoceptors: are there families of P2X and P2Y purinoceptors? Pharmacol Ther 64(3):445–475

Aherne CM, Saeedi B, Collins CB, Masterson JC, McNamee EN, Perrenoud L, Rapp CR, Curtis VF, Bayless A, Fletcher A, Glover LE, Evans CM, Jedlicka P, Furuta GT, de Zoeten EF, Colgan SP, Eltzschig HK (2015) Epithelial-specific A2B adenosine receptor signaling protects the colonic epithelial barrier during acute colitis. Mucosal Immunol 8(3):699

Altwegg R, Vincent T (2014) TNF blocking therapies and immunomonitoring in patients with inflammatory bowel disease. Mediators Inflamm 2014:172821

Amar J, Chabo C, Waget A, Klopp P, Vachoux C, Bermudez-Humaran LG, Smirnova N, Berge M, Sulpice T, Lahtinen S, Ouwehand A, Langella P, Rautonen N, Sansonetti PJ, Burcelin R (2011) Intestinal mucosal adherence and translocation of commensal bacteria at the early onset of type 2 diabetes: molecular mechanisms and probiotic treatment. EMBO Mol Med 3(9):559–572

Aparicio-Domingo P, Romera-Hernandez M, Karrich JJ, Cornelissen F, Papazian N, Lindenbergh-Kortleve DJ, Butler JA, Boon L, Coles MC, Samsom JN, Cupedo T (2015) Type 3 innate lymphoid cells maintain intestinal epithelial stem cells after tissue damage. J Exp Med 212 (11):1783–1791

Atarashi K, Nishimura J, Shima T, Umesaki Y, Yamamoto M, Onoue M, Yagita H, Ishii N, Evans R, Honda K, Takeda K (2008) ATP drives lamina propria T(H)17 cell differentiation. Nature 455(7214):808–812

Atarashi K, Tanoue T, Shima T, Imaoka A, Kuwahara T, Momose Y, Cheng G, Yamasaki S, Saito T, Ohba Y, Taniguchi T, Takeda K, Hori S, Ivanov II, Umesaki Y, Itoh K, Honda K (2011) Induction of colonic regulatory T cells by indigenous Clostridium species. Science 331 (6015):337–341

Barrett JC, Hansoul S, Nicolae DL, Cho JH, Duerr RH, Rioux JD, Brant SR, Silverberg MS, Taylor KD, Barmada MM, Bitton A, Dassopoulos T, Datta LW, Green T, Griffiths AM, Kistner EO, Murtha MT, Regueiro MD, Rotter JI, Schumm LP, Steinhart AH, Targan SR, Xavier RJ, Consortium NIG, Libioulle C, Sandor C, Lathrop M, Belaiche J, Dewit O, Gut I, Heath S, Laukens D, Mni M, Rutgeerts P, Van Gossum A, Zelenika D, Franchimont D, Hugot JP, de Vos M, Vermeire S, Louis E, Belgian-French IBDC, Wellcome Trust Case Control C, Cardon LR, Anderson CA, Drummond H, Nimmo E, Ahmad T, Prescott NJ, Onnie CM, Fisher SA, Marchini J, Ghori J, Bumpstead S, Gwilliam R, Tremelling M, Deloukas P, Mansfield J, Jewell D, Satsangi J, Mathew CG, Parkes M, Georges M, Daly MJ (2008) Genome-wide association defines more than 30 distinct susceptibility loci for Crohn's disease. Nat Genet 40 (8):955–962

Battat R, Kopylov U, Szilagyi A, Saxena A, Rosenblatt DS, Warner M, Bessissow T, Seidman E, Bitton A (2014) Vitamin B12 deficiency in inflammatory bowel disease: prevalence, risk factors, evaluation, and management. Inflamm Bowel Dis 20(6):1120–1128

Bibiloni R, Fedorak RN, Tannock GW, Madsen KL, Gionchetti P, Campieri M, De Simone C, Sartor RB (2005) VSL#3 probiotic-mixture induces remission in patients with active ulcerative colitis. Am J Gastroenterol 100(7):1539–1546

Brandt LJ (2013) American Journal of Gastroenterology Lecture: Intestinal microbiota and the role of fecal microbiota transplant (FMT) in treatment of C. difficile infection. Am J Gastroenterol 108(2):177–185

Bravo JA, Forsythe P, Chew MV, Escaravage E, Savignac HM, Dinan TG, Bienenstock J, Cryan JF (2011) Ingestion of Lactobacillus strain regulates emotional behavior and central GABA receptor expression in a mouse via the vagus nerve. Proc Natl Acad Sci U S A 108 (38):16050–16055

Brumfield S, Matasi JJ, Tulshian D, Czarniecki M, Greenlee W, Garlisi C, Qiu H, Devito K, Chen SC, Sun Y, Bertorelli R, Ansell J, Geiss W, Le VD, Martin GS, Vellekoop SA, Haber J, Allard ML (2011) Synthesis and SAR development of novel P2X7 receptor antagonists for the treatment of pain: part 2. Bioorg Med Chem Lett 21(24):7287–7290

Bry L, Falk PG, Midtvedt T, Gordon JI (1996) A model of host-microbial interactions in an open mammalian ecosystem. Science 273(5280):1380–1383

Cammarota G, Ianiro G, Gasbarrini A (2014) Fecal microbiota transplantation for the treatment of Clostridium difficile infection: a systematic review. J Clin Gastroenterol 48(8):693–702

Cario E, Gerken G, Podolsky DK (2007) Toll-like receptor 2 controls mucosal inflammation by regulating epithelial barrier function. Gastroenterology 132(4):1359–1374

Chassaing B, Koren O, Goodrich JK, Poole AC, Srinivasan S, Ley RE, Gewirtz AT (2015) Dietary emulsifiers impact the mouse gut microbiota promoting colitis and metabolic syndrome. Nature 519(7541):92–96

Chekeni FB, Elliott MR, Sandilos JK, Walk SF, Kinchen JM, Lazarowski ER, Armstrong AJ, Penuela S, Laird DW, Salvesen GS, Isakson BE, Bayliss DA, Ravichandran KS (2010) Pannexin 1 channels mediate 'find-me' signal release and membrane permeability during apoptosis. Nature 467(7317):863–867

Chow J, Mazmanian SK (2010) A pathobiont of the microbiota balances host colonization and intestinal inflammation. Cell Host Microbe 7(4):265–276

Clevers HC, Bevins CL (2013) Paneth cells: maestros of the small intestinal crypts. Annu Rev Physiol 75:289–311

Consortium UIG, Barrett JC, Lee JC, Lees CW, Prescott NJ, Anderson CA, Phillips A, Wesley E, Parnell K, Zhang H, Drummond H, Nimmo ER, Massey D, Blaszczyk K, Elliott T, Cotterill L, Dallal H, Lobo AJ, Mowat C, Sanderson JD, Jewell DP, Newman WG, Edwards C, Ahmad T, Mansfield JC, Satsangi J, Parkes M, Mathew CG, C. Wellcome Trust Case Control, Donnelly P, Peltonen L, Blackwell JM, Bramon E, Brown MA, Casas JP, Corvin A, Craddock N, Deloukas P, Duncanson A, Jankowski J, Markus HS, Mathew CG, McCarthy MI, Palmer CN, Plomin R, Rautanen A, Sawcer SJ, Samani N, Trembath RC, Viswanathan AC, Wood N, Spencer CC, Barrett JC, Bellenguez C, Davison D, Freeman C, Strange A, Donnelly P, Langford C, Hunt SE, Edkins S, Gwilliam R, Blackburn H, Bumpstead SJ, Dronov S, Gillman M, Gray E, Hammond N, Jayakumar A, McCann OT, Liddle J, Perez ML, Potter SC, Ravindrarajah R, Ricketts M, Waller M, Weston P, Widaa S, Whittaker P, Deloukas P, Peltonen L, Mathew CG, Blackwell JM, Brown MA, Corvin A, McCarthy MI, Spencer CC, Attwood AP, Stephens J, Sambrook J, Ouwehand WH, McArdle WL, Ring SM, Strachan DP (2009) Genome-wide association study of ulcerative colitis identifies three new susceptibility loci, including the HNF4A region. Nat Genet 41(12):1330–1334

Coyne MJ, Reinap B, Lee MM, Comstock LE (2005) Human symbionts use a host-like pathway for surface fucosylation. Science 307(5716):1778–1781

Di Virgilio F (2007) Liaisons dangereuses: P2X(7) and the inflammasome. Trends Pharmacol Sci 28(9):465–472

Diehl GE, Longman RS, Zhang JX, Breart B, Galan C, Cuesta A, Schwab SR, Littman DR (2013) Microbiota restricts trafficking of bacteria to mesenteric lymph nodes by CX(3)CR1(hi) cells. Nature 494(7435):116–120

Dogan B, Suzuki H, Herlekar D, Sartor RB, Campbell BJ, Roberts CL, Stewart K, Scherl EJ, Araz Y, Bitar PP, Lefebure T, Chandler B, Schukken YH, Stanhope MJ, Simpson KW (2014) Inflammation-associated adherent-invasive Escherichia coli are enriched in pathways for use of propanediol and iron and M-cell translocation. Inflamm Bowel Dis 20(11):1919–1932

Du J, Chen Y, Shi Y, Liu T, Cao Y, Tang Y, Ge X, Nie H, Zheng C, Li YC (2015) 1,25-Dihydroxyvitamin D Protects Intestinal Epithelial Barrier by Regulating the Myosin Light Chain Kinase Signaling Pathway. Inflamm Bowel Dis 21(11):2495–506

Dutilh BE, Cassman N, McNair K, Sanchez SE, Silva GG, Boling L, Barr JJ, Speth DR, Seguritan V, Aziz RK, Felts B, Dinsdale EA, Mokili JL, Edwards RA (2014) A highly abundant bacteriophage discovered in the unknown sequences of human faecal metagenomes. Nat Commun 5:4498

Dvorak AM, Monahan RA, Osage JE, Dickersin GR (1978) Mast-cell degranulation in Crohn's disease. Lancet 1(8062):498

Eltzschig HK, Eckle T, Mager A, Kuper N, Karcher C, Weissmuller T, Boengler K, Schulz R, Robson SC, Colgan SP (2006) ATP release from activated neutrophils occurs via connexin 43 and modulates adenosine-dependent endothelial cell function. Circ Res 99(10):1100–1108

Eltzschig HK, Sitkovsky MV, Robson SC (2012) Purinergic signaling during inflammation. N Engl J Med 367(24):2322–2333

Eser A, Colombel JF, Rutgeerts P, Vermeire S, Vogelsang H, Braddock M, Persson T, Reinisch W (2015) Safety and Efficacy of an Oral Inhibitor of the Purinergic Receptor P2X7 in Adult Patients with Moderately to Severely Active Crohn's Disease: A Randomized Placebo-controlled, Double-blind, Phase IIa Study. Inflamm Bowel Dis 21(10):2247–53

Florholmen J (2015) Mucosal healing in the era of biologic agents in treatment of inflammatory bowel disease. Scand J Gastroenterol 50(1):43–52

Frantz AL, Rogier EW, Weber CR, Shen L, Cohen DA, Fenton LA, Bruno ME, Kaetzel CS (2012) Targeted deletion of MyD88 in intestinal epithelial cells results in compromised antibacterial immunity associated with downregulation of polymeric immunoglobulin receptor, mucin-2, and antibacterial peptides. Mucosal Immunol 5(5):501–512

Friedman DJ, Kunzli BM, A-Rahim YI, Sevigny J, Berberat PO, Enjyoji K, Csizmadia E, Friess H, Robson SC (2009) From the Cover: CD39 deletion exacerbates experimental murine colitis and human polymorphisms increase susceptibility to inflammatory bowel disease. Proc Natl Acad Sci U S A 106(39):16788–16793

Fries W, Belvedere A, Vetrano S (2013) Sealing the broken barrier in IBD: intestinal permeability, epithelial cells and junctions. Curr Drug Targets 14(12):1460–1470

Gentile D, Natale M, Lazzerini PE, Capecchi PL, Laghi-Pasini F (2015) The role of P2X receptors in tissue fibrosis: a brief review. Signal, Purinergic

Giannasca PJ, Giannasca KT, Leichtner AM, Neutra MR (1999) Human intestinal M cells display the sialyl Lewis A antigen. Infect Immun 67(2):946–953

Goto Y, Kiyono H (2012) Epithelial barrier: an interface for the cross-communication between gut flora and immune system. Immunol Rev 245(1):147–163

Goto Y, Obata T, Kunisawa J, Sato S, Ivanov II, Lamichhane A, Takeyama N, Kamioka M, Sakamoto M, Matsuki T, Setoyama H, Imaoka A, Uematsu S, Akira S, Domino SE, Kulig P, Becher B, Renauld JC, Sasakawa C, Umesaki Y, Benno Y, Kiyono H (2014) Innate lymphoid cells regulate intestinal epithelial cell glycosylation. Science 345(6202):1254009

Gulbransen BD, Bashashati M, Hirota SA, Gui X, Roberts JA, MacDonald JA, Muruve DA, McKay DM, Beck PL, Mawe GM, Thompson RJ, Sharkey KA (2012) Activation of neuronal P2X7 receptor pannexin-1 mediates death of enteric neurons during colitis. Nat Med 18 (4):600–604

Hamilton MJ, Sinnamon MJ, Lyng GD, Glickman JN, Wang X, Xing W, Krilis SA, Blumberg RS, Adachi R, Lee DM, Stevens RL (2011) Essential role for mast cell tryptase in acute experimental colitis. Proc Natl Acad Sci U S A 108(1):290–295

Hase K, Kawano K, Nochi T, Pontes GS, Fukuda S, Ebisawa M, Kadokura K, Tobe T, Fujimura Y, Kawano S, Yabashi A, Waguri S, Nakato G, Kimura S, Murakami T, Iimura M, Hamura K, Fukuoka S, Lowe AW, Itoh K, Kiyono H, Ohno H (2009) Uptake through glycoprotein 2 of FimH(+) bacteria by M cells initiates mucosal immune response. Nature 462(7270):226–230

Hepworth MR, Fung TC, Masur SH, Kelsen JR, McConnell FM, Dubrot J, Withers DR, Hugues S, Farrar MA, Reith W, Eberl G, Baldassano RN, Laufer TM, Elson CO, Sonnenberg GF (2015) Immune tolerance. Group 3 innate lymphoid cells mediate intestinal selection of commensal bacteria-specific CD4(+) T cells. Science 348(6238):1031–1035

Hironaka I, Iwase T, Sugimoto S, Okuda K, Tajima A, Yanaga K, Mizunoe Y (2013) Glucose triggers ATP secretion from bacteria in a growth-phase-dependent manner. Appl Environ Microbiol 79(7):2328–2335

Hirota SA, Ng J, Lueng A, Khajah M, Parhar K, Li Y, Lam V, Potentier MS, Ng K, Bawa M, McCafferty DM, Rioux KP, Ghosh S, Xavier RJ, Colgan SP, Tschopp J, Muruve D, MacDonald JA, Beck PL (2011) NLRP3 inflammasome plays a key role in the regulation of intestinal homeostasis. Inflamm Bowel Dis 17(6):1359–1372

Hofman P, Cherfils-Vicini J, Bazin M, Ilie M, Juhel T, Hebuterne X, Gilson E, Schmid-Alliana A, Boyer O, Adriouch S, Vouret-Craviari V (2015) Genetic and pharmacological inactivation of the purinergic P2RX7 receptor dampens inflammation but increases tumor incidence in a mouse model of colitis-associated cancer. Cancer Res 75(5):835–845

Hormannsperger G, von Schillde MA, Haller D (2013) Lactocepin as a protective microbial structure in the context of IBD. Gut Microbes 4(2):152–157

Huttenhower C, Kostic AD, Xavier RJ (2014) Inflammatory bowel disease as a model for translating the microbiome. Immunity 40(6):843–854. doi:10.1016/j.immuni.2014.05.013

Idzko M, Ferrari D, Eltzschig HK (2014) Nucleotide signalling during inflammation. Nature 509 (7500):310–317

Iglesias R, Locovei S, Roque A, Alberto AP, Dahl G, Spray DC, Scemes E (2008) P2X7 receptor-Pannexin1 complex: pharmacology and signaling. Am J Physiol Cell Physiol 295(3):C752–C760

Iwase T, Shinji H, Tajima A, Sato F, Tamura T, Iwamoto T, Yoneda M, Mizunoe Y (2010) Isolation and identification of ATP-secreting bacteria from mice and humans. J Clin Microbiol 48(5):1949–1951

Jang MH, Kweon MN, Iwatani K, Yamamoto M, Terahara K, Sasakawa C, Suzuki T, Nochi T, Yokota Y, Rennert PD, Hiroi T, Tamagawa H, Iijima H, Kunisawa J, Yuki Y, Kiyono H (2004) Intestinal villous M cells: an antigen entry site in the mucosal epithelium. Proc Natl Acad Sci U S A 101(16):6110–6115

Jeon SG, Kayama H, Ueda Y, Takahashi T, Asahara T, Tsuji H, Tsuji NM, Kiyono H, Ma JS, Kusu T, Okumura R, Hara H, Yoshida H, Yamamoto M, Nomoto K, Takeda K (2012) Probiotic Bifidobacterium breve induces IL-10-producing Tr1 cells in the colon. PLoS Pathog 8(5), e1002714

Johansson ME, Phillipson M, Petersson J, Velcich A, Holm L, Hansson GC (2008) The inner of the two Muc2 mucin-dependent mucus layers in colon is devoid of bacteria. Proc Natl Acad Sci U S A 105(39):15064–15069

Kaser A, Zeissig S, Blumberg RS (2010) Inflammatory bowel disease. Annu Rev Immunol 28:573–621

Kau AL, Ahern PP, Griffin NW, Goodman AL, Gordon JI (2011) Human nutrition, the gut microbiome and the immune system. Nature 474(7351):327–336

Keating C, Pelegrin P, Martinez CM, Grundy D (2011) P2X7 receptor-dependent intestinal afferent hypersensitivity in a mouse model of postinfectious irritable bowel syndrome. J Immunol 187(3):1467–1474

Kiyono H, McGhee JR, Michalek SM (1980) Lipopolysaccharide regulation of the immune response: comparison of responses to LPS in germfree, Escherichia coli-monoassociated and conventional mice. J Immunol 124(1):36–41

Klotz U, Schwab M (2005) Topical delivery of therapeutic agents in the treatment of inflammatory bowel disease. Adv Drug Deliv Rev 57(2):267–279

Kunisawa J, Kiyono H (2015) Vitamins mediate immunological homeostasis and diseases at the surface of the body. Endocr Metab Immune Disord Drug Targets 15(1):25–30

Kurashima Y, Kiyono H (2014) New era for mucosal mast cells: their roles in inflammation, allergic immune responses and adjuvant development. Exp Mol Med 46, e83

Kurashima Y, Amiya T, Nochi T, Fujisawa K, Haraguchi T, Iba H, Tsutsui H, Sato S, Nakajima S, Iijima H, Kubo M, Kunisawa J, Kiyono H (2012) Extracellular ATP mediates mast cell-dependent intestinal inflammation through P2X7 purinoceptors. Nat Commun 3:1034

Kurashima Y, Kiyono H, Kunisawa J (2015) Pathophysiological role of extracellular purinergic mediators in the control of intestinal inflammation. Mediators Inflamm 2015:427125

Kurtz CC, Drygiannakis I, Naganuma M, Feldman SH, Bekiaris V, Linden J, Ware CF, Ernst PB (2014) Extracellular Adenosine Regulates Colitis through Effects on Lymphoid and Non-Lymphoid Cells. J Physiol Gastrointest Liver Physiol, Am

Kwon HK, Lee CG, So JS, Chae CS, Hwang JS, Sahoo A, Nam JH, Rhee JH, Hwang KC, Im SH (2010) Generation of regulatory dendritic cells and CD4 + Foxp3 + T cells by probiotics administration suppresses immune disorders. Proc Natl Acad Sci U S A 107(5):2159–2164

Liu L, Guo X, Rao JN, Zou T, Xiao L, Yu T, Timmons JA, Turner DJ, Wang JY (2009) Polyamines regulate E-cadherin transcription through c-Myc modulating intestinal epithelial barrier function. Am J Physiol Cell Physiol 296(4):C801–C810

Longman RS, Diehl GE, Victorio DA, Huh JR, Galan C, Miraldi ER, Swaminath A, Bonneau R, Scherl EJ, Littman DR (2014) CX(3)CR1(+) mononuclear phagocytes support colitis-associated innate lymphoid cell production of IL-22. J Exp Med 211(8):1571–1583

Louis NA, Robinson AM, MacManus CF, Karhausen J, Scully M, Colgan SP (2008) Control of IFN-alphaA by CD73: implications for mucosal inflammation. J Immunol 180(6):4246–4255

Macho Fernandez E, Pot B, Grangette C (2011) Beneficial effect of probiotics in IBD: are peptidogycan and NOD2 the molecular key effectors? Gut Microbes 2(5):280–286

Macpherson AJ, Gatto D, Sainsbury E, Harriman GR, Hengartner H, Zinkernagel RM (2000) A primitive T cell-independent mechanism of intestinal mucosal IgA responses to commensal bacteria. Science 288(5474):2222–2226

Mar JS, Nagalingam NA, Song Y, Onizawa M, Lee JW, Lynch SV (2014) Amelioration of DSS-induced murine colitis by VSL#3 supplementation is primarily associated with changes in ileal microbiota composition. Gut Microbes 5(4):494–503

Marques CC, Castelo-Branco MT, Pacheco RG, Buongusto F, do Rosario Jr, Schanaider A, Coutinho-Silva R, de Souza HS (2014) Prophylactic systemic P2X7 receptor blockade prevents experimental colitis. Biochim Biophys Acta 1842(1):65–78

Mascanfroni ID, Yeste A, Vieira SM, Burns EJ, Patel B, Sloma I, Wu Y, Mayo L, Ben-Hamo R, Efroni S, Kuchroo VK, Robson SC, Quintana FJ (2013) IL-27 acts on DCs to suppress the T cell response and autoimmunity by inducing expression of the immunoregulatory molecule CD39. Nat Immunol 14(10):1054–1063

Mashimo H, Wu DC, Podolsky DK, Fishman MC (1996) Impaired defense of intestinal mucosa in mice lacking intestinal trefoil factor. Science 274(5285):262–265

Matsukawa T, Izawa K, Isobe M, Takahashi M, Maehara A, Yamanishi Y, Kaitani A, Okumura K, Teshima T, Kitamura T, Kitaura J (2015) Ceramide-CD300f binding suppresses experimental colitis by inhibiting ATP-mediated mast cell activation. Gut 65(5):777–787

McGovern, D. P., M. R. Jones, K. D. Taylor, K. Marciante, X. Yan, M. Dubinsky, A. Ippoliti, E. Vasiliauskas, D. Berel, C. Derkowski, D. Dutridge, P. Fleshner, D. Q. Shih, G. Melmed, E. Mengesha, L. King, S. Pressman, T. Haritunians, X. Guo, S. R. Targan, J. I. Rotter and I. B. D. G. C. International (2010) Fucosyltransferase 2 (FUT2) non-secretor status is associated with Crohn's disease. Hum Mol Genet 19(17):3468–3476

McGuckin MA, Linden SK, Sutton P, Florin TH (2011) Mucin dynamics and enteric pathogens. Nat Rev Microbiol 9(4):265–278

Miele E, Pascarella F, Giannetti E, Quaglietta L, Baldassano RN, Staiano A (2009) Effect of a probiotic preparation (VSL#3) on induction and maintenance of remission in children with ulcerative colitis. Am J Gastroenterol 104(2):437–443

Mifflin RC, Pinchuk IV, Saada JI, Powell DW (2011) Intestinal myofibroblasts: targets for stem cell therapy. Am J Physiol Gastrointest Liver Physiol 300(5):G684–G696

Millet YA, Alvarez D, Ringgaard S, von Andrian UH, Davis BM, Waldor MK (2014) Insights into Vibrio cholerae intestinal colonization from monitoring fluorescently labeled bacteria. PLoS Pathog 10(10), e1004405

Muise AM, Walters TD, Glowacka WK, Griffiths AM, Ngan BY, Lan H, Xu W, Silverberg MS, Rotin D (2009) Polymorphisms in E-cadherin (CDH1) result in a mis-localised cytoplasmic protein that is associated with Crohn's disease. Gut 58(8):1121–1127

Nenci A, Becker C, Wullaert A, Gareus R, van Loo G, Danese S, Huth M, Nikolaev A, Neufert C, Madison B, Gumucio D, Neurath MF, Pasparakis M (2007) Epithelial NEMO links innate immunity to chronic intestinal inflammation. Nature 446(7135):557–561

Neves AR, Castelo-Branco MT, Figliuolo VR, Bernardazzi C, Buongusto F, Yoshimoto A, Nanini HF, Coutinho CM, Carneiro AJ, Coutinho-Silva R, de Souza HS (2014) Overexpression of ATP-activated P2X7 receptors in the intestinal mucosa is implicated in the pathogenesis of Crohn's disease. Inflamm Bowel Dis 20(3):444–457

Nochi T, Yuki Y, Matsumura A, Mejima M, Terahara K, Kim DY, Fukuyama S, Iwatsuki-Horimoto K, Kawaoka Y, Kohda T, Kozaki S, Igarashi O, Kiyono H (2007) A novel M cell-specific carbohydrate-targeted mucosal vaccine effectively induces antigen-specific immune responses. J Exp Med 204(12):2789–2796

Obata T, Goto Y, Kunisawa J, Sato S, Sakamoto M, Setoyama H, Matsuki T, Nonaka K, Shibata N, Gohda M, Kagiyama Y, Nochi T, Yuki Y, Fukuyama Y, Mukai A, Shinzaki S, Fujihashi K, Sasakawa C, Iijima H, Goto M, Umesaki Y, Benno Y, Kiyono H (2010) Indigenous opportunistic bacteria inhabit mammalian gut-associated lymphoid tissues and share a mucosal antibody-mediated symbiosis. Proc Natl Acad Sci U S A 107(16):7419–7424

Ochoa-Cortes F, Linan-Rico A, Jacobson KA, Christofi FL (2014) Potential for developing purinergic drugs for gastrointestinal diseases. Inflamm Bowel Dis 20(7):1259–1287

Ochoa-Reparaz J, Mielcarz DW, Begum-Haque S, Kasper LH (2011) Gut, bugs, and brain: role of commensal bacteria in the control of central nervous system disease. Ann Neurol 69 (2):240–247

Odashima M, Bamias G, Rivera-Nieves J, Linden J, Nast CC, Moskaluk CA, Marini M, Sugawara K, Kozaiwa K, Otaka M, Watanabe S, Cominelli F (2005) Activation of A2A adenosine receptor attenuates intestinal inflammation in animal models of inflammatory bowel disease. Gastroenterology 129(1):26–33

Ogura Y, Bonen DK, Inohara N, Nicolae DL, Chen FF, Ramos R, Britton H, Moran T, Karaliuskas R, Duerr RH, Achkar JP, Brant SR, Bayless TM, Kirschner BS, Hanauer SB, Nunez G, Cho JH (2001) A frameshift mutation in NOD2 associated with susceptibility to Crohn's disease. Nature 411(6837):603–606

Ohman L, Simren M (2010) Pathogenesis of IBS: role of inflammation, immunity and neuroimmune interactions. Nat Rev Gastroenterol Hepatol 7(3):163–173

Papi C, Fasci-Spurio F, Rogai F, Settesoldi A, Margagnoni G, Annese V (2013) Mucosal healing in inflammatory bowel disease: treatment efficacy and predictive factors. Dig Liver Dis 45 (12):978–985

Petersen C, Round JL (2014) Defining dysbiosis and its influence on host immunity and disease. Cell Microbiol 16(7):1024–1033

Pham TA, Clare S, Goulding D, Arasteh JM, Stares MD, Browne HP, Keane JA, Page AJ, Kumasaka N, Kane L, Mottram L, Harcourt K, Hale C, Arends MJ, Gaffney DJ, Sanger Mouse Genetics Project, Dougan G, Lawley TD (2014) Epithelial IL-22RA1-mediated fucosylation promotes intestinal colonization resistance to an opportunistic pathogen. Cell Host Microbe 16(4):504–516

Pickard JM, Maurice CF, Kinnebrew MA, Abt MC, Schenten D, Golovkina TV, Bogatyrev SR, Ismagilov RF, Pamer EG, Turnbaugh PJ, Chervonsky AV (2014) Rapid fucosylation of intestinal epithelium sustains host-commensal symbiosis in sickness. Nature 514 (7524):638–641

Podolsky DK, Gerken G, Eyking A, Cario E (2009) Colitis-associated variant of TLR2 causes impaired mucosal repair because of TFF3 deficiency. Gastroenterology 137(1):209–220

Proietti M, Cornacchione V, Rezzonico Jost T, Romagnani A, Faliti CE, Perruzza L, Rigoni R, Radaelli E, Caprioli F, Preziuso S, Brannetti B, Thelen M, McCoy KD, Slack E, Traggiai E, Grassi F (2014) ATP-gated ionotropic P2X7 receptor controls follicular T helper cell numbers in Peyer's patches to promote host-microbiota mutualism. Immunity 41(5):789–801

Rakoff-Nahoum S, Paglino J, Eslami-Varzaneh F, Edberg S, Medzhitov R (2004) Recognition of commensal microflora by toll-like receptors is required for intestinal homeostasis. Cell 118 (2):229–241

Ralevic V, Burnstock G (1998) Receptors for purines and pyrimidines. Pharmacol Rev 50 (3):413–492

Randall TD, Carragher DM, Rangel-Moreno J (2008) Development of secondary lymphoid organs. Annu Rev Immunol 26:627–650

Rausch P, Rehman A, Kunzel S, Hasler R, Ott SJ, Schreiber S, Rosenstiel P, Franke A, Baines JF (2011) Colonic mucosa-associated microbiota is influenced by an interaction of Crohn disease and FUT2 (Secretor) genotype. Proc Natl Acad Sci U S A 108(47):19030–19035

Reyes A, Haynes M, Hanson N, Angly FE, Heath AC, Rohwer F, Gordon JI (2010) Viruses in the faecal microbiota of monozygotic twins and their mothers. Nature 466(7304):334–338

Rieder F, Fiocchi C (2009) Intestinal fibrosis in IBD--a dynamic, multifactorial process. Nat Rev Gastroenterol Hepatol 6(4):228–235

Rivera J, Fierro NA, Olivera A, Suzuki R (2008) New insights on mast cell activation via the high affinity receptor for IgE. Adv Immunol 98:85–120

Roggenbuck D, Hausdorf G, Martinez-Gamboa L, Reinhold D, Buttner T, Jungblut PR, Porstmann T, Laass MW, Henker J, Buning C, Feist E, Conrad K (2009) Identification of GP2, the major zymogen granule membrane glycoprotein, as the autoantigen of pancreatic antibodies in Crohn's disease. Gut 58(12):1620–1628

Roggenbuck D, Reinhold D, Werner L, Schierack P, Bogdanos DP, Conrad K (2013) Glycoprotein 2 antibodies in Crohn's disease. Adv Clin Chem 60:187–208

Sartor RB (2004) Therapeutic manipulation of the enteric microflora in inflammatory bowel diseases: antibiotics, probiotics, and prebiotics. Gastroenterology 126(6):1620–1633

Sato S, Kaneto S, Shibata N, Takahashi Y, Okura H, Yuki Y, Kunisawa J, Kiyono H (2013) Transcription factor Spi-B-dependent and -independent pathways for the development of Peyer's patch M cells. Mucosal Immunol 6(4):838–846

Scaldaferri F, Gerardi V, Lopetuso LR, Del Zompo F, Mangiola F, Boskoski I, Bruno G, Petito V, Laterza L, Cammarota G, Gaetani E, Sgambato A, Gasbarrini A (2013) Gut microbial flora, prebiotics, and probiotics in IBD: their current usage and utility. Biomed Res Int 2013:435268

Sears CL, Geis AL, Housseau F (2014) Bacteroides fragilis subverts mucosal biology: from symbiont to colon carcinogenesis. J Clin Invest 124(10):4166–4172

Simpson HL, Campbell BJ, Rhodes JM (2014) IBD: microbiota manipulation through diet and modified bacteria. Dig Dis 32(Suppl 1):18–25

Sonnenberg GF, Monticelli LA, Alenghat T, Fung TC, Hutnick NA, Kunisawa J, Shibata N, Grunberg S, Sinha R, Zahm AM, Tardif MR, Sathaliyawala T, Kubota M, Farber DL, Collman RG, Shaked A, Fouser LA, Weiner DB, Tessier PA, Friedman JR, Kiyono H, Bushman FD, Chang KM, Artis D (2012) Innate lymphoid cells promote anatomical containment of lymphoid resident commensal bacteria. Science 336(6086):1321–1325

Sun M, He C, Cong Y, Liu Z (2015) Regulatory immune cells in regulation of intestinal inflammatory response to microbiota. Mucosal Immunol 8(5):969–978

Terahara K, Yoshida M, Igarashi O, Nochi T, Pontes GS, Hase K, Ohno H, Kurokawa S, Mejima M, Takayama N, Yuki Y, Lowe AW, Kiyono H (2008) Comprehensive gene expression profiling of Peyer's patch M cells, villous M-like cells, and intestinal epithelial cells. J Immunol 180(12):7840–7846

Thabane M, Kottachchi DT, Marshall JK (2007) Systematic review and meta-analysis: The incidence and prognosis of post-infectious irritable bowel syndrome. Aliment Pharmacol Ther 26(4):535–544

Thompson JA, Oliveira RA, Djukovic A, Ubeda C, Xavier KB (2015) Manipulation of the quorum sensing signal AI-2 affects the antibiotic-treated gut microbiota. Cell Rep 10(11):1861–1871

Tsai SH, Kinoshita M, Kusu T, Kayama H, Okumura R, Ikeda K, Shimada Y, Takeda A, Yoshikawa S, Obata-Ninomiya K, Kurashima Y, Sato S, Umemoto E, Kiyono H, Karasuyama H, Takeda K (2015) The ectoenzyme E-NPP3 negatively regulates ATP-dependent chronic allergic responses by basophils and mast cells. Immunity 42 (2):279–293

Umesaki Y, Setoyama H, Matsumoto S, Okada Y (1993) Expansion of alpha beta T-cell receptor-bearing intestinal intraepithelial lymphocytes after microbial colonization in germ-free mice and its independence from thymus. Immunology 79(1):32–37

Uo M, Hisamatsu T, Miyoshi J, Kaito D, Yoneno K, Kitazume MT, Mori M, Sugita A, Koganei K, Matsuoka K, Kanai T, Hibi T (2013) Mucosal CXCR4+ IgG plasma cells contribute to the pathogenesis of human ulcerative colitis through FcgammaR-mediated CD14 macrophage activation. Gut 62(12):1734–1744

Vaishnava S, Behrendt CL, Ismail AS, Eckmann L, Hooper LV (2008) Paneth cells directly sense gut commensals and maintain homeostasis at the intestinal host-microbial interface. Proc Natl Acad Sci U S A 105(52):20858–20863

Van der Sluis M, De Koning BA, De Bruijn AC, Velcich A, Meijerink JP, Van Goudoever JB, Buller HA, Dekker J, Van Seuningen I, Renes IB, Einerhand AW (2006) Muc2-deficient mice spontaneously develop colitis, indicating that MUC2 is critical for colonic protection. Gastroenterology 131(1):117–129

Van Praet JT, Donovan E, Vanassche I, Drennan MB, Windels F, Dendooven A, Allais L, Cuvelier CA, van de Loo F, Norris PS, Kruglov AA, Nedospasov SA, Rabot S, Tito R, Raes J, Gaboriau-Routhiau V, Cerf-Bensussan N, Van de Wiele T, Eberl G, Ware CF, Elewaut D (2015) Commensal microbiota influence systemic autoimmune responses. EMBO J 34(4):466–474

Velcich A, Yang W, Heyer J, Fragale A, Nicholas C, Viani S, Kucherlapati R, Lipkin M, Yang K, Augenlicht L (2002) Colorectal cancer in mice genetically deficient in the mucin Muc2. Science 295(5560):1726–1729

Verbeke KA, Boesmans L, Boets E (2014) Modulating the microbiota in inflammatory bowel diseases: prebiotics, probiotics or faecal transplantation? Proc Nutr Soc 73(4):490–497

Vora P, Youdim A, Thomas LS, Fukata M, Tesfay SY, Lukasek K, Michelsen KS, Wada A, Hirayama T, Arditi M, Abreu MT (2004) Beta-defensin-2 expression is regulated by TLR signaling in intestinal epithelial cells. J Immunol 173(9):5398–5405

Wehkamp J, Harder J, Weichenthal M, Schwab M, Schaffeler E, Schlee M, Herrlinger KR, Stallmach A, Noack F, Fritz P, Schroder JM, Bevins CL, Fellermann K, Stange EF (2004) NOD2 (CARD15) mutations in Crohn's disease are associated with diminished mucosal alpha-defensin expression. Gut 53(11):1658–1664

Weinstein DL, O'Neill BL, Hone DM, Metcalf ES (1998) Differential early interactions between Salmonella enterica serovar Typhi and two other pathogenic Salmonella serovars with intestinal epithelial cells. Infect Immun 66(5):2310–2318

Wu HJ, Ivanov II, Darce J, Hattori K, Shima T, Umesaki Y, Littman DR, Benoist C, Mathis D (2010) Gut-residing segmented filamentous bacteria drive autoimmune arthritis via T helper 17 cells. Immunity 32(6):815–827

Xu MQ, Cao HL, Wang WQ, Wang S, Cao XC, Yan F, Wang BM (2015) Fecal microbiota transplantation broadening its application beyond intestinal disorders. World J Gastroenterol 21(1):102–111

Yao Y, Levings MK, Steiner TS (2012) ATP conditions intestinal epithelial cells to an inflammatory state that promotes components of DC maturation. Eur J Immunol 42(12):3310–3321

Ye JH, Rajendran VM (2009) Adenosine: an immune modulator of inflammatory bowel diseases. World J Gastroenterol 15(36):4491–4498

Zaki MH, Boyd KL, Vogel P, Kastan MB, Lamkanfi M, Kanneganti TD (2010) The NLRP3 inflammasome protects against loss of epithelial integrity and mortality during experimental colitis. Immunity 32(3):379–391

Zhou Y, Kawasaki H, Hsu SC, Lee RT, Yao X, Plunkett B, Fu J, Yang K, Lee YC, Huang SK (2010) Oral tolerance to food-induced systemic anaphylaxis mediated by the C-type lectin SIGNR1. Nat Med 16(10):1128–1133

Chapter 44
Control of Intestinal Regulatory T Cells by Human Commensal Bacteria

Koji Atarashi

Abstract The human gut harbours diverse bacteria that play a fundamental role in our health and well-being. Recent studies have shown that the gut microbiota plays an important role in the development of the intestinal immune system, and that an altered microbiota composition is associated to some inflammatory diseases. Therefore, the microbiota is increasingly being recognised as a potential therapeutic target in several inflammatory disorders. Here, recent findings are presented on human microbiota-mediated colonic regulatory T (Treg) cell induction. We identified $CD4^+$ $FoxP3^+$ Treg cell-inducing bacteria derived from a healthy human fecal sample. These bacteria included 17 strains belonging to *Clostridium* cluster IV, XIVa, and XVIII. The 17 strains were found to increase markedly the number of colonic Treg cells and enhance their immunosuppressive capacity. Oral administration of the 17 strains to specific-pathogen–free mice resulted in significant attenuation of intestinal inflammation and allergic diarrhoea, suggesting that the isolated 17 strains may have therapeutic potential in the treatment of immune disorders.

Keywords Microbiota • Dysbiosis • Treg • IBD • Clostridium • SCFAs

44.1 Introduction

The human gastrointestinal tract harbours a diverse and complex microbial community that comprises thousands of different species of bacteria and 100 trillion bacterial cells. Humans have coevolved with their microbes and coexist as 'superorganisms', in which the gut microbiota performs essential physiological functions that help maintain human health. For instance, human enzymes cannot degrade most complex carbohydrates and polysaccharides, such as cellulose, xylans, starch, and inulin. However, commensal bacteria can metabolise indigestible

K. Atarashi (✉)
Department of Microbiology and Immunology, Keio University School of Medicine, 35 Shinanomachi, Shinjuku-ku, Tokyo 160-8582, Japan

PRESTO, Japan Science and Technology Agency, Kawaguchi 332-0012, Japan
e-mail: kojiatarashi@keio.jp

M. Miyasaka, K. Takatsu (eds.), *Chronic Inflammation*,
DOI 10.1007/978-4-431-56068-5_44

polysaccharides to readily absorbable monosaccharides and short-chain fatty acids (SCFAs), mainly acetate, butyrate, and propionate (Macfarlane and Macfarlane 2003). The microbiota is also capable of metabolising bile acids and sterols, and synthesising many essential vitamins and amino acids. Many of these metabolites derived from microbiota have beneficial effects on host health. In addition, gut microbiota can provide increased resistance to pathogens by occupying their required niche, producing antagonistic compounds, and activating the immune system (Guarner and Malagelada 2003).

The intestinal immune system is composed of unique lymphoid tissues and many cell types. Studies using germ-free (GF) animals raised in the absence of all microorganisms, established that gut microbiota is essential for the development and functions of the intestinal immune system. The development of gut-associated lymphoid tissues (GALT), including Peyer's patches, isolated lymphoid follicles, and mesenteric lymph nodes (MLN), are defective in GF mice. GF mice have fewer and smaller Peyer's patches and isolated lymphoid follicles, and smaller MLN with reduced cellularity compared to those in mice housed under specific-pathogen–free (SPF) conditions (Macpherson and Harris 2004). In relation to defective lymphoid development, GF mice show decreased immunoglobulin A (IgA)-producing plasma cells in the lamina propria as well as decreased luminal IgA levels. The induction and activation of T helper (Th) 17 cells, interleukin-17 (IL-17) producing $CD4^+$ T cells, and regulatory T (Treg) cells in lamina propria also depend on the presence of the microbiota (Ivanov et al. 2009; Atarashi et al. 2011). In addition to adaptive immune cells, the microbiota influences the development and function of innate immune cells such as innate lymphoid cells (ILCs), natural killer (NK) cells, and monocytes. Consequently, GF animals are more susceptible to oral infection with various pathogens and are defective in oral tolerance. Taken together, it is clear that the microbiota is essential for the development, maturation, and maintenance of the intestinal immune system.

44.2 Dysbiosis in Disease

The gut microbiota is composed of thousands of bacterial species that include symbionts (beneficial), commensals (neither pathogenic nor beneficial), and pathobionts (potentially pathogenic). Firmicutes, Bacteroidetes, Actinobacteria, and Proteobacteria are the predominant phyla in human gut. However, there is no apparent consensus on the composition of a healthy microbiota, aside from the fact that a more complex and diverse community is better. Two decades ago, culture-based studies were performed in an attempt to assess intestinal microbial diversity. These were insufficient in identifying the composition of the microbiota, because a large fraction of the intestinal bacteria is difficult to cultivate and only a small number of samples can be tested. To address this limitation, culture-independent methods for analysing microbiota composition have been developed in the past few decades. For instance, bacterial 16S ribosomal RNA (rRNA) gene sequences can be profiled by temperature or denaturing gradient gel electrophoresis (TGGE or

DGGE) and terminal-restriction fragment length polymorphism (T-RFLP). Although these techniques provided a more comprehensive analysis of the microbiota composition, they also possess a resolution limit with respect to detecting minor species. In recent years, technological advances in next-generation sequencing (NGS) have made it possible to identify the entire microbial community at the species level even for scarce members. Furthermore, metagenomic analysis provides access to the functional gene composition of microbial communities.

Using NGS-based 16S rRNA gene and metagenomic sequence analyses, a number of studies have shown the link between gut microbiota and human health. The composition of the microbiota in patients with inflammatory bowel diseases (IBDs), cancers, obesity, diabetes, asthma, allergies, and even autism differs significantly from that of healthy subjects (Petersen and Round 2014). It is widely accepted that an altered microbial composition, termed dysbiosis, is one of the important factors contributing to the initiation, chronicity, and persistence of many of these diseases, although it remains unclear whether dysbiosis is a cause or consequence of these diseases. Dysbiosis is commonly characterised by decreased diversity, a significant reduction in some beneficial bacteria, and an increase in potentially pathogenic bacteria. The main factors causing dysbiosis are the use of antibiotics, poor diet, infection, and inflammation.

IBDs, such as Crohn's disease (CD) and ulcerative colitis (UC), are chronic relapsing inflammatory disorders of the gastrointestinal tract, resulting from altered interactions between the gut microbiota and the intestinal immune system. Many studies have observed dysbiosis in IBD patients, and suggested that dysbiosis leads to dysregulation of immune homeostasis and immune tolerance to gut microbiota. The consistent features of dysbiosis in IBD patients are low bacterial diversity, a decrease in the proportions of the phyla Firmicutes and Bacteroidetes, and an increase in the proportions of Proteobacteria and Fusobacteria (Frank et al. 2007). Among Firmicutes, *Clostridium* cluster XIVa and IV (e.g., *Faecalibacterium prausnitzii*) are especially decreased in IBD patients (Sokol et al. 2009). However, the mechanisms by which dysbiosis contributes to intestinal inflammation are poorly understood because the specific effects of individual bacteria on the host immune system remain unclear. Elucidating the role of individual bacterial species will be very important for the development of microbe-based therapeutic approaches for patients with dysbiosis-associated diseases.

44.3 Intestinal Treg cells

Intestinal surfaces are continuously exposed to a large number of bacterial antigens. Reduced immune tolerance and an overactivation of immune responses to intestinal bacteria are known to lead to the development of intestinal inflammation, as evidenced by the attenuation of the spontaneous development of colitis in many mouse models housed under GF conditions. Treg cells play an important role in maintaining immune tolerance and in suppressing excessive immune responses in the intestine as well as the whole body. Indeed, Treg cells are particularly abundant in the intestine, where they prevent aberrant inflammatory reactions to the microbiota. The transcription factor forkhead box P3 (FoxP3) is essential for the development and suppressive function of Treg cells (Sakaguchi et al. 2008). Treg cells can suppress the activation of various immune cells including effector $CD4^+$ T (Th1, Th2, and Th17) cells, $CD8^+$ T cells, γδT cells, B cells, NK cells, NKT cells, monocytes, and dendritic cells (DCs). The mechanisms behind Treg-mediated suppression of effector T cells are mainly considered cell contact-dependent based on in vitro suppression assays. Contact-dependent suppression is mediated by cell-surface inhibitory molecules such as cytotoxic T-lymphocyte–associated protein 4 (CTLA-4), lymphocyte activation gene-3 (LAG-3), and perforin/granzyme B. CTLA-4 binds to CD80/CD86 costimulatory molecules with high avidity and causes downregulation of surface CD80/CD86 on antigen-presenting cells (APCs). LAG-3 binds to MHC class II on APCs and inhibits DC activation. Although CTLA-4 and LAG-3 indirectly suppress effector T cells, Treg cells directly induce apoptosis of T cells via perforin/granzyme B. Another mechanism for Treg-mediated suppression of effector T cells is the consumption of IL-2, which is essential for their proliferation and survival. Treg cells express constitutively high levels of IL-2 receptor α subunit (CD25) that sequesters IL-2 and causes effector T-cell apoptosis. Treg cells also secrete immunosuppressive cytokines such as transforming growth factor-β (TGF-β), IL-10, and IL-35. IL-10 is well known as an important mediator in preventing intestinal inflammation (Rubtsov et al. 2008).

Treg cells are classified into two main populations: thymus-derived Treg (tTreg) or peripherally induced Treg (pTreg) cells (Shevach and Thornton 2014). tTreg cells develop in the thymus through high-affinity T-cell receptor (TCR) interactions with self-peptides in the presence of TGF-β and IL-2. In the periphery, TGF-β is also essential for the induction of FoxP3 expression in naïve $CD4^+$ $FoxP3^-$ T cells, and IL-2 is required for expansion, activity, and survival of pTreg cells. It has been difficult to distinguish between tTreg and pTreg cells because they share Treg-specific signatures, such as high expression of FoxP3, CD25, CTLA-4, and GITR (glucocorticoid-induced tumour necrosis factor family receptor). Recently, several studies have suggested phenotypic differences between tTreg and pTreg cells. Helios (Ikzf2), a member of the Ikaros transcription factor family, and neuropilin-1 (Nrp-1), a receptor for semaphorins and VEGF family members, are solely expressed by tTreg cells and thus, could be used as a marker for distinguishing tTreg cells from pTreg cells (Thornton et al. 2010; Weiss et al. 2012; Yadav et al. 2012). Epigenetic differences have also been observed between tTreg and pTreg cells. The CpG-rich Treg-specific demethylated region (TSDR) of the FoxP3 locus, also known as conserved noncoding sequence 2 (CNS2), is completely demethylated in tTreg cells but partially methylated in pTreg cells (Floess et al. 2007). CNS2 demethylation is critical for stable FoxP3 expression. Although functional differences between tTreg and pTreg subsets are not fully understood, these differences in Helios and Nrp-1 expression and epigenetic modifications may contribute to their different functionality and stability.

44.4 Isolation of Treg-Inducing Human Microbiota

Recently, several studies have shown that the number and function of colonic Treg cells are influenced by gut microbiota. For instance, *Bacteroides fragilis* is a human commensal bacterium with several immunoregulatory functions. Polysaccharide A (PSA) of *B. fragilis* directly acts on Treg cells through their TLR2 to induce IL-10 expression (Round and Mazmanian 2010). Similarly, the altered Schaedler flora (ASF), a cocktail of 8 defined bacterial strains, predominantly induces Treg cells in the colon via a TLR-dependent mechanism (Geuking et al. 2011). Additional evidence for the link between microbiota and Treg cells stems from previous work indicating that Treg cells are markedly decreased in the colon of GF mice and that 46 strains of Clostridia indigenous to mice strongly induce the accumulation and differentiation of colonic Treg cells (Atarashi et al. 2011). Furthermore, our recent studies revealed that 17 strains of Clostridia from human gut microbiota are potent Treg-inducing bacteria (Atarashi et al. 2013). Our recent results are briefly introduced below.

First, to determine whether Treg-inducing bacteria were present in the human intestine, we orally administered GF mice with human feces from a healthy adult. Treg cells were significantly increased in the colon of the mice colonised with human microbiota compared with that observed in the colon of GF mice. Moreover, human feces treated with chloroform, which is generally used to select for spore-forming bacteria including *Clostridium* and *Bacillus* species, significantly increased the number of Treg cells as compared to that observed in GF mice inoculated with untreated human feces. Interestingly, compared with untreated human feces, chloroform-treated human feces poorly induced Th17 differentiation, indicating that spore-forming bacteria in the human intestine are potent Treg-inducers, and weak Th17-inducers.

Next, we serially diluted the cecal contents of colonised mice and inoculated GF mice to reduce the bacterial composition and isolate Treg-inducing bacteria. After two repetitions, we cultured the cecal contents of these mice, collected 31 bacterial strains, and selected 23 strains based on less than 99% 16S rRNA gene sequence similarity. Colonic Treg cells were fully induced when GF mice were inoculated with the 23 strains isolated. Meta-analysis of the 16S rRNA gene sequences from the gut microbiota of the mice inoculated with the 23 strains revealed that 17 out of 23 strains could colonise in the intestine. Full genome analysis of these 17 strains revealed that two strains belonged to the *Clostridium* cluster IV, 12 strains to cluster XIVa, and 3 strains to cluster XVIII (Fig. 44.1). Furthermore, colonisation of these 17 strains could induce Treg cells, particularly Helios-negative pTreg cells, in mice with various genetic backgrounds (IQI, C57BL/6, and BALB/c), as well as in F344 rats, suggesting common mechanisms for Treg induction by these 17 strains.

Strain No.	Closest relatives	Database	Similarity (%)	*Clostridium* cluster
Strain 1	*Clostridium saccharogumia* (T) SDG-Mt85-3Db *Clostridium spiroforme* DSM 1552	RDP GenomeDB	99.46 95.89	cluster XVIII
Strain 3	*Lachnospiraceae* bacterium 7 1 58FAA *Flavonifractor plautii* ATCC 29863	RDP GenomeDB	100 99.84	cluster IV
Strain 4	*Clostridium hathewayi* *Clostridium hathewayi* DSM 13479	RDP GenomeDB	99.06 98.11	cluster XIVa
Strain 6	*Blautia producta* JCM 1471 *Lachnospiraceae* bacterium 6_1_63FAA	RDP GenomeDB	99.79 96.52	cluster XIVa
Strain 7	*Clostridium bolteae* (T) type strain 16351 *Clostridium bolteae* ATCC BAA 613	RDP GenomeDB	99.53 99.53	cluster XIVa
Strain 8	*Clostridiaceae* bacterium JC13 *Erysipelotrichaceae* bacterium 2 2 44A	RDP GenomeDB	99.16 92.92	cluster XVIII
Strain 9	*Clostridium indolis* CM971 *Anaerostipes caccae* DSM 14662	RDP GenomeDB	99.26 97.67	cluster XIVa
Strain 13	*Anaerotruncus colihominis* HKU19 *Anaerotruncus colihominis* DSM 17241	RDP GenomeDB	100 100	cluster IV
Strain 14	*Ruminococcus* sp. ID8 *Coprococcus comes* ATCC 27758	RDP GenomeDB	98.59 93.23	cluster XIVa
Strain 15	*Clostridium lavalense* CCRI-9929 *Clostridium asparagiforme* DSM 15981	RDP GenomeDB	99.52 99.73	cluster XIVa
Strain 16	*Clostridium* sp. 7 3 54FAA *Clostridium symbiosum* WAL 14673	GenomeDB GenomeDB	100 99.87	cluster XIVa
Strain 18	*Clostridium ramosum* JCM 5234 *Clostridium ramosum* DSM 1402	RDP GenomeDB	100 100	cluster XVIII
Strain 21	*Eubacterium contortum* type strain DSM 3982 *Eubacterium fissicatena*	RDP GenomeDB	99.58 100	cluster XIVa
Strain 26	*Clostridium scindens* VPI 12708 *Clostridium scindens* ATCC 35704	RDP GenomeDB	99.72 99.58	cluster XIVa
Strain 27	*Lachnospiraceae* bacterium 3 1 57FAA CT1	RDP	97.54	cluster XIVa
Strain 28	*Clostridium aldenense* (T) RMA 9741 *Clostridiales* bacterium 1_7_47FAA	RDP GenomeDB	98.41 99.73	cluster XIVa
Strain 29	*Lachnospiraceae* bacterium 3 1 57FAA CT1	GenomeDB	99.6	cluster XIVa

Fig. 44.1 16S rRNA gene analysis for each of the isolated 17 strains (adapted from Atarashi et al. 2013). The full-length or part of the 16S rRNA gene was amplified by PCR using 16S rRNA gene-specific primer pairs. Each amplified 16S rRNA gene was sequenced and compared to the Ribosomal Database Project (RDP) database and genome database (GenomeDB) constructed from

44.5 Bacterial Treg Induction Mechanism

We hypothesised that the 17 strains identified previously acted on intestinal epithelial cells to produce TGF-β, because TGF-β is critical for the development and differentiation of Treg cells. Indeed, cecal contents from the mice colonised with the 17 strains stimulate epithelial cells to reduce TGF-β production; however, it is still unknown if this is the mechanism underlying colonic Treg cell induction. An important mediator of TGF-β production and Treg cell induction, which has been proposed previously, may involve SCFAs. These organic fatty acids are produced by bacterial fermentation of carbohydrates, and are abundantly present in the cecal content of mice colonised with the 17 strains identified. Moreover, this idea is supported by the fact that the increase in TGF-β levels in the colon of inoculated mice was not inhibited by pretreatment of the cecal contents with a protease, nuclease, or heat. Interestingly, epithelial cells actually secrete high amounts of active TGF-β following the addition of SCFAs to the culture media. In addition, SCFAs, particularly butyrate, have been shown to act on DCs through GPR109a to induce anti-inflammatory genes and promote Treg differentiation (Arpaia et al. 2013; Singh et al. 2014). SCFAs, butyrate, and propionate can also directly induce Treg cell differentiation, and Treg cell proliferation, and IL-10 production via GPR43 by inhibition of histone deacetylase (Furusawa et al. 2013; Smith et al. 2013).

In vitro production of SCFAs by each of the 17 strains was analysed using liquid chromatography–electrospray ionisation–tandem mass spectrometry (LC-ESI-MS/MS). Most of the 17 strains, mainly belonging to cluster XIVa, produced acetate and/or butyrate at high concentrations. Among these bacteria, strains 9, 16, 27, and 29 showed very high butyrate production. Comparative genome analysis revealed that most of the 17 strains contain abundant genes involved in acetate and butyrate-synthesis pathways such as butyryl CoA:acetate CoA-transferase genes. Importantly, when GF mice were monocolonised with one of each of the 17 strains, the percentage and the absolute number of Treg cells were significantly lower in the colon than in the mice colonised with the mixture of the 17 strains. These data suggest that synergistic or mutualistic interactions among the 17 strains identified contribute to effective Treg induction (Fig. 44.2).

Fig. 44.1 (continued) publicly available genome sequences in NCBI and HMP databases. Close relative species/strains, % similarity, and *Clostridium* cluster are shown

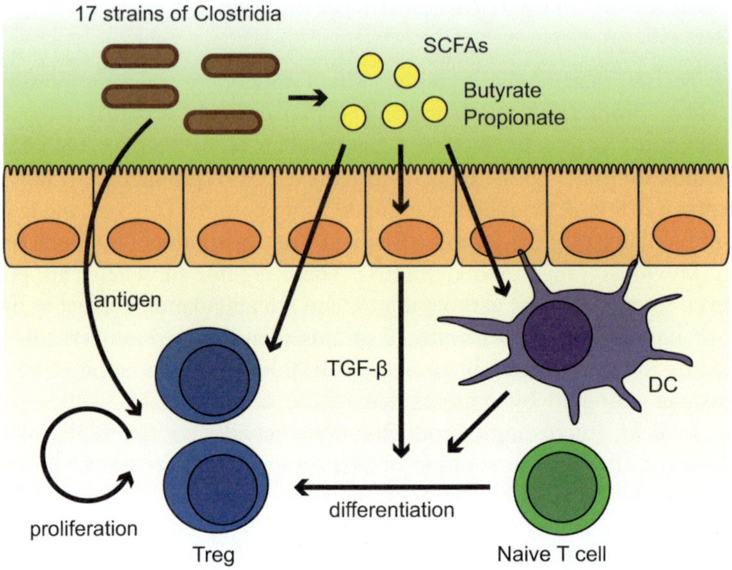

Fig. 44.2 Treg induction by 17 strains of Clostridia from the human microbiota
Commensal microbial metabolism controls regulatory T-cell (Treg) homeostasis in the colon. Colonisation of GF mice with 17 strains of Clostridia isolated from the human microbiota induces the accumulation of Treg cells in the colon. Short-chain fatty acids (SCFAs), especially butyrate and propionate, promote Treg differentiation and proliferation directly or through activation of epithelial dendritic cell (DC)

44.6 Suppression of Intestinal Inflammation

Colonic Treg cells from GF mice show immature phenotypes as well as decreased cell number. Although Treg cells produce large amounts of IL-10, and express high levels of CTLA-4 and inducible T cell costimulator (ICOS) in the colon of SPF mice, all of these suppressive effector molecules are decreased in the GF condition. Colonisation of GF mice with the Treg-inducing 17 strains led to increased expression of IL-10, CTLA-4, and ICOS in colonic Treg cells equivalent to that observed in SPF mice. Thus, the 17 strains promote both the accumulation and functional maturation of Treg cells in the colonic lamina propria. It is also known that Treg-mediated suppression depends on the strength of the antigen stimulus. To address whether the 17 strains induced Ag-specific Treg cells, we have analysed the suppressive activity of Treg cells from mice colonised with the 17 strains. OT-I $CD8^+$ T cells proliferated when stimulated with DCs pulsed with the MHC class I $OVA_{257–264}$ peptide. The addition of colonic Treg cells suppressed OT-I proliferation in a dose-dependent manner. Furthermore, this Treg-mediated suppression was markedly enhanced in the presence of the 17 strains that had been killed by heat. The TCR repertoire of colonic Treg cells recognised antigens derived from the gut microbiota (Lathrop et al. 2011). These results are consistent with a previous report (Lathrop et al. 2011) and suggest that colonic Treg cells induced by the 17 strains have mainly TCRs recognising the derived antigens.

Treg cells are critical for intestinal immune homeostasis, and the 17 strains identified could provide therapeutic benefit in patients with IBD or diarrhoea by inducing the development of Treg cells in the colon. Oral administration of the 17 strains could increase the number of Treg cells in the colon even under SPF conditions. As expected, SPF mice colonised by the 17 strains showed lower mortality and decreased severity of 2,4,6-trinitrobenzene sulfonic acid (TNBS)-induced colitis. The occurrence and severity of OVA-induced allergic diarrhoea were significantly lower in the SPF mice colonised by the 17 strains as compared to control SPF mice. These findings suggest that supplementation with the 17 strains may induce Treg cells and be effective in the treatment of allergic and inflammatory diseases in humans.

44.7 Discussion

An altered microbiota composition, termed dysbiosis, is associated with many human diseases. Several studies have reported decreased levels of Clostridia in fecal samples from patients with IBD, and atopic dermatitis (Candela et al. 2012; Frank et al. 2007). Animal studies have also shown that bacterial dysbiosis leads to spontaneous development of ulcerative colitis-like inflammation (Garrett et al. 2007). Therefore, manipulation of the gut microbiota holds the promise of important advances in the treatment of a number of inflammatory diseases. Fecal microbiota transplantation (FMT) is a targeted microbiota-based therapy that has garnered a great deal of attention in recent years. FMT is the introduction of commensal bacteria derived from stool samples of a healthy donor into the gastrointestinal tract of the patient through colonoscopy, nasoduodenal tubes, or enemas. It has shown remarkable clinical effectiveness in patients with pseudomembranous colitis induced by *Clostridium difficile* infection (van Nood et al. 2013). There are a number of studies evaluating the efficacy of FMT in other diseases associated with dysbiosis, including IBD, irritable bowel syndrome, diabetes, obesity, multiple sclerosis, and autism. Despite the increasing application of FMT, many questions remain regarding the selection and screening of donors, the sample processing, and the risks of transmission of pathogens. Therefore, treatment with a composite of well-characterised effective microorganisms may be preferable.

In this context, our Treg-inducing 17 strains can form the basis for new live biotherapeutic candidates for the treatment of IBD, allergy, and other immune-inflammatory diseases. However, there is a need for further studies to assess the benefits and the potential risks of the 17 strains and to optimise the therapeutic mixture by eliminating components of the 17 strains that are nonessential to Treg cell induction. In addition, it is necessary to identify and isolate other bacterial strains from the human intestine that can affect the development and activation of other immune cell subsets, such as Th17 cells or Th1 cells. It is also possible, sometime in the future, to develop a necessary and sufficient bacterial consortium as a universal panacea against diseases associated with dysbiosis.

References

Arpaia N, Campbell C, Fan X, Dikiy S, van der Veeken J, deRoos P, Liu H, Cross JR, Pfeffer K, Coffer PJ, Rudensky AY (2013) Metabolites produced by commensal bacteria promote peripheral regulatory T-cell generation. Nature 504(7480):451–455. doi:10.1038/nature12726

Atarashi K, Tanoue T, Oshima K, Suda W, Nagano Y, Nishikawa H, Fukuda S, Saito T, Narushima S, Hase K, Kim S, Fritz JV, Wilmes P, Ueha S, Matsushima K, Ohno H, Olle B, Sakaguchi S, Taniguchi T, Morita H, Hattori M, Honda K (2013) Treg induction by a rationally selected mixture of Clostridia strains from the human microbiota. Nature 500(7461):232–236. doi:10.1038/nature12331

Atarashi K, Tanoue T, Shima T, Imaoka A, Kuwahara T, Momose Y, Cheng G, Yamasaki S, Saito T, Ohba Y, Taniguchi T, Takeda K, Hori S, Ivanov II, Umesaki Y, Itoh K, Honda K (2011) Induction of colonic regulatory T cells by indigenous Clostridium species. Science 331 (6015):337–341. doi:10.1126/science.1198469

Candela M, Rampelli S, Turroni S, Severgnini M, Consolandi C, De Bellis G, Masetti R, Ricci G, Pession A, Brigidi P (2012) Unbalance of intestinal microbiota in atopic children. BMC Microbiol 12:95. doi:10.1186/1471-2180-12-95

Floess S, Freyer J, Siewert C, Baron U, Olek S, Polansky J, Schlawe K, Chang HD, Bopp T, Schmitt E, Klein-Hessling S, Serfling E, Hamann A, Huehn J (2007) Epigenetic control of the foxp3 locus in regulatory T cells. PLoS Biol 5(2), e38. doi:10.1371/journal.pbio.0050038

Frank DN, St Amand AL, Feldman RA, Boedeker EC, Harpaz N, Pace NR (2007) Molecular-phylogenetic characterization of microbial community imbalances in human inflammatory bowel diseases. Proc Natl Acad Sci U S A 104(34):13780–13785. doi:10.1073/pnas.0706625104

Furusawa Y, Obata Y, Fukuda S, Endo TA, Nakato G, Takahashi D, Nakanishi Y, Uetake C, Kato K, Kato T, Takahashi M, Fukuda NN, Murakami S, Miyauchi E, Hino S, Atarashi K, Onawa S, Fujimura Y, Lockett T, Clarke JM, Topping DL, Tomita M, Hori S, Ohara O, Morita T, Koseki H, Kikuchi J, Honda K, Hase K, Ohno H (2013) Commensal microbe-derived butyrate induces the differentiation of colonic regulatory T cells. Nature 504(7480):446–450. doi:10.1038/nature12721

Garrett WS, Lord GM, Punit S, Lugo-Villarino G, Mazmanian SK, Ito S, Glickman JN, Glimcher LH (2007) Communicable ulcerative colitis induced by T-bet deficiency in the innate immune system. Cell 131(1):33–45. doi:10.1016/j.cell.2007.08.017

Geuking MB, Cahenzli J, Lawson MA, Ng DC, Slack E, Hapfelmeier S, McCoy KD, Macpherson AJ (2011) Intestinal bacterial colonization induces mutualistic regulatory T cell responses. Immunity 34(5):794–806. doi:10.1016/j.immuni.2011.03.021

Guarner F, Malagelada JR (2003) Gut flora in health and disease. Lancet 361(9356):512–519. doi:10.1016/S0140-6736(03)12489-0

Ivanov II, Atarashi K, Manel N, Brodie EL, Shima T, Karaoz U, Wei D, Goldfarb KC, Santee CA, Lynch SV, Tanoue T, Imaoka A, Itoh K, Takeda K, Umesaki Y, Honda K, Littman DR (2009) Induction of intestinal Th17 cells by segmented filamentous bacteria. Cell 139(3):485–498. doi:10.1016/j.cell.2009.09.033

Lathrop SK, Bloom SM, Rao SM, Nutsch K, Lio CW, Santacruz N, Peterson DA, Stappenbeck TS, Hsieh CS (2011) Peripheral education of the immune system by colonic commensal microbiota. Nature 478(7368):250–254. doi:10.1038/nature10434

Macfarlane S, Macfarlane GT (2003) Regulation of short-chain fatty acid production. Proc Nutr Soc 62(1):67–72. doi:10.1079/PNS2002207

Macpherson AJ, Harris NL (2004) Interactions between commensal intestinal bacteria and the immune system. Nat Rev Immunol 4(6):478–485. doi:10.1038/nri1373

Petersen C, Round JL (2014) Defining dysbiosis and its influence on host immunity and disease. Cell Microbiol 16(7):1024–1033. doi:10.1111/cmi.12308

Round JL, Mazmanian SK (2010) Inducible Foxp3+ regulatory T-cell development by a commensal bacterium of the intestinal microbiota. Proc Natl Acad Sci U S A 107 (27):12204–12209. doi:10.1073/pnas.0909122107

Rubtsov YP, Rasmussen JP, Chi EY, Fontenot J, Castelli L, Ye X, Treuting P, Siewe L, Roers A, Henderson WR Jr, Muller W, Rudensky AY (2008) Regulatory T cell-derived interleukin-10 limits inflammation at environmental interfaces. Immunity 28(4):546–558. doi:10.1016/j.immuni.2008.02.017

Sakaguchi S, Yamaguchi T, Nomura T, Ono M (2008) Regulatory T cells and immune tolerance. Cell 133(5):775–787. doi:10.1016/j.cell.2008.05.009

Shevach EM, Thornton AM (2014) tTregs, pTregs, and iTregs: similarities and differences. Immunol Rev 259(1):88–102. doi:10.1111/imr.12160

Singh N, Gurav A, Sivaprakasam S, Brady E, Padia R, Shi H, Thangaraju M, Prasad PD, Manicassamy S, Munn DH, Lee JR, Offermanns S, Ganapathy V (2014) Activation of Gpr109a, receptor for niacin and the commensal metabolite butyrate, suppresses colonic inflammation and carcinogenesis. Immunity 40(1):128–139. doi:10.1016/j.immuni.2013.12.007

Smith PM, Howitt MR, Panikov N, Michaud M, Gallini CA, Bohlooly YM, Glickman JN, Garrett WS (2013) The microbial metabolites, short-chain fatty acids, regulate colonic Treg cell homeostasis. Science 341(6145):569–573. doi:10.1126/science.1241165

Sokol H, Seksik P, Furet JP, Firmesse O, Nion-Larmurier I, Beaugerie L, Cosnes J, Corthier G, Marteau P, Dore J (2009) Low counts of Faecalibacterium prausnitzii in colitis microbiota. Inflamm Bowel Dis 15(8):1183–1189. doi:10.1002/ibd.20903

Thornton AM, Korty PE, Tran DQ, Wohlfert EA, Murray PE, Belkaid Y, Shevach EM (2010) Expression of Helios, an Ikaros transcription factor family member, differentiates thymic-derived from peripherally induced Foxp3+ T regulatory cells. J Immunol 184(7):3433–3441. doi:10.4049/jimmunol.0904028

van Nood E, Vrieze A, Nieuwdorp M, Fuentes S, Zoetendal EG, de Vos WM, Visser CE, Kuijper EJ, Bartelsman JF, Tijssen JG, Speelman P, Dijkgraaf MG, Keller JJ (2013) Duodenal infusion of donor feces for recurrent Clostridium difficile. N Engl J Med 368(5):407–415. doi:10.1056/NEJMoa1205037

Weiss JM, Bilate AM, Gobert M, Ding Y, Curotto de Lafaille MA, Parkhurst CN, Xiong H, Dolpady J, Frey AB, Ruocco MG, Yang Y, Floess S, Huehn J, Oh S, Li MO, Niec RE, Rudensky AY, Dustin ML, Littman DR, Lafaille JJ (2012) Neuropilin 1 is expressed on thymus-derived natural regulatory T cells, but not mucosa-generated induced Foxp3+ T reg cells. J Exp Med 209(10):1723–1742. doi:10.1084/jem.20120914, S1721

Yadav M, Louvet C, Davini D, Gardner JM, Martinez-Llordella M, Bailey-Bucktrout S, Anthony BA, Sverdrup FM, Head R, Kuster DJ, Ruminski P, Weiss D, Von Schack D, Bluestone JA (2012) Neuropilin-1 distinguishes natural and inducible regulatory T cells among regulatory T cell subsets in vivo. The Journal of experimental medicine 209 (10):1713–1722, S1711-1719. doi:10.1084/jem.20120822

Chapter 45
Roles of the Epithelial Autophagy in the Intestinal Mucosal Barrier

Koji Aoki and Manabu Sugai

Abstract Intestines are a part of the gastrointestinal tract and essential for digestion and absorbtion of food. It is also home to a dense community of resident bacteria and exposed to a very large number of microorganisms associated with food and from the external environment. The epithelial surface of the intestines plays a critical role in host protection by forming the intestinal mucosal layer that functions as a barrier to microorganisms and by producing a diverse repertoire of antimicrobial proteins that directly kill or hinder the growth of microorganisms. Autophagy in the intestinal epithelial cells plays key roles in the mucosal immune defence, whereas its deregulation is one of the major mechanisms that cause intestinal chronic inflammation and subsequent development of intestinal bowel disease such as Crohn's disease and intestinal ulcerative colitis. Here we review the roles of autophagy in the intestinal epithelial barrier functions and consider the mechanisms by which deregulated autophagy causes intestinal bowel disease.

Keywords Autophagy • Intestinal bowel disease • Crohn's disease • ATG16L1 • Intestinal epithelial cells • Enterocytes • Goblet cells • Paneth cells

---

K. Aoki (✉)
The Aoki Laboratory, Life Science Unit, Tenure-Track Program for Innovative Research, University of Fukui, 23-3 Matsuokashimoaizuki, Eiheiji-cho, Yoshida-gun, Fukui 910-1193, Japan

Division of Pharmacology, Department of Biochemistry and Bioinformative Sciences, School of Medicine, University of Fukui, 23-3 Matsuokashimoaizuki, Eiheiji-cho, Yoshida-gun, Fukui 910-1193, Japan

PREST, Japan Science and Technology Agency, Kawaguchi 332-0012, Japan
e-mail: aokik@u-fukui.ac.jp

M. Sugai
Division of Molecular Genetics, Department of Biochemistry and Bioinformative Sciences, School of Medicine, University of Fukui, 23-3 Matsuokashimoaizuki, Eiheiji-cho, Yoshida-gun, Fukui 910-1193, Japan
e-mail: msugai@u-fukui.ac.jp

M. Miyasaka, K. Takatsu (eds.), *Chronic Inflammation*,
DOI 10.1007/978-4-431-56068-5_45

45.1 Introduction

Intestines are essential for absorption of nutrients. To digest foodstuffs and absorb nutrients and water efficiently, the intestinal mucosa is covered by a very large number of epithelial cells such as absorptive, goblet, Paneth, and enteroendocrine cells (Clevers 2013). The intestinal mucosa is exposed not only to various food antigens but also to a large number of microorganisms that contaminate the eaten food (Palm et al. 2015; Lozupone et al. 2012). In addition, there are a huge number of resident microbacteria in the intestines (Lozupone et al. 2012). Thus, the intestinal mucosa is a key platform that provides a crosstalk between antigens or microbacteria and the immune system of the body (Kagnoff 2014). In addition, some microbacteria including pathogenic ones pass through the intestinal epithelial cells to establish infection, whereas the epithelial cells defend the intestines from infection and produce signals that activate immune reactions (Kagnoff 2014; Peterson and Artis 2014). Accordingly, the intestinal epithelial cells are considered as the leading edge of the immune defence system of the intestines.

The dysfunctions in the intestinal immune system can induce inflammatory bowel diseases (IBD) such as Crohn's disease (CD) and intestinal ulcerative colitis (UC) (Maloy and Powrie 2011; Khor et al. 2011; Boyapati et al. 2015; Baumgart and Sandborn 2012). The CD and UC are chronic, refractory, and relapsing immune-mediated inflammatory diseases, accompanied with various symptoms including diarrhoea and also with complications outside the gastrointestinal tract including rheumatologic disease. In CD, inflammation can be induced at any part of the digestive tract from mouth to anus. The prevalence of IBD is increased in most advanced countries, and it has been more common in the developing countries since the 1970s. Although the epidemiologic data suggest that incidence of IBD is influenced by environmental factors, recent studies indicate that it is also affected by genetic predisposition. Genome-wide association studies have highlighted autophagy as one of key pathways whose deregulation contributes to IBD (Khor et al. 2011; Stappenbeck et al. 2011). Consistently, a number of studies have indicated that autophagy plays key roles in the intestinal immune functions of both intestinal epithelial (IECs) and immune cells. In this review, we discuss the roles of autophagy in intestinal epithelial defence, and the mechanisms by which suppression of autophagy in IECs contributes to intestinal inflammation and subsequent IBD development.

45.2 Inflammatory Bowel Diseases, Crohn's Disease, and Ulcerative Colitis

Inflammatory bowel disease (IBD), such as Crohn's disease (CD) and ulcerative colitis (UC), is the disorder that develops chronic inflammation and mucosal ulceration anywhere along the gastrointestinal tract (Maloy and Powrie 2011;

Khor et al. 2011; Boyapati et al. 2015; Strowig et al. 2012; Sartor 2006). In Japan, CD and UC affect nearly 40,000 and 17,00,000, respectively, and those numbers have increased continuously for these several decades as well as other western countries (Boyapati et al. 2015; Sartor 2006).

Whereas the precise mechanisms underlying IBD are still unclear, significant progress has been made to understand its pathogenesis in the past 10–20 years through using advanced technologies such as genome-wide association studies. It has been suggested that IBD is caused by an inappropriate and uncontrolled immune response and by subsequent chronic inflammation (Maloy and Powrie 2011; Khor et al. 2011; Strowig et al. 2012; Sartor 2006). Thus, current therapeutic drugs include anti-inflammatory agents such as corticosteroids, immunomodulators such as thiopurines and methotrexate, and biological agents such as antibodies against tumour necrosis factor (TNF), although IBD can be cured completely by none of these drugs. Accordingly, further drug discovery is necessary for IBD.

The onset and development of chronic inflammation can be affected by the genetic alterations and environmental factors that perturb the composition and balance of the commensal microbes, the intestinal mucosal barrier, and immune reaction (Maloy and Powrie 2011). The environmental factors include meal antigens, smoking, and so on, whereas the relationships among such factors and chronic inflammation are poorly understood.

45.3 Genetic Alterations of Autophagy Genes in Crohn's Disease

Recently, the number of genetic alterations has been found by genome-wide association studies for CD and UC (Franke et al. 2010; Anderson et al. 2011; Hampe et al. 2007; Rioux et al. 2007; Jostins et al. 2012). Notably, these analyses identify single nucleotide polymorphisms (SNPs) in the *ATG16L1* gene as one of the susceptibility loci (Hampe et al. 2007; Rioux et al. 2007) as well as *NOD2* and *IL23R,* and deletion polymorphism for the upstream region of *IRGM* (Hugot et al. 2001; Ogura et al. 2001; Duerr et al. 2006; McCarroll et al. 2008; Parkes et al. 2007; Brest et al. 2011). The protein product of the *ATG16L1* gene, *autophagy related 16-like 1*, is one of the components essential to form the autophagosome, a core machinery of autophagy (Nakatogawa et al. 2009; Xie and Klionsky 2007) (Fig. 45.1). The protein product of the *NOD2* gene, *nucleotide-binding oligomerization domain containing 2*, carries two caspase recruitment (CARD) domains and six leucine-rich repeats (LRRs) (Caruso et al. 2014). It functions as a sensor of intracellular bacteria and is involved in host defence. In addition, the *Immunity-Related GTPase M* (*IRGM*) gene belongs to the Immunity-Related GTPase family, one of IFN-inducible GTPase subfamilies (Kim et al. 2012). IRGM plays a key role in xenophagy that targets microbial pathogens through autophagy (McCarroll et al. 2008; Brest et al. 2011; Kim et al. 2012). These results strongly suggest that

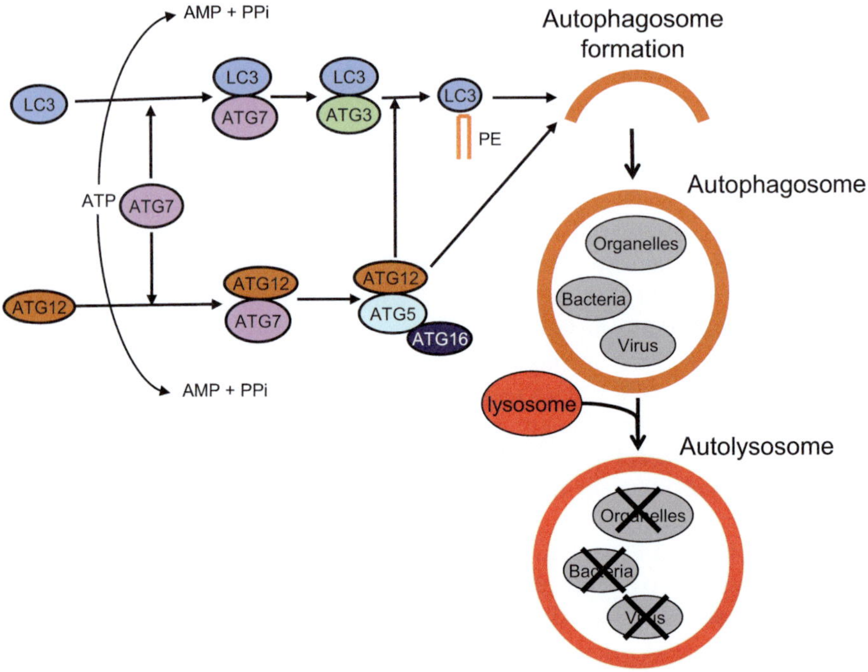

Fig. 45.1 Scheme of two ubiquitin-like conjugation systems. ATG7 mediates conjugation of ATG12 with ATG5, and ATG12-ATG5 conjugate further forms a complex with ATG16. It also mediates conjugation of LC3B with ATG3, and LC3 is finally conjugated with phosphatidylethanol amine (PE). Both of the above conjugates are required for autophagosome expansion. The isolation membranes expand, and sequester a portion of the cytosol, organelles, and microorganisms, forming autophagosomes. The degradation occurs in autophagolysosomes that are fusions of the autophagosomes with lysosomes

deregulation of autophagy contributes to Crohn's disease. It is noted that the genomic alterations in the *IRGM* gene are associated also with UC (Jostins et al. 2012), suggesting that deregulation of autophagy contributes to ulcerative colitis as well.

45.4 Autophagy

Autophagy is an evolutionally conserved cellular pathway in eukaryotes that degrades cytoplasmic components including proteins, organelles, and invaded bacteria, through the lysosomal and vacuolar system (Nakatogawa et al. 2009; Mizushima et al. 2008; Fig. 45.1). Autophagy plays important roles not only in cell survival in eukaryotes under nutrient-limited conditions but also in a wide variety of development and physiology in mammals (Mizushima and Levine 2010; Mizushima and Komatsu 2011). In addition, deregulation of autophagy is implicated to be involved in a number of diseases such as neurodegenerative disease, liver disease, myodegenerative disease, cardiac disease, and cancer (Levine and Kroemer 2008).

There are various types of autophagy that include micro- and macroautophagy, and chaperon-mediated autophagy (CMA) (Nakatogawa et al. 2009). Macroautophagy, referred as autophagy, is the main pathway to degrade cytoplasmic components, and involves the formation of a double membrane called autophagosomes. On the other hand, microautophagy involves the direct engulfment of cytoplasmic components into the lysosome. Similarly, cytosolic proteins are translocated directly across the lysosomal membrane during chaperone-mediated autophagy. The formation of autophagosomes requires isolation membranes that are single membrane-bound structures. The isolation membranes appear and expand, and sequester a portion of the cytosol and organelles, forming autophagosomes. The degradation through autophagy occurs in autophagolysosomes that are fusions of the autophagosomes with lysosomes (Fig. 45.1).

ATG molecules and two ubiquitin-like conjugation systems contribute to the formation of autophagosomes. Among ATGs, ATG12, and ATG8/LC3 are ubiquitin-like proteins that play essential roles in the autophagosome formation through the ubiquitin-like conjugation systems (Mizushima et al. 1998; Ichimura et al. 2000; Fig. 45.1). ATG12 is conjugated with ATG5 and the ATG12-ATG5 conjugate further forms a multimeric complex with ATG16, whereas LC3B is finally conjugated with phosphatidyl-ethanol amine (PE). Both of the above conjugates are required for autophagosome expansion (Ichimura et al. 2000; Tanida et al. 1999).

45.5 The Roles of Intestinal Epithelial Cells in the Intestinal Mucosal Barrier

One of main physiological functions of intestines is to digest and absorb nutrients (Clevers 2013; Barker 2014; Gardner et al. 1970a, b, c), whereas the mucosal barrier is an important boundary that protects the surface of the intestinal tract from adhesion and invasion of the microorganisms (Peterson and Artis 2014; Antoni et al. 2014). In the mucosal barrier, central roles are played by the columnar epithelial cells that form a sheet covering the intestinal mucosal layer. The intestinal epithelial cells are composed of absorptive enterocytes and secretory goblet, enteroendocrine, and Paneth cells (Clevers 2013). The enterocytes, goblet, and enteroendocrine cells migrate toward the top of the villus, whereas Paneth cells migrate to the base of the crypt of Lieberkühn. The enterocytes uptake nutrients such as sugar, amino acids, and lipids as well as water (Gardner et al. 1970a, b, c). The goblet cells secrete mucins that form mucus that protects the intestinal mucosal layer, as the first line of defence of the mucosal barrier (McCauley and Guasch

2015; McGuckin et al. 2011). On the other hand, Paneth cells secrete proteins that are known to play an another key roles in host defence, including antimicrobial proteins and peptides such as defensins (Stappenbeck 2009, 2010). In addition, the enteroendocrine cells secrete peptides that are involved in the control of appetite and digestive responses (Harrison et al. 2013). Autophagy is reported to play essential roles in the mucosal barrier functions of enterocytes, goblet, and Paneth cells as follows.

45.6 Paneth Cells and Autophagy

Paneth cells, located in the small intestine, produce and secrete a wide variety of antimicrobial proteins and peptides, so-called AMPs, which kill or inactivate microorganisms as a part of the intestinal mucosal immune defense (Stappenbeck 2009, 2010; Mukherjee and Hooper 2015; Table 45.1). These AMPs include defensins, lysozyme, and secretory phospholipase A2 (sPLA2). Lysozyme and sPLA2 kill bacteria through an enzymatic mechanism, attacking critical structures of the bacterial cell wall and disrupting bacterial membrane integrity, respectively (Mukherjee and Hooper 2015; Ganz 2004a; Brogden 2005). On the other hand, defensins suppress synthesis of the bacterial cell wall (Sass et al. 2010; Ganz 2004b). Through secretion of these AMPs, Paneth cells contribute to the innate immune response to the microorganisms.

The genetic alterations in the *ATG16L1* gene are found to be associated with Crohn's disease as described above (Franke et al. 2010; Anderson et al. 2011; Hampe et al. 2007; Rioux et al. 2007; Jostins et al. 2012). Upon suppression of autophagy through lowering the *Atg16l1* level by gene-trap–mediated disruptions, abnormalities in morphology and the granule exocytosis pathway are found in Paneth cells (Cadwell et al. 2008; Table 45.1). Likewise, knockout of other Atg proteins Atg5 or Atg7 also induce abnormalities in the granule exocytosis pathway and increase expression level of genes involved in inflammatory reactions such as peroxisome proliferator-activated receptor (PPAR) signaling (Cadwell et al. 2008, 2009). In addition, Paneth cell abnormalities and increased susceptibility to intestinal inflammation are found in mice deficient for the *Irgm1* gene as well (Liu et al. 2013). These results collectively suggest that deregulated autophagy results in the dysfunction of Paneth cells. Notably, the morphological defects of Paneth cells such as the granule abnormalities are found also in the Crohn's disease patients who carry the homozygous *ATG16L1* risk allele (Cadwell et al. 2008) as well as in the knock-in mice which carry Atg16L1 T300A, a common polymorphism of Crohn's disease (Lassen et al. 2014). Note that the high proportions of abnormal Paneth cells were associated with shorter time to disease recurrence after surgery (VanDussen et al. 2014), suggesting that the defects of autophagy function in Paneth cells contribute to Crohn's disease. Consistent with these results, autophagy dysfunction in the Paneth cells causes ER-stress–induced intestinal inflammation, NF-kB overactivation, and intestinal epithelial cell death (Adolph et al. 2013).

Table 45.1 Roles of autophagy in the intestinal epithelial barrier functions

Epithelial cell types	Roles in intestinal mucosal immunity	Mice analyzed	Phenotypes of mice deficient for *Atg* or its related genes	References
Enterocyte	Killing bacteria that invade into enterocytes through xenophagy	*Atg5* and *Atg7* knockout mouse	The increased number of bacteria in enterocytes and enhanced systemic translocation of bacteria from the intestine to spleen and liver	Benjamin et al. (2013); Conway et al. (2013); Inoue et al. (2012)
Goblet cell	Formation of the intestinal mucus layer through generation of mucin glycoproteins	*Atg5* knockout mouse	The morphological defects and goblet cell dysfunctions including granule secretion	Lassen et al. (2014); Patel et al. (2013); Wlodarska et al. (2014)
		Mice carrying *Atg16l1* T300A		
		NLRP6 knockout mouse		
			Reduced autophagy activity in goblet cells and abrogated mucus secretion	Wlodarska et al. (2014)
Paneth cell	Killing or inactivation of microorganisms through generation of antimicrobial proteins and peptides such as lysozome, sPLA2 and defensins	*Atg5*, *Atg7* and *Atg16l1* knockout mouse	Abnormalities in Paneth cell morphology, secretion of antimicrobial proteins and peptides and expression level of genes involved in inflammatory reactions	Cadwell et al. (2008); Cadwell et al. (2009); Liu et al. (2013); Lassen et al. (2014)
		Mice carrying Atg16l1 hypomorphic allele		
		Mice carrying *Atg16l1* T300A		
		Irgm1 knockout mouse		
		Mice carrying Atg16l1 hypomorphic allele	Abnormal response of inflammatory transcription pathway by virus infection	Adolph et al. (2013)

In the low Atg16L1 mice, virus infection also causes abnormal response of the inflammatory transcription pathway in Paneth cells (Cadwell et al. 2010). Upon treatment with dextran sodium sulfate (DSS), virus-infected low Atg16l1 mice develop phenotypes resembling those of Crohn's disease, such as increased inflammation in the muscularis and associated mesenteric fat and blood vessels, increases in lymphoid aggregates, and hyperplasia of the colonic epithelial cells (Cadwell et al. 2010). These results suggest that dysfunction of autophagy increases susceptibility of intestinal inflammation caused by multiple stimuli. On the other hand, it is unclear how secretion of granules and cell morphology in Paneth cells are controlled by autophagy.

Notably, deficiency of the *Atg16L1* gene in the haematopoietic cells increases susceptibility of intestinal inflammation induced by DSS through inflammatory signals such as IL1-β and IL-18 (Saitoh et al. 2008), suggesting that autophagy in the immune cells also contributes to suppression of intestinal inflammation.

45.7 Goblet Cells and Autophagy

The mucus layer of the intestinal tract is the first barrier as an innate host defence against intestinal pathogenic microorganisms (Peterson and Artis 2014; Antoni et al. 2014; Ashida et al. 2012; Table 45.1). Because goblet cells secrete mucin glycoproteins and other proteins that mainly constitute the mucus layer, they play key roles in providing an antimicrobial protective mechanism at the interface between the intestinal luminal side and mucosa. Both autophagy proteins and endosomes are shown to be required for generation of reactive oxygen species (ROS) that is essential to control mucin granule accumulation in colonic goblet cells (Patel et al. 2013). Consistent with these results, goblet cell dysfunctions and its morphological defects are found for the knock-in mice carrying Atg16l1 T300A and the mice deficient in Atg5 (Lassen et al. 2014; Wlodarska et al. 2014).

Interestingly, mice deficient in NLRP6, one of innate sensors of stress or damage-associated molecular patterns, exhibit reduced autophagy activity in goblet cells and abrogated mucus secretion (Wlodarska et al. 2014; Table 45.1). Importantly, the NLRP6-deficient mice fail to clear enteric pathogens from the mucosal layer, which results in increased susceptibility to persistent infection (Wlodarska et al. 2014), indicating a critical role of goblet cells in intestinal mucosal immunity. Because autophagy regulates the level of IL1-β production (Lassen et al. 2014; Saitoh et al. 2008), these results also suggest that autophagy and inflammasome signal regulate their functions each other's functions whereas the mechanism by which NLRP6 controls autophagy activity remains unclear.

45.8 Enterocytes and Autophagy

The intestinal epithelial cells are one of the routes where the bacterial pathogens invade the intestinal mucosal layer (Haraga et al. 2008; Barnich et al. 2007). Upon bacterial infection in the small intestine, bacterial invasion of the cells activates autophagy in the intestinal epithelial cells and the bacteria are targeted to autophagosomes (Benjamin et al. 2013; Conway et al. 2013). Suppression of autophagy by introducing the *Atg5* knockout allele into the intestinal epithelial cells results in the increased number of bacteria in the intestinal epithelial cells and enhanced systemic translocation of bacteria from the intestine to spleen and liver

(Benjamin et al. 2013; Conway et al. 2013; Inoue et al. 2012; Table 45.1). In contrast, xenophagy is shown to be proficient in macrophages or dendritic cells deficient for the *Atg5* gene (Benjamin et al. 2013; Homer et al. 2010). Thus, it is conceivable that some additional factors specific to the intestinal epithelial cells play a role in xenophagy. These results also suggest that autophagy in the enterocytes contributes to the intestinal mucosal barrier.

45.9 Activation of Autophagy in Intestinal Epithelial Cells

In the intestine, autophagy is less active in differentiated epithelial cells than undifferentiated ones (Groulx et al. 2012), suggesting some mechanisms that activate autophagy in differentiated cells such as enterocytes, and Paneth and goblet cells. NOD1 and NOD2, members of the cytosolic Nod-like receptors (NLRs), are sensors of pathogens as the innate immune system, detecting pathogen-associated molecular patterns (Caruso et al. 2014; Boyle et al. 2014). Genomic alterations in the *NOD2* gene are associated with Crohn's disease, including frameshift mutations and missense mutations (Hugot et al. 2001; Inoue et al. 2012). NOD1 and NOD2 are shown to activate autophagy by recruiting ATG16L1 to the plasma membrane at the site of bacterial entry, linking bacterial invasion with intracellular pathogen defence through autophagy (Travassos et al. 2010; Cooney et al. 2010). On the other hand, autophagy is not activated by the protein product of the *NOD2* L1007insC polymorphism, the most prevalent frameshift mutation in Crohn's disease, which generates truncated *NOD2* protein (Travassos et al. 2010), highlighting the contribution of autophagy to the intracellular pathogen defence mediated by NOD2.

Interestingly, autophagy-dependent suppression of *Salmonella* survival is abrogated by the ATG16L1 T300A variant that is associated with Crohn's disease, whereas NOD2-dependent signalling is unaffected by ATG16L1 T300A (Homer et al. 2010), suggesting that deregulation of the autophagy pathway contributes to Crohn's disease other than NOD2-dependent signalling pathway. Notably, interaction between ATG16L1 and NOD2 is shown not only in dendritic cells and mesenchymal embryonic fibroblast, but also in intestinal epithelial cells (Homer et al. 2010; Travassos et al. 2010; Cooney et al. 2010). These results suggest that *NOD2* polymorphisms contribute to Crohn's disease through deregulated autophagy activity in intestinal epithelial cells as well as macrophages and dendritic cells.

Interaction between ATG16L1 and NOD2 also suppresses inflammatory cytokine production stimulated by *NOD2* through the autophagy-independent function (Sorbara et al. 2013). On the other hand, ATG16L1 T300A fails to suppress NOD2-driven cytokine production (Sorbara et al. 2013), suggesting that ATG16L1 contributes to suppression of inflammation independent of autophagy as well.

In addition, we have recently found that autophagy activity was stimulated by CDX2 (our unpublished data), a homeobox transcription factor expressed specifically in the intestinal epithelial cells (Aoki et al. 2011; Aoki et al. 2003). We also found that CDX2 played some roles in the intestinal mucosal barrier (our unpublished data). These results also suggest that the intestinal mucosal barrier through autophagy is controlled by mechanisms specific to the intestinal epithelial cells.

45.10 Suppression of Autophagy by Bacteria in the Intestinal Epithelial Cells

The adherent-invasive *Escherichia coli* (AIEC) colonises the ileal mucosa of patients with Crohn's disease, and adheres to and invades intestinal epithelial cells. AIEC infection increases the levels of microRNA MIR30C and MIR130A in T84 cells, derived from human intestinal epithelial cells (Nguyen et al. 2014). The MIR 30C and MIR30A target *ATG5* and *ATG16L1*, respectively, and inhibit autophagy, leading to the increased number of intracellular bacteria and stimulation of inflammatory response (Nguyen et al. 2014). These results suggest that invasive bacteria survive through suppression of autophagy in the target cells.

45.11 Autophagy and Inflammatory Signals in the Intestinal Epithelial Cells

As described above, the amounts of IL1β and IL18 are increased in macrophages deficient in the *Atg16L1* gene, which results in colonic inflammation upon treatment with DSS (Lassen et al. 2014; Saitoh et al. 2008). On the other hand, IL18 production in the intestinal epithelial cells was unaffected by inhibiting autophagy through expression of shRNA for *ATG7* (our preliminary observation), suggesting that autophagy played different roles in control of the inflammatory signals between the stromal cells and intestinal epithelial ones. Surprisingly, however, we recently found that expression of CDX2 reduced the amount of IL18 in the intestinal epithelial cells through suppression of the caspase 1 activity (our unpublished data). These results suggested that CDX2 contributed to suppression of intestinal inflammation through control of inflammatory signals in the intestinal epithelial cells.

45.12 Summary

Intestinal epithelial cells play essential roles in the mucosal immune defense in the intestines. Recent genome-wide association studies identify genomic alterations in the autophagy or autophagy-related genes such as *ATG16L1* and *IRGM1*, highlighting the importance of autophagy in the barrier functions of the intestinal epithelial cells. Through autophagy, enterocytes suppress survival and dissemination of microoganisms that invade the intestinal mucosal layer. Goblet cells secrete mucin that functions as a first line of barrier of the intestinal mucosal layer, whereas Paneth cells secrete antimicrobial peptides that directly target microorganisms. Autophagy is also critical to the secretion pathway in both goblet and Paneth cells. Thus, deregulated autophagy weakens intestinal mucosal barrier functions and leads to subsequent chronic inflammation. In order to understand the pathogenesis of inflammatory bowel disease, it is essential to reveal precise molecular mechanisms by which autophagy contributes to the mucosal defense of the intestinal epithelial cells.

Acknowledgments The research programs in our laboratory have been supported by grants from PREST, Japan, Tenure-Track Program, and Research Program for Life Science, University of Fukui, Japan. We also thank all members of our laboratory.

References

Adolph TE, Tomczak MF, Niederreiter L, Ko HJ, Bock J, Martinez-Naves E et al (2013) Paneth cells as a site of origin for intestinal inflammation. Nature 503(7475):272–276

Anderson CA, Boucher G, Lees CW, Franke A, D'Amato M, Taylor KD et al (2011) Meta-analysis identifies 29 additional ulcerative colitis risk loci, increasing the number of confirmed associations to 47. Nat Genet 43(3):246–252

Antoni L, Nuding S, Wehkamp J, Stange EF (2014) Intestinal barrier in inflammatory bowel disease. World J Gastroenterol 20(5):1165–1179

Aoki K, Tamai Y, Horiike S, Oshima M, Taketo MM (2003) Colonic polyposis caused by mTOR-mediated chromosomal instability in *Apc*$^{+/\Delta716}$*Cdx2*$^{+/-}$ compound mutant mice. Nat Genet 35 (1):323–330

Aoki K, Kakizaki F, Sakashita H, Manabe T, Aoki M, Taketo MM (2011) Suppression of colonic polyposis by homeoprotein CDX2 through its nontranscriptional function that stabilizes p27 ^{Kip1}. Cancer Res 71(2):593–602

Ashida H, Ogawa M, Kim M, Mimuro H, Sasakawa C (2012) Bacteria and host interactions in the gut epithelial barrier. Nat Chem Biol 8(1):36–45

Barker N (2014) Adult intestinal stem cells: critical drivers of epithelial homeostasis and regeneration. Nat Rev Mol Cell Biol 15(1):19–33

Barnich N, Carvalho FA, Glasser AL, Darcha C, Jantscheff P, Allez M et al (2007) CEACAM6 acts as a receptor for adherent-invasive E. coli, supporting ileal mucosa colonization in Crohn disease. J Clin Invest 117(6):1566–1574

Baumgart DC, Sandborn WJ (2012) Crohn's disease. Lancet 380(9853):1590–1605

Benjamin JL, Sumpter R Jr, Levine B, Hooper LV (2013) Intestinal epithelial autophagy is essential for host defense against invasive bacteria. Cell Host Microbe 13(6):723–734

Boyapati R, Satsangi J, Ho GT (2015) Pathogenesis of Crohn's disease. F1000Prime Rep **7**:44

Boyle JP, Parkhouse R, Monie TP (2014) Insights into the molecular basis of the NOD2 signalling pathway. Open Biol 4(12):140178

Brest P, Lapaquette P, Souidi M, Lebrigand K, Cesaro A, Vouret-Craviari V et al (2011) A synonymous variant in IRGM alters a binding site for miR-196 and causes deregulation of IRGM-dependent xenophagy in Crohn's disease. Nat Genet 43(3):242–245

Brogden KA (2005) Antimicrobial peptides: pore formers or metabolic inhibitors in bacteria? Nat Rev Microbiol 3(3):238–250

Cadwell K, Liu JY, Brown SL, Miyoshi H, Loh J, Lennerz JK et al (2008) A key role for autophagy and the autophagy gene Atg16l1 in mouse and human intestinal Paneth cells. Nature 456 (7219):259–263

Cadwell K, Patel KK, Komatsu M, Virgin HW, Stappenbeck TS (2009) A common role for Atg16L1, Atg5 and Atg7 in small intestinal Paneth cells and Crohn disease. Autophagy 5 (2):250–252

Cadwell K, Patel KK, Maloney NS, Liu TC, Ng AC, Storer CE et al (2010) Virus-plus-susceptibility gene interaction determines Crohn's disease gene Atg16L1 phenotypes in intestine. Cell 141(7):1135–1145

Caruso R, Warner N, Inohara N, Nunez G (2014) NOD1 and NOD2: signaling, host defense, and inflammatory disease. Immunity 41(6):898–908

Clevers H (2013) The intestinal crypt, a prototype stem cell compartment. Cell 154(2):274–284

Conway KL, Kuballa P, Song JH, Patel KK, Castoreno AB, Yilmaz OH, Jijon HB et al (2013) Atg16l1 is required for autophagy in intestinal epithelial cells and protection of mice from Salmonella infection. Gastroenterology 145(6):1347–1357

Cooney R, Baker J, Brain O, Danis B, Pichulik T, Allan P et al (2010) NOD2 stimulation induces autophagy in dendritic cells influencing bacterial handling and antigen presentation. Nat Med 16(1):90–97

Duerr RH, Taylor KD, Brant SR, Rioux JD, Silverberg MS, Daly MJ et al (2006) A genome-wide association study identifies IL23R as an inflammatory bowel disease gene. Science 314 (5804):1461–1463

Franke A, McGovern DP, Barrett JC, Wang K, Radford-Smith GL, Ahmad T et al (2010) Genome-wide meta-analysis increases to 71 the number of confirmed Crohn's disease susceptibility loci. Nat Genet 42(12):1118–1125

Ganz T (2004a) Antimicrobial polypeptides. J Leukoc Biol 75(1):34–38

Ganz T (2004b) Defensins: antimicrobial peptides of vertebrates. C R Biol 327(6):539–549

Gardner JD, Brown MS, Laster L (1970a) The columnr epithelial cell of the small intestine: digestion and transport. 3. N Engl J Med 283(24):1317–1324

Gardner JD, Brown MS, Laster L (1970b) The columnar epithelial cell of the small intestine: digestion and transport II. N Engl J Med 283(23):1264–1271

Gardner JD, Brown MS, Laster L (1970c) The columnar epithelial cell of the small intestine: digestion and transport. (First of three parts). N Engl J Med 283(22):1196–1202

Groulx JF, Khalfaoui T, Benoit YD, Bernatchez G, Carrier JC, Basora N et al (2012) Autophagy is active in normal colon mucosa. Autophagy 8(6):893–902

Hampe J, Franke A, Rosenstiel P, Till A, Teuber M, Huse K et al (2007) A genome-wide association scan of nonsynonymous SNPs identifies a susceptibility variant for Crohn disease in ATG16L1. Nat Genet 39(2):207–211

Haraga A, Ohlson MB, Miller SI (2008) Salmonellae interplay with host cells. Nat Rev Microbiol 6(1):53–66

Harrison E, Lal S, McLaughlin JT (2013) Enteroendocrine cells in gastrointestinal pathophysiology. Curr Opin Pharmacol 13(6):941–945

Homer CR, Richmond AL, Rebert NA, Achkar JP, McDonald C (2010) ATG16L1 and NOD2 interact in an autophagy-dependent antibacterial pathway implicated in Crohn's disease pathogenesis. Gastroenterology 139(5):1630–1641

Hugot JP, Chamaillard M, Zouali H, Lesage S, Cezard JP, Belaiche J et al (2001) Association of NOD2 leucine-rich repeat variants with susceptibility to Crohn's disease. Nature 411 (6837):599–603

Ichimura Y, Kirisako T, Takao T, Satomi Y, Shimonishi Y, Ishihara N et al (2000) A ubiquitin-like system mediates protein lipidation. Nature 408(6811):488–492

Inoue J, Nishiumi S, Fujishima Y, Masuda A, Shiomi H, Yamamoto K et al (2012) Autophagy in the intestinal epithelium regulates Citrobacter rodentium infection. Arch Biochem Biophys 521(1–2):95–101
Jostins L, Ripke S, Weersma RK, Duerr RH, McGovern DP, Hui KY et al (2012) Host-microbe interactions have shaped the genetic architecture of inflammatory bowel disease. Nature 491 (7422):119–124
Kagnoff MF (2014) The intestinal epithelium is an integral component of a communications network. J Clin Invest 124(7):2841–2843
Khor B, Gardet A, Xavier RJ (2011) Genetics and pathogenesis of inflammatory bowel disease. Nature 474(7351):307–317
Kim BH, Shenoy AR, Kumar P, Bradfield CJ, MacMicking JD (2012) IFN-inducible GTPases in host cell defense. Cell Host Microbe 12(4):432–444
Lassen KG, Kuballa P, Conway KL, Patel KK, Becker CE, Peloquin JM et al (2014) Atg16L1 T300A variant decreases selective autophagy resulting in altered cytokine signaling and decreased antibacterial defense. Proc Natl Acad Sci U S A 111(21):7741–7746
Levine B, Kroemer G (2008) Autophagy in the pathogenesis of disease. Cell 132(1):27–42
Liu B, Gulati AS, Cantillana V, Henry SC, Schmidt EA, Daniell X et al (2013) Irgm1-deficient mice exhibit Paneth cell abnormalities and increased susceptibility to acute intestinal inflammation. Am J Physiol Gastrointest Liver Physiol 305(8):G573–G584
Lozupone CA, Stombaugh JI, Gordon JI, Jansson JK, Knight R (2012) Diversity, stability and resilience of the human gut microbiota. Nature 489(7415):220–230
Maloy KJ, Powrie F (2011) Intestinal homeostasis and its breakdown in inflammatory bowel disease. Nature 474(7351):298–306
McCarroll SA, Huett A, Kuballa P, Chilewski SD, Landry A, Goyette P et al (2008) Deletion polymorphism upstream of IRGM associated with altered IRGM expression and Crohn's disease. Nat Genet 40(9):1107–1112
McCauley HA, Guasch G (2015) Three cheers for the goblet cell: maintaining homeostasis in mucosal epithelia. Trends Mol Med 21(8):492–503
McGuckin MA, Linden SK, Sutton P, Florin TH (2011) Mucin dynamics and enteric pathogens. Nat Rev Microbiol 9(4):265–278
Mizushima N, Komatsu M (2011) Autophagy: renovation of cells and tissues. Cell 147 (4):728–741
Mizushima N, Levine B (2010) Autophagy in mammalian development and differentiation. Nat Cell Biol 12(9):823–830
Mizushima N, Sugita H, Yoshimori T, Ohsumi Y (1998) A new protein conjugation system in human. The counterpart of the yeast Apg12p conjugation system essential for autophagy. J Biol Chem 273(51):33889–33892
Mizushima N, Levine B, Cuervo AM, Klionsky DJ (2008) Autophagy fights disease through cellular self-digestion. Nature 451(7182):1069–1075
Mukherjee S, Hooper LV (2015) Antimicrobial defense of the intestine. Immunity 42(1):28–39
Nakatogawa H, Suzuki K, Kamada Y, Ohsumi Y (2009) Dynamics and diversity in autophagy mechanisms: lessons from yeast. Nat Rev Mol Cell Biol 10(7):458–467
Nguyen HT, Dalmasso G, Muller S, Carriere J, Seibold F, Darfeuille-Michaud A (2014) Crohn's disease-associated adherent invasive Escherichia coli modulate levels of microRNAs in intestinal epithelial cells to reduce autophagy. Gastroenterology 146(2):508–519
Ogura Y, Bonen DK, Inohara N, Nicolae DL, Chen FF, Ramos R et al (2001) A frameshift mutation in NOD2 associated with susceptibility to Crohn's disease. Nature 411 (6837):603–606
Palm NW, de Zoete MR, Flavell RA (2015) Immune-microbiota interactions in health and disease. Clin Immunol. doi:10.1016/j.clim.2015.05.014
Parkes M, Barrett JC, Prescott NJ, Tremelling M, Anderson CA, Fisher SA et al (2007) Sequence variants in the autophagy gene IRGM and multiple other replicating loci contribute to Crohn's disease susceptibility. Nat Genet 39(7):830–832

Patel KK, Miyoshi H, Beatty WL, Head RD, Malvin NP, Cadwell K et al (2013) Autophagy proteins control goblet cell function by potentiating reactive oxygen species production. EMBO J 32(24):3130–3144

Peterson LW, Artis D (2014) Intestinal epithelial cells: regulators of barrier function and immune homeostasis. Nat Rev Immunol 14(3):141–153

Rioux JD, Xavier RJ, Taylor KD, Silverberg MS, Goyette P, Huett A et al (2007) Genome-wide association study identifies new susceptibility loci for Crohn disease and implicates autophagy in disease pathogenesis. Nat Genet 39(5):596–604

Saitoh T, Fujita N, Jang MH, Uematsu S, Yang BG, Satoh T et al (2008) Loss of the autophagy protein Atg16L1 enhances endotoxin-induced IL-1beta production. Nature 456 (7219):264–268

Sartor RB (2006) Mechanisms of disease: pathogenesis of Crohn's disease and ulcerative colitis. Nat Clin Pract Gastroenterol Hepatol 3(7):390–407

Sass V, Schneider T, Wilmes M, Korner C, Tossi A, Novikova N et al (2010) Human beta-defensin 3 inhibits cell wall biosynthesis in Staphylococci. Infect Immun 78(6):2793–2800

Sorbara MT, Ellison LK, Ramjeet M, Travassos LH, Jones NL, Girardin SE et al (2013) The protein ATG16L1 suppresses inflammatory cytokines induced by the intracellular sensors Nod1 and Nod2 in an autophagy-independent manner. Immunity 39(5):858–873

Stappenbeck TS (2009) Paneth cell development, differentiation, and function: new molecular cues. Gastroenterology 137(1):30–33

Stappenbeck TS (2010) The role of autophagy in Paneth cell differentiation and secretion. Mucosal Immunol 3(1):8–10

Stappenbeck TS, Rioux JD, Mizoguchi A, Saitoh T, Huett A, Darfeuille-Michaud A et al (2011) Crohn disease: a current perspective on genetics, autophagy and immunity. Autophagy 7 (4):355–374

Strowig T, Henao-Mejia J, Elinav E, Flavell R (2012) Inflammasomes in health and disease. Nature 481(7381):278–286

Tanida I, Mizushima N, Kiyooka M, Ohsumi M, Ueno T, Ohsumi Y et al (1999) Apg7p/Cvt2p: a novel protein-activating enzyme essential for autophagy. Mol Biol Cell 10(5):1367–1379

Travassos LH, Carneiro LA, Ramjeet M, Hussey S, Kim YG (2010) Nod1 and Nod2 direct autophagy by recruiting ATG16L1 to the plasma membrane at the site of bacterial entry. Nat Immunol 11(1):55–62

VanDussen KL, Liu TC, Li D, Towfic F, Modiano N, Winter R et al (2014) Genetic variants synthesize to produce paneth cell phenotypes that define subtypes of Crohn's disease. Gastroenterology 146(1):200–209

Wlodarska M, Thaiss CA, Nowarski R, Henao-Mejia J, Zhang JP, Brown EM et al (2014) NLRP6 inflammasome orchestrates the colonic host-microbial interface by regulating goblet cell mucus secretion. Cell 156(5):1045–1059

Xie Z, Klionsky DJ (2007) Autophagosome formation: core machinery and adaptations. Nat Cell Biol 9(10):1102–1109

Chapter 46
Development of Sentinel-Cell Targeted Therapy for Inflammatory Bowel Diseases

Kenichi Asano and Masato Tanaka

Abstract The primary goal of the immune system is to distinguish self and non-self, and expel the latter from our body. The lumen of the gastrointestinal tract is always exposed to exogenous antigens. The intestinal immune response is strictly regulated not to overreact to them without compromising defenses against hazardous antigens. Many researchers have tried to describe how our mucosal immune system discriminate beneficial commensals and harmful pathogens, without much success.

Innate immune cells in the lamina propria (LP) are the first line of cells that counter invading pathogens. Although macrophages and DCs are considered crucial for the maintenance of mucosal homeostasis, their precise roles in the induction of tolerance and protective immunity are not totally understood. $CD169^{+}$ macrophages constitute 30 % of $CX3CR1^{hi}$ resident macrophages. They do not reside at the villus tip; instead, their localisation is biased toward muscularis mucosa. In response to epithelial injury, the $CD169^{+}$ macrophages produce CCL8 that recruits $Ly6C^{hi}$ inflammatory monocytes. Depletion of $CD169^{+}$ macrophages or inhibition of CCL8 activity in vivo suppress the symptoms of experimentally induced colitis in mouse. These findings suggest that $CD169^{+}$ macrophage-derived CCL8 is a promising therapeutic target for the treatment of mucosal injury and subsequent inflammation.

Keywords Inflammatory bowel disease • DSS • Macrophage • CD169 • Sialoadhesin • CCL8 • MCP-2

K. Asano (✉)
Laboratory of Immune Regulation, School of Life Science, Tokyo University of Pharmacy and Life Sciences, Tokyo, Japan

PRESTO, Japan Science and Technology Agency, Kawaguchi 332-0012, Japan
e-mail: asanok@toyaku.ac.jp

M. Tanaka
Laboratory of Immune Regulation, School of Life Science, Tokyo University of Pharmacy and Life Sciences, Tokyo, Japan

M. Miyasaka, K. Takatsu (eds.), *Chronic Inflammation*,
DOI 10.1007/978-4-431-56068-5_46

46.1 Introduction

The surface of the intestinal tract is constantly exposed to both self and nonself antigens. In order to avoid excess inflammation, the mucosal immune system remains unresponsive to harmless antigens such as commensal bacteria or diet. One of the unanswered questions in the field of mucosal immunology is how the immune system distinguishes hazardous agents from potentially beneficial ones (Pedron and Sansonetti 2008; Vance et al. 2009). Lamina propria (LP)-resident mononuclear phagocytes, mainly macrophages and dendritic cells (DCs), play a major role in the orchestration of mucosal immune balance (Denning et al. 2007; Varol et al. 2010). They comprise heterogeneous subsets (Gordon and Taylor 2005; Rivollier et al. 2012) in terms of surface phenotype and function. Among them, LP-resident CX3CR1^{+} macrophages are significant because they maintain tolerance toward commensal bacteria in an interleukin (IL)-10–dependent fashion (Ueda et al. 2010). Unlike DCs, a consensus regarding how to characterise different macrophage subsets in the LP was lacking until the proposal of the so-called 'monocyte-waterfall' model (Tamoutounour et al. 2012). In the intestine, CD64 is expressed only by resident macrophages, and its expression level is positively correlated with that of pan-macrophage marker CX3CR1 and negatively correlated with Ly6C (Tamoutounour et al. 2012; Auffray et al. 2007). Macrophages in the LP are constantly replenished by blood-borne monocytes (Bain et al. 2014). Mucosal inflammation induces the robust infiltration of blood-borne monocytes in mice model of IBD. Those infiltrating cells are characterised by high expression of Ly6C and the intermediate to low expression of CX3CR1 and CD64 (Tamoutounour et al. 2012; Bain et al. 2014; Farache et al. 2013; Zigmond et al. 2012). As they enter the site of inflammation, they acquire inflammatory phenotypes by producing IL-6 and IL-23, to further activate Th17 cells and innate lymphoid cells. LyC6^{hi} monocytes that migrate from the bloodstream gradually lose Ly6C expression, and in turn gain CD64 expression as they mature into the resident phenotype. Other myeloid cells, such as eosinophils, neutrophils, and DCs, can readily be discriminated from macrophages because they do not express CD64. It is likely that each subset plays a distinct role in the maintenance of gut homeostasis. However, the roles played by different subsets during mucosal immune responses remain largely unknown.

46.2 CD169^{+} Macrophages in the Intestine

CD169, also known as 'Sialoadhesin', was originally identified as a receptor for unopsonised sheep erythrocyte (Crocker and Gordon 1986). Its expression is restricted on the surface of a certain subset of macrophages (Kraal and Janse 1986). CD169^{+} macrophages have a unique localisation at the boundary between the lymphoid organ and bloodstream or lymphatic flow. In the marginal zone of the spleen, they capture apoptotic cells in the bloodstream and induce cell-associated antigen-specific tolerance (Miyake et al. 2007). A $CD169^+$ counterpart in the lymph node sinus engulfs dead tumour cells that flow into the draining lymph node, and activates tumour antigen-specific CD8 T cells (Asano et al. 2011). They also capture lymph-borne immune complexes and relay them to follicular B cells (Phan et al. 2007). Accumulation of NK cells in response to viral infection or viral-based vaccine is dependent on subcapsular sinus $CD169^+$ macrophages in the lymph node (Garcia et al. 2012). This evidence indicates that $CD169^+$ macrophages act as sentinels that capture particulate antigens and confer their information to other immune cells.

Recently, it was reported that $CD169^+$ macrophages also localise in the intestine, the largest barrier organ of our body (Asano et al. 2015; Hiemstra et al. 2014). $CD169^+$ macrophages constitute roughly 30 % and 15 % of $CX3CR1^{hi}$, $CD64^{hi}$, and $Ly6C^{lo}$ resident macrophages in the colon and the small intestine, respectively. Monocytes, DCs, eosinophils, and neutrophils do not express CD169 under both physiological and inflammatory conditions (Fig. 46.1a). $CD169^+$ macrophages in the colon express other macrophage markers such as CD11c, F4/80, MHC II, CD172α, and CD206. In contrast to $CX3CR1^+$ total macrophages that distribute evenly within the LP, $CD169^+$ macrophages are absent from the area adjacent to the epithelial border but are abundant in the area that surrounds crypts (Fig. 46.1b). The biased localisation of $CD169^+$ macrophages may facilitate them to avoid constant exposure to bacterial- and dead epithelial cell-derived antigens. Collectively, these findings indicate that $CD169^+$ macrophages play some unique role in the regulation of gut homeostasis.

46.3 Depletion of CD169 Macrophages Ameliorates DSS-Induced Colitis in Mice

The administration of DSS in drinking water causes epithelial injury and subsequent intestinal inflammation in mice (Okayasu et al. 1990). This model is useful to explore therapies that modulate innate immune response because DSS-induced colitis progresses even in lymphocyte-deficient mice (Kim et al. 2006; Tlaskalova-Hogenova et al. 2005). DSS-fed WT mice exhibit transient weight loss and hemorrhagic diarrhoea. Pathologically, DSS administration induces loss of goblet cells, distortion of crypts, and infiltration of mononuclear cells into the LP. CD169-DTR mice express human DTR under the control of CD169 gene promoter. Injection of DT into CD169 DTR mice leads to selective and transient depletion of $CD169^+$ cells (Miyake et al. 2007). Surprisingly, DSS-induced colitis is almost totally suppressed in the absence of $CD169^+$ cells. In WT mice, DSS-induced epithelial injury promotes the accumulation of inflammatory cells, such as monocytes, eosinophils, and neutrophils. The numbers of $CD64^-$, Siglec-F^+ eosinophils, $CD64^-$, $Ly6G^+$ neutrophils, and $CD64^-$, $CD103^+$ DCs are similar

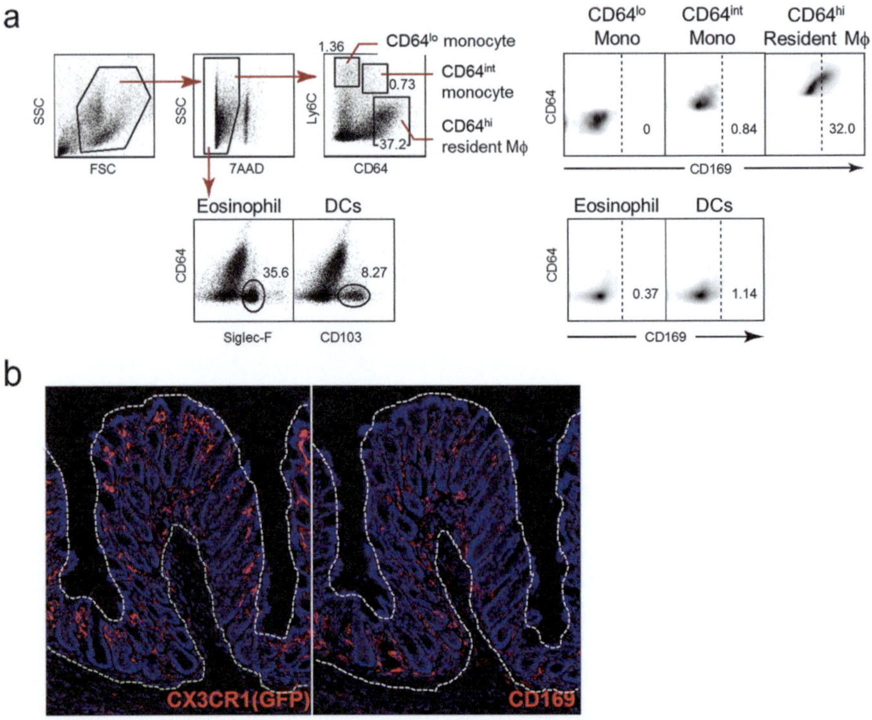

Fig. 46.1 **(a)** CD169 expression is restricted to CD64^{hi} resident macrophage compartment. CD11b^{+} and/or CD11c^{+} cells were enriched by magnetic sorting and stained for surface markers for monocytes (Ly6C), resident macrophages (CD64), eosinophils (Siglec-F) and DCs (CD103) (*left*). Frequencies of CD169^{+} cells among each compartment are shown (*right*). *Dashed lines* represent isotype control. **(b)** CD169^{+} macrophages are located distantly from epithelial border. Consecutive colon sections from CX3CR1^{gfp} mouse were stained for GFP (CX3CR1, *left*, *red*) or CD169 (*right, red*) and nucleus (*blue*). *Dashed white lines* represent epithelial border or muscularis mucosa. Original magnification, ×20. Modified from Asano, K. et al. Nat. Commun. 6:7802 doi: 10.1038/ncomms8802 (2015), licensed under a Creative Commons Attribution 4.0 International License

between WT and CD169-DTR mice although the clinical symptoms of colitis were suppressed in the latter. In contrast, the number of monocytes is significantly reduced in the absence of CD169^{+} macrophages. These findings suggest that following mucosal injury CD169^{+} macrophages recruit inflammatory monocytes that further exacerbate inflammation.

The adoptive transfer of naïve CD4 T cells into lymphocyte-deficient mice causes chronic inflammation in the small intestine and colon that mimic the human disease of IBD (Morrissey et al. 1993). RAG1-deficient mice that were crossed with CD169-DTR mice were transferred with CD45Rb^{hi}, CD4 T cells to induce colitis. Unlike DSS-induced colitis, the depletion of CD169^{+} macrophages does not ameliorate the clinical symptoms of T cell–transfer colitis. This result highlights the importance of CD169$^+$ macrophages in the pathology of the acute, innate immunity-dominant phase of colitis but not of chronic, CD4 T cell–dependent colitis.

46.4 CCL8/MCP-2 is Produced by CD169 Macrophages in the Colon

What are the monocyte-recruiting signals released from CD169$^+$ macrophages? Microarray analysis revealed cytokine mRNAs that are increased only in CD169$^+$ macrophages under the inflammatory condition. Among those cytokines, CCL8 mRNA is upregulated in CD169$^+$ macrophages but not in CD169$^-$ cell types after DSS administration. CCL8 is upregulated as early as 5 days after DSS administration, and gradually decreases after the removal of DSS. Selective reduction of inflammatory monocytes in DSS-fed CD169-DTR mice suggests that following epithelial injury, CD169$^+$ macrophages produce CCL8 in order to recruit CD169$^-$, Ly6C^{hi} monocytes.

The physiological and pathological role of CCL8 is not fully understood. CCL8-deficient mice develop normally and are fertile. It is generally accepted that the recruitment of monocytes from bone marrow to inflammatory sites is controlled by the CCL2-CCR2 pathway. Unlike several other CCL chemokines, mouse CCL8 does not bind to CCR2 but has its own receptor, CCR8 (Zlotnik and Yoshie 2012). Thus, it is not surprising if CCL2 and CCL8 play nonredundant roles in the maintenance of the LP macrophage pool. The serum concentration of CCL8 positively correlates with the severity of graft-versus-host disease (GVHD) both in humans and mice (Hori et al. 2008; Ota et al. 2009; Yamamoto et al. 2011). In one report, CCL8 expression in the colon biopsy specimen was upregulated in IBD patients (Banks et al. 2003). Future study is needed to determine the role of CCL8 in the progression of human IBD.

Endogenous ligands that trigger CCL8 production remain to be identified. Bone marrow-derived macrophages (BMDMs) gain CD169 expression as they differentiate in the presence of M-CSF. In contrast, BMDCs differentiated in the presence of GM-CSF do not express CD169 (Karasawa et al. 2015). CD169$^+$ BMDMs, but not CD169$^-$ BMDCs, produced CCL8 when they were stimulated in vitro with LPS or HMGB1. IL-6 and TNFα production by BMDMs and BMDCs to the same stimuli is similar. These results raise the possibility that CD169$^+$ macrophages in the colon produce CCL8 in response not only to commensal bacterium-derived components but also to damaged cell-derived antigens.

46.5 Anti-CCL8 Antibody Suppresses DSS-Induced Colitis in Mice

Upregulation of CCL8 after epithelial injury suggests that this cytokine acts as a molecular link between epithelial damage and monocyte recruitment. Indeed, monocytic cell line WEHI-3 cells migrate toward CCL8 in vitro. Moreover, intravenous injection of anti-CCL8 antibody after the DSS-administration ameliorates the clinical symptoms of colitis in WT mice (Asano et al. 2015). Tissue injury was also reduced in the anti-CCL8 antibody injected mice. These results highlight the clinical importance of $CD169^+$ macrophage-derived CCL8 in the progression of colitis, and its potential as a novel therapeutic target for epithelial injury and the subsequent inflammation.

46.6 Discussion

The Intestine is a major route of nonself-antigen entry. In addition to exogenous antigens, immune cells in the intestine are constantly exposed to dead cell-derived self antigens due to rapid turnover of epithelial cells. Gut immune cells are tolerant to harmless nonself and self antigens yet respond to invading pathogens. Various innate immune cells are associated with the orchestration of mucosal immune balance. For instance, tolerance toward bacterial antigens is maintained by $CX3CR1^{hi}$ resident macrophages (Ueda et al. 2010). Epithelial injury and subsequent invasion of commensals induce $Ly6C^{hi}$ inflammatory monocyte accumulation in mice. However, a cellular and molecular link between the breach of the epithelial barrier and monocyte recruitment has been missing. $CD169^+$ cells are a subset of $CX3CR1^{hi}$ resident macrophages. They are strategically located distantly from the epithelium-LP border. Unlike $CX3CR1^{hi}$, $CD169^-$ resident macrophages, $CD169^+$ macrophages might stay reactive to commensal bacterium and dead cell-derived antigens because they are not constantly exposed to those antigens under physiological conditions. $CD169^+$ macrophage-derived CCL8 recruits inflammatory monocytes to the LP, which further promote mucosal injury (Fig. 46.2). Neutralisation of CCL8 in vivo by monoclonal antibody ameliorates DSS-induced colitis. These findings demonstrate the clinical importance of the $CD169^+$ macrophage-CCL8 axis in the development of mucosal inflammation. It is important to identify a human counterpart of mouse CCL8 and determine its relevance in the development of IBD.

Stimulation of BM-derived $CD169^+$ macrophages but not $CD169^-$ DCs with LPS or HMGB1 *in vitro* induces CCL8 production. This result suggests that $CD169^+$ macrophages in the intestine also respond to PAMPs and/or DAMPs *in vivo*. Future study is needed to identify endogenous stimuli that trigger CCL8 production.

Inflammatory bowel disease (IBD) is characterized by the chronic inflammation of the gastrointestinal tract (Kaser et al. 2010). The abnormal activation of innate immune cells toward commensal bacteria or dietary antigen is considered key event to the initiation of mucosal inflammation. In human, hereditary background as well as an imbalance in the composition of microbiota are also associated with IBD (Maloy and Powrie 2011). Conventional treatments of IBD comprise aminosalicylates, immunomodulators such as azathioprine, and corticosteroids. While they are useful to induce remission, their applications are limited by concerns of side effects. Anti-TNFα antibody dramatically improved the management of Crohn's disease (Derkx et al. 1993; Rutgeerts et al. 1999). However, anti-TNFα therapy is still under optimization in the management of ulcerative colitis (Danese et al. 2013; Laharie et al. 2012). Blockade of TNFα is effective in the mainteinance of mucosal healing and clinical remission in human patients. In contrast, the treatment with anti-TNFα rather leads to aggravation in DSS-induced acute colitis in mice (Kojouharoff et al. 1997). This study indicates reciprocal roles of TNFα during acute and chronic phase of inflammation. Considering critical role of TNFα in host defense against infectious agents, it is desired to explore more cell type-specific target for IBD treatment.

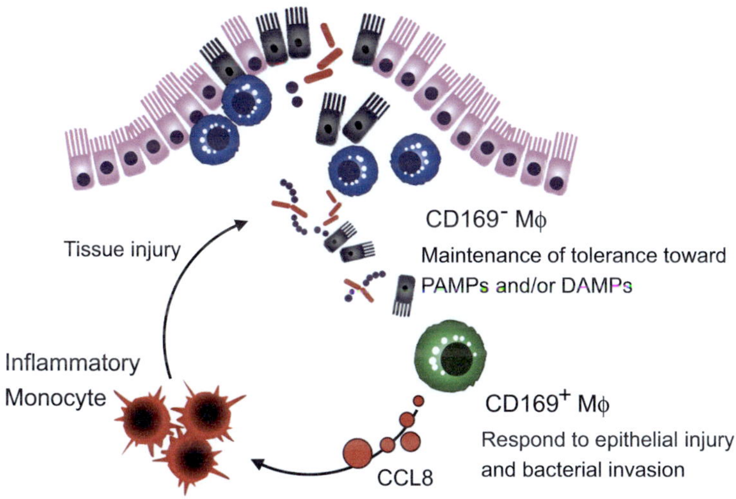

Fig. 46.2 Predicted roles of CD169^{+} macrophages and CCL8 in the development of mucosal injury. CX3CR1^{+}, CD169^{-} macrophages adjacent to the epithelial border induce tolerance toward bacterial components in an IL-10-dependent manner. Profound barrier injury may allow CD169^{+} macrophages to come into contact with DAMPs and/or PAMPs. CD169^{+} macrophages produce CCL8 that recruits inflammatory monocytes from the blood stream. Infiltrated monocytes further promote mucosal injury

46.7 Conclusion

CCL8 is a signature cytokine of CD169$^+$ macrophage. The cytokine may promote the accumulation of inflammatory monocytes to the LP in response to profound epithelial damage. Current data raise the hope that targeting CD169$^+$ macrophage-CCL8 axis could provide new therapeutic approaches to mucosal injury and inflammation.

References

Asano K, Nabeyama A, Miyake Y, Qiu CH, Kurita A, Tomura M, Kanagawa O, Fujii S, Tanaka M (2011) CD169-positive macrophages dominate antitumor immunity by crosspresenting dead cell-associated antigens. Immunity 34(1):85–95. doi:10.1016/j.immuni.2010.12.011

Asano K, Takahashi N, Ushiki M, Monya M, Aihara F, Kuboki E, Moriyama S, Iida M, Kitamura H, Qiu CH, Watanabe T, Tanaka M (2015) Intestinal CD169(+) macrophages initiate mucosal inflammation by secreting CCL8 that recruits inflammatory monocytes. Nat Commun 6:7802. doi:10.1038/ncomms8802

Auffray C, Fogg D, Garfa M, Elain G, Join-Lambert O, Kayal S, Sarnacki S, Cumano A, Lauvau G, Geissmann F (2007) Monitoring of blood vessels and tissues by a population of monocytes with patrolling behavior. Science 317(5838):666–670. doi:10.1126/science.1142883

Bain CC, Bravo-Blas A, Scott CL, Gomez Perdiguero E, Geissmann F, Henri S, Malissen B, Osborne LC, Artis D, Mowat AM (2014) Constant replenishment from circulating monocytes maintains the macrophage pool in the intestine of adult mice. Nat Immunol 15(10):929–937. doi:10.1038/ni.2967

Banks C, Bateman A, Payne R, Johnson P, Sheron N (2003) Chemokine expression in IBD. Mucosal chemokine expression is unselectively increased in both ulcerative colitis and Crohn's disease. J Pathol 199(1):28–35. doi:10.1002/path.1245

Crocker PR, Gordon S (1986) Properties and distribution of a lectin-like hemagglutinin differentially expressed by murine stromal tissue macrophages. J Exp Med 164(6):1862–1875

Danese S, Colombel JF, Peyrin-Biroulet L, Rutgeerts P, Reinisch W (2013) Review article: the role of anti-TNF in the management of ulcerative colitis – past, present and future. Aliment Pharmacol Ther 37(9):855–866. doi:10.1111/apt.12284

Denning TL, Wang YC, Patel SR, Williams IR, Pulendran B (2007) Lamina propria macrophages and dendritic cells differentially induce regulatory and interleukin 17-producing T cell responses. Nat Immunol 8(10):1086–1094. doi:10.1038/ni1511

Derkx B, Taminiau J, Radema S, Stronkhorst A, Wortel C, Tytgat G, van Deventer S (1993) Tumour-necrosis-factor antibody treatment in Crohn's disease. Lancet 342(8864):173–174

Farache J, Zigmond E, Shakhar G, Jung S (2013) Contributions of dendritic cells and macrophages to intestinal homeostasis and immune defense. Immunol Cell Biol 91(3):232–239. doi:10.1038/icb.2012.79

Garcia Z, Lemaitre F, van Rooijen N, Albert ML, Levy Y, Schwartz O, Bousso P (2012) Subcapsular sinus macrophages promote NK cell accumulation and activation in response to lymph-borne viral particles. Blood 120(24):4744–4750. doi:10.1182/blood-2012-02-408179

Gordon S, Taylor PR (2005) Monocyte and macrophage heterogeneity. Nat Rev Immunol 5(12):953–964. doi:10.1038/nri1733

Hiemstra IH, Beijer MR, Veninga H, Vrijland K, Borg EG, Olivier BJ, Mebius RE, Kraal G, den Haan JM (2014) The identification and developmental requirements of colonic CD169(+) macrophages. Immunology 142(2):269–278

Hori T, Naishiro Y, Sohma H, Suzuki N, Hatakeyama N, Yamamoto M, Sonoda T, Mizue Y, Imai K, Tsutsumi H, Kokai Y (2008) CCL8 is a potential molecular candidate for the diagnosis of graft-versus-host disease. Blood 111(8):4403–4412. doi:10.1182/blood-2007-06-097287

Karasawa K, Asano K, Moriyama S, Ushiki M, Monya M, Iida M, Kuboki E, Yagita H, Uchida K, Nitta K, Tanaka M (2015) Vascular-resident CD169-positive monocytes and macrophages control neutrophil accumulation in the kidney with ischemia-reperfusion injury. J Am Soc Nephrol 26(4):896–906. doi:10.1681/ASN.2014020195

Kaser A, Zeissig S, Blumberg RS (2010) Inflammatory bowel disease. Annu Rev Immunol 28:573–621. doi:10.1146/annurev-immunol-030409-101225

Kim TW, Seo JN, Suh YH, Park HJ, Kim JH, Kim JY, Oh KI (2006) Involvement of lymphocytes in dextran sulfate sodium-induced experimental colitis. World J Gastroenterol 12(2):302–305

Kojouharoff G, Hans W, Obermeier F, Mannel DN, Andus T, Scholmerich J, Gross V, Falk W (1997) Neutralization of tumour necrosis factor (TNF) but not of IL-1 reduces inflammation in chronic dextran sulphate sodium-induced colitis in mice. Clin Exp Immunol 107(2):353–358

Kraal G, Janse M (1986) Marginal metallophilic cells of the mouse spleen identified by a monoclonal antibody. Immunology 58(4):665–669

Laharie D, Bourreille A, Branche J, Allez M, Bouhnik Y, Filippi J, Zerbib F, Savoye G, Nachury M, Moreau J, Delchier JC, Cosnes J, Ricart E, Dewit O, Lopez-Sanroman A, Dupas JL, Carbonnel F, Bommelaer G, Coffin B, Roblin X, Van Assche G, Esteve M, Farkkila M, Gisbert JP, Marteau P, Nahon S, de Vos M, Franchimont D, Mary JY, Colombel JF, Lemann M, Groupe d'Etudes Therapeutiques des Affections Inflammatoires D (2012) Ciclosporin versus infliximab in patients with severe ulcerative colitis refractory to intravenous steroids: a parallel, open-label randomised controlled trial. Lancet 380(9857):1909–1915. doi:10.1016/S0140-6736(12)61084-8

Maloy KJ, Powrie F (2011) Intestinal homeostasis and its breakdown in inflammatory bowel disease. Nature 474(7351):298–306. doi:10.1038/nature10208

Miyake Y, Asano K, Kaise H, Uemura M, Nakayama M, Tanaka M (2007) Critical role of macrophages in the marginal zone in the suppression of immune responses to apoptotic cell-associated antigens. J Clin Invest 117(8):2268–2278. doi:10.1172/JCI31990

Morrissey PJ, Charrier K, Braddy S, Liggitt D, Watson JD (1993) CD4+ T cells that express high levels of CD45RB induce wasting disease when transferred into congenic severe combined immunodeficient mice. Disease development is prevented by cotransfer of purified CD4+ T cells. J Exp Med 178(1):237–244

Okayasu I, Hatakeyama S, Yamada M, Ohkusa T, Inagaki Y, Nakaya R (1990) A novel method in the induction of reliable experimental acute and chronic ulcerative colitis in mice. Gastroenterology 98(3):694–702

Ota A, Yamamoto M, Hori T, Miyai S, Naishiro Y, Sohma H, Maeda M, Kokai Y (2009) Upregulation of plasma CCL8 in mouse model of graft-vs-host disease. Exp Hematol 37 (4):525–531. doi:10.1016/j.exphem.2008.12.006

Pedron T, Sansonetti P (2008) Commensals, bacterial pathogens and intestinal inflammation: an intriguing menage a trois. Cell Host Microbe 3(6):344–347. doi:10.1016/j.chom.2008.05.010

Phan TG, Grigorova I, Okada T, Cyster JG (2007) Subcapsular encounter and complement-dependent transport of immune complexes by lymph node B cells. Nat Immunol 8 (9):992–1000. doi:10.1038/ni1494

Rivollier A, He J, Kole A, Valatas V, Kelsall BL (2012) Inflammation switches the differentiation program of Ly6Chi monocytes from antiinflammatory macrophages to inflammatory dendritic cells in the colon. J Exp Med 209(1):139–155. doi:10.1084/jem.20101387

Rutgeerts P, D'Haens G, Targan S, Vasiliauskas E, Hanauer SB, Present DH, Mayer L, Van Hogezand RA, Braakman T, DeWoody KL, Schaible TF, Van Deventer SJ (1999) Efficacy and safety of retreatment with anti-tumor necrosis factor antibody (infliximab) to maintain remission in Crohn's disease. Gastroenterology 117(4):761–769

Tamoutounour S, Henri S, Lelouard H, de Bovis B, de Haar C, van der Woude CJ, Woltman AM, Reyal Y, Bonnet D, Sichien D, Bain CC, Mowat AM, Reis e Sousa C, Poulin LF, Malissen B,

Guilliams M (2012) CD64 distinguishes macrophages from dendritic cells in the gut and reveals the Th1-inducing role of mesenteric lymph node macrophages during colitis. Eur J Immunol 42(12):3150–3166. doi:10.1002/eji.201242847

Tlaskalova-Hogenova H, Tuckova L, Stepankova R, Hudcovic T, Palova-Jelinkova L, Kozakova H, Rossmann P, Sanchez D, Cinova J, Hrncir T, Kverka M, Frolova L, Uhlig H, Powrie F, Bland P (2005) Involvement of innate immunity in the development of inflammatory and autoimmune diseases. Ann N Y Acad Sci 1051:787–798. doi:10.1196/annals.1361.122

Ueda Y, Kayama H, Jeon SG, Kusu T, Isaka Y, Rakugi H, Yamamoto M, Takeda K (2010) Commensal microbiota induce LPS hyporesponsiveness in colonic macrophages via the production of IL-10. Int Immunol 22(12):953–962. doi:10.1093/intimm/dxq449

Vance RE, Isberg RR, Portnoy DA (2009) Patterns of pathogenesis: discrimination of pathogenic and nonpathogenic microbes by the innate immune system. Cell Host Microbe 6(1):10–21. doi:10.1016/j.chom.2009.06.007

Varol C, Zigmond E, Jung S (2010) Securing the immune tightrope: mononuclear phagocytes in the intestinal lamina propria. Nat Rev Immunol 10(6):415–426. doi:10.1038/nri2778

Yamamoto M, Ota A, Hori T, Imai S, Sohma H, Suzuki N, Hatakeyama N, Inazawa N, Ito YM, Kimura H, Tsutsumi H, Kokai Y (2011) Early expression of plasma CCL8 closely correlates with survival rate of acute graft-vs.-host disease in mice. Exp Hematol 39(11):1101–1112. doi:10.1016/j.exphem.2011.07.006

Zigmond E, Varol C, Farache J, Elmaliah E, Satpathy AT, Friedlander G, Mack M, Shpigel N, Boneca IG, Murphy KM, Shakhar G, Halpern Z, Jung S (2012) Ly6C hi monocytes in the inflamed colon give rise to proinflammatory effector cells and migratory antigen-presenting cells. Immunity 37(6):1076–1090. doi:10.1016/j.immuni.2012.08.026

Zlotnik A, Yoshie O (2012) The chemokine superfamily revisited. Immunity 36(5):705–716. doi:10.1016/j.immuni.2012.05.008

Chapter 47
Identification of Long Non-Coding RNAs Involved in Chronic Inflammation in *Helicobacter Pylori* Infection and Associated Gastric Carcinogenesis

Reo Maruyama

Abstract *Helicobacter pylori* infection and associated chronic gastritis increase the risk for development of gastric cancer (GC). It is desirable to ascertain the precise underlying molecular mechanisms and pathophysiological implications of chronic inflammation and associated carcinogenesis. There is growing evidence that long non-coding RNAs (lncRNAs) play important roles in a wide repertoire of biological processes. It is very likely that lncRNAs also play critical roles in chronic inflammation and carcinogenesis. However, few studies have reported the involvement of lncRNAs in chronic inflammation. In this study, we aimed comprehensively to identify novel lncRNAs critically involved in chronic inflammation and associated carcinogenesis. To search for novel lncRNAs, we used ChIP-seq technology to obtain global histone methylation profiles in gastric epithelial cells. We first compared global histone H3 lysine 4 trimethylation profiles between gastric mucosae with and without active *H. pylori* infection and then identified lncRNAs associated with *H. pylori* infection. Next, we compared gastric mucosa from healthy individuals with background mucosa from patients with GC. *TCONS_00027118* was identified as a downregulated lncRNA gene in the background mucosa; it was silenced by DNA methylation. Moreover, the methylation of this region was universally observed in various tumours in other organs. In contrast, *TCONS_00006264* was identified as an upregulated lncRNA gene. Knockdown of the lncRNA substantially reduced gastric cancer cell viability, suggesting that it has important functions. Here we provide lists of lncRNAs potentially involved in *H. pylori* infection and gastric carcinogenesis. These lists will be useful resources for further functional and clinical studies.

Keywords Long non-coding RNA • Gastric cancer • Chronic gastritis • Helicobacter pylori • Epigenetics • DNA methylation • ChIP-seq

---

R. Maruyama (✉)
Department of Molecular Biology, Sapporo Medical University School of Medicine, Sapporo, Japan

PRESTO, Japan Science and Technology Agency, Kawaguchi 332-0012, Japan
e-mail: reomaru@sapmed.ac.jp

M. Miyasaka, K. Takatsu (eds.), *Chronic Inflammation*,
DOI 10.1007/978-4-431-56068-5_47

47.1 Introduction

Chronic inflammation promotes cancer development. It is estimated that up to 25 % of all cancers are associated with chronic inflammation (Mantovani et al. 2008). For example, chronic gastritis with *Helicobacter pylori* infection is a risk factor for gastric cancer (GC). Not only *H. pylori* infection itself as a definite carcinogen but also long-term inflammation is thought to exert pronounced effects on gastric carcinogenesis (Chiba et al. 2012). A wide range of cytokines released from immune, mesenchymal, and epithelial cells work in concert to promote cancer development, wherein proinflammatory cytokines and reactive oxygen and nitrogen species are thought to induce genetic and epigenetic changes. However, precise underlying molecular mechanisms and pathophysiological significance of chronic inflammation in carcinogenesis are not fully understood.

Long noncoding RNAs (lncRNAs) have emerged as key players in a wide repertoire of biological processes. The Encyclopedia of DNA Elements (ENCODE) project revealed that more than 60 % of the human genome is pervasively transcribed (Consortium 2012), and given that exonic regions of protein-coding genes represent only 2 % of the genome, the majority is considered to consist of lncRNAs. A number of studies have shown that lncRNAs have distinct molecular functions, and appear to be involved in all aspects of gene regulation (Geisler and Coller 2013). Intriguingly, a class of lncRNAs represented by *HOTAIR* seemingly regulates hundreds of specific target genes on the genome by recruiting and guiding appropriate chromatin regulators to specific sites on the genome at the right times (Chu et al. 2011). Although it is evident that transcription factors are the most important players in the direct regulation of gene expression, it is possible that lncRNAs also play critical roles in the fine tuning of gene regulation as transcription factors. Therefore, studying chronic inflammation from a different perspective focusing on lncRNAs would reveal new insights. A number of studies have reported novel lncRNAs functionally involved in the innate and adaptive immune systems (Fitzgerald and Caffrey 2014). However, in most cases, immune-related lncRNAs are identified by differential expression analysis in response to activation of immune cells, and few studies have reported the involvement of lncRNAs in chronic inflammation or their association with carcinogenesis.

In this study, we aimed to identify novel lncRNAs comprehensively that are critically involved in chronic inflammation and related carcinogenesis. In particular, we analysed *H. pylori*-associated gastritis and GC as a model.

47.2 Results

47.2.1 Genome-Wide Analysis of Histone Methylation Profiles in the Gastric Mucosa

To identify novel lncRNAs, we analysed histone H3 lysine 4 trimethylation (H3K4me3) profiles. Given that H3K4me3 is enriched at transcription start sites (TSS) of actively transcribed genes, H3K4me3 marks can be used as a clue to the presence of transcripts (Suzuki et al. 2011), whether mRNA or lncRNA. For sample preparation, the crypt isolation technique was employed to isolate epithelial cells from specimens collected by upper gastrointestinal endoscopy. Then, chromatin immunoprecipitation (ChIP) and ChIP followed by high-throughput sequencing (ChIP-seq) protocols were optimised to obtain unbiased data from small amounts of starting materials. Eventually, we were able to establish a workflow to perform ChIP-seq experiments of clinical samples. A total of 36 ChIP-seq experiments (22 for H3K4me3 and 14 for H3K27me3) were performed.

47.2.2 Identification of lncRNAs Involved in H. Pylori *Infection*

To identify novel lncRNAs directly involved in *H. pylori* infection, we compared H3K4me3 profiles between one sample with active *H. pylori* infection and two samples without active *H. pylori* infection (posteradication samples). Around 21,000 H3K4me3 peaks were detected in the genome of each sample. Approximately three fourths (15,553) of the peaks were common to the two groups, whereas 2383 peaks were specific for the sample with active *H. pylori* infection (Fig. 47.1a). Representative examples of ChIP-seq data are shown in Fig. 47.1b. Of the 2383 *H. pylori*-specific peaks, 437 were annotated as TSSs of RefSeq genes, whereas 41 were associated with lncRNA genes (Fig. 47.1c). Intriguingly, gene ontology (GO) term enrichment analysis of 437 RefSeq genes revealed that biological functions associated with immune response were significantly enriched for these genes (Fig. 47.1d). For example, *IL2RA*, *ICAM1*, *IL21R*, *PAX5*, and *CXCR5* are included in the list. On the basis of this result, it can be reasonably expected that 41 (functionally unknown) lncRNAs identified in this analysis have similar functions, such as immune response. In principle, this finding should reflect changes in gastric epithelial cells with *H. pylori* infection, although we cannot exclude the possibility of infiltration by leukocytes.

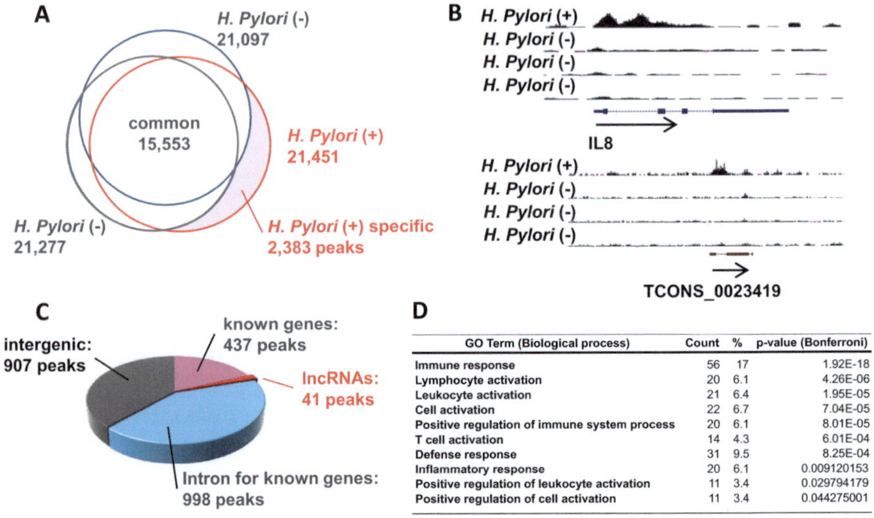

GO Term (Biological process)	Count	%	p-value (Bonferroni)
Immune response	56	17	1.92E-18
Lymphocyte activation	20	6.1	4.26E-06
Leukocyte activation	21	6.4	1.95E-05
Cell activation	22	6.7	7.04E-05
Positive regulation of immune system process	20	6.1	8.01E-05
T cell activation	14	4.3	6.01E-04
Defense response	31	9.5	8.25E-04
Inflammatory response	20	6.1	0.009120153
Positive regulation of leukocyte activation	11	3.4	0.029794179
Positive regulation of cell activation	11	3.4	0.044275001

Fig. 47.1 lncRNAs involved in *H. pylori* infection.
(**a**) Venn diagram showing the number of H3K4me3 peaks detected specifically in the *H. pylori*-positive sample
(**b**) Representative examples of ChIP-seq data for *IL8* and lncRNA gene
(**c**) Pie chart showing gene annotations of 2383 *H. pylori*-positive sample specific peaks
(**d**) GO terms (biological functions) enriched for 437 RefSeq genes with *H. pylori*-positive specific H3K4me3 peaks

47.2.3 Identification of lncRNAs Potentially Involved in Gastric Carcinogenesis Associated with Chronic Inflammation

We next sought to identify lncRNAs that exhibit a change in activity over the course of long-term inflammation and could affect gastric carcinogenesis. To this end, we compared global H3K4me3 profiles between the normal gastric mucosa from healthy individuals with the background noncancerous mucosa from patients with GC (hereafter referred to as 'background mucosa' and assumed to contain precancerous lesions). As seen in the heat map (Fig. 47.2a), unsupervised hierarchical clustering of seven samples based on H3K4me3 peaks showed a clear separation into two groups (healthy individuals and cancer patients), suggesting that global H3K4me3 profiles are distinct between the two groups and reflected molecular alterations in epithelial cells due to chronic inflammation. On the basis of clustering of H3K4me3 peaks, we arbitrarily defined three groups of peaks: unchanged, downregulated, and upregulated peaks in the background mucosa (Fig. 47.2b). Approximately one-third (11,398) of the peaks were detected in all seven samples, indicating that these genomic regions show no changes over the course of long-term inflammation. In contrast, 3140 peaks appeared to be downregulated (defined as positive in all three healthy individuals and negative in

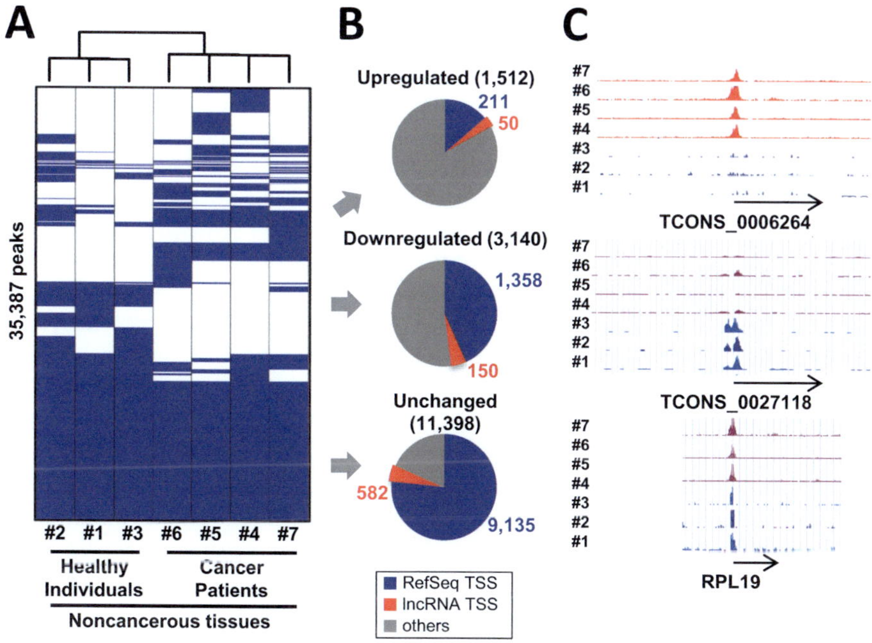

Fig. 47.2 Global H3K4me3 profiles in the gastric mucosa from healthy individuals and patients with gastric cancer.
(**a**) Heat map showing H3K4me3 patterns and hierarchical clustering of seven samples
(**b**) Proportion of gene annotations for H3K4me3 peaks in the three categories
(**c**) Representative example of ChIP-seq results for each category

more than two background mucosae), whereas 1512 peaks appeared to be upregulated (defined as negative in all healthy individuals and positive in all background mucosae), indicating that these genomic regions are repressed or activated in the background mucosa for some reason. This observation of these regions probably reflects changes in the molecular characteristics of gastric epithelial cells over the course of long-term inflammation. Incorporating annotation data of lncRNAs, 150 lncRNA genes were identified as downregulated and 50 lncRNA as upregulated genes in the background mucosa (Fig. 47.2b, c).

47.2.4 Gene Downregulation in the Background Mucosa from Patients with Cancer may Reflect Global Epigenomic Changes

We next focused on 150 lncRNA genes identified as downregulated in the background mucosa of patients with cancer. To narrow the list, we analysed RNA-seq data of 175 patients with GC (148 cancerous and 27 noncancerous samples) available in the Cancer Genome Atlas (TCGA) database. Of the 150 lncRNAs, expression levels of eight lncRNAs were significantly lower in cancer than those in noncancerous samples (Fig. 47.3a). We selected the *TCONS_0027118* gene for further analysis. As shown in Fig. 47.2c, H3K4me3 peaks in the *TCONS_0027118* promoter region were reduced or absent in the background mucosa from patients with GC. Bisulfite pyrosequencing of the same samples showed that the promoter region was methylated in various degrees in the background mucosa of patients with GC but not methylated in healthy individuals (data not shown), suggesting the existence of epigenetic field defect (Chiba et al. 2012). Examination of GC cell lines and gastric normal tissues revealed that *TCONS_0027118* was weakly expressed in normal tissues, whereas its expression was suppressed by DNA methylation in all GC cell lines examined (Fig. 47.3b).

We next assessed the methylation status in a series of clinical samples. We analysed samples of endoscopically-obtained background noncancerous mucosa from 62 patients with GC and samples of the normal gastric mucosa from 50 healthy individuals (43 individuals with active *H. pylori* infection and 7 without). Bisulfite pyrosequencing revealed that the levels of *TCONS_0027118* promoter methylation were significantly higher in the background mucosa from patients with GC than those in the mucosa of patients without cancer and with active *H. pylori* infection (median 13.8 % vs 26.8 %, $p < 0.05$) (Fig. 47.3c). The high frequency of *TCONS_0027118* methylation in the background mucosa suggests that it could be a biomarker for risk prediction of gastric cancer among *H. pylori*-positive patients.

Colorectal cancer cell lines showed the same DNA methylation pattern (data not shown). We accordingly hypothesised that the methylation of this region is a universal phenomenon in precancerous lesions and early stages of cancer. To test this hypothesis, we analysed DNA methylation levels of the *TCONS_0027118* promoter in 20 different tumour types in the TCGA database. The analysis revealed that not only gastric or colorectal cancers but also other types of cancers, such as head and neck, esophageal, cervical squamous, pancreatic, and lung squamous carcinomas, showed high frequencies of methylation in tumour samples but low frequencies in corresponding normal tissues (Fig. 47.3d). This result suggests that similar epigenomic changes occur between the background mucosa from patients with GC and cancerous tissues from these organs. The same mechanism may be involved in both tissues. To identify the mechanism, we investigated the TCGA methylation data in more detail. We calculated the correlation coefficients of the beta values of the probe corresponding to the *TCONS_0027118* promoter with those of each of the remaining probes to identify probes that showed the highest correlation with *TCONS_0027118* in all TCGA samples (red dots in Fig. 47.3e). Probes showing patterns similar (with high correlation) to those of *TCONS_0027118* formed several clusters on the genome, and notably, most were located close to coding regions for zinc finger proteins (ZNFs) as well as unknown lncRNAs. This observation may have some biological meaning, although further study will be needed.

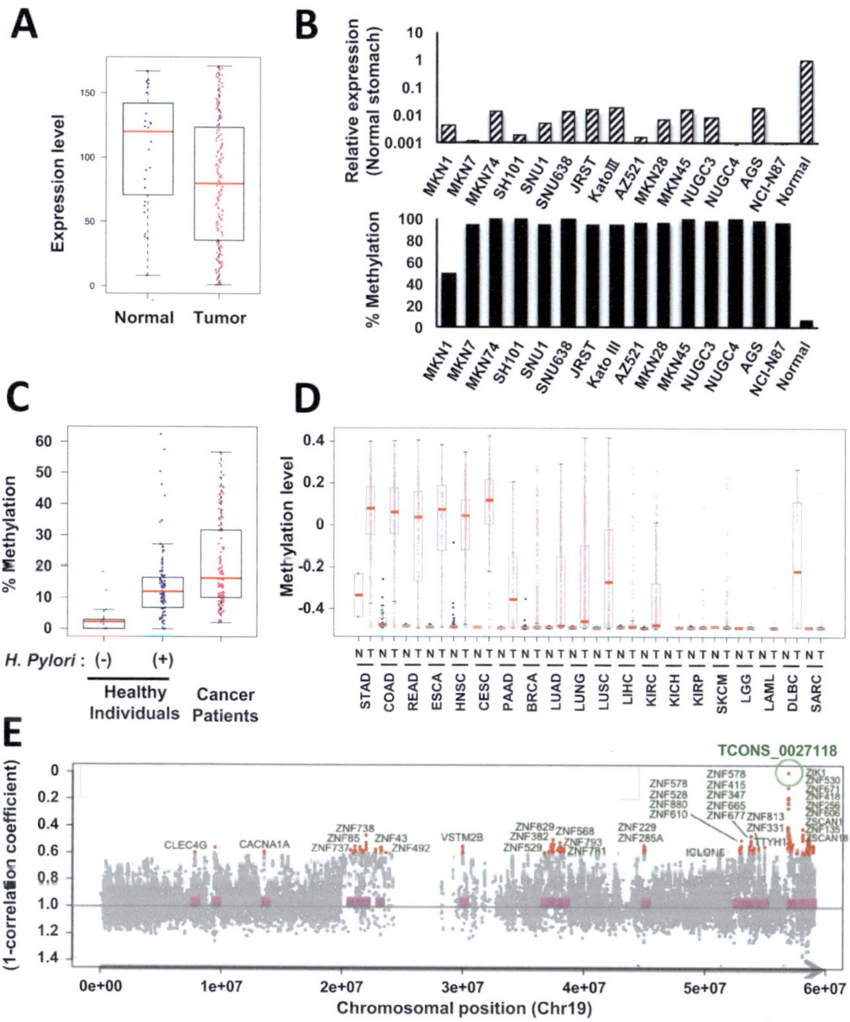

Fig. 47.3 Downregulated lncRNA in the background mucosa from patients with gastric cancer.
(**a**) Expression levels of *TCONS_0027118* in clinical samples of GCs in the TCGA database
(**b**) Expression and DNA methylation levels of *TCONS_0027118* in GC cell lines
(**c**) DNA methylation levels of the lncRNA in the normal gastric mucosa from healthy individuals and the background mucosa from GC patients
(**d**) Levels of DNA methylation of the lncRNA in indicated tumours (T) and corresponding normal tissues (N) in the TCGA database
(**e**) Correlation between methylation level of the lncRNA and those of each probe on chromosome 19 in all samples in the TCGA database. Each *dot* represents each probe in HumanMethylation 450 k data. *X*-axis represents chromosomal location. *Y*-axis represents distance (1-correlation coefficient) between the probe of the lncRNA and each probe

47.2.5 Genes Upregulated in the Background Mucosa from Patients with GC

Lastly, we focused on lncRNA genes upregulated in the background mucosa from patients with GC. ChIP-seq profiles revealed 50 lncRNAs as upregulated lncRNA genes (Fig. 47.2b). Quantitative reverse transcription PCR (qRT-PCR) screening experiments revealed that six of the 16 selected lncRNAs showed an increased expression in various GC cell lines compared with the normal gastric mucosa (Fig. 47.4a). We selected the *TCONS_0006264* gene, one of the six lncRNAs, for further study because there is a divergently transcribed mRNA neighbour whose protein is thought to be involved in cancer and immunity (Fig. 47.4b). qRT-PCR experiments showed that the expressions of both lncRNA and mRNA were upregulated in approximately two-thirds of GC cell lines compared with those in the normal gastric mucosa (Fig. 47.4c).

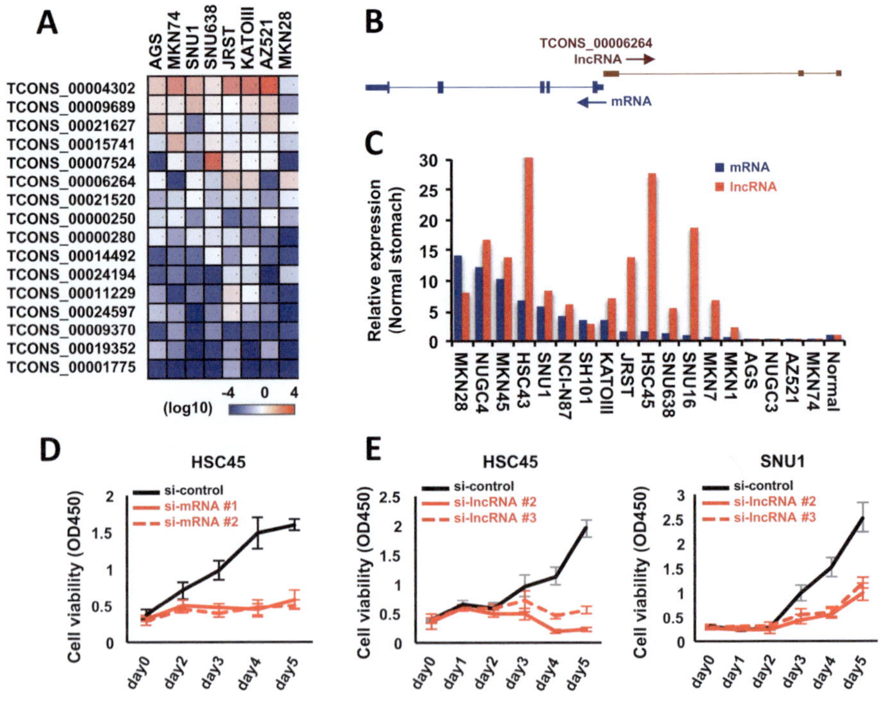

Fig. 47.4 Upregulated lncRNAs in the background mucosa from patients with gastric cancer. (**a**) qRT-PCR screening of selected lncRNAs. Heat map showing relative expression levels of lncRNAs in GC cell lines compared with the normal stomach
(**b**) Genomic organization of *TCONS_0006264* and divergently transcribed mRNA
(**c**) Relative expression of genes compared with the normal stomach
(**d, e**) Knockdown of the mRNA or the lncRNA attenuated gastric cancer cell viability

We next designed siRNAs specifically targeting the mRNA or the lncRNA to evaluate the effect of genes on cancer cells. Knockdown of the mRNA reduced the cell viability of HSC45 GC cells (Fig. 47.4d), a result that was somewhat expected because the mRNA is known to be overexpressed in several cancers and is speculated to have oncogenic properties. However, knockdown of *TCONS_0006264* also significantly decreased the cell viability of HSC45 cells (Fig. 47.4e). This effect was confirmed in an additional three cell lines (JRST, MKN45, and SNU1), in which the lncRNA was overexpressed. These results suggest that not only the protein but also the divergently-transcribed lncRNA itself has certain biological functions in GC cells. We did not observe a *cis*-effect of the lncRNA on the protein expression (data not shown). Further mechanistic study of the lncRNA is ongoing and will be reported (manuscript under preparation).

47.3 Discussion

In this study, we sought to identify novel lncRNAs that are deeply involved in *H. pylori* infection and related carcinogenesis. Our eventual goal is to clarify some of the molecular mechanisms of chronic inflammation and its association with carcinogenesis. We have provided lists of lncRNAs that may be involved in these processes, although we are still conducting mechanistic studies of these lncRNAs.

We first compared gastric epithelial tissues with and without active *H. pylori* infection. We identified a list of candidate lncRNAs that are potentially involved in *H. pylori* infection. Overall, these results suggested that changes of molecular profiles occur in gastric epithelial cells during active *H. pylori* infection. However, it should be noted that the results are potentially affected by infiltration of immune cells in the *H. pylori*-positive sample, although we used the crypt isolation technique to isolate only epithelial cells as much as possible. It might be better to consider the gene list as describing a mixed population with gastric epithelial cells and infiltrating immune cells. In any case, it is reasonable to infer that these lncRNAs have some functions associated with immune response, according to the result of GO term enrichment analysis of the corresponding RefSeq genes. This list should be useful for identifying novel lncRNAs with functions associated with immune response, although further functional studies will be required.

To the best of our knowledge, this is the first study of global H3K4me3 profiles in the gastric mucosa from healthy individuals and patients with GC. The global histone methylation patterns clearly differed more than expected between the two groups. The unsupervised hierarchical clustering result shows that their molecular cellular properties were distinct. Certain changes (probably epigenomic changes) occur over the course of long-term inflammation that contribute to the formation of precancerous lesions. It is uncertain whether the changes result from very long-term inflammation or are results of some specific event that occurs only in patients with GC (such as point mutations in stem cells). Because only 1–3 % of *H. pylori*-associated gastritis patients develop gastric cancer, they might have some specific changes distinguishable from those in other noncancerous individuals. For further investigation, the background mucosa from chronic gastritis patients with or without gastric cancer should be compared.

We also attempted to obtain an overview of the epigenetic field defect by analysing publicly-available DNA methylation data. Intriguingly, correlation analysis of each probe revealed that probes in certain clusters in chromosome 19 always showed patterns similar to that of the TCONS_00027118 probe (meaning that if the lncRNA is methylated in one sample, the other probes in the clusters are also methylated). Thus, these cluster regions will always be methylated in the background mucosa from patients with GC, suggesting that three-dimensional chromatin structures are involved in this epigenomic change. Although further studies will be required, we have shown that analysing large public datasets is a promising approach.

In summary, we examined global H3K4me3 profiles from various samples and identified a list of lncRNA genes potentially involved in *H. pylori* infection and gastric carcinogenesis. This study provides resources useful to a wide range of researchers in the field of chronic inflammation, RNA biology, and cancer biology.

47.4 Experimental Procedures

47.4.1 Sample Collection

Gastric mucosa specimens were obtained by endoscopic biopsy. For ChIP-seq experiments, the crypt isolation technique (Nakamura et al. 1994) was used to isolate the epithelial cell-enriched fraction. Informed consent was obtained from all patients before collection of specimens.

47.4.2 ChIP and ChIPseq Experiment

ChIP-seq for the H3K27me3 mark and data analysis were performed essentially as previously described (Maruyama et al. 2011), but samples were sequenced using the SOLiD platform and the size-selection step was omitted to minimize sample loss.

47.4.3 Data Analysis

An annotation dataset for human lncRNA genes was obtained from the Human Body Map of lincRNAs at the Broad Institute. H3K4me3 peaks were defined using

MACS1.4 software with the default setting. Genome-wide DNA methylation and expression data in TCGA datasets were obtained from the Cancer Genomics Browser. Abbreviations used in Fig. 47.3e are listed in the browser. GO term enrichment analysis was conducted using the DAVID Bioinformatics resources 6.7.

47.4.4 Expression and DNA Methylation Analysis

Total RNA from cell lines was obtained using an RNeasy Mini kit (QIAGEN) and total RNA from normal gastric mucosa was purchased from Ambion. Single-stranded cDNA was prepared using PrimeScript RT Master Mix Perfect Realtime (TaKaRa). qRT-PCR was performed using SYBR Select Master Mix (Applied Biosystems) and a 7500 Fast Real-Time PCR System (Applied Biosystems). Relative expression levels of target lncRNAs were determined using the endogenous housekeeping gene ACTB as internal controls. For DNA methylation analysis, genomic DNA (1 μg) was modified with sodium bisulfite using an EpiTect Bisulfite kit (Qiagen). Pyrosequencing reaction was performed using a PSQ96 system with a PyroGold reagent Kit (Qiagen).

47.4.5 Cell Viability Assay

Predesigned and custom-designed siRNAs were purchased from Sigma Genosys (Japan). siRNAs (20 nM) were transfected into GC cells using Lipofectamin RNAiMax (Life Technologies). Sufficient knockdown of target genes was confirmed by qRT-PCR for all siRNAs. GC cells were seeded in 96-well plates at a density of 2000 or 5000 cells/100 μl 1 day before siRNA transfection. Cell numbers at each time point were determined indirectly by measuring absorbance at 450 nm 2 h after adding Cell Counting Kit-8 (Dojindo).

References

Chiba T, Marusawa H, Ushijima T (2012) Inflammation-associated cancer development in digestive organs: mechanisms and roles for genetic and epigenetic modulation. Gastroenterology 143(3):550–563. doi:10.1053/j.gastro.2012.07.009

Consortium EP (2012) An integrated encyclopedia of DNA elements in the human genome. Nature 489(7414):57–74. doi:10.1038/nature11247

Chu C, Qu K, Zhong FL, Artandi SE, Chang HY (2011) Genomic maps of long noncoding RNA occupancy reveal principles of RNA-chromatin interactions. Mol Cell 44(4):667–78. doi:10.1016/j.molcel.2011.08.027

Fitzgerald KA, Caffrey DR (2014) Long non coding RNAs in innate and adaptive immunity. Curr Opin Immunol 26:140–146. doi:10.1016/j.coi.2013.12.001

Geisler S, Coller J (2013) RNA in unexpected places: long non-coding RNA functions in diverse cellular contexts. Nat Rev Mol Cell Biol 14(11):699–712. doi:10.1038/nrm3679

Mantovani A, Allavena P, Sica A, Balkwill F (2008) Cancer-related inflammation. Nature 454 (7203):436–444. doi:10.1038/nature07205

Maruyama R, Choudhury S, Kowalczyk A, Bessarabova M, Beresford-Smith B, Conway T, Kaspi A, Wu Z, Nikolskaya T, Merino VF, Lo PK, Liu XS, Nikolsky Y, Sukumar S, Haviv I, Polyak K (2011) Epigenetic regulation of cell type-specific expression patterns in the human mammary epithelium. PLoS Genet 7(4):e1001369. doi:10.1371/journal.pgen.1001369

Nakamura S, Goto J, Kitayama M, Kino I (1994) Application of the crypt-isolation technique to flow-cytometric analysis of DNA content in colorectal neoplasms. Gastroenterology 106 (1):100–107

Suzuki H, Takatsuka S, Akashi H, Yamamoto E, Nojima M, Maruyama R, Kai M, Yamano HO, Sasaki Y, Tokino T, Shinomura Y, Imai K, Toyota M (2011) Genome-wide profiling of chromatin signatures reveals epigenetic regulation of MicroRNA genes in colorectal cancer. Cancer Res 71(17):5646–5658. doi:10.1158/0008-5472.can-11-1076

Part X
Chronic Inflammation and Central Nervous System Disease

Chapter 48
The Research for the Mechanism of Chronically Intractable Pain Based on the Functions of Microglia as Brain Immunocompetent Cell

Kazuhide Inoue and Makoto Tsuda

Abstract Injury to the nervous system often causes a debilitating chronic pain syndrome, termed neuropathic pain. Neuropathic pain is refractory to currently available analgesics. Accumulating evidence indicates that spinal microglia react and undergo a series of changes that influence directly establishing the pain. P2X4 receptor (P2X4R), a subtype of ionotropic ATP receptors, is upregulated in spinal microglia after nerve injury by several factors including the CC chemokine CCL21 derived from damaged neurons, CC chemokine receptor CCR2, the extracellular matrix protein fibronectin in the spinal cord, and the transcription factor interferon regulatory factor 8 and 5 expressed in microglia. The inhibition of the function of P2X4R and P2X4R-regulating molecules suppresses the excitability of dorsal horn neurons and neuropathic pain. These findings indicate that microglia overexpressing P2X4R are a central player in neuropathic pain.

Keywords P2X4 receptor • Microglia • Neuropathic pain • Spinal cord

K. Inoue (✉)
Department of Molecular and System Pharmacology, Graduate School of Pharmaceutical Sciences, Kyushu University, 3-1-1 Maidashi, Higashi-ku, Fukuoka 812-8582, Japan

AMED-CREST, Fukuoka 812-8582, Japan
e-mail: inoue@phar.kyushu u.ac.jp

M. Tsuda
Department of Life Innovation, Graduate School of Pharmaceutical Sciences, Kyushu University, 3-1-1 Maidashi, Higashi-ku, Fukuoka 812-8582, Japan

M. Miyasaka, K. Takatsu (eds.), *Chronic Inflammation*,
DOI 10.1007/978-4-431-56068-5_48

48.1 Introduction

Injury to the nervous system arising from bone compression in cancer, diabetes mellitus, infection, autoimmune diseases, or traumatic injury often causes debilitating chronic pain states, namely neuropathic pain. Characteristic symptoms of neuropathic pain include tactile allodynia (pain hypersensitivity to normally innocuous stimuli), spontaneous pain, and hyperalgesia (increased pain perception of noxious stimuli). Neuropathic pain is refractory to currently available treatments, such as nonsteroidal anti-inflammatory drugs and opioids (Costigan et al. 2009). The evidence from basic pain research indicated that neuropathic pain is a reflection of the aberrant excitability of dorsal horn neurons evoked by peripheral sensory inputs (Woolf and Salter 2000; Costigan et al. 2009). Recently, many studies have shown that glial cells including microglia activated in the spinal cord in response to peripheral nerve injury (PNI) changes pathologically the neurotransmission of the pain signalling (Watkins et al. 2001; Tsuda et al. 2005; McMahon and Malcangio 2009; Ren and Dubner 2010; Tsuda et al. 2013).

Microglia are thought to derive from primitive macrophages in the yolk sac (Ginhoux et al. 2010). In normal condition, microglia are ubiquitously distributed throughout CNS and have small cell body bearing branched and motile processes, which monitor the local environment (Davalos et al. 2005; Nimmerjahn et al. 2005). In response to PNI, microglia activation in the spinal cord progresses through hypertrophic morphology, an increase in cell number, and alters the expression of genes including neurotransmitter receptors such as purinergic P2 receptors (Tsuda et al. 2005; Suter et al. 2007; Tsuda et al. 2009b; Pocock and Kettenmann 2007). Extracellular stimuli such as ATP activate microglia to evoke various cellular responses such as the production and release of bioactive factors including cytokines and neurotrophic factors (Inoue 2006), which in turn lead to neuropathic pain.

Among purinergic P2 receptors [ionotropic receptors (P2XRs) and metabotropic receptors (P2YRs)], activated microglia express several subtypes of P2XRs and P2YRs, and these receptors play a key role in establishing and maintaining neuropathic pain states (Tsuda et al. 2012; 2013). In this chapter, we describe recent advances to understand the mechanism of neuropathic pain with a specific focus on P2X4R in spinal microglia after PNI.

48.2 P2X4R in Activated Spinal Microglia and Neuropathic Pain

We first observed that tactile allodynia after PNI was reversed by a pharmacological blocker of P2X4Rs in the spinal cord (Tsuda et al. 2003). We also revealed that expression of P2X4Rs in the spinal cord was upregulated exclusively in microglia by immunohistochemical studies. These results indicated that PNI-induced allodynia depended on signalling via microglial P2X4Rs. Furthermore, several papers reported that animals with P2X4R knockdown or knockout in the spinal

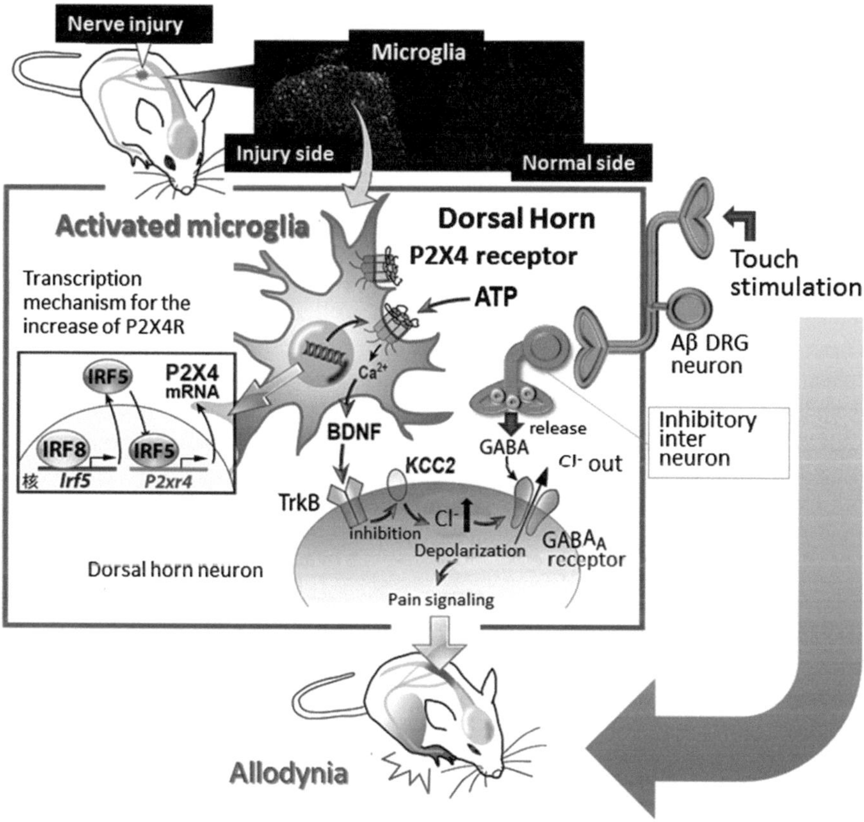

Fig. 48.1 The mechanisms of allodynia via P2X4R stimulation in activated microglia after peripheral nerve injury. Nerve injury activates spinal cord microglia in the dorsal horn. In activated microglia increased P2X4R is stimulated by ATP to cause the release of BDNF. BDNF downregulates the potassium-chloride–transporter KCC2 via TrkB, causes an increase in intracellular [Cl^-], and leads to the collapse of the transmembrane anion gradient in dorsal horn neurons. In the meanwhile, touch stimulation causes GABA release from inhibitory interneurons. GABA opens Cl^- channels of the dorsal horn neurons and leads Cl^- outflow, resulting in the depolarisation of these neurons. The spikes of these neurons reach the cortex through synaptic transmissions and evoke pain sensation. In this way, touch stimulation evokes allodynia. This might be the mechanism of mechanical allodynia of neuropathic pain

cord were resistant to PNI-induced tactile allodynia (Tsuda et al. 2003; Ulmann et al. 2008; Tsuda et al. 2009a), indicating a necessity for P2X4Rs. It was also demonstrated that activation of microglial P2X4Rs stimulated the synthesis and release of brain-derived neurotrophic factor (BDNF; Trang et al. 2009; Ulmann et al. 2008). BDNF may cause downregulation of the neuronal chloride transporter KCC2 to alter the transmembrane anion gradient in a subpopulation of dorsal horn lamina I neurons, which in turn renders GABA and glycine effects depolarising, rather than hyperpolarising, in these neurons (Coull et al. 2005). Thus, microglial P2X4Rs are central players in the pathogenesis of neuropathic pain (Fig. 48.1).

48.3 Regulation of P2X4R Expression in Microglia

Overexpression of P2X4R in microglia is a key process in evoking neuropathic pain. Recently, it was reported that the chemokine CCL21 (chemokine C-C motif ligand 21) was induced in injured dorsal root ganglion (DRG) neurons and transported through the neurons to the central terminals of the dorsal horn (Biber et al. 2011). The treatment with CCL21-neutralizing antibody and deficiency of CCL21 showed attenuation of microglial P2X4R upregulation and tactile allodynia. In cultured microglia, CCL21 increased the expression of P2X4R. Intrathecal supply of CCL21 caused tactile allodynia in CCL21-deficient mice after PNI. These results indicate that CCL21 derived from injured DRG neurons directly contributes to the P2X4R expression in microglia and neuropathic pain (Biber et al. 2011).

It was well known that blood–spinal cord barrier functions collapse after PNI (Beggs et al. 2010; Echeverry et al. 2011); then there is the possibility that proteins leaked from the blood might change P2X4R expression in microglia. The extracellular matrix protein fibronectin might be one of the proteins. The protein level of fibronectin was elevated in the dorsal horn after PNI (Nasu-Tada et al. 2006; Echeverry et al. 2011). Fibronectin also induces upregulation of mRNA and protein of P2X4R in primary cultured microglial cells (Nasu-Tada et al. 2006). It was shown in in vitro and in vivo experiments using integrin blockers that fibronectin/integrin signalling was crucial for the overexpression of P2X4R and tactile allodynia (Tsuda et al. 2008a). Furthermore, intrathecal injection of fibronectin to naïve animals produced tactile allodynia which was not observed in P2X4R-deficient mice after the administration of fibronectin (Tsuda et al. 2008a). Microglial Lyn tyrosine kinase, a member of Src-family kinases (SFKs) that belong to the nonreceptor protein tyrosine kinase family, is an important molecule because fibronectin failed to cause the upregulation of P2X4R gene expression in microglial cells lacking Lyn (Tsuda et al. 2008b). Lyn was the predominant SFK in spinal cord microglia (Tsuda et al. 2008b) amongst the five members (Src, Fyn, Lck, Yes, and Lyn) expressed in the CNS (Salter and Kalia 2004). Lyn expression in the spinal cord in vivo is highly restricted to microglia, and the level of Lyn increased after PNI (Tsuda et al. 2008b) in interferon-γ dependently (Tsuda et al. 2009b). Tactile allodynia and the upregulation of spinal P2X4R expression after PNI was suppressed in mice lacking Lyn (Tsuda et al. 2008b).

Two intracellular signalling cascades are distinctly activated after Lyn tyrosine kinase activation. One is a pathway through phosphatidylinositol 3-kinase (PI3K)-Akt and the other is through mitogen-activated protein kinase kinase (MAPK kinase, MEK)-extracellular signal-regulated kinase (ERK; Tsuda et al. 2009c). In the PI3K-Akt pathway, the degradation of p53 was occurred in a proteasome-dependent manner. The attenuation of the repressive effect of p53 may cause an enhancement of P2X4 gene expression. Activated MEK-ERK signalling in microglia exposed to fibronectin enhanced eukaryotic translation initiation factor 4E (eIF4E) phosphorylation through activated MAPK-interacting protein kinase-1, which may play a role in regulating P2X4 expression at translational levels. Because pharmacological inhibition of SFK effectively suppressed ERK activity in spinal microglia (Katsura et al. 2006), the Lyn-ERK signalling pathway seems likely to be active in spinal microglia after PNI. Therefore, Lyn might be a key kinase in the molecular mechanism of the P2X4R upregulation in microglia.

48.4 Transcriptional Factors of Microglia

PNI activates microglia and converts them to reactive phenotypes through the activation of gene transcription. Recent studies identified interferon regulatory factor 8 (IRF8) as a transcription factor in microglia (Masuda et al. 2012; Minten et al. 2012; Horiuchi et al. 2012; Kierdorf et al. 2013) that was critical for their activation and neuropathic pain (Masuda et al. 2012). IRF8 is a member of the IRF family (IRF1–9), and is expressed in immune cells such as lymphocytes and dendritic cells (Tamura et al. 2008). IRF8 expression was markedly enhanced in microglia, but not in neurons or astrocytes, in the spinal cord after PNI (Masuda et al. 2012). The upregulation of IRF8 expression occurs as early as day 1, peaks on day 3, and persists for at least several weeks after the injury. IRF8-deficient mice showed a reduction of PNI-induced tactile allodynia without any change in basal mechanical sensitivity. Intrathecal administration of a small interfering RNA (siRNA) targeting IRF8 in wild-type mice suppressed the upregulation of spinal IRF8 and allodynia after PNI, indicating an ongoing activation of IRF8 in spinal microglia after nerve injury. In vitro and in vivo studies demonstrated that IRF8 promoted the transcription of P2X4R as well as innate immune response (TLR2), P2Y12R, the chemokine receptor CX3CR1, interleukin 1β, cathepsin S, and BDNF that are related to the pain. The next question is how IRF8 expression is induced in microglia after PNI.

Recently we have shown that interferon regulatory factor-5 (IRF5) directly controls the transcription of P2X4R in microglia after PNI. Upon stimulation of microglia by fibronectin, IRF5 induced de novo expression of P2X4R by directly binding to the promoter region of the P2rx4 gene. Mice lacking Irf5 did not upregulate spinal P2X4R after PNI, and also exhibited substantial resistance to pain hypersensitivity. It is very important that microglial IRF5 is an IRF8-regulated gene, and required for IRF8-induced P2X4R expression. Therefore, a transcriptional axis from IRF8 to IRF5 might contribute to the activation of spinal microglia to overexpress P2X4R after PNI (Fig. 48.1 small box).

48.5 Ending Remarks

Pharmacological, molecular, and genetic studies on P2X4Rs mentioned above provide results that P2X4R in activated microglia are a central player in the mechanisms of neuropathic pain and the targets of new medicines for the pain. It is noted that activated microglia are necessary but not enough to cause the pain as mentioned before; that is, PNI caused microglial activation but not the pain in CCL21-deficient mice. After the intrathecal supply of CCL21 into the spinal cord, these mice with overexpressed P2X4R showed the pain.

The role of upregulated P2X4R in microglia has also been reported not only in neuropathic pain but also in stroke (Cavaliere et al. 2003), brain tumour (Guo et al. 2004), traumatic brain injury (Guo and Schluesener 2005; Zhang et al. 2007), spinal cord injury (Schwab et al. 2005), and epilepsy (Ulmann et al. 2013) of animal models, and acute inflammatory demyelinating polyradiculoneuropathy in humans (Zhang et al. 2008), as typical examples of chronic inflammation and central nervous system disease. More recently, it was reported that P2X4R-positive microglia are essential for morphine-induced hyperalgesia through a P2X4R-BDNF-KCC2 disinhibition cascade in the relation between microglia and dorsal horn neurons (Ferrini et al. 2013). The increased evidence of the functions of P2X4R in microglia might provide us exciting insights into the mechanisms of the neuropathic pain and other CNS diseases.

References

Beggs S, Liu XJ, Kwan C et al (2010) Peripheral nerve injury and TRPV1-expressing primary afferent C-fibers cause opening of the blood-brain barrier. Mol Pain 6:74

Biber K, Tsuda M, Tozaki-Saitoh H et al (2011) Neuronal CCL21 up-regulates microglia P2X4 expression and initiates neuropathic pain development. EMBO J 30:1864–1873

Cavaliere F, Florenzano F, Amadio S et al (2003) Up-regulation of P2X2, P2X4 receptor and ischemic cell death: prevention by P2 antagonists. Neuroscience 120:85–98

Costigan M, Scholz J, Woolf CJ (2009) Neuropathic pain: a maladaptive response of the nervous system to damage. Annu Rev Neurosci 32:1–32

Coull JA, Beggs S, Boudreau D et al (2005) BDNF from microglia causes the shift in neuronal anion gradient underlying neuropathic pain. Nature 438:1017–1021

Davalos D, Grutzendler J, Yang G et al (2005) ATP mediates rapid microglial response to local brain injury in vivo. Nat Neurosci 8:752–758

Echeverry S, Shi XQ, Rivest S et al (2011) Peripheral nerve injury alters blood-spinal cord barrier functional and molecular integrity through a selective inflammatory pathway. J Neurosci 31:10819–10828

Ferrini F, Trang T, Mattioli TA et al (2013) Morphine hyperalgesia gated through microglia-mediated disruption of neuronal Cl(−) homeostasis. Nat Neurosci 16:183–192

Ginhoux F, Greter M, Leboeuf M et al (2010) Fate mapping analysis reveals that adult microglia derive from primitive macrophages. Science 330:841–845

Guo LH, Schluesener HJ (2005) Lesional accumulation of P2X(4) receptor(+) macrophages in rat CNS during experimental autoimmune encephalomyelitis. Neuroscience 134:199–205

Guo LH, Trautmann K, Schluesener HJ (2004) Expression of P2X4 receptor in rat C6 glioma by tumor-associated macrophages and activated microglia. J Neuroimmunol 152:67–72
Horiuchi M, Wakayama K, Itoh A et al (2012) Interferon regulatory factor 8/interferon consensus sequence binding protein is a critical transcription factor for the physiological phenotype of microglia. J Neuroinflammation 9:227
Inoue K (2006) The function of microglia through purinergic receptors: neuropathic pain and cytokine release. Pharmacol Ther 109:210–226
Katsura H, Obata K, Mizushima T et al (2006) Activation of Src-family kinases in spinal microglia contributes to mechanical hypersensitivity after nerve injury. J Neurosci 26:8680–8690
Kierdorf K, Erny D, Goldmann T, Sander V et al (2013) Microglia emerge from erythromyeloid precursors via Pu.1- and Irf8-dependent pathways. Nat Neurosci 16:273–280
Masuda T, Tsuda M, Yoshinaga R et al (2012) IRF8 is a critical transcription factor for transforming microglia into a reactive phenotype. Cell Rep 1:334–340
McMahon SB, Malcangio M (2009) Current challenges in glia-pain biology. Neuron 64:46–54
Minten C, Terry R, Deffrasnes C et al (2012) IFN regulatory factor 8 is a key constitutive determinant of the morphological and molecular properties of microglia in the CNS. PLoS One 7, e49851
Nasu-Tada K, Koizumi S, Tsuda M et al (2006) Possible involvement of increase in spinal fibronectin following peripheral nerve injury in upregulation of microglial P2X(4), a key molecule for mechanical allodynia. Glia 53:769–775
Nimmerjahn A, Kirchhoff F, Helmchen F (2005) Resting microglial cells are highly dynamic surveillants of brain parenchyma in vivo. Science 308:1314–1318
Pocock JM, Kettenmann H (2007) Neurotransmitter receptors on microglia. Trends Neurosci 30:527–535
Ren K, Dubner R (2010) Interactions between the immune and nervous systems in pain. Nat Med 16:1267–1276
Salter MW, Kalia LV (2004) Src kinases: a hub for NMDA receptor regulation. Nat Rev Neurosci 5:317–328
Schwab JM, Guo L, Schluesener HJ (2005) Spinal cord injury induces early and persistent lesional P2X4 receptor expression. J Neuroimmunol 163:185–189
Suter MR, Wen YR, Decosterd I et al (2007) Do glial cells control pain? Neuron Glia Biol 3:255–268
Tamura T, Yanai H, Savitsky D et al (2008) The IRF family transcription factors in immunity and oncogenesis. Annu Rev Immunol 26:535–584
Trang T, Beggs S, Wan X et al (2009) P2X4-receptor-mediated synthesis and release of brain-derived neurotrophic factor in microglia is dependent on calcium and p38-mitogen-activated protein kinase activation. J Neurosci 29:3518–3528
Tsuda M, Shigemoto-Mogami Y, Koizumi S et al (2003) P2X4 receptors induced in spinal microglia gate tactile allodynia after nerve injury. Nature 424:778–783
Tsuda M, Inoue K, Salter MW (2005) Neuropathic pain and spinal microglia: a big problem from molecules in "small" glia. Trends Neurosci 28:101–107
Tsuda M, Toyomitsu E, Komatsu T et al (2008a) Fibronectin/integrin system is involved in P2X (4) receptor upregulation in the spinal cord and neuropathic pain after nerve injury. Glia 56:579–585
Tsuda M, Tozaki-Saitoh H, Masuda T et al (2008b) Lyn tyrosine kinase is required for P2X (4) receptor upregulation and neuropathic pain after peripheral nerve injury. Glia 56:50–58
Tsuda M, Kuboyama K, Inoue T et al (2009a) Behavioral phenotypes of mice lacking purinergic P2X4 receptors in acute and chronic pain assays. Mol Pain 5:28
Tsuda M, Masuda T, Kitano J et al (2009b) IFN-gamma receptor signaling mediates spinal microglia activation driving neuropathic pain. Proc Natl Acad Sci U S A 106:8032–8037
Tsuda M, Toyomitsu E, Kometani M et al (2009c) Mechanisms underlying fibronectin-induced upregulation of P2XR expression in microglia: distinct roles of PI3K-Akt and MEK-ERK signaling pathways. J Cell Mol Med 13:3251–3259

Tsuda M, Tozaki-Saitoh H, Inoue K (2012) Purinergic system, microglia and neuropathic pain. Curr Opin Pharmacol 12:74–79

Tsuda M, Masuda T, Tozaki-Saitoh H et al (2013) Microglial regulation of neuropathic pain. J Pharmacol Sci 121:89–94

Ulmann L, Hatcher JP, Hughes JP et al (2008) Up-regulation of P2X4 receptors in spinal microglia after peripheral nerve injury mediates BDNF release and neuropathic pain. J Neurosci 28:11263–11268

Ulmann L, Levavasseur F, Avignone E et al (2013) Involvement of P2X4 receptors in hippocampal microglial activation after status epilepticus. Glia 61:1306–1319

Watkins LR, Milligan ED, Maier SF (2001) Glial activation: a driving force for pathological pain. Trends Neurosci 24:450–455

Woolf CJ, Salter MW (2000) Neuronal plasticity: increasing the gain in pain. Science 288:1765–1769

Zhang Z, Zhang Z, Artelt M et al (2007) Dexamethasone attenuates early expression of three molecules associated with microglia/macrophages activation following rat traumatic brain injury. Acta Neuropathol (Berl) 113(6):675–682

Zhang Z, Zhang Z, Fauser U et al (2008) Mechanical allodynia and spinal up-regulation of P2X4 receptor in experimental autoimmune neuritis rats. Neuroscience 152:495–501

Chapter 49
The Role of Innate Immunity in Ischemic Stroke

Takashi Shichita, Minako Ito, Rimpei Morita, and Akihiko Yoshimura

Abstract Inflammation is an essential step in the progression of ischemic stroke pathology. Various mechanisms of innate immunity are implicated in the acute sterile inflammation following ischemic brain injury. Ischemic insults induce the necrotic death of brain cells, which leads to the extracellular release of danger-associated molecular patterns (DAMPs). DAMPs are recognised by pattern recognition receptors (PRRs) in innate immune cells. IL-1β and various neurotoxic mediators are produced by the activation of PRRs. Caspase-1 activation and the formation of inflammasome complex are essential for the production of the active form of IL-1β in ischemic stroke. Recent evidence has supported the use of new therapeutic opportunities for ischemic stroke by targeting DAMPs and inflammasome. This review examines the evidence regarding the essential roles of innate immunity in the creation of novel treatments for ischemic stroke.

Keywords Ischemic stroke • Acute sterile inflammation • DAMPs • Pattern recognition receptors • Inflammasome

49.1 Introduction

Ischemic stroke is a leading cause of serious disability all over the world. Although significant effort has gone toward the development of an effective therapeutic method, there are few established strategies other than reperfusion therapy with tissue plasminogen activator (t-PA), the only United States Federal Drug Administration (FDA)-approved treatment. Considering the increase in the aged

T. Shichita (✉)
Department of Microbiology and Immunology, School of Medicine, Keio University, 35 Shinanomachi, Shinjuku-ku, Tokyo 160-8582, Japan

PRESTO, Japan Science and Technology Agency , Kawaguchi 332-0012, Japan
e-mail: to.7-chin@mba.ocn.ne.jp

M. Ito • R. Morita • A. Yoshimura
Department of Microbiology and Immunology, School of Medicine, Keio University, 35 Shinanomachi, Shinjuku-ku, Tokyo 160-8582, Japan

M. Miyasaka, K. Takatsu (eds.), *Chronic Inflammation*,
DOI 10.1007/978-4-431-56068-5_49

population in developed and emerging countries, it is very important to develop a new-generation therapy for ischemic stroke in order to improve the prognosis of ischemic stroke patients.

Ischemic stroke involves the death (mostly through necrosis) of cerebral tissue as a result of severe ischemia. This process is so rapid that neuroprotective agents cannot work sufficiently. Reperfusion therapy using t-PA offers significant neuroprotection against ischemic brain damage if provided within 4.5 h after stroke onset. In this context, early treatment is a fundamental principle in ischemic stroke treatment, as emphasised by the phrase ‘time is brain’. In order to broaden this therapeutic time window, inflammation following ischemic brain injury has recently been studied (Macrez et al. 2011; Iadecola and Anrather 2011).

Postischemic inflammation, which lasts for several days after stroke onset, induces the secondary progression of ischemic stroke pathology. Inflammation keeps blood vessels leaky and increases extracellular fluids in brain tissue. Edematous brain tissue compresses normal tissue around the infarct area, which leads to secondary ischemia that causes further neuronal death and enlarges infarct volume. Severe brain oedema is the cause of exaggerated neurological deficits and is often fatal for stroke patients.

Inflammation enhances the infiltration of circulating immune cells into the brain through the production of various chemokines and the formation of leaky microvasculature (Eltzschig and Eckle 2011). Neutrophils and monocytes are major players in postischemic inflammation (Benakis et al. 2015). T and B lymphocytes and other types of blood cells also play unexpected roles in the ischemic brain (Hurn et al. 2007; Shichita et al. 2009). Although these immune cells produce various inflammatory mediators for several days after stroke onset, the resolution of inflammation turns these immune cells into neuroprotective cells for tissue repair (Fig. 49.1) (Shichita et al. 2014). Regulating the postischemic inflammation appropriately (i.e., suppressing the neurotoxic inflammation but promoting the resolution) is considered important for ischemic stroke treatment. In this review, we introduce the role of innate immunity and its mechanisms that provide opportunities to develop novel ischemic stroke treatment.

49.2 Acute Sterile Inflammation After Ischemic Stroke

Innate immunity is essential for host defence against microbes. As such, innate immune cells can distinguish nonself molecules from self (endogenous) molecules (Kawai and Akira 2011). Nonself molecules, which include nucleic acids, lipoproteins, and lipopolysaccharide (LPS) derived from microbes, are recognised by the pattern recognition receptor (PRR) on immune cells. To date, various types of PRRs have been identified. Toll-like receptor (TLR), which is the most well-known PRR, is essential for host defence against various microbes. Compounds of microbes activate immune cells via TLR and induce the inflammation that eliminates microbes and enhances tissue repair.

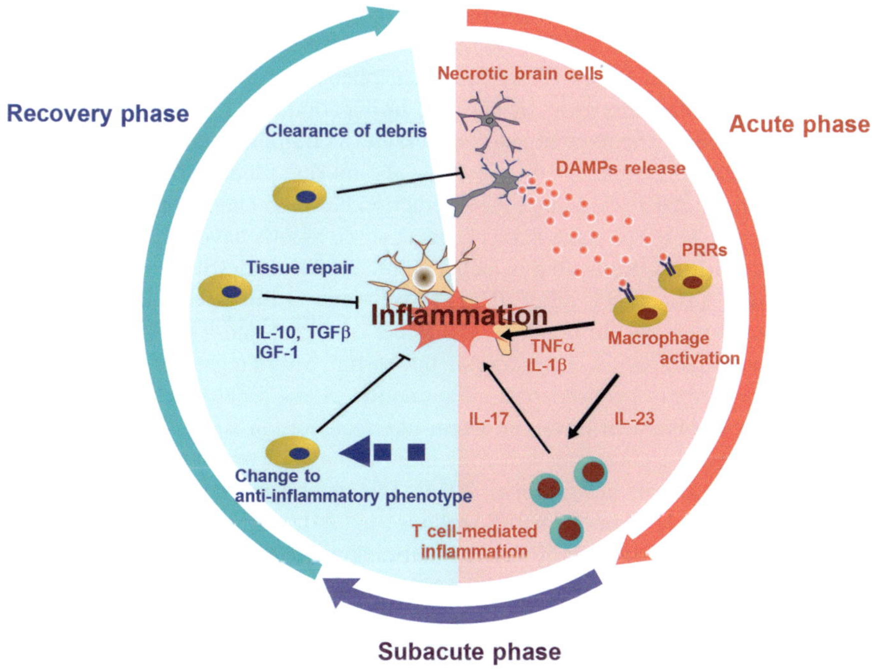

Fig. 49.1 Mechanisms of inflammatory responses after ischemic stroke. Ischemic insults cause the necrotic death of brain cells. DAMPs are released extracellularly from dying cells; they activate infiltrating macrophages and neutrophils through PRRs signalling. Various inflammatory mediators and T-cell–mediated inflammation promote the secondary progression of ischemic brain damage. In the recovery phase, immune cells, which acquire the anti-inflammatory phenotype, promote tissue repair and debris removal in infarcts

The brain is a sterile organ in which there are no pathogens. Inflammation after stroke is thus sterile inflammation that is induced by activated immune cells without microbial compounds. In this situation, immune cells are believed to be activated by endogenous molecules, which are called danger-associated molecular patterns (DAMPs) (Shichita et al. 2014; Kono and Rock 2008). There are two types of DAMPs: one is an intracellular compound released from dying cells due to the loss of membrane integrity and the other is an extracellular compound modified by the enzymes from dying cells. Various types of molecules have been identified as DAMPs that are recognised by PRRs. These danger signals from dying cells can activate immune cells via PRRs, and thus induce acute sterile inflammation.

The major signalling pathways to recognise DAMPs differ by disease pathology. In ischemic stroke, it appears that TLR2 and TLR4 play an essential role, given the many reports on the importance of TLR2 and TLR4 in postischemic inflammation (Suzuki et al. 2012; Ziegler et al. 2011; Marsh et al. 2009). A deficiency in either or both reduces infarct volume and decreases the inflammatory cytokine expression in infiltrating immune cells (Shichita et al. 2012). Neuronal TLR2 and TLR4 are also implicated in ischemic brain injury (Tang et al. 2007). Although Myd88, which is a critical adaptor molecule for most TLR signalling, is critical for the inflammatory cytokine expression of infiltrating immune cells in the ischemic brain, a deficiency of Myd88 is reported to reveal an equivocal phenotype in the murine model of ischemic stroke (Famakin et al. 2011). The implication of TLRs other than TLR2 and TLR4 has been investigated using knockout (KO) mice. A deficiency in TLR3 or TLR9 shows no significant difference in the murine model of ischemic stroke (Hyakkoku et al. 2010). Although the interaction between endogenous ligands and other TLR3 and TLR9 remains to be clarified, some reports have suggested that the activation of TLR9 is implicated in ischemic preconditioning (Stevens et al. 2008). It is possible that TLRs other than TLR2 and TLR4 have unknown anti-inflammatory functions. In addition, the effect of TLR2 and TLR4 deficiency in postischemic inflammation is remarkable but partial. This suggests that other PRRs are also implicated in postischemic inflammation. Some recent reports have demonstrated the involvement of c-type lectin-like receptors in ischemic stroke. Mannose binding lectin exaggerates ischemic brain damage by promoting intravascular thrombus formation (de la Rosa et al. 2014). Mincle and Syk signalling is reported to enhance ischemic brain injury (Suzukio et al. 2013). Thus, various PRRs are associated with the induction of sterile inflammation after ischemic stroke. Future investigations are expected to unveil the previously unknown function of PRRs in postischemic inflammation.

49.3 DAMPs in Ischemic Stroke

There are many reports about the role of DAMPs in sterile inflammation. The molecules that function as DAMPs include nucleic acids, lipids, proteins, and so on; they are hidden intracellularly or extracellularly from immune cells in normal conditions (Shichita et al. 2014; Kono and Rock 2008). Cells dying due to tissue injury expose them to immune cells. Intracellular compounds (self DNA, RNA, lipids, and proteins) are released into the extracellular space, and some of them are recognised by PRRs on immune cells (Fig. 49.2). Tissue injury modifies some lipids and proteins (e.g., the oxidisation of lipids, glycation of proteins, and degradation of the extracellular matrix). The necrotic death of ischemic brain cells induces the release of intracellular compounds and results in drastic changes in the phospholipids and extracellular matrix in the brain. These molecules function as DAMPs in postischemic sterile inflammation.

The compounds in the extracellular matrix are well known to function as DAMPs (Frey et al. 2013). Fibrinogen and proteoglycans (hyaluronan, biglycan, and versican) are reported to activate TLRs and initiate inflammatory responses (Petrey and de la Motte 2014; Barbelova et al. 2009). Although they are generated after ischemic stroke, the functions of these molecules in the ischemic brain have not yet been elucidated sufficiently (Al'Qteishat et al. 2006; Tang et al. 2014). Some reports have suggested the neuroprotective roles of these proteoglycans. The administration of small synthetic hyaluronan disaccharides reduces infarct volume

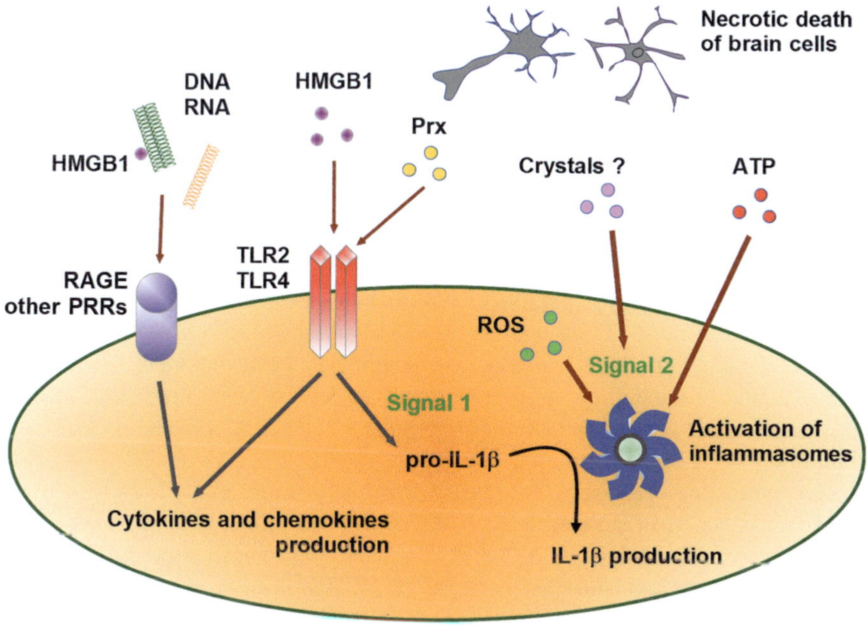

Fig. 49.2 DAMPs and pattern recognition receptors in postischemic inflammation. Various DAMPs are released into the extracellular space from dying cells. These DAMPs are recognised by pattern recognition receptors (PRRs) and induce the production of inflammatory mediators. Pro-IL-1β (inactive form) is produced by DAMPs-PRRs signalling (Signal 1). Pro-IL-1β is cleaved by activated caspase-1 to mature IL-1β (active form), which needs the activation of inflammasomes induced by stimuli (Signal 2) other than PRRs signalling

(Eqea et al. 2014). Biglycan is reported to have neurotrophic activity (Junghans et al. 1995). Versican and other peptidoglycans are important for neuronal migration in the developing brain (Maeda 2015). Further investigation is needed to determine the precise functions of the compounds in the cerebral extracellular matrix.

Many reports about intracellular DAMPs have been summarised previously elsewhere (Shichita et al. 2014). In ischemic stroke, high mobility group box 1 (HMGB1) and peroxiredoxin (Prx) family proteins have been reported to be DAMPs. There are differences in the roles and the phases between these two DAMPs (Shichita et al. 2012). HMGB1 is included in the nucleus in normal brain cells. After ischemic injury, HMGB1 is translocated into cytoplasm and released into the extracellular space by necrotic cell death (Zhang et al. 2011). This release of HMGB1 occurs within 2–4 h after the onset of ischemic stroke and exaggerates the blood–brain barrier breakdown (Zhang et al. 2011; Qiu et al. 2008). Extracellular HMGB1 is recognised by TLRs and receptor for advanced glycation end-products (RAGE). Both HMGB1-TLR and HMGB1-RAGE pathways have a neurotoxic effect in ischemic brain injury (Muhammad et al. 2008). The inhibition of HMGB1 by antibody or small interfering RNA (siRNA) attenuates ischemic brain damage (Kim et al. 2012). The intravenous administration of immunoglobulin also reveals a neuroprotective effect by modulating HMGB1-induced TLR and RAGE signalling pathways (Lok et al. 2015). Furthermore, serum levels of HMGB1 and soluble RAGE are associated with poor functional outcome in ischemic stroke patients (Sapojnikova et al. 2014).

Recently, the link between HMGB1 and systemic responses in ischemic stroke has been suggested. After ischemic stroke, immunosuppression and severe splenic atrophy has been reported (Offner et al. 2006). The RAGE activation by circulating HMGB1 induces the functional exhaustion of mature monocytes and lymphopenia (Liesz et al. 2015). The depletion of lymphocytes decreases the serum level of HMGB1 (Gu et al. 2013). In addition to these inflammatory effects of HMGB1, HMGB1 production from reactive astrocytes has been reported to promote neurovascular remodelling and functional recovery after ischemic stroke (Hayakawa et al. 2012). Thus, broad functions of HMGB1 released by ischemic brain injury have been identified.

The extracellular release of Prx family proteins functions as DAMPs, which has been demonstrated in several articles (Shichita et al. 2012; Mullen et al. 2015; Salzano et al. 2014; Riddell et al. 2010). Classically, Prx family proteins have been investigated as antioxidants that catalyse the reduction of hydrogen peroxide to water. Prx family proteins, which have six isoforms (Prx1-6), are almost ubiquitously expressed in the body, and relatively high amounts of Prxs are involved in brain tissue (Seo et al. 2000). Various injuries (e.g., infection, hypoxia, degeneration) increase the intracellular expression of Prxs, which has protective effects against oxidative stress. Intracellular Prxs have a neuroprotective role in neurodegenerative disease and ischemic stroke (Hu et al. 2011; Hwang et al. 2010). However, the necrotic death of brain cells releases Prxs into the extracellular space. These extracellular Prxs are recognised by TLR2 and TLR4 in immune cells (Shichita et al. 2012). This is one of the mechanisms that initiate postischemic inflammation. A recent article has demonstrated that ligustilide has neuroprotective effects in ischemic stroke by inhibiting the extracellular release of Prx6 and decreasing TLR4 activation (Kuang et al. 2014). In addition, the active secretion of oxidised Prx2 and thioredoxin (Trx) from LPS-activated macrophages has been reported to induce inflammatory responses (Salzano et al. 2014). Thus, multiple functions of Prxs in ischemic stroke have been identified.

Another difference between HMGB1 and Prx family proteins acting as DAMPs has recently been pointed out. Although circulating HMGB1 can be detected in the serum of human ischemic stroke patients, the levels of circulating Prx family proteins do not correlate with the severity of human ischemic stroke (Kunze et al. 2014). An increased level of Prxs has been demonstrated in the extracellular fluid of ischemic brain tissue and cerebrospinal fluid (Kuang et al. 2014; Dayon et al. 2011). These observations may indicate the unique mechanisms for scavenging DAMPs in the ischemic brain.

49.4 IL-1β, Inflammasome and Brain Injury

Ischemic damage results in leaky microvasculature which enhances the infiltration of immune cells. Neutrophils and macrophages are activated by DAMPs signalling and produce neurotoxic mediators (cytokines, elastase, and oxygen metabolites). Among these, IL-1β is one of the most important mediators in sterile inflammation (Chen et al. 2007). IL-1β is a neurotoxic cytokine that directly induces neuronal cell death and induces the production of inflammatory mediators in microglia and astrocytes (Allan et al. 2005). Although a deficiency in the IL-1β gene does not result in a significant attenuation in ischemic brain damage, both IL-1α and IL-1β are essential for ischemic stroke (Boutin et al. 2001). The secretion of IL-1β from immune cells needs both PRR activation and caspase-1 (IL-1β converting enzyme) activation, which cleaves pro-IL-1β (inactivated form) to process it into an active 17 kDa form (IL-1β). The deficiency or inhibition of caspase-1 significantly attenuates ischemic brain damage (Hara et al. 1997). These observations suggest the IL-1β–independent neurotoxic effect of caspase-1, although the discrepancy in phenotype between IL-1β–deficient mice and caspase-1–deficient mice in ischemic stroke should be clarified by future investigations.

The activation of inflammasome in ischemic brain injury has been examined as a novel therapeutic target. Inflammasome comprises the large protein complexes (~700 kDa) realised by recruiting ASC (an apoptosis-associated speck-like protein containing a caspase recruitment domain) and caspase-1 (Martinon et al. 2002). Activated inflammasome triggers the autocatalytic cleavage of caspase-1 to process pro-IL-1β and pro-IL-18 into their active form (Keller et al. 2008). Although the role of IL-18 in ischemic stroke is marginal, inflammasome is also implicated in a form of cell death called pyroptosis (Wheeler et al. 2003). The induction of pyroptosis in brain cells may be another neurotoxic mechanism by inflammasome activation.

49.5 Mechanism of Inflammasome Activation

There are two categories of inflammasomes, the NLR (nucleotide binding and oligomerisation domain and leucine-rich repeat containing) family and the PYHIN (pyrin domain and HIN domain containing) family. NLRP1, NLRP3, and NLRC4 inflammasomes, belonging to the NLR family, are related to the induction of inflammation (Fann et al. 2013; Yang et al. 2014). NLRP1 is mainly expressed in neurons and NLRP1 inflammasome activation causes the pyroptosis of neurons. NLRP3 is mainly expressed in immune cells and is activated to produce mature IL-1β. By a recent study, AIM2 (Absent in melanoma 2) inflammasome, belonging to the PYHIN family, has been also demonstrated to be implicated in ischemic brain injury (Denes et al. 2015; Hornung et al. 2009). Thus, inflammasome activation is important for sterile inflammation after ischemic stroke.

NLRP3 inflammasome is activated by various stimuli, such as the rupture of lysosome membranes by the phagocytosis of crystals (e.g., monosodium urate [MSU]), alum, silica, and cholesterol) (Hornung et al. 2008; Martinon et al. 2006; Duewell et al. 2010), the disturbance of cellular ion balance (K^+ or Ca^{2+}) by ATP or nigericin, and the release of mitochondrial reactive oxygen species (ROS) into the cytoplasm (Fig. 49.2) (Mariathasan et al. 2006; Lee et al. 2012). Furthermore, two recent studies have demonstrated that the oligomeric complex of NLRP3 inflammasome is extracellularly released and functions as a danger signal (Franklin et al. 2014; Baroja-Mazo et al. 2014). Although the specific trigger of inflammasome activation in the ischemic brain has not yet been identified, it is known that drastic changes in cerebral circumstance by ischemic insults activate various inflammasomes.

49.6 Involvement of BTK in the Inflammasome Activation and Stroke

Various protein kinases are also required for the activation of inflammasomes. PKR, PKC, DAPK, IRAK, Syk, and JNK have been reported to be important for regulating inflammasome activation (Lu et al. 2012; Qu et al. 2012; Chuang et al. 2011; Lin et al. 2014; Gross et al. 2009; Okada et al. 2014). Recently, Bruton's tyrosine kinase (BTK) was identified as a previously unknown modulator of NLRP3 inflammasome (Ito et al. 2015). BTK, a member of the Tec family of nonreceptor tyrosine kinases, is mainly expressed in B cells and myeloid cells. BTK physically interacts with both NLRP3 and ASC, resulting in the promotion of ASC oligomerisation in a kinase activity-dependent manner. BTK functions as a platform protein for the interaction between NLRP3 and ASC, which promotes caspase-1 activation and IL-1β secretion (Fig. 49.3). BTK is activated in infiltrating macrophages/neutrophils 24 h after the onset of ischemic stroke, coincident with the activation of NLRP3 and caspase-1. Ibrutinib is a potent covalent inhibitor of BTK that was recently approved by the United States Food and Drug Administration (FDA) for the treatment of mantle cell lymphoma (MCL) and chronic lymphocytic leukemia (CLL). The administration of ibrutinib, which is therapeutic up to 12 h after stroke onset, significantly reduces infarct volume and improves neurological deficits by suppressing caspase-1 activation and IL-1β secretion. These results may enhance our understanding of the mechanism underlying NLRP3 inflammasome activation. Ibrutinib, a BTK inhibitor, has been shown to be well tolerated by patients and to have few adverse effects. As such, ibrutinib could also be effective against inflammasome-dependent diseases, including ischemic stroke.

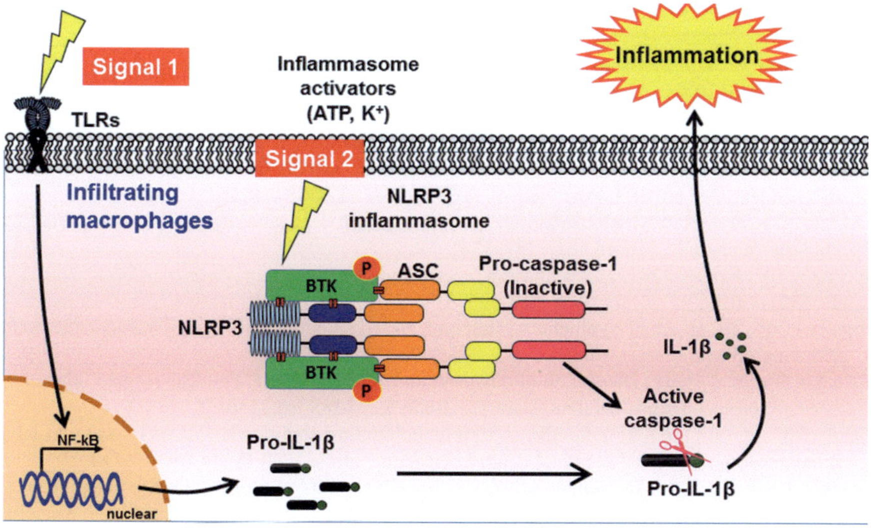

Fig. 49.3 Significance of BTK in NLRP3 inflammasome cascade. BTK, which is activated by Signal 1 (DAMPs-PRRs signaling), interacts with ASC. Activated BTK enhances both the oligomerisation of ASC and the formation of the ASC-NLRP3 complex, and this strengthens the function of the NLRP3 inflammasome cascade

49.7 Conclusion

The cellular processes underlying acute sterile inflammation after ischemic stroke have only recently been coming to light. Future investigation could identify previously unknown therapeutic targets in postischemic inflammation and provide a broader therapeutic time window for the treatment of ischemic stroke.

References

Al'Qteishat A et al (2006) Changes in hyaluronan production and metabolism following ischaemic stroke in man. Brain 129:2158–2176

Allan SM et al (2005) Interleukin-1 and neuronal injury. Nat Rev Immunol 5:629–640

Barbelova A et al (2009) Biglycan, a danger signal that activates the NLRP3 inflammasome via Toll and P2X receptors. J Biol Chem 284:24035–24048

Baroja-Mazo A et al (2014) The NLRP3 inflammasome is released as a particulate danger signal that amplifies the inflammatory response. Nat Immunol 15:738–748

Benakis C et al (2015) The role of microglia and myeloid immune cells in acute cerebral ischemia. Front Cell Neurosci 8:461

Boutin H et al (2001) Role of Il-1alpha and IL-1beta in ischemic brain damage. J Neurosci 21:5528–5534

Chen CJ et al (2007) Identification of a key pathway required for the sterile inflammatory response triggered by dying cells. Nat Med 13:851–856

Chuang YT et al (2011) Tumor suppressor death-associated protein kinase is required for full IL-1beta production. Blood 117:960–970
Dayon L et al (2011) Brain extracellular fluid protein changes in acute stroke patients. J Proteome Res 10:1043–1051
de la Rosa X et al (2014) Mannose-binding lectin promotes local microvascular thrombosis after transient brain ischemia in mice. Stroke 45:1453–1459
Denes A et al (2015) AIM2 and NLRC4 inflammasomes contribute with ASC to acute brain injury independently of NLRP3. Proc Natl Acad Sci U S A 112:4050–4055
Duewell P et al (2010) NLRP3 inflammasomes are required for atherogenesis and activated by cholesterol crystals. Nature 464:1357–1361
Eltzschig HK, Eckle T (2011) Ischemia and reperfusion—from mechanism to translation. Nat Med 17:1391–1401
Eqea J et al (2014) Small synthetic hyaluronan disaccharides afford neuroprotection in brain ischemia-related models. Neuroscience 265:313–322
Famakin BM et al (2011) Disruption of downstream MyD88 or TRIF Toll-like receptor signaling does not protect against cerebral ischemia. Brain Res 1388:148–156
Fann DY et al (2013) Intravenous immunoglobulin suppresses NLRP1 and NLRP3 inflammasome-mediated neuronal death in ischemic stroke. Cell Death Dis 4, e790
Franklin BS et al (2014) The adaptor ASC has extracellular and 'prionoid' activities that propagate inflammation. Nat Immunol 15:727–737
Frey H et al (2013) Biological interplay between proteoglicans and their innate immune receptors in inflammation. FEBS J 280:2165–2179
Gross O et al (2009) Syk kinase signalling couples to the Nlrp3 inflammasome for anti-fungal host defence. Nature 459:433–436
Gu L et al (2013) T cells contribute to stroke-induced lymphopenia in rats. PLoS One 8, e59602
Hara H et al (1997) Inhibition of interleukin 1beta converting enzyme family proteases reduces ischemic and excitotoxic neuronal damage. Proc Natl Acad Sci U S A 94:2007–2012
Hayakawa K et al (2012) Astrocytic high-mobility group box 1 promotes endothelial progenitor cell-mediated neurovascular remodeling during stroke recovery. Proc Natl Acad Sci U S A 109:7505–7510
Hornung V et al (2008) Silica crystals and aluminum salts activate the NALP3 inflammasome through phagosomal destabilization. Nat Immunol 9:847–856
Hornung V et al (2009) AIM2 recognizes cytosolic dsDNA and forms a caspase-1-activating inflammasome with ASC. Nature 458:514–518
Hu X et al (2011) Peroxiredoxin-2 protects against 6-hydroxydopamine-induced dopaminergic neurodegeneration via attenuation of the apoptosis signal-regulating kinase (ASK1) signaling cascade. J Neurosci 31:247–261
Hurn PD et al (2007) T- and B-cell-deficient mice with experimental stroke have reduced lesion size and inflammation. J Cereb Blood Flow Metab 27:1798–1805
Hwang IK et al (2010) Changes in the expression of mitochondrial peroxiredoxin and thioredoxin in neurons and glia and their protective effects in the experimental cerebral ischemic damage. Free Radic Biol Med 48:1242–1251
Hyakkoku K et al (2010) Toll-like receptor 4 (TLR4), but not TLR3 or TLR9, knock-out mice have neuroprotective effects against focal cerebral ischemia. Neuroscience 171:258–267
Iadecola C, Anrather J (2011) The immunology of stroke: from mechanisms to translation. Nat Med 17:796–808
Ito M et al (2015) Bruton's tyrosine kinase (BTK) is essential for NLRP3 inflammasome activation and therapeutic for ischemic brain injury. Nat Commun 6:7360
Junghans U et al (1995) Purification of a meningeal cell-derived chondroitin sulphate proteoglycan with neurotrophic activity for brain neurons and its identification as biglycan. Eur J Neurosci 7:2341–2350
Kawai T, Akira S (2011) The role of pattern-recognition receptors in innate immunity: update on Toll-like receptors. Nat Immunol 11:373–384

Keller M et al (2008) Active caspase-1 is a regulator of unconventioNLRProtein secretion. Cell 132:818–831

Kim ID et al (2012) Intranasal delivery of HMGB1 siRNA confers target gene knockdown and robust neuroprotection in the post-ischemic brain. Mol Ther 20:829–839

Kono H, Rock KL (2008) How dying cells alert the immune system to danger. Nat Rev Immunol 8:279–289

Kuang X et al (2014) Ligustilide ameliorates neuroinflammation and brain injury in focal cerebral ischemia/reperfusion rats: involvement of inhibition of TLR4/peroxiredoxin 6 signaling. Free Radic Biol Med 71:165–175

Kunze A et al (2014) Peroxiredoxin 5 (PRX5) is correlated inversely to systemic markers of inflammation in acute stroke. Stroke 45:608–610

Lee GS et al (2012) The calcium-sensing receptor regulates the NLRP3 inflammasome through Ca2+ and cAMP. Nature 492:123–127

Liesz A et al (2015) DAMP signaling is a key pathway inducing immune modulation after brain injury. J Neurosci 35:583–598

Lin KM et al (2014) IRAK-1 bypasses priming and directly links TLRs to rapid NLRP3 inflammasome activation. Proc Natl Acad Sci U S A 111:775–780

Lok KZ et al (2015) Intravenous immunoglobulin (IVIg) dampens neuronal toll-like receptor-mediated responses in ischemia. J Neuroinflammation 12:73

Lu B et al (2012) Novel role of PKR in inflammasome activation and HMGB1 release. Nature 488:670–674

Macrez R et al (2011) Stroke and the immune system: from pathophysiology to new therapeutic strategies. Lancet Neurol 10:471–480

Maeda N (2015) Proteoglycans and neuronal migration in the cerebral cortex during development and disease. Front Neurosci 9:98

Mariathasan S et al (2006) Cryopyrin activates the inflammasome in response to toxins and ATP. Nature 440:228–232

Marsh BJ et al (2009) Toll-like receptor signaling in endogenous neuroprotection and stroke. Neuroscience 158:1007–1020

Martinon F et al (2002) The inflammasome: a molecular platform triggering activation of inflammatory caspases and processing of proIL-beta. Mol Cell 10:417–426

Martinon F et al (2006) Gout-associated uric acid crystals activate the NALP3 inflammasome. Nature 440:237–241

Muhammad S et al (2008) The HMGB1 receptor RAGE mediates ischemic brain damage. J Neurosci 28:12023–12031

Mullen L et al (2015) Cysteine oxidation targets peroxiredoxins 1 and 2 for exosomal release through a novel mechanism of redox-dependent secretion. Mol Med. doi:10.2119/molmed.2015.00033

Offner H et al (2006) Splenic atrophy in experimental stroke is accompanied by increased regulatory T cells and circulating macrophages. J Immunol 176:6523–6531

Okada M et al (2014) Lysosome rupture-activated TAK1 JNK pathway regulates NLRP3 inflammasome activation. J Biol Chem 289:32926–32936

Petrey AC, de la Motte CA (2014) Hyaluronan, a crucial regulator of inflammation. Front Immunol 5:101

Qiu J et al (2008) Early release of HMGB-1 from neurons after the onset of brain ischemia. J Cereb Blood Flow Metab 28:927–938

Qu Y et al (2012) Phosphorylation of NLRC4 is critical for inflammasome activation. Nature 490:539–542

Riddell JR et al (2010) Peroxiredoxin 1 stimulates secretion of proinflammatory cytokines by binding to TLR4. J Immunol 184:1022–1030

Salzano S et al (2014) Linkage of inflammation and oxidative stress via release of glutathionylated peroxiredoxin-2, which acts as a danger signal. Proc Natl Acad Sci U S A 111:12157–12162

Sapojnikova N et al (2014) Correlation between MMP-9 and extracellular cytokine HMGB1 in prediction of human ischemic stroke outcome. Biochim Biophys Acta 1842:1379–1384
Seo MS et al (2000) Identification of a new type of mammalian peroxiredoxin that forms an intramolecular disulfide as a reaction intermediate. J Biol Chem 275:20346–20354
Shichita T et al (2009) Pivotal role of cerebral interleukin-17-producing gammadeltaT cells in the delayed phase of ischemic brain injury. Nat Med 15:946–950
Shichita T et al (2012) Peroxiredoxin family proteins are key initiators of post-ischemic inflammation in the brain. Nat Med 18:911–917
Shichita T et al (2014) Post-ischemic inflammation regulates neural damage and protection. Front Cell Neurosci 8:319
Stevens SL et al (2008) Toll-like receptor 9: a new target of ischemic preconditioning in the brain. J Cereb Blood Flow Metab 28:1040–1047
Suzuki Y et al (2012) Pharmacological inhibition of TLR4-NOX4 signal protects against neuronal death in transient focal ischemia. Sci Rep 2:896
Suzukio Y et al (2013) Involvement of Mincle and Syk in the changes to innate immunity after ischemic stroke. Sci Rep 3:3177
Tang SC et al (2007) Pivotal role for neuronal Toll-like receptors in ischemic brain injury and functional deficits. Proc Natl Acad Sci U S A 104:13798–13803
Tang SC et al (2014) Association between plasma levels of hyaluronic acid and functional outcome in acute stroke patients. J Neuroinflammation 11:101
Wheeler RD et al (2003) No role for interleukin-18 in acute murine stroke-induced brain injury. J Cereb Blood Flow Metab 23:531–535
Yang F et al (2014) NLRP3 deficiency ameliorates neurovascular damage in experimental ischemic stroke. J Cereb Blood Flow Metab 34:660–667
Zhang J et al (2011) Anti-high mobility group box-1 monoclonal antibody protects the blood-brain barrier from ischemia-induced disruption in rats. Stroke 42:1420–1428
Ziegler G et al (2011) Blocking TLR2 in vivo protects against accumulation of inflammatory cells and neuronal injury in experimental stroke. J Cereb Blood Flow Metab 31:757–766

Chapter 50
Chronic Neuroinflammation Underlying Pathogenesis of Alzheimer's Disease

Takashi Saito

Abstract Neuroinflammation is involved in neurodegenerative disorders such as amyotrophic lateral sclerosis, Huntington's disease, multiple sclerosis, Parkinson's disease, and Alzheimer disease (AD). AD, the most common form of dementia in the elderly, is characterized by specific pathologies in the brain. Currently, it is thought that longitudinal pathophysiological processes cause AD patients' brains to be in a state of chronic neuroinflammation; additionally, reactive glial cells, the innate immune system in the brain, contribute to AD pathogenesis. However, the detailed molecular and cellular mechanisms underlying this pathogenesis are still unclear because of the disease complexities. In addition, because the existing animal models of AD are not accurate and reproducible, the development of therapeutic strategies for the disease is slow. To overcome these obstacles, we have to revalidate the previously obtained results whether they are beneficial or detrimental, and promote the elucidation of the pathogenic mechanisms of AD using relevant animal models. This review reconsiders insights of the current situation of AD research and shares perspectives for understanding AD pathogenesis when considered as a whole body disorder.

Keywords Alzheimer's disease (AD) • Neuroinflammation • Amyloid cascade hypothesis • Amyloid-β peptide (Aβ) • Tau • Glial cells • Astrocyte • Microglia • Animal model

50.1 Introduction

Alzheimer disease (AD), the primary cause of dementia in the elderly, imposes a tremendous social and economic burden on modern society. The current number of patients with dementia worldwide is estimated to be >35 million and is expected to reach 77 million by 2030 and >110 million by 2050. However, four drugs, three

T. Saito (✉)
Laboratory for Proteolytic Neuroscience, RIKEN Brain Science Institute, 2-1 Hirosawa, Wako-shi 351-0198, Saitama, JAPAN

PRESTO, Japan Science and Technology Agency, Kawaguchi 332-0012, Japan
e-mail: takasai@brain.riken.jp

M. Miyasaka, K. Takatsu (eds.), *Chronic Inflammation*,
DOI 10.1007/978-4-431-56068-5_50

comprising choline esterase inhibitors and one N-methyl-D-aspartate (NMDA) receptor antagonist, have been approved worldwide for clinical use, although their therapeutic effect is a temporary stopgap for a few years. Unfortunately, development of drugs that are capable of ameliorating this disease has been difficult. To overcome these difficulties, elucidation of the mechanism underlying AD pathogenesis and then finding druggable candidates, which modulate the key molecules involved in AD pathogenesis, are necessary. On the other hand, epidemiological observation of people with rheumatoid arthritis has revealed that non-steroidal anti-inflammatory drugs (NSAIDs) may decrease the risk of AD (McGeer et al. 1996). Furthermore, in a cohort study of approximately 7000 subjects without dementia (age $\geq$ 55 years), the long-term use of NSAIDs potentially protected subjects against AD but not against vascular dementia (in't Veld et al. 2001). However the protective mechanism by such NSAID treatment is still unclear; inflammation in the brain or in the peripheral body might be involved in AD pathogenesis.

50.2 Neuroinflammation in Brains with AD

Neuroinflammation is caused by infectious insults, injury, or disease, particularly in the acute phase, and its function is to protect the central nervous system (CNS). Because the CNS is segregated from the peripheral region by the blood–brain barrier (BBB), glial cells, such as astrocytes and microglia, act as immune cells in the brain. Astrocytes have multiple functions, including supporting the endothelial cells that form neurovascular units as the BBB, supplying nutrients to CNS, maintaining the extracellular ion/water balance, and repairing or remodelling tissue during the process of neuroinflammation. Microglia perform macrophage-like phagocytic actions to remove pathogens in the brain and to protect the neurons from these toxic species. However, microglia as the *Janus* generate and release low-molecular–weight, toxic species such as reactive oxygen species (ROS) or nitric oxide (NO) that result in microglia-mediated neurotoxicity (Block et al. 2007). These glial cells generate proinflammatory cytokines and chemokines to respond to danger signals and communicate thoroughly with each other, thus serving as a guardian of the brain environment (Burda and Sofroniew 2014).

Two pathological hallmarks of AD in the brain are senile plaques and neurofibrillary tangles (NFTs), which are formed during the process of AD pathogenesis. Senile plaques and NFTs start appearing >25 years and approximately 15 years before the onset of AD, respectively (Bateman et al. 2012: Fig. 50.1). Because these protein aggregates are thought to be a danger signal, accumulation of reactive microglia (microgliosis) and reactive astrocyte (astrocytosis) occur around the senile plaques or NFTs. Thus, glial cells are chronically activated in the brain even before the onset of AD (Maeda et al. 2011). Chronic activation of glial cells, particularly microglias, contribute to a detrimental environment in healthy brain tissues, that is, chronic neuroinflammation elicits AD pathogenesis.

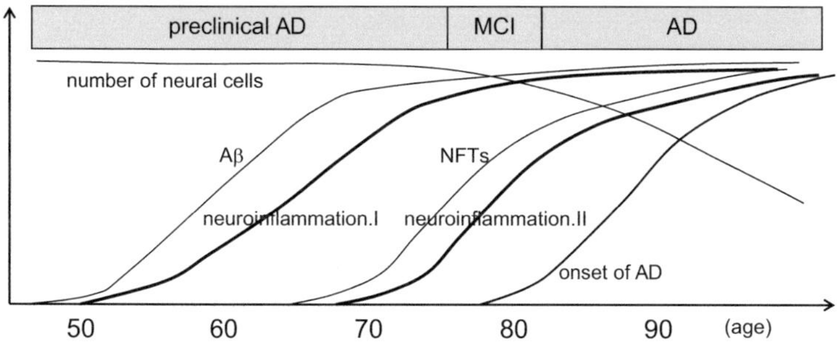

Fig. 50.1 Chronological progression of AD pathogenesis. Amyloid deposition begins in the preclinical stage, followed by neurofibrillary tangle (NFT) formation, which leads to neurodegeneration and neuronal cell death and direct triggering of the onset of AD via the stage of mild cognitive impairment (MCI). Neuroinflammation (I and II) may facilitate AD pathogenesis in the brain

50.3 AD Pathogenesis and Chronic Neuroinflammation

Amyloid-β peptide (Aβ), a major component of senile plaques, are proteolytically generated from amyloid precursor protein (APP;De Strooper et al. 2010), and serve as the primary trigger not only for the onset of AD, but also for the normal ageing of the brain (Jonsson et al. 2012). Four major forms of the C-terminal variants of Aβ exist, Aβx-38, Aβx-40, Aβx-42, and Aβx-43, with Aβx-42 (and Aβx-43) being more neurotoxic due to its higher hydrophobicity, which leads to faster aggregation (Blennow et al. 2006; Saito et al. 2011). Mutations associated with early onset familial AD (FAD) have been identified in the APP and presenilin genes. These mutations lead to an accelerated production both of Aβx-42 and Aβx-43, and an increase in the Aβx-42/Aβx-38 and Aβx-43/Aβx-40 ratio, respectively. Since steady-state levels of Aβx-42 are much higher than Aβx-43 in the brain, these findings indicate that Aβx-42 plays an important role in AD pathogenesis. Another hallmark of AD, NFTs comprises tau proteins, which are microtubule-associated proteins that stabilise microtubules in neurons. Pathological tau proteins are hyperphosphorylated biochemically. A number of mutations of the tau gene, for example, P301L, P301S, and other mutants, present as frontotemporal dementia with Parkinsonism linked to chromosome 17 (FTDP-17), which is also a progressive neurodegenerative disorder that affects behavior, language, and movement, although it is different from dementia associated with AD (Goedert et al. 2000). In AD, extracellular Aβ deposition precedes intracellular wild-type tau deposition with hyperphosphorylation at the cortical region in the brain followed by irreversible brain atrophy caused by longitudinal processes of neurodegeneration and neuronal cell death. These chronological and pathological progressions of AD established a working hypothesis that is the amyloid cascade hypothesis (Hardy and Higgins 1992; Fig. 50.2). Although the pathological molecular mechanisms connecting Aβ deposition to NFT formation or NFTs formation to neurodegeneration and neural cell death remain unclear, neuroinflammation is

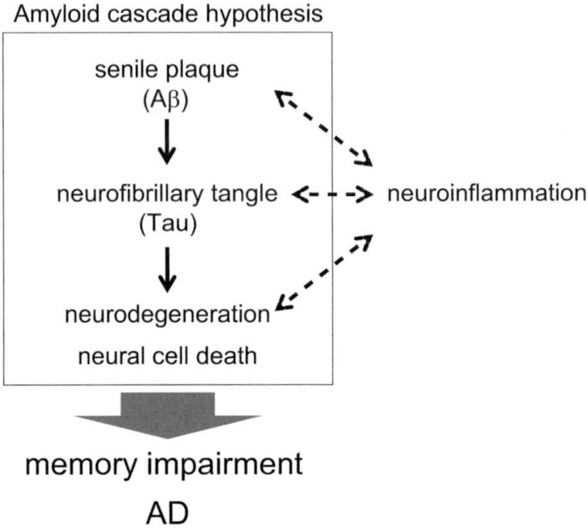

Fig. 50.2 Amyloid cascade hypothesis and neuroinflammation. Based on genetic studies, APP and PS1 mutations accelerate toxic Aβ production, leading to neurodegeneration and neuronal cell death via the formation of NFTs. However, detailed molecular mechanisms remain unclear; neuroinflammation may be a key player in the development of AD pathology

currently recognised as an important pathophysiological event in AD pathogenesis (Fig. 50.2).

When, where, and why does neuroinflammation appear in the brain? Aβ and tau deposition begins in the presymptomatic stage, recently termed preclinical AD, before the onset of AD (Fig. 50.1). Because in vivo imaging of inflammation is difficult in the human brain due to technical limitations, the researchers performed immunohistochemical analyses using postmortem AD brain tissues. Such postmortem AD brain cells exhibit astrocytosis and microgliosis around the senile plaques and NFTs. These pathological signs were detected as the endpoint markers of AD in the brain. Although recent advances in research have enabled cyclooxigenase-2 (COX-2)-positron emission tomography (PET) and translocator protein (TSPO)-PET *in vivo* imaging of neuroinflammation in the mouse brain (Ji et al. 2013; Yasuno et al. 2012), immunohistochemical analyses using the brain from AD mouse models are necessary due to the limitations of spatiotemporal resolution of live imaging. APP transgenic (Tg) model mice with Aβ amyloidosis but without NFTs exhibited amyloid plaque formation followed by gliosis in response to the amyloid depositions (Games et al. 1995). On the other hand, tau Tg mice with the FTDP-17 mutations, for example, P301L-tau Tg mice and mThy-1 3R tau mice, formed NFT-like tau aggregates but not amyloid plaques followed by the accumulation of reactive glial cells around the aggregates (Lewis et al. 2000; Rockenstein et al. 2015). This evidence suggests that activated glial cells remove abnormal aggregates in the brain as part of its defence system. However, it is still unclear which and how glial cells, astrocyte or microgliamicroglia, recognize such abnormal aggregates as danger signals, especially considering that the communication between astrocyte and microglia is strictly regulated by cytokines and chemokines. The author suggests that the receptivity and responsiveness of glial cells against amyloid plaques and NFTs may be different. Therefore, neuroinflammation I and II have been included in Fig. 50.1. Furthermore, dysregulation of glial communication at the stage when both amyloid and tau pathologies exist in the brain along with long-lasting abnormal activation of glial cells may facilitate AD pathogenesis, including neuroinflammation II. Interestingly, P301S-tau Tg mice exhibit reactive gliosis before tau aggregation, and immunosuppression of P301S-tau Tg mice with FK506 attenuates tau pathology (Yoshiyama et al. 2007), which suggests that neuroinflammation, particularly neuroinflammation I as shown in Fig. 50.1, could possibly connect amyloid pathology and NFTs formation. Taken together, chronic neuroinflammation is thought to be a key player in the process of AD pathogenesis.

50.4 Inflammatory Regulators in AD Pathogenesis

Inflammatory response in the brain is distinguished by the activation of the glial cells and the release of proinflammatory mediators such as cytokines and chemokines, as well as the innate immune system in the peripheral region. During the acute stage of inflammation, cytokines (in particular), interleukin (IL)-1β, tumour necrosis factor-α, and interferon-γ, work as major effectors. During brain ageing, secreted IL-6 is a key component of senescence-associated secretory phenotype (SASP), which may be an important link between cellular/tissue senescence and age-related diseases in the brain, including AD. Genetic manipulation of inflammatory factors in AD mouse models has been performed (summarised in Birch et al. 2014). These results showed that the modulation of inflammatory factors alters amyloid pathology in APP Tg mice. Interestingly, inflammasome, a key inflammatory signalling platform that activates IL-1β and IL-18 (Latz et al. 2013), may contribute to AD pathogenesis (Heneka et al. 2013; Liu and Chan 2014).

Modulation of glial functions through manipulation of chemokines and their receptors has also been performed (Birch et al. 2014; Liu et al. 2014). The results from these studies showed that microglial fractalkine receptor (CX3CR1) could potentially contribute to the acceleration of tau pathology and neuronal loss, thus supporting the notion that reactive microglial neuroinflammation drives AD pathogenesis, particularly connecting tau pathology and neuronal cell death (Fuhrmann et al. 2010; Maphis et al. 2015). This evidence strongly suggests that neuroinflammation and its modulating factors might contribute to AD pathogenesis. However, most of these studies were based on mouse models that have a tendency to overexpress particular genes, thus these results may not be entirely reliable.

50.5 Animal Models for AD Research

Mouse models have been developed to reproduce the pathologies and elucidate the molecular mechanisms of AD. However, almost all existing mouse models are transgenic mouse models which overexpress the genes with FAD mutations and/or FTDP-17 mutations, leading to the nonphysiological production of particular proteins. Therefore, it is common for the results obtained from these mouse models to be misinterpreted. APP Tg mice were developed to test the amyloid cascade hypothesis, and provided extensive knowledge regarding AD. Although APP Tg mice, such as Tg2576 and APP23, develop senile plaque-like aggregates with neuroinflammation (Benzing et al. 1999; Bornemann et al. 2001), the size and composition (Aβx-40 dominant plaques) of these aggregates were different from the senile plaques in the brain of AD patients (Iwatusbo et al. 1994; Saito et al. 2011, 2014; Fig. 50.3). APP overexpression ensures high production of Aβ and also induces the generation of high levels of the by-products that are the proteolytic fragments of APP; therefore, such by-products disrupt the physiological functions of the brain (Chasseigneaux and Allinquant 2012; Mitani et al. 2012;

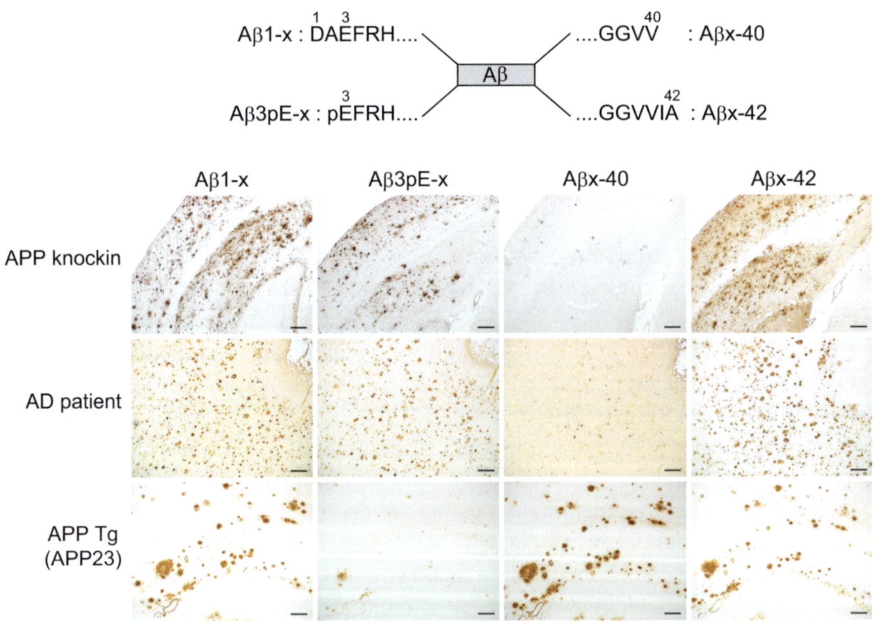

Fig. 50.3 Relevance of amyloid pathology in the model mice. N- and C-terminal structures of Aβ subspecies are schematically illustrated, and pE represents pyloglutamate. APP knockin mouse models exhibit amyloid pathology that is relevant to AD patients compared with the existing overexpression-based mouse models of AD (APP knockin mice and AD patients exhibit Aβ42 dominant amyloid depositions including Aβ3pE-x deposition. On the other hand, APP Tg mice exhibit Aβ40 dominant amyloid deposition and lower level of Aβ3pE-x, compared to APP knockin mice and AD patients.). *Scale bars* represent 200 μm

Pardossi-Piquard and Checler 2012). Moreover, APP Tg mice exhibit no NFTs, neurodegeneration, nor neuronal cell death; nevertheless, APP Tg mice usually exhibit aggressive nature in the cage, and often die suddenly of unknown reasons. Tau Tg mice crossbred with APP Tg mice and the triple-transgenic model (3× Tg-AD) both exhibit senile plaque-like aggregates and NFT-like aggregates caused by the effect of the FTDP-17 'non-AD' mutation of tau. Consequently, the pathological process of AD underlying the amyloid cascade hypothesis is disregarded. Previous characterisation of the neuroinflammatory process in AD has also been demonstrated by genetic manipulation of several immune and inflammatory pathways in overexpression-based AD mouse models (Wyss-Coray 2006; Birch et al. 2014). Our unpublished results also suggest that different responses of glial cells to the Aβ subspecies, including Aβ3pE-x (Saido et al. 1995; Saito et al. 2014), could be detected in the mouse brain. This evidence leads to uncertainty as to whether the responsiveness of glial cells against the non-AD type aggregates at the artificial pathological stage in the brain of transgenic mouse models is relevant. Additionally, it also reveals that the generation of more relevant AD mouse models is necessary to understand and reliably reproduce the AD pathogenesis process.

To address these issues, novel AD mouse models were successfully generated by manipulating the APP gene using gene knockin without overexpression (Saito et al. 2014; Nilsson et al. 2014). The APP knockin mice exhibited Aβx-42-dominant amyloid plaques including Aβ3pE-x depositions, similar to those in the AD brain (Fig. 50.3). In addition, APP knockin mice showed normal expression levels and processing of APP as well as normal lifespans with normal nature in the cage. Unfortunately, the APP knockin mice did not exhibit NFTs, neurodegeneration, and neuronal cell death as previous APP Tg mice; however, hyperphosphorylation of tau, dystrophic neurite, and neuritic plaques were detected. Regarding neuroinflammation in the APP knockin mice, gliosis occurred around the amyloid plaques in particular (Fig. 50.4), which is similar to that in AD patients. At this stage of amyloidosis in the knockin mice, the specific cytokines/chemokines, which are different from cytokines/chemokines found in APP Tg-based studies (Birch et al. 2014), may play important roles in AD pathogenesis (unpublished results). Our single-locus APP knockin mouse models will be valuable tools for distinguishing between facts and artefacts observed in existing mouse models. Furthermore, knockin mice crossbred with other mutant mice could be powerful mouse models in AD research.

To understand the relationship between tau pathology and neuroinflammation in AD, we need to establish improved tauopathy mouse models, because almost all the existing tauopathy mouse models overexpress a single isoform of the tau protein with or without FTDP-17 mutations (http://www.alzforum.org/research-models). In normal human adult brains, the tau protein has six distinct isoforms controlled by alternative splicing. Such tau isoforms are classified as 3R tau and 4R-tau depending on whether there are three or four repeated microtubule-binding domains, respectively. Note that NFTs in AD comprise a mix of both 3R-tau and 4R-tau in all six isoforms, although 4R-tau aggregates may cause corticobasal

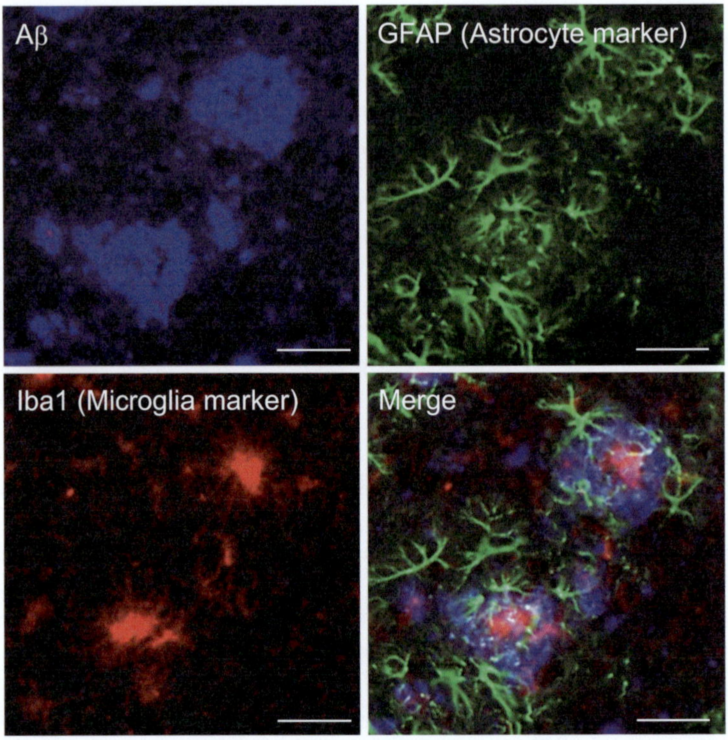

Fig. 50.4 Amyloidosis and gliosis. Reactive glial cells surround amyloid plaques in the brain. *Scale bars* represent 25 μm

degeneration or progressive supranuclear palsy; in addition, Pick's disease is characterised by the deposition of 3R-tau aggregates in the brain (Buée et al. 2000). Furthermore, only 4R-tau isoforms, but not 3R-tau, exist in normal mouse adult brains, suggesting that human and mouse brains contain distinct isoforms of tau. Nevertheless, almost all the existing tauopathy mouse models contain one of six isoforms of the tau overexpressing model. These single isoforms of tau-expressing mice with or without the FTDP-17 mutation are not relevant to the AD model. In addition, because mouse tau may hamper human tau aggregation, a mouse tau deleted model would be better (Ando et al. 2010). Ideally, the tau model would preferably be based on the knockin model, because overexpression of tau may disturb synaptic functions. Overall, relevant animal models are key to solving the AD pathogenesis and to finding a promising method of preventing, treating, and diagnosing AD.

50.6 Conclusion

A number of studies in AD research have revealed that chronological factors and complex pathological systems in the brain make it difficult to elucidate the pathophysiological mechanisms of AD. In addition, current animal models of AD are not accurate or reproducible, and thus, the development of therapeutic strategies for this disease is slow. As highlighted in this chapter, neuroinflammation is currently recognised as an important process in AD pathogenesis, and involves a rate-limiting phase potentially connecting Aβ amyloidosis with neurodegeneration and neuronal cell death through tauopathy. Thus, understanding the glial functions associated with AD or other neurodegenerative disorders, as a new doctrine: *gliostasis (glial cells homeostasis)*, may lead to a promising solution. In order to promote such investigations, development of a relevant animal model is necessary.

The relationship between long-term NSAID use and neuroinflammation is still unclear. A recent preliminary meta-analysis on the treatment effect of NSAIDs suggested no beneficial effect on AD (Niguel-Álvarez et al. 2015), in contrast to the epidemiological evidence. This gap may indicate that peripheral inflammation, such as that caused by rheumatoid arthritis, affects the brain pathologically. Interestingly, gut microbiota may influence brain disorders (Hsiao et al. 2013; Mayer et al. 2015) and modulate the immune system in the individual host (Atarashi et al. 2013). As chronic inflammation accelerates ageing, it is possible that NSAIDs ameliorate ageing signs in the peripheral region (Jurk et al. 2014), thus suggesting that NSAIDs may affect AD not by modulating the neuroinflammatory pathway directly but by affecting the brain indirectly. This evidence brings to light a new perspective that AD pathogenesis may be linked not only to the condition of the CNS, but also to peripheral conditions, as a whole body disorder.

Finally, to prevent, treat, and delay the onset of AD, the detailed process of AD pathogenesis must first be elucidated. Furthermore, it is imperative to identify surrogate markers that correctly and timely indicate the pathological stage, for example, Aβ amyloidosis, neuroinflammation, tauopathy, and neuronal damage; it is hoped that they are comparable with other neurodegenerative disorders. Such indicators in serum or plasma reflecting a specific pathological stage in the brain and body would be useful for the therapeutic intervention of the disease.

Acknowledgment I thank Takaomi C. Saido (RIKEN BSI) for supervision and leading of the research, and also thank Yukio Matsuba, Naomi Mihira, and Yuya Tomita (RIKEN BSI) for technical assistance. I am grateful for Per Nilsson (RIKEN BSI) for helpful discussion during the preparation of this manuscript. This work was supported by research grants from JST PRESTO, RIKEN BSI, MEXT Kakenhi (grants-in-aid for Young Scientist (A):19689009, Young Scientist (A):23680039 and Scientific research (B):26290019) and the Cell Science Research Foundation.

References

Ando K, Leroy K, Heraud C et al (2010) Deletion of murine tau gene increases tau aggregation in a human mutant tau transgenic mouse model. Biochem Soc Trans 38:1001–1005

Atarashi K, Tanoue T, Oshima K et al (2013) Treg induction by a rationally selected mixture of chostridia strains from the human microbiota. Nature 500:232–236

Bateman RJ, Xiong C, Benzinger TLS et al (2012) Clinical and biomarker changes in dominantly inherited Alzheimer's disease. N Engl J Med 367:795–804

Benzing WC, Wujek JR, Ward EK et al (1999) Evidence for glial-mediated inflammation in aged APPSW transgenic mice. Neurobiol Aging 20:581–589

Birch AM, Katsouri L, Sastre M (2014) Modulation of inflammation in transgenic models of Alzheimer's disease. J Neuroinflam 11:25

Blennow K, De Leon MJ, Zetterberg H (2006) Alzheimer's disease. Lancet 368:387–403

Block ML, Zecca L, Hong JS (2007) Microglia-mediated neurotoxicity: uncovering the molecular mechanism. Nat Rev Neurosci 8:57–69

Bornemann KD, Wiederhold KH, Pauli C et al (2001) Abeta-induced inflammatory processes in microglia cells of APP23 transgenic mice. Am J Pathol 158:63–73

Buée L, Bussière T, Buée-Scherrer V et al (2000) Tau protein isoforms, phosphorylation and role in neurodegenerative disorder. Brain Res Rev 33:95–130

Burda JE, Sofroniew M (2014) Reactive gliosis and the multicellular response to CNS damage and disease. Neuron 81:229–248

Chasseigneaux S, Allinquant B (2012) Functions of Aβ, sAPPα and sAPPβ: similarities and differences. J Neurochem 120:99–108

De Strooper B, Vassar R, Golde T (2010) The secretase: enzymes with therapeutic potential in Alzheimer disease. Nat Rev Neurol 6:99–107

Fuhrmann M, Bittner T, Jung CKE et al (2010) Microglial Cx3cr1 knockout prevents neuron loss in a mouse model of Alzheimer's disease. Nat Neurosci 13:411–413

Games D, Adams D, Alessandrini R et al (1995) Alzheimer-type neuropathology in transgenic mice overexpressing V717F β-amyloid precursor protein. Nature 373:523–527

Goedert M, Ghetti B, Spillantini MG (2000) Tau gene mutations in frontotemporal dementia and parkinsonism linked to chormosome 17 (FTDP-17). Their relevance for understanding the neurogenerative process. Ann N Y Acad Sci 20:74–83

Hardy JA, Higgins GA (1992) Alzheimer's disease: the amyloid cascade hypothesis. Science 256:184–185

Heneka MT, Kummer MP, Stutz A et al (2013) NLRP3 is activated in Alzheimer's disease and contributes to pathology in APP/PS1 mice. Nature 493:674–678

Hsiao EY, McBride SW, Hsien S et al (2013) Microbiota modulate behavioral and physiological abnormalities associated with neurodevelopmental disorders. Cell 155:1451–1463

in't Veld BA, Ruitenberg A, Hofman A et al (2001) Nonsteroidal antiinflammatory drugs and the risk of Alzheimer's disease. N Engl J Med 345:1515–1521

Iwatusbo T, Odaka A, Suzuki N et al (1994) Visualization of Aβ42(43) and Aβ40 in senile plaques with end-specific Aβ monoclonals: evidence that an initially deposited species is Aβ42(43). Neuron 13:45–53

Ji B, Kumata K, Onoe H et al (2013) Assessment of redioligands for PET imaging of cyclooxygenease-2 in an ischemic neuronal injury model. Brain Res 1533:152–162

Jonsson T, Atwal JK, Steinberg S et al (2012) A mutation in APP protects against Alzheimer's disease and age-related cognitive decline. Nature 488:96–99

Jurk D, Wilson C, Passos JF et al (2014) Chronic inflammation induces telomer dysfunction and accelerates ageing in mice. Nat Commun 2:4172

Latz E, Xiao TS, Stutz A (2013) Activation and regulation of the inflammasomes. Nat Rev 13:397–411

Lewis J, McGowan E, Rockwood J et al (2000) Neurofibrillary tangles, amyotrophy and progressive motor disturbance in mice expressing mutant (P301L) tau protein. Nat Genet 25:402–405

Liu L, Chan C (2014) The role of inflammasome in Alzheimer's disease. Ageing Res Rev 2:6–15
Liu C, Cui G, Zhu M et al (2014) Neuroinflammation in Alzheimer's disease: chemokines produced by astrocytes and chemokine receptors. Int J Clin Exp Pathol 7:8342–8355
Maeda J, Zhang MR, Okauchi T et al (2011) In vivo positron emission tomographic imaging of glial responses to amyloid-beta and tau pathologies in mouse models of Alzheimer's disease and related disorders. J Neurosci 31:4720–4730
Maphis N, Xu G, Kokiko-Cochran ON et al (2015) Reactive microglia drive tau pathology and contribute to the spreading of pathological tau in the brain. Brain (in press). doi:10.1093/brain/awv081
Mayer EA, Tillisch K, Gupta A (2015) Gut/brain axis and the microbiota. J Clin Invest 125:926–938
McGeer PL, Schulzer M, McGeer EG (1996) Arthritis and anti-inflammatory agents as possible protective factors for Alzheimer's disease: a review of 17 epidemiologic studies. Neurology 47:425–432
Mitani Y, Yarimizu J, Saita K et al (2012) Differential effects between γ-secretase inhibitors and modulators on cognitive function in amyloid precursor protein-transgenic and nontransgenic mice. J Neurosci 32:2037–2050
Niguel-Álvarez M, Santos-Lozano A, Sanchis-Gomar F et al (2015) Non-steroidal anti-inflammatory drugs as a treatment for Alzheimer's disease: a systematic review and meta-analysis of treatment effect. Drugs Aging 32:139–147
Nilsson P, Saito T, Saido TC (2014) New mouse model of Alzheimer's. ACS Chem Neurosci 5:499–502
Pardossi-Piquard R, Checler F (2012) The physiology of the β-amyloid precursor protein intracellular domain AICD. J Neurochem 120:109–124
Rockenstein E, Overk CR, Ubhi K et al (2015) A novel triple repeat mutant tau trasngenic model that mimics aspects of Pick's disease and fronto-temporal tauopathies. Plos One 10:e0121570
Saido TC, Iwatsubo T, Mann DM et al (1995) Dominant and differential deposition of distinct β-amyloid peptide species, AβN3(pE), in senile plaques. Neuron 14:457–466
Saito T, Suemoto T, Brouwers N et al (2011) Potent amyloidogenecity and pathogenecity of Aβ43. Nat Neurosci 11:1023–1032
Saito T, Matsuba Y, Mihira N et al (2014) Single *App* knock-in mouse models of Alzheimer's disease. Nat Neurosci 17:661–663
Wyss-Coray T (2006) Inflammation in Alzheimer disease: driving force, bystander or beneficial response ? Nat Med 12:1005–1015
Yasuno F, Kosaka J, Ota M et al (2012) Increased binding of peripheral benzodiazepine receptor in mild cognitive impairment-dementia converters measured by positron emission tomography with [11C]DAA1106. Psychiatry Res 203:67–74
Yoshiyama Y, Higuchi M, Zhang B et al (2007) Synapse loss and microglial activation precede tangles in a P301S tauopathy model mouse. Neuron 53:337–351

Part XI
Chronic Inflammation and Cardiovascular Diseases

Chapter 51
The Roles of Hypoxic Responses During the Pathogenesis of Cardiovascular Diseases

Norihiko Takeda

Abstract Chronic activation of the inflammatory processes contributes to the pathogenesis of cardiovascular diseases. Accumulation of macrophages accelerates atherosclerotic plaque formation. In addition, serum levels of inflammatory cytokines are significantly elevated in patients with heart failure. Therefore, it is critically important to understand the molecular mechanisms by which macrophage activation and its resolution are regulated.

Accumulating evidence showed that macrophages can be broadly classified into two types; those include proinflammatory (M1) and anti-inflammatory (M2) macrophages. An M1 macrophage expresses inducible nitric oxide synthase (*iNOS*), and synthesises nitric oxide, one of the inflammatory mediators. In contrast, a M2 macrophage expresses *Arginase1* (*Arg1*), which is known as an M2 marker gene. Intriguingly, both iNOS and Arg1 compete for the same substrate, *l*-arginine, for each reaction, therefore iNOS and Arg1 act antagonistically with regard to nitric oxide synthesis. Monocyte-derived macrophage extravasates from blood vessels into the inflammatory area, thus it encounters a gradual decrease in oxygen availability. In hypoxic condition, hypoxia-inducible factor-α (HIF-α) plays an essential role in its transcriptional responses. We have analysed the roles of HIF-1α and HIF-2α in macrophage activation, and elucidated that HIF-1α and HIF-2α are acting a critical role in M1 and M2 macrophage activation, respectively. Intriguingly, HIF-1α induces the expression of *iNOS*, whereas *Arg1* expression is mainly mediated by HIF-2α. Based on these results, we identified that the balance between HIF-1α and HIF-2α, namely HIF-α switching, plays a critical roles in both activation and resolution of the inflammation processes. Moreover, HIF-α switching in keratinocytes also contributes to the regulation of blood pressure through the release of nitric oxide in the skin.

Keywords Inflammation • Hypoxia • Cardiovascular remodelling • Nitric oxide • HIF-α • Blood pressure • Macrophage

---

N. Takeda (✉)
Department of Cardiovascular Medicine, Graduate School of Medicine, The University of Tokyo, 7-3-1 Hongo, Bunkyo-ku, Tokyo 113-8655, Japan

PRESTO, Japan Science and Technology Agency, Kawaguchi 332-0012, Japan
e-mail: ntakeda-tky@umin.ac.jp

M. Miyasaka, K. Takatsu (eds.), *Chronic Inflammation*,
DOI 10.1007/978-4-431-56068-5_51

51.1 Introduction

Hypertension, diabetes mellitus, dyslipidemia, smoking, and age are well known as risk factors for the development of atherosclerosis (Wilson et al. 1998). As an underlying mechanism, the inflammatory process plays an integral part in the development of atherosclerosis. Inflammation not only promotes plaque progression, but also triggers the thrombotic events resulting in the occurrence of acute coronary syndrome (Libby and Hansson 2015). The surface expression of adhesion molecules in vascular endothelial cells is increased at the plaque area (Bevilacqua et al. 1987). Intriguingly, dyslipidemia seems to induce the inflammation. Mice with hypercholesterolemia had an elevated number of the Ly6c positive inflammatory monocytes in the peripheral blood (Swirski et al. 2007; Tacke et al. 2007). Subsequently, inflammatory cells including monocytes or macrophages accumulate at the plaque area of the atheroma (Aikawa et al. 1998). These mononuclear phagocytes become a major source of the form cells that accumulate in the atheroma plaques (Aqel et al. 1985; Jonasson et al. 1986). In addition, inflammatory mediators released by these monocytes or macrophages are considered to elicit smooth muscle cell apoptosis, which induces plaque instability (Geng et al. 1996). Moreover, the impairment of dead cell clearance, efferocytosis, also accelerates the progression of unstable plaques (Tabas 2010).

Heart failure is a syndrome in which the heart cannot acquire a suffifcient amount of cardiac output in order to meet the demand of the body. The underlying pathological processes of heart failure include Ca handling, metabolic disorders , and the alterations in the neuroendocrine or immune system. It has also been reported that monocytes or macrophages accumulate in the cardiac remodelling processes (Nahrendorf and Swirski 2013). In addition, peripheral monocyte levels are considered as an independent risk factor for readmission with heart failure or myocardial infarction (Maekawa et al. 2002). These results support the hypothesis that inflammation also plays an important role in the pathophysiology of heart failure. Indeed, the serum level of inflammatory cytokines, including tumour necrosis factor (TNF)-α or interleukin1-β (IL-1β) are elevated in patients with heart failure (Bozkurt et al. 2010). Moreover, the serum level of TNF-α correlated with the severity of the heart failure (Levine et al. 1990). On the other hand, clinical studies trying to modulate inflammation processes were mostly disappointing for the management of patients with heart failure, which emphasises the importance of understanding the roles of inflammatory processes in heart failure.

51.2 Antagonistic Function of M1 and M2 Macrophages in Nitric Oxide Synthesis

Monocytes and macrophages constitute a major part of the innate immune response which is activated as a host defence mechanism. In addition, macrophages have a homeostatic function including tissue remodelling or metabolic regulation (Gordon and Martinez 2010).

Macrophages were originally considered to promote inflammation which takes place in tissue injury or infection (Bonecini-Almeida et al. 1998). However, recent studies revealed that macrophages consist of a heterogeneous cell population, which includes proinflammatory and anti-inflammatory macrophages (Bonecini-Almeida et al. 1998; Gordon 2003; Mantovani et al. 2004; Mosser and Edwards 2008). A classically-activated macrophage, which highly expresses proinflammatory cytokines, reactive oxygen or nitrogen species, is now called a M1 macrophage. One of the M1 marker genes is inducible nitric oxide synthase (*iNOS*), which produces nitric oxide (NO), one of the proinflammatory mediators. On the other hand, an M2 macrophage displays several types of activation profiles, which is distinct from M1. Th2 cytokines such as interleukin-4 (IL-4) or IL-13 induce M2-type macrophage activation (Gordon 2003), which is involved in host defence to parasites. Arginase1 *(Arg1)* or mannose receptor expression is specifically increased in M2 macrophages, and thus are considered M2 marker genes. IL-10, glucocorticoid hormones, and immune complexes also influence the activation status of macrophages, and are termed M2-like (Biswas and Mantovani 2010). M2 macrophages also include a tissue resident macrophage, which has a homeostatic function in each tissue (Mantovani et al. 2004).

The two signature enzymes, iNOS and Arg1 share the same metabolic substrate, *l*-arginine, and thus act antagonistically with regard to NO synthesis. In M1 macrophages, iNOS synthesises NO, and thus strikingly promotes the inflammatory processes. In contrast, Arg1 sequesters *l*-arginine in M2 macrophages, and thus prevents the excessive production of NO (El Kasmi et al. 2008). Therefore, the balance between iNOS and Arg1 critically regulates the production of NO.

Although extensive studies have been carried out on the molecular mechanisms of acute inflammation, little is known about the molecular mechanisms by which acute inflammation resolves. Therefore, understanding the processes of macrophage activation or its resolution will greatly help to elucidate the molecular link between chronic inflammation and organ dysfunction.

51.3 Roles of HIF-α in the Cellular Responses to Hypoxia

Under hypoxic condition, each cell exhibits several types of adaptive or maladaptive responses. Although gene expression is mostly suppressed in the hypoxic condition, the abundance of some genes is significantly induced, which have been

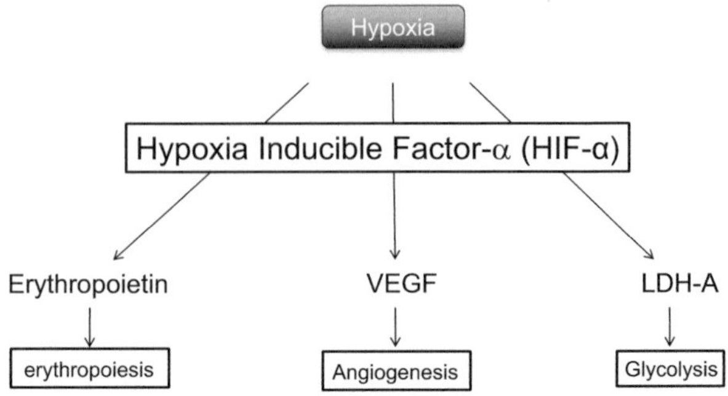

Fig. 51.1 HIF-α mediates cellular responses to hypoxia. In hypoxic condition, the transcript levels of hypoxia-inducible genes are elevated in a HIF-α dependent manner. Hypoxia-inducible genes include erythropoietin, vascular endothelial growth factor, and lactate dehydrogenase-a

termed hypoxia-inducible genes. Hypoxia-inducible genes include genes related to inflammation (*iNOS*) (Melillo et al. 1995), angiogenesis (vascular endothelial growth factor-a, Vegf-a) (Tuder et al. 1995), erythropoiesis (erythropoietin, Epo), or cellular metabolism (pyruvate dehydrogenase kinase, isoform1, Pdk1, or lactate dehydrogenase-a, Ldh-a) (Kim et al. 2006; McClelland 1985). Note that most of the transcriptional responses of these hypoxia-responsive genes are mediated through a group of transcription factors, hypoxia inducible factor-1α (HIF-1α) and HIF-2α (Fig. 51.1) (Tian et al. 1997; Wang and Semenza 1993; Weidemann and Johnson 2008). The abundance of both HIF-1α and HIF-2α proteins is dramatically increased in hypoxia. Under normoxic condition, the proline residues of HIF-α are hydroxylated oxygen dependently through prolyl hydroxylase domain containing proteins. Hydroxylated HIF-α is degraded via thevon Hippel–Lindau protein mediated ubiquitin-proteasomal system (Maxwell et al. 1999). On the other hand, HIF-α protein is stabilised in hypoxic condition, and translocates into the nucleus, where it dimerises with its binding partner, HIF-1β. A heterodimer of HIF-α and HIF-1β binds to the hypoxia-responsive element (HRE) upstream of hypoxia-inducible genes, resulting in the activation of its gene expression. Several HREs have been identified upstream or downstream of the hypoxia inducible genes, including Vegf-a, Epo, and Pdk1.

51.4 The Roles of HIF-α Switching in Macrophage Activation and Its Resolution

Monocyte-derived macrophages accumulate at the inflammatory area, and accelerate or modulate the inflammatory processes. It is well known that the inflammatory area is in an hypoxic state (Murdoch et al. 2004). In addition to the reduced oxygen delivery, the rise of the local oxygen consumption makes the inflammatory area hypoxic. More important, proinflammatory macrophages are known to accumulate in the hypoxic area (Murdoch et al. 2004).

The roles of HIF-1α in M1 macrophage activation have been well characterized. Lipopolysaccharide (LPS) or interferon-γ (IFNγ) upregulates the expression of *iNOS*, a classical M1 marker gene, and thus is well known to elicit M1 macrophage activation (Bonecini-Almeida et al. 1998; Mantovani et al. 2004). As an underlying mechanism by which HIF-1α signal promotes M1 macrophage activation, we and other groups has revealed that LPS or IFNγ induces HIF-1α protein accumulation even in normoxic condition, which transactivates the promoter activity of the *iNOS* gene. Consistent with this, the severity of the septic shock, which is induced by lipopolysaccharide (LPS), was attenuated in myeloid-specific HIF-1α deficient (LysM/HIF-1α) mice (Peyssonnaux et al. 2007; Takeda et al. 2010). The severity of chemically induced cutaneous inflammation was also attenuated in LysM/HIF-1α mice (Cramer et al. 2003). In addition, the development of experimental arthritis was dependent on HIF-1α activation in myeloid cells.

In contrast to the roles of HIF-1α in an M1 macrophage, the roles of HIF-2α in macrophage activation have not been fully elucidated. We checked the gene expression of HIF-1α and HIF-2α in M1 or M2 macrophages, and identified that HIF-1α and HIF-2α are specifically expressed in M1 and M2 macrophages, respectively (Takeda et al. 2010). LPS or IFNγ significantly upregulates the HIF-1α mRNA level, resulting in the increased HIF-1α protein accumulation in hypoxia. On the other hand, LPS or IFNγ strikingly suppresses HIF-2α mRNA level, therefore no HIF-2α protein was observed in the M1 macrophage. Note that IL-4 or IL-13, one of the Th2 cytokines, significantly increased HIF-2α mRNA level.

Intriguingly, we identified that HIF-2α accelerates the transcription of *Arg1* in M2 macrophages. Both *iNOS* and *Arg1* gene expression were induced in hypoxia, and HIF-1α plays a critical role in the induction of *iNOS*. On the other hand, *Arg1* expression was significantly decreased in HIF-2α deficient macrophages. These results showed that *iNOS* and *Arg1* utilise different HIF-α isoforms for their hypoxic induction.

Consistent with this, *iNOS* expression and NO synthesis were decreased in HIF-1α deficient macrophages, whereas NO production was significantly increased in HIF-2α deficient macrophages. Finally, LPS-elicited production of NO is significantly suppressed in myeloid-specific HIF-1α deficient mice, whereas serum $NO_{(x)}$ level was significantly elevated in myeloid specific HIF-2α deficient mice. Collectively, these results showed that the balance between HIF-1α and HIF-2α critically regulates serum NO level (Fig. 51.2). Therefore, the antiparallel expression of HIF-1α and HIF-1α in M1 and M2 macrophages, called HIF-α switching, could contribute to the temporal control of initiation and resolution in septic shock.

51.5 The Roles of Skin HIF-α Switching in Cardiovascular Remodelling

Skin contains an extensive series of arterial plexus, and some of them are involved in thermal control or blood pressure control. Skin circulation is altered in several disorders including heart failure, hypercholesterolemia, and hypertension (Green

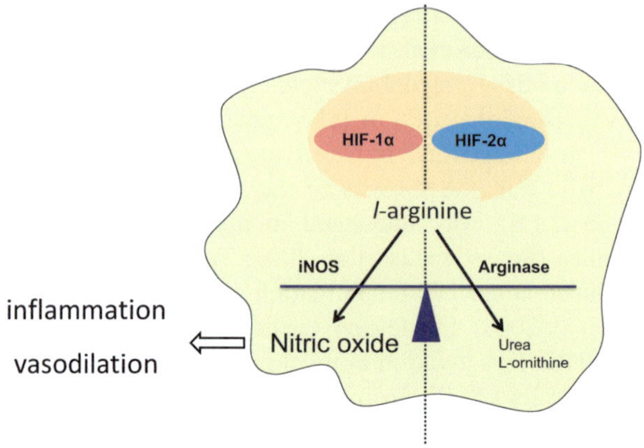

Fig. 51.2 HIF-α switching regulates nitric oxide level in macrophages and keratinocytes Nitric oxide (NO) is synthesized via an inducible nitric oxide synthase (*iNOS*). HIF-1α activates the transcription of *iNOS* gene, resulting in the NO synthesis. In contrast, HIF-2α augments the expression of *Arg1*, which suppresses the NO production in macrophages or keratinocytes

et al. 2006; Khan et al. 1999). Local regulation seems to play an essential role in controlling tissue blood flow or oxygenation, thus it has been a scientific topic how the skin or keratinocytes control vascular tone.

We have investigated the expression of HIF-1α and HIF-2α in human skin epithelial cells, and found that hypertensive patients had smaller number of HIF-1α positive cells, whereas the number of HIF-2α positive cells is higher in hypertensive patients. Similar to the macrophages, HIF-1α in skin keratinocytes also activates *iNOS* expression, whereas *Arg1* expression was induced by HIF-2α (Cowburn et al. 2013). To further examine the roles of HIF-α in the regulation of local perfusion, we generated keratinocyte-specific HIF-1α deficient (K14cre-HIF-1α) or HIF-2α deficient (K14cre- HIF-2α) mice. Skin $NO_{(x)}$ was higher in K14cre-HIF-2α mice than wild-type mice. Moreover, K14cre-HIF-2α mice quickly become hypothermic a in cold environment, which seems to be related to the excessive cutaneous vasodilation. In contrast, the cutaneous vessels in K14cre-HIF-1α mice were less dilated than those in wild-type mice. Consistent with this, K14cre-HIF-1α mice showed a more significant elevation of core temperature after physical exercise than wild-type mice. These results showed that HIF-α switching in keratinocytes is essential in the homeostasis of body temperature regulation.

Skin HIF-α switching also regulates systemic blood pressure. Systolic or diastolic blood pressure of K14cre-HIF-1α mice was higher than those of wild-type mice, resulting in the increased extent of cardiac fibrosis. Angiotensin II (AngII) is one of the most potent vasoconstrictors, which elevates blood pressure and induces cardiac fibrosis. Intriguingly, AngII induced the elevation of blood pressure and was attenuated in K14cre-HIF-2α mice compared with wild-type mice. Consistent with this, cardiac fibrosis was less prominent in K14cre-HIF-2α mice. Collectively, these results illuminated that HIF-α switching in the skin critically regulates the systemic blood pressure, which results in the modulation of cardiovascular remodelling.

51.6 Conclusion

Although extensive studies have been carried out on the initiation processes of acute inflammation, little is known about the molecular mechanisms by which inflammation resolves or sustains as a chronic inflammation. Transient activation of the inflammatory processes seems to be beneficial in tissue remodelling. Chronic inflammation, however, significantly deteriorates each organ function. Understanding the macrophage heterogeneity in cardiovascular remodelling seems to offer new mechanistic insight into the disease progression, and provide a novel avenue for immune cell-mediated manipulation of cardiovascular remodelling.

We have identified HIF-α switching in macrophages or keratinocytes, and characterised its roles in tissue remodelling. HIF-α switching seems to play an essential role in the initiation and resolution of inflammatory processes. Moreover, HIF-α switching also critically regulates cardiovascular remodelling. Further studies regarding the roles of HIF-α in other cells will elucidate the biological processes by which tissue or organ homeostasis is maintained.

Acknowledgment N.T. was supported by a grant from JST-PRESTO program, and a Grant-in-Aid for Scientific Research on Innovative Areas 'Oxygen Biology: a new criterion for integrated understanding of life' (No. 26111003) of MEXT, Japan, the Banyu Life Science Foundation International and the Takeda Science Foundation.

References

Aikawa M, Rabkin E, Okada Y, Voglic SJ, Clinton SK, Brinckerhoff CE, Sukhova GK, Libby P (1998) Lipid lowering by diet reduces matrix metalloproteinase activity and increases collagen content of rabbit atheroma: a potential mechanism of lesion stabilization. Circulation 97 (24):2433–2444

Aqel NM, Ball RY, Waldmann H, Mitchinson MJ (1985) Identification of macrophages and smooth muscle cells in human atherosclerosis using monoclonal antibodies. J Pathol 146 (3):197–204. doi:10.1002/path.1711460306

Bevilacqua MP, Pober JS, Mendrick DL, Cotran RS, Gimbrone MA Jr (1987) Identification of an inducible endothelial-leukocyte adhesion molecule. Proc Natl Acad Sci U S A 84 (24):9238–9242

Biswas SK, Mantovani A (2010) Macrophage plasticity and interaction with lymphocyte subsets: cancer as a paradigm. Nat Immunol 11(10):889–896. doi:10.1038/ni.1937

Bonecini-Almeida MG, Chitale S, Boutsikakis I, Geng J, Doo H, He S, Ho JL (1998) Induction of in vitro human macrophage anti-Mycobacterium tuberculosis activity: requirement for IFN-gamma and primed lymphocytes. J Immunol 160(9):4490–4499

Bozkurt B, Mann DL, Deswal A (2010) Biomarkers of inflammation in heart failure. Heart Fail Rev 15(4):331–341. doi:10.1007/s10741-009-9140-3

Cowburn AS, Takeda N, Boutin AT, Kim JW, Sterling JC, Nakasaki M, Southwood M, Goldrath AW, Jamora C, Nizet V, Chilvers ER, Johnson RS (2013) HIF isoforms in the skin differentially regulate systemic arterial pressure. Proc Natl Acad Sci U S A 110(43):17570–17575. doi:10.1073/pnas.1306942110

Cramer T, Yamanishi Y, Clausen BE, Forster I, Pawlinski R, Mackman N, Haase VH, Jaenisch R, Corr M, Nizet V, Firestein GS, Gerber HP, Ferrara N, Johnson RS (2003) HIF-1alpha is essential for myeloid cell-mediated inflammation. Cell 112(5):645–657

El Kasmi KC, Qualls JE, Pesce JT, Smith AM, Thompson RW, Henao-Tamayo M, Basaraba RJ, Konig T, Schleicher U, Koo MS, Kaplan G, Fitzgerald KA, Tuomanen EI, Orme IM, Kanneganti TD, Bogdan C, Wynn TA, Murray PJ (2008) Toll-like receptor-induced arginase 1 in macrophages thwarts effective immunity against intracellular pathogens. Nat Immunol 9 (12):1399–1406. doi:10.1038/ni.1671

Geng YJ, Wu Q, Muszynski M, Hansson GK, Libby P (1996) Apoptosis of vascular smooth muscle cells induced by in vitro stimulation with interferon-gamma, tumor necrosis factor-alpha, and interleukin-1 beta. Arterioscler Thromb Vasc Biol 16(1):19–27

Gordon S (2003) Alternative activation of macrophages. Nat Rev Immunol 3(1):23–35. doi:10.1038/nri978

Gordon S, Martinez FO (2010) Alternative activation of macrophages: mechanism and functions. Immunity 32(5):593–604. doi:10.1016/j.immuni.2010.05.007

Green DJ, Maiorana AJ, Siong JH, Burke V, Erickson M, Minson CT, Bilsborough W, O'Driscoll G (2006) Impaired skin blood flow response to environmental heating in chronic heart failure. Eur Heart J 27(3):338–343. doi:10.1093/eurheartj/ehi655

Jonasson L, Holm J, Skalli O, Bondjers G, Hansson GK (1986) Regional accumulations of T cells, macrophages, and smooth muscle cells in the human atherosclerotic plaque. Arteriosclerosis 6 (2):131–138

Khan F, Litchfield SJ, Stonebridge PA, Belch JJ (1999) Lipid-lowering and skin vascular responses in patients with hypercholesterolaemia and peripheral arterial obstructive disease. Vasc Med 4(4):233–238

Kim JW, Tchernyshyov I, Semenza GL, Dang CV (2006) HIF-1-mediated expression of pyruvate dehydrogenase kinase: a metabolic switch required for cellular adaptation to hypoxia. Cell Metab 3(3):177–185. doi:10.1016/j.cmet.2006.02.002

Levine B, Kalman J, Mayer L, Fillit HM, Packer M (1990) Elevated circulating levels of tumor necrosis factor in severe chronic heart failure. N Engl J Med 323(4):236–241. doi:10.1056/NEJM199007263230405

Libby P, Hansson GK (2015) Inflammation and immunity in diseases of the arterial tree: players and layers. Circ Res 116(2):307–311. doi:10.1161/CIRCRESAHA.116.301313

Maekawa Y, Anzai T, Yoshikawa T, Asakura Y, Takahashi T, Ishikawa S, Mitamura H, Ogawa S (2002) Prognostic significance of peripheral monocytosis after reperfused acute myocardial infarction:a possible role for left ventricular remodeling. J Am Coll Cardiol 39(2):241–246

Mantovani A, Sica A, Sozzani S, Allavena P, Vecchi A, Locati M (2004) The chemokine system in diverse forms of macrophage activation and polarization. Trends Immunol 25(12):677–686. doi:10.1016/j.it.2004.09.015

Maxwell PH, Wiesener MS, Chang GW, Clifford SC, Vaux EC, Cockman ME, Wykoff CC, Pugh CW, Maher ER, Ratcliffe PJ (1999) The tumour suppressor protein VHL targets hypoxia-inducible factors for oxygen-dependent proteolysis. Nature 399(6733):271–275. doi:10.1038/20459

McClelland GB (1985) Brooks GA (2002) Changes in MCT 1, MCT 4, and LDH expression are tissue specific in rats after long-term hypobaric hypoxia. J Appl Physiol 92(4):1573–1584. doi:10.1152/japplphysiol.01069.2001

Melillo G, Musso T, Sica A, Taylor LS, Cox GW, Varesio L (1995) A hypoxia-responsive element mediates a novel pathway of activation of the inducible nitric oxide synthase promoter. J Exp Med 182(6):1683–1693

Mosser DM, Edwards JP (2008) Exploring the full spectrum of macrophage activation. Nat Rev Immunol 8(12):958–969. doi:10.1038/nri2448

Murdoch C, Giannoudis A, Lewis CE (2004) Mechanisms regulating the recruitment of macrophages into hypoxic areas of tumors and other ischemic tissues. Blood 104(8):2224–2234. doi:10.1182/blood-2004-03-1109

Nahrendorf M, Swirski FK (2013) Monocyte and macrophage heterogeneity in the heart. Circ Res 112(12):1624–1633. doi:10.1161/CIRCRESAHA.113.300890

Peyssonnaux C, Cejudo-Martin P, Doedens A, Zinkernagel AS, Johnson RS, Nizet V (2007) Cutting edge: Essential role of hypoxia inducible factor-1alpha in development of lipopolysaccharide-induced sepsis. J Immunol 178(12):7516–7519

Swirski FK, Libby P, Aikawa E, Alcaide P, Luscinskas FW, Weissleder R, Pittet MJ (2007) Ly-6Chi monocytes dominate hypercholesterolemia-associated monocytosis and give rise to macrophages in atheromata. J Clin Invest 117(1):195–205. doi:10.1172/JCI29950

Tabas I (2010) Macrophage death and defective inflammation resolution in atherosclerosis. Nat Rev Immunol 10(1):36–46. doi:10.1038/nri2675

Tacke F, Alvarez D, Kaplan TJ, Jakubzick C, Spanbroek R, Llodra J, Garin A, Liu J, Mack M, van Rooijen N, Lira SA, Habenicht AJ, Randolph GJ (2007) Monocyte subsets differentially employ CCR2, CCR5, and CX3CR1 to accumulate within atherosclerotic plaques. J Clin Invest 117(1):185–194. doi:10.1172/JCI28549

Takeda N, O'Dea EL, Doedens A, Kim JW, Weidemann A, Stockmann C, Asagiri M, Simon MC, Hoffmann A, Johnson RS (2010) Differential activation and antagonistic function of HIF-{alpha} isoforms in macrophages are essential for NO homeostasis. Genes Dev 24 (5):491–501. doi:10.1101/gad.1881410

Tian H, McKnight SL, Russell DW (1997) Endothelial PAS domain protein 1 (EPAS1), a transcription factor selectively expressed in endothelial cells. Genes Dev 11(1):72–82

Tuder RM, Flook BE, Voelkel NF (1995) Increased gene expression for VEGF and the VEGF receptors KDR/Flk and Flt in lungs exposed to acute or to chronic hypoxia. Modulation of gene expression by nitric oxide. J Clin Invest 95(4):1798–1807. doi:10.1172/JCI117858

Wang GL, Semenza GL (1993) General involvement of hypoxia-inducible factor 1 in transcriptional response to hypoxia. Proc Natl Acad Sci U S A 90(9):4304–4308

Weidemann A, Johnson RS (2008) Biology of HIF-1alpha. Cell Death Differ 15(4):621–627. doi:10.1038/cdd.2008.12

Wilson PW, D'Agostino RB, Levy D, Belanger AM, Silbershatz H, Kannel WB (1998) Prediction of coronary heart disease using risk factor categories. Circulation 97(18):1837–1847

Chapter 52
Prevention and Treatment of Heart Failure Based on the Control of Inflammation

Motoaki Sano

Abstract Heart failure is a complex clinical syndrome that results from any structural or functional impairment of ventricular filling or ejection of blood. For decades, there have been three different concepts of heart failure and their corresponding treatment strategies: diuretics for volume retention, cardiotonics, vasodilators for abnormal hemodynamic state, neurohormonal antagonists for abnormalities in interorgan communication. These conventional treatments consistently prolongs survival in patients with heart failure but never cure the underlying heart diseases. Various risk factors induce heart injury and aortic damages, eventually leading to a final stage of cardiovascular disease, namely heart failure. It has become widely accepted that pathogenesis of heart injury and aortic damage is considered to be the result of a chronic inflammatory process. Therefore, chronic inflammation can be regarded as a new target that may treat and/or prevent heart failure. This chapter describes the recent advances in understanding of the mechanisms of chronic inflammation of cardiovascular diseases, especially focused on our accomplishments.

Keywords Myocardial infarction • Pulmonary oedema • Aortic dissection • Natural killer cells • Neutrophils • γδT cells • Cardiomyocytes • Fibrosis

52.1 Introduction

Medical devices have been developed because of advances in medical engineering technology. Catheter-based treatment of coronary artery disease, valvular disease, structural heart disease, and peripheral vascular disease are about to enter a hay-day. These noninvasive treatments will achieve further progress due to advances in devices. It can be said that 'device therapy of cardiovascular diseases' has matured. By contrast, medical treatment of heart failure is immature and a 'paradigm shift' is required.

M. Sano (✉)
Department of Cardiology, Keio University School of Medicine, 35 Shinanomachi Shinjuku-ku, Tokyo, Japan

PRESTO, Japan Science and Technology Agency, Kawaguchi 332-0012, Japan
e-mail: msano@a8.keio.jp

M. Miyasaka, K. Takatsu (eds.), *Chronic Inflammation*,
DOI 10.1007/978-4-431-56068-5_52

Heart failure is a complex clinical syndrome that results from any structural or functional impairment of ventricular filling or ejection of blood. The cardinal manifestations of heart failure are dyspnea and fatigue, which may limit exercise tolerance, and fluid retention, which may lead to pulmonary and/or splanchnic congestion and/or peripheral edema (Yancy et al. 2013). There are more than two million patients with heart failure in Japan. Two hundred thousand patients die each year. Five-year survival of patients with heart failure is worse than for most cancers. They repeatedly get hospitalised. As a matter of fact, 35 % of patients are readmitted to hospital within 1 year. For decades, there have been three different concepts of heart failure and their corresponding treatment strategies: diuretics for volume retention, cardiotonics, vasodilators for abnormal hemodynamic state, neurohormonal antagonists for abnormalities in interorgan communication. These conventional treatments consistently prolongs survival in patients with heart failure but never cure underlying heart disease. There are many inherited and environmental factors that play a role in the pathogenesis of heart failure. Recent advances in our understanding of the molecular mechanisms of various heart diseases hold great promise for the development of molecular target therapy in a disease subtype-specific fashion. Rather than conventional heart failure treatment, targeted therapy may be more therapeutically effective in patients whose heart disease has a specific molecular target. Molecular subtype-based stratification to guide therapeutic decisions is a promising new strategy.

This review discusses recent advances regarding the regulation of inflammation in various cardiovascular diseases that can lead to heart failure, and the potential molecular targets for anti-inflammatory therapy.

52.2 Left Ventricular Remodelling After MI

Myocardial infarction (MI) is usually caused by a blood clot, which stops the blood flowing to a part of heart muscle. Despite the introduction of current gold-standard cardioprotective therapies including β-blockers, renin–angiotensin–aldosterone system antagonists, antiplatelet agents, and statins, prognosis remains poor in post-MI patients, who often display adverse left ventricular (LV) remodelling after MI (Krum and Teerlink 2011) (Fig. 52.1). LV remodelling leads to heart failure and is a main determinant of morbidity and mortality after MI (Lewis et al. 2003). At the present time, therapeutic options to prevent LV remodelling are limited.

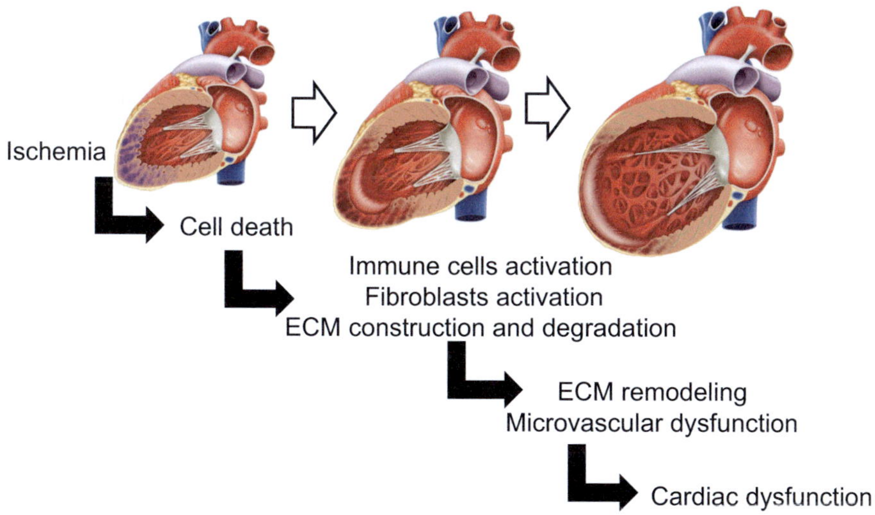

Fig. 52.1 Postmyocardial infarction wound healing and adverse cardiac remodelling. MI is a predominant cause of congestive heart failure

52.3 Temporal Dynamics of Immune Cell Accumulation Following MI

In this setting, myocyte loss, inflammation, and ventricular remodelling are the principal causes of heart failure. MI causes inflammation, which is characterised by the recruitment and activation of immune cells of the innate and adaptive immune systems. These cells may have a cell type-specific function in the time course after MI that involves clearance of dead tissues, the reparative response, and adverse remodelling (Anzai et al. 2012; Nahrendorf et al. 2007, 2010; Leuschner et al. 2012; Arslan et al. 2011; Frangogiannis 2012). Therefore, immunomodulatory therapies may harbour a promising potential for accelerating cardiac repair and ameliorating LV remodelling after MI. We performed a comprehensive characterisation of the temporal dynamics of immune cell accumulation following MI by both flow cytometry and immunohistochemistry (Yan et al. 2013). The total number of infiltrating leukocytes gradually increased after MI to a peak on day 7. Neutrophils accumulated in the infarcted heart, peaking at 3 days after MI and then notably, continuing to accumulate in the infarcted myocardium over 7–14 days after the MI onset. Numerically, macrophages were the predominant cells infiltrating the infarcted myocardium, and these cells showed a biphasic pattern of activation. M1 macrophages dominated on 1–3 days post-MI, whereas M2 macrophages increased more gradually and represented the predominant macrophage subset after 5 days post-MI. DC accumulation reached a maximum on day 5–7 after MI. Infiltrating CD4 + −αßT cells, CD8 + −αßT, cells and γδT cells, and B cells started to increase gradually to a peak on day 7 after MI. NK cells and NKT cells started to increase on day 3 and peaked on day 7 after MI.

52.4 Cell-Type Specific Function of Immune Cells in the Infarcted Myocardium After MI

Disruption of myeloperoxidase (MPO), released predominantly by neutrophils, decreases leukocyte infiltration and LV dilation, enhances ventricular function, and delays early death attributable to myocardial rupture (Nahrendorf et al. 2007), whereas macrophage depletion impairs wound healing and increases LV remodelling after MI (Krum and Teerlink 2011; Leuschner et al. 2012). Dendritic cells are a potent immunoprotective regulator during the postinfarction healing process via their control of monocyte/macrophage homeostasis (Arslan et al. 2011). Treg cells serve to protect against adverse ventricular remodelling and contribute to improved cardiac function after MI via inhibition of inflammation and direct protection of cardiomyocytes (Weirather et al. 2014). γδT cells are the major source of interleukin (IL)-17 in the infarcted myocardium and function specifically in the late remodelling stages by promoting sustained infiltration of neutrophils and macrophages, stimulating macrophages to produce proinflammatory cytokines, aggravating cardiomyocyte death, and enhancing fibroblast proliferation and profibrotic gene expression via IL-17 production (Yan et al. 2012). NKT cells play a protective role against post-MI LV remodelling and failure through the enhanced expression of cardioprotective cytokines such as IL-10 (Sobirin et al. 2012).

52.5 The Pathogenesis of Cardiogenic Pulmonary Oedema After MI

Acute cardiogenic pulmonary oedema after MI is thought to arise when abnormally high pulmonary capillary pressure induces the characteristic accumulation of low-protein fluid in the interstitial and alveolar spaces of the lung associated with this disorder. However, murine MI causes protein-rich alveolar oedema accompanied by neutrophil-predominant infiltration. We found that IL-1β plays a key role in the inflammatory mechanism behind these severe abnormalities in respiratory gas exchange after cardiogenic pulmonary oedema (Yan et al. 2014). IL-1β sensitizes lung-endothelial cells (ECs) to neutrophil adhesion, resulting in an increased chance of neutrophil trafficking from the intravascular environment into the interstitial and alveolar compartments. Experiments in small MI model [no left ventricular endodiastolic pressure (LVEDP) elevation/moderate myocardial inflammation] and TAC model (LVEDP elevation with mild myocardial inflammation) supported both the infarcted myocardium and pulmonary capillary ECs exposed to high microvascular pressure, which are the source of IL-1β.

Interestingly, depletion of NK cells from mice had little effect on LV remodelling after MI; however, these mice exhibited severe respiratory distress associated with protein-rich, high-permeability alveolar oedema accompanied by neutrophil infiltration (Yan et al. 2014). There were 20-fold more NK cells in the mouse lungs than in the heart, and these cells were accumulated around the vasculature. The unique pulmonary environment promotes the development of NK cells with a lung-specific phenotype (Vivier et al. 2008; Shi et al. 2011). The pulmonary NK cells, which have a $CD11b^{high}$ $CD27^{low}$ mature phenotype and express IL-10, translocate to the site of vascular inflammation after MI. NK cells are a major IL-10 source in the lung and IL-10 secreted from lung NK cells alleviates the increased permeability of the inflamed alveolar–capillary barrier after MI. Beneficial effects of systemic IL-10 administration on pulmonary neutrophil accumulation and lung oedema after injury have been reported in various models. Intravenous administration of IL-10 protects against hepatic ischemia-reperfusion–induced lung injury by inhibiting lung nuclear factor-kB activation and the resulting pulmonary neutrophil accumulation and lung oedema (Yoshidome et al. 1999). In a porcine model of acute bacterial pneumonia, local expression of IL-10 suppressed lung oedema and neutrophil invasion, resulting in significantly reduced lung damage (Morrison et al. 2000). Together, these findings indicate that NK cells play a counterregulatory role against an inflammatory change in microvascular permeability in the lung associated with MI (Fig. 52.2).

52.5.1 *Identification of Novel AntiFibrotic Lipid Mediator 18 HEPE*

The beneficial effects of n-3 polyunsaturated fatty acids (PUFAs), primarily eicosapentaenoic acid (EPA) and docosahexaenoic acid (DHA), were first recognised in the late 1960s with epidemiological evidence of the Inuit population, who consumed an n-3 PUFA-rich diet, having a low incidence of myocardial infarction (Mozaffarian and Wu 2011). Subsequently, large-scale randomised clinical trials confirmed that dietary supplementation of n-3 PUFAs prevented cardiovascular events in patients with recent myocardial infarction (Yokoyama et al. 2007), reduced mortality in patients with symptomatic chronic heart failure who were receiving standard treatment (Gissi et al. 2008), and ameliorated left ventricular (LV) functional capacity and decreased the circulating concentrations of inflammatory cytokines such as TNF, IL-1β, and IL-6 in nonischemic dilated cardiomyopathy (Nodari et al. 2011). Thus, PUFAs have potential cardiovascular benefit, although the mechanisms underlying such effects remain poorly understood.

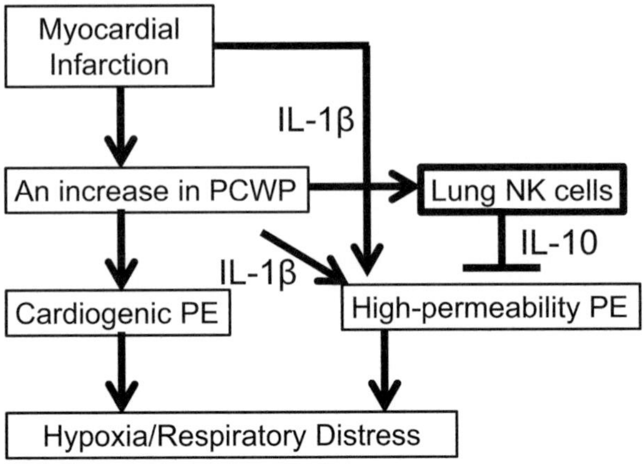

Fig. 52.2 Interleukin-10 secreted from lung NK cells alleviates the increased permeability of the inflamed alveolar–capillary barrier after MI. Impaired gas exchange after myocardial infarction is not solely because of haemodynamic changes but is in part attributable to neutrophil infiltration of the lung that induces an inflammatory response and, hence, increased endothelial–alveolar permeability. Interleukin-1β is the primary initiator of pulmonary inflammation after MI in mice. Lung NK cells play a protective role against cardiogenic pulmonary oedema via a paracrine secretion of interleukin-10

Mammals cannot naturally produce n-3 fatty acids, so they must rely on a dietary supply. Recently, Kang et al. (2004) developed a transgenic mouse expressing the C. elegans fat-1 gene encoding n-3 desaturase, which converts n-6 to n-3 PUFAs. These fat-1 mice show enrichment of n-3 PUFAs in almost all organs and tissues.

To examine the impact of elevated tissue n-3 PUFA levels in the setting of maladaptive cardiac remodelling, we subjected both fat-1 transgenic mice and WT mice to pressure overload by transverse aortic constriction (TAC) (Endo et al. 2014). The pressure overload induced a similar degree of cardiomyocyte hypertrophy. However, the fibroblast activation and macrophage infiltration were less in fat-1 mice compared with WT mice. Consequently, the pressure overload-induced decline in cardiac systolic function observed in the WT mice was significantly alleviated in fat-1 mice.

Bone marrow (BM) transplantation studies revealed that the fat-1 transgenic BM cells accounted for the anti-remodelling effect of the fat-1 transgene, and the contribution of the fat-1 transgenic cardiomyocytes to favourable changes in fat-1 mice was minor, if any. Lipidomic analysis revealed selective enrichment of EPA in fat-1 transgenic BM cells and EPA-metabolite 18-hydroxyeicosapentaenoic acid (18-HEPE) in fat-1 transgenic macrophages. Interestingly, the increased expression of IL-6, Ccl2, and Tgfb1 mRNA in cardiac fibroblasts with WT macrophages conditioned media was significantly attenuated by 18-HEPE. Cardiac fibroblasts can be activated directly by pressure overload or secondarily by inflammatory mediators released from activated inflammatory cells (Nicoletti et al. 1996; Koyanagi et al. 2000; Kuwahara et al. 2004; Kai et al. 2006). This profibrotic feedforward loop accelerated the progression of cardiac fibrosis under pressure overload. EPA-enriched fat-1 transgenic macrophages generated an 18-HEPE-rich milieu in the heart, thereby counterregulating the profibrotic feedforward loop established under pressure overload conditions (Endo et al. 2014). Notably, dietary intake of EPA can increase the plasma level of 18-HEPE in humans and 18-HEPE administration effectively suppressed perivascular and interstitial fibrosis and macrophage infiltration, thereby preventing the development of heart failure under pressure overload.

52.5.2 *Elucidation of Underlying Mechanisms of Vascular Inflammation After Acute Aortic Dissection*

Acute aortic dissection (AAD) is a tear in the inner layer of the aortic wall, which allows blood to enter into the wall of the aorta, creating a new passage for blood, known as the 'false lumen'. Blood flow into the false lumen may decrease blood flow to vital organs or may lead to rupture, which may be fatal, or to formation of a balloon-like expansion of the aorta, known as an aneurysm. To delineate treatment, the Stanford classification divides AAD into 2 types, type A and type B. Type A AAD, which affects the ascending aorta, is highly lethal and surgical repair is the first choice of treatment. Type B AAD, which does not affect the ascending aorta, is generally more benign and can be medically managed (blood pressure is reduced as much as possible) in most circumstances. However, a substantial proportion of medically treated patients encounter catastrophic events within a month including aortic expansion and rupture. Furthermore, lung oxygen impairment happens always happens even if there is a difference in degree. Thus, better understanding of the molecular and cellular mechanisms involved in the AAD associated processes and the search for a novel therapeutic approach are matters of great importance.

We established a novel mouse model wherein AAD develops in 100 % of cases within 24 h after Ang II administration in mice (Kurihara et al. 2012). This model is dependent on preconditioning with the lysyl oxidase inhibitor β-aminopropionitrile monofumarate (BAPN) to create a pre-AAD status in immature mice. Collagen and elastin crosslinks, which are critical for maintaining vessel wall integrity, are disrupted by BAPN administration, leading to the generation of a mechanically fragile aorta. This type of pre-AAD status is typically seen in human connective tissue diseases, such as Ehlers–Danlos syndrome, but also arises secondary to ageing and atherosclerosis (Schlatmann and Becker 1977; Carlson et al. 1970). As such, our mouse model would seem to recapitulate a similar state in which suitable triggers, such as Ang II, precipitate the transition from a pre-AAD status to AAD. Therefore, our mouse AAD model recapitulates key features of human aortic dissection and facilitates in vivo investigation of the natural disease course after its onset.

The dissection was initiated at the proximal site of the descending thoracic aorta and propagated distally into an abdominal site. Dissection of the aorta caused dilatation, and $\approx 70\,\%$ of the mice died of aortic rupture (Anzai et al. 2015). We found that AAD promotes rapid mobilisation of neutrophils to the aortic tunica adventitia. AAD triggered the gene expression levels of CXCL1, CXCL2, and granulocyte-colony stimulating factor (G-CSF) in the tunica adventitia of the dissected aorta, which are known to be the major neutrophil chemoattractants. In parallel with temporal dynamics of gene expression in the dissected aorta, serum CXCL1 and G-CSF concentrations rapidly increased, whereas serum CXCL2 concentrations did not increase in response to AAD. The percentage of peripheral blood $CD11b^+$ $Ly6G^+$ neutrophils in response to AAD. Because bone marrow (BM) serves as a reservoir for neutrophils, we examined the BM environment in response to AAD. Interestingly, $CD45^+$ $CD11b^+$ $Ly6G^+$ mature neutrophils in BM were transiently reduced after AAD. Neutrophil release from the BM to the blood depends on the balance between the CXCL1/2-CXCR2 and CXCL12-CXCR4 axes (Sadik et al. 2011; Summers et al. 2010). The CXCR2 ligands, CXCL1 and CXCL2, promote neutrophil egress, whereas the function of the CXCR4 ligand, CXCL12, retains neutrophils in the BM. G-CSF stimulates proliferation of myeloid precursors and promotes neutrophil release from the BM by reducing the BM fluid levels of CXCL12 produced by BM stromal cells (Christopher and Link 2007; Martin et al. 2003). The fluorescent intensity of CXCR2 was significantly increased in BM CD45+ CD11b + Ly6G+ neutrophils in response to AAD. In contrast, the protein concentration of CXCL12 in BM fluids was significantly reduced in response to AAD. Bone marrow CXCL12 was reduced. These chemokine changes facilitated neutrophil egress from bone marrow and infiltration into the aortic adventitia.

To determine whether neutrophil mobilisation from BM plays a causative role in lethal aortic rupture, BAPN/Ang II–treated mice were administered intraperitoneally with either a neutralising anti-CXCR2 antibody or control IgG. Interference of CXCL1 function using an anti-CXCR2 antibody reduced neutrophil accumulation and limited aortic rupture post AAD. The tunica adventitia of the expanded dissected aorta demonstrated high levels of interleukin-6 (IL-6) expression. Neutrophils were the major sources of IL-6, and CXCR2 neutralisation significantly reduced local and systemic levels of IL-6. These results prompted us to speculate that blockade of neutrophil migration from BM by neutralising an anti-CXCR2 antibody prevents progression of aortic dissection and lethal aortic rupture, at least in part via the inhibition of IL-6–mediated adventitial inflammation. To determine whether IL-6 plays a causative role in the lethal complications after AAD onset, IL-$6^{-/-}$ mice were subjected to BAPN and Ang II treatment. Disruption of IL-6 effectively suppressed dilatation and rupture of the dissected aorta without any influence on the incidence of AAD and neutrophil mobilisation.

To the best of our knowledge, this study is the first to examine aortic behaviour after AAD onset and has revealed as yet unrecognised cellular and molecular pathways leading to aortic expansion and subsequent rupture (Anzai et al. 2015). Our preliminary human data suggest that the response to dissection may be similar to that in the murine model and that there is a common pathway leading to subsequent rupture. Although a prospective study is needed that focuses on serum levels of CXCL8, G-CSF, and IL-6 in early and late complications of type B aortic dissection, our study implicates CXCR2- or interleukin-6 neutralisation as novel therapeutic strategies to prevent large-artery complications, including aneurysm formation and rupture, in patients with type B AAD.

52.6 Conclusions

Various risk factors induce heart injury (e.g. myocardial infarction) and aortic damages (e.g. acute aortic dissection), leading to a final stage of cardiovascular diseases, namely heart failure (Fig. 52.3). Chronic inflammation is caused by a series of stress responses and leaves 'memory' in the organ. To elucidate the cellular mechanism of chronic inflammation in the cardiovascular system in a temporally and spatially discrete manner facilitates the discovery of novel therapeutic targets to prevent and treat heart failure.

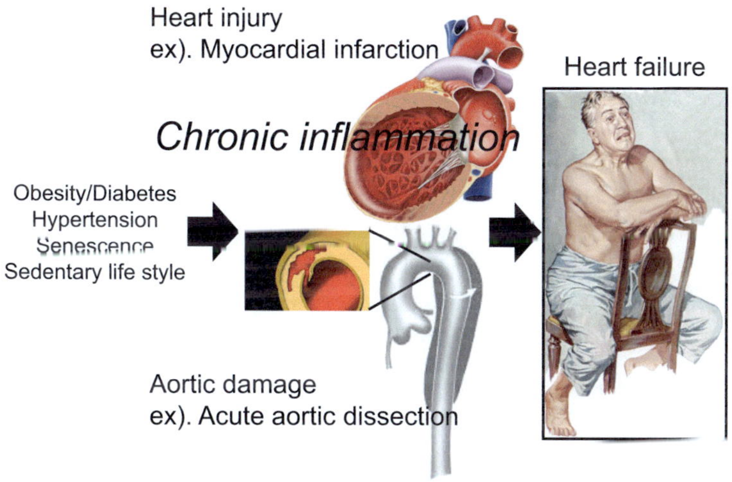

Fig. 52.3 Chronic inflammation is caused by a series of stress responses and leaves 'memory' in organ. Various risk factors induce heart injury (e.g. myocardial infarction) and aortic damages (e.g. acute aortic dissection), leading to a final stage of cardiovascular diseases, namely heart failure. To elucidate the cellular mechanism of chronic inflammation in cardiovascular system in a temporally and spatially discrete manner facilitates the discovery of novel therapeutic targets to prevent and treat heart failure

References

Anzai A, Anzai T, Nagai S, Maekawa Y, Naito K, Kaneko H et al (2012) Regulatory role of dendritic cells in postinfarction healing and left ventricular remodeling. Circulation 125:1234–1245

Anzai A, Shimoda M, Endo J, Kohno T, Katsumata Y, Matsuhashi T et al (2015) Adventitial CXCL1/G-CSF expression in response to acute aortic dissection triggers local neutrophil recruitment and activation leading to aortic rupture. Circ Res 116:612–623

Arslan F, de Kleijn DP, Pasterkamp G (2011) Innate immune signaling in cardiac ischemia. Nat Rev Cardiol 8:292–300

Carlson RG, Lillehei CW, Edwards JE (1970) Cystic medial necrosis of the ascending aorta in relation to age and hypertension. Am J Cardiol 25:411–415

Christopher MJ, Link DC (2007) Regulation of neutrophil homeostasis. Curr Opin Hematol 14:3–8

Endo J, Sano M, Isobe Y, Fukuda K, Kang JX, Arai H et al (2014) 18-HEPE, an n-3 fatty acid metabolite released by macrophages, prevents pressure overload-induced maladaptive cardiac remodeling. J Exp Med 211:1673–1687

Frangogiannis NG (2012) Regulation of the inflammatory response in cardiac repair. Circ Res 110:159–173

Gissi HFI, Tavazzi L, Maggioni AP, Marchioli R, Barlera S, Franzosi MG et al (2008) Effect of n-3 polyunsaturated fatty acids in patients with chronic heart failure (the GISSI-HF trial): a randomised, double-blind, placebo-controlled trial. Lancet 372:1223–1230

Kai H, Mori T, Tokuda K, Takayama N, Tahara N, Takemiya K et al (2006) Pressure overload-induced transient oxidative stress mediates perivascular inflammation and cardiac fibrosis through angiotensin II. Hypertens Res 29:711–718

Kang JX, Wang J, Wu L, Kang ZB (2004) Transgenic mice: fat-1 mice convert n-6 to n-3 fatty acids. Nature 427:504

Koyanagi M, Egashira K, Kitamoto S, Ni W, Shimokawa H, Takeya M et al (2000) Role of monocyte chemoattractant protein-1 in cardiovascular remodeling induced by chronic blockade of nitric oxide synthesis. Circulation 102:2243–2248

Krum H, Teerlink JR (2011) Medical therapy for chronic heart failure. Lancet 378:713–721

Kurihara T, Shimizu-Hirota R, Shimoda M, Adachi T, Shimizu H, Weiss SJ et al (2012) Neutrophil-derived matrix metalloproteinase 9 triggers acute aortic dissection. Circulation 126:3070–3080

Kuwahara F, Kai H, Tokuda K, Takeya M, Takeshita A, Egashira K et al (2004) Hypertensive myocardial fibrosis and diastolic dysfunction: another model of inflammation? Hypertension 43:739–745

Leuschner F, Rauch PJ, Ueno T, Gorbatov R, Marinelli B, Lee WW et al (2012) Rapid monocyte kinetics in acute myocardial infarction are sustained by extramedullary monocytopoiesis. J Exp Med 209:123–137

Lewis EF, Moye LA, Rouleau JL, Sacks FM, Arnold JM, Warnica JW et al (2003) Predictors of late development of heart failure in stable survivors of myocardial infarction: the CARE study. J Am Coll Cardiol 42:1446–1453

Martin C, Burdon PC, Bridger G, Gutierrez-Ramos JC, Williams TJ, Rankin SM (2003) Chemokines acting via CXCR2 and CXCR4 control the release of neutrophils from the bone marrow and their return following senescence. Immunity 19:583–593

Morrison DF, Foss DL, Murtaugh MP (2000) Interleukin-10 gene therapy-mediated amelioration of bacterial pneumonia. Infect Immun 68:4752–4758

Mozaffarian D, Wu JH (2011) Omega-3 fatty acids and cardiovascular disease: effects on risk factors, molecular pathways, and clinical events. J Am Coll Cardiol 58:2047–2067

Nahrendorf M, Swirski FK, Aikawa E, Stangenberg L, Wurdinger T, Figueiredo JL et al (2007) The healing myocardium sequentially mobilizes two monocyte subsets with divergent and complementary functions. J Exp Med 204:3037–3047

Nahrendorf M, Pittet MJ, Swirski FK (2010) Monocytes: protagonists of infarct inflammation and repair after myocardial infarction. Circulation 121:2437–2445
Nicoletti A, Heudes D, Mandet C, Hinglais N, Bariety J, Michel JB (1996) Inflammatory cells and myocardial fibrosis: spatial and temporal distribution in renovascular hypertensive rats. Cardiovasc Res 32:1096–1107
Nodari S, Triggiani M, Campia U, Manerba A, Milesi G, Cesana BM et al (2011) Effects of n-3 polyunsaturated fatty acids on left ventricular function and functional capacity in patients with dilated cardiomyopathy. J Am Coll Cardiol 57:870–879
Sadik CD, Kim ND, Luster AD (2011) Neutrophils cascading their way to inflammation. Trends Immunol 32:452–460
Schlatmann TJ, Becker AE (1977) Histologic changes in the normal aging aorta: implications for dissecting aortic aneurysm. Am J Cardiol 39:13–20
Shi FD, Ljunggren HG, La Cava A, Van Kaer L (2011) Organ-specific features of natural killer cells. Nat Rev Immunol 11:658–671
Sobirin MA, Kinugawa S, Takahashi M, Fukushima A, Homma T, Ono T et al (2012) Activation of natural killer T cells ameliorates postinfarct cardiac remodeling and failure in mice. Circ Res 111:1037–1047
Summers C, Rankin SM, Condliffe AM, Singh N, Peters AM, Chilvers ER (2010) Neutrophil kinetics in health and disease. Trends Immunol 31:318–324
Vivier E, Tomasello E, Baratin M, Walzer T, Ugolini S (2008) Functions of natural killer cells. Nat Immunol 9:503–510
Weirather J, Hofmann UD, Beyersdorf N, Ramos GC, Vogel B, Frey A et al (2014) Foxp3+ CD4+ T cells improve healing after myocardial infarction by modulating monocyte/macrophage differentiation. Circ Res 115:55–67
Yan X, Shichita T, Katsumata Y, Matsuhashi T, Ito H, Ito K et al (2012) Deleterious effect of the IL-23/IL-17A axis and gammadeltaT cells on left ventricular remodeling after myocardial infarction. J Am Heart Assoc 1:e004408
Yan X, Anzai A, Katsumata Y, Matsuhashi T, Ito K, Endo J et al (2013) Temporal dynamics of cardiac immune cell accumulation following acute myocardial infarction. J Mol Cell Cardiol 62:24–35
Yan X, Hegab AE, Endo J, Anzai A, Matsuhashi T, Katsumata Y et al (2014) Lung natural killer cells play a major counter-regulatory role in pulmonary vascular hyperpermeability after myocardial infarction. Circ Res 114:637–649
Yancy CW, Jessup M, Bozkurt B, Butler J, Casey DE Jr, Drazner MH et al (2013) 2013 ACCF/AHA guideline for the management of heart failure: a report of the American College of Cardiology Foundation/American Heart Association Task Force on Practice Guidelines. J Am Coll Cardiol 62:e147–e239
Yokoyama M, Origasa H, Matsuzaki M, Matsuzawa Y, Saito Y, Ishikawa Y et al (2007) Effects of eicosapentaenoic acid on major coronary events in hypercholesterolaemic patients (JELIS): a randomised open-label, blinded endpoint analysis. Lancet 369:1090–1098
Yoshidome H, Kato A, Edwards MJ, Lentsch AB (1999) Interleukin-10 inhibits pulmonary NF-kappaB activation and lung injury induced by hepatic ischemia-reperfusion. Am J Physiol 277:L919–L923

Index

A

Acetylcholine, 431
Activation, 9, 22, 38, 56, 65, 93, 97–105, 120, 135, 176, 189, 202, 212, 254, 278, 303, 311, 320, 331, 347, 358, 371, 380, 406, 419, 432, 442, 462, 477, 490, 502, 512–611, 622, 628, 642, 652, 662, 677, 687
Acute sterile inflammation, 650–652, 657
Adaptive immunity, 109, 320, 339, 358, 418, 431, 543
Adenine- and uridine(AU)-rich elements (AREs), 265, 320–324
Adipocytes, 71, 72, 103, 141–143, 294–296, 382, 383
Adoptive immunity, 434
Adrenaline, 429
Adrenergic nerve, 430–432, 435, 436
Aicardi–Goutières syndrome (AGS), 512, 515–520
Aire, 490, 492–496
Air pollutant, 261–263, 265, 267
Airway hyperreactivity (AhR), 263–264, 267
Allergic rhinitis, 42–43
Alzheimer disease (AD), 661–669
Amyloid-β peptide (Aβ), 663
Amyloid cascade hypothesis, 663, 664, 666, 667
Anemia of chronic inflammation (ACI), 83–85
Angiogenesis, 4, 11, 12, 37, 201, 228, 294, 460, 678
Animal model, 9, 13, 71, 72, 228, 349, 402, 434, 494–496, 559–560, 646, 666–669
Antibody responses, 163, 165
Antioxidant, 265, 266, 309, 310, 313, 314, 422, 560
Aortic dissection, 691–693
Arthritis, 5, 52, 55, 56, 67, 91, 322, 347, 350, 449, 468, 477–480, 482, 572, 580, 679
Asthma, 11, 41–43, 95, 252, 253, 255, 258, 259, 261, 262, 266, 351, 402, 404, 405, 572, 593
Astrocytes, 445, 452, 645, 654, 655, 662, 664
ATG16L1, 605, 608–612
Atopic dermatitis, 42–43, 250–267, 599
Autoimmune disease, 4, 8, 40, 43, 56, 64, 100, 113, 116, 118, 164, 165, 320, 363, 402, 412, 413, 422, 430, 442, 448–449, 451–453, 476, 478–480, 483, 489–497, 511 520, 526 531, 555, 612
Autoimmunity, 39, 324, 325, 328, 346, 349, 418, 475, 476, 480, 481, 490, 491, 494–496, 516, 520, 530, 575
Autoinflammatory disorders, 65
Autophagy, 122, 302, 303, 391, 604–613
Autotaxin (ATX), 78–82, 153, 155, 464, 465, 467–468

B

Bach2, 420, 422–425
BART clusters, 236
β_2ARs, 430–436
BHRF1, 236, 239, 240, 242
Bleomycin, 21, 23–25, 27–31, 469
Blood pressure, 292, 549, 551, 679, 680, 691

M. Miyasaka, K. Takatsu (eds.), *Chronic Inflammation*,
DOI 10.1007/978-4-431-56068-5

C

C2 domain-containing protein, 373, 376
Calcitonin gene-related peptide, 436
Calcium, 374, 375, 377, 526–529, 531–532, 573, 579
Cancer, 9
Carbohydrates, 333, 334, 337, 576, 591, 597
Cardiomyocytes, 312, 688, 690
Cardiovascular remodeling, 679–681
CCL8, 621, 623, 624
CCL20, 436
CCR4, 254
CCR7, 432
CD1d, 346
CD169, 618–621
Cell–cell communication, 212, 219
Cellular senescence, 188–197, 215–219, 310–314, 420, 421, 425, 555–562
Chemokine receptor, 206, 254, 413, 432–433, 435, 462, 527, 645
ChIP followed by high-throughput sequencing (ChIP-seq), 629–631, 634, 636
Chronic gastritis, 223, 240, 628, 636
Chronic inflammation, 4–13, 20–31, 37–45, 64, 72, 78–86, 98–105, 134–143, 174–185, 188–197, 223–227, 230, 240, 255, 279, 287, 292–296, 320, 325, 328, 345–352, 371, 380–393, 402, 410, 418–425, 442, 445, 450, 451, 464, 469, 475–483, 489–497, 530–532, 542–551, 555, 558, 604, 605, 613, 620, 622, 627–637, 677, 681, 693
Chronic rhinosinusitis (CRS), 410
Clostridium, 571, 572, 593, 595, 597, 599
Colitis, 7, 8, 39, 43, 67, 91, 223, 225, 336, 391, 453, 464, 479, 480, 483, 571, 573, 576, 578, 579, 581, 593, 599, 604, 606, 619–623
Collagen-induced arthritis (CIA), 8
Colorectal cancer (CRC), 9, 10, 12, 195, 225, 391, 410, 632
Commensal mutualism, 569–573, 578
Corticotropin-releasing hormone, 436
Crohn's disease (CD), 8, 43, 113, 225, 336, 479, 593, 604–606, 608, 609, 611, 612
CRTH2, 254
C-type lectin, 293, 294, 320, 333–341, 652
Cutaneous leukocyte-associated antigen, 253
CXCR4, 432
Cyclooxygenase (COX), 4, 5, 90, 179
Cyclooxygenase 2 (COX-2), 5, 8–10, 12, 13, 52, 174–178, 183, 184, 321, 322, 559–561, 664
Cysteine persulfides, 309–314
Cytokine, 4–7, 10, 26, 28, 29, 40, 42, 43, 55–57, 67, 72, 83, 91, 94, 98, 101, 103, 110, 114, 116, 117, 125, 135, 162, 177, 179, 180, 184, 195, 208, 215, 231, 240–242, 253, 254, 256–259, 264, 266, 292, 294, 295, 302, 320–328, 346, 349, 358, 360, 364–366, 381, 382, 384, 389, 391, 402, 404, 407, 410, 411, 413, 418, 419, 435, 442, 444, 445, 448, 450, 452, 477, 529, 542, 543, 556–558, 568, 569, 594, 624, 628, 642, 655, 662, 664, 665, 667, 677, 679

D

Damage-associated molecular patterns (DAMPs), 4, 6, 278, 300, 303, 304, 306, 339, 380, 577, 622, 623, 651–655
Delayed-type hypersensitivity, 434
Dendritic cells (DCs), 6, 30, 38, 39, 55, 57, 67, 114–117, 162, 251, 258, 262, 320, 346, 348, 357–366, 370, 373, 434, 435, 444, 462, 468, 490, 491, 526, 594, 611, 618, 645, 688
Deoxycholic acid (DCA), 195, 196
Dextran sodium sulfate (DSS), 609, 610, 612, 619, 621, 622
Diabetes, 103, 142, 194, 227, 280, 282, 349, 448, 449, 451, 452, 477–478, 480, 495, 496, 572, 593, 599
DNA methylation, 134–137, 419, 632, 633, 636, 637
Dnmt3a, 136, 137
Double-stranded RNA (dsRNA), 277–287, 513
Drosophila, 120–127, 212, 214, 216–219, 528
Dysbiosis, 569–573, 576, 578, 592–593, 599, 600

E

E3 ligase, 123, 124, 126, 326
Endothelial cell, 26, 40, 468
Enterocytes, 83, 607, 610–611
Eosinophilic CRS (ECRS), 410–412
Epigenetics, 419, 542
Epithelial cells, 20, 21, 42, 94, 177, 178, 182, 183, 185, 212, 225, 239, 262, 338, 404, 410, 411, 445, 447, 451, 468, 569, 573–576, 597, 604, 607, 609, 622, 629–631, 635, 680
Epstein–Barr virus (EBV), 235–243, 347
Exocytosis, 370–377, 608
Experimental autoimmune encephalomyelitis (EAE), 7

F

Fecal transplantation, 570, 599
Ferroportin, 83–84
Fibrinogen, 202, 206, 207, 652
Fibroblast, 5, 10, 12, 13, 20–26, 28, 31, 116, 189, 191, 193, 211, 218, 241, 312, 561, 582, 688, 690
Fibrosis, 4, 11–13, 20–31, 82, 193, 226, 294–296, 402, 469, 544, 549, 578, 582, 680, 690
Follicular helper T (Tfh) cells, 162–164, 324, 444, 577, 578
Forkhead box P3 (FoxP3), 443, 448, 476, 477, 479, 482, 483, 594

G

γδT cells, 594, 687, 688
Gastric cancer (GC), 177–180, 182, 184, 223, 225, 240, 241, 451, 628, 631–635
Gastritis, 449, 478, 628, 635
Glial cells, 642, 662, 664, 665, 667, 668
Glycolipid, 346, 347
Glycyrrhiza uralensis, 381, 384–388
Glycyrrhizin, 381, 384–386
Goblet cell, 43, 573, 610
G protein-coupled receptors (GPCRs), 4, 78, 432, 460, 465, 502, 503

H

Helicobacter pylori, 10, 175, 180, 184, 225, 628–637
Helper T cell (Th) subsets, 6
Hepatitis, 223, 240
Hepcidin, 83–86
HSP70, 115, 116
Hypoxia, 654, 677–679
Hypoxia-inducible factor-α (HIF-α), 677–681

I

Idiopathic pulmonary fibrosis (IPF), 24, 25, 30, 310, 469
IkB kinase (IKK), 56, 121, 278, 323, 326, 358, 381, 387
IL 5, 12, 253, 259, 402, 405–408, 410, 412
IL-17, 6, 8, 56, 110, 402, 403, 413, 452, 530
IL-33, 26, 256–259, 403–408, 410–412
Imaging, 137–143, 148, 149, 151–155, 446
Imaging mass spectrometry, 148–157
Immune cell trafficking, 460–469
Immune checkpoint, 442
Immune network, 496
Immunodeficiency, 236, 374, 526, 527, 529–530, 533, 534
Immunological synapse, 419, 446, 447
Immunoproteasome, 65–72
Immunosenescence, 418, 419, 556, 557
Inflammaging, 418, 542, 543, 555–557, 559
Inflammasome, 65, 300, 303–306, 334, 380–393, 610, 653, 655–656, 665
Inflammation, 676, 677, 679, 681, 685–693
Inflammatory bowel diseases (IBDs), 43, 45, 64, 110, 142, 336, 452, 479, 568–573, 575, 578, 580–582, 593, 599, 604, 605, 618–624
Inflammatory cytokine, 55, 56, 67, 84, 125, 141, 188, 191, 192, 195, 213–214, 216, 219, 256, 257, 322, 334, 339, 375, 376, 384, 389, 418, 506, 512, 515, 516, 576, 579, 581, 611, 651, 676, 689
Inflammatory diseases, 38, 45, 50, 54, 56, 57, 64, 65, 67–68, 72, 110, 133, 143, 148, 263–266, 310, 321, 325, 327, 328, 389, 434, 516, 599
Inflammatory mediator, 5, 20, 26–28, 83, 148, 157, 204, 280, 380, 501–507, 543, 650, 651, 653, 655, 676, 690
Innate immune, 94, 109, 113, 120–127, 183, 184, 206, 208, 279, 281, 300, 302–303, 306, 320, 321, 323, 327, 337, 338, 371, 381–383, 389–392, 515, 520, 542, 543, 569, 576, 592, 608, 611, 619, 622, 650, 665, 677
Innate immunity, 4–6, 98, 182–184, 381, 384, 551, 649–657
Innate lymphoid cells (ILCs), 259, 350, 568, 572, 592, 618
Interferon (IFN), 98, 242, 444, 511, 512, 514–520
Interleukin-6 (IL-6), 5, 7, 27, 55, 67, 69–71, 84, 98, 101, 103, 110, 134, 180, 192–195, 312, 320–322, 324, 325, 327, 331, 336, 338, 360, 365, 380, 384, 420, 421, 424, 435, 464, 468, 512, 515, 516, 555, 558, 562, 568, 577, 579, 581, 618, 621, 665, 689, 690, 692
Intestinal epithelial (IECs), 180, 182–183, 452, 570, 573, 576, 577, 597, 604, 607–608, 610–612
Intracranial aneurysm (IA), 8–10
Intratracheal transfer, 24
Invariant natural killer T (iNKT) cells, 345–352, 532
In vivo imaging, 664

Iron homeostasis, 526–534
Ischemic stroke, 391, 649–657
Isoliquiritigenin (ILG), 381, 385–388

J
c-Jun N-terminal kinase (JNK), 214, 215, 217–219, 278, 280–282, 285, 286, 381, 385, 408, 656

K
Kelch-like ECH-associated protein 1 (Keap1), 265

L
LIM protein, 111, 116–118
Lipid mediator, 78, 83, 90–94, 138, 153, 163, 460, 464–465, 501, 689–691
Lipidomics, 91, 92
Lipodystrophy, 64, 65, 68, 71
Lipoxygenases (LOX), 90
Liquid chromatography tandem mass spectrometry (LC-MS/MS), 91, 92, 502
Liver cancer, 188, 194–196, 223–225, 228
Long non-coding RNAs, 628–637
Lymph-node egress, 432–434
Lymphocytes, 4, 43, 55, 94, 110, 138, 148, 155, 241, 403, 409, 419, 430–433, 435, 452, 461–463, 467, 468, 527, 557, 569, 571, 575, 645, 650, 654
Lymphocyte trafficking, 155, 431, 433–435, 463, 467
Lymphoma, 51, 228, 235, 236, 241, 242, 451
Lysophosphatidic acid (LPA), 78–82, 153–157, 460, 464–469, 501, 502, 504, 505
Lysophosphatidylserine (LysoPS), 82, 501–507
Lysophospholipid (LPL), 78, 460–469, 501, 503
Lysosomal digestive enzymes, 370
Lysosome, 44, 122, 370–372, 374, 375, 606, 607
Lysosome-related disease, 370

M
Macrophage, 5, 8, 676–681
Mammalian target of rapamycin (mTOR), 38, 45, 278, 448, 481
Mast cell, 42, 101, 259, 336, 370, 372, 405, 406, 445, 463, 468, 502, 503, 526, 527, 568, 577–581
MCP-1, 9
MCP-2, 621
Membrane fusion, 301, 372–374
Membrane trafficking, 372
Memory Th2 cell, 250, 253–254, 256–259, 266, 267, 402–410, 412, 413
Menin, 419–425
Metabolic disorder, 278, 281, 285, 380, 382, 383, 478, 676
Metabolic syndrome, 4, 91, 292–293, 380, 478, 542, 558
Microbiota, 162, 184, 195, 196, 568–572, 576, 591–595, 598, 599, 623, 669
Microcluster, 446, 447
Microglia, 452, 642–646, 655, 662
MicroRNA (miRNA), 50, 54–57, 98, 99, 223–231, 236–243, 285–287, 324, 612
miR-140, 53, 54
miR-146a, 54–57, 325
miR-155, 54–57
Mitochondrial dysfunction, 212–213, 215–217, 219, 305, 306, 560
Mitophagy, 301–304
Monocyte, 27, 30, 41, 42, 55, 56, 89, 91, 134, 138, 139, 142, 305, 336, 337, 348, 444, 450, 451, 550, 592, 594, 618–624, 650, 654, 676, 677, 688
mRNA stability, 321–324
Multiple sclerosis (MS), 8, 40–42, 164, 165, 264, 266, 349, 434, 449, 452, 502, 515, 572, 599
Myeloid cells, 21, 57, 101, 335, 336, 338, 452, 568, 579, 581, 618, 656, 679
Myocardial infarction (MI), 310, 676, 686, 689, 690, 693
Myofibroblast, 20–24, 31, 294, 582

N
NADPH oxidase 1 (NOX1), 182
Natural killer (NK) cells, 30, 31, 192, 242, 337, 346, 347, 349, 370, 462, 468, 526, 530, 592, 594, 619, 688–690
Natural products, 380–393
Neuroinflammation, 661–669
Neuropathic pain, 642–646
Neutrophils, 10, 26, 30, 138, 303, 336, 338, 370, 372, 380, 409, 442, 468, 503, 577, 578, 618, 619, 650, 651, 655, 656, 687–690, 692
Nicotinic acetylcholine receptor subunit α7, 431
Nitric oxide (NO), 98, 101, 135, 310, 662, 677, 680
NLRP3, 65, 300, 303–306, 381, 383–392, 655–657

Noradrenaline, 430
Nrf2, 265–267, 311, 422
Nuclear factor kB (NFkB), 9, 10, 56, 67, 70, 120, 121, 125–127, 204, 406, 408, 464
Nuclear factor of activated T-cell (NFAT), 526, 528, 529, 532, 533

O

Obesity, 103–104, 134, 141–143, 188, 194–196, 277–287, 292–296, 380–393, 478, 480, 593, 599
Omega-3 fatty acid, 90, 91
Organ-specific autoimmunity, 489–497
Osteoarthritis, 52–53, 56, 468
Osteoclasts, 101, 135, 141
Oxidative stress, 44, 45, 214, 215, 217, 240, 265, 284, 285, 309–314, 322, 557, 560, 654

P

p21, 216, 217, 557, 560, 561
P2X4R, 642–646
Paneth cells, 569, 573–575, 607–610, 613
Pathogen-associated molecular pattern (PAMP), 4–6, 278, 303, 304, 320, 611, 622, 623
Pathogenic Th2 cells (Tpath2 cells), 402–405, 407–411, 413
Pattern recognition receptor (PRR), 121, 122, 320, 357, 380, 381, 392, 650–653, 655
PD-1, 91, 162, 442–453
PDLIM1, 116, 117
PDLIM2, 110–117
Permeability, 91, 155, 202–207, 573, 580, 581, 689, 690
PGF2α, 4, 13
Phosphatase and tensin homologue (PTEN), 38, 39, 51, 99–102
Phosphatases, 370
Phospholipase, 464, 505
PKR, 278–287, 512, 514, 515, 656
Polarization, 98–105
Polycyclic aromatic hydrocarbons, 262
Positive feedback loop, 8–10
Postmortem degradation of metabolites, 149–152
Premetastatic, 202, 203, 206, 208
Preterm birth, 555–562
Probiotics, 570, 571
Programmed cell death, 577
Programmed cell death-1 (PD-1), 420, 442
Prostaglandin, 4–13, 78, 90, 149
Prostaglandin E_2 (PGE_2), 4, 6–11, 174–178, 183, 184, 384
Pruritus, 255
Psychoimmunology, 430
Psychoneuroimmunology, 430
PtdIns(3,4,5)P_3, 99–105
Pulmonary oedema, 688–693
Purinergic signalling, 569, 570, 576–581

R

Reactive oxygen species (ROS), 26, 135, 182, 189, 191, 193, 214, 240, 266, 284, 301, 304, 305, 309–313, 334, 338, 392, 610, 656, 662
Receptor activator of nuclear factor kB ligand (RANKL), 5, 135, 140–141
Regnase-1, 323–328
Regulatory T (Treg) cells, 28, 39, 264, 350, 443, 444, 448, 475, 483, 530, 532, 591–600
Resolution of inflammation, 57, 89–95, 320–328, 503, 650
Retinal pigment epithelium, 43
Retinitis pigmentosa, 43
Retinosis pigmentosa, 506
Rheumatoid arthritis (RA), 54–56, 110, 118, 141, 164, 328, 418, 419, 422, 434, 468, 478, 482, 555, 662, 669
RIG-I-like receptors (RLRs), 320, 511–515, 517, 520
RNA modification, 286, 323
Roquin, 164, 323–328

S

S1P receptor, 80, 139, 140, 460, 461, 463, 468
S100A8, 141–143, 206, 207
S100A9, 207
SAA3, 206
Sema4A, 37–45
Semaphorins, 37, 38, 594
Senescence-associated secretary phenotype (SASP), 188, 192, 194, 196, 215–219, 312, 418, 420–422, 557, 562, 665
Senescence-associated secretome (SAS), 191–197
Senescence-messaging secretome (SMS), 188, 215, 515
Sensitisation, 251–262

SHIP1, 57, 99–105
Short-chain fatty acids (SCFAs), 568, 592, 597, 598
Sialoadhesin, 618
Signal transducer and activator of transcription 3 (STAT3), 101, 110, 113, 183, 445, 464
Signal transducer and activator of transcription 4 (STAT4), 110–112, 117
Silica, 30, 390, 656
Sphingosine-1-phosphate (S1P), 138, 140, 150, 163, 432, 460–461, 501
Spinal cord, 22, 41, 642–646
ST2, 258, 259, 403–408, 410
Sterile inflammation, 303, 652, 655
Stress, 8, 9, 435
Structural biology, 78–86
Sympathetic nervous system, 435
Systemic lupus erythematosus (SLE), 56, 363, 449, 480, 481, 512, 515–518

T
TAR-RNA binding protein (TRBP), 281, 284–287
Tau, 663–665, 667, 668
T-cell-mediated inflammatory disease, 434
T-cell receptor (TCR), 251, 254, 323, 325, 327, 346, 423, 443–448, 452, 478, 479, 481, 483, 526, 527, 531, 594, 599
T-cell regulatory, 43, 336
T cell signaling, 446
T follicular helper (Tfh), 350, 477
Th1, 6, 7
Th2 cells, 253
Th17, 6–8, 41, 42, 68, 110, 113, 117, 140–141, 262, 264, 325, 327, 335, 338, 339, 350, 402, 403, 412, 413, 453, 478, 479, 482, 577, 579, 581, 594, 595, 600, 618
Thymic epithelial cells (TECs), 490
Thymic stromal lymphopoietin (TSLP), 256–258, 262, 266
Thymus, 350, 418, 447, 462, 475, 490–494, 503, 530–532, 594
Tissue remodeling, 5, 11–13
Tolerance, 39, 280, 285, 445, 448, 452, 482, 490, 492, 494–496, 532, 533, 571, 592, 593, 618, 619, 622, 623, 686
Toll-like receptor (TLR), 54, 98, 109, 120, 164, 278, 320, 323, 325, 326, 339, 358, 380–382, 511
Tumor immunotherapy, 450–451
Tumor necrosis factor α (TNFα), 7, 10, 11, 27, 55, 56, 67, 101, 103, 121, 134, 142, 176, 178–182, 184, 202, 229, 231, 240, 292, 294, 295, 323, 365, 380–382, 384, 388, 445, 463, 464, 478, 512, 555, 568, 579, 621, 623, 676
Tumor progression, 37, 98, 211–220

U
Ubiquitin E3 ligase, 110–113, 116, 117
Ubiquitination, 84, 114, 124, 125, 515

V
Vagus nerve, 431
Vasoactive intestinal peptide, 436
Virus, 66, 122, 235–243, 281, 305, 322, 347, 351, 370, 442, 450, 511–515, 520, 530, 573, 609

MIX
Papier aus verantwortungsvollen Quellen
Paper from responsible sources
FSC® C105338

If you have any concerns about our products,
you can contact us on
ProductSafety@springernature.com

In case Publisher is established outside the EU,
the EU authorized representative is:
Springer Nature Customer Service Center GmbH
Europaplatz 3, 69115 Heidelberg, Germany

Printed by Libri Plureos GmbH
in Hamburg, Germany